Lecture Notes in Computer Science   12140

More information about this series at http://www.springer.com/series/7407

Valeria V. Krzhizhanovskaya ·
Gábor Závodszky · Michael H. Lees ·
Jack J. Dongarra · Peter M. A. Sloot ·
Sérgio Brissos · João Teixeira (Eds.)

Computational Science – ICCS 2020

20th International Conference
Amsterdam, The Netherlands, June 3–5, 2020
Proceedings, Part IV

 Springer

Editors
Valeria V. Krzhizhanovskaya ⓘ
University of Amsterdam
Amsterdam, The Netherlands

Michael H. Lees
University of Amsterdam
Amsterdam, The Netherlands

Peter M. A. Sloot ⓘ
University of Amsterdam
Amsterdam, The Netherlands

ITMO University
Saint Petersburg, Russia

Nanyang Technological University
Singapore, Singapore

João Teixeira
Intellegibilis
Setúbal, Portugal

Gábor Závodszky ⓘ
University of Amsterdam
Amsterdam, The Netherlands

Jack J. Dongarra ⓘ
University of Tennessee
Knoxville, TN, USA

Sérgio Brissos
Intellegibilis
Setúbal, Portugal

ISSN 0302-9743   ISSN 1611-3349 (electronic)
Lecture Notes in Computer Science
ISBN 978-3-030-50422-9   ISBN 978-3-030-50423-6 (eBook)
https://doi.org/10.1007/978-3-030-50423-6

LNCS Sublibrary: SL1 – Theoretical Computer Science and General Issues

This Springer imprint is published by the registered company Springer Nature Switzerland AG
The registered company address is: Gewerbestrasse 11, 6330 Cham, Switzerland

Preface

Twenty Years of Computational Science

Welcome to the 20th Annual International Conference on Computational Science (ICCS – https://www.iccs-meeting.org/iccs2020/).

During the preparation for this 20th edition of ICCS we were considering all kinds of nice ways to celebrate two decennia of computational science. Afterall when we started this international conference series, we never expected it to be so successful and running for so long at so many different locations across the globe! So we worked on a mind-blowing line up of renowned keynotes, music by scientists, awards, a play written by and performed by computational scientists, press attendance, a lovely venue… you name it, we had it all in place. Then corona hit us.

After many long debates and considerations, we decided to cancel the physical event but still support our scientists and allow for publication of their accepted peer-reviewed work. We are proud to present the proceedings you are reading as a result of that.

ICCS 2020 is jointly organized by the University of Amsterdam, NTU Singapore, and the University of Tennessee.

The International Conference on Computational Science is an annual conference that brings together researchers and scientists from mathematics and computer science as basic computing disciplines, as well as researchers from various application areas who are pioneering computational methods in sciences such as physics, chemistry, life sciences, engineering, arts and humanitarian fields, to discuss problems and solutions in the area, to identify new issues, and to shape future directions for research.

Since its inception in 2001, ICCS has attracted increasingly higher quality and numbers of attendees and papers, and 2020 was no exception, with over 350 papers accepted for publication. The proceedings series have become a major intellectual resource for computational science researchers, defining and advancing the state of the art in this field.

The theme for ICCS 2020, "Twenty Years of Computational Science", highlights the role of Computational Science over the last 20 years, its numerous achievements, and its future challenges. This conference was a unique event focusing on recent developments in: scalable scientific algorithms, advanced software tools, computational grids, advanced numerical methods, and novel application areas. These innovative novel models, algorithms, and tools drive new science through efficient application in areas such as physical systems, computational and systems biology, environmental systems, finance, and others.

This year we had 719 submissions (230 submissions to the main track and 489 to the thematic tracks). In the main track, 101 full papers were accepted (44%). In the thematic tracks, 249 full papers were accepted (51%). A high acceptance rate in the thematic tracks is explained by the nature of these, where many experts in a particular field are personally invited by track organizers to participate in their sessions.

ICCS relies strongly on the vital contributions of our thematic track organizers to attract high-quality papers in many subject areas. We would like to thank all committee members from the main and thematic tracks for their contribution to ensure a high standard for the accepted papers. We would also like to thank Springer, Elsevier, the Informatics Institute of the University of Amsterdam, the Institute for Advanced Study of the University of Amsterdam, the SURFsara Supercomputing Centre, the Netherlands eScience Center, the VECMA Project, and Intellegibilis for their support. Finally, we very much appreciate all the Local Organizing Committee members for their hard work to prepare this conference.

We are proud to note that ICCS is an A-rank conference in the CORE classification.

We wish you good health in these troubled times and hope to see you next year for ICCS 2021.

June 2020

Valeria V. Krzhizhanovskaya
Gábor Závodszky
Michael Lees
Jack Dongarra
Peter M. A. Sloot
Sérgio Brissos
João Teixeira

Organization

d Simulations, Adaptive Algorithms and Solvers – ABS-AAS

aszynski
Pardo
r Calo
obert Schaefer
Quanling Deng

Applications of Computational Methods in Artificial Intelligence and Machine Learning – ACMAIML

Kourosh Modarresi
Raja Velu
Paul Hofmann

Biomedical and Bioinformatics Challenges for Computer Science – BBC

Mario Cannataro
Giuseppe Agapito
Mauro Castelli
Riccardo Dondi
Rodrigo Weber dos Santos
Italo Zoppis

Classifier Learning from Difficult Data – CLD^2

Michał Woźniak
Bartosz Krawczyk
Paweł Ksieniewicz

Complex Social Systems through the Lens of Computational Science – CSOC

Debraj Roy
Michael Lees
Tatiana Filatova

Computational Health – CompHealth

Sergey Kovalchuk
Stefan Thurner
Georgiy Bobashev

Computational Methods for Emerging Problems in (dis-)Information Analysis – DisA

Michal Choras
Konstantinos Demestichas

Computational Optimization, Modelling and Simulation – COMS

Xin-She Yang
Slawomir Koziel
Leifur Leifsson

Computational Science in IoT and Smart Systems – IoTSS

Vaidy Sunderam
Dariusz Mrozek

Computer Graphics, Image Processing and Artificial Intelligence – CGIPAI

Andres Iglesias
Lihua You
Alexander Malyshev
Hassan Ugail

Data-Driven Computational Sciences – DDCS

Craig C. Douglas
Ana Cortes
Hiroshi Fujiwara
Robert Lodder
Abani Patra
Han Yu

Machine Learning and Data Assimilation for Dynamical Systems – MLDADS

Rossella Arcucci
Yi-Ke Guo

Meshfree Methods in Computational Sciences – MESHFREE

Vaclav Skala
Samsul Ariffin Abdul Karim
Marco Evangelos Biancolini
Robert Schaback

Rongjiang Pan
Edward J. Kansa

Multiscale Modelling and Simulation – MMS

Derek Groen
Stefano Casarin
Alfons Hoekstra
Bartosz Bosak
Diana Suleimenova

Quantum Computing Workshop – QCW

Katarzyna Rycerz
Marian Bubak

Simulations of Flow and Transport: Modeling, Algorithms and Computation – SOFTMAC

Shuyu Sun
Jingfa Li
James Liu

Smart Systems: Bringing Together Computer Vision, Sensor Networks and Machine Learning – SmartSys

Pedro J. S. Cardoso
João M. F. Rodrigues
Roberto Lam
Janio Monteiro

Software Engineering for Computational Science – SE4Science

Jeffrey Carver
Neil Chue Hong
Carlos Martinez-Ortiz

Solving Problems with Uncertainties – SPU

Vassil Alexandrov
Aneta Karaivanova

Teaching Computational Science – WTCS

Angela Shiflet
Alfredo Tirado-Ramos
Evguenia Alexandrova

Uncertainty Quantificatior⟩   ⟨ Models – UNEQUIvOCAL

Wouter Edeling
Anna Nikishova
Peter Coveney

Program Committe ₁ation

Ahmad Abdelfattah
Samsul Ariffin
   Abdul Karim
Evgenia Adamopoulou
Jaime Afonso Martins
Giuseppe Agapito
Ram Akella
Elisabete Alberdi Celaya
Luis Alexandre
Vassil Alexandrov
Evguenia Alexandrova
Hesham H. Ali
Julen Alvarez-Aramberri
Domingos Alves
Julio Amador Diaz Lopez
Stanislaw
   Ambroszkiewicz
Tomasz Andrysiak
Michael Antolovich
Hartwig Anzt
Hideo Aochi
Hamid Arabnejad
Rossella Arcucci
Khurshid Asghar
Marina Balakhontceva
Bartosz Balis
Krzysztof Banas
João Barroso
Dominik Bartuschat
Nuno Basurto
Pouria Behnoudfar
Joern Behrens
Adrian Bekasiewicz
Gebrai Bekdas
Stefano Beretta
Benjamin Berkels
Martino Bernard

Ba
Mari.
Jérémy
Robert Bu⟨
Michael Bur⟨₍ᵢ₎
Allah Bux
Aleksander Byrski
Cristiano Cabrita
Xing Cai
Barbara Calabrese
Jose Camata
Mario Cannataro
Alberto Cano
Pedro Jorge Sequeira
   Cardoso
Jeffrey Carver
Stefano Casarin
Manuel Castañón-Puga
Mauro Castelli
Eduardo Cesar
Nicholas Chancellor
Patrikakis Charalampos
Ehtzaz Chaudhry
Chuanfa Chen
Siew Ann Cheong
Andrey Chernykh
Lock-Yue Chew
Su Fong Chien
Marta Chinnici
Sung-Bae Cho
Michal Choras
Loo Chu Kiong

Neil Chue Hong
Svetlana Chuprina
Paola Cinnella
Noélia Correia
Adriano Cortes
Ana Cortes
Enrique
   Costa-Montenegro
David Coster
Helene Coullon
Peter Coveney
Attila Csikasz-Nagy
Loïc Cudennec
Javier Cuenca
Yifeng Cui
António Cunha
Ben Czaja
Pawel Czarnul
Flávio Martins
Bhaskar Dasgupta
Konstantinos Demestichas
Quanling Deng
Nilanjan Dey
Khaldoon Dhou
Jamie Diner
Jacek Dlugopolski
Simona Domesová
Riccardo Dondi
Craig C. Douglas
Linda Douw
Rafal Drezewski
Hans du Buf
Vitor Duarte
Richard Dwight
Wouter Edeling
Waleed Ejaz
Dina El-Reedy

Amgad Elsayed
Nahid Emad
Chriatian Engelmann
Gökhan Ertaylan
Alex Fedoseyev
Luis Manuel Fernández
Antonino Fiannaca
Christos
  Filelis-Papadopoulos
Rupert Ford
Piotr Frackiewicz
Martin Frank
Ruy Freitas Reis
Karl Frinkle
Haibin Fu
Kohei Fujita
Hiroshi Fujiwara
Takeshi Fukaya
Wlodzimierz Funika
Takashi Furumura
Ernst Fusch
Mohamed Gaber
David Gal
Marco Gallieri
Teresa Galvao
Akemi Galvez
Salvador García
Bartlomiej Gardas
Delia Garijo
Frédéric Gava
Piotr Gawron
Bernhard Geiger
Alex Gerbessiotis
Ivo Goncalves
Antonio Gonzalez Pardo
Jorge
  González-Domínguez
Yuriy Gorbachev
Pawel Gorecki
Michael Gowanlock
Manuel Grana
George Gravvanis
Derek Groen
Lutz Gross
Sophia
  Grundner-Culemann

Pedro Guerreiro
Tobias Guggemos
Xiaohu Guo
Piotr Gurgul
Filip Guzy
Pietro Hiram Guzzi
Zulfiqar Habib
Panagiotis Hadjidoukas
Masatoshi Hanai
John Hanley
Erik Hanson
Habibollah Haron
Carina Haupt
Claire Heaney
Alexander Heinecke
Jurjen Rienk Helmus
Álvaro Herrero
Bogumila Hnatkowska
Maximilian Höb
Erlend Hodneland
Olivier Hoenen
Paul Hofmann
Che-Lun Hung
Andres Iglesias
Takeshi Iwashita
Alireza Jahani
Momin Jamil
Vytautas Jancauskas
João Janeiro
Peter Janku
Fredrik Jansson
Jirí Jaroš
Caroline Jay
Shalu Jhanwar
Zhigang Jia
Chao Jin
Zhong Jin
David Johnson
Guido Juckeland
Maria Juliano
Edward J. Kansa
Aneta Karaivanova
Takahiro Katagiri
Timo Kehrer
Wayne Kelly
Christoph Kessler

Jakub Klikowski
Harald Koestler
Ivana Kolingerova
Georgy Kopanitsa
Gregor Kosec
Sotiris Kotsiantis
Ilias Kotsireas
Sergey Kovalchuk
Michal Koziarski
Slawomir Koziel
Rafal Kozik
Bartosz Krawczyk
Elisabeth Krueger
Valeria Krzhizhanovskaya
Pawel Ksieniewicz
Marek Kubalcík
Sebastian Kuckuk
Eileen Kuehn
Michael Kuhn
Michal Kulczewski
Krzysztof Kurowski
Massimo La Rosa
Yu-Kun Lai
Jalal Lakhlili
Roberto Lam
Anna-Lena Lamprecht
Rubin Landau
Johannes Langguth
Elisabeth Larsson
Michael Lees
Leifur Leifsson
Kenneth Leiter
Roy Lettieri
Andrew Lewis
Jingfa Li
Khang-Jie Liew
Hong Liu
Hui Liu
Yen-Chen Liu
Zhao Liu
Pengcheng Liu
James Liu
Marcelo Lobosco
Robert Lodder
Marcin Los
Stephane Louise

Frederic Loulergue
Paul Lu
Stefan Luding
Onnie Luk
Scott MacLachlan
Luca Magri
Imran Mahmood
Zuzana Majdisova
Alexander Malyshev
Muazzam Maqsood
Livia Marcellino
Tomas Margalef
Tiziana Margaria
Svetozar Margenov
Urszula
  Markowska-Kaczmar
Osni Marques
Carmen Marquez
Carlos Martinez-Ortiz
Paula Martins
Flávio Martins
Luke Mason
Pawel Matuszyk
Valerie Maxville
Wagner Meira Jr.
Roderick Melnik
Valentin Melnikov
Ivan Merelli
Choras Michal
Leandro Minku
Jaroslaw Miszczak
Janio Monteiro
Kourosh Modarresi
Fernando Monteiro
James Montgomery
Andrew Moore
Dariusz Mrozek
Peter Mueller
Khan Muhammad
Judit Muñoz
Philip Nadler
Hiromichi Nagao
Jethro Nagawkar
Kengo Nakajima
Ionel Michael Navon
Philipp Neumann

Mai Nguyen
Hoang Nguyen
Nancy Nichols
Anna Nikishova
Hitoshi Nishizawa
Brayton Noll
Algirdas Noreika
Enrique Onieva
Kenji Ono
Eneko Osaba
Aziz Ouaarab
Serban Ovidiu
Raymond Padmos
Wojciech Palacz
Ivan Palomares
Rongjiang Pan
Joao Papa
Nikela Papadopoulou
Marcin Paprzycki
David Pardo
Anna Paszynska
Maciej Paszynski
Abani Patra
Dana Petcu
Serge Petiton
Bernhard Pfahringer
Frank Phillipson
Juan C. Pichel
Anna
  Pietrenko-Dabrowska
Laércio L. Pilla
Armando Pinho
Tomasz Piontek
Yuri Pirola
Igor Podolak
Cristina Portales
Simon Portegies Zwart
Roland Potthast
Ela Pustulka-Hunt
Vladimir Puzyrev
Alexander Pyayt
Rick Quax
Cesar Quilodran Casas
Barbara Quintela
Ajaykumar Rajasekharan
Celia Ramos

Lukasz Rauch
Vishal Raul
Robin Richardson
Heike Riel
Sophie Robert
Luis M. Rocha
Joao Rodrigues
Daniel Rodriguez
Albert Romkes
Debraj Roy
Katarzyna Rycerz
Alberto Sanchez
Gabriele Santin
Alex Savio
Robert Schaback
Robert Schaefer
Rafal Scherer
Ulf D. Schiller
Bertil Schmidt
Martin Schreiber
Alexander Schug
Gabriela Schütz
Marinella Sciortino
Diego Sevilla
Angela Shiflet
Takashi Shimokawabe
Marcin Sieniek
Nazareen Sikkandar
  Basha
Anna Sikora
Janaína De Andrade Silva
Diana Sima
Robert Sinkovits
Haozhen Situ
Leszek Siwik
Vaclav Skala
Peter Sloot
Renata Slota
Grazyna Slusarczyk
Sucha Smanchat
Marek Smieja
Maciej Smolka
Bartlomiej Sniezynski
Isabel Sofia Brito
Katarzyna Stapor
Bogdan Staszewski

Jerzy Stefanowski
Dennis Stevenson
Tomasz Stopa
Achim Streit
Barbara Strug
Pawel Strumillo
Dante Suarez
Vishwas H. V. Subba Rao
Bongwon Suh
Diana Suleimenova
Ray Sun
Shuyu Sun
Vaidy Sunderam
Martin Swain
Alessandro Taberna
Ryszard Tadeusiewicz
Daisuke Takahashi
Zaid Tashman
Osamu Tatebe
Carlos Tavares Calafate
Kasim Tersic
Yonatan Afework
   Tesfahunegn
Jannis Teunissen
Stefan Thurner

Nestor Tiglao
Alfredo Tirado-Ramos
Arkadiusz Tomczyk
Mariusz Topolski
Paolo Trunfio
Ka-Wai Tsang
Hassan Ugail
Eirik Valseth
Pavel Varacha
Pierangelo Veltri
Raja Velu
Colin Venters
Gytis Vilutis
Peng Wang
Jianwu Wang
Shuangbu Wang
Rodrigo Weber
   dos Santos
Katarzyna
   Wegrzyn-Wolska
Mei Wen
Lars Wienbrandt
Mark Wijzenbroek
Peter Woehrmann
Szymon Wojciechowski

Maciej Woloszyn
Michal Wozniak
Maciej Wozniak
Yu Xia
Dunhui Xiao
Huilin Xing
Miguel Xochicale
Feng Xu
Wei Xue
Yoshifumi Yamamoto
Dongjia Yan
Xin-She Yang
Dongwei Ye
Wee Ping Yeo
Lihua You
Han Yu
Gábor Závodszky
Yao Zhang
H. Zhang
Jinghui Zhong
Sotirios Ziavras
Italo Zoppis
Chiara Zucco
Pawel Zyblewski
Karol Zyczkowski

Contents – Part IV

Classifier Learning from Difficult Data

Different Strategies of Fitting Logistic Regression for Positive
and Unlabelled Data 3
   Paweł Teisseyre, Jan Mielniczuk, and Małgorzata Łazęcka

Branch-and-Bound Search for Training Cascades of Classifiers........... 18
   Dariusz Sychel, Przemysław Klęsk, and Aneta Bera

Application of the Stochastic Gradient Method in the Construction
of the Main Components of PCA in the Task Diagnosis of Multiple
Sclerosis in Children... 35
   Mariusz Topolski

Grammatical Inference by Answer Set Programming.................. 45
   *Wojciech Wieczorek, Łukasz Strąk, Arkadiusz Nowakowski,
   and Olgierd Unold*

Dynamic Classifier Selection for Data with Skewed Class Distribution
Using Imbalance Ratio and Euclidean Distance..................... 59
   Paweł Zyblewski and Michał Woźniak

On Model Evaluation Under Non-constant Class Imbalance 74
   Jan Brabec, Tomáš Komárek, Vojtěch Franc, and Lukáš Machlica

A Correction Method of a Base Classifier Applied to Imbalanced
Data Classification 88
   Pawel Trajdos and Marek Kurzynski

Standard Decision Boundary in a Support-Domain of Fuzzy Classifier
Prediction for the Task of Imbalanced Data Classification 103
   Pawel Ksieniewicz

Employing One-Class SVM Classifier Ensemble for Imbalanced Data
Stream Classification... 117
   Jakub Klikowski and Michał Woźniak

Clustering and Weighted Scoring in Geometric Space Support Vector
Machine Ensemble for Highly Imbalanced Data Classification 128
   Paweł Ksieniewicz and Robert Burduk

Performance Analysis of Binarization Strategies for Multi-class Imbalanced
Data Classification   141
    Michał Żak and Michał Woźniak

Towards Network Anomaly Detection Using Graph Embedding   156
    Qingsai Xiao, Jian Liu, Quiyun Wang, Zhengwei Jiang, Xuren Wang,
    and Yepeng Yao

Maintenance and Security System for PLC Railway LED Sign
Communication Infrastructure.   170
    Tomasz Andrysiak and Łukasz Saganowski

Fingerprinting of URL Logs: Continuous User Authentication
from Behavioural Patterns   184
    Jakub Nowak, Taras Holotyak, Marcin Korytkowski, Rafał Scherer,
    and Slava Voloshynovskiy

On the Impact of Network Data Balancing in Cybersecurity Applications . . .   196
    Marek Pawlicki, Michał Choraś, Rafał Kozik, and Witold Hołubowicz

Pattern Recognition Model to Aid the Optimization of Dynamic
Spectrally-Spatially Flexible Optical Networks   211
    Paweł Ksieniewicz, Róża Goścień, Mirosław Klinkowski,
    and Krzysztof Walkowiak

Missing Features Reconstruction Using a Wasserstein Generative
Adversarial Imputation Network   225
    Magda Friedjungová, Daniel Vašata, Maksym Balatsko,
    and Marcel Jiřina

Complex Social Systems Through the Lens of Computational Science

Cooperation for Public Goods Under Uncertainty   243
    Jeroen Bruggeman and Rudolf Sprik

An Information-Theoretic and Dissipative Systems Approach to the Study
of Knowledge Diffusion and Emerging Complexity in Innovation Systems. . .   252
    Guillem Achermann, Gabriele De Luca, and Michele Simoni

Mapping the Port Influence Diffusion Patterns: A Case Study of Rotterdam,
Antwerp and Singapore   266
    Peng Peng and Feng Lu

Entropy-Based Measure for Influence Maximization in Temporal Networks . . .   277
    Radosław Michalski, Jarosław Jankowski, and Patryk Pazura

Evaluation of the Costs of Delayed Campaigns for Limiting the Spread
of Negative Content, Panic and Rumours in Complex Networks   291
    Jaroslaw Jankowski, Piotr Bartkow, Patryk Pazura, and Kamil Bortko

From Generality to Specificity: On Matter of Scale in Social Media
Topic Communities   305
    Danila Vaganov, Mariia Bardina, and Valentina Guleva

Computational Health

Hybrid Text Feature Modeling for Disease Group Prediction Using
Unstructured Physician Notes   321
    Gokul S. Krishnan and S. Sowmya Kamath

Early Signs of Critical Slowing Down in Heart Surface Electrograms
of Ventricular Fibrillation Victims   334
    Berend Nannes, Rick Quax, Hiroshi Ashikaga, Mélèze Hocini,
    Remi Dubois, Olivier Bernus, and Michel Haïssaguerre

A Comparison of Generalized Stochastic Milevsky-Promislov Mortality
Models with Continuous Non-Gaussian Filters   348
    Piotr Śliwka and Leslaw Socha

Ontology-Based Inference for Supporting Clinical Decisions
in Mental Health   363
    Diego Bettiol Yamada, Filipe Andrade Bernardi,
    Newton Shydeo Brandão Miyoshi, Inácia Bezerra de Lima,
    André Luiz Teixeira Vinci, Vinicius Tohoru Yoshiura,
    and Domingos Alves

Towards Prediction of Heart Arrhythmia Onset Using Machine Learning   376
    Agnieszka Kitlas Golińska, Wojciech Lesiński, Andrzej Przybylski,
    and Witold R. Rudnicki

Stroke ICU Patient Mortality Day Prediction   390
    Oleg Metsker, Vozniuk Igor, Georgy Kopanitsa, Elena Morozova,
    and Prohorova Maria

Universal Measure for Medical Image Quality Evaluation Based
on Gradient Approach   406
    Marzena Bielecka, Andrzej Bielecki, Rafał Obuchowicz,
    and Adam Piórkowski

Constructing Holistic Patient Flow Simulation Using System Approach   418
    Tesfamariam M. Abuhay, Oleg G. Metsker, Aleksey N. Yakovlev,
    and Sergey V. Kovalchuk

Investigating Coordination of Hospital Departments in Delivering
Healthcare for Acute Coronary Syndrome Patients Using Data-Driven
Network Analysis 430
  Tesfamariam M. Abuhay, Yemisrach G. Nigatie, Oleg G. Metsker,
  Aleksey N. Yakovlev, and Sergey V. Kovalchuk

A Machine Learning Approach to Short-Term Body Weight Prediction
in a Dietary Intervention Program 441
  Oladapo Babajide, Tawfik Hissam, Palczewska Anna,
  Gorbenko Anatoliy, Arne Astrup, J. Alfredo Martinez,
  Jean-Michel Oppert, and Thorkild I. A. Sørensen

An Analysis of Demographic Data in Irish Healthcare Domain
to Support Semantic Uplift. 456
  Kris McGlinn and Pamela Hussey

From Population to Subject-Specific Reference Intervals 468
  Murih Pusparum, Gökhan Ertaylan, and Olivier Thas

Analyzing the Spatial Distribution of Acute Coronary Syndrome Cases
Using Synthesized Data on Arterial Hypertension Prevalence 483
  Vasiliy N. Leonenko

The Atrial Fibrillation Risk Score for Hyperthyroidism Patients 495
  Ilya V. Derevitskii, Daria A. Savitskaya, Alina Y. Babenko,
  and Sergey V. Kovalchuk

Applicability of Machine Learning Methods to Multi-label Medical
Text Classification. 509
  Iuliia Lenivtceva, Evgenia Slasten, Mariya Kashina,
  and Georgy Kopanitsa

Machine Learning Approach for the Early Prediction
of the Risk of Overweight and Obesity in Young People 523
  Balbir Singh and Hissam Tawfik

Gait Abnormality Detection in People with Cerebral Palsy Using
an Uncertainty-Based State-Space Model 536
  Saikat Chakraborty, Noble Thomas, and Anup Nandy

Analyses of Public Health Databases via Clinical Pathway
Modelling: TBWEB 550
  Anderson C. Apunike, Lívia Oliveira-Ciabati, Tiago L. M. Sanches,
  Lariza L. de Oliveira, Mauro N. Sanchez, Rafael M. Galliez,
  and Domingos Alves

Preliminary Results on Pulmonary Tuberculosis Detection in Chest X-Ray
Using Convolutional Neural Networks.   563
    Márcio Eloi Colombo Filho, Rafael Mello Galliez,
    Filipe Andrade Bernardi, Lariza Laura de Oliveira, Afrânio Kritski,
    Marcel Koenigkam Santos, and Domingos Alves

Risk-Based AED Placement - Singapore Case   577
    Ivan Derevitskii, Nikita Kogtikov, Michael H. Lees, Wentong Cai,
    and Marcus E. H. Ong

Time Expressions Identification Without Human-Labeled Corpus
for Clinical Text Mining in Russian   591
    Anastasia A. Funkner and Sergey V. Kovalchuk

Experiencer Detection and Automated Extraction of a Family Disease
Tree from Medical Texts in Russian Language.   603
    Ksenia Balabaeva and Sergey Kovalchuk

Computational Methods for Emerging Problems in (dis-)Information Analysis

Machine Learning – The Results Are Not the only Thing that Matters!
What About Security, Explainability and Fairness?   615
    Michał Choraś, Marek Pawlicki, Damian Puchalski, and Rafał Kozik

Syntactic and Semantic Bias Detection and Countermeasures   629
    Roman Englert and Jörg Muschiol

Detecting Rumours in Disasters: An Imbalanced Learning Approach.   639
    Amir Ebrahimi Fard, Majid Mohammadi, and Bartel van de Walle

Sentiment Analysis for Fake News Detection by Means
of Neural Networks.   653
    Sebastian Kula, Michał Choraś, Rafał Kozik, Paweł Ksieniewicz,
    and Michał Woźniak

Author Index   667

Classifier Learning from Difficult Data

Different Strategies of Fitting Logistic Regression for Positive and Unlabelled Data

Paweł Teisseyre[1]([✉]) [ID], Jan Mielniczuk[1,2] [ID], and Małgorzata Łazęcka[1,2] [ID]

[1] Institute of Computer Science, Polish Academy of Sciences, Warsaw, Poland
{teisseyrep,miel,malgorzata.lazecka}@ipipan.waw.pl
[2] Faculty of Mathematics and Information Sciences, Warsaw University of Technology, Warsaw, Poland

Abstract. In the paper we revisit the problem of fitting logistic regression to positive and unlabelled data. There are two key contributions. First, a new light is shed on the properties of frequently used naive method (in which unlabelled examples are treated as negative). In particular we show that naive method is related to incorrect specification of the logistic model and consequently the parameters in naive method are shrunk towards zero. An interesting relationship between shrinkage parameter and label frequency is established. Second, we introduce a novel method of fitting logistic model based on simultaneous estimation of vector of coefficients and label frequency. Importantly, the proposed method does not require prior estimation, which is a major obstacle in positive unlabelled learning. The method is superior in predicting posterior probability to both naive method and weighted likelihood method for several benchmark data sets. Moreover, it yields consistently better estimator of label frequency than other two known methods. We also introduce simple but powerful representation of positive and unlabelled data under Selected Completely at Random assumption which yields straightforwardly most properties of such model.

Keywords: Positive unlabelled learning · Logistic regression · Empirical risk minimization · Misspecification

1 Introduction

Learning from positive and unlabelled data (PU learning) has attracted much interest within the machine learning literature as this type of data naturally arises in many applications (see e.g. [1]). In the case of PU data, we have an access to positive examples and unlabeled examples. Unlabeled examples can be either positive or negative. In this setting the true class label $Y \in \{0, 1\}$ is not observed directly. We only observe surrogate variable $S \in \{0, 1\}$, which indicates whether an example is labeled (and thus positive; $S = 1$) or unlabeled ($S = 0$). PU problem naturally occurs in under-reporting [2] which frequently happens in survey data, and it

© Springer Nature Switzerland AG 2020
V. V. Krzhizhanovskaya et al. (Eds.): ICCS 2020, LNCS 12140, pp. 3–17, 2020.
https://doi.org/10.1007/978-3-030-50423-6_1

refers to situation when some respondents fail to the answer a question truthfully. For example, imagine that we are interested in predicting an occurrence of some disease ($Y = 1$ denotes presence of disease and $Y = 0$ its absence) using some feature vector X. In some cases we only have an access to self-reported data [3], i.e. respondents answer to the question concerning the occurrence of the disease. Some of them admit to the disease truthfully ($S = 1 \implies Y = 1$) and the other group reports no disease ($S = 0$). The second group consists of respondents who suffer from disease but do not report it ($Y = 1, S = 0$) and those who really do not have a disease ($Y = 0, S = 0$). Under-reporting occurs due to a perceived social stigma concerning e.g. alcoholism, HIV disease or socially dangerous behaviours such as talking on the phone frequently while driving. PU data occur frequently in text classification problems [4–6]. When classifying user's web page preferences, some pages can be bookmarked as positive ($S = 1$) whereas all other pages are treated as unlabelled ($S = 0$). Among unlabelled pages, one can find pages that users visit ($Y = 1, S = 0$) as well as those which are avoided by users ($Y = 0, S = 0$). The third important example is a problem of disease gene identification which aims to find which genes from the human genome are causative for diseases [7,8]. In this case all the known disease genes are positive examples ($S = 1$), while all other candidates, generated by traditional linkage analysis, are unlabelled ($S = 0$). Several approaches exist to learn with PU data. A simplest approach is to treat S as a class label (this approach is called naive method or non-traditional classification) [9]. To organize terminology, learning with true class label Y will be called oracle method. Although this approach cannot be used in practice, it may serve as a benchmark method with which all considered methods are compared.

In this paper we focus on logistic regression. Despite its popularity, there is a lack of thorough analysis of different learning methods based on logistic regression for PU data. We present the following novel contributions. First, we analyse theoretically the naive method and its relationship with oracle method. We show that naive method is related to incorrect specification of the logistic model and we establish the connection between risk minimizers corresponding to naive and oracle methods, for certain relatively large class of distributions. Moreover, we show that parameters in naive method are shrunk towards zero and the amount of shrinkage depends on label frequency $c = P(S = 1|Y = 1)$. Secondly, we propose an intuitive method of parameter estimation in which we simultaneously estimate parameter vector and label frequency c (called joint method hereafter). The method does not require prior estimation which is a difficult task in PU learning [10,11]. Finally, we compare empirically the proposed method with two existing methods (naive method and the method based on optimizing weighted empirical risk, called briefly weighted method) with respect to estimation errors.

Finally, the popular taxonomy used in PU learning [1] differentiates between three categories of methods. The first group are postprocessing methods which first use naive method and then modify output probabilities using label frequency [9]. The second group are preprocessing methods that weigh the examples using label frequency [12–14]. We refer to [1] (Sect. 5.3.2) for a description of general empirical risk minimization framework in which the weights of observations depending on label frequency c, for any loss function are determined. The last group are methods incorporating label frequency into learning algorithms. A representative algorithm from this group is POSC4.5 [15], which is PU tree learning method. The three methods considered in this paper (naive, weighted and joint method) represent the above three categories, respectively.

This paper is organized as follows. In Sect. 2, we state the problem and discuss its variants and assumptions. In Sect. 3, we analyse three learning methods based on logistic regression in detail. Section 4 discusses the relationship between naive and oracle methods. We report the results of experiments in Sect. 5 and conclude the paper in Sect. 6. Technical details are stated in Sect. 7. Some additional experiments are described in Supplement[1].

2 Assumptions and Useful Representation for PU Data

In this work we consider single training data (STD) scenario, which can be described as follows. Let X be feature vector, $Y \in \{0, 1\}$ be a true class label and $S \in \{0, 1\}$ an indicator of whether an example is labelled ($S = 1$) or not ($S = 0$). We assume that there is some unknown distribution $P(Y, X, S)$ such that $(y_i, x_i, s_i), i = 1, \ldots, n$ is iid sample drawn from it and data $(x_i, s_i), i = 1, \ldots, n$, is observed. Thus, instead of a sample (x_i, y_i) which corresponds to classical classification task, we observe only sample (x_i, s_i), where s_i depends on (x_i, y_i). Only positive examples ($Y = 1$) can be labelled, i.e. $P(S = 1|X, Y = 0) = 0$. The true class label is observed only partially, i.e. when $S = 1$ we know that $Y = 1$, but when $S = 0$, then Y can be either 1 or 0. A commonly used assumption is SCAR (Selected Completely At Random) assumption which states that labelled examples are selected randomly from a set of positives examples, independently from X, i.e.

$$P(S = 1|Y = 1, X) = P(S = 1|Y = 1).$$

Note that this is equivalent to X and S being independent given Y (denoted $X \perp S|Y$) as $P(S = 1|Y = 0, X) = P(S = 1|Y = 0) = 0$. Parameter $c := P(S = 1|Y = 1)$ is called label frequency and plays an important role in PU learning. In the paper we introduce a useful representation of variable (X, S) under SCAR assumption. Namely, we show that S can be represented as

$$S = Y \cdot \varepsilon, \text{ where } \varepsilon \perp (X, Y) \text{ and } \varepsilon \sim Bern(1, p), \tag{1}$$

for a certain $0 < p < 1$ and $Bern(1, p)$ stands for Bernoulli distribution. Indeed, we have $S = Y\varepsilon \perp X$ given Y, as $\varepsilon \perp (X, Y)$ implies that $\varepsilon \perp X$ given Y. Moreover,

$$P(S = 1|Y = 1) = P(Y\varepsilon = 1|Y = 1) = P(\varepsilon = 1) = p.$$

Thus probability of success $P(\varepsilon = 1)$ coincides with c. Under SCAR assumption we have

$$P(Y = 1|X) = c^{-1}P(S = 1|X), \tag{2}$$

---

[1] https://github.com/teisseyrep/PUlogistic.

$$P(Y = 1|S = 0, X) = \frac{1-c}{c} \frac{P(S = 1|X)}{P(S = 0|X)} \qquad (3)$$

[9] and

$$P(X = x|Y = 1) = P(X = x|S = 1). \qquad (4)$$

[2]. Properties (2)–(4) are easily derivable when (1) is applied (see Sect. 7).

We also note that the assumed STD scenario should be distinguished from case-control scenario when two independent samples are observed: labeled sample consisting of independent observations drawn from distribution of X given $Y = 1$ and the second drawn from distribution of X. This is carefully discussed in [1]. Both PU scenarios should be also distinguished from semi-supervised scenario when besides fully observable sample from distribution of (X, Y) we also have at our disposal sample from distribution of X [16] or, in extreme case, we have full knowledge of distribution of X, see [17] and references therein. One of the main goals of PU learning is to estimate the posterior probability $f(x) := P(Y = 1|X = x)$. The problem is discussed in the following sections.

3   Logistic Regression for PU Data

In this section we present three different methods of estimating $f(x) := P(Y = 1|X = x)$ using logistic loss. When data is fully observed the natural way to learn a model is to consider risk for logistic loss

$$R(b) = -E_{X,Y}[Y \log(\sigma(X^T b)) + (1 - Y) \log(1 - \sigma(X^T b))], \qquad (5)$$

where $\sigma(s) = 1/(1 + \exp(-s))$ and minimize its empirical version. This will be called oracle method. Note that using logistic loss function in the definition of $R(b)$ above corresponds to fitting logistic regression using Maximum Likelihood (ML) method. Obviously, for PU data, this approach is not feasible as we do not observe Y and inferential procedures have to be based on (S, X). The simplest approach (called naive estimation or non-traditional estimation) is thus to consider risk

$$R_1(b) = -E_{X,S}[S \log(\sigma(X^T b)) + (1 - S) \log(1 - \sigma(X^T b))] \qquad (6)$$

and the corresponding empirical risk

$$\hat{R}_1(b) = -\frac{1}{n} \sum_{i=1}^{n} [s_i \log(\sigma(x_i^T b)) + (1 - s_i) \log(1 - \sigma(x_i^T b))],$$

which can be directly optimized. In Sect. 4 we study the relationship between minimizers of $R(b)$ and $R_1(b)$

$$b^* = \arg\min_b R(b), \quad b_1^* = \arg\min_b R_1(b).$$

It turns out that for certain, relatively large, class of distributions of X, $b_1^* = \eta b^*$, for some $\eta \in R$ (i.e. b_1^* and b^* are collinear). Moreover, when predictors X are normal and when (Y, X) corresponds to logistic model, we establish the relationship between η and label frequency c which shows that $\eta < 1$ and thus naive approach leads to shrinking of vector b^*. To estimate the posterior $f(x) = P(Y = 1|X = x)$ using naive estimation, we perform a two-step procedure, i.e. we first estimate $\hat{b}_{naive} = \arg\min_b \hat{R}_1(b)$ and then let $\hat{f}_{naive}(x) := c^{-1}\sigma(x^T \hat{b}_{naive})$, where unknown c has to be estimated using some external procedure. Note that even when (Y, X) corresponds to logistic regression model, b^* and whence posterior probability is not consistently estimated by naive method.

The second approach is based on weighted empirical risk minimization. As mentioned before, the empirical counterpart of risk $R(b)$ cannot be directly optimized as we do not observe Y. However it can be shown [1] that

$$R(b) = -P(S = 1)E_{X|S=1}\left[\frac{1}{c}\log\sigma(X^T b) + \left(1 - \frac{1}{c}\right)\log(1 - \sigma(X^T b)))\right]$$
$$+ P(S = 0)E_{X|S=0}\log(1 - \sigma(X^T b)).$$

The risk above is approximated by

$$\hat{R}(b) = -\frac{1}{n}\sum_{i:s_i=1}\left[\frac{1}{c}\log\sigma(x_i^T b) + \left(1 - \frac{1}{c}\right)\log(1 - \sigma(x_i^T b))\right]$$
$$+ \frac{1}{n}\sum_{i:s_i=0}\log(1 - \sigma(x_i^T b)).$$

This means that all unlabelled examples are assigned weight 1, whereas each labelled example is treated as a combination of positive example with weight $1/c$ and negative example with weight $(1 - 1/c)$. The posterior estimator is defined as $\hat{f}_{weighted}(x) = \sigma(x^T \hat{b}_{weighted})$, where $\hat{b}_{weighted} = \arg\min_b \hat{R}(b)$. The above idea of weighted empirical risk minimization was used in case-control scenario for which the above formulas have slightly different forms, see [12,13].

In the paper we propose a novel, intuitive approach, called joint method (name refers to joint estimation of b and c). In this method we avail ourselves of an important feature of logistic regression, namely that posterior probability is directly parametrized. This in turn allows to directly plug in the equation (2) into the risk function

$$R_2(b, c) = -E_{X,S}[S\log(c\sigma(X^T b)) + (1 - S)\log(1 - c\sigma(X^T b))].$$

The empirical counterpart of the above risk is

$$\hat{R}_2(b, c) = -\frac{1}{n}\sum_{i=1}^{n}[s_i\log(c\sigma(x_i^T b)) + (1 - s_i)\log(1 - c\sigma(x_i^T b))].$$

The empirical risk $\hat{R}_2(b, c)$ can be optimized with respect to b if c is assumed to be known or can be optimized simultaneously with respect to both b and c.

In the latter case the posterior estimator is $\hat{f}_{\text{joint}}(x) := \sigma(x^T \hat{b}_{\text{joint}})$ where $(\hat{b}_{\text{joint}}, \hat{c}_{\text{joint}}) = \arg\min_{b,c} \hat{R}_2(b,c)$. Note that when conditional distribution of Y given X is governed by logistic model i.e. $P(Y = 1|X = x) = \sigma(\beta^T x)$, for some unknown vector β, then in view of (2) $P(S = 1|X = x) = c\sigma(\beta^T x)$ and $\hat{R}_2(b,c)$ is log-likelihood for observed sample (x_i, s_i). Whence under regularity conditions, maximisation of $\hat{R}_2(b,c)$ yields consistent estimator of (β, c) in view of known results in consistency of maximum likelihood method. To optimize function $\hat{R}_2$ we use BFGS algorithm, which requires the knowledge of functional form of gradient. The partial derivatives of $\hat{R}_2$ are given by

$$\frac{\partial \hat{R}_2(b,c)}{\partial b} = -\frac{1}{n}\sum_{i=1}^{n} x_i \sigma(x_i^T b)(1 - \sigma(x_i^T b)) \left[\frac{s_i - c\sigma(x_i^T b)}{\sigma(x_i^T b)(1 - c\sigma(x_i^T b))} \right],$$

$$\frac{\partial \hat{R}_2(b,c)}{\partial c} = -\frac{1}{n}\sum_{i=1}^{n} \left[\frac{s_i}{c} - \frac{(1 - s_i)\sigma(x_i^T b)}{1 - c\sigma(x_i^T b)} \right].$$

For $c = 1$, the first equation above reduces to well-known formula for gradient of the maximum likelihood function for standard logistic regression. In general we observe quick convergence of BFGS algorithm. The proposed method is described by the following scheme.

Algorithm 1. Joint method for posterior estimation

Input   : Observed data (x_i, s_i), $i = 1, \ldots, n$; new instance x
$(\hat{b}_{\text{joint}}, \hat{c}_{\text{joint}}) = \arg\min_{b,c} -\frac{1}{n}\sum_{i=1}^{n}[s_i \log(c\sigma(x_i^T b)) + (1 - s_i)\log(1 - c\sigma(x_i^T b))]$
Compute $\hat{f}_{\text{joint}}(x) := \sigma(x^T \hat{b}_{\text{joint}})$
Output   : $\hat{f}_{\text{joint}}(x)$

Finally, we note that the joint method above is loosely related to non-linear regression fit in dose-response analysis when generalized logistic curve is fitted [18].

4   Naive Method as an Incorrect Specification of Logistic Regression

In this Section we show that naive method is related to incorrect specification of the logistic model and that the corresponding parameter vector will be shrunk towards zero for relatively large class of distributions of X. Moreover, we establish the relationship between the amount of shrinkage and label frequency.

Assume for simplicity of exposition that components of X are non-constant random variables (in the case when one of predictors is a dummy variable which allows for the intercept in the model, collinearity in (9) corresponds to vector of predictors with dummy variable omitted) and assume that regression function of Y given X has the following form

$$P(Y = 1|X = x) = q(\beta^T X), \tag{7}$$

for a certain response function q taking its values in $(0, 1)$ and a certain $\beta \in R^p$. We note that when oracle method (5) is correctly specified, i.e. $q(\cdot) = \sigma(\cdot)$, then $\beta = b^*$ (cf [19]). Here we consider more general situation in which we may have $q(\cdot) \neq \sigma(\cdot)$. Under SCAR assumption, $P(S = 1|X = x) = cq(\beta^T X)$ and thus when $cq(\cdot) \neq \sigma(\cdot)$ then maximising $\widehat{R}_1(b)$ corresponds to fitting misspecified logistic model to (X, S). Importantly, this model is misspecified even if the oracle model is correctly specified. Observe that in this case shrinking of parameters is intuitive as they have to move towards 0 to account for diminished ($c < 1$) aposteriori probability. We explain in the following why misspecified fit, which occurs frequently in practice may still lead to reasonable results. Assume namely that distribution of X satisfies linear regression condition (LRC)

$$E(X|\beta^T X = x) = wx + w_0 \tag{8}$$

for a certain $w_0, w \in R^p$. Note that (8) has to be satisfied for a true β only. LRC is fulfilled (for all β) by normal distribution, and more generally, by a larger class of elliptically contoured distributions (multivariate t-Student distribution is a representative example). Then it follows (see e.g. [20])

$$b_1^* = \eta\beta \tag{9}$$

and $\eta \neq 0$ provided $\text{Cov}(Y, X) \neq 0$. In this case true vector β and its projection on a logistic model are collinear which partly explains why logistic classification works even when data does not follow logistic model. When oracle method (5) is correctly specified, i.e. $q(\cdot) = \sigma(\cdot)$, then (9) can be written as

$$b_1^* = \eta b^* = \eta\beta, \tag{10}$$

i.e. risk minimizers corresponding to naive and oracle methods are collinear. In the following we investigate the relationship between label frequency c and collinearity factor η. Intuition suggests that small c should result in shrinking of estimators towards zero. First, we have a general formula (see [19] for derivation) describing the relationship between c and η when (7) holds

$$\frac{1}{c} = \frac{E_X[\sigma(X^T\beta)X_j]}{E_X[\sigma(X^T b_1^*)X_j]} = \frac{E_X[\sigma(X^T\beta)X_j]}{E_X[\sigma(X^T\eta\beta)X_j]},$$

for any j, where X_j is j-th coordinate of $X = (X_1, \ldots, X_p)$. Unfortunately, the above formula does not yield simple relationship between c and η. Some additional assumptions are needed to find more revealing one. In the case when X has normal distribution $N(0, \Sigma)$ it follows from [20] together with (2) that the following equality holds

$$\frac{E\sigma'(\beta^T X)}{E\sigma'(\eta\beta^T X)} = \frac{\eta}{c}, \tag{11}$$

where $\sigma'(s)$ denotes derivative of $\sigma(s)$ wrt to s. This is easily seen to be a corollary of Stein's lemma stating that $\text{Cov}(h(Z_1), Z_2) = \text{Cov}(Z_1, Z_2)Eh'(Z_1)$ for bivariate normal (Z_1, Z_2). Equation (11) can be used to find upper and lower bounds for η. Namely, we prove the following Theorem.

Theorem 1. *Assume that X follows normal distribution $N(0, \Sigma)$ and that linear regression condition holds (8). Then*

$$4cE\sigma'(\beta^T X) \leq \eta \leq c\frac{E\sigma'(\beta^T X)}{E\sigma'(c\beta^T X)} \leq c. \qquad (12)$$

Note that RHS inequality in (1) yields the lower bound on the amount of shrinkage of true vector β^* whereas LHS gives a lower bound on this amount.

Proof. Let $Z = \beta^T X$ and note that Z has normal distribution $N(0, a^2)$ with $a^2 = \beta^T \Sigma \beta$. It follows from the fact that $\sigma'(s) = \sigma(s)(1 - \sigma(s))$ is nonincreasing for $s > 0$ that function $h(\lambda) = E\sigma'(\lambda Z)$ is non-increasing. This justifies the last equality on the right as $c \leq 1$. Define $g(\lambda) = h(1) - (\lambda/c)h(\lambda)$ and note that $g(0) = h(1) > 0$, $g(c) \leq 0$ and g is continuous. Thus for a certain $\lambda_0 \in [0, c]$ it holds that $g(\lambda_0) = 0$ and it follows from (11) and uniqueness of projection that $\eta = \lambda_0$. In order to prove the RHS inequality it is enough to prove that $g(\lambda)$ is convex as then $\lambda_0 \leq \lambda^*$, where λ^* is a point at which a line $h(1) - \lambda h(c)/c$ joining points $(0, g(0))$ and $(c, g(c))$ crosses x-axis. As $\lambda^* = (h(1)/h(c))c$ the inequality follows. Convexity of g follows from concavity of $\lambda h(\lambda)$ which is proved in Supplement. In order to prove the left inequality it is enough to observe that $\sigma'(x) \leq 1/4$ and use (11) again.

Note for $c \to 0$ the ratio of the lower and upper bound tends to 1 as $E\sigma'(c\beta^T X) \to 1/4$. To illustrate the above theoretical result we performed simulation experiment in which we artificially generated a sample of size $n = 10^6$ in such a way that X followed 3-dimensional standard normal distribution and Y was generated from (7) with $q(\cdot) = \sigma(\cdot)$, with known β. Then $Z = \beta^T X$ has $N(0, ||\beta||^2)$ distribution and the bounds in (12) depend only on c and $||\beta||$. Figure 1 shows how collinearity parameter η and the corresponding bounds depend on c, for three different norms $||\beta||$. Note that the bounds become tighter for smaller $||\beta||$ and smaller c. Secondly, for small c, the lower bound is nearly optimal.

5 Experiments

5.1 Datasets

We use 9 popular benchmark datasets from UCI repository[2]. To create PU datasets from the completely labelled datasets, the positive examples are selected to be labelled with label frequencies $c = 0.1, 0.2, \ldots, 0.9$. For each label frequency c

---

[2] https://archive.ics.uci.edu/ml/datasets.php.

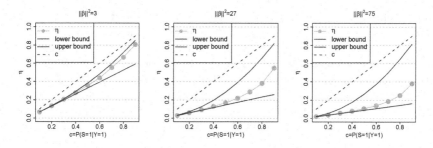

Fig. 1. Shrinkage parameter η wrt c for simulated dataset for $n = 10^6$.

we generated 100 PU datasets labelling randomly elements having $Y = 1$ with probability c and then averaged the results over 100 repetitions.

In addition, we consider one artificial dataset having n observations, generated as follows. Feature vector X was drawn from 3-dimensional standard normal distribution and Y was simulated from (7) with $q(\cdot) = \sigma(\cdot)$, with known $\beta = (1, 1, 1)$. This corresponds to correct specification of the oracle method. The observed variable S was labelled as 1 for elements having $Y = 1$ with probability c. Note however, that in view of discussion in Sect. 4, the naive model is incorrectly specified. Moreover, recall that in this case $\beta = b^* = \arg \min R(b)$. The main advantage of using artificial data is that β (and thus also b^*) is known and thus we can analyse the estimation error for the considered methods. For artificial dataset, we experimented with different values of c and n.

5.2 Methods and Evaluation Measures

The aim of the experiments is to compare the three methods of learning parameters in logistic regression: naive, weighted and joint. Our implementation of the discussed methods is available at https://github.com/teisseyrep/PUlogistic. Our main goal is to investigate how the considered methods relate to the oracle method, corresponding to idealized situation in which we have an access to Y. In view of this, as an evaluation measure we use approximation error for posterior defined as $AE = n^{-1} \sum_{i=1}^{n} |\hat{f}_{\text{oracle}}(x_i) - \hat{f}_{\text{method}}(x_i)|$, where 'method' corresponds to one of the considered methods (naive, weighted or joint), i.e. $\hat{f}_{\text{naive}}(x) := c^{-1} \sigma(x^T \hat{b}_{\text{naive}})$, $\hat{f}_{\text{weighted}}(x_i) := \sigma(x_i^T \hat{b}_{\text{weighted}})$ or $\hat{f}_{\text{joint}}(x_i) := \sigma(x_i^T \hat{b}_{\text{joint}})$. The oracle classifier is defined as $\hat{f}_{\text{oracle}}(x_i) := \sigma(x_i^T \hat{b}_{\text{oracle}})$, where $\hat{b}_{\text{oracle}}$ is minimizer of empirical version of (5). Estimation error for posterior, defined above, measures how accurate we can approximate the oracle classifier when using S instead of true class label Y. We consider two scenarios. In the first one we assume that c is known and we only estimate parameters corresponding to vector X. This setting corresponds to known prior probability $P(Y = 1)$ (c can be estimated accurately when prior is known via equation $c = P(S = 1)/P(Y = 1)$ by plugging-in corresponding fraction for $P(S = 1)$). In the second more realistic scenario, c is unknown and is estimated from data. For joint method we jointly minimize empirical risk $\widehat{R}_2(b, c)$ with respect to b and c. For two remaining methods (naive and weighted) we use external methods of estimation of c. We employ two methods; the first one was proposed by Elkan and Noto [9] (called EN) is based on averaging predictions of naive classifier over labeled examples for validation data. The second method, described in recent paper [11], is based on optimizing a lower bound of c via top-down decision tree induction (this method will be called TI). In order to analyse prediction performance of the proposed methods, we calculate AUC (Area Under ROC curve) of classifiers based on $\hat{f}_{\text{method}}$ on independent test set.

For artificial datasets, the true parameter β is known so we can analyse mean estimation error defined as $EE = p^{-1} \sum_{j=1}^{p} |\hat{b}_j - \beta_j|$, where $\hat{b}$ corresponds to one of the considered methods. Moreover, we consider an angle between β and $\hat{b}$. In view of property (9) the angle should be small, for sufficiently large sample size. Finally, let us note, that some real datasets may contain large number of features, so to make the estimation procedures more stable, we first performed feature selection. We used filter method recommended in [21] based on mutual information and select top $t = 3, 5, 10$ features for each dataset (we present the results for $t = 5$, the results for other t are similar and are presented in Supplement). This step is common for all considered methods.

5.3   Results

First, we analyse how the approximation errors for posterior depend on c, for real datasets (Fig. 2). We show the results for unknown c, the results for known c are presented in Supplement https://github.com/teisseyrep/PUlogistic. For unknown c, estimation of label frequency plays an important role. We observe that the performance curves vary depending on the method used. For most datasets, TI method outperforms EN, which is consistent with experiments described in [11], an exception is spambase for which TI works poorly. Importantly, joint method is a clear winner for most of the datasets, what suggests that simultaneous estimation of c and b is more effective than performing these two steps separately. Its superiority is frequently quite dramatic (see diabetes, credit-g and spambase). For most datasets, we observe the deterioration in posterior approximation when c becomes smaller. This is concordant with expectations, as for small c, the level of noise in observed variable S increases (cf Eq. (1)) and thus the gap between oracle and naive methods increases.

Tables 1 and 2 show values of AUC, for cases of known and unknown c, respectively. The results are averaged over 100 repetitions. In each repetition, we randomly chose $c \in (0, 1)$, then generate PU dataset and finally split it into training and testing subsets. For naive and weighted methods, c is estimated using TI algorithm (the performance for EN algorithm is generally worse and thus not presented in the Table). The last row contains averaged ranks, the larger the rank for AUC the better. The best method from three (naive, weighted and joint method) is in bold. As expected, the oracle method is an overall winner. The differences between the remaining methods are not very pronounced. Surprisingly, naive and joint methods work in most cases on par, whereas weighted

Table 1. AUC, known c

	Oracle	Joint	Naive	Weighted
Breastc	0.993	0.981	**0.987**	0.974
Diabetes	0.821	0.805	**0.808**	0.805
Heart-c	0.879	0.847	0.849	**0.850**
Credit-a	0.914	0.875	**0.899**	0.891
Credit-g	0.740	0.726	**0.727**	0.725
Adult	0.874	**0.874**	0.869	0.874
Vote	0.973	**0.974**	0.968	0.970
Wdbc	0.987	**0.981**	0.971	0.970
Spambase	0.911	**0.914**	0.892	0.899
Rank	3.8	**2.4**	2.1	1.7

Table 2. AUC (est. c)

Oracle	Joint	Naive	Weighted
0.993	0.983	**0.988**	0.977
0.821	0.798	**0.805**	0.796
0.879	0.843	0.850	**0.853**
0.914	0.889	**0.899**	0.897
0.740	0.724	**0.730**	0.718
0.874	**0.872**	0.869	0.863
0.973	0.972	0.968	**0.977**
0.987	**0.981**	0.969	0.973
0.911	**0.913**	0.893	0.856
3.8	**2.2**	**2.2**	1.8

Table 3. $|c - \hat{c}|$

EN	TI	Joint
0.060	0.064	**0.030**
0.234	0.169	**0.071**
0.138	0.121	**0.043**
0.125	0.130	0.317
0.287	0.261	**0.143**
0.244	0.214	**0.059**
0.044	0.088	**0.024**
0.099	0.068	**0.033**
0.189	0.267	**0.033**
2.4	2.3	**1.2**

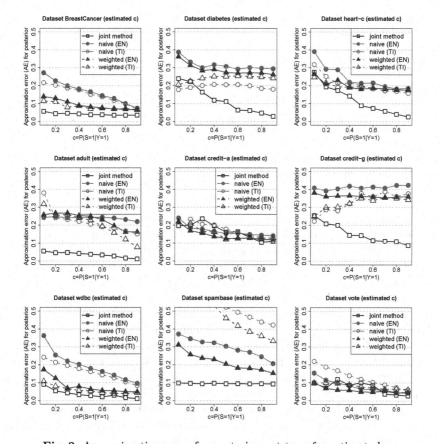

Fig. 2. Approximation error for posterior wrt to c, for estimated c.

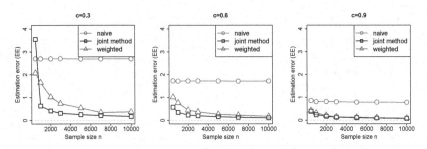

Fig. 3. Mean absolute error $p^{-1} \sum_{j=1}^{p} |\hat{b}_j - \beta|$ wrt to sample size n, where $\hat{b}$ corresponds to one of the methods: naive, weighted and joint method.

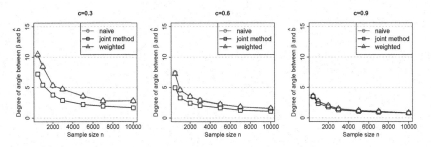

Fig. 4. Degree of angle between β and $\hat{b}$ wrt to sample size n, where $\hat{b}$ corresponds to one of the methods: naive, weighted and joint.

method performs slightly worse. The advantage of joint method is the most pronounced for spambase, for which we also observed superior performance of the joint method wrt approximation error (Fig. 2, bottom panel). Finally, joint method turns out to be effective for estimating c (Table 3)- the estimation errors for joint method are smaller than for TI and EN, for almost all datasets.

Figures 3 and 4 show results for artificial data, for $c = 0.3, 0.6, 0.9$, respectively. Mean estimation error converges to zero with sample size for weighted and joint methods (Fig. 3) and the convergence for joint method is faster. As expected, the estimation error for naive method is much larger than for joint and weighted methods, which is due to incorrect specification of the logistic regression. Note that weighted and joint methods account for wrong specification and therefore both methods perform better. Next we analysed an angle between true β (or equivalently b^*) and $\hat{b}$. Although the naive method does not recover the true signal β, it is able to consistently estimate the direction of β. Indeed the angle for naive method converges to zero with sample size (Fig. 4), which is in line with property (9). Interestingly the speed of converge for weighted method is nearly the same as for naive method, whereas the convergence for joint method is a bit faster.

6   Conclusions

We analysed three different approaches to fitting logistic regression model for PU data. We study theoretically the naive method. Although it does not estimate the true signal β consistently, it is able to consistently estimate the direction of β. This property can be particularly useful in the context of feature selection, where consistent estimation of the direction allows to discover the true significant features - this issue is left for future research. We have shown that under mild assumptions, risk minimizers corresponding to naive and oracle methods are collinear and the collinearity factor η is related to label frequency c. Moreover, we proposed novel method that allows to estimate parameter vector and label frequency c simultaneously. The proposed joint method achieves the smallest approximation error, which indicates that it is the closest to the oracle method among considered methods. Secondly, the joint method, unlike weighted and naive methods, does not require using external procedures to estimate c. Importantly, it outperforms the two existing methods (EN and TI) wrt to estimation error for c. In view of above, joint method can be recommended in practice, especially for estimating posterior probability and c; the differences in AUC for classifiers between the considered methods are not very pronounced.

7   Proofs

Equation (2) follows from

$$P(S = 1|X = x) = P(Y\varepsilon = 1|X = x) = P(Y = 1, \varepsilon = 1|X = x)$$
$$= P(Y = 1|X = x)P(\varepsilon = 1|X = x) = P(Y = 1|X = x)P(\varepsilon = 1)$$
$$= P(Y = 1|X = x)P(S = 1|Y = 1).$$

The third equality follows from conditional independence of Y and ε given X.
    To prove (3), note that $P(Y = 1|S = 0, X)$ can be written as

$$\frac{P(Y = 1, \varepsilon = 0, X)}{P(S = 0, X)} = \frac{P(\varepsilon = 0)}{P(\varepsilon = 1)} \frac{P(Y = 1, X)P(\varepsilon = 1)}{P(S = 0, X)}$$
$$= \frac{P(\varepsilon = 0)}{P(\varepsilon = 1)} \frac{P(Y = 1, \varepsilon = 1, X)}{P(S = 0, X)} \frac{1 - c}{c} \frac{P(S = 1, X)}{P(S = 0, X)} = \frac{1 - c}{c} \frac{P(S = 1|X)}{P(S = 0|X)},$$

where the second to last equality follows from $P(\varepsilon = 0)/P(\varepsilon = 1) = (1 - c)/c$.
To prove (4) we write

$$P(X = x|S = 1) = P(X = x|Y = 1, \varepsilon = 1) = P(X = x|Y = 1).$$

The third equality follows from conditional independence of X and ε given Y.

References

1. Bekker, J., Davis, J.: Learning from positive and unlabeled data: a survey (2018)
2. Sechidis, K., Sperrin, M., Petherick, E.S., Luján, M., Brown, G.: Dealing with under-reported variables: an information theoretic solution. Int. J. Approx. Reason. **85**, 159–177 (2017)
3. Onur, I., Velamuri, M.: The gap between self-reported and objective measures of disease status in India. PLOS ONE **13**(8), 1–18 (2018)
4. Liu, B., Dai, Y., Li, X., Lee, W.S., Yu, P.S.: Building text classifiers using positive and unlabeled examples. In: Proceedings of the Third IEEE International Conference on Data Mining, ICDM 2003, p. 179 (2003)
5. Fung, G.P.C., Yu, J.X., Lu, H., Yu, P.S.: Text classification without negative examples revisit. IEEE Trans. Knowl. Data Eng. **18**(1), 6–20 (2006)
6. Li, X., Liu, B.: Learning to classify texts using positive and unlabeled data. In: Proceedings of the 18th International Joint Conference on Artificial Intelligence, pp. 587–592 (2003)
7. Mordelet, F., Vert, J.-P.: ProDiGe: prioritization of disease genes with multi-task machine learning from positive and unlabeled examples. BMC Bioinformatics **12**(1), 389 (2011)
8. Cerulo, L., Elkan, C., Ceccarelli, M.: Learning gene regulatory networks from only positive and unlabeled data. BMC Bioinformatics **11**, 228 (2010)
9. Elkan, C., Noto, K.: Learning classifiers from only positive and unlabeled data. In: Proceedings of the 14th ACM SIGKDD International Conference on Knowledge Discovery and Data Mining, KDD 2008, pp. 213–220 (2008)
10. du Plessis, M.C., Niu, G., Sugiyama, M.: Class-prior estimation for learning from positive and unlabeled data. Mach. Learn. **106**(4), 463–492 (2016). https://doi.org/10.1007/s10994-016-5604-6
11. Bekker, J., Davis, J.: Estimating the class prior in positive and unlabeled data through decision tree induction. In: Proceedings of the 32th AAAI Conference on Artificial Intelligence, February 2018
12. Steinberg, D., Cardell, N.S.: Estimating logistic regression models when the dependent variable has no variance. Commun. Stat. Theory Methods **21**(2), 423–450 (1992)
13. Lancaster, T., Imbens, G.: Case-control studies with contaminated controls. J. Econom. **71**(1), 145–160 (1996)
14. Kiryo, R., Niu, G., du Plessis, M.C., Sugiyama, M.: Positive-unlabeled learning with non-negative risk estimator. In: Proceedings of the 31st International Conference on Neural Information Processing Systems, NIPS 2017, pp. 1674–1684 (2017)
15. Denis, F., Gilleron, R., Letouzey, F.: Learning from positive and unlabeled examples. Theoret. Comput. Sci. **348**(1), 70–83 (2005)
16. Chapelle, O., Schölkopf, B., Zien, A.: Semi-Supervised Learning. The MIT Press, Cambridge (2010)
17. Candès, E., Fan, Y., Janson, L., Lv, J.: Panning for gold: model-x knockoffs for high-dimensional controlled variable selection. Manuscript (2018)
18. Gottschalk, P.G., Dunn, J.R.: The five-parameter logistic: a characterization and comparison with the four-parameter logistic. Anal. Biochem. **343**(1), 54–65 (2005)

19. Mielniczuk, J., Teisseyre, P.: What do we choose when we err? Model selection and testing for misspecified logistic regression revisited. In: Matwin, S., Mielniczuk, J. (eds.) Challenges in Computational Statistics and Data Mining. SCI, vol. 605, pp. 271–296. Springer, Cham (2016). https://doi.org/10.1007/978-3-319-18781-5_15
20. Kubkowski, M., Mielniczuk, J.: Active set of predictors for misspecified logistic regression. Statistics **51**, 1023–1045 (2017)
21. Sechidis, K., Brown, G.: Simple strategies for semi-supervised feature selection. Mach. Learn. **107**(2), 357–395 (2017). https://doi.org/10.1007/s10994-017-5648-2

Branch-and-Bound Search for Training Cascades of Classifiers

Dariusz Sychel$^{(\boxtimes)}$ ⓘ, Przemysław Klęsk ⓘ, and Aneta Bera ⓘ

Faculty of Computer Science and Information Technology, West Pomeranian
University of Technology, ul. Żołnierska 49, 71-210 Szczecin, Poland
{dsychel,pklesk,abera}@wi.zut.edu.pl

Abstract. We propose a general algorithm that treats cascade training
as a tree search process working according to the *branch-and-bound* technique. The algorithm allows to reduce the *expected number of features*
used by an operating cascade—a key quantity we focus on in the paper.
While searching, we observe suitable lower bounds on partial expectations and prune tree branches that cannot improve the best-so-far result.
Both exact and approximate variants of the approach are formulated.
Experiments pertain to cascades trained to be face or letter detectors
with Haar-like features or Zernike moments being the input information, respectively. Results confirm shorter operating times of cascades
obtained owing to the reduction in the number of extracted features.

Keywords: Cascade of classifiers · Branch-and-bound tree search ·
Expected number of features

1 Introduction

Branch-and-bound technique is a useful tool in computer science. Multiple application examples can be named—let us mention DNA regulatory motif finding [8]
and α-β pruning in games, just to give two examples from quite remote fields.
In this paper we adopt the technique to train cascades of classifiers.

Cascades were in principle designed to work as classifying systems operating
under the following two conditions: (1) very large number of incoming requests,
(2) significant classes imbalance. The second condition should not be seen as a
difficulty but rather a favorable setting that makes the whole idea viable. Namely,
a cascade should vary its computational efforts depending on the contents of an
object to be classified. Objects that are obvious negatives (non-targets) should
be recognized fast, using only a few features extracted. Targets, or objects resem-
bling, them are allowed to employ more features and time for computations.

Despite the development of deep learning, recent literature shows that cascades of classifiers are still widely applied in detection systems or batch classification jobs. Let us list a few examples: crowd analysis and people counting [1],

This work was financed by the National Science Centre, Poland. Research project
no.: 2016/21/B/ST6/01495.

V. V. Krzhizhanovskaya et al. (Eds.): ICCS 2020, LNCS 12140, pp. 18–34, 2020.
https://doi.org/10.1007/978-3-030-50423-6_2

human detection in thermal images [6], localization of white blood cells [4], eye tracking [9,11], detection of birds near high power electric lines [7].

There exist a certain *average* value of the computational cost incurred by an operating cascade. It can be mathematically defined as an *expected value* and, in fact, calculated explicitly for a given cascade (we do this in Sect. 2.3) in terms of: the number of features applied on successive stages, false alarm and detection rates on successive stages, probability distribution from which the data is drawn. Since the true distribution underlying the data is typically unknown in practice, the exact expected value cannot be determined. Interestingly though, it can be accurately approximated using just the feature counts and false alarm rates.

Training procedures for cascades are time-consuming, taking days or even weeks. As Viola and Jones noted in their pionieering work [18], cascade training is a difficult combinatorial optimization involving many parameters: number of stages, number of features on successive stages, selection of those features, and finally decision thresholds. The problem has not been ultimately solved yet. Viola and Jones tackled it by imposing the final requirements the whole cascade should meet in order to be accepted, defined by a pair of numbers (A, D), where A denotes the largest allowed false alarm rate (FAR), and D the smallest allowed detection rate (sensitivity). Due to probabilistic properties of cascade structure, one can translate final requirements onto per-stage requirements as geometric means: $a_{max} = A^{1/K}$ and $d_{min} = D^{1/K}$, where K is the fixed number of stages.

Many modifications to cascade training have been introduced over the years. Most of them try out different: feature selection approaches, subsampling methods, or are simply tailored to a particular type of features [3,10,12,17] (e.g. Haar, HOG, LBP, etc.). Some authors obtain modified cascades by designing new boosting algorithms that underlie the training [14,15], but due to mathematical difficulties, the expected number of features is seldom the main optimization criterion. One of few exceptions is an elegant work by Saberian and Vasconcelos [14]. The authors use gradient descent to optimize explicitly a Lagrangian representing the trade-off between cascade's error rate and the operating cost (expected value). They use a trick that translates non-differentiable recursive formulas to smooth ones using hyperbolic tangent approximations. The approach is analytically tractable but expensive, because all cascade stages are kept open while training. In every step one has to check *all* variational derivatives based on features at disposal for *all* open stages.

The main contribution of this paper is an algorithm—or in fact a general framework—for training cascades of classifiers via a tree search approach and the *branch-and-bound* technique. Successive tree levels correspond to successive cascade stages. Sibling nodes represent variants of the same stage with different number of features applied. We provide suitable formulas for lower bounds on the expected value that we optimize. During an ongoing search, we observe the lower bounds, and whenever a bound for some tree branch is greater than (or equal to) the best-so-far expectation, the branch becomes pruned. Once the search is finished, one of the paths from the root to some terminal node indicates the cascade with the smallest expected number of features. Apart from the exact approach to pruning, we additionally propose an approximate one, using suitable predictions of expected values.

2 Preliminaries

2.1 Notation

Throughout this paper we use the following notation:

- K — number of cascade stages,
- $n = (n_1, n_2, \ldots, n_K)$—numbers of features used on successive stages,
- $(a_1, a_2, \ldots, a_K)$—FAR values on successive stages (false alarm rates),
- $(d_1, d_2, \ldots, d_K)$—sensitivities on successive stages (detection rates),
- A—required FAR for the whole cascade,
- D—required detection rate (sensitivity) for the whole cascade,
- $F = (F_1, F_2, \ldots, F_K)$—ensemble classifiers on successive stages (the cascade),
- A_k—FAR observed up to k-th stage of cascade ($A_k = \prod_{1 \leqslant i \leqslant k} a_i$),
- D_k—sensitivity observed up to k-th stage of cascade ($D_k = \prod_{1 \leqslant i \leqslant k} d_i$),
- $(p, 1 - p)$—true probability distribution of classes (unknown in practice),
- $\mathcal{D}, \mathcal{V}$—training and validation data sets,
- $\#$—set size operator (cardinality of a set),
- $\|$—concatenation operator (to concatenate cascade stages).

The probabilistic meaning of relevant quantities is as follows. The final requirements (A, D) demand that: $P\left(F(\mathbf{x}) = + \,|y=-\right) \leqslant A$ and $P\left(F(\mathbf{x}) = + \,|y=+\right) \geqslant D$, whereas false alarm and detection rates observed on particular stages are, respectively, equal to:

$$a_k = P\left(F_k(\mathbf{x}) = + \,|y=-, F_1(\mathbf{x}) = \cdots = F_{k-1}(\mathbf{x}) = +\right),$$
$$d_k = P\left(F_k(\mathbf{x}) = + \,|y=+, F_1(\mathbf{x}) = \cdots = F_{k-1}(\mathbf{x}) = +\right). \tag{1}$$

2.2 Classical Cascade Training Algorithm (Viola-Jones Style)

The classical cascade training algorithm given below (Algorithm 1) can be treated as a reference for new algorithms we propose.

Please note, in the final line of the pseudocode, that we return $(F_1, F_2, \ldots, F_k)$ rather than $(F_1, F_2, \ldots, F_K)$. This is because the training procedure can potentially stop earlier, when $k < K$, provided that the final requirements (A, D) for the entire cascade are already satisfied i.e. $A_k \leqslant A$ and $D_k \geqslant D$.

The step "Adjust decision threshold" requires a more detailed explanation. The real-valued response of any stage can be suitably thresholded to obtain either some wanted sensitivity or FAR. Hence, the resulting $\{-1, +1\}$-decision of a stage is, in fact, calculated as the sign of expression

$$F_k(\mathbf{x}) - \theta_k,$$

where θ_k represents the decision threshold. Suppose $(v_1, v_2, \ldots, v_{\#\mathcal{P}})$ denotes a sequence of sorted, $v_i \leqslant v_{i+1}$, real-valued responses of a new cascade stage F_{k+1} obtained on positive examples (subset $\mathcal{P}$). Then, the $d_{\min}$ per-stage requirement can be satisfied by simply choosing: $\theta_{k+1} = v_{\lfloor (1-d_{\min}) \cdot \#\mathcal{P} \rfloor}$.

Algorithm 1. VJ-style training algorithm for cascade of classifiers

---

procedure TRAINVJCASCADE($\mathcal{D}$, A, D, K, $\mathcal{V}$)

  From $\mathcal{D}$ take subsets $\mathcal{P}$, $\mathcal{N}$ with positive and negative examples, respectively.

  $F := ()$ $\triangleright$ initial cascade — empty sequence

  $a_{\max} := A^{1/K}$, $d_{\min} := D^{1/K}$, $A_0 := 1$, $D_0 := 1$, $k := 0$.

  while $A_k > A$ **do**

    $n_{k+1} := 0$, $F_{k+1} := 0$, $A_{k+1} := A_k$, $a_{k+1} := A_{k+1}/A_k$.

    while $a_{k+1} > a_{\max}$ **do**

      $n_{k+1} := n_{k+1} + 1$.

      Train new weak classifier f using $\mathcal{P}$ and $\mathcal{N}$

      $F_{k+1} := F_{k+1} + f$.

      Adjust decision threshold θ_{k+1} for F_{k+1} to satisfy $d_{\min}$ requirement.

      Use cascade $F\|F_{k+1}$ on validation set $\mathcal{V}$ to measure A_{k+1} and D_{k+1}.

      $a_{k+1} := A_{k+1}/A_k$.

    $F := F\|F_{k+1}$.

    if $A_{k+1} > A$ **then**

      $\mathcal{N} := \emptyset$.

    Use cascade F to populate set $\mathcal{N}$ with false detections

        sampled from non-target images.

    $k := k + 1$

  return $F = (F_1, F_2, \ldots, F_k)$.

2.3 Expected Number of Extracted Features

Definition-Based Formula. A cascade stops operating after a certain number of stages. It does not stop in the middle of a stage. Therefore the possible outcomes of the random variable of interest, describing the disjoint events, are: n_1, $n_1 + n_2$, ..., $n_1 + n_2 + \cdots + n_K$. Hence, by the definition of expected value, the expected number of features can be calculated as follows:

$$E(n) = \sum_{1 \leqslant k \leqslant K} \left(\sum_{1 \leqslant i \leqslant k} n_i \right) \left(p \Big(\prod_{1 \leqslant i < k} d_i \Big)(1 - d_k)^{[k < K]} + (1 - p)\Big(\prod_{1 \leqslant i < k} a_i \Big)(1 - a_k)^{[k < K]} \right),$$
$$(2)$$

where $[\cdot]$ is an indicator function.

Incremental Formula and Its Approximation. By grouping the terms in (2) with respect to n_k the following alternative formula can be derived:

$$E(n) = \sum_{1 \leqslant k \leqslant K} n_k \left(p \prod_{1 \leqslant i < k} d_i + (1 - p) \prod_{1 \leqslant i < k} a_i \right). \qquad (3)$$

Obviously, in practical applications the true probability distribution underlying the data is unknown. Since the probability p of the positive class is very small (typically $p < 10^{-4}$), the expected value can be accurately approximated using only the summands related to the negative class as follows:

$$\widehat{E}(n) = \sum_{1\leqslant k\leqslant K} n_k \prod_{1\leqslant i<k} a_i \approx E(n). \tag{4}$$

It is also interesting to remark that in the original Viola and Jones' paper [18] the authors proposed an incorrect formula to estimate the expected number of features, namely:

$$E_{\mathrm{VJ}}(n) = \sum_{k=1}^{K} n_k \prod_{i=1}^{k-1} r_i, \tag{5}$$

where r_i represents the "positive rate" of i-th stage. This is equivalent to

$$E_{\mathrm{VJ}}(n) = \sum_{k=1}^{K} n_k \prod_{i=1}^{k-1} (pd_i + (1-p)a_i). \tag{6}$$

Please note that by multiplying positive rates of stages, one obtains mixed terms of form $d_i \cdot a_j$ that do not have any probabilistic sense. For example for $k = 3$ the product under summation becomes $(pd_1 + (1-p)a_1)(pd_2 + (1-p)a_2)$, with the terms $d_1 a_2$ and $a_1 d_2$ having no sense, because a fixed data point does not change its class label while traveling along the cascade.

3   Cascade Training as a Tree Search

In stage-wise training procedures, each stage, once fixed, must not be altered. The paper [14], discussed in the introduction, represents an opposite approach, where all stages can be extended with a weak classifier at any time. The approach we propose is in-between the two mentioned above. It provides more flexibility than stage-wise training and simultaneously avoids high complexity of [14].

We treat cascade training as a tree search process. The root of the tree represents an empty cascade. Successive tree levels correspond to successive cascade stages. Each non-terminal tree node is going to have an odd number of children nodes. They will represent variants of a subsequent stage with slightly different number of features. The children will be processed recursively from left to right until the stop condition is met. It should be understood that the nodes are not simply generated mechanically but, in fact, trained as ensemble classifiers.

The size of the tree shall be controlled by two integer parameters L and C, predefined by the user. To keep the tree fairly small, the branching of variants shall take place only at L top-most levels, e.g. $L = 2$. At those levels the branching factor will be equal to C, an odd number, e.g. $C = 5$. At deeper levels the branching factor will be one. Therefore, the actual branching shall affect only initial stages having the largest impact on the expected number of features. Once the tree search is finished, one of the paths from the root to some terminal node shall indicate the best cascade i.e. having the smallest expectation.

For notation purposes, children nodes being variants of the same stage use an additional subindex. For example, the classifier $F_{1,0}$ denotes the main variant of the first stage (using a certain number of features) and is graphically represented as the *middle* child. Its *left* siblings $F_{1,-1}, F_{1,-2}, \ldots$ denote classifiers using fewer features (one less, two less, etc.). The *right* siblings $F_{1,+1}, F_{1,+2}, \ldots$ use more features than the middle child (one more, two more, etc.). This notation will be used only locally within single recursive calls (due to global ambiguity).

3.1 Pruning Search Tree Using Current Partial Expectations—Exact Branch-and-bound

During an ongoing tree search (combined with cascade training) one can observe *partial* values for the expected value of interest — formula (4). Suppose a new $(k+1)$-th stage has been completed, revealing n_{k+1} features. The formula

$$\widehat{E}\Big((n_1,\ldots,n_{k+1})\Big) = \sum_{1\leqslant j\leqslant k} n_j \prod_{1\leqslant i<j} a_i + n_{k+1}\prod_{1\leqslant i<k+1} a_i = \widehat{E}\Big((n_1,\ldots,n_k)\Big) + n_{k+1}\prod_{1\leqslant i<k+1} a_i.$$
(7)

expresses the partial expectation for the extended cascade in an incremental manner. It should be clear that whenever a partial expectation for some tree branch is greater than (or equal to) the best-so-far exact expectation, say $\widehat{E}((n_1,\ldots,n_{k+1})) \geqslant \widehat{E}^*$, then there is no point in pursuing that branch further down the tree. In other words, pruning can be applied because formula (7) provides a lower bound on the final unknown expectation.

Figure 1 provides a symbolic illustration of a search tree with pruning. In the figure, the subindexes $E_1, E_2, \ldots$ are meant to indicate chronologically the partial expected values observed on the successive branches as the tree is being traversed from left to right. Crossed-out lines represent the pruned branches.

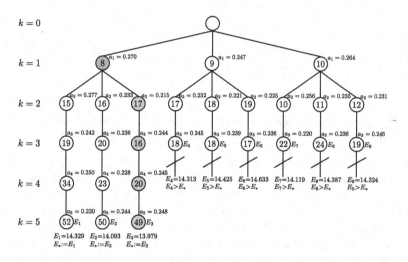

Fig. 1. Cascade training as a tree search with pruning—example illustration.

Algorithm 2 stated as a recursion. A single recursive call can be summarized as follows. It takes as input a partial cascade F with k stages and trains the new $(k+1)$-th stage in its main variant $F_{k+1,0}$. We refer to it as the *middle child*. Then, the algorithm "branches" the stage (if level not greater than L) by creating clones of the middle child with fewer features: $F_{k+1,-1}, F_{k+1,-2}, \ldots$ (*left children*), and with more features: $F_{k+1,+1}, F_{k+1,+2}, \ldots$ (*right children*). The algorithm iterates over all children and performs recursive calls to train their subsequent stages provided that the lower bound (7) on the final expectation is not worse than the best expectation $\widehat{E}^*$ so far. A recursion path, representing some cascade, reaches its stopping point when final requirements (A, D) are satisfied and when its expected value is strictly less than $\widehat{E}^*$ (initially, set to ∞). The outermost recursion call is

$$\textsc{TrainTreeCascade} \left(\mathcal{D}, A, D, K, 0, \mathcal{V}, (), L, C, \text{null}, \infty \right)$$

yielding a pair of results: the best cascade F^* and its expectation $\widehat{E}^*$.

Inside the subroutine $\textsc{TrainStage}$ we train a single ensemble using per-stage requirements. They can be calculated a standard geometric means (classical VJ-style), leading to constant per-stage requirements for the whole training, or as updated geometric means (UGM): uniform or greedy. The formulas below represent the three options.

$$\text{VJ}: \quad a_{\max,k+1} = A^{1/K}, \qquad\qquad d_{\min,k+1} = D^{1/K}. \qquad (8)$$

$$\text{UGM}: \quad a_{\max,k+1} = \left(A \Big/ \prod_{1 \leqslant i \leqslant k} a_i \right)^{1/(K-k)}, \quad d_{\min,k+1} = \left(D \Big/ \prod_{1 \leqslant i \leqslant k} d_i \right)^{1/(K-k)}. \quad (9)$$

$$\text{UGM-G}: \quad a_{\max,k+1} = A^{(k+1)/K} \Big/ \prod_{1 \leqslant i \leqslant k} a_i, \quad d_{\min,k+1} = D^{(k+1)/K} \Big/ \prod_{1 \leqslant i \leqslant k} d_i. \qquad (10)$$

3.2 Pruning Search Tree Using Expectation Predictions— Approximate Branch-and-bound

Suppose we have completed the training of stage $k+1$ and would like to make a prediction about the partial expectation for stage $k+2$ without training it. Obviously, the training of any stage is time-consuming, hence a significant gain would be benefited by not wasting time on a stage that is not going to improve the best-so-far expectation. Observe that when the stage $k+1$ is completed, we get to know two new pieces of information: n_{k+1} and a_{k+1}. That second piece is not needed to calculate formula (7) for stage $k+1$, but it is needed for stage $k+2$. Therefore, the only unknown preventing us from calculating the exact partial expectation for stage $k+2$ is n_{k+2}. We are going to approximate it.

Algorithm 2 Training cascade of classifiers via tree search with exact pruning

---

procedure TRAINTREECASCADE($\mathcal{D}$, A, D, K, k, $\mathcal{V}$, F, C, L, F^*, $\widehat{E}^*$)

From $\mathcal{D}$ take subset $\mathcal{P}$ with all positive examples, and subset $\mathcal{N}$ with all negative examples.

Train stage for middle child: $F_{k+1,0} :=$ TRAINSTAGE($\mathcal{P}$, $\mathcal{N}$, K, k, $\mathcal{V}$, F).

Use cascade $F\|F_{k+1,0}$ on validation set $\mathcal{V}$ to measure $A_{k+1,0}$ and $D_{k+1,0}$.

if $k > L$ **then**

$C := 1$.

for $c := -1, -2, \ldots, -\lfloor C/2 \rfloor$ **do**   ▷ left children

Create $F_{k+1,c}$ by cloning $F_{k+1,c+1}$.

Remove last weak classifier from $F_{k+1,c}$.

Adjust decision threshold $\theta_{k+1,c}$ for $F_{k+1,c}$ to satisfy $d_{\min,k+1}$ requirement.

Use cascade $F\|F_{k+1,c}$ on validation set $\mathcal{V}$ to measure $A_{k+1,c}$ and $D_{k+1,c}$.

for $c := 1, 2, \ldots, \lfloor C/2 \rfloor$ **do**   ▷ right children

Create $F_{k+1,c}$ by cloning $F_{k+1,c-1}$.

Train new weak classifier f using $\mathcal{P}$ and $\mathcal{N}$

$F_{k+1,c} := F_{k+1,c} + f$.

Adjust decision threshold $\theta_{k+1,c}$ for $F_{k+1,c}$ to satisfy $d_{\min,k+1}$ requirement.

Use cascade $F\|F_{k+1,c}$ on validation set $\mathcal{V}$ to measure $A_{k+1,c}$ and $D_{k+1,c}$.

for $c := -\lfloor C/2 \rfloor, \ldots, 0, \ldots, \lfloor C/2 \rfloor$ **do**   ▷ all children

Calculate expectation $\widehat{E}$ for cascade $F\|F_{k+1,c}$ using (7).

if $A_{k+1,c} > A$ and $\widehat{E} < \widehat{E}^*$ **then**

Prepare new training set $\mathcal{D}_{k+1,c}$ and new validation set $\mathcal{V}_{k+1,c}$.

$(F^*, \widehat{E}^*) :=$ TRAINTREECASCADE($\mathcal{D}_{k+1,c}$, A, D, K, $k+1$, $\mathcal{V}_{k+1,c}$, $F\|F_{k+1,c}$,

L, C, E^*, F^*)

else if $A_{k+1,c} \leqslant A$ and $\widehat{E} < \widehat{E}^*$ **then**

$\widehat{E}^* := \widehat{E}$, $F^* := F\|F_{k+1,c}$.

return $(F^*, \widehat{E}^*)$.

return (F^*, E^*).

procedure TRAINSTAGE($\mathcal{P}$, $\mathcal{N}$, K, k, $\mathcal{V}$, F)

$n_{k+1} := 0$, $F_{k+1} := 0$, $A_{k+1} := A_k$.

Calculate per-stage requirements

$(a_{\max,k+1}, d_{\min,k+1})$ using (8) or (9) or (10).

$a_{k+1} := A_{k+1}/A_k$.

while $a_{k+1} > a_{\max,k+1}$ **do**

$n_{k+1} := n_{k+1} + 1$.

Train new weak classifier f using $\mathcal{P}$ and $\mathcal{N}$.

$F_{k+1} := F_{k+1} + f$.

Adjust decision threshold θ_{k+1} for F_{k+1}

to satisfy $d_{\min,k+1}$ requirement.

Use cascade $F\|F_{k+1}$ on validation

set $\mathcal{V}$ to measure A_{k+1} and D_{k+1}.

$a_{k+1} := A_{k+1}/A_k$.

return F_k.

As cascade experiments on real-data show, the counts of features $(n_k)_{k=1,\ldots,K}$ typically form a non-decreasing sequence. There exist counter-examples, but in the vast majority of cases it is true that $n_{k+1} \geqslant n_k$. Therefore, to build our prediction it could potentially be sufficient to lowerbound n_{k+2} by n_{k+1}. Instead, we prefer to propose a safer parameterized approach — by assuming:

$$n_{k+2} \geqslant \alpha \, n_{k+1}, \tag{11}$$

where parameter α could be selected e.g. from $[0.5, 1.5]$ interval. The following lines demonstrate explicitly the prediction we are going to apply:

$$\widehat{E}\Big((n_1, \ldots, n_{k+2})\Big) = \widehat{E}\Big((n_1, \ldots, n_k)\Big) + n_{k+1} \prod_{1 \leqslant i < k+1} a_i + n_{k+2} \prod_{1 \leqslant i < k+2} a_i$$

$$\approx \widehat{E}\Big((n_1, \ldots, n_k)\Big) + n_{k+1} \prod_{1 \leqslant i < k+1} a_i + \alpha \, n_{k+1} \Big(\prod_{1 \leqslant i < k+1} a_i\Big) a_{k+1} \equiv \widehat{E}_\alpha \tag{12}$$

The influence of parameter α can be described as follows. By lowering α, one decreases the risk of pruning a branch incorrectly, but simultaneously one strengthens the underestimation of the expected value, which can lead to training continuation despite a negligible chance of improvement. In contrast, higher α values lead to more pruning but with some risk of missing the optimum solution. Additionally, it is worth to remark that the prediction we make is only for *one* stage ahead, ignoring all subsequent stages. Since those stages shall too contribute their summands to the final expectation then this suggests that high α values should still be safe, especially for initial levels.

Algorithm 3 represents the described approach for cascade training based on tree search and approximate pruning.

Algorithm 3 Training cascade of classifiers algorithm via tree search with approximate pruning

---

procedure TRAINTREECASCADEAPPROX($\mathcal{D}$, A, D, K, k, $\mathcal{V}$, F, C, L, F^*, $\widehat{E}^*$, α)

$\quad\vdots$ ▷ initial steps same as in TRAINTREECASCADE

    for $c := -\lfloor C/2 \rfloor, \ldots, 0, \ldots, \lfloor C/2 \rfloor$ **do**   ▷ all children
      if $A_{k+1,c} > A$ **then**
          Calculate expectation prediction $\widehat{E}_\alpha$ for $F\|F_{k+1,c}\|F_{k+2,\cdot}$ using (12).
          if $\widehat{E}_\alpha < \widehat{E}^*$ **then**
              Prepare new training set $\mathcal{D}_{k+1,c}$ and new validation set $\mathcal{V}_{k+1,c}$.
              $(F^*, \widehat{E}^*) :=$ TRAINTREECASCADEAPPROX($\mathcal{D}_{k+1,c}$, A, D, K, $k+1$, $\mathcal{V}_{k+1,c}$, $F\|F_{k+1,c}$, C, L, E^*, F^*, α)
      else
          Calculate expectation $\widehat{E}$ for cascade $F\|F_{k+1,c}$ using (7).
          if $\widehat{E} < \widehat{E}^*$ **then**
              $\widehat{E}^* := \widehat{E}, \quad F^* := F\|F_{k+1,c}$.
              return $(F^*, \widehat{E}^*)$.
    return (F^*, E^*).

---

4   Experiments

In all experiments we apply *RealBoost+bins* [13] as the main learning algorithm, producing ensembles of weak classifiers as successive cascade stages. Each weak classifier is based on a single selected feature.

Experiments on two collections of images are carried out. Firstly, we test the proposed approach in face detection task, using Haar-like features (HFs) as input information. Secondly, we experiment with synthetic images representing letters (computer fonts originally prepared by T.E. de Campos et al. [5]) and we treat the 'A' letter as our target object. In that experiment we expect to detect our targets regardless of their rotation. To do so, we apply rotationally invariant features based on Zernike moments (ZMs) [2]. In both cases, feature extraction is backed with integral images (complex-valued for ZMs).

In experiments we used a machine with Intel Core i7-4790K 4/8 c/t, 8MB cache. For clear interpretation of time measurements, we report detection times using only a single thread [ST]. The software has been programmed in C#, with key computational procedures implemented in C++ as a dll library.

Experiment: "Faces" (Haar-like features). Training faces were cropped from 3 000 images, looked up using *Google Images*, yielding 7 258 face examples described by 14 406 HFs. The test set contained 3 014 examples from *Essex Face Data* [16]. Validation sets contained 1 000 examples. The number of negatives in the test set was constant and equal to 1 000 000. To reduce training time, the number of negatives in training and validation sets was gradually reduced for successive stages, as described in Table 1. Detection times, reported later, were determined as averages from 200 executions of the detection procedure.

Table 1. "Faces": experimental setup.

Train data		Validation data		Test data		Detection procedure	
qty./parameter	value	qty./parameter	value	qty./parameter	value	qty./parameter	value
no. of positives	7 258	no. of positives	1000	no. of positives	3014	no. of repetitions	200
no. of negatives	139 373	no. of negatives	40 000	no. of negatives	1 000 000	image resolution	600 × 480
" 2nd stage	42 742	" other stages	24 000	total set size	1 003 014	no. of detection scales	5
" other stages	27 742	total set size	41 000			window growing coef.	1.2
total set size	146 631	" other stages	25 000			smallest window	48 × 48
" 2nd stage	50 000					largest window size	100 × 100
" other stages	35 000					window jumping coef.	0.05

We start reporting results by showing some visual examples of detection outcomes obtained by two best detectors (in terms of expected number of features) trained to satisfy $A = 10^{-4}$ and $A = 10^{-5}$ requirements, respectively, see Fig. 2.

Table 2 provides detailed information about cascades trained with $A = 10^{-3}$ requirement. Every row contains a cascade, represented by two sequences: a sequence of features counts n_k on successive stages (top), and a sequence of false alarm rates a_k (bottom). The third column reports the expected value $\widehat{E}(n)$ calculated according to (4). The right-most columns provide information about the

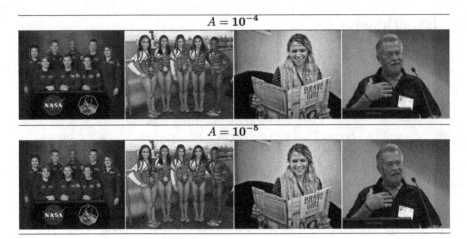

Fig. 2. "Faces": detection examples (false alarms marked in yellow).

effectiveness of tree pruning, showing how many nodes were in fact trained with respect to the potential total. We allow ourselves to report approximate pruning (for both $\alpha = 0.8$ and $\alpha = 1.2$) in the same row as exact pruning, because in all experiments the approximate pruning has never led to a suboptimal solution. The table shows clearly that in general the greater the "bushiness" of the tree the better the expected value we try to minimize — an increase in either C or L parameter lead to an improvement. Additionally, owing to pruning, the time needed to train cascades involving wider trees did not increase proportionally to the overall number of nodes. One should realize that nodes (stages) lying deeper in the tree, with low effective FAR resulting from chain multiplication of a_k rates, require much time for resampling, since only a small fraction of negative examples reaches those stages. That is why it is so important to prune redundant nodes. In particular, for TREE-C3-L2-UGM-G an exhaustive search would require 84 nodes, exact pruning reduces this number to 60, whereas approximate pruning cuts it further down to 57 (for $\alpha = 0.8$) and 55 (for $\alpha = 1.2$).

Table 3 compares cascades trained traditionally (VJ) against selected best cascades trained via tree search. The comparison pertains to accuracy and detection times. This time we show three variants of A requirement: 10^{-3}, 10^{-4} and 10^{-5} (that last setting only for cascades with 10 stages). In addition, the theoretical expected value for cascades can be compared against an average observed on the test set (column $\bar{n}$). We remark that the tree-based approach combined with greedy per-stage requirements — TREE-C3-L1-UGM-G — produced the best cascades (marked with dark gray) having the smallest expected values. Savings in detection times per image with respect to VJ approach are at the level of ≈ 7.5 ms (about 8% per thread). This may seems not large but we remind that the measurements are for single-threaded executions [ST]. For example, if 8 threads are used this implies a reduction of ≈ 4 FPS.

Table 2. "Faces": cascades trained for $A = 10^{-3}$ (pruning information in last columns).

Training alghorithm	Cascade	$E(n)$	Validation FAR	sensitivity	Trained nodes pruning exact	pruning approximate $\alpha = 0.8$	$\alpha = 1.2$
			Requirements: $10^{-3} = 0.9500$				
TREE-C3-L1 VJ	8 16 20 23 50 0.2703, 0.2329, 0.2357, 0.2284, 0.2439	14.0932	0.000827	0.9520	11/15	10/15	10/15
TREE-C3-L1 UGM	8 16 20 22 32 0.2703, 0.2329, 0.2357, 0.2324, 0.2781	14.0193	0.000959	0.9510	11/15	10/15	10/15
TREE-C3-L1 UGM-G	8 16 20 22 32 0.2703, 0.2329, 0.2357, 0.2324, 0.2781	14.0193	0.000959	0.9510	11/15	11/15	10/15
TREE-C3-L2 VJ	8 17 16 20 49 0.2703, 0.2152, 0.2442, 0.2445, 0.2483	13.9789	0.000862	0.9520	27/39	24/39	23/39
TREE-C3-L2 UGM	8 17 16 20 46 0.2703, 0.2152, 0.2442, 0.2434, 0.2834	13.9678	0.000980	0.9510	27/39	25/39	23/39
TREE-C3-L2 UGM-G	8 17 16 19 41 0.2703, 0.2152, 0.2442, 0.2776, 0.2406	13.9562	0.000949	0.9510	27/39	26/39	24/39
TREE-C5-L1 VJ	7 18 20 35 40 0.2770, 0.2134, 0.2358, 0.2402, 0.2328	13.7886	0.000779	0.9520	17/25	15/25	14/25
TREE-C5-L1 UGM	7 18 20 31 34 0.2770, 0.2134, 0.2358, 0.2564, 0.2685	13.7204	0.000959	0.9510	17/25	15/25	14/25
TREE-C5-L1 UGM-G	7 18 18 27 53 0.2770, 0.2134, 0.2612, 0.2492, 0.2483	13.6694	0.000955	0.9510	17/25	15/25	13/25
TREE-C3-L1 VJ	3 5 12 4 12 11 16 25 39 38 0.7595, 0.4615, 0.4471, 0.4997, 0.4532, 0.4714, 0.4763, 0.4879, 0.4639, 0.4877	13.6463	0.000880	0.9550	22/30	21/30	19/30
TREE-C3-L1 UGM	4 13 7 9 12 13 12 25 16 17 0.4521, 0.4820, 0.3889, 0.5291, 0.4676, 0.4938, 0.5189, 0.5688, 0.5490, 0.5581	13.3128	0.000987	0.9500	24/30	23/30	22/20
TREE-C3-L1 UGM-G	4 9 10 8 11 12 23 19 33 24 0.4521, 0.5445, 0.4980, 0.4787, 0.5053, 0.5089, 0.4976, 0.4954, 0.4451, 0.5970	13.1655	0.000989	0.9510	23/30	23/30	23/30
TREE-C3-L2 VJ	3 5 12 4 12 11 16 25 39 38 0.7595, 0.4615, 0.4471, 0.4997, 0.4532, 0.4714, 0.4763, 0.4879, 0.4639, 0.4877	13.6463	0.000880	0.9550	51/84	45/84	40/84
TREE-C3-L2 UGM	4 13 7 9 12 13 12 25 16 17 0.4521, 0.4820, 0.3889, 0.5291, 0.4676, 0.4938, 0.5189, 0.5688, 0.5490, 0.5581	13.3128	0.000987	0.9500	52/84	44/84	39/84
TREE-C3-L2 UGM-G	4 10 9 6 13 22 15 14 22 21 0.4521, 0.5038, 0.4470, 0.6043, 0.4975, 0.5153, 0.5436, 0.5666, 0.4644, 0.5352	13.1160	0.000985	0.9510	60/84	57/84	55/84

Table 3. "Faces": VJ vs tree-based cascades with $K=5$ (left) and $K=10$ (right) stages.

$K=5$ (left):

Training algorithm	$E(n)$	Validation		Test			Detection time	
		FAR	sensitivity	FAR	sensitivity	$\bar{n}$	image [ST][ms]	window [ST][µs]
		Requirements: 10^{-3}	**0.9500**				(windows per image: 130 971)	
VJ	14.97	0.000792	0.9520	0.000815	0.9549	15.88	83	0.64
TREE C5-L1 UGM-G	13.67	0.000955	0.9510	0.001078	0.9668	14.59	73	0.56
		Requirements: 10^{-4}	**0.9500**				(windows per image: 130 971)	
VJ	23.88	0.000091	0.9520	0.000107	0.9482	25.80	125	0.95
TREE C3-L1 UGM-G	22.21	0.000097	0.9510	0.000099	0.9542	23.88	115	0.87

$K=10$ (right):

Training algorithm	$E(n)$	Validation		Test			Detection time	
		FAR	sensitivity	FAR	sensitivity	$\bar{n}$	image [ST][ms]	window [ST][µs]
		Requirements: 10^{-3}	**0.9500**				(windows per image: 130 971)	
VJ	13.78	0.000950	0.9600	0.001048	0.9569	15.78	75	0.57
TREE C3-L2 UGM-G	13.12	0.000985	0.9510	0.001155	0.9545	14.71	71	0.55
		Requirements: 10^{-4}	**0.9500**				(windows per image: 130 971)	
VJ	15.14	0.000077	0.9550	0.000073	0.9562	16.45	89	0.68
TREE C3-L1 UGM-G	14.38	0.000099	0.9510	0.000108	0.9552	15.29	83	0.63
		Requirements: 10^{-5}	**0.9500**				(windows per image: 130 971)	
VJ	17.52	0.000006	0.9550	0.000010	0.9482	19.45	100	0.77
TREE C3-L1 UGM-G	17.28	0.000010	0.9510	0.000010	0.9317	19.01	92	0.70

Experiment: "Synthetic A letters" (Zernike Moments). Table 4 lists details of the experimental setup for this experiment. In train images, only objects with limited rotations were allowed ($\pm 45°$ with respect to their upright positions). In contrast, in test images, rotations within the full range of $360°$ were allowed. During the training 540 features were at disposal [2].

Table 4. "Synthetic A letters": experimental setup.

Train data		Validation data		Test data		Detection procedure	
qty./parameter	value	qty./parameter	value	qty./parameter	value	qty./parameter	value
no. of positives	20 384	no. of positives	1 000	no. of positives	20 000	no. of repetitions	200
no. of negatives	50 546	no. of negatives	10 000	no. of negatives	1 000 000	image resolution	600 × 480
total set size	70 930	total set size	11 000	total set size	1 020 000	no. of detection scales	5
						window growing coef.	1.2
						smallest window	100 × 100
						largest window size	208 × 208
						window jumping coef.	0.05

Figure 3 presents examples of detection outcomes obtained by best detectors trained to satisfy 10^{-3} and 10^{-4} FAR requirements. As it turned out for this data, the cascades did not need many stages nor features. Table 5 compares VJ against tree-based cascades. One can note that despite small feature counts (comparing to the previous experiment), the proposed method still allows to reduce the expectations. The smallest were achieved by the TREE-C3-L1-UGM-G variant, yielding 2.5682 and 2.9910, respectively for $A = 10^{-3}$ and $A = 10^{-4}$.

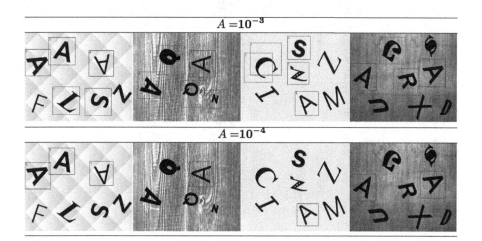

Fig. 3. "Synthetic A letters": detection examples.

Table 5. "Synthetic A letters": VJ vs tree-based cascades with $K = 5$ stages.

Training alghorithm	Cascade	Expected value	Validation		Test			Detection time	
			FAR	sensiti-vity	FAR	sensiti-vity	$\bar{n}$	image [ST][ms]	window [ST][μs]
			Requirements: 10^{-3}	0.9500				(windows per image: 18 752)	
VJ	2 2 3 5 7 / 0.2115, 0.2128, 0.1893, 0.2180, 0.2098	2.6136	0.000389	0.9540	0.000869	0.9313	2.56	127	6.77
TREE-C3-L1-UGM-G	2 2 3 5 / 0.2115, 0.2128, 0.2954, 0.2313, 0.2812	2.5682	0.000865	0.9500	0.002016	0.9351	2.50	124	6.61
			Requirements: 10^{-4}	0.9500				(windows per image: 18 752)	
VJ	3 4 5 8 10 / 0.1497, 0.1460, 0.1444, 0.1368, 0.1206	3.7393	0.000052	0.9530	0.000163	0.9411	3.84	157	8.37
TREE-C3-L1-UGM-G	2 4 5 6 11 / 0.2115, 0.1137, 0.1284, 0.1824, 0.1535	2.9910	0.000086	0.9510	0.000203	0.9418	2.99	133	7.10

5   Conclusion

Training a cascade of classifiers is a difficult optimization problem that, in our opinion, should be always carried out with a primary focus on the *expected number of extracted features*. This quantity reflects directly how fast an operating cascade is. Our proposition of the tree search-based training allows to 'track' more than one variant of a cascade. Potentially, this approach can be computationally expensive, but we have managed to reduce it with suitable branch-and-bound techniques. Being able to prune some of the subtrees, we save both the training and resampling time needed by later cascade stages. To our knowledge, no such proposition regarding the cascade structure has been tried out before. In our future research we plan to investigate more the approximate variant, trying to predict partial expectations for more than one stage ahead.

References

1. Abbas, S., et al.: Crowd detection and management using cascade classifier on ARMv8 and OpenCV-Python. In: 2017 International Conference on Innovations in Information, Embedded and Communication Systems (ICIIECS), pp. 1–6, March 2017
2. Bera, A., Klęsk, P., Sychel, D.: Constant-time calculation of zernike moments for detection with rotational invariance. IEEE Trans. Pattern Anal. Mach. Intell. **41**(3), 537–551 (2019)
3. Bourdev, L., Brandt, J.: Robust object detection via soft cascade. In: Proceedings of the 2005 IEEE Computer Society Conference on Computer Vision and Pattern Recognition (CVPR 2005) - Volume 2, pp. 236–243. IEEE Computer Society (2005)
4. Budiman, R.A.M., Achmad, B., Faridah, Arif, A., Nopriadi, Zharif, L.: Localization of white blood cell images using Haar cascade classifiers. In: 2016 1st International Conference on Biomedical Engineering (IBIOMED), pp. 1–5, October 2016
5. de Campos, T.E., et al.: Character recognition in natural images. In: International Conference on Computer Vision Theory and Applications, Portugal, pp. 273–280 (2009)
6. Setjo, C.H., et al.: Thermal image human detection using Haar-cascade classifier. In: 2017 7th International Annual Engineering Seminar (InAES), pp. 1–6 (2017)
7. Lu, J., et al.: Detection of bird's nest in high power lines in the vicinity of remote campus based on combination features and cascade classifier. IEEE Access **6**, 39063–39071 (2018)
8. Jones, N., Pevzner, P.: An Introduction to Bioinformatics Algorithms. MIT Press, Cambridge (2002)
9. Cuimei, L., et al.: Human face detection algorithm via Haar cascade classifier combined with three additional classifiers. In: 2017 13th IEEE International Conference on Electronic Measurement Instruments (ICEMI), pp. 483–487 (2017)
10. Li, J., Zhang, Y.: Learning SURF cascade for fast and accurate object detection. In: Proceedings of the 2013 IEEE Conference on Computer Vision and Pattern Recognition, pp. 3468–3475. CVPR 2013. IEEE Computer Society (2013)
11. Li, Y., Xu, X., Mu, N., Chen, L.: Eye-gaze tracking system by Haar cascade classifier. In: 2016 IEEE 11th Conference on Industrial Electronics and Applications (ICIEA), pp. 564–567, June 2016

12. Pham, M., Cham, T.: Fast training and selection of Haar features using statistics in boosting-based face detection. In: IEEE 11th International Conference on Computer Vision, ICCV 2007, pp. 1–7 (2007)
13. Rasolzadeh, B., et al.: Response binning: improved weak classifiers for boosting. In: IEEE Intelligent Vehicles Symposium, pp. 344–349 (2006)
14. Saberian, M., Vasconcelos, N.: Boosting algorithms for detector cascade learning. J. Mach. Learn. Res. **15**, 2569–2605 (2014)
15. Shen, C., Wang, P., Paisitkriangkrai, S., van den Hengel, A.: Training effective node classifiers for cascade classification. Int. J. Comput. Vis. **103**(3), 326–347 (2013). https://doi.org/10.1007/s11263-013-0608-1
16. University of Essex: Face Recognition Data. https://cswww.essex.ac.uk/mv/allfaces/faces96.html (1997). Accessed 11 May 2019
17. Vallez, N., Deniz, O., Bueno, G.: Sample selection for training cascade detectors. PLos ONE **10**, e0133059 (2015)
18. Viola, P., Jones, M.: Robust real-time face detection. Int. J. Comput. Vis. **57**(2), 137–154 (2004). https://doi.org/10.1023/B:VISI.0000013087.49260.fb

Application of the Stochastic Gradient Method in the Construction of the Main Components of PCA in the Task Diagnosis of Multiple Sclerosis in Children

Mariusz Topolski$^{(\boxtimes)}$ (iD)

Department of Systems and Computer Networks, Faculty of Electronics,
Wrocław University of Science and Technology,
Wybrzeże Wyspiańskiego 27, 50-370 Wrocław, Poland
mariusz.topolski@pwr.edu.pl

Abstract. Many different medical problems are characterized by quite large spatial dimensions, which causes the task of recognizing patterns to become troublesome. This is a well-known phenomenon called curse of dimensionality. These problems force the creation of various methods of reducing dimensionality. These methods are based on selection and extraction of features. The most commonly used method in literature, regarding the later, is the analysis of the main components of pca. The natural problem of this method is the possibility of applying it to linear space. It is a natural problem to develop the pca concept for cases of nonlinear feature spaces, optimization of feature selection for principal components and the inclusion of classes in the task of supervised learning. An important problem in the perspective of machine learning is not only a reduction of features and attributes but also separation of classes. The developed method was tested in two computer experiments using real data of multiple sclerosis in children. The discussed problem, even from the very nature of the data itself, is important because it can contribute to practical implementations in medical diagnostics. The purpose of the research is to develop a method of extracting features with the application of the stochastic gradient method in the task diagnosis of multiple sclerosis in children. This solution could contribute to the increasing quality of classification and thus may be the basis for building systems that support the medical diagnostics in recognition of multiple sclerosis in children.

Keywords: Principal components analysis · Stochastic gradient · Recognition of returns · Multiple sclerosis

1 Introduction

Nowadays machine learning techniques are being used in ever more fields, such as broadly understood medicine, neuroimaging, image classification and detection

© Springer Nature Switzerland AG 2020
V. V. Krzhizhanovskaya et al. (Eds.): ICCS 2020, LNCS 12140, pp. 35–44, 2020.
https://doi.org/10.1007/978-3-030-50423-6_3

of network attacks. They produce huge amounts of data with many attributes. Such a large dose of information, paradoxically, does not improve the quality of algorithms, and the data itself is expensive to acquire and store. This resulted in the need for methods to reduce the size of the data, without degrading (or even improving) the quality of classifiers. The reason why more information does not mean better classification is the so-called *curse of dimensionality*, described for the first time by Richard Bellman [1]. When adding dimensions to collections, the distances between specific points are constantly increasing. The number of objects needed for proper generalization is also increasing. It is estimated that in the case of linear classifiers this number increases linearly with dimensionality, and squarely in the case of quadratic algorithms. Even worse is the case of non-parametric classifiers, such as neural networks or those using radial base functions, where the number of objects needed for proper generalization increases exponentially [2]. Sometimes the problem of the curse of dimensionality is called *small n large p"* [4].

The curse of dimensionality results in the *Hughes phenomenon* [3]. For a fixed number of samples, recognition accuracy may first increase algorithms increase, but decreases when the number of attributes exceeds a certain optimal value. In addition to the distance between the samples, this is also caused by the noise in the data or insignificant features. *Selection and extraction* (reduction) features are used to reduce the dimensionality of the data. Feature selection is designed to select a subset of the features used for classification, while *feature extraction* is used to transform (e.g., linear) feature space.

2 Methods

Principal Component Analysis belongs to projection methods. The goal of projection methods is to find a mapping from original space with d dimensions for a new one $(k <= d)$ space, to minimize information loss [5].

It is an unsupervised learning method, which means it doesn't need class labels. In the case of PCA, the new attributes are created in a way that maximises their variance. The algorithm aims to create new features (the so-called principal components) that will be uncorrelated (orthogonal) and ordered according to decreasing variance. In order for the algorithm to give correct results, the input data should be normalized first. The principal components are eigenvectors of the input attribute covariance matrix. Because the direction is important in them, these lengths are selected 1. Assuming that λ_i is the eigenvalue of the ith eigenvector, after ordering the proportion of total variance is descending derived from the first k vectors can be calculated using the formula:

$$\frac{\lambda_1 + \lambda_2 + \ldots + \lambda_k}{\lambda_1 + \lambda_2 + \ldots + \lambda_k + \ldots + \lambda n} \tag{1}$$

If the original dimensions of the input data are strongly correlated with each other, we get a small number of eigenvectors with large eigenvalues. A large reduction in dimensions is then possible. However, if the dimensions are not strongly correlated, k will be similar to n and it is not possible to reduce the dimensions without losing the initial part of the set variance [5]. If the number of attributes exceeds the number of objects, it is possible to reduce the dimensions to at most to the number of samples [6].

One of the disadvantages of PCA is that it uses a linear transformation, which makes it unsuitable for more complex spaces. The solution to this problem may be to develop a basic algorithm with the so-called *kernel trick*, getting KPCA (*Kernel Principal Component Analysis*).

In order to solve a non-linear problem, one would first have to transform the input space X as a certain highly-dimensional space F using the function $\phi(x)$, and then e.g. calculate the scalar product $<\phi(x), \phi(x')>$. However, it would be computationally complicated. Therefore, choose the $k(x, x') = <\phi(x), \phi(x')>$ for some transformation ϕ [7]. One of the models using this trick is e.g. SVM classifier.

Another idea for developing PCA is, for example, using class labels as in the development of Karhunen-Loève or carrying out selection of features in the space obtained by PCA [8]. In addition to using the standard PCA, new versions are often created to suit specific problems. One such variation of PCA method is *SuperPCA* [12]. It is used in the classification problem related to *hyperspectral imagining* [17]. The method combines PCA with a segmentation algorithm by means of super pixelization.

Another interesting development of PCA is the DiPCA (*Dynamic Inner PCA*), method, also used in process monitoring, but focusing on the aspect of data dynamics [13]. Its goal is to maximize covariance between components and their earlier values. It accomplishes this by extracting a model of dynamic hidden variables on which standard PCA is then performed.

When it comes to supervised methods, LDA is also still widely used. An example of the use of linear discriminant analysis is the already mentioned feature extraction for the task of cancer recognition based on microscopic tissue images [11]. A team from India used a different approach to diagnose lung cancer [14], that used computed tomography images as input. In the study, LDA was used to reduce the size of the data (*Optimal Deep Neural Network*). The results showed an improvement in quality compared to previously used classifiers.

Another proposed method is factor-rotation-modified CCPCA analysis. The authors [15] proposed factor rotation in terms of decision-making centroids. The method was used to assess the risk of *lymphocytic leukaemia*.

The article presents a new concept of GPCA for building main components in the pca method. For this purpose, the *stochastic-gradient-optimization* method was used [16].

In the case of GPCA properties and eigenvectors we are looking for a K matrix such that:

$$K_{i,j} = L(Z_i, Z_j), \tag{2}$$

where L is a function of the goal, Z is a standardized variable, k is e.g. the kernel:

$$L(Z_i, Z_j) = \sum_{i=1}^{n} \left(x_i - \omega^T Z_j \right)^2, \tag{3}$$

where: $L(Z_i, Z_j)$ is a overall error on the training set, ω^T is a gradient.

By minimizing the function $L(Z_i, Z_j)$ it starts with the selected start-up solution $\omega_0 = 0$. Then the gradient is determined at the point $\omega_{k-1}, \alpha_k \nabla L (\omega_{k-1})$. The step along the negative gradient is determined one by one:

$$\omega_k = \omega_{k-1} - \alpha_k \nabla L (\omega_{k-1}), \tag{4}$$

where α_k is the step length determined before the linear search. We calculate the gradient ∇_L using the difference:

$$\frac{\partial \left(Z_i - \omega^T Z_j \right)^2}{\partial \omega_j} = -2 \left(Z_i - \omega^T Z_j \right) Z_i j \tag{5}$$

Finally

$$\nabla_L (\omega) = -2 \left(x_i - \omega^T Z_j \right) Z_j. \tag{6}$$

The number of principal components can now be represented as a linear combination of original variables Z

$$G_{k_{ij}} = \sum_{i=1}^{k} \sum_{j=1}^{m} a_{k_{ij},j} Z_j, \tag{7}$$

where m is the number of primary variables in the training set, w is the number of main components, Z_j is the j-th standardized variable, $G_{k_{ij}}$ is the i-th main component, $a_{k_{ij},j}$ are factor loads.

The developed GPCA method can be used in non-linear feature spaces. Other kernel functions may be proposed depending on the class the problem. In the article we consider a linear case.

3   Experimental Set-Up

The aim of the research is to build a feature extraction method that will allow more accurate classification of children with multiple sclerosis. The problem is important because the prognosis for the development of the disease is an extremely difficult process. Often, only appropriately selected variables allow for accurate classification of children to certain risk groups. The developed method gives a chance to build a tool that will support the physician in diagnostics and thus can contribute to the correct diagnosis and treatment of children. Because multiple sclerosis does not give initial clear-cut symptoms, well-chosen variables and risk groups can improve the quality of classification. This goal has become the most important reason for undertaking research on the construction of the extraction model, which will form the basis for classification using known algorithms. Similar studies have already been conducted and the developed CCPCA method [15] has found real application in the classification people with lymphocytic leukaemia. Particular attention was paid to the newly developed GPCA concept focusing on the optimization of factor rotation axes using the gradient method.

The real-world dataset was used in own research. Actual data relate to prognosis of multiple sclerosis in children. The data contained 230 instances and 20 features and two classes: 1 – poor prognosis, 2 – good prognosis. The number of respondents in the classes is 110, 120 instances. So we have balanced data.

In the experiments, several methods of extracting features known from the literature have been compared. Including: PCA (*Principal Component Analysis*) [5], KPCA (*Kernel Principal Component Analysis*) [7], CCPCA (Centroid Class Principal Component Analysis) [15], FA (*Factor Analysis*) [9], ICA (*Independent Component Analysis*) [10], GPCA (*Gradient Component Analysis*), which is the proposed proprietary method in this article.

Two experiments were performed in the tests, in which the accuracy score for three classifiers was verified in succession: SVM (*Support Vector Machine*, RF (*Random Forest*) and k-NN (*k-Nearest Neighbours*).

The *accuracy score metric* was used to assess the quality of the classification. Wilcoxon signed rank test at statistical significance level $\alpha = 0.05$, was used to assess the differences between accuracy for different methods and algorithms. A five-stratified cross-validation was used in all experiments.

4   Experimental Evaluation

The conducted research was divided into two experiments. The results of the second experiment depend on the first experiment. In the first experiment, the number of principal components were determined experimentally for the PCA, CCPCA and GPCA methods, which explain the set threshold of total variance. Thanks to this approach, we control the selection of main components, and thus the number of features that will form the basis of the classification. The thresholds for which the best algorithm classifications were obtained were included in the second experiment.

4.1   Experiment 1 - Determining the Quality of the Classification Depending on the Threshold of Total Explained Variance

Experiment 1 was carried out for three PCA, CCPCA and GPCA methods. The thresholds of explained total variance were adopted by 1 to 100. The study was conducted on three algorithms SVM, RF and k-NN. The results are presented in the chart Figs. 1 and 2.

The results of the tests in Experiment 1 show that for each PCA, CCPCA and GPCA method there is a threshold of total variance at which the quality of all classifiers is the highest. As you can see, these thresholds are consistent and the

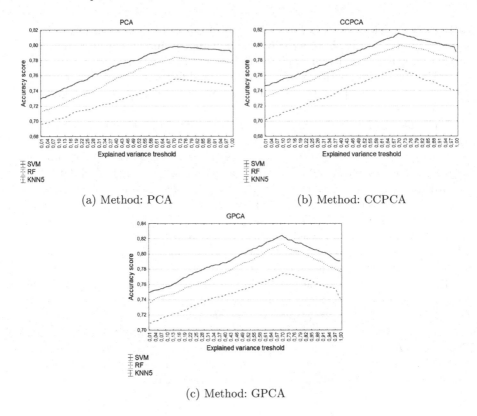

(a) Method: PCA   (b) Method: CCPCA

(c) Method: GPCA

Fig. 1. The plot of the dependence of the classification accuracy on the applied thresholds of the total explained variance for the methods of extracting the PCA, CCPCA and GPCA features on 230 teaching standards.

best results of correct classifications with each PCA method and classification algorithm are within 68–72%. It should be noted that for threshold 1 all features are taken for classification. In the case of 0.01, we have a situation where there is only one main component that combines are one to three attributes. For the 0.7 threshold, there are 3 main components. Also note that there is a slight data drift for different and near thresholds. However, as you can see, matching attributes to principal components is getting better. Therefore, there is a very interesting conclusion that as the total variance is threshold, the quality of matching attributes to these components increases. Figure 2 shows the results showing which features were assigned to a given principal component. The basis for classification of features into main components was the factor load value $\lambda > 0.6$. The results indicate that we will get a better fit for decision class 2 of the problem for component 1, and class 2 will be better classified by the set of features in components 2 and 3. Based on the GPCA method, the features $Z7$, $Z8$, $Z10$, $Z12$, $Z14$ and $Z18$ were rejected, which do not make a significant contribution to explaining object classes.

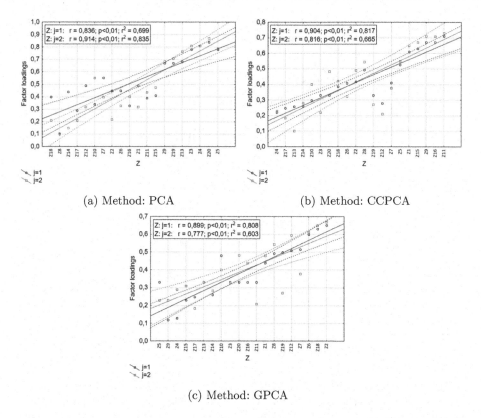

(a) Method: PCA   (b) Method: CCPCA

(c) Method: GPCA

Fig. 2. Plot of the relationship between the selection of object features for each of the three main components and the factor load values

4.2   Experiment 2. Determining the Quality of Classification for Various Methods of Feature Extraction

The purpose of the experiment is to verify how proprietary CCPCA and GPCA algorithms perform in the task of extracting features against other methods, i.e. PCA, KPCA, FA and ICA. The goal was achieved by checking the quality of real data classification using three algorithms: SVM, RF and k-NN. Based on the results obtained in experiment 1, 70% of the total explained variance for the PCA, CCPCA and GPCA methods was selected for the training data set. The Accuracy score obtained and Wilcoxon signed rank test is shown in Table 1. The first measuring points with the names of the algorithms relate to the case without using the feature extraction method. The next results, i.e. PCA, CCPCA, GPCA, KPCA, FA and ICA relate to the classification for a given algorithm after the extraction of features by a given method.

Table 1. The results of the experiments for the binary case with application of *accurace-score* metrics. In the columns the algorithms are presented, where NO means lack of extraction of an object's features.

Method	SVM	RF	KNN
1 NO	0.791 –	0.740 –	0.750 –
2 PCA	0.798 1	0.755 1	0.770 1
3 CCPCA	0.823 1,2,5,6,7	0.769 1,2,5,6,7	0.828 1,2,5,6,7
4 GPCA	0.826 1,2,5,6,7	0.771 1,2,5,6,7	0.833 1,2,5,6,7
5 KPCA	0.810 1,2	0.764 1,2,7	0.802 1,2,7
6 FA	0.806 1,2	0.759 1,2	0.797 1,2
7 ICA	0.806 1,2	0.757 1,2	0.793 1,2

The first significant conclusion from the research is that after extraction with any of the methods, the quality of classification with each of the three algorithms increased statistically significantly ($p < 0.05$). In the task of feature extraction, the best results are obtained by using the GPCA and CCPCA methods. Classification quality after application of GPCA and CCPCA were statistically comparable. Methods KPCA and FA don't differ significantly from each other. Method ICA for algorithms RF and KNN gave better results than in the case of extraction with the ica method ICA.

5 Conclusions

The purpose of the work was to develop a feature extraction method based on updating the property matrix and eigenvector values. In this task, the stochastic gradients method was used, where the function of the goal was the regression function. The study was conducted on a balanced set describing prognosis of children with multiple sclerosis. In during the analysis, it was possible to create a model that gives promising results for such a task. Two experiments were carried out in the work. The first assumed estimation of the GPCA model parameters, i.e. the threshold of the greedy explained variance giving the best quality of classification, estimation of the belonging of variables to the main components. In experiment 2, the quality of SVM, RF and k-NN algorithm classification was tested for various methods of feature extraction. The obtained results showed that the best extraction method is GPCA and CCPCA. The method of stochastic gradients used in the task of minimizing the error in estimating the matrix of eigenvector values proved to be a good approach. The estimation of GPCA components was also carried out for each decision class. In this way, although the same sets of characteristics for each class in each component were obtained, but different matching attributes of the teaching set, which in turn contributed to improving the quality of classification. The GPCA algorithm proved comparable to CCPCA method which was based on *Varimax* rotation normalized with respect to decision-making centroids. The elaborated method was, as already mentioned, tested on real data with MS disease in children. However, it can be used for other learning collections. In further research, the developed method will be tested on other learning sets, which will confirm the ability to handle various types of data. The biggest problem that can be encountered in using the stochastic gradient approach is the algorithm step.

References

1. Bellman, R.E.: Adaptive Control Processes: A Guided Tour, vol. 2045. Princeton University Press, Princeton (2015)
2. Jimenez, L.O., Landgrebe, D.A.: Hyperspectral data analysis and supervised feature reduction via projection pursuit. IEEE Trans. Geosci. Remote Sens. **37**(6), 2653–2667 (1999)
3. Hughes, G.: On the mean accuracy of statistical pattern recognizers. IEEE Trans. Inf. Theory **14**(1), 55–63 (1968)
4. Fort, G., Lambert-Lacroix, S.: Classification using partial least squares with penalized logistic regression. Bioinformatics **21**(7), 1104–1111 (2004)
5. Alpaydin, E.: Introduction to Machine Learning. MIT Press, Cambridge (2009)
6. Ringnér, M.: What is principal component analysis? Nat. Biotechnol. **26**(3), 303 (2008)
7. Schölkopf, B.: The kernel trick for distances. In: Advances in Neural Information Processing Systems, pp. 301–307 (2001)
8. Mao, K.Z.: Identifying critical variables of principal components for unsupervised feature selection. IEEE Trans. Systems Man Cybern. Part B (Cybern.) **35**(2), 339–344 (2005)
9. Jain, P.M., Shandliya, V.K.: A survey paper on comparative study between principal component analysis (PCA) and exploratory factor analysis (EFA). Int. J. Comput. Sci. Appl. **6**(2), 373–375 (2013)
10. Hyvärinen, A., Oja, E.: Independent component analysis: algorithms and applications. Neural Netw. **13**(4–5), 411–430 (2000)
11. Kaznowska, E., et al.: The classification of lung cancers and their degree of malignancy by FTIR, PCA-LDA analysis, and a physics-based computational model. Talanta **186**, 337–345 (2018)
12. Jiang, J., Ma, J., Chen, C., Wang, Z., Cai, Z., Wang, L.: SuperPCA: a super pixel-wise PCA approach for unsupervised feature extraction of hyperspectral imagery. IEEE Trans. Geosci. Remote Sens. **56**(8), 4581–4593 (2018)
13. Dong, Y., Qin, S.J.: A novel dynamic PCA algorithm for dynamic data modelling and process monitoring. J. Process Control **67**, 1–11 (2018)
14. Lakshmanaprabu, S.K., Mohanty, S.N., Shankar, K., Arunkumar, N., Ramirez, G.: Optimal deep learning model for classification of lung cancer on CT images. Future Gener. Comput. Syst. **92**, 374–382 (2019)

15. Topolski, M., Topolska, K.: Algorithm for constructing a classifier team using a modified PCA (Principal Component Analysis) in the task of diagnosis of acute lymphocytic leukaemia type B-CLL. In: Pérez García, H., Sánchez González, L., Castejón Limas, M., Quintián Pardo, H., Corchado Rodríguez, E. (eds.) HAIS 2019. LNCS (LNAI), vol. 11734, pp. 614–624. Springer, Cham (2019). https://doi.org/10.1007/978-3-030-29859-3_52
16. Bootou, L.: Large-scale machine learning with stochastic gradient descent. In: Proceedings of COMPSTAT' 2010, pp. 177–186 (2010)
17. Krawczyk, B., Ksieniewicz, P., Woźniak, M.: Hyperspectral image analysis based on color channels and ensemble classifier. In: Polycarpou, M., de Carvalho, A.C.P.L.F., Pan, J.-S., Woźniak, M., Quintian, H., Corchado, E. (eds.) HAIS 2014. LNCS (LNAI), vol. 8480, pp. 274–284. Springer, Cham (2014). https://doi.org/10.1007/978-3-319-07617-1_25

Grammatical Inference by Answer Set Programming

Wojciech Wieczorek[1]([✉]) [ID], Łukasz Strąk[1] [ID], Arkadiusz Nowakowski[1] [ID], and Olgierd Unold[2] [ID]

[1] Institute of Computer Science, University of Silesia in Katowice, Sosnowiec, Poland
{wojciech.wieczorek,lukasz.strak,arkadiusz.nowakowski}@us.edu.pl
[2] Department of Computer Engineering,
Wrocław University of Science and Technology, Wrocław, Poland
olgierd.unold@pwr.edu.pl

Abstract. In this paper, the identification of context-free grammars based on the presentation of samples is investigated. The main idea of solving this problem proposed in the literature is reformulated in two different ways: in terms of general constrains and as an answer set program. In a series of experiments, we showed that our answer set programming approach is much faster than our alternative method and the original SAT encoding method. Similarly to a pioneer work, some well-known context-free grammars have been induced correctly, and we also followed its test procedure with randomly generated grammars, making it clear that using our answer set programs increases computational efficiency. The research can be regarded as another evidence that solutions based on the stable model (answer set) semantics of logic programming may be a right choice for complex problems.

Keywords: Grammatical inference · Answer set programming · Constraint satisfaction problem

1 Introduction

In grammatical inference [9], a learning algorithm LA takes a finite sequence (usually strings) of examples as input and outputs a language description (usually grammars). There are two main types of presentations: (i) A *text* for a language L is an infinite sequence of strings $x_1, x_2, \ldots$ from L such that every string of L occurs at least once in the text; (ii) An *informant* for a language L is an infinite sequence of pairs $(x_1, d_1), (x_2, d_2), \ldots$ in $\Sigma^* \times \mathbb{B}$ such that every string of Σ^* occurs at least once in the sequence and $d_i = \text{true} \iff x_i \in L$. The inference algorithms that use type (ii) of information are said to learn from *positive and negative examples*. From the Gold's results [7], we know that the class of context-free languages (and even regular languages) cannot be identified from

This research was supported by National Science Center (Poland), grant number 2016/21/B/ST6/02158.

V. V. Krzhizhanovskaya et al. (Eds.): ICCS 2020, LNCS 12140, pp. 45–58, 2020.
https://doi.org/10.1007/978-3-030-50423-6_4

presentation (i), but can be identified using presentation (ii). However, de la Higuera [8] showed that it is computationally hard.

In this work, the following informant learning environment is exploited. Suppose that the inferring process is based on the existence of an *Oracle*, which can be seen as a device that:

1. Knows the language and has to answer correctly.
2. Can answer *equivalence queries*. They are made by proposing some hypothesis to the Oracle. The hypothesis is a grammar representing the unknown language. The Oracle answers YES in the positive case. In the negative case, the Oracle has to return the shortest string in the symmetric difference between the target language and the submitted hypothesis.

Then the following procedure can be applied. Start from a small[1] sample S and $k = 1$. The parameter k denotes the number of non-terminal symbols in the target grammar. Run an answer set program (or another exact method). Every time it turns out that there is no solution that satisfies all of the constraints, increase k by 1. As long as the Oracle returns a pair (x, d) in response to an equivalent query, add (x, d) to S and run the answer set program again (or respectively another exact method). Stop after the answer is YES. Unfortunately, there is no guarantee that the procedure will terminate in a polynomial number of steps, even when the target language is regular [1]. The equivalence checking may be done by random sampling. The positive answer could be incorrect, but this probability decreases if the sampling is repeated.

A very similar procedure for the induction of context-free grammars was proposed by Imada and Nakamura [11]. However, for the exact searching of k-variable grammar, they used Boolean formulas and applied an SAT solver. We took over their main Boolean variables, treating them as predicates, and then constructed a new encoding founded on answer set programming. In an alternative approach, we used general constraints of Gurobi Optimizer[2] instead of ASP.

1.1 Related Work

The most closely related work to CFG identification is by Imada and Nakamura [11]. They proposed a way to synthesize CFGs from positive and negative samples based on solving a Boolean satisfiability problem (SAT). They translated the learning problem for a CFG into a SAT, which is then solved by a SAT solver. The result of the SAT solver satisfying the SAT contains a minimal set of rules (it can be easily changed to a minimal set of variables) that derives all positive samples and no negative samples.

They used one *derivation constraint* and two main types of Boolean variables:

---

[1] We are aware of this imprecision. The number of words and their lengths should allow of executing a program in a reasonable amount of time. In experiments, we took two words: one example and one counter-example.

[2] https://www.gurobi.com/.

Derivation variables. A set of derivation variables represents a relation between nonterminal symbols and substrings (in other words, derivation or parse tree) of each (positive or negative) sample w as follows: for any substring x of w and $p \in V$, the derivation variable T_x^p represents that the nonterminal p derives the string x.

Rule variables. A set of rule variables represents a rule set as follows: for any $p, q, r \in V$, $a \in \Sigma$, a variable R_{qr}^p (or R_a^p) determines whether the production rule $p \to q\,r$ (or $p \to a$) is a member of the set of rules or not.

The derivation constraint is a set of following Boolean expressions for any string $a_1 \cdots a_n$ $(n > 1)$ and nonterminal $p \in V$.

$$T_{a_1 \cdots a_n}^p \leftrightarrow \bigvee_{i=1}^{n-1} \bigvee_{q \in V} \bigvee_{r \in V} \left(R_{qr}^p \wedge T_{a_1 \cdots a_i}^q \wedge T_{a_{i+1} \cdots a_n}^r \right).$$

Nakamura et al. have been working on another approach for incremental learning of CFGs implemented in the Synapse system [15]. This approach is based on rule generation by analyzing the results of bottom-up parsing for positive samples and searching for rule sets. Their system can also learn similar CFGs but does it only from positive samples. Both methods synthesized similar rule sets for each language in their experiments. They reported that the computation time by the SAT-based approach is rather shorter than Synapse in most languages.

1.2   Our Contribution

The purpose of the present proposal is to investigate to what extent the power of an ASP solver makes it possible to tackle the context-free inference problem for large-size instances and to compare our approach with the original one. Because of the possibility of future comparisons with other methods, the Python implementation[3] of our winning method is given via GitLab.

The main original scientific contributions are as follows:

- the formulation of the induction of a k-variable context-free grammar in terms of logical rules with answer set semantics;
- the formulation of the induction of a k-variable context-free grammar in terms of general constraints;
- the construction of an informant learning algorithm based on ASP, CSP, and SAT solvers;
- the conduct of an appropriate statistical test in order to determine the fastest CFG inference method.

This paper is organized into five sections. In Sect. 2, we present necessary definitions and facts originating from formal languages and declarative

---

[3] The Python scripting language is used only for generating appropriate AnsProlog facts.

problem-solving. Section 3 describes our inference algorithms: (a) based on solving an answer set program, and (b) based on solving a constraint satisfaction program, including general constraints such as AND/OR. Section 4 shows the experimental results of our approaches in comparison with the original one. Concluding comments are made in Sect. 5.

2   Preliminaries

We assume the reader to be familiar with basic context-free languages theory, e.g., from [10], so that we introduce only some notations and notions used later in the paper.

2.1   Words and Languages

An *alphabet* is a finite, non-empty set of symbols. We use the symbol Σ for the alphabet. A *word* is a finite sequence of symbols chosen from the alphabet. We denote the length of the word w by $|w|$. The *empty word* ε is the word with zero occurrences of symbols. Let x and y be words. Then xy denotes the *catenation* of x and y, that is, the word formed by making a copy of x and following it by a copy of y. As usual, Σ^* denotes the set of words over Σ. The word w is called a *prefix* of the word u if there is a word x such that $u = wx$. We call it a *proper* prefix if $x \neq \varepsilon$. The word w is called a *suffix* of the word u if there is a word x such that $u = xw$. It is a *proper* suffix if $x \neq \varepsilon$. A *factor* (or *subword*) is a prefix of a suffix. A set of words, all of which are chosen from some Σ^*, where Σ is a particular alphabet, is called a *language*.

2.2   Context-Free Grammars

A *context-free grammar* (CFG) is defined by a quadruple $G = (V, \Sigma, P, v_0)$, where V is an alphabet of *variables* (or sometimes *non-terminal symbols*), Σ is an alphabet of *terminal symbols* such that $V \cap \Sigma = \emptyset$, P is a finite set of *production rules* in the form $A \to \alpha$ for $A \in V$ and $\alpha \in (V \cup \Sigma)^*$, and v_0 is a special non-terminal symbol called the *start symbol*. For simplicity's sake, we write $A \to \alpha_1 \mid \alpha_2 \mid \cdots \mid \alpha_k$ instead of $A \to \alpha_1, A \to \alpha_1, \ldots, A \to \alpha_k$. We call the word $x \in (V \cup \Sigma)^*$ a *sentential form*. Let u, v be two words in $(V \cup \Sigma)^*$ and $A \in V$. Then, we write $uAv \Rightarrow uxv$, if $A \to x$ is a rule in P. That is, we can substitute the word x for symbol A in a sentential form if $A \to x$ is a rule in P. We call this rewriting a *derivation*. For any two sentential forms x and y, we write $x \Rightarrow^* y$, if there exists a sequence $x = x_0, x_1, x_2, \ldots, x_n = y$ of sentential forms such that $x_i \Rightarrow x_{i+1}$ for all $i = 0, 1, \ldots, n - 1$. The language $L(G)$ generated by G is the set of all words over Σ that are generated by G; that is, $L(G) = \{x \in \Sigma^* \mid v_0 \Rightarrow^* x\}$. A language is called a *context-free language* if it is generated by a context-free grammar. Assume that G is the unknown (target) CFG to be identified. An *example* (a *positive word*) of G is a word in $L(G)$, and a *counter-example* (a *negative word*) of G is a word not in $L(G)$.

A *normal form* for context-free grammars is a form, for which any grammar can be converted to the respective normal form version. Amongst all normal forms for context-free grammars, the most useful and the most well-known one is the Chomsky normal form (CNF). A grammar is said to be in *Chomsky normal form* if each of its rules is in one of two possible forms:

(a) $X \to x$, $x \in \Sigma$, $X \in V$, or
(b) $X \to Y Z$, $X, Y, Z \in V$.

2.3 Answer Set Programming

We will briefly introduce the idea of answer set programming (ASP). Those who are interested in a more detailed description of the topic, alternative definitions, and the formal specification of this kind of logic programming are referred to handbooks [3,6], and [12].

A variable or constant is a *term*. An *atom* is $a(t_1, \ldots, t_n)$, where a is a *predicate* of arity n and $t_1, \ldots, t_n$ are terms. A *literal* is either a *positive literal* p or a *negative literal* $\neg p$, where p is an atom.

A *rule* r is a clause of the form

$$a_0 \leftarrow a_1 \wedge \cdots \wedge a_k \wedge \neg a_{k+1} \wedge \cdots \wedge \neg a_m \quad m \geq 0, \tag{1}$$

where $a_0, \ldots, a_m$ are atoms. The atom a_0 is the *head* or r, while the conjunction $a_1 \wedge \cdots \wedge a_k \wedge \neg a_{k+1} \wedge \cdots \wedge \neg a_m$ is the *body* of r. By $H(r)$, we denote the head atom, and by $B(r)$ the set $\{a_1, \ldots, a_k, \neg a_{k+1}, \ldots, \neg a_m\}$ of the body literals. $B^+(r)$ ($B^-(r)$, resp.) denotes the set of atoms occurring positively (negatively, resp.) in $B(r)$. A *program* (also called ASP program) is a finite set of rules. A $\neg$-free program is called *positive*. A term, atom, literal, rule, or a program is *ground* if no variables appear in it.

Let $\mathcal{P}$ be a program. Let r be a rule in $\mathcal{P}$, a *ground instance* of r is a rule obtained from r by replacing[4] every variable X in r by constants occurring in $\mathcal{P}$. We denote the set of all the ground instances of the rules occurring in $\mathcal{P}$ by $\mathrm{ground}(\mathcal{P})$.

An *interpretation* I for $\mathcal{P}$ is a set of ground atoms. A ground positive literal A is *true* (*false*, resp.) w.r.t. I if $A \in I$ ($A \notin I$, resp.). A ground negative literal $\neg A$ is *true* (*false*, resp.) w.r.t. I if $A \notin I$ ($A \in I$, resp.).

Let r be a ground rule in $\mathrm{ground}(\mathcal{P})$. The head of r is *true* w.r.t. I if $H(r) \in I$. The body of r is *true* w.r.t. I if all body literals of r are true w.r.t. I (i.e., $B^+(r) \subseteq I$ and $B^-(r) \cap I = \emptyset$) and is *false* w.r.t. I otherwise. The rule r is *satisfied* (or *true*) w.r.t. I if r head is true w.r.t. I or r body is false w.r.t. I.

A *model* for $\mathcal{P}$ is an interpretation M for $\mathcal{P}$ such that every rule $r \in \mathrm{ground}(\mathcal{P})$ is true w.r.t. M.

---

[4] This process can be done efficiently, because many ground instances can be discarded; see Chapter 4 of [6].

Given a program $\mathcal{P}$ and an interpretation I, the *reduct* $\mathcal{P}^I$ is the set of positive rules defined as follows:

$$\mathcal{P}^I = \{H(r) \leftarrow \bigwedge B^+(r) \mid r \in \text{ground}(\mathcal{P}) \text{ and } B^-(r) \cap I = \emptyset\}. \qquad (2)$$

I is an *answer set* of $\mathcal{P}$ if I is the $\subseteq$-smallest model for $\mathcal{P}^I$.

Over the last years, answer set programming has emerged as a declarative problem-solving paradigm. It is a programming methodology rooted in research on artificial intelligence and computational logic, and researchers use it in many areas of science and technology. For experiments we took advantages of CLINGO—one of the most efficient and widely used answer set programming system available[5] today. In addition to standard definitions, CLINGO allows to define *constraints*, i.e., rules with the empty head, for instance

$$\leftarrow a(t) \qquad (3)$$

By adding this constraint to a program, we eliminate its answer sets that contain $a(t)$. Adding the 'opposite' constraint

$$\leftarrow \neg a(t) \qquad (4)$$

eliminates those answers that do not contain $a(t)$. A constraint can be translated into a normal rule. To this end, the constraint

$$\leftarrow a_1 \wedge \cdots \wedge a_k \wedge \neg a_{k+1} \wedge \cdots \wedge \neg a_m \qquad (5)$$

is mapped onto the rule

$$x \leftarrow a_1 \wedge \cdots \wedge a_k \wedge \neg a_{k+1} \wedge \cdots \wedge \neg a_m \wedge \neg x \qquad (6)$$

where x is a new atom.

Example. Suppose we have three numbered urns and two distinguishable balls. Every ball has been put to an urn, maybe to the same. An ASP program to code this knowledge is as follows:

$$\text{urn}(1) \leftarrow \qquad (7)$$
$$\text{urn}(2) \leftarrow \qquad (8)$$
$$\text{urn}(3) \leftarrow \qquad (9)$$
$$\text{ball}(q) \leftarrow \qquad (10)$$
$$\text{ball}(r) \leftarrow \qquad (11)$$
$$\text{contains}(U, B) \leftarrow \text{urn}(U) \wedge \text{ball}(B) \wedge \neg\text{not\_in}(U, B) \qquad (12)$$
$$\text{not\_in}(U, B) \leftarrow \text{urn}(U) \wedge \text{urn}(V) \wedge U \neq V \wedge \text{ball}(B) \wedge \text{contains}(V, B) \qquad (13)$$
$$\text{in}(B) \leftarrow \text{urn}(U) \wedge \text{ball}(B) \wedge \text{contains}(U, B) \qquad (14)$$
$$\leftarrow \text{ball}(B) \wedge \neg\text{in}(B) \qquad (15)$$

---

[5] https://potassco.org/.

Please notice that as usual in logic programming, identifiers with initial upper-case letters are assigned to variables. Rules 7–11 are simple facts concerning urns and balls. Rules 12 and 13 define predicates that tell whether a ball is inside in a particular urn. Inequality $U \neq V$ is only used during grounding to eliminate some ground instances of rule 13. It is worth mentioning that grounding systems do not make unnecessary replacements, for example, 1 for U. Rules 14 and 15 ensure that every ball is exactly in one urn.

Suppose now that we have discovered that urn 2 is empty and we want to know possible configurations. It is enough to add two facts:

$$\text{not\_in}(2, q) \leftarrow \tag{16}$$

$$\text{not\_in}(2, r) \leftarrow \tag{17}$$

and find all answer sets. A possible answer set is: ball(q), ball(r), urn(1), urn(2), in(r), not_in(2, q), not_in(2, r), not_in(3, q), not_in(3, r), contains(1, q), in(q), urn(3), contains(1, r), which describes the placement of both balls into the first urn.

CLINGO also allows using choice constructions, for instance:

$$\{\text{p}(U, B)\colon \text{urn}(U)\} = 2 \leftarrow \text{ball}(B) \tag{18}$$

describes all possible ways to choose which two of the atoms p(1, q), p(2, q), p(3, q) and which two of the atoms p(1, r), p(2, r), p(3, r) are included in the resultant model. Before and after an expression in braces, we can put integers, which express bounds on the cardinality of the stable models described by the rule. The number on the left is the lower bound (0 is default), and the number on the right is the upper bound (unbounded is default).

3   Proposed Encodings for the Induction of CFGs

Our translation converts CFG identification into an ASP program (the main approach) and CSP model (an alternative approach, constraint satisfaction problem). Suppose we are given a sample composed of examples, S_+, and counter-examples, S_-, over an alphabet Σ, and a positive integer k. We want to find a k-variable CFG $G = (V, \Sigma, P, v_0)$ such that $S_+ \subseteq L(G)$ and $S_- \cap L(G) = \emptyset$.

3.1   Using Logic Programming with Answer Set Semantics

Let F be the set of all factors (excluding the empty word) of $S_+ \cup S_-$. Let us now see how to describe the rules for the relationship between a grammar G and a sample $S_+ \cup S_-$ in terms of ASP. There are three main predicates: $y(I, J, L)$, which indicates the presence of $I \rightarrow JL$ in P; $w(I, Q)$, which indicates that $I \Rightarrow^* Q$, where Q represents a factor; and $z(I, A)$, which indicates the presence of $I \rightarrow A$.

1. We have the following domain specification, our facts.

$$\text{variable}(i) \leftarrow \qquad\qquad \text{for } i = 0, 1, \ldots, k - 1 \qquad\qquad (19)$$

$$\text{factor}(f) \leftarrow \qquad\qquad \text{for all } f \in F \qquad\qquad (20)$$

$$\text{terminal}(a) \leftarrow \qquad\qquad \text{for all } a \in \Sigma \qquad\qquad (21)$$

$$\text{positive}(s) \leftarrow \qquad\qquad \text{for all } s \in S_+ \qquad\qquad (22)$$

$$\text{negative}(s) \leftarrow \qquad\qquad \text{for all } s \in S_- \qquad\qquad (23)$$

$$\text{compose}(f, b, c) \leftarrow \qquad\qquad \text{for such } f, b, c \in F \text{ that } f = bc \qquad\qquad (24)$$

2. The next rules ensure that in a grammar G a factor can or cannot be derived from a specific variable and ensure that in the grammar there is a subset of all possible productions.

$$\{\text{w}(I, F)\} \leftarrow \text{variable}(I) \wedge \text{factor}(F) \qquad\qquad (25)$$

$$\{\text{z}(I, A)\} \leftarrow \text{variable}(I) \wedge \text{terminal}(A) \qquad\qquad (26)$$

$$\{\text{y}(I, J, L)\} \leftarrow \text{variable}(I) \wedge \text{variable}(J) \wedge \text{variable}(L) \qquad\qquad (27)$$

$$\text{w}(I, A) \leftarrow \text{variable}(I) \wedge \text{terminal}(A) \wedge \text{z}(I, A) \qquad\qquad (28)$$

$$\text{z}(I, A) \leftarrow \text{variable}(I) \wedge \text{terminal}(A) \wedge \text{w}(I, A) \qquad\qquad (29)$$

3. All examples should be accepted, and no counter-example can be accepted.

$$\leftarrow \text{positive}(F) \wedge \neg\text{w}(0, F) \qquad\qquad (30)$$

$$\leftarrow \text{negative}(F) \wedge \text{w}(0, F) \qquad\qquad (31)$$

4. For every $f \in F$ for which $|f| \geq 2$ and for every pair (b, c) $(b, c \in F)$ of such factors that $bc = f$, f can be derived from a non-terminal I if there are two non-terminals, J and L, such that b can be derived from J, c can be derived from L, and there is a production $I \rightarrow J L$.

$$\text{w}(I, F) \leftarrow \text{variable}(I) \wedge \text{variable}(J) \wedge \text{variable}(L)$$
$$\wedge \text{compose}(F, B, C) \wedge \text{y}(I, J, L) \wedge \text{w}(J, B) \wedge \text{w}(L, C) \qquad (32)$$

5. On the other hand, if $I \Rightarrow^* f$, then at least one such pair (J, L) should exist, that $I \rightarrow J L$ is in P and $J \Rightarrow^* b$ and $L \Rightarrow^* c$.

$$\leftarrow \text{variable}(I) \wedge \text{factor}(F) \wedge \neg\text{terminal}(F) \wedge \text{w}(I, F)$$
$$\wedge \{\text{y}(I, J, L)\colon \text{variable}(J) \wedge \text{variable}(L)$$
$$\wedge \text{compose}(F, B, C) \wedge \text{w}(J, B) \wedge \text{w}(L, C)\} = 0 \qquad (33)$$

3.2   Using General Constraints

This time, instead of predicates, w, y, and z are binary variables. We use the following constraints

$$w_{0s} = 1 \qquad\qquad \text{for all } s \in S_+ \qquad\qquad (34)$$

$$w_{0s} = 0 \qquad\qquad \text{for all } s \in S_- \qquad\qquad (35)$$

and

$$w_{if} \leftrightarrow \sum_{j,l \in K,\, bc=f} y_{ijl} \wedge w_{jb} \wedge w_{lc} + (z_{if} \text{ if } f \in \Sigma) \qquad (36)$$

for each $(i, f) \in K \times F$, where $\alpha \leftrightarrow \beta$ means if $\alpha = 0$ then $\beta = 0$ and if $\alpha = 1$ then $\beta \geq 1$, and $K = \{0, 1, \ldots, k-1\}$.

4 Experimental Results

In this section, we describe some experiments comparing the performance of our approaches implemented[6] in Python, using CLINGO (ASP) and using Gurobi Optimizer, with our implementation of Imada et al. algorithm [11] using the PicoSAT solver (SAT), when positive and negative words are given. For these experiments, we use a set of 40 samples: partly based on randomly generated grammars (33 samples) and partly based on the set of fundamental CFGs appearing in grammatical inference research (the last 7 samples).

4.1 Benchmarks

For testing the learning power for general CFGs, we randomly generated 33 CFGs and prepared positive and negative samples with lengths no longer than 14 exhaustively enumerated for them. The grammars are in Chomsky normal form with 6 to 12 rules on the alphabet $\{a, b\}$. In every sample, positive words constitute not less than 20% of the total.

The last seven samples are also with lengths no longer than 14 exhaustively enumerated, but they were generated based on the following descriptions:

(a) The set of palindromes over $\{a, b\}$.
(b) The parentheses language: the set of strings consisting of equal numbers of a's and b's such that every prefix does not have more b's than a's.
(c) The set of strings consisting of b's twice as many as a's.
(d) The set of strings of a's and b's not of the form ww.
(e) The complement of the language (b).
(f) $\{a^n b^n \mid n \geq 1\}$.
(g) The set of strings consisting of equal numbers of a's and b's.

4.2 Performance Comparison

In all experiments, we used Intel Xeon CPU E5-2650 v2, 2.6 GHz (single-core out of eight), under Ubuntu 18.04 operating system with 60 GB available RAM. Algorithm 1 shows the process for synthesizing a grammar (the set of production rules with v_0 being always the start symbol) from positive and negative words. In the algorithm, S_+ and S_- represent the set of positive and negative words

---

[6] https://gitlab.com/answer-set-programming/asp4cfg.

Algorithm 1. Synthesize CFG G from examples and counter-examples

---

Require: S_+ positive words, S_- negative words, Σ a set of terminal symbols
Ensure: G a context-free grammar consistent with S_+ and S_-
   $S'_+ \leftarrow \{$the shortest word from $S_+\}$
   $S'_- \leftarrow \{$the shortest word from $S_-\}$
   $k \leftarrow 1$
   loop
      $R \leftarrow \text{Convert}(S'_+, S'_-, \Sigma, k)$
      find a stable model M for R by the solver
      while R has no stable model M **do**
         $k \leftarrow k + 1$
         $R \leftarrow \text{Convert}(S'_+, S'_-, \Sigma, k)$
         find a stable model M for R by the solver
      end while
      $P \leftarrow \text{Extract}(M)$
      $G \leftarrow (\{v_0, v_1, \ldots, v_{k-1}\}, \Sigma, P, v_0)$
      $X \leftarrow S_+ \setminus L(G)$
      $Y \leftarrow S_- \cap L(G)$
      if $X = \emptyset$ and $Y = \emptyset$ **then**
         return G
      else
         add appropriately the shortest word from $X \cup Y$ to S'_+ or to S'_-
      end if
   end loop

---

as an input. The variables S'_+ and S'_- hold sets of samples to be covered in the next loop iteration. The algorithm picks up a word from S_+ or S_- that is not covered by the inferred grammar G, and add it to S'_+ or S'_-. The function *Convert* translates the problem into a set of ASP rules R (or Gurobi general constraints or a Boolean expression). If the ASP solver (or Gurobi Optimizer or the SAT solver) finds a stable model M, the function *Extract* returns a set of production rules by analyzing the presence of particular $y(i, j, l)$ and $z(i, a)$ atoms. The algorithm repeats this process—increasing k to relaxe the limit on the number of non-terminals—until G covers the all given S_+ and S_-.

The results are listed in Table 1. In order to determine whether the observed CPU time differences between ASP's runs and the remaining methods' runs did not occur by chance, we use the Wilcoxon signed-rank test [17, pp. 915–916] for ASP vs SAT and ASP vs Gurobi. The *null hypothesis* to be tested is that the median of the paired differences is negative (against the alternative that it is positive). As we can see from Table 2, p-value is high in both cases, so the null hypothesis cannot be rejected, and we may conclude that using our ASP encoding is likely to improve CPU time performance for most of this kind of benchmarks.

Table 1. Execution times of exact solving CFG identification in seconds

Language	$|V|$	ASP	SAT	Gurobi
1	3	51.70	48.65	56.42
2	6	646.39	21049.22	>21050
3	4	74.31	189.85	143.76
4	5	75.90	347.84	>2000
5	4	27.91	64.82	18.36
6	5	75.96	335.98	10.33
7	4	68.35	61.87	>2000
8	4	57.14	118.25	28.85
9	3	45.17	94.86	73.03
10	5	211.33	568.12	568.06
11	5	62.48	166.65	>2000
12	3	21.50	58.12	33.28
13	6	112.69	705.80	>2000
14	6	943.02	4807.32	>4808
15	7	19358.09	252290.70	>252291
16	4	49.01	111.22	103.05
17	7	2921.44	8035.44	>8036
18	5	361.52	1369.22	>2000
19	5	63.47	238.71	186.10
20	2	12.96	5.64	3.88
21	5	96.68	512.83	671.62
22	2	11.38	12.02	10.54
23	3	11.84	43.03	9.92
24	4	109.98	159.73	176.49
25	3	22.65	22.40	29.65
26	5	38.74	271.30	420.11
27	5	94.76	295.81	>2000
28	5	216.61	625.07	>2000
29	5	271.88	324.43	>2000
30	6	228.98	412.16	>2000
31	2	10.97	15.29	19.84
32	5	62.17	293.98	105.74
33	3	10.42	18.30	13.15
34	5	31.13	49.28	32.83
35	3	12.84	20.97	12.86
36	4	118.17	76.98	73.74
37	6	173.66	191.42	>2000
38	4	29.33	54.63	36.71
39	4	4.12	21.00	9.02
40	3	66.71	50.65	40.40

Table 2. Obtained p-values from the Wilcoxon signed-rank test

ASP vs SAT	ASP vs Gurobi
0.999999647	0.999987068

4.3 ASP-Based CFG Induction on Bioinformatics Datasets

Our induction method can also be applied to other data, that are not taken from context-free infinite languages. We tried its classification quality on two bioinformatics datasets: WALTZ-DB database [4], composed by 116 hexapeptides known to induce amyloidosis (S_+) and by 161 hexapeptides that do not induce amyloidosis (S_-) and Maurer-Stroh et al. database from the same domain [14], where the ratio of S_+/S_- is 240/836.

We chose a few standard machine learning methods for comparison: BNB (Naive Bayes classifier for multivariate Bernoulli models [13, pp. 234–265]), DTC (Decision Trees Classifier, CART method [5]), MLP (Multi-layer Perceptron [16]), and SVM (Support Vector Machine classifier with the linear kernel [18]). In all methods except ASP and BNB, an unsupervised data-driven distributed representation, called ProtVec [2], was applied in order to convert words (protein representations) to numerical vectors. For using BNB, we represented words as binary-valued feature vectors that indicated the presence or absence of every pair of protein letters. In case of ASP, the training set was partitioned randomly into n parts, and the following process was being performed m times. Choosing one part for synthesizing a CFG and use rest $n - 1$ parts for validating it. The best of all m grammars—in terms of higher F-measure—was then confronted with the test set. For WALTZ-DB n and m have been set to 20, for Maurer-Stroh n has been set to 10 and m to 30. These values were selected experimentally based on the size of databases and the running time of the ASP solver.

To estimate the ASP's and compared approaches' ability to classify unseen hexapeptides repeated 10-fold cross-validation (cv) strategy was used. It means splitting the data randomly into 10 mutually exclusive folds, building a model on all but one fold, and evaluating the model on the skipped fold. The procedure was repeated 10 times and the overall assessment of the model was based on the mean of those 10 individual evaluations. Table 3 summarizes the performances of the compared methods on WALTZ-DB and Maurer-Stroh databases. It is noticable that the ASP approach achieved best F-score for smaller dataset (Maurer-Stroh) and an average F-score for the bigger one (WALTZ-DB), hence it can be used with a high reliability to recognize amyloid proteins. BNB is outstanding for the WALTZ-DB and almost as good as ASP for Maurer-Stroh database.

Table 3. Performance of compared methods on WALTZ-DB and Maurer-Stroh databases in terms of Precision (P), Recall (R), and F-score (F1)

Method	WALTZ-DB			Maurer-Stroh		
	P	R	F1	P	R	F1
ASP	0.38 ± 0.09	0.58 ± 0.12	0.45 ± 0.07	0.58 ± 0.12	0.66 ± 0.18	0.61 ± 0.12
BNB	0.51 ± 0.09	0.69 ± 0.14	0.59 ± 0.10	0.61 ± 0.11	0.60 ± 0.13	0.60 ± 0.11
DTC	0.43 ± 0.11	0.59 ± 0.26	0.46 ± 0.11	0.36 ± 0.20	0.74 ± 0.39	0.48 ± 0.26
MLP	0.49 ± 0.20	0.57 ± 0.27	0.46 ± 0.10	0.43 ± 0.09	0.90 ± 0.07	0.58 ± 0.10
SVM	0.37 ± 0.06	0.69 ± 0.07	0.48 ± 0.06	0.24 ± 0.21	0.51 ± 0.44	0.32 ± 0.28

5 Conclusion

In this paper, we proposed an approach for learning context-free grammars from positive and negative samples by using logic programming. We encode the set of samples, together with limits on the number of non-terminals to be synthesized as an answer set program. A stable model (an answer set) for the program contains a set of grammar rules that derives all positive samples and no negative samples. A feature of this approach is that we can synthesize a compact set of rules in Chomsky normal form. The other feature is that our learning method reflects future improvements on ASP solvers. We present experimental results on learning CFGs for fundamental context-free languages, including a set of strings composed of the equal numbers of a's and b's and the set of strings over $\{a, b\}$ not of the form ww. Another series of experiments on random languages shows that our encoding can speed up computations in comparison with SAT and CSP encodings.

References

1. Angluin, D.: Negative results for equivalence queries. Mach. Learn. **5**(2), 121–150 (1990). https://doi.org/10.1007/BF00116034
2. Asgari, E., Mofrad, M.R.K.: Continuous distributed representation of biological sequences for deep proteomics and genomics. PLoS ONE **10**(11), 1–15 (2015). https://doi.org/10.1371/journal.pone.0141287
3. Baral, C.: Knowledge Representation, Reasoning, and Declarative Problem Solving. Cambridge University Press, New York (2003)
4. Beerten, J., et al.: WALTZ-DB: a benchmark database of amyloidogenic hexapeptides. Bioinformatics **31**(10), 1698–1700 (2015)
5. Breiman, L., Friedman, J.H., Olshen, R.A., Stone, C.J.: Classification and Regression Trees. Wadsworth and Brooks, Monterey (1984)
6. Gebser, M., Kaminski, R., Kaufmann, B., Schaub, T.: Answer Set Solving in Practice. Morgan & Claypool Publishers, San Rafael (2012)
7. Gold, E.M.: Language identification in the limit. Inf. Control **10**, 447–474 (1967)
8. de la Higuera, C.: Characteristic sets for polynomial grammatical inference. Mach. Learn. **27**(2), 125–138 (1997). https://doi.org/10.1023/A:1007353007695

9. de la Higuera, C.: Grammatical Inference: Learning Automata and Grammars. Cambridge University Press, New York (2010)
10. Hopcroft, J.E., Motwani, R., Ullman, J.D.: Introduction to Automata Theory, Languages, and Computation, 2nd edn. Addison-Wesley, Reading (2001)
11. Imada, K., Nakamura, K.: Learning context free grammars by using SAT solvers. In: Proceedings of the 2009 International Conference on Machine Learning and Applications, pp. 267–272. IEEE Computer Society (2009)
12. Lifschitz, V.: Answer Set Programming. Springer, Cham (2019). https://doi.org/10.1007/978-3-030-24658-7
13. Manning, C.D., Raghavan, P., Schütze, H.: Introduction to Information Retrieval. Cambridge University Press, Cambridge (2008)
14. Maurer-Stroh, S., et al.: Exploring the sequence determinants of amyloid structure using position-specific scoring matrices. Nat. Methods **7**, 237–242 (2010)
15. Nakamura, K., Matsumoto, M.: Incremental learning of context free grammars based on bottom-up parsing and search. Pattern Recognint. **38**(9), 1384–1392 (2005). https://doi.org/10.1016/j.patcog.2005.01.004
16. Pedregosa, F., et al.: Scikit-learn: machine learning in Python. J. Mach. Learn. Res. **12**, 2825–2830 (2011)
17. Salkind, N.J.: Encyclopedia of Research Design. SAGE Publications Inc., London (2010)
18. Wu, T.F., Lin, C.J., Weng, R.C.: Probability estimates for multi-class classification by pairwise coupling. J. Mach. Learn. Res. **5**, 975–1005 (2004)

Dynamic Classifier Selection for Data with Skewed Class Distribution Using Imbalance Ratio and Euclidean Distance

Paweł Zyblewski$^{(\boxtimes)}$ ⓘ and Michał Woźniak ⓘ

Department of Systems and Computer Networks, Faculty of Electronics,
Wrocław University of Science and Technology,
Wybrzeże Wyspiańskiego 27, 50-370 Wrocław, Poland
{pawel.zyblewski,michal.wozniak}@pwr.edu.pl

Abstract. Imbalanced data analysis remains one of the critical challenges in machine learning. This work aims to adapt the concept of *Dynamic Classifier Selection* (DCS) to the pattern classification task with the skewed class distribution. Two methods, using the similarity (distance) to the reference instances and class imbalance ratio to select the most confident classifier for a given observation, have been proposed. Both approaches come in two modes, one based on the k-Nearest Oracles (KNORA) and the other also considering those cases where the classifier makes a mistake. The proposed methods were evaluated based on computer experiments carried out on 41 datasets with a high imbalance ratio. The obtained results and statistical analysis confirm the usefulness of the proposed solutions.

Keywords: Classifier ensemble · Dynamic Classifier Selection · Imbalanced data

1 Introduction

Traditional machine learning algorithms assume that the number of instances belonging to problem classes is relatively similar. However, it is worth noting that in many real problems the size of one class (*majority class*) may significantly exceed the size of the second one (*minority class*). This makes the algorithms biased towards the majority class, although the correct recognition of less common class is often more important. This research trend is known as learning from imbalanced data [8] and it is still widely discussed in scientific works.

There are three main approaches to dealing with the imbalanced data classification:

- *Data-level methods* focusing on modifying the training set in such a way that it becomes suitable for classic learning algorithms (e.g., *oversampling* and *undersampling*).

© Springer Nature Switzerland AG 2020
V. V. Krzhizhanovskaya et al. (Eds.): ICCS 2020, LNCS 12140, pp. 59–73, 2020.
https://doi.org/10.1007/978-3-030-50423-6_5

- *Algorithm-level methods* that modify existing classification algorithms to off-set their bias towards the majority class.
- *Hybrid methods* combining the strengths of the previously mentioned approaches.

Many works on imbalanced data classification employ classifier ensembles [16]. One of the more promising directions is the *Dynamic Ensemble Selection* (DES) [5]. Dynamic selection (DS) methods select a single classifier or an ensemble (from an available classifier pool) to predict the decision for each unknown query. This is based on the assumption that each of the base classifiers is an expert in a different region of the feature space. The classification of each unknown sample by DES involves three steps:

- Definition of the region of competence; that is, how to define the local region surrounding the unknown sample, in which the competence level of the base models is estimated. This local region of competence is found in the dynamic selection dataset (DSEL), which is usually part of the training set.
- Defining the selection criterion later used to assess the competence of the base classifiers in the local region of competence (e.g., accuracy or diversity).
- Determination of the selection mechanism deciding whether we choose a single classifier or an ensemble.

Previous work related to the imbalanced data classification using classifier ensembles and DES involves various approaches. Ksieniewicz in [9] proposed an *Undersampled Majority Class Ensemble* (UMCE) employing different combination methods and pruning, based on a k-fold division of the majority class to divide an imbalanced problem into many balanced ones. Chen et al. [4] presented the *Dynamic Ensemble Selection Decision-making* (DESD) algorithm to select the most appropriate classifiers using a weighting mechanism to highlight the base models that are better suited for recognizing the minority class. Zyblewski et al. in [17] proposed the *Minority Driven Ensemble* (MDE) for highly imbalanced data streams classification and Roy et al. in [14] combined preprocessing with dynamic ensemble selection to classify both binary and multiclass stationary imbalanced datasets.

The main contributions of this work are as follows:

- The proposition of the new dynamic selection methods adapted for the classification of highly imbalanced data.
- Experimental evaluation of the proposed algorithms based on a high number of diverse benchmark datasets and a detailed comparison with the *state-of-art* approaches.

2   Dynamic Ensemble Selection Based on Imbalance Ratio and Euclidean Distance

This paper proposes two algorithms for dynamic classifier selection for the imbalanced data classification problem. These are respectively the Dynamic Ensemble

Selection using Euclidean distance (DESE) and the Dynamic Ensemble Selection using Imbalance Ratio and Euclidean distance (DESIRE).

The generation of the classifier pool is based on the *Bagging* approach [2], and more specifically on the *Stratified Bagging*, in which the samples are drawn with replacement from the minority and majority class separately in such a way that each bootstrap maintains the original training set class proportion. This is necessary due to the high imbalance, which in the case of standard bagging can lead to the generation of training sets containing only the majority class.

Both proposed methods are derived in part from algorithms based on local oracles, and more specifically on KNORA-U [7], which gives base classifiers weights based on the number of correctly classified instances in the local region of competence and then combines them by weighted majority voting. The computational cost in this type of method is mainly related to the size of the classifier pool and the DSEL size, as the k-nearest neighbors technique is used to define local competence regions, which can be costly for large datasets. Instead of hard voting, DESE and DESIRE are based on the probabilities returned by the base models and they calculate weights for each classifier for both the minority and majority classes separately.

Proposed methods come in two variants: *Correct* (denoted as C), where weights are modified only in the case of correct classification, and *All* (denoted as A), where, in addition to correct decisions, weights are also affected by incorrect ones. The exact way of weights calculation is presented in Algorithm 1.

For each instance, the proposed algorithms perform the following steps:

- In step 2, the k-nearest neighbors of a given instance are found in DSEL, which form the local region of competence LRC.
- In step 4, each classifier Ψ_j from the pool classifies all samples belonging to LRC.
- In steps 5–13, the classifier weights are modified separately for the minority and majority class, starting from the value of 0. The *All* variant uses all four conditions, while the *Correct* variant is based only on the conditions in lines 6 and 8. In the case of DESE, the modifications are based on the Euclidean distance between the classified sample and its neighbor from the local competence region, and in the case of DESIRE, the Euclidean distance is additionally scaled by a percentage of the minority or majority class in such a way that more emphasis is placed on the minority class.

Finally, the weights obtained from DESE or DESIRE are normalized to the $[0, 1]$ range and multiplied by the ensemble support matrix. The combination is carried out according to the maximum rule [6], which chooses the classifier that is most confident of itself. The choice of this combination rule was dictated by a small number of instances in the datasets, which significantly reduces the risk of base classifiers overfitting.

Algorithm 1: Pseudocode of the proposed weight calculation methods assuming that the minority class is positive and the majority class is negative.

Input:

E, classifier pool,

D_t, test dataset,

$DSEL$, Dynamic Selection Dataset,

k, number of nearest neighbors,

min, maj, respectively the percentage of minority and majority classes in the training set,

$W \leftarrow \varnothing$, empty weights array of shape (n_classifiers, n_samples, 2).

Output:

W, weights array of shape (n_classifiers, n_samples, 2).

1: **for each** sample x_i in D_t **do**

2:      $LRC \leftarrow$ the k nearest neighbors of x_i in $DSEL$

3:      **for each** Classifier Ψ_j in E **do**

4:          $Predict \leftarrow predict(LRC, \Psi_j)$

5:          **for each** $neighbor$ in $len(LRC)$ **do**

6:              **if** $Predict[neighbor] =$ True negative **then**

7:                  $W[j, i, 0]+ = \{ {ED[x_i, neighbor] \text{ for DESE} \atop ED[x_i, neighbor]*min \text{ for DESIRE}}$

8:              **else if** $Predict[neighbor] =$ True positive **then**

9:                  $W[j, i, 1]+ = \{ {ED[x_i, neighbor] \text{ for DESE} \atop ED[x_i, neighbor]*maj \text{ for DESIRE}}$

10:             **else if** $Predict[neighbor] =$ False negative **then**

11:                 $W[j, i, 1]- = \{ {ED[x_i, neighbor] \text{ for DESE} \atop ED[x_i, neighbor]*min \text{ for DESIRE}}$

12:             **else if** $Predict[neighbor] =$ False positive **then**

13:                 $W[j, i, 0]- = \{ {ED[x_i, neighbor] \text{ for DESE} \atop ED[x_i, neighbor]*maj \text{ for DESIRE}}$

14:          **end for**

15:      **end for**

16: **end for**

(lines 6–9 braced as **Correct**; lines 6–13 braced as **All**)

---

3   Experimental Evaluation

This section presents the details of the experimental study, the datasets used and the results that the proposed approaches have achieved compared to the *state-of-art* methods.

3.1   Experimental Set-Up

The main goal of the following experiments was to compare the performance of proposed dynamic selection methods, designed specifically for the task of imbalanced data classification, with the *state-of-art* ensemble methods paired with preprocessing. The evaluation in each of the experiments is based on 5 metrics commonly used to assess the quality of classification for imbalanced problems. These are *F1 score* [15], *precision* and *recall* [13], *G-mean* [11] and *balanced accuracy score* (BAC) [3] according to the *stream-learn* [10] implementation. All experiments have been implanted in *Python* and can be repeated using the code on *Github*[1].

As the base models three popular classifiers, according to the *scikit-learn* [12] implementation, were selected, i.e. *Gaussian Naive Bayes* (GNB), *Classification and Regression Trees* (CART) and *k-Nearest Neighbors* classifier (KNN). The fixed size of the classifier pool has been determined successively as 5, 15, 30 and 50 base models. The evaluation was carried out using 10 times repeated 5-fold cross-validation. Due to the small number of instances in the datasets, DSEL is defined as the entire training set.

The experiments were carried out on 41 datasets from the KEEL repository [1], which contain binary problems created through various combinations of class merging. All datasets have a high imbalance ratio of at least 9. Problems characteristics are presented in Table 1.

Table 1. Datasets characteristics.

Dataset	Instances	Features	IR	Dataset	Instances	Features	IR
ecoli-0-1_vs_2-3-5	244	7	9	glass2	214	9	12
ecoli-0-1_vs_5	240	6	11	glass4	214	9	15
ecoli-0-1-3-7_vs_2-6	281	7	39	glass5	214	9	23
ecoli-0-1-4-6_vs_5	280	6	13	led7digit-0-2-4-5-6-7-8-9_vs_1	443	7	11
ecoli-0-1-4-7_vs_2-3-5-6	336	7	11	page-blocks-1-3_vs_4	472	10	16
ecoli-0-1-4-7_vs_5-6	332	6	12	shuttle-c0-vs-c4	1829	9	14
ecoli-0-2-3-4_vs_5	202	7	9	shuttle-c2-vs-c4	129	9	20
ecoli-0-2-6-7_vs_3-5	224	7	9	vowel0	988	13	10
ecoli-0-3-4_vs_5	200	7	9	yeast-0-2-5-6_vs_3-7-8-9	1004	8	9
ecoli-0-3-4-6_vs_5	205	7	9	yeast-0-2-5-7-9_vs_3-6-8	1004	8	9
ecoli-0-3-4-7_vs_5-6	257	7	9	yeast-0-3-5-9_vs_7-8	506	8	9
ecoli-0-4-6_vs_5	203	6	9	yeast-0-5-6-7-9_vs_4	528	8	9
ecoli-0-6-7_vs_3-5	222	7	9	yeast-1_vs_7	459	7	14
ecoli-0-6-7_vs_5	220	6	10	yeast-1-2-8-9_vs_7	947	8	31
ecoli4	336	7	16	yeast-1-4-5-8_vs_7	693	8	22
glass-0-1-4-6_vs_2	205	9	11	yeast-2_vs_4	514	8	9
glass-0-1-5_vs_2	172	9	9	yeast-2_vs_8	482	8	23
glass-0-1-6_vs_2	192	9	10	yeast4	1484	8	28
glass-0-1-6_vs_5	184	9	19	yeast5	1484	8	33
glass-0-4_vs_5	92	9	9	yeast6	1484	8	41
glass-0-6_vs_5	108	9	11				

Subsections 3.2 and 3.3 present the results of experiments comparing the presented methods, DESE in experiment 1 and DESIRE in experiment 2, with *state-of-art* ensemble algorithms used for the imbalanced data classification.

Both proposed and reference methods occur in versions with preprocessing (in the form of *random oversampling*) and without it, the use of oversampling is denoted by the letter *O* found before the acronym of the method. As a reference method, a single classifier, as well as stratified bagging and dynamic selection in the form of the KNORA-U algorithm were selected.

---

[1] https://github.com/w4k2/iccs20-desire.

The radar diagrams show the average global ranks achieved by each of the tested algorithms in terms of each of the 5 evaluation metrics, while the tables show the results of the Wilcoxon signed-rank ($p = 0.05$) statistical test for a pool size of 5 base classifiers. The numbers under the average rank of each method indicate the algorithms which are statistically significantly worse than the one in question. The complete results for each of the 41 datasets and the full statistical analysis can be found on the *Github*[2].

3.2 Experiment 1 – Euclidean Distance-Based Approach

In Fig. 1 we can see how the average ranks for DESE and reference methods changed in terms of different metrics depending on the ensemble size. We can see that the proposed methods (especially ODESE-C) for 5 base models achieve higher rankings in terms of each metric with an exception of *recall*. While the single classifier and bagging are preferring *recall*, ODESE-C and DESE-C prefer *precision*. As the number of base classifiers increases, BAC and *G-mean*-based rankings deteriorate to KNORA-U level, while the *F1 score* remains high due to high *precision*.

Table 2 presents the results of the statistical analysis, which shows that the ODESE-C method performs statistically significantly better than all reference methods in terms of each metric except for *recall*.

When the base classifier is CART, as seen in Fig. 2, for the smallest pool, DESE-C (both without and with oversampling) achieves higher ranks than the reference methods in terms of each of the five metrics. Along with the increase in the number of classifiers, we can observe that while OKNORA-U and OSB stand out in terms of *precision*, ODESE-C performs better in terms of other metrics, and ODESE-A, despite the low *F1 score* and *precision*, achieves the highest average ranks in terms of BAC, *G-mean* and *recall*. Table 3 confirms that for the five base classifiers, ODESE-C is statistically significantly better than all reference methods, while ODESE-A performs statistically significantly better than ODESE-C in terms of *recall*, *G-mean* and BAC.

Table 2. Statistical tests on mean ranks for GNB with pool size = 5.

	GNB (1)	OSB (2)	OKNORA-U (3)	DESE-C (4)	ODESE-C (5)	DESE-A (6)	ODESE-A (7)
F1 score	2.146	2.085	3.500	5.549	5.963	4.159	4.598
	–	–	1,2	1,2,3,6,7	1,2,3,6,7	1,2,3	1,2,3
precision	1.829	1.756	3.220	6.256	5.866	4.720	4.354
	–	–	1,2	all	1,2,3,6,7	1,2,3	1,2,3
recall	4.207	5.159	4.902	2.134	3.744	3.329	4.524
	4	4,5,6	4,5,6	–	4	4	4,5,6
G-mean	2.341	2.695	4.183	4.695	5.890	3.622	4.573
	–	–	1,2	1,2,6	all	1	1,2,6
BAC	2.317	2.634	3.963	4.720	5.976	3.671	4.720
	–	–	1,2	1,2,6	all	1,2	1,2,6

---

[2] https://github.com/w4k2/iccs20-desire/tree/master/article_results.

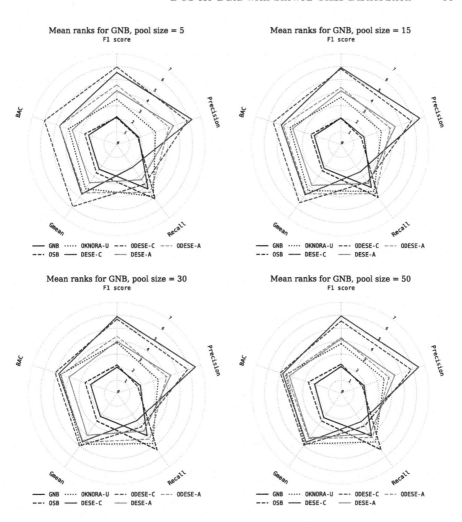

Fig. 1. Mean ranks for GNB classifier.

Table 3. Statistical tests on mean ranks for CART with pool size = 5.

	CART (1)	OSB (2)	OKNORA-U (3)	DESE-C (4)	ODESE-C (5)	DESE-A (6)	ODESE-A (7)
F1 score	2.683	2.841	2.988	5.329	5.561	4.256	4.341
	–	–	–	1,2,3,6,7	1,2,3,6,7	1,2,3	1,2,3
precision	2.634	3.976	4.195	5.695	5.134	3.195	3.171
	–	1	1,6,7	*all*	1,2,3,6,7	–	–
recall	3.293	2.622	2.695	3.890	4.463	5.366	5.671
	2,3	–	–	2,3	1,2,3,4	1,2,3,4,5	1,2,3,4,5
G-mean	3.098	2.671	2.817	4.061	4.634	5.232	5.488
	–	–	–	2,3	1,2,3,4	1,2,3,4	1,2,3,4,5
BAC	3.098	2.585	2.732	4.280	4.829	5.085	5.390
	–	–	–	1,2,3	1,2,3,4	1,2,3,4	1,2,3,4,5

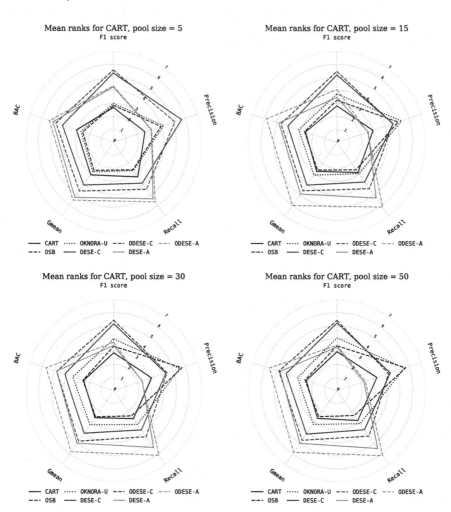

Fig. 2. Mean ranks for CART classifier.

Table 4. Statistical tests on mean ranks for KNN with pool size = 5.

	KNN (1)	OSB (2)	OKNORA-U (3)	DESE-C (4)	ODESE-C (5)	DESE-A (6)	ODESE-A (7)
F1 score	3.585	4.305	3.476	4.549	4.390	3.744	3.951
	–	3	–	1,6	–	–	–
precision	5.317	3.963	3.049	4.976	3.659	3.878	3.159
	3,5,6,7	3,7	–	2,3,5,6,7	–	7	–
recall	1.427	5.232	5.366	2.463	4.939	3.305	5.268
	–	1,4,6	1,4,6	1	1,4,6	1,4	1,4,6
G-mean	1.537	5.061	4.866	2.720	5.110	3.427	5.280
	–	1,4,6	1,4,6	1	1,4,6	1,4	1,4,6
BAC	1.659	5.012	4.841	2.780	5.024	3.415	5.268
	–	1,4,6	1,4,6	1	1,4,6	1,4	1,4,6

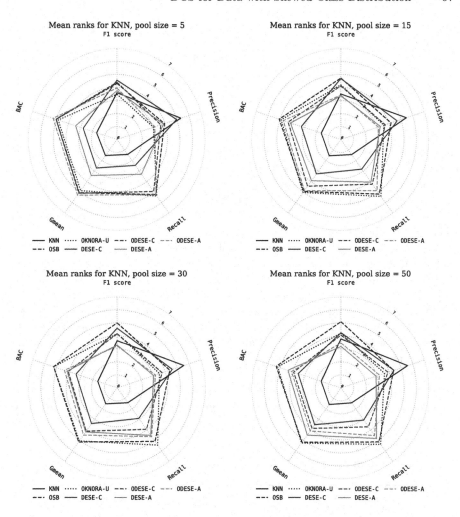

Fig. 3. Mean ranks for KNN classifier.

In Fig. 3 and Table 4 we can see that the proposed methods using oversampling do not differ statistically from the reference methods, except for a single classifier, which is characterized by high *precision* but at the same time achieves the worst mean ranks based on the remaining metrics. Together with the increase in the base classifier number, KNORA-U and OSB achieve higher average ranks than ODESE-C and ODESE-A.

3.3   Experiment 2 – Scaled Euclidean Distance-Based Approach

The results below show the average ranks for the proposed DESIRE method, which calculates weights based on Euclidean distances scaled by the percentages of the minority and majority classes in the training set.

In the case of GNB as the base model (Fig. 4), the ODESIRE-C method achieves the best results compared to reference methods in terms of mean ranks based on *F1 score*, *precision*, *G-mean* and BAC. When the ensemble size increases, the proposed method is equal to OKNORA-U in terms of BAC and *G-mean* but retains the advantage in terms of *F1 score* and *precision*. Also, the more base classifiers the smaller the differences between DESIRE using preprocessing and the version without it. Table 5 presents the results of the statistical analysis, which shows that ODESIRE-C is statistically better than all reference methods when the number of base classifiers is low.

Figure 5 shows that for a small classifier pool, ODESIRE-C achieves higher ranks than reference methods in terms of each evaluation metric, and as the classifier number increases, it loses significantly in *precision* compared to OSB and OKNORA-U. ODESIRE-A has a high *recall*, which unfortunately is reflected by the lowest *precision* and *F1 score*. In Table 6 we see that for 5 base classifiers, DSIRE-C both with and without preprocessing is statistically significantly better than reference methods in terms of all metrics except one, *G-mean* in the case DESIRE-C and *recall* for ODESIRE-C.

When the base classifier is KNN (Fig. 6), as in the case of DESE, ODESIRE-C is not statistically worse than OSB and OKNORA-U (Table 7) and as the number of classifiers in the pool increases, the average global ranks significantly deteriorate compared to reference methods.

Table 5. Statistical tests on mean ranks for GNB with pool size = 5.

	GNB (1)	OSB (2)	OKNORA-U (3)	DESIRE-C (4)	ODESIRE-C (5)	DESIRE-A (6)	ODESIRE-A (7)
F1 score	2.341	2.280	4.159	5.634	6.098	3.878	3.610
	–	–	1,2	1,2,3,6,7	1,2,3,6,7	1,2	1,2
precision	2.244	2.098	3.902	6.341	6.098	3.976	3.341
	–	–	1,2	1,2,3,6,7	1,2,3,6,7	1,2,7	1,2
recall	4.037	4.890	4.427	1.939	3.305	4.183	5.220
	4	4,5	4,5	–	4	4,5	1,3,4,5,6
G-mean	2.341	2.793	4.622	4.829	5.976	3.610	3.829
	–	–	1,2,6	1,2,6,7	all	1	1,2
BAC	2.341	2.634	4.427	4.829	6.061	3.610	4.098
	–	–	1,2,6	1,2,6	all	1,2	1,2

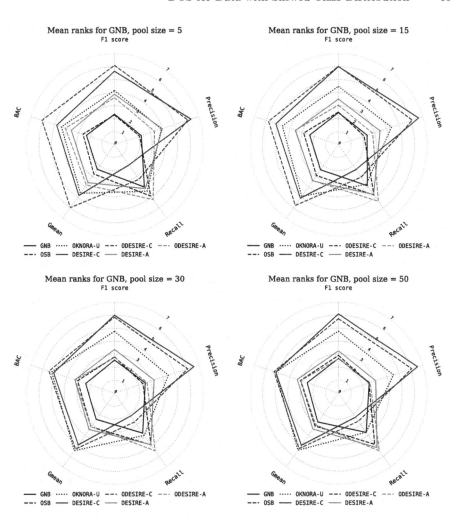

Fig. 4. Mean ranks for GNB classifier.

Table 6. Statistical tests on mean ranks for CART with pool size = 5.

	CART (1)	OSB (2)	OKNORA-U (3)	DESIRE-C (4)	ODESIRE-C (5)	DESIRE-A (6)	ODESIRE-A (7)
F1 score	3.415	3.768	3.915	5.622	5.768	2.524	2.988
	6	6	6,7	1,2,3,6,7	1,2,3,6,7	–	–
precision	3.683	4.659	4.878	5.793	5.256	1.793	1.939
	6,7	1,6,7	1,6,7	*all*	1,6,7	–	–
recall	3.146	2.488	2.561	3.793	4.110	5.817	6.085
	2,3	–	–	2,3	1,2,3	1,2,3,4,5	1,2,3,4,5
G-mean	3.049	2.598	2.744	4.280	4.817	5.183	5.329
	–	–	–	1,2,3	1,2,3,4	1,2,3,4	1,2,3,4
BAC	3.073	2.537	2.683	4.744	5.110	4.695	5.159
	–	–	–	1,2,3	1,2,3	1,2,3	1,2,3

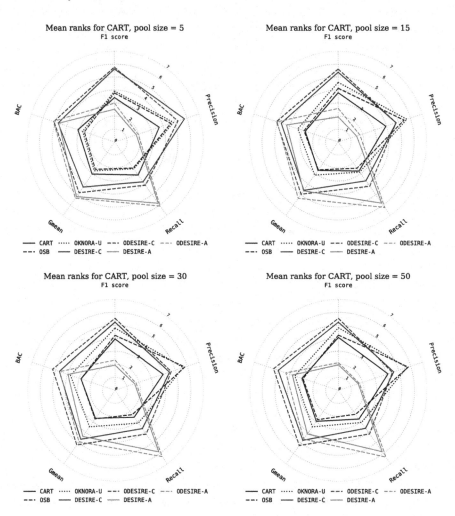

Fig. 5. Mean ranks for CART classifier.

Table 7. Statistical tests on mean ranks for KNN with pool size = 5.

	KNN (1)	OSB (2)	OKNORA-U (3)	DESIRE-C (4)	ODESIRE-C (5)	DESIRE-A (6)	ODESIRE-A (7)
F1 score	3.902	4.963	4.134	4.780	4.878	2.878	2.463
precision	6,7 5.354	1,3,6,7 4.695	6,7 3.854	6,7 5.207	6,7 4.293	– 2.732	– 1.866
recall	5,6,7 1.354	3,6,7 4.695	6,7 4.841	3,5,6,7 2.341	6,7 4.146	7 4.500	– 6.122
G-mean	– 1.451	1,4 4.866	1,4 4.500	1 2.683	1,4 4.610	1,4 4.524	*all* 5.366
BAC	– 1.561	1,4 4.841	1,4 4.573	1 2.768	1,4 4.744	1,4 4.354	1,3,4,5,6 5.159
	–	1,4	1,4	1	1,4	1,4	1,4,6

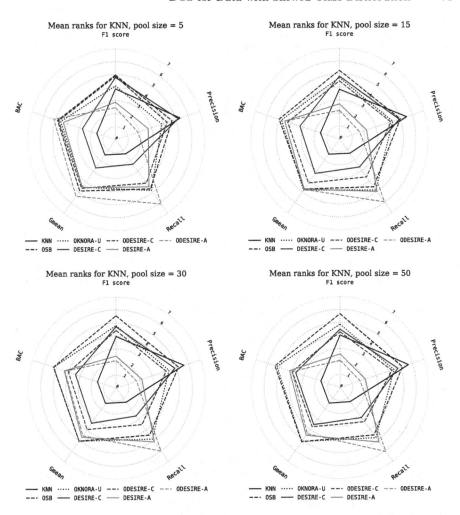

Fig. 6. Mean ranks for KNN classifier.

3.4 Lessons Learned

The presented results confirmed that dynamic selection methods adapted specifically for the imbalanced data classification can achieve statistically better results than *state-of-art* ensemble methods coupled with preprocessing, especially when the pool of base classifiers is relatively small. This may be due to the fact that *bagging* has not yet stabilized, while the proposed method chooses the best single classifier. The *Correct* approach in which the weights of the models were changed only if the instances belonging to the local competence region were correctly classified, proved to be more balanced in terms of all 5 evaluation measures. This may indicate too high weight penalties with incorrect classification in the *All* approach. When KNN is used as the base classifier, with a small pool the proposed methods performed statistically similar to KNORA-U, and with a larger number of classifiers, achieved statistically inferior rank compared to the reference methods. This may be probably due to the support calculation method in the KNN, which is not suitable for the algorithms proposed in this work. For GNB and CART, DESE-C and DESIRE-C achieved results which are statistically better than or similar to the reference methods, often without the use of preprocessing, since it has a built-in mechanism to deal with the imbalance.

4   Conclusions

The main purpose of this work was to propose a novel solution based on dynamic classifier selection for imbalanced data classification problem. Two methods were proposed, namely DESE and DESIRE, which use the Euclidean distance and imbalance ratio in the training set to select the most appropriate model for the classification of each new sample. Research conducted on benchmark datasets and statistical analysis confirmed the usefulness of proposed methods, especially when there is a need to maintain a relatively low number of classifiers.

Future work may involve the exploration of different approaches to the base classifiers' weighting, as well as using different combination methods and the use of proposed methods for the imbalanced data stream classification.

Acknowledgments. This work was supported by the Polish National Science Centre under the grant No. 2017/27/B/ST6/01325.

References

1. Alcala-Fdez, J., et al.: Keel data-mining software tool: data set repository, integration of algorithms and experimental analysis framework. J. Mult. Valued Log. Soft Comput. **17**, 255–287 (2010)
2. Breiman, L.: Bagging predictors. Mach. Learn. **24**(2), 123–140 (1996). https://doi.org/10.1007/BF00058655
3. Brodersen, K.H., Ong, C.S., Stephan, K.E., Buhmann, J.M.: The balanced accuracy and its posterior distribution. In: Proceedings of the 2010 20th International Conference on Pattern Recognition, ICPR 2010, Washington, DC, USA, pp. 3121–3124. IEEE Computer Society (2010)
4. Chen, D., Wang, X.-J., Wang, B.: A dynamic decision-making method based on ensemble methods for complex unbalanced data. In: Cheng, R., Mamoulis, N., Sun, Y., Huang, X. (eds.) WISE 2020. LNCS, vol. 11881, pp. 359–372. Springer, Cham (2019). https://doi.org/10.1007/978-3-030-34223-4_23
5. Cruz, R.M.O., Sabourin, R., Cavalcanti, G.D.C.: Dynamic classifier selection: recent advances and perspectives. Inf. Fus. **41**, 195–216 (2018)
6. Duin, R.P.W.: The combining classifier: to train or not to train? In: Object Recognition Supported by User Interaction for Service Robots, vol. 2, pp. 765–770, August 2002
7. Ko, A.H., Sabourin, R., Alceu Souza Britto, J.: From dynamic classifier selection to dynamic ensemble selection. Pattern Recogn. **41**(5), 1718–1731 (2008)

8. Krawczyk, B.: Learning from imbalanced data: open challenges and future directions. Progress Artif. Intell. **5**(4), 221–232 (2016). https://doi.org/10.1007/s13748-016-0094-0
9. Ksieniewicz, P.: Undersampled majority class ensemble for highly imbalanced binary classification. In: Proceedings of the Second International Workshop on Learning with Imbalanced Domains: Theory and Applications. Proceedings of Machine Learning Research, Dublin, Ireland, vol. 94, pp. 82–94. PMLR, ECML-PKDD, 10 September 2018
10. Ksieniewicz, P., Zyblewski, P.: Stream-learn-open-source Python library for difficult data stream batch analysis. arXiv preprint arXiv:2001.11077 (2020)
11. Kubat, M., Matwin, S.: Addressing the curse of imbalanced training sets: one-sided selection. In: ICML (1997)
12. Pedregosa, F., et al.: Scikit-learn: machine learning in Python. J. Mach. Learn. Res. **12**, 2825–2830 (2011)
13. Powers, D.: Evaluation: from precision, recall and F-measure to ROC, informedness, markedness & correlation. J. Mach. Learn. Technol **2**, 2229–3981 (2011)
14. Roy, A., Cruz, R.M., Sabourin, R., Cavalcanti, G.D.: A study on combining dynamic selection and data preprocessing for imbalance learning. Neurocomputing **286**, 179–192 (2018)
15. Sasaki, Y.: The truth of the F-measure. Teach Tutor Mater, January 2007
16. Woźniak, M., Graña, M., Corchado, E.: A survey of multiple classifier systems as hybrid systems. Inf. Fus. **16**, 3–17 (2014). Special Issue on Information Fusion in Hybrid Intelligent Fusion Systems
17. Zyblewski, P., Ksieniewicz, P., Woźniak, M.: Classifier selection for highly imbalanced data streams with *minority driven ensemble*. In: Rutkowski, L., Scherer, R., Korytkowski, M., Pedrycz, W., Tadeusiewicz, R., Zurada, J.M. (eds.) ICAISC 2019. LNCS (LNAI), vol. 11508, pp. 626–635. Springer, Cham (2019). https://doi.org/10.1007/978-3-030-20912-4_57

On Model Evaluation Under Non-constant Class Imbalance

Jan Brabec[1,2(✉)], Tomáš Komárek[1,2], Vojtěch Franc[2], and Lukáš Machlica[3]

[1] Cisco Systems, Inc., Karlovo Namesti 10 Street, Prague, Czech Republic
{janbrabe,tomkomar}@cisco.com

[2] Faculty of Electrical Engineering, Czech Technical University in Prague, Prague, Czech Republic
xfrancv@cmp.felk.cvut.cz

[3] Resistant.AI, Prague, Czech Republic
lukas.machlica@resistant.ai

Abstract. Many real-world classification problems are significantly class-imbalanced to detriment of the class of interest. The standard set of proper evaluation metrics is well-known but the usual assumption is that the test dataset imbalance equals the real-world imbalance. In practice, this assumption is often broken for various reasons. The reported results are then often too optimistic and may lead to wrong conclusions about industrial impact and suitability of proposed techniques. We introduce methods (Supplementary code related to techniques described in this paper is available at: https://github.com/CiscoCTA/nci_eval) focusing on evaluation under non-constant class imbalance. We show that not only the absolute values of commonly used metrics, but even the order of classifiers in relation to the evaluation metric used is affected by the change of the imbalance rate. Finally, we demonstrate that using sub-sampling in order to get a test dataset with class imbalance equal to the one observed in the wild is not necessary, and eventually can lead to significant errors in classifier's performance estimate.

Keywords: Evaluation metrics · Imbalanced data · Precision · ROC

1 Introduction

Class-imbalanced problems arise if number of samples in one of the classes, often in the class of interest, is significantly lower than in the other class, often the background class. Such problems are present in variety of different domains such as medicine [16], finance [15,20,21], cybersecurity [1,3,5] and many others.

In highly imbalanced problems it is essential to use suitable evaluation metrics to correctly assess the merit of pursued algorithms and realistically judge

V. Franc—Was supported by OP VVV project CZ.02.1.01\0.0\0.0\16_019\0000765 Research Center for Informatics.

V. V. Krzhizhanovskaya et al. (Eds.): ICCS 2020, LNCS 12140, pp. 74–87, 2020.
https://doi.org/10.1007/978-3-030-50423-6_6

their impact before they are deployed into the wild. Methods for evaluation of classifiers on class-imbalanced datasets are well known and have been thoroughly described in the past [4,9,11,19].

It is usually assumed that the imbalance of the test dataset is the same as in the real distribution on which the model will operate once deployed into production environment. However, this assumption is often broken, because of different reasons ranging from selection bias when constructing the test dataset, high costs of acquiring large dataset mainly in situations when the imbalance is high (e.g. $1:10^4$), to the fact that often not a single general distribution exists (e.g. disease classifier may face different priors depending on the location).

Discrepancy between imbalances in test datasets and real world is often the root cause of too optimistic results leading to wrong expectations of the impact in industrial applications. This is detrimental to the research community, because it creates confusion about which problems are still open and which are solved. It might discourage groups from working on such problems, and make it harder for researchers still investigating the field to convince the community that in the light of the too optimistic prior work their results have still impact.

Throughout this paper, we frame and investigate the problem of classifier evaluation dropping the assumption of constant class imbalance. We focus on precision related metrics as one of the most popular metrics for imbalanced problems [4,9]. We show how these metrics can be computed for arbitrary class imbalances and any test dataset without the need to re-sample the data. We also inspect their behavior as a function of the imbalance rate. We show that Precision-Recall (PR) curves have little value without stating the corresponding imbalance ratio which can dramatically affect the results and their assessment.

We demonstrate that change in imbalance rate, maybe surprisingly, affects also the ranking of classifiers under these metrics. We argue that instead of tabulating the results for a single dataset, it is beneficial to plot the dependence on the class imbalance rate whenever possible. Such plots provide considerably more information for wider audience.

We also describe how errors in measurements can be assessed and that they can significantly affect the reliability of measured precision mainly in cases when low regions of false positive rate are of interest. This can be primarily attributed to the fact that the test dataset is finite. Therefore, we further elaborate how the class imbalance increases the demands on the size of test dataset.

Most importantly, *we refute the common understanding that the best practice is to alter the test dataset so that class imbalance matches the imbalance of the pursued distribution* as is suggested e.g. in [14]. We show how re-sampling of a dataset may lead to significant errors in measurements. We stress that the test dataset should be constructed in a way to allow measurements of false-positive and true-positive rates with errors as small as possible. We show that the crucial entity to focus on is the coefficient of variation related to both true-positive and false-positive rates.

2   Preliminaries

Throughout this paper we are concerned with the binary classification task. Let $x \in \mathcal{X}$ be an input and $y \in \mathcal{Y} = \{-1, 1\}$ be a target. We call the class $y = -1$ negative class and the class $y = 1$ positive class. The positive class is assumed to be the minority class and the negative class is the majority class. We do not assume that there exists a single real-world joint-probability distribution $p(x, y)$ but instead consider a parametric family:

$$p(x, y; \eta) = p(x|y) \cdot P(y; \eta), \text{ where } P(y; \eta) = \begin{cases} 1 - \eta & y = -1 \\ \eta & y = 1 \end{cases}. \tag{1}$$

Parameter $\eta \in [0, 1]$ specifies the positive class prevalence. If we consider a classifier $h : \mathcal{X} \mapsto \mathcal{Y}$ then the following classifier evaluation metrics can be expressed as probabilities:

$$\text{TPR} = \text{Recall} = P(h(x) = 1|y = 1) = \mathbb{E}_{x \sim p(x|y=1)}[h(x) = 1] \tag{2}$$

$$\text{FPR} = P(h(x) = 1|y = -1) = \mathbb{E}_{x \sim p(x|y=-1)}[h(x) = 1] \tag{3}$$

$$\text{Prec}(\eta) = P(y = 1|h(x) = 1) = \frac{\text{TPR} \cdot \eta}{\text{TPR} \cdot \eta + \text{FPR} \cdot (1 - \eta)} \tag{4}$$

TPR stands for true-positive-rate (also called recall or sensitivity), FPR for false-positive-rate and Prec for precision. Formula (4) is derived using Bayes' theorem. We can observe that both TPR and FPR are not affected by the positive class prevalence but precision is. This observation is very important for the rest of this paper.

To estimate the above-mentioned metrics we need to evaluate the classifier on a test dataset. We assume that the test dataset is sampled i.i.d. from $p(x, y; \eta_{test})$ where η_{test} may or may not correspond to a positive class prevalence connected to some real-world application of the classifier. TP, FP, TN, FN denote the number of true positives, false positives, true negatives and false negatives, respectively and $N = TP + FP + TN + FN$ equals the size of the test set.

Prevalence of the positive class in the test dataset p_+ and imbalance ratio (IR) are defined as (one can be computed from the other easily):

$$p_+ = \frac{TP + FN}{N}, \quad IR = \frac{TP + FN}{TN + FP}. \tag{5}$$

$\widehat{\text{TPR}}$ is defined as the fraction of positive samples that were classified correctly:

$$\widehat{\text{TPR}} = \frac{1}{|\mathcal{X}^+|} \sum_{x \in \mathcal{X}^+} [\![h(x) = 1]\!] = \frac{\text{TP}}{|\mathcal{X}^+|} = \frac{\text{TP}}{\text{TP} + \text{FN}}, \tag{6}$$

where $[\![\cdot]\!]$ is the indicator function. $\widehat{\text{FPR}}$ is defined as the fraction of negatives samples that were classified incorrectly:

$$\widehat{\text{FPR}} = \frac{1}{|\mathcal{X}^-|} \sum_{x \in \mathcal{X}^-} [\![h(x) = 1]\!] = \frac{\text{FP}}{|\mathcal{X}^-|} = \frac{\text{FP}}{\text{FP} + \text{TN}}. \tag{7}$$

$\widehat{\text{Prec}}$ is the number of true positives out of all the positive predictions:

$$\widehat{\text{Prec}}(\eta) = \frac{\widehat{\text{TPR}} \cdot \eta}{\widehat{\text{TPR}} \cdot \eta + \widehat{\text{FPR}} \cdot (1 - \eta)} \tag{8}$$

It can be easily shown that $\widehat{\text{Prec}}(p_+) = \text{TP}/(\text{TP} + \text{FP})$ resolves to the standard formula used to compute precision. It holds that the metrics measured on the test dataset approach their true values originating from the distribution $p(x, y; \eta)$ as the size of the dataset grows. In other words $p_+ \rightarrow \eta_{test}$, $\widehat{\text{TPR}} \rightarrow$ TPR, $\widehat{\text{FPR}} \rightarrow$ FPR and $\widehat{\text{Prec}} \rightarrow$ Prec as N approaches infinity, but the errors in estimation caused by limited size of test dataset are often significant enough to deserve consideration, particularly during classifier evaluation in settings that are heavily class-imbalanced. We elaborate on this in Sect. 5.

3 Precision in the Light of Different Class Imbalance Ratios

Equation (8) in Sect. 2 shows that the class imbalance ratio of the test dataset directly impacts the measured precision. As such, the test dataset class imbalance must be considered when interpreting the results to assess viability of the classifier for a given application.

Fortunately, it is not necessary for a test dataset's imbalance ratio to be equivalent to the real-world imbalance. Equation (8) shows how to estimate precision ($\widehat{\text{Prec}}$), that corresponds to any class imbalance, from $\widehat{\text{TPR}}$ and $\widehat{\text{FPR}}$ which are estimated from the test dataset and are unaffected by it's imbalance.

In Sect. 5 we provide rationale and show that matching the real-world class imbalance is often sub-optimal and not desirable for correct evaluation.

3.1 Positive-Prevalence Precision Curve

Positive prevalence adjusted precision computed by Equation (8) is a linear rational function of the positive class prevalence η. As such, it can be plotted over an interval of positive prevalence values. We call such plot Positive-Prevalence Precision (P^3) curve. The curve should be plotted with log-scaled x-axis (lin-log P^3 curve) to easily distinguish between different orders of magnitude of the positive prevalence as demonstrated in Fig. 1.

P^3 curve is a useful instrument when evaluating a classifier to determine it's performance beyond a particular dataset. The downside of the plot is that contrary to ROC or PR curves, it captures the performance only for a single operating point of the classifier. Each point on an ROC curve thus has it's corresponding P^3 curve.

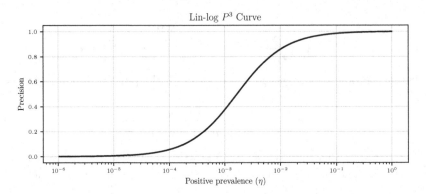

Fig. 1. Positive-Prevalence Precision (P³) curve for a hypothetical classifier with TPR = 0.6 and FPR = 0.001. The graph is plotted in logarithmic scale of the x-axis.

Given a particular ROC curve, each point on the curve corresponds to a different value of TPR. Instead of saying that P³ curve corresponds to a particular point on the ROC curve, it can also be said that it corresponds to a fixed value of TPR. For example, P³ curve in Fig. 1 corresponds to a classifier with TPR fixed at 60%.

P³ curve answers the question "How does precision of a given classifier evolve when changing the class imbalance-ratio?" and allows to quickly visually assess some of the conditions under which the classifier is suitable for production environment. Also, even if P³ curve may not be used in a particular evaluation of a classifier it is still important to possess intuition about it's general shape.

3.2   Precision-Recall Curves

PR curve is a very popular method to evaluate classifiers on imbalanced datasets. It captures the relationship between recall (TPR) on the x-axis and precision on the y-axis. As is the case with ROC curve, PR curve is usually created by applying different thresholds on the raw output of a classifier. While ROC is a strictly increasing function, PR curves do not have to be monotonous because it is possible for precision to both increase or decrease for different threshold values.

As discussed in Sect. 3.1, contrary to the ROC curve, *PR curve is affected by the imbalance ratio present in the test dataset.* This behavior is demonstrated in Fig. 2. PR curves can immediately reveal poor performance on class-imbalanced datasets that might not be obvious when inspecting ROC curves alone [18]. Because of this property PR curves are well suited and popular choice for evaluation of classifiers on class-imbalanced sets.

We suggest that the particular imbalance ratio present in a test dataset for which the PR curve was created should always be reported and considered when interpreting the impact of the results. When different research teams perform their experiments on different test sets while solving the same problem, and

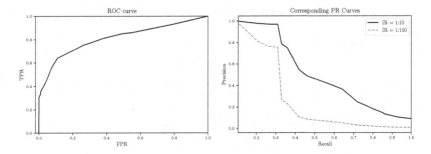

Fig. 2. Example of how a single ROC curve can correspond to two different PR curves given different imbalance ratios. The solid PR curve was created from the ROC curve with assumption that the IR was 1:10 while the dashed PR curve corresponds to IR equal to 1:100.

even if the data originate from the same source, the resulting PR curves will not be comparable if different imbalance ratios are present. For example, in computer security the datasets of downloaded files might originate from the VirusTotal[1] service, but different teams may work with different subsets that have different imbalance ratios.

Another danger is that the class imbalance ratio in a particular test dataset is often not representative of the imbalance ratios encountered once the classifier is deployed in real environment. It is often the case that the imbalance ratios experienced in the wild are lower than the ratio in the test dataset (not rarely the test datasets are even not imbalanced at all). In such situations, too optimistic estimates of the classifier's performance will be obtained if evaluation based on PR curve computed directly on the test dataset is used.

To remedy these risks, often test datasets with the same class imbalance ratios that would be encountered in the real environment are created. In Sect. 5 we demonstrate that this should not be the goal. Rather, a test dataset should be assembled that allows estimation of TPR and FPR with low enough variance and (8) should be used to compute Precision-Recall curves for different class imbalance ratios of interest.

4   Comparing Performance of Classifiers

When comparing performance of classifiers that need to deal with imbalanced data, the area under PR-curve (PR-AUC) or F1 score ($F_1 = 2 \cdot \frac{\text{Prec} \cdot \text{Recall}}{\text{Prec} + \text{Recall}}$) are often used out of convenience because they can be expressed as a single number [8]. In this section, we show that not only the values of these metrics dramatically depend on the imbalance rate in the selected test dataset, but the rate has notable influence even on the order of classifiers related to their efficacy. That is, based on these metrics two classifiers can switch places given different

---

[1] https://www.virustotal.com.

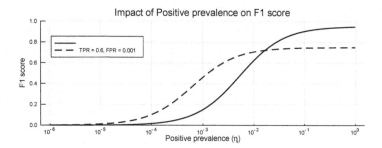

Fig. 3. The graph is similar to Positive Prevalence-Precision plot in Fig. 1 but instead of precision it plots F1 score of two distinct classifiers computed on the same dataset but assuming different imbalance rates. It can be seen that not only the absolute value of the score but even the order of the classifiers depends on the positive class prevalence.

imbalance rates. This can lead to incorrect conclusions about performance of classifiers on real data. The fact can be also misused for cherry-picking of an imbalance rate to pick the one where a classifier achieves better results than any other method it competes with.

4.1 Affecting Ordering of Classifiers: F1 Score

F1 score is defined as harmonic mean of precision and recall. The comparison of F1 scores of two classifiers is therefore affected by the selected imbalanced rate since precision depends on the rate while recall does not. Figure 3 demonstrates how the F1 score of two classifiers depends on the imbalance rate present in a test dataset.

Therefore, *we suggest to plot F1 scores in relation to imbalance rates, such as seen in Fig. 3 instead of tabulated F1 scores in any applied research papers.* The plot contains a superset of information, it is easily interpretable, space-efficient and conveys an overall better picture about performance of classifiers independent of the particular imbalance rate in the selected test dataset. The imbalance rate of the particular test dataset can be easily highlighted on the x-axis.

4.2 Affecting Ordering of Classifiers: PR-AUC

Firstly, it is proven that if a classifier *dominates* in ROC space it also dominates in PR space [6], but dominance is not linked to the area under ROC curve (ROC-AUC). It is easily possible for a classifier to have greater ROC-AUC than another but smaller area under PR curve (PR-AUC) on the same test dataset.

A convenient property of evaluating classifier by ROC-AUC is that it's value is invariant to class imbalance. On the other hand, the value of ROC-AUC can be dominated by insignificant regions in the ROC space, e.g. high values of FPR,

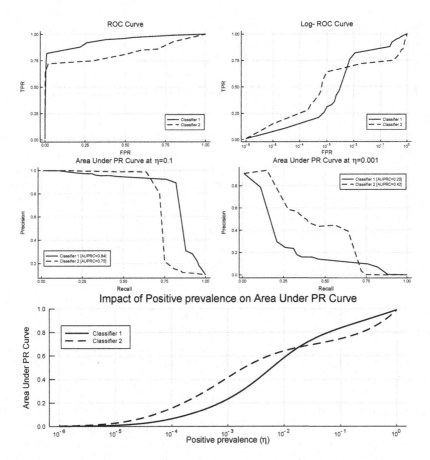

Fig. 4. The top-left plot is an example plot of two classifier ROC curves. In the top-right plot the same ROC curves are displayed with logarithmically scaled x-axis. The middle row displays corresponding PR curves for the ROC curves under different positive class prevalences (namely 10^{-1} and 10^{-3}). The bottom plot shows how PR-AUC of the classifiers depends on the class imbalance rate and that the order of the classifiers can easily switch for two different prevalences.

which are in practice of no importance. If the problem is heavily class imbalanced it is usually not an appropriate method for evaluation of classifiers [2] and PR-AUC should be considered.

However, it is often not realized that PR-AUC values depend on class imbalance and notably that also the order of classifiers under this metric depends on the imbalance rate as demonstrated in Fig. 4. It may be more surprising than in the case of F1 score computed only at a single operating point, while PR-AUC is evaluated over the whole range of operating points. Therefore, one might wrongly expect the metric to preserve ordering of classifiers across different imbalance rates.

We offer similar advice as with F1 score about the need to report the dataset imbalance rate together with PR-AUC values and to ideally use plots as in Fig. 4 instead of tabulated values for a single imbalance ratio.

5   Impact of Errors on Estimates of TPR and FPR

Class-imbalanced problems have increased demands on the test dataset size. It is often ignored that $\widehat{\text{TPR}}$ and $\widehat{\text{FPR}}$ computed on test dataset are just point estimates of the real TPR and FPR, given in (2) and (3), respectively, and as such they may be affected by uncertainty related to insufficient amount of samples of the minority class. In this section, we investigate how this uncertainty impacts the measured precision and how to correctly design experiments in presence of imbalanced data to suppress the uncertainty in the outcome.

A common approach to quantify the uncertainty of estimates based on finite samples is to use the interval estimates. We say that $\mathcal{I}_{\text{TPR}} = (\widehat{\text{TPR}} - \sigma_{\text{TPR}}, \widehat{\text{TPR}} + \sigma_{\text{TPR}})$ is the α-confidence interval of TPR if it holds that

$$\text{Prob}(\text{TPR} \in \mathcal{I}_{\text{TPR}}) \geq \alpha, \tag{9}$$

where the probability is w.r.t. randomly generated positive test samples $\mathcal{X}^+$ which are used to compute $\widehat{\text{TPR}}$ by (6). The interval (half-)width σ_{TPR}, the number of samples $|\mathcal{X}^+|$ and the confidence level $\alpha \in (0, 1)$ are dependent variables the exact relation of which is characterized by numerous concentration bounds like the Hoeffding's inequality. For example, by fixing σ_{TPR} and α we can compute the minimal number of samples in $\mathcal{X}^+$ which guarantee that $\mathcal{I}_{\text{TPR}}$ is the α-confidence interval. In the sequel we assume that the interval width σ_{TPR} is not greater than $\widehat{\text{TPR}}$. Note that this formalisation does not introduce any specific constraints on the shape of TPR distribution. The confidence interval $\mathcal{I}_{\text{TPR}}$ can be characterized by a single number, the coefficient of variation, defined as

$$\text{CV}_{\text{TPR}} = \frac{\sigma_{\text{TPR}}}{\widehat{\text{TPR}}}. \tag{10}$$

Analogously, we can define $\mathcal{I}_{\text{FPR}} = (\widehat{\text{FPR}} - \sigma_{\text{FPR}}, \widehat{\text{FPR}} + \sigma_{\text{FPR}})$, $\text{CV}_{\text{FPR}} = \frac{\sigma_{\text{FPR}}}{\widehat{\text{FPR}}}$, and we also assume that $\sigma_{\text{FPR}} < \widehat{\text{FPR}}$.

Let us define the precision as a function of the positive class prevalence η, TPR and FPR[2]:

$$\text{Prec}(\eta, \text{TPR}, \text{FPR}) = \frac{\eta \cdot \text{TPR}}{\eta \cdot \text{TPR} + (1 - \eta) \cdot \text{FPR}}. \tag{11}$$

Given $\text{TPR} \in \mathcal{I}_{\text{TPR}}$ and $\text{FPR} \in \mathcal{I}_{\text{FPR}}$, the value of $\text{Prec}(\eta, \text{TPR}, \text{FPR})$ has to be for any fixed $\eta \in (0, 1)$ inside the interval $(\text{LB}(\eta), \text{UB}(\eta))$ where

$$\text{LB}(\eta) = \min_{\substack{\text{TPR} \in \mathcal{I}_{\text{TPR}} \\ \text{FPR} \in \mathcal{I}_{\text{FPR}}}} \text{Prec}(\eta, \text{TPR}, \text{FPR}), \tag{12}$$

$$\text{UB}(\eta) = \max_{\substack{\text{TPR} \in \mathcal{I}_{\text{TPR}} \\ \text{FPR} \in \mathcal{I}_{\text{FPR}}}} \text{Prec}(\eta, \text{TPR}, \text{FPR}). \tag{13}$$

---

[2] In (8) we used $\text{Prec}(\eta)$ since the values of TPR and FPR were assumed to be fixed.

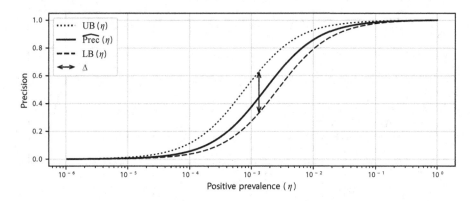

Fig. 5. The figure visualizes the uncertainty band containing the value of $\mathrm{Prec}(\eta, \mathrm{TPR}, \mathrm{FPR}) \in (\mathrm{UB}(\eta), \mathrm{LB}(\eta))$ when TPR and FPR are bound to intervals $\mathcal{I}_{\mathrm{TPR}} = (\widehat{\mathrm{TPR}} - \sigma_{\mathrm{TPR}}, \widehat{\mathrm{TPR}} + \sigma_{\mathrm{TPR}})$ and $\mathcal{I}_{\mathrm{FPR}} = (\widehat{\mathrm{FPR}} - \sigma_{\mathrm{FPR}}, \widehat{\mathrm{FPR}} + \sigma_{\mathrm{FPR}})$, respectively. The value $\Delta = \max_{\eta \in (0,1)} (\mathrm{UB}(\eta) - \mathrm{LB}(\eta))$ corresponds to the maximal width of the uncertainty band. The solid line corresponds to the point estimate $\widehat{\mathrm{Prec}}(\eta) = \mathrm{Prec}(\eta, \widehat{\mathrm{TPR}}, \widehat{\mathrm{FPR}})$.

Let Δ be the maximal width of the interval $(\mathrm{LB}(\eta), \mathrm{UB}(\eta))$ w.r.t. η, that is,

$$\Delta = \max_{\eta \in (0,1)} (\mathrm{UB}(\eta) - \mathrm{LB}(\eta)) . \tag{14}$$

The number Δ can be interpreted as the maximal uncertainty in measurements of precision when the exact values of TPR and FPR are replaced by their confidence intervals $\mathcal{I}_{\mathrm{TPR}}$ and $\mathcal{I}_{\mathrm{FPR}}$, respectively. It is easy to see that $\mathrm{TPR} \in \mathcal{I}_{\mathrm{TPR}}$ and $\mathrm{FPR} \in \mathcal{I}_{\mathrm{FPR}}$ imply

$$\mathrm{Prec}(\eta, \mathrm{TPR}, \mathrm{FPR}) \in (\widehat{\mathrm{Prec}}(\eta) - \Delta, \widehat{\mathrm{Prec}}(\eta) + \Delta). \tag{15}$$

The concepts of $\mathrm{UB}(\eta)$, $\mathrm{LB}(\eta)$ and Δ as well as their relation to $\mathrm{Prec}(\eta, \mathrm{TPR}, \mathrm{FPR})$ are illustrated in Fig. 5. The following theorem relates the maximal uncertainty Δ and the coefficients of variation $\mathrm{CV}_{\mathrm{TPR}}$ and $\mathrm{CV}_{\mathrm{FPR}}$, which characterize the confidence intervals $\mathcal{I}_{\mathrm{TPR}}$ and $\mathcal{I}_{\mathrm{FPR}}$, respectively.

Theorem 1. *Let* $TPR \in (\widehat{TPR} - \sigma_{TPR}, \widehat{TPR} + \sigma_{TPR})$ *and* $FPR \in (\widehat{FPR} - \sigma_{FPR}, \widehat{FPR} + \sigma_{FPR})$. *Let further* $\widehat{TPR} > \sigma_{TPR}$ *and* $\widehat{FPR} > \sigma_{FPR}$. *Then*

$$\Delta \le \max\{CV_{TPR}, CV_{FPR}\}$$

and the equality is attained iff $CV_{TPR} = CV_{FPR}$.[3]

---

[3] The proof for Theorem 1 is available in the appendix of this paper at: https://arxiv.org/pdf/2001.05571.pdf.

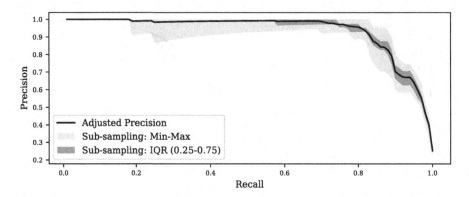

Fig. 6. PR curves for $\eta = 10^{-2}$. The black PR curve is computed from full dataset with $p_+ = 10^{-3}$ and adjusted to $\eta = 10^{-2}$ using (8), whereas the gray areas indicate IQR and min-max range of PR curves computed on 30 datasets with randomly sub-sampled negative class to match $p_+ = 10^{-2}$. Note that some PR curves are inside of IQR only partially.

Corollary 1. *Let $\mathcal{I}_{TPR}$ and $\mathcal{I}_{FPR}$ be α-confidence intervals of the true TPR and FPR, respectively, and let CV_{TPR} and CV_{FPR} be their corresponding coefficients of variation. Let further $\Delta = \max\{CV_{TPR}, CV_{FPR}\}$. Then $\mathcal{I}_{Prec} = (\widehat{Prec}(\eta) - \Delta, \widehat{Prec}(\eta) + \Delta)$ is the α^2-confidence interval of $Prec(\eta, TPR, FPR)$, i.e.*

$$Prec(\eta,\ TPR,\ FPR) \in \mathcal{I}_{Prec}$$

holds with probability α^2 at least.

The α^2-confidence level stems from the fact that $TPR \in \mathcal{I}_{TPR}$ and $FPR \in \mathcal{I}_{FPR}$ are two independent random events with probability not less than α.

Theorem 1 shows the relationship between confidence intervals for precision, widths of these intervals and point estimates of TPR, FPR. That is, coefficients of variation for TPR and FPR are the crucial quantities to consider when designing test dataset. If a test set is constructed we first need to manually fix both σ_{TPR} and σ_{FPR} at reasonable values based on the purpose of the dataset, and then ensure sufficient number of testing samples necessary to estimate TPR, FPR with desired Δ. If, for example, one is interested in $FPR = 10^{-3}$ on a dataset having only 10,000 negative samples, the estimate around this working point may become extremely noisy. Since such low FPR corresponds to only 10 FP samples ($10,000 * 10^{-3}$), just a small increase or decrease in number of FPs suffice to significantly alter the relative value of the FPR. Therefore, if such low values of FPR are of interest, one should increase the amount of negatives. Different methods exist that can quantify the concentration bounds. For example, Hoeffding's inequality can be used, which states that the upper bound on the number of required samples is proportional to $\frac{1}{\sigma_{FPR}^2}$, but Hoeffding's bound is very loose and usually less samples are required.

On the other hand, given a test dataset, in order to find Δ we need to estimate $\sigma_{\mathrm{TPR}}, \sigma_{\mathrm{FPR}}$ to get $\mathrm{CV}_{\mathrm{TPR}}, \mathrm{CV}_{\mathrm{FPR}}$. For that purpose cross-validation or bootstrapping can be used. For example, a classifier with $\widehat{\mathrm{TPR}} = 0.6, \sigma_{\mathrm{TPR}} = 0.06, \widehat{\mathrm{FPR}} = 10^{-3}, \sigma_{\mathrm{FPR}} = 10^{-4}$ has $\mathrm{CV}_{\mathrm{TPR}} = \mathrm{CV}_{\mathrm{FPR}} = \Delta = 0.1$, which might be reasonable width of the precision's confidence interval (i.e. $\pm 10\%$ change). But, if we increase $\sigma_{\mathrm{FPR}} = 5 * 10^{-4}$ then even though the number might seem small and it may be not indicative of the impact on estimate of the precision, the bound for precision becomes $\Delta = 0.5$ (i.e. $\pm 50\%$ change), which will immediately shed light on the reliability of estimates of the precision.[4]

5.1   Example of Errors Caused by Sub-sampling

To illustrate the error of sub-sampling we used ResNet-50 [10] on the ImageNet validation dataset [17] to detect images of 'agama' in a one-vs-all manner. The p_+ in such dataset is 10^{-3}.

To plot PR curves for $\eta = 10^{-2}$ we can either use the full dataset and then apply (8) to adjust the precision, or sub-sample the dataset to $p_+ = 10^{-2}$. Figure 6 compares these two approaches, where we repeated the sub-sampling 30 times to estimate the variance introduced by random reduction of the negative class. The results show that PR curves measured on the sub-sampled datasets are encumbered by a considerable measurement errors even though each one has 5000 samples, which might otherwise be a reasonable number for evaluation on balanced problems. Moreover $\eta = 10^{-2}$ is not as drastic imbalance as is often encountered in applications and the errors could be even more pronounced if η was lower.

Unlike the common practice of sub-sampling of the test dataset to the desired imbalance rate [14], we recommend to use a bigger dataset (to decrease the coefficients of variation) and adjust the metrics to the desired imbalance rate instead.

6   Related Work

Several comprehensive papers about methodology of evaluation on imbalanced datasets were written [4,7,9,11,19]. They focus on measuring the performance on the test dataset and do not address the problem of mismatch between class imbalances in test and application datasets.

In [5] authors use a plot with area under PR curve on the y-axis and a quantity related to the imbalance ratio on the x-axis. The plot is similar to Fig. 4, it is used because it is useful in the context of the paper but it's properties and impacts are not discussed.

In [2] authors discuss several bad practices in handling of class-imbalanced problems. Apart from other causes, they discuss the importance of addressing

---

[4] In this example, $\Delta \approx 0.31$ for $\eta \approx 1.45 \cdot 10^{-3}$. Computation can be found in the supplementary code to this paper.

the real imbalance ratios that can be different from the test dataset. They also present a formula for adjusting the precision to different imbalance ratios but do not explore this formula in greater detail neither inspect the impact of uncertainty originating from the finite size of the test dataset on precision.

Paper [12] introduces measure based on area under PR curve, which is further integrated across different class imbalances yielding a single evaluation number. The idea is based on the relationship between PR and ROC given in (8). No additional investigations related to multiple working points, ordering of classifiers according to the score, nor errors in measurements are carried out.

In [14] authors raise the issue of experimental results in cybersecurity often not being reproducible in real applications. They mention the problem that the class imbalance is often different in test dataset and in practice. They do not address the issue analytically but instead choose to re-sample the test dataset to desired imbalance ratios. This goes directly against our observations in Sect. 5 and applying such method leads to results heavily affected by noise.

It should be mentioned that other evaluation metrics well-suited for evaluation of class-imbalanced problems were proposed. A notable example is Matthews Correlation Coefficient (MCC) [13], but is not in the scope of this paper. MCC is not as widely used as PR [8] and it's values are not that easily interpretable as values of precision and recall.

7   Conclusion

This paper addressed evaluation of classifiers under consideration that the class imbalance ratio encountered in real world is different from imbalance present in the test dataset or is suspect to change. We focused on precision as one of the most popular evaluation metrics for imbalanced problems.

We stress that it is of significant importance to report also the imbalance ratio under which the classifier was developed and is aimed for, because assuming different imbalance ratios may easily lead to swapping of places of classifiers. This holds also for both PR-AUC and F1 score.

We have shown that even very small absolute values of σ_{FPR} can result in large variance in measured precision. The larger the class imbalance, the greater are the demands on the amount of negative samples present in the test dataset. Therefore, rather than sub-sampling a dataset to reach desired imbalance rate, all the samples should be kept to decrease the coefficients of variation, and the evaluation metrics should be computed given the presented formulas.

References

1. Axelsson, S., Sands, D.: The base-rate fallacy and the difficulty of intrusion detection. Understanding Intrusion Detection Through Visualization, pp. 31–47 (2006)
2. Brabec, J., Machlica, L.: Bad practices in evaluation methodology relevant to class-imbalanced problems. Critiquing and correcting trends in machine learning workshop at NeurIPS abs/1812.01388 (2018). http://arxiv.org/abs/1812.01388

3. Brabec, J., Machlica, L.: Decision-forest voting scheme for classification of rare classes in network intrusion detection. In: 2018 IEEE International Conference on Systems, Man, and Cybernetics (SMC), pp. 3325–3330, October 2018. https://doi.org/10.1109/SMC.2018.00563

4. Chawla, N.V.: Data mining for imbalanced datasets: an overview. In: Maimon, O., Rokach, L. (eds.) Data Mining and Knowledge Discovery Handbook, pp. 875–886. Springer, Boston (2009). https://doi.org/10.1007/978-0-387-09823-4_45

5. Damodaran, A., Di Troia, F., Visaggio, C.A., Austin, T.H., Stamp, M.: A comparison of static, dynamic, and hybrid analysis for malware detection. J. Comput. Virol. Hacking Tech. **13**(1), 1–12 (2015). https://doi.org/10.1007/s11416-015-0261-z

6. Davis, J., Goadrich, M.: The relationship between precision-recall and ROC curves. In: Proceedings of the 23rd International Conference on Machine Learning, pp. 233–240. ACM (2006)

7. Fawcett, T.: An introduction to ROC analysis. Pattern Recogn. Lett. **27**(8), 861–874 (2006)

8. Haixiang, G., Yijing, L., Shang, J., Mingyun, G., Yuanyue, H., Bing, G.: Learning from class-imbalanced data: review of methods and applications. Expert Syst. Appl. **73**, 220–239 (2017)

9. He, H., Garcia, E.A.: Learning from imbalanced data. IEEE Trans. Knowl. Data Eng. **21**(9), 1263–1284 (2009)

10. He, K., Zhang, X., Ren, S., Sun, J.: Deep residual learning for image recognition. In: The IEEE Conference on Computer Vision and Pattern Recognition (CVPR), June 2016

11. Kotsiantis, S., Kanellopoulos, D., Pintelas, P., et al.: Handling imbalanced datasets: a review. GESTS Int. Trans. Comput. Sci. Eng. **30**(1), 25–36 (2006)

12. Landgrebe, T.C., Paclik, P., Duin, R.P.: Precision-recall operating characteristic (P-ROC) curves in imprecise environments. In: 2006 18th International Conference on Pattern Recognition, ICPR 2006, vol. 4, pp. 123–127. IEEE (2006)

13. Matthews, B.W.: Comparison of the predicted and observed secondary structure of T4 phage lysozyme. Biochimica et Biophysica Acta (BBA) Protein Struct. **405**(2), 442–451 (1975)

14. Pendlebury, F., Pierazzi, F., Jordaney, R., Kinder, J., Cavallaro, L.: TESSERACT: eliminating experimental bias in malware classification across space and time. In: 28th USENIX Security Symposium (USENIX Security 2019), pp. 729–746 (2019)

15. Phua, C., Alahakoon, D., Lee, V.: Minority report in fraud detection: classification of skewed data. SIGKDD Explor. Newsl. **6**(1), 50–59 (2004). https://doi.org/10.1145/1007730.1007738. http://doi.acm.org/10.1145/1007730.1007738

16. Rahman, M.M., Davis, D.: Addressing the class imbalance problem in medical datasets. Int. J. Mach. Learn. Comput. **3**(2), 224 (2013)

17. Russakovsky, O., et al.: Imagenet large scale visual recognition challenge. Int. J. Comput. Vis. **115**(3), 211–252 (2015)

18. Saito, T., Rehmsmeier, M.: The precision-recall plot is more informative than the ROC plot when evaluating binary classifiers on imbalanced datasets. PLoS ONE **10**(3), e0118432 (2015)

19. Sokolova, M., Lapalme, G.: A systematic analysis of performance measures for classification tasks. Inf. Process. Manage. **45**(4), 427–437 (2009)

20. Wei, W., Li, J., Cao, L., Ou, Y., Chen, J.: Effective detection of sophisticated online banking fraud on extremely imbalanced data. World Wide Web **16**(4), 449–475 (2013)

21. Yu, L., Wang, S., Lai, K.K., Wen, F.: A multiscale neural network learning paradigm for financial crisis forecasting. Neurocomputing **73**(4–6), 716–725 (2010)

A Correction Method of a Base Classifier Applied to Imbalanced Data Classification

Pawel Trajdos$^{(\boxtimes)}$ and Marek Kurzynski

Wroclaw University of Science and Technology, Wroclaw, Poland
{pawel.trajdos,marek.kurzynski}@pwr.edu.pl

Abstract. In this paper, the issue of tailoring the soft confusion matrix classifier to deal with imbalanced data is addressed. This is done by changing the definition of the soft neighbourhood of the classified object. The first approach is to change the neighbourhood to be more local by changing the Gaussian potential function approach to the nearest neighbour rule. The second one is to weight the instances that are included in the neighbourhood. The instances are weighted inversely proportional to the a priori class probability. The experimental results show that for one of the investigated base classifiers, the usage of the KNN neighbourhood significantly improves the classification results. What is more, the application of the weighting schema also offers a significant improvement.

Keywords: Classification · Probabilistic model · Randomized reference classifier · Soft confusion matrix · Imbalanced data

1 Introduction

Imbalanced dataset, denoting the case when there is a significant difference between the prior probabilities for different classes, is a difficult problem for classification. It results from the fact that – on the one hand – for most such problems it is desirable to build classifiers with good performance on the minority class being the class of interest, but – on the other hand – in highly imbalanced datasets, the minority class is mostly sensitive to singular classification errors. Let's cite two practical classification problems as examples of such situation. The first example concerns fraud detection in online monetary transactions. Although fraud is becoming more common and this is a growing problem for banking systems, the number of fraudulent transactions is typically a small fraction of all financial transactions. So, we have here an imbalanced classification problem in which the classifier should correctly recognize objects from the minority class, i.e. detect all fraud transactions and at the same time it should not give false alarms. A similar situation is in the second example regarding computer-aided medical diagnosis. In the simple task of medical screening tests we have two classes: healthy people (majority class) and people suffering from a rare disease (minority class). Requirements for the diagnostic algorithm are the same as before: to successfully detect ill people.

© Springer Nature Switzerland AG 2020
V. V. Krzhizhanovskaya et al. (Eds.): ICCS 2020, LNCS 12140, pp. 88–102, 2020.
https://doi.org/10.1007/978-3-030-50423-6_7

There are more negative consequences of imbalanced dataset that hinder correct classification. We can mention here [28]: overlapping classes (clusters of minority class are heavily contaminated with majority class), lack of density (learners do not have enough data to make generalization about the distribution of minority samples), noisy data (the presence of noise degrades the information capacity of minority class samples) and dataset shift (training and testing data follow the different distribution).

The difficulty in classifying imbalanced datasets has caused great interest among the pattern recognition research community in methods and algorithms that would effectively solve this problem. The proposed methods of classification of imbalanced datasets can be divided into two following categories [1,23]:

1. **Data level approach** (or external techniques) involves manipulating instances of the learning set to obtain a more balanced class distribution. This goal can be achieved through undersampling and/or oversampling procedures. In the first approach, instances are removed from the majority class, while in the second technique new artificial instances are added to the minority class. Different specified algorithms for both methods define the way of removing (adding) instances from the majority (to the minority) class. Random undersampling [17], ACOSampling [41], EUSBoost [10,19] for undersampling approach and SMOTE [3], ADASYN [14], SNOCC [42] for oversampling procedures are exemplary algorithms for this category of methods.

2. **Algorithm level approach** (or internal techniques) denotes classifiers which directly learn class characteristics from the imbalanced data. The leading approaches in this category of methods are:
   - **Improved algorithms** denote classifiers that are modified (improved) to fit their properties to the specifics of imbalanced classification. Support vector machines [15], artificial neural networks [8], k-nearest neighbours [25], decision tree [24], fuzzy inference system [7] and random forest [40] are the most popular methods which have been adapted to classification of imbalanced data.
   - **One-class learning** algorithms for imbalanced problem are trained on the representation of the minority class [32].
   - **Cost-sensitive learning** is based on a very-well known classification scheme in which the cost of misclassification depends on the kind of error made. For example, in the Bayes decision theory this cost is modeled by loss function (loss matrix), which practically can have any values [6]. Application of this scheme to the classification of imbalanced data denotes that first we define cost of misclassification of objects from the minority (class of interest) and majority class (e.g. using domain expert opinion) and then we build a classifier (learner) which takes into account different costs for different classification errors (e.g. minimizing the expected cost or risk) [18,20,31].
   - **Ensemble learning** – in this approach several base classifiers are trained and their predictions are combined to produce the final classification decision [9]. Ensemble methods applied to the imbalanced data classification

combine ensemble learning algorithms and techniques dedicated to imbalanced problems, e.g. undersampling/oversampling procedures [29,37] or cost-sensitive learning [31].

This paper is devoted to the new classifier for imbalanced data which belongs to the algorithm level category of methods. The algorithm developed is based on the author's method of improving the operation of any classifier called base classifier. In the method first the local class-dependent probabilities of misclassification and correct classification are determined. For this purpose two original concepts of randomized reference classifier (RRC) [39] and soft confusion matrix (SCM) [34] are used. Then, the determined probabilities are used for correction of the decision of the base classifier to increase the chance of correct classification of the recognized object. The developed method has already been successfully applied for the construction of multi-classifier systems [34], in multi-label recognition [35,36] and in the recognition of biosignals [22]. However, the algorithm is sensitive to imbalanced data distribution. In other words, its correction ability is lower when the class imbalance ratio is higher. To make the developed approach more practical, it is necessary to provide a mechanism of dealing with imbalanced class distribution. And this paper is aimed at dealing with this issue. In the proposed algorithm for imbalanced data, the classification functions have additional factors inversely proportional to the class size with the parameter experimentally tuned. This mechanism allows a controlled change in the degree of correction of the base classifier to highlight minority classes.

The paper is organized as follows. Section 2 introduces the formal notation used in the paper and provides a description of the proposed approach. The experimental setup is given in Sect. 3. In Sect. 4 experimental results are given and discussed. Section 5 concludes the paper.

2   Proposed Method

2.1   Preliminaries

Let be given pattern recognition problem in which x denotes d-dimensional feature vector of an object and j is its class number. Feature vector x belongs to the feature space $\mathcal{X} = \Re^d$ and class number j takes value in a finite set $\mathcal{M} = \{1, 2, 3, ..., M\}$. Let ψ be a trained classifier which assigns a class number to the recognized object. In other words, ψ_n maps the feature space to the set of class labels, viz. $\psi : \mathcal{X} \rightarrow \mathcal{M}$. Classifier ψ will be called base classifier. We suppose that ψ is described by the canonical model, i.e. for given object x it first produces values of normalized classification functions (supports) $g_i(x), i \in \mathcal{M}$ $(g_i(x) \in [0,1], \sum g_i(x) = 1)$ and then classifies object according to the maximum support rule:

$$\psi(x) = i \Leftrightarrow g_i(x) = \max_{k \in \mathcal{M}} g_k(x). \tag{1}$$

However, the base classifier ψ and formula (1) will not be used directly for classification. To classify object x a decision scheme will be used, which indirectly takes into account classification result of ψ and additionally uses the local (relative to x) properties of base classifier for correction of its decision to increase the chance of correct classification of the recognized object. The proposed decision scheme will be further modified in terms of imbalanced data classification. Source of information about the properties of the base classifier used in the correction procedure of ψ is a validation set:

$$\mathcal{V} = \{(x_1, j_1), (x_2, j_2), \ldots, (x_N, j_N)\}; \quad x_k \in \mathcal{X}, \ j_k \in \mathcal{M} \tag{2}$$

containing pairs of feature vectors and their corresponding class labels.

The basis for the proposed method of classification is the probabilistic model meaning the assumption that x and j are observed values of random variables X and J, respectively.

2.2   Correction of Base Classifier

The corrected base classifier $\psi^{(Corr)}$, using the probabilistic model of the recognition task, acts according to the known Bayes scheme:

$$\psi^{(Corr)}(x) = i \Leftrightarrow P(i|x) = \max_{k \in \mathcal{M}} P(k|x). \tag{3}$$

Now, however, we will express *a posteriori* probabilities $P(j|x), j \in \mathcal{M}$ in a different way than in the classic Bayesian formula, making them dependent on the probabilistic properties of the base classifier, namely:

$$P(j|x) = \sum_{i \in \mathcal{M}} P(i, j|x) = \sum_{i \in \mathcal{M}} P(i|x)P(j|i, x), \tag{4}$$

where $P(i|x) = P(\psi(x) = i)$ and $P(j|i, x)$ denotes the probability that x belongs to the j-th class given that $\psi(x) = i$. Unfortunately, it should be noted that with both probabilities there is a serious problem. First, for the deterministic base classifier ψ probabilities $p(i|x), i \in \mathcal{M}$ are equal to 0 or 1. Secondly, probabilities $P(j|i, x)$ are class-dependent probabilities of the correct classification (for $i = j$) and the misclassification (for $i \neq j$) of ψ at the point x and estimating these probabilities would require many validation objects at this point.

To give the formula (4) a constructive character and calculate both probabilities we will use two concepts: the randomized reference classifier (RRC) and the soft confusion matrix (SCM). The RRC is randomized model of classifier ψ and with its help the probabilities $p(\psi(x) = i) \in [0, 1]$ will be calculated. In turn, the SCM will be used to determine the assessment of correct and misclassification of ψ at the point x, i.e. probabilities $P(j|\psi(x) = i), i, j \in \mathcal{M}$. The method defines the surrounding of the point x containing validation objects in terms of fuzzy sets allowing for flexible selection of membership functions and taking into account the case of imbalanced classes.

2.3   Randomized Reference Classifier

RRC is a probabilistic classifier which is defined by a probability distribution over the set of class labels $\mathcal{M}$. Its classifying functions $\{\delta_j(x)\}_{j \in \mathcal{M}}$ are observed values of random variables $\{\Delta_j(x)\}_{j \in \mathcal{M}}$ fulfilling the following conditions:

$$\Delta_i(x) \in [0,1], \tag{5}$$

$$\sum_{i \in \mathcal{M}} \Delta_i(x) = 1, \tag{6}$$

$$\mathbf{E}\left[\Delta_i(x)\right] = g_i(x), \ i \in \{0,1\}, \tag{7}$$

where $\mathbf{E}$ is the expected value operator. Conditions (5) and (6) follow from the normalization properties of class supports, whereas condition (7) provides the equivalence of the randomized model $\psi^{(RRC)}$ and base classifier ψ. Based on the latter condition, the RRC can be used to provide a randomized model of any classifier that returns a vector of class-specific supports $g(x)$.

It is obvious, that the probability of classifying an object x into the class i using the RRC is as follows:

$$P(\psi^{(RRC)}(x) = i) = P[\Delta_i(x) > \Delta_k(x), k \in \mathcal{M} \setminus i]. \tag{8}$$

The probability on the right side of (8) can be easily determined if we assume – as in the original work of Woloszynski and Kurzynski [39] – that $\Delta_i(x)$ have the beta distribution.

Since $\psi^{(RRC)}$ acts – on average – as the modeled base classifier, the following approximation is fully justified:

$$P(\psi(x) = i) \approx P[\Delta_i(x) > \Delta_k(x), k \in \mathcal{M} \setminus i], \ x \in \mathcal{X}, i \in \mathcal{M}. \tag{9}$$

2.4   Soft Confusion Matrix

Classically, the confusion matrix is in the form of two-dimensional table, in which the rows correspond to the true classes while the columns match the outcomes of the classifier ψ, as it shown in Table 1.

Table 1. The multiclass confusion matrix of classifier ψ

		Classification by ψ			
		1	2	...	M
	1	$\varepsilon_{1,1}$	$\varepsilon_{1,2}$	...	$\varepsilon_{1,M}$
True	2	$\varepsilon_{2,1}$	$\varepsilon_{2,2}$	...	$\varepsilon_{2,M}$
Class	:	:	:		:
	M	$\varepsilon_{M,1}$	$\varepsilon_{M,2}$	...	$\varepsilon_{M,M}$

The value $\varepsilon_{i,j}$ is determined from validation set (2) as the following ratio ($|\cdot|$ is the cardinality of a set):

$$\varepsilon_{i,j} = \frac{|\bar{\mathcal{V}}_j \cap \bar{\mathcal{D}}_i|}{|\bar{\mathcal{V}}_j|}, \tag{10}$$

where $\bar{\mathcal{V}}_j = \{x_k \in \mathcal{V} : j_k = j\}$ (class set) denotes the set of validation objects from the j-th class and $\bar{\mathcal{D}}_i = \{x_k \in \mathcal{V} : \psi(x_k) = i\}$ (decision set) is the set of validation objects assigned by ψ to the i-th class.

The confusion matrix (10) gives an overall (for the whole feature space) image of the classifier properties, while our purpose is to assess the local probabilities $P(j|i,x)$. For this reason, we will generalize the term of confusion matrix enabling free shaping of the concept of "locality" and assigning weights to individual validation objects. Generalized confusion matrix, called the soft confusion matrix (SCM), referred to the recognized object $x \in \mathcal{X}$ is defined as follows:

$$\varepsilon_{i,j}(x) = \frac{|\mathcal{V}_j \cap \mathcal{D}_i \cap \mathcal{N}(x)|}{|\mathcal{V}_j \cap \mathcal{N}(x)|}, \tag{11}$$

where $\mathcal{V}_j, \mathcal{D}_i$ and $\mathcal{N}(x)$ are fuzzy sets specified in the validation set $\mathcal{V}$ and $|\cdot|$ denotes the cardinality of a fuzzy set [5].

Now we will define and give a practical interpretation of fuzzy sets that create the proposed SCM concept (11).

The Class Set $\mathcal{V}_j$. Identically as in (10), this set denotes the set of validation objects from the j-th class. Formulating the set $\mathcal{V}_j$ in terms of fuzzy sets theory it can be assumed that the grade of membership of validation object x_k to $\mathcal{V}_j$ is the class indicator which leads to the following definition of $\mathcal{V}_j$ as the fuzzy set:

$$\mathcal{V}_j = \{(x_k, \mu_{\mathcal{V}_j}(x_k))\}, \quad \text{where} \quad \mu_{\mathcal{V}_j}(x_k) = \begin{cases} 1 & \text{if } j_k = j, \\ 0 & \text{elsewhere.} \end{cases} \tag{12}$$

The Decision Set $\mathcal{D}_i$. For the confusion matrix (10) the crisp decision set $\bar{\mathcal{D}}_i$ includes validation objects x_k for which $\psi(x_k) = i$. The original concept of fuzzy decision set $\mathcal{D}_j$ is defined as follows:

$$\mathcal{D}_i = \{(x_k, \mu_{\mathcal{D}_i}(x_k)) : x_k \in \mathcal{V}, \mu_{\mathcal{D}_i}(x_k) = P(i|x_k)\}, \tag{13}$$

where $P(i|j, x_k)$ is calculated according to (9). Formula (13) demonstrates that now the membership of validation object x_k to the set $\mathcal{D}_i$ is not determined by the decision of classifier ψ. The grade of membership of validation object x_k to $\mathcal{D}_i$ depends on the potential chance of classifying object x_k to the i-th class by the base classifier. We assume, that this potential chance is equal to the probability $P(i|x) = P(\psi(x) = i)$ calculated using the randomized model RRC of base classifier ψ.

The Neighbourhood Set $\mathcal{N}(x)$. As it seems, this set play the crucial role in the proposed concept of SCM, because it decide which validation objects x_k and with which weights will be taken into account in the procedure of determining the local properties of $\psi(x)$. Formally, $\mathcal{N}(x)$ is also a fuzzy set:

$$\mathcal{N}(x) = \{(x_k, \mu_{\mathcal{N}(x)}(x_k)) : x_k \in \mathcal{V}\}, \tag{14}$$

but its membership function is not defined univocally because it depends on the adopted concept of "locality" (relative to x). There are two typical methods of determining the set $\mathcal{V}(x)$. In the first approach the neighbourhood of x is precisely defined and only validation objects belonging to this neighbourhood are used to calculate (11). In the second method all validation points are members of the set $\mathcal{N}(x)$ and its membership functions is equal to 1 for $x_k = x$ and decreases with increasing the distance between x_k and x. In the further experimental investigations two forms of the fuzzy set $\mathcal{N}(x)$ were used as representative of both approaches.

1. KNN **Neighborhood.** Let first define the K-neighbourhood of the test object x as the set of K nearest validation objects, viz.

$$\mathcal{K}_K(x) = \{x_{n1}, \ldots, x_{nK} \in \mathcal{V} : \max_{l=1,2,\ldots,K} \text{dist}(x_{nl}, x)^2 \leq \min_{x_k \notin \mathcal{K}_K(x)} \text{dist}(x_k, x)^2\},$$

(15)

where $\text{dist}(x_k, x)^2$ denotes the Euclidean distance in the feature space $\mathcal{X}$. The KNN-related membership function of $\mathcal{N}(x)$ is defined as follows:

$$\mu_{\mathcal{N}(x)}^{(K)}(x_k) = \begin{cases} 1 \text{ if } x_k \in \mathcal{K}(x), \\ 0 \text{ otherwise.} \end{cases}$$

(16)

This kind of neighbourhood should be more fragile to the local properties of the data since it completely ignores the instances that are not in $\mathcal{K}$.

2. **Gaussian Neighborhood.** In this method the Gaussian membership function was applied for defining the set $\mathcal{N}(x)$:

$$\mu_{\mathcal{N}(x)}^{(G)}(x_k) = \exp(-\beta \text{dist}(x, x_k)^2),$$

(17)

where $\beta \in \mathbb{R}_+$ is parameter of μ. The Gaussian-based neighbourhood was originally proposed to use with the SCM classifier in [34].

2.5   Dealing with Imbalanced Data

In this paper, the issue of imbalanced class distribution is dealt with via modification of the membership function of the neighbourhood set $\mathcal{N}(x)$. We propose to add a new factor (weight) to original membership function $\mu_{\mathcal{N}(x)}(x_k)$ which is inversely proportional to the *a priori* probability of class $P(j), j \in \mathcal{M}$. Assuming that the minority class is the class of interest, such a method relatively enhances class proportionally to its importance. The proposed approach also means, that the neighbourhood set $\mathcal{N}(x)$ is now dependent on the class j to which the validation objects used to calculate $\varepsilon_{i,j}(x)$ in (11) belong. Thus, the membership function of the neighbourhood set (14) that includes imbalanced classes is as follows:

$$\mu_{\mathcal{N}_j(x)}(x_k) = c\mu_{\mathcal{N}(x)}(x_k)P(j)^{-\gamma}.$$

(18)

$\gamma \in \mathbb{R}_+$ is the coefficient that controls weighting intensity and c is normalized coefficient.

Finally, from (14) and modification (18) we get the following approximation:

$$P(j|i, x) \approx \frac{\varepsilon_{i,j}(x)}{\sum_{j \in \mathcal{M}} \varepsilon_{i,j}(x)}, \qquad (19)$$

which together with (9), (4) and (3) give the corrected base classifier $\psi^{(Corr)}(x)$ in the version tailored for the case of imbalanced data.

3   Experimental Setup

To validate the classification quality obtained by the proposed approaches the experimental evaluation, which setup is described below, is performed.

The following base classifiers were employed:

- ψ_{NB} – Naive Bayes classifier with kernel density estimation [13].
- ψ_{J48} – Weka version of the C4.5 algorithm [27] with Laplace smoothing [26].
- ψ_{NC} – nearest centroid (Nearest Prototype) [21].

The classifiers implemented in WEKA framework [12] were used. If not stated otherwise, the classifier parameters were set to their defaults. For each base classifier, the training dataset is resampled with weights inversely proportional to the *a priori* probability of instance-specific class. This is to make base classifiers robust against imbalanced data.

During the experimental evaluation the following classifiers were compared:

1. ψ_{R} – unmodified base classifier,
2. ψ_{G} – SCM classifier with unmodified Gaussian neighbourhood,
3. ψ_{Gw} – SCM classifier with weighted Gaussian neighbourhood,
4. ψ_{K} – SCM classifier with unmodified KNN neighbourhood,
5. ψ_{Kw} – SCM classifier with weighted KNN neighbourhood.

The size of the neighbourhood, expressed as β coefficient, the number of nearest neighbours K and the weighting coefficient γ, were chosen using a fivefold cross-validation procedure and the grid search technique. The following values of β, K and γ were considered: $\beta \in \{2^{-2}, 2^{-1}, 2^1, \cdots, 2^6\}$, $K \in \{1, 3, 5, 7, \cdots, 15\}$, $\gamma \in \{0, 2^{-6}, 2^{-5}, 2^{-4}, \cdots 2^2\}$. The values were chosen in such a way that minimizes macro-averaged kappa coefficient.

The experimental code was implemented using WEKA framework. The source code of the algorithms is available online[1].

To evaluate the proposed methods the following classification-loss criteria are used [30]: Macro-averaged FDR (1-precision), FNR (1-recall), F_1, Matthews correlation coefficient (MCC); Micro-averaged F_1, MCC. More quality measures from the macro-averaging group are considered because this kind of measures is more sensitive to the performance for minority classes.

---

[1] https://github.com/ptrajdos/rrcBasedClassifiers/tree/develop.

Following the recommendations of [4,11], the statistical significance of the obtained results was assessed using the two-step procedure. The first step is to perform the Friedman test [4] for each quality criterion separately. Since the multiple criteria were employed, the familywise errors (FWER) should be controlled [2]. To do so, the Bergman-Hommel [2] procedure of controlling FWER of the conducted Friedman tests was employed. When the Friedman test shows that there is a significant difference within the group of classifiers, the pairwise tests using the Wilcoxon signed-rank test [4,38] were employed. To control FWER of the Wilcoxon-testing procedure, the Bergman-Hommel approach was employed [2]. For all tests the significance level was set to $\alpha = 0.05$.

The experimental evaluation was conducted on the collection of the 78 benchmark datasets taken from the Keel repository containing imbalanced datasets with imbalance ratio higher than 9^2.

During the preprocessing stage, the datasets underwent a few transformations. First, all nominal attributes were converted into a set of binary variables. The transformation is necessary whenever the distance-based algorithms are employed [33]. To reduce the computational burden and remove irrelevant information, the PCA procedure with the variance threshold set to 95% was applied [16]. The features were also normalized to have zero mean value and zero unit variance.

4   Results and Discussion

To compare multiple algorithms on multiple benchmark sets the average ranks approach [4] is used. To provide a visualization of the average ranks, the radar plots are employed. In the plots, the data is visualized in such way that the lowest ranks are closer to the centre of the graph. The radar plots related to the experimental results are shown in Figs. 1a–c.

Due to the page limit, the full results are published online[3].

The numerical results are given in Tables 2, 3 and 4. Each table is structured as follows. The first row of each section contains names of the investigated algorithms. Then the table is divided into six sections – one section is related to a single evaluation criterion. The first row of each section is the name of the quality criterion investigated in the section. The second row shows the p-value of the Friedman test. The third one shows the average ranks achieved by algorithms. The following rows show p-values resulting from pairwise Wilcoxon test. The p-value equal to 0.00 informs that the p-values are lower than 10^{-3} and p-value equal to 1.00 informs that the value is higher than 0.999.

4.1   Macro Averaged Criteria

Let us begin with the analysis of the results related to KNN neighbourhood. For the Naive Bayes and J48 classifiers, there are no significant differences between the Gaussian neighbourhood and KNN neighbourhood. For the Nearest Centroid classifier, on the other hand, the KNN neighbourhood gives better results in terms of FNR, F_1 and MCC. For the FDR criterion, there is no significant difference. It means that for ψ_{NC} classifier applying KNN neighbourhood improves recall without affecting precision what results in better overall performance. What is more, for the J48 classifier, only the classifiers based on the KNN neighbourhood offers a significant improvement in terms of F_1 criterion.

Now the impact of applying the weighting scheme is assessed. Generally speaking, the application of the weighting scheme results in improving recall at the cost of reducing precision. However, in general, the reduction of precision is not significant (except for ψ_{NC} classifier and KNN approach). As a consequence, the overall classification quality, measured in terms of F_1 criterion remains unchanged (no significant difference). This kind of change is the expected consequence of applying the weighting scheme. On the other hand, in cases of J48 and NC classifiers, there are significant improvements in terms of MCC criterion. What is more, for the J48 classifier, only the classifiers based on the weighted neighbourhood offers a significant improvement in terms of MCC criterion.

Now the correction ability of the SCM classifier is investigated. As it was said above, for the J48 base classifier, the overall correction ability depends on the type of the neighbourhood applied. For ψ_{NB} and ψ_{NC} classifiers, on the other hand, there is always a significant improvement in terms of F_1 and MCC criteria. In general, the application of SCM classifier, when compared to the base classifier, improves the precision at the cost of decreasing recall. The recall-decrease is lower for the SCM classifiers using the weighted neighbourhood approach. So, applying the weighting scheme eliminates the main drawback of the SCM classifier used to the imbalanced data.

4.2   Micro Averaged Criteria

For the micro-averaged criteria, the statistical tests show that all differences are significant. Consequently, all investigated approaches improve the overall majority-class-performance in comparison to the unmodified base classifier. However, the classifiers with weighted neighbourhood show lower classification quality compared with classifiers that use no weights. This is an obvious consequence of trying to improve the performance for the minority class.

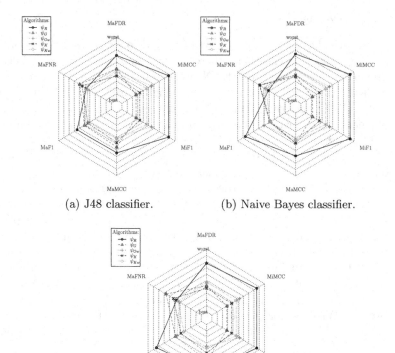

(a) J48 classifier.   (b) Naive Bayes classifier.

(c) Nearest Centroid classifier.

Fig. 1. Radar plots for the investigated classifiers.

Table 2. Statistical evaluation. Wilcoxon test results for J48 classifier.

	Ψ_R	Ψ_G	Ψ_{Gw}	Ψ_K	Ψ_{Kw}	Ψ_R	Ψ_G	Ψ_{Gw}	Ψ_K	Ψ_{Kw}	Ψ_R	Ψ_G	Ψ_{Gw}	Ψ_K	Ψ_{Kw}
Nam.	MaFDR					MaFNR					MaF1				
Frd.	4.079e−06					3.984e−02					3.189e−03				
Rank	3.846	2.981	2.897	2.532	2.744	2.865	3.385	2.737	3.192	2.821	3.590	3.013	2.987	2.750	2.660
Ψ_R		.002	.000	.000	.000		.128	1.00	.128	1.00		.217	.217	.009	.003
Ψ_G			.933	.491	.491			.003	.952	.007			.985	.572	.206
Ψ_{Gw}				.491	.491				.022	1.00				.572	.217
Ψ_K					.491					.000					.217
Nam.	MaMCC					MiF1					MiMCC				
Frd.	1.494e−03					1.412e−24					1.412e−24				
Rank	3.564	3.205	2.622	2.910	2.699	4.506	2.064	3.205	2.346	2.878	4.506	2.064	3.205	2.346	2.878
Ψ_R		.603	.009	.129	.026		.000	.004	.000	.000		.000	.004	.000	.000
Ψ_G			.010	.535	.045			.000	.005	.000			.000	.005	.000
Ψ_{Gw}				.173	.717				.000	.003				.000	.003
Ψ_K					.026					.000					.000

Table 3. Statistical evaluation. Wilcoxon test results for NB classifier.

	Ψ_R	Ψ_G	Ψ_{Gw}	Ψ_K	Ψ_{Kw}	Ψ_R	Ψ_G	Ψ_{Gw}	Ψ_K	Ψ_{Kw}	Ψ_R	Ψ_G	Ψ_{Gw}	Ψ_K	Ψ_{Kw}
Nam.	MaFDR					MaFNR					MaF1				
Frd.	2.116e−08					1.697e−02					7.976e−18				
Rank	3.955	2.494	2.987	2.609	2.955	2.654	3.308	2.872	3.327	2.840	4.417	2.494	2.994	2.564	2.532
Ψ_R		.000	.000	.000	.000		.005	.392	.020	.392		.000	.000	.000	.000
Ψ_G			.103	.612	.501			.001	.511	.006			.197	.664	.664
Ψ_{Gw}				.096	.501				.019	.833				.091	.041
Ψ_K					.074					.006					.749
Nam.	MaMCC					MiF1					MiMCC				
Frd.	1.224e−04					1.226e−31					1.226e−31				
Rank	3.737	2.929	2.744	2.923	2.667	4.699	1.878	3.154	2.391	2.878	4.699	1.878	3.154	2.391	2.878
Ψ_R		.004	.000	.000	.000		.000	.000	.000	.000		.000	.000	.000	.000
Ψ_G			.145	.894	.300			.000	.002	.000			.000	.002	.000
Ψ_{Gw}				.397	.939				.000	.015				.000	.015
Ψ_K					.397					.000					.000

Table 4. Statistical evaluation. Wilcoxon test results for NC classifier.

	Ψ_R	Ψ_G	Ψ_{Gw}	Ψ_K	Ψ_{Kw}	Ψ_R	Ψ_G	Ψ_{Gw}	Ψ_K	Ψ_{Kw}	Ψ_R	Ψ_G	Ψ_{Gw}	Ψ_K	Ψ_{Kw}
Nam.	MaFDR					MaFNR					MaF1				
Frd.	3.149e−10					2.684e−04					4.529e−17				
Rank	4.090	2.667	2.827	2.506	2.910	2.865	3.654	2.647	3.096	2.737	4.378	2.942	2.827	2.532	2.321
Ψ_R		.000	.000	.000	.000		.057	.651	1.00	.651		.000	.000	.000	.000
Ψ_G			1.00	.470	1.00			.000	.004	.000			1.00	.043	.044
Ψ_{Gw}				.171	1.00				.225	1.00				.044	.043
Ψ_K					.043					.056					1.00
Nam.	MaMCC					MiF1					MiMCC				
Frd.	5.340e−11					4.793e−20					4.001e−20				
Rank	3.994	3.282	2.397	2.821	2.506	4.404	2.147	3.096	2.500	2.853	4.410	2.147	3.096	2.500	2.846
Ψ_R		.000	.000	.000	.000		.000	.000	.000	.000		.000	.000	.000	.000
Ψ_G			.000	.028	.010			.000	.010	.000			.000	.010	.000
Ψ_{Gw}				.310	.979				.001	.010				.001	.010
Ψ_K					.310					.001					.001

5 Conclusions

This paper addresses the issue of tailoring the soft confusion matrix classifier to dealing with imbalanced data. Two concepts based on the change of the neighbourhood were proposed. The experimental results show that, in some circumstances, these approaches can improve the obtained classification quality. It shows that classifiers based on the RRC concept and SCM concept, in particular, are robust tools that can deal with various types of data. The other way of tailoring the SCM classifier to imbalanced data may be the modification of the $P(i|x)$ probability distribution. This aspect should be studied carefully.

Acknowledgments. This work was supported by the statutory funds of the Department of Systems and Computer Networks, Wroclaw University of Science and Technology.

References

1. Ali, A., Shamsuddin, S.M., Ralescu, A.L., et al.: Classification with class imbalance problem: a review. Int. J. Adv. Soft Comput. Appl. **7**(3), 176–204 (2015)
2. Bergmann, B., Hommel, G.: Improvements of general multiple test procedures for redundant systems of hypotheses. In: Bauer, P., Hommel, G., Sonnemann, E. (eds.) Multiple Hypothesenprüfung/Multiple Hypotheses Testing. Medizinische Informatik und Statistik, vol. 70, pp. 100–115. Springer, Heidelberg (1988). https://doi.org/10.1007/978-3-642-52307-6_8
3. Chawla, N.V., Bowyer, K.W., Hall, L.O., Kegelmeyer, W.P.: SMOTE: synthetic minority over-sampling technique. JAIR **16**, 321–357 (2002). https://doi.org/10.1613/jair.953
4. Demšar, J.: Statistical comparisons of classifiers over multiple data sets. J. Mach. Learn. Res. **7**, 1–30 (2006)
5. Dhar, M.: On cardinality of fuzzy sets. IJISA **5**(6), 47–52 (2013). https://doi.org/10.5815/ijisa.2013.06.06
6. Duda, R.: Pattern Classification. Wiley, New York (2001)
7. Fernández, A., del Jesus, M.J., Herrera, F.: Hierarchical fuzzy rule based classification systems with genetic rule selection for imbalanced data-sets. Int. J. Approximate Reasoning **50**(3), 561–577 (2009). https://doi.org/10.1016/j.ijar.2008.11.004
8. Fu, K., Cheng, D., Tu, Y., Zhang, L.: Credit card fraud detection using convolutional neural networks. In: Hirose, A., Ozawa, S., Doya, K., Ikeda, K., Lee, M., Liu, D. (eds.) ICONIP 2016. LNCS, vol. 9949, pp. 483–490. Springer, Cham (2016). https://doi.org/10.1007/978-3-319-46675-0_53
9. Galar, M., Fernandez, A., Barrenechea, E., Bustince, H., Herrera, F.: A review on ensembles for the class imbalance problem: bagging-, boosting-, and hybrid-based approaches. IEEE Trans. Syst. Man Cybern. C **42**(4), 463–484 (2012). https://doi.org/10.1109/tsmcc.2011.2161285
10. Galar, M., Fernández, A., Barrenechea, E., Herrera, F.: EUSBoost: enhancing ensembles for highly imbalanced data-sets by evolutionary undersampling. Pattern Recognit. **46**(12), 3460–3471 (2013). https://doi.org/10.1016/j.patcog.2013.05.006
11. Garcia, S., Herrera, F.: An extension on "statistical comparisons of classifiers over multiple data sets" for all pairwise comparisons. J. Mach. Learn. Res. **9**, 2677–2694 (2008)
12. Hall, M., Frank, E., Holmes, G., Pfahringer, B., Reutemann, P., Witten, I.H.: The WEKA data mining software. SIGKDD Explor. Newsl. **11**(1), 10 (2009). https://doi.org/10.1145/1656274.1656278
13. Hand, D.J., Yu, K.: Idiot's Bayes: not so stupid after all? Int. Stat. Rev./Revue Internationale de Statistique **69**(3), 385 (2001). https://doi.org/10.2307/1403452
14. He, H., Bai, Y., Garcia, E.A., Li, S.: ADASYN: adaptive synthetic sampling approach for imbalanced learning. In: 2008 IEEE International Joint Conference on Neural Networks (IEEE World Congress on Computational Intelligence), pp. 1322–1328. IEEE, June 2008. https://doi.org/10.1109/ijcnn.2008.4633969

15. Hwang, J.P., Park, S., Kim, E.: A new weighted approach to imbalanced data classification problem via support vector machine with quadratic cost function. Expert Syst. Appl. **38**(7), 8580–8585 (2011). https://doi.org/10.1016/j.eswa.2011.01.061
16. Jolliffe, I.T., Cadima, J.: Principal component analysis: a review and recent developments. Philos. Trans. R. Soc. A **374**(2065), 20150202 (2016). https://doi.org/10.1098/rsta.2015.0202
17. Kaur, H., Pannu, H.S., Malhi, A.K.: A systematic review on imbalanced data challenges in machine learning. CSUR **52**(4), 1–36 (2019). https://doi.org/10.1145/3343440
18. Khan, S.H., Hayat, M., Bennamoun, M., Sohel, F.A., Togneri, R.: Cost-sensitive learning of deep feature representations from imbalanced data. IEEE Trans. Neural Netw. Learn. Syst. **29**(8), 3573–3587 (2018). https://doi.org/10.1109/tnnls.2017.2732482
19. Krawczyk, B., Galar, M., Jelen, L., Herrera, F.: Evolutionary undersampling boosting for imbalanced classification of breast cancer malignancy. Appl. Soft Comput. **38**, 714–726 (2016). https://doi.org/10.1016/j.asoc.2015.08.060
20. Krawczyk, B., Woźniak, M., Schaefer, G.: Cost-sensitive decision tree ensembles for effective imbalanced classification. Appl. Soft Comput. **14**, 554–562 (2014). https://doi.org/10.1016/j.asoc.2013.08.014
21. Kuncheva, L., Bezdek, J.: Nearest prototype classification: clustering, genetic algorithms, or random search? IEEE Trans. Syst. Man Cybern. Part C (Appl. Rev.) **28**(1), 160–164 (1998). https://doi.org/10.1109/5326.661099
22. Kurzynski, M., Krysmann, M., Trajdos, P., Wolczowski, A.: Multiclassifier system with hybrid learning applied to the control of bioprosthetic hand. Comput. Biol. Med. **69**, 286–297 (2016). https://doi.org/10.1016/j.compbiomed.2015.04.023
23. López, V., Fernández, A., García, S., Palade, V., Herrera, F.: An insight into classification with imbalanced data: empirical results and current trends on using data intrinsic characteristics. Inf. Sci. **250**, 113–141 (2013). https://doi.org/10.1016/j.ins.2013.07.007
24. Park, Y., Ghosh, J.: Ensembles of (α)-trees for imbalanced classification problems. IEEE Trans. Knowl. Data Eng. **26**(1), 131–143 (2014). https://doi.org/10.1109/tkde.2012.255
25. Patel, H., Thakur, G.: A hybrid weighted nearest neighbor approach to mine imbalanced data. In: Proceedings of the International Conference on Data Mining (DMIN), pp. 106–110. The Steering Committee of The World Congress in Computer Science, Computer... (2016)
26. Provost, F., Domingos, P.: Tree induction for probability-based ranking. Mach. Learn. **52**(3), 199–215 (2003). https://doi.org/10.1023/a:1024099825458
27. Quinlan, J.R.: C4.5: Programs for Machine Learning. Morgan Kaufmann Publishers Inc., San Francisco (1993)
28. Ramyachitra, D., Manikandan, P.: Imbalanced dataset classification and solutions: a review. Int. J. Comput. Bus. Res. (IJCBR) **5**(4), 186–194 (2014)
29. Seiffert, C., Khoshgoftaar, T.M., Van Hulse, J., Napolitano, A.: RUSBoost: a hybrid approach to alleviating class imbalance. IEEE Trans. Syst. Man Cybern. A **40**(1), 185–197 (2010). https://doi.org/10.1109/tsmca.2009.2029559
30. Sokolova, M., Lapalme, G.: A systematic analysis of performance measures for classification tasks. Inf. Process. Manage. **45**(4) (2009). https://doi.org/10.1016/j.ipm.2009.03.002

31. Sun, Y., Kamel, M.S., Wong, A.K., Wang, Y.: Cost-sensitive boosting for classification of imbalanced data. Pattern Recognit. **40**(12), 3358–3378 (2007). https://doi.org/10.1016/j.patcog.2007.04.009

32. Sun, Y., Wong, A.K.C., Kamel, M.S.: Classification of imbalanced data: a review. Int. J. Pattern Recogn. Artif. Intell. **23**(04), 687–719 (2009). https://doi.org/10.1142/s0218001409007326

33. Tian, Y., Deng, N.: Support vector classification with nominal attributes. In: Hao, Y., et al. (eds.) CIS 2005. LNCS (LNAI), vol. 3801, pp. 586–591. Springer, Heidelberg (2005). https://doi.org/10.1007/11596448_86

34. Trajdos, P., Kurzynski, M.: A dynamic model of classifier competence based on the local fuzzy confusion matrix and the random reference classifier. Int. J. Appl. Math. Comput. Sci. **26**(1) (2016). https://doi.org/10.1515/amcs-2016-0012

35. Trajdos, P., Kurzynski, M.: A correction method of a binary classifier applied to multi-label pairwise models. Int. J. Neural Syst. **28**(09), 1750062 (2018). https://doi.org/10.1142/s0129065717500629

36. Trajdos, P., Kurzynski, M.: Weighting scheme for a pairwise multi-label classifier based on the fuzzy confusion matrix. Pattern Recognit. Lett. **103**, 60–67 (2018). https://doi.org/10.1016/j.patrec.2018.01.012

37. Wang, S., Yao, X.: Diversity analysis on imbalanced data sets by using ensemble models. In: 2009 IEEE Symposium on Computational Intelligence and Data Mining, pp. 324–331. IEEE, March 2009. https://doi.org/10.1109/cidm.2009.4938667

38. Wilcoxon, F.: Individual comparisons by ranking methods. Biometrics Bull. **1**(6), 80 (1945). https://doi.org/10.2307/3001968

39. Woloszynski, T., Kurzynski, M.: A probabilistic model of classifier competence for dynamic ensemble selection. Pattern Recogn. **44**(10–11), 2656–2668 (2011). https://doi.org/10.1016/j.patcog.2011.03.020

40. Wu, Q., Ye, Y., Zhang, H., Ng, M.K., Ho, S.S.: ForesTexter: an efficient random forest algorithm for imbalanced text categorization. Knowl. Based Syst. **67**, 105–116 (2014). https://doi.org/10.1016/j.knosys.2014.06.004

41. Yu, H., Ni, J., Zhao, J.: ACOSampling: an ant colony optimization-based undersampling method for classifying imbalanced DNA microarray data. Neurocomputing **101**, 309–318 (2013). https://doi.org/10.1016/j.neucom.2012.08.018

42. Zheng, Z., Cai, Y., Li, Y.: Oversampling method for imbalanced classification. Comput. Inform. **34**(5), 1017–1037 (2016)

Standard Decision Boundary in a Support-Domain of Fuzzy Classifier Prediction for the Task of Imbalanced Data Classification

Pawel Ksieniewicz$^{(\boxtimes)}$ (iD)

Department of Systems and Computer Networks,
Wroclaw University of Science and Technology, Wrocław, Poland
pawel.ksieniewicz@pwr.edu.pl

Abstract. Many real classification problems are characterized by a strong disturbance in a prior probability, which for the most of classification algorithms leads to favoring majority classes. The action most often used to deal with this problem is oversampling of the minority class by the SMOTE algorithm. Following work proposes to employ a modification of an individual binary classifier support-domain decision boundary, similar to the fusion of classifier ensembles done by the *Fuzzy Templates* method to deal with imbalanced data classification without introducing any repeated or artificial patterns into the training set. The proposed solution has been tested in computer experiments, which results shows its potential in the *imbalanced data classification*.

Keywords: Pattern recognition · Classification · Imbalanced data · Fuzzy classifiers · Standard normalization

1 Introduction

The base and the most important element of any *artificial intelligence* application is the decision module, most often being a trained *pattern recognition* model [4]. The development of such a solution requires the use of an algorithm capable of building knowledge around the specific type of training data.

In the case where training samples are only a set of non-described patterns, for example, to gather groups of objects based on *cluster analysis*, we are dealing with the problem of *unsupervised learning*. In most situations, however, we are not interested in identifying groups in the data set. The goal is preferably in assigning new objects, seen for the first time, to classes that we already have known there is a possibility to learn about their properties on the example of existing patterns. This type of learning is called *supervised learning*, and this specific task is *classification* [17].

In real classification problems, it is relatively rare for each class of a training set to be represented evenly. A significant disturbance in the proportions between

© Springer Nature Switzerland AG 2020
V. V. Krzhizhanovskaya et al. (Eds.): ICCS 2020, LNCS 12140, pp. 103–116, 2020.
https://doi.org/10.1007/978-3-030-50423-6_8

classes is widely studied in the literature under the name of *imbalanced data classification* [5,7].

Solutions for such problems are usually divided into three groups [9]. The first are *built-in methods* that try to modify the algorithm's principles or its decision process to take into consideration the disturbed prior probability of the problem [19,24]. The second group, which is also the most popular in literature and applications, is based on data preprocessing aiming to balance the class counts in the training set. The most common solutions of this type are *under-* [20] and *oversampling* [18] together with methods for generating synthetic patterns such as SMOTE [6,21,22] or ADASYN [1,8,25]. The third group consists of *hybrid methods* [23], mainly feasting on achievements of *ensemble learning*, using a pool of diversified base classifiers [12,13] and a properly constructed, imbalanced decision principle [10,11].

Following work tries to propose a practical method from the *built-in methods* group of solutions, modifying the support-domain decision boundary of the *fuzzy classifier*. It is done using the knowledge acquired on the basis of support vectors obtained on the training set by the already built model, similarly to the propositions of *Fuzzy templates* [15,16]. The second section describes how to adapt them to work with a single classification model and how to modify this approach to the proposed *Standard Decision Boundary* algorithm. The third chapter contains the design of computer experiments carried out and summarized in the fourth chapter, and the fifth one focuses on the overall conclusions drawn from the research.

2   Methods

The *feature space* of a problem in which the decision boundary of the classifier is drawn is the most often undertaken area of considering the construction of a classification method. However, its modification may also take place in the space of supports obtained by the model, which is the subject of the method proposed in this article.

*Regular Decision Boundary (*RDB*).* A fitting algorithm of every *fuzzy classifier* does not only provide bare prediction but also calculates the complementary (adding up to one) probability of belonging to each of the problem classes, which constructs the *support vector* of a predicted sample [3]. The classifier's decision, in the most popular approach, is taken in a favor of the class for which the highest support was obtained [14].

By simplifying the classification problem only for binary tasks, one may determine such a decision rule by the most straightforward equation of a straight line:

$$y = x, \tag{1}$$

where the x-axis represents support for the negative class and the y-axis is positive support. For the following paper, this rule will state as *Regular Decision Boundary* (RDB), and it is illustrated in Fig. 1.

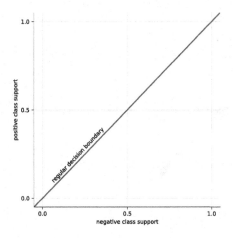

Fig. 1. Illustration of a *Regular Decision Boundary* (RDB).

Fuzzy Templates Decision Boundary (FTDB). A commonly perceived phenomenon that occurs in classification models build on an imbalanced training set is the general tendency to favor the majority class [5]. The support obtained for it receives a particular bonus, caused directly by the increased prior probability.

One of the possible counteractions to this phenomenon may be the modification of a decision rule in the support domain. Solutions of this type are quite common in the construction of *fusers* for the needs of classifier ensembles [15]. One of such approaches is the proposition of *Fuzzy Templates*, introducing the *Decision Profile*, being the matrix of *support vectors* obtained for all patterns from the training set by each classifier from the available pool [16]. To produce a prediction, algorithm determines class centroids of obtained supports, and the final decision is based on the *Nearest Mean* principle.

In the case of a single *fuzzy classifier*, in contrast to the ensemble products of *Decision Profiles*, each of the complementary support vectors obtained for the training set, by definition, must be on a diagonal of a support space perpendicular to the *Regular Decision Rule*. An attempt to employ the *Fuzzy Templates* approach in a single classification model may be described by the equation of a straight line parallel to the *Regular Decision Boundary*, but passing through a point determined by the mean support values calculated separately for the patterns of both the training set classes:

$$y = x + \mu_2 - \mu_1, \tag{2}$$

where μ_1 and μ_2 are mean supports of each class. For the purpose of the following paper this rule will state as *Fuzzy Templates Decision Boundary* (FTDB), and its example is illustrated in Fig. 2a.

Standard Decision Boundary (SDB). The *Fuzzy Templates* method, is an additional, simple classifier, supplementing any fuzzy classification algorithm with

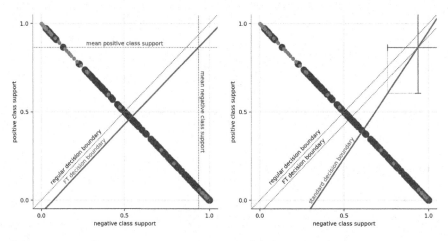

(a) *Fuzzy Templates Decision Boundary*   (b) *Standard Decision Boundary*

Fig. 2. Illustration of trainable decision boundaries.

the model learned from its answers. It is based on the calculation of the basic statistical measure (*mean value*) and its inclusion in the final prediction of the hierarchical ensemble. The following work proposes an enhancement of this approach by including into the decision process also the basic knowledge about the distribution of supports obtained by the base classifier, using a *standard deviation* measure.

This approach still assumes that the decision boundary goes through the intersection of mean supports, but its gradient is further modified. It depends directly on the ratio between standard deviations, so it also goes through the point designated as the difference between the expected values of the distribution and the standard deviations vector. The formula may represent the equation of the proposed decision boundary:

$$y = \frac{\sigma_2(x - \mu_1)}{\sigma_1} + \mu_2, \tag{3}$$

where σ_1 and σ_2 are standard deviations of both classes. Due to the employment of both statistical measures calculated for the needs of a standard normalization, this rule will state as *Standard Decision Boundary* (SDB), and its example is illustrated in Fig. 2b.

Supposition. Intuition suggests that changes in the prediction method implemented both by the FTDB and SDB models should increase the *precision* of the obtained decisions, although the linear nature of the used decision boundary in a presence of a such tendency must simultaneously lead to a worsening of the results of the *recall* metric. Using aggregate information about class distributions in a decision rule, ignoring the prior probabilities of the training set, may result in an increase in the overall quality of predictions in imbalanced data.

So if the proposed method will obtain significantly better results in aggregate metrics, such as *F1-score, balanced accuracy score* or *geometric mean score*, it will be considered as promising.

3   Design of Experiments

Datasets. The problems considered in research during experiments are directly expressed by the selection of datasets that meets specific conditions. For the purposes of conducted experimental evaluation, it was decided to use data with a high degree of imbalance, exceeding the 1:9 ratio, with relatively low dimensionality (up to 20 features). The appropriate collection is contained in the KEEL data repository [2]. A summary of the datasets selected for testing, supplemented with information on the *imbalance ratio*, the count of features and patterns is presented in Table 1.

Compared Approaches. The basis of considerations taken in this work are the differences between approaches to draw a decision boundary in the *support space* and the effectiveness of this type of solutions in *imbalanced data classification* problems. For the purposes of evaluation, the three methods presented in Sect. 2 have been supplemented with the preprocessing method, being a *state-of-art* solution for this type of problems. Due to the very large *imbalance ratio*, it is often impossible to apply the SMOTE algorithm (with default parameterization it requires at least 5 minority class examples in the learning set), therefore *random oversampling* was chosen. The full list of compared algorithms presents as follows:

1. **RDB**—*Regular Decision Boundary* used in *Gaussian Naive Bayes* classifier,
2. **ROS-RDB**—*Regular Decision Boundary* used in *Gaussian Naive Bayes* classifier trained on datasets with *randomly oversampled* minority class,
3. **FTDB**—*Fuzzy Templates Decision boundary* used in *Gaussian Naive Bayes* classifier,
4. **SDB**—*Standard Decision boundary* used in *Gaussian Naive Bayes* classifier.

Evaluation Methodology and Metrics Used. During the experimental evaluation, a *stratified 5-fold cross validation* was used, for the non-deterministic ROS-RDB algorithm by performing an additional ten-time replication of the results. Both pair tests between the quality of classifiers for individual data sets and ranking tests, used for general assessment of the relations between them, were carried out using the Wilcoxon test using 5% significance level. Due to the imbalanced nature of the considered problems, in assessing the quality of solutions it was decided to use *precision* and *recall* metrics, supplemented with aggregated *F1-score, balanced accuracy score* and *geometric-mean-score* metrics. Full source code of the performed tests, along with the method implementations and a full report of results, are located on the publicly available Git repository[1].

---

[1] http://github.com/w4k2/sdb.

Table 1. Overview of imbalanced classification datasets selected for experimental evaluation.

Dataset	Samples	Features	IR
ecoli-0-3-4-vs-5	200	7	1:9
yeast-2-vs-4	514	8	1:9
ecoli-0-6-7-vs-3-5	222	7	1:9
ecoli-0-2-3-4-vs-5	202	7	1:9
glass-0-1-5-vs-2	172	9	1:9
yeast-0-3-5-9-vs-7-8	506	8	1:9
yeast-0-2-5-6-vs-3-7-8-9	1004	8	1:9
yeast-0-2-5-7-9-vs-3-6-8	1004	8	1:9
ecoli-0-4-6-vs-5	203	6	1:9
ecoli-0-1-vs-2-3-5	244	7	1:9
ecoli-0-2-6-7-vs-3-5	224	7	1:9
glass-0-4-vs-5	92	9	1:9
ecoli-0-3-4-6-vs-5	205	7	1:9
ecoli-0-3-4-7-vs-5-6	257	7	1:9
yeast-0-5-6-7-9-vs-4	528	8	1:9
vowel0	988	13	1:10
ecoli-0-6-7-vs-5	220	6	1:10
glass-0-1-6-vs-2	192	9	1:10
ecoli-0-1-4-7-vs-2-3-5-6	336	7	1:11
led7digit-0-2-4-5-6-7-8-9-vs-1	443	7	1:11
glass-0-6-vs-5	108	9	1:11
ecoli-0-1-vs-5	240	6	1:11
glass-0-1-4-6-vs-2	205	9	1:11
glass2	214	9	1:12
ecoli-0-1-4-7-vs-5-6	332	6	1:12
ecoli-0-1-4-6-vs-5	280	6	1:13
shuttle-c0-vs-c4	1829	9	1:14
yeast-1-vs-7	459	7	1:14
glass4	214	9	1:15
ecoli4	336	7	1:16
page-blocks-1-3-vs-4	472	10	1:16
glass-0-1-6-vs-5	184	9	1:19
shuttle-c2-vs-c4	129	9	1:20
yeast-1-4-5-8-vs-7	693	8	1:22
glass5	214	9	1:23
yeast-2-vs-8	482	8	1:23
yeast4	1484	8	1:28
yeast-1-2-8-9-vs-7	947	8	1:31
yeast5	1484	8	1:33
ecoli-0-1-3-7-vs-2-6	281	7	1:39
yeast6	1484	8	1:41

4   Experimental Evaluation

4.1   Results

Scores and Paired Tests. Table 2 contains the results achieved by each of the considered algorithms for the aggregate, *F1-score* metric. The ROS-RDB method, being a typical approach to deal with imbalanced data using single model, looks the worst in the pool, which not only does not improve RDB results, but also often leads to statistically significant worse results. The FTDB method, although sporadically, leads to a significant improvement over RDB, never achieving results significantly inferior to it. Definitely the best in this competition is the SDB method proposed in this paper, which in eleven cases is statistically significantly better than each of the other methods, and in fourteen cases better than RDB.

For both the *precision metric* and the other aggregate measures (*balanced accuracy score* and *geometric mean score*), the observations are identical to those drawn from the *F1-score*, so the relevant result tables are not attached directly to the article, while still being public in the repository indicated in the previous section.

The aggregate metrics, such as *F1-score*, allows to draw some binding conclusions, but does not give a full picture of interpretation. As expected, with the *recall* metric (Table 3), the FTDB and RDB algorithms give some deterioration relative to both the base method and the ROS-RDB approach. Statistical significance occurs in this difference, however, only once for DTDB and twice for RDB.

Rank Tests. The final comparison of the considered solutions was carried out by ranking tests, included in Table 4. The ROS-RDB method obtains a small, but statistically significant advantage in the ranking over all other methods for the *recall* metric, but in all other measures it stands out very negatively, which leads to suggestions about its overall uselessness in the considered task of highly imbalanced data classification. If the goal of counteracting the tendency of favoring in the prediction of the majority class (which was stated as the basic problem in the classification of imbalanced data) is to equalize the impact of both classes, on the example of the considered data sets, the ROS method must be rejected because it leads to the reverse tendency. In the case of *precision* and each of the aggregate metrics the same statistically significant relation is observed. The RDB method is better than ROS-RDB, the FTDB method is better than both RDB methods, and the SDB proposed in this paper is significantly better than all the competitors in the considered pool of solutions.

Table 2. Results achieved by analyzed methods for all considered datasets with *F1-score* metric. Bold values shows dependency to the best classifier in a competition and the numbers below scores show classifier significantly worse than the one in the column.

Dataset	1 RDB	2 ROS-RDB	3 FTDB	4 SDB
ecoli-0-3-4-vs-5	0.340 2	0.268 —	0.396 2	**0.670** all
yeast-2-vs-4	**0.295** —	0.269 —	**0.334** 2	**0.454** 2
ecoli-0-6-7-vs-3-5	**0.190** —	**0.298** —	**0.218** —	**0.338** —
ecoli-0-2-3-4-vs-5	0.332 —	0.260 —	0.383 —	**0.659** all
glass-0-1-5-vs-2	**0.218** —	**0.183** —	**0.232** —	**0.239** —
yeast-0-3-5-9-vs-7-8	**0.269** —	**0.212** —	**0.252** —	**0.229** —
yeast-0-2-5-6-vs-3-7-8-9	**0.262** —	**0.478** 3	0.401 —	**0.469** —
yeast-0-2-5-7-9-vs-3-6-8	0.201 2	0.165 —	0.272 1, 2	**0.381** all
ecoli-0-4-6-vs-5	**0.629** —	**0.584** —	**0.629** —	**0.736** —
ecoli-0-1-vs-2-3-5	**0.217** —	**0.244** —	**0.217** —	**0.387** —
ecoli-0-2-6-7-vs-3-5	**0.208** —	**0.186** —	**0.208** —	**0.256** —
glass-0-4-vs-5	**0.960** —	**0.960** —	**0.960** —	**0.760** —
ecoli-0-3-4-6-vs-5	0.312 2	0.247 —	0.350 2	**0.669** all
ecoli-0-3-4-7-vs-5-6	0.356 —	0.251 —	0.489 —	**0.665** all
yeast-0-5-6-7-9-vs-4	0.174 —	0.173 —	0.195 —	**0.362** all
vowel0	**0.709** 2	0.562 —	**0.697** 2	**0.676** 2
ecoli-0-6-7-vs-5	**0.633** —	**0.663** —	**0.660** —	**0.688** —
glass-0-1-6-vs-2	0.199 —	**0.231** —	**0.218** —	**0.236** 1
ecoli-0-1-4-7-vs-2-3-5-6	**0.324** —	**0.384** —	**0.357** —	**0.384** —
led7digit-0-2-4-5-6-7-8-9-vs-1	**0.640** —	**0.622** —	**0.640** —	**0.646** —

(continued)

Table 2. (*continued*)

Dataset	1 RDB	2 ROS-RDB	3 FTDB	4 SDB
glass-0-6-vs-5	**0.867**	**0.867**	**0.867**	0.733
	—	—	—	—
ecoli-0-1-vs-5	0.632	0.582	0.638	**0.823**
	—	—	—	—
glass-0-1-4-6-vs-2	0.229	**0.260**	0.230	0.240
	—	—	—	—
glass2	0.169	**0.195**	0.179	0.187
	—	—	—	1
ecoli-0-1-4-7-vs-5-6	0.538	0.662	0.570	**0.688**
	—	—	—	—
ecoli-0-1-4-6-vs-5	0.709	0.664	0.723	**0.764**
	—	—	—	—
shuttle-c0-vs-c4	**0.980**	**0.980**	**0.980**	**0.980**
	—	—	—	—
yeast-1-vs-7	0.141	0.136	0.153	**0.223**
	—	—	1, 2	*all*
glass4	0.190	**0.481**	0.233	0.233
	—	—	—	—
ecoli4	**0.696**	0.602	**0.696**	0.787
	—	—	—	2
page-blocks-1-3-vs-4	0.511	0.521	0.524	**0.540**
	—	—	—	—
glass-0-1-6-vs-5	**0.760**	**0.760**	**0.760**	0.667
	—	—	—	—
shuttle-c2-vs-c4	0.813	0.800	0.813	**1.000**
	—	—	—	—
yeast-1-4-5-8-vs-7	0.086	**0.088**	0.085	0.103
	—	—	—	1
glass5	**0.768**	**0.768**	**0.768**	0.693
	—	—	—	—
yeast-2-vs-8	0.254	0.190	**0.262**	0.202
	—	—	2	—
yeast4	0.073	0.071	0.086	**0.117**
	—	—	1, 2	*all*
yeast-1-2-8-9-vs-7	0.067	0.066	0.068	**0.098**
	—	—	—	*all*
yeast5	0.154	0.120	0.165	**0.642**
	2	—	1, 2	*all*
ecoli-0-1-3-7-vs-2-6	0.434	0.388	0.434	**0.490**
	—	—	—	—
yeast6	0.066	0.060	0.066	**0.169**
	2	—	2	*all*

Table 3. Results achieved by analyzed methods for all considered datasets with *recall* metric. Bold values shows dependency to the best classifier in a competition and the numbers below scores show classifier significantly worse than the one in the column.

Dataset	1 RDB	2 ROS-RDB	3 FTDB	4 SDB
ecoli-0-3-4-vs-5	**0.850**	**0.850**	**0.850**	**0.850**
	—	—	—	—
yeast-2-vs-4	**0.902**	**0.922**	**0.902**	0.825
	—	—	—	—
ecoli-0-6-7-vs-3-5	0.170	0.260	0.210	**0.360**
	—	—	—	—
ecoli-0-2-3-4-vs-5	**0.850**	**0.850**	**0.850**	**0.850**
	—	—	—	—
glass-0-1-5-vs-2	0.633	**0.733**	0.633	0.633
	—	—	—	—
yeast-0-3-5-9-vs-7-8	**0.880**	**0.880**	0.800	0.760
	—	—	—	—
yeast-0-2-5-6-vs-3-7-8-9	0.307	**0.557**	0.436	0.505
	—	3	—	—
yeast-0-2-5-7-9-vs-3-6-8	**0.917**	0.854	0.897	0.897
	—	—	—	—
ecoli-0-4-6-vs-5	0.650	0.650	0.650	**0.850**
	—	—	—	—
ecoli-0-1-vs-2-3-5	0.160	0.200	0.160	**0.440**
	—	—	—	—
ecoli-0-2-6-7-vs-3-5	0.190	0.190	0.190	**0.310**
	—	—	—	—
glass-0-4-vs-5	**1.000**	**1.000**	**1.000**	0.800
	—	—	—	—
ecoli-0-3-4-6-vs-5	**0.850**	**0.850**	**0.850**	**0.850**
	—	—	—	—
ecoli-0-3-4-7-vs-5-6	0.760	0.760	0.760	**0.920**
	—	—	—	—
yeast-0-5-6-7-9-vs-4	**0.960**	**0.960**	0.920	0.864
	—	—	—	—
vowel0	0.811	**0.844**	0.811	0.811
	—	—	—	—
ecoli-0-6-7-vs-5	0.700	**0.850**	0.750	0.800
	—	—	—	—
glass-0-1-6-vs-2	0.683	**0.800**	0.683	0.683
	—	—	—	—
ecoli-0-1-4-7-vs-2-3-5-6	0.267	**0.333**	0.300	**0.333**
	—	—	—	—
led7digit-0-2-4-5-6-7-8-9-vs-1	0.757	**0.832**	0.757	0.786
	—	—	—	—

(*continued*)

Table 3. (*continued*)

Dataset	1 RDB	2 ROS-RDB	3 FTDB	4 SDB
glass-0-6-vs-5	0.900	0.900	0.900	0.800
	—	—	—	—
ecoli-0-1-vs-5	0.600	0.650	0.650	0.850
	—	—	—	—
glass-0-1-4-6-vs-2	0.650	0.700	0.650	0.650
	—	—	—	—
glass2	0.733	0.833	0.733	0.733
	—	—	—	—
ecoli-0-1-4-7-vs-5-6	0.560	0.760	0.600	0.760
	—	—	—	—
ecoli-0-1-4-6-vs-5	0.800	0.850	0.850	0.850
	—	—	—	—
shuttle-c0-vs-c4	0.984	0.984	0.984	0.984
	—	—	—	—
yeast-1-vs-7	0.933	0.933	0.933	0.833
	—	—	—	—
glass4	0.200	0.567	0.267	0.267
	—	—	—	—
ecoli4	0.950	0.950	0.950	0.900
	—	—	—	—
page-blocks-1-3-vs-4	0.593	0.667	0.627	0.667
	—	—	—	—
glass-0-1-6-vs-5	0.900	0.900	0.900	0.800
	—	—	—	—
shuttle-c2-vs-c4	1.000	1.000	1.000	1.000
	—	—	—	—
yeast-1-4-5-8-vs-7	0.967	1.000	0.933	0.800
	4	4	4	—
glass5	0.900	0.900	0.900	0.800
	—	—	—	—
yeast-2-vs-8	0.950	1.000	0.900	0.650
	—	—	—	—
yeast4	0.962	0.982	0.962	0.904
	—	—	—	—
yeast-1-2-8-9-vs-7	1.000	1.000	1.000	0.733
	4	4	4	—
yeast5	1.000	1.000	1.000	0.886
	—	—	—	—
ecoli-0-1-3-7-vs-2-6	0.800	0.800	0.800	0.800
	—	—	—	—
yeast6	1.000	0.971	0.971	0.914
	—	—	—	—

Table 4. Results for mean ranks according to all considered metrics.

Metric	1 RDB	2 ROS-RDB	3 FTDB	4 SDB
F1-score	2.215	1.927	2.607	3.251
	2	–	1,2	*all*
Precision	2.271	1.837	2.646	3.246
	2	–	1,2	*all*
Recall	2.463	2.800	2.446	2.290
	–	*all*	–	–
Balanced accuracy	2.224	1.963	2.595	3.217
	2	–	1,2	*all*
Geometric mean score	2.205	1.868	2.641	3.285
	2	–	1,2	*all*

5   Conclusions

Following paper, considering the binary classification of imbalanced data, proposed the application of the *Fuzzy Templates* method in the construction of the support-domain decision boundary for a single model in order to balance the impact of classes of different counts on the prediction of the decision system. The proposal was further developed to use both *standard normalization* metrics, introducing the *Standard Decision Boundary* method. Both solutions were tested in computer experiments on the example of a highly imbalanced dataset collection and compared to both the base method and the *state-of-art* preprocessing method.

Both proposed solutions seem to improve the quality of imbalanced data classification in relation to the regular support-domain decision boundary, in contrast to oversampling, without leading to overweight of the predictive towards the minority class. Modification of the use of *Fuzzy Templates* in the form of *Standard Decision Boundary* is also more effective than the simple use of a class support prototype and may be considered a recommendable solution to the problem of binary classification of imbalanced data. Due to the promising results achieved for individual models, the next works will attempt to generalize the SDB method for *classifier ensembles*.

Acknowledgements. This work was supported by the Polish National Science Centre under the grant No. 2017/27/B/ST6/01325 as well as by the statutory funds of the Department of Systems and Computer Networks, Faculty of Electronics, Wroclaw University of Science and Technology.

References

1. Aditsania, A., Adiwijaya, Saonard, A.L.: Handling imbalanced data in churn prediction using ADASYN and backpropagation algorithm. In: Proceeding - 2017 3rd International Conference on Science in Information Technology: Theory and Application of IT for Education, Industry and Society in Big Data Era, ICSITech 2017 (2017)
2. Alcalá-Fdez, J., et al.: KEEL data-mining software tool: data set repository, integration of algorithms and experimental analysis framework. J. Multiple-Valued Logic Soft Comput. **17**, 255–287 (2011)
3. del Amo, A., Montero, J., Cutello, V.: On the principles of fuzzy classification. In: Annual Conference of the North American Fuzzy Information Processing Society - NAFIPS (1999)
4. Bishop, C.M.: Pattern Recognition and Machine Learning. Springer, New York (2006)
5. Fernández, A., García, S., Galar, M., Prati, R.C., Krawczyk, B., Herrera, F.: Learning from Imbalanced Data Sets (2018)
6. Fernández, A., García, S., Herrera, F., Chawla, N.V.: SMOTE for learning from imbalanced data: progress and challenges, marking the 15-year anniversary. J. Artif. Intell. Res. **61**, 863–905 (2018)
7. Ganganwar, V.: An overview of classification algorithms for imbalanced datasets. Int. J. Emerg. Technol. Adv. Eng. **2**(4), 42–47 (2012)
8. He, H., Bai, Y., Garcia, E.A., Li, S.: ADASYN: adaptive synthetic sampling approach for imbalanced learning. In: Proceedings of the International Joint Conference on Neural Networks (2008)
9. Krawczyk, B.: Learning from imbalanced data: open challenges and future directions. Prog. Artif. Intell. **5**(4), 221–232 (2016)
10. Ksieniewicz, P.: Undersampled majority class ensemble for highly imbalanced binary classification. In: Second International Workshop on Learning with Imbalanced Domains: Theory and Applications, pp. 82–94 (2018)
11. Ksieniewicz, P.: Combining *Random Subspace* approach with SMOTE oversampling for imbalanced data classification. In: Pérez García, H., Sánchez González, L., Castejón Limas, M., Quintián Pardo, H., Corchado Rodríguez, E. (eds.) HAIS 2019. LNCS (LNAI), vol. 11734, pp. 660–673. Springer, Cham (2019). https://doi.org/10.1007/978-3-030-29859-3_56
12. Ksieniewicz, P., Woźniak, M.: Imbalanced data classification based on feature selection techniques. In: Yin, H., Camacho, D., Novais, P., Tallón-Ballesteros, A.J. (eds.) IDEAL 2018. LNCS, vol. 11315, pp. 296–303. Springer, Cham (2018). https://doi.org/10.1007/978-3-030-03496-2_33
13. Ksieniewicz, P., Wozniak, M., Torgo, L., Krawczyk, B., Branco, P., Moniz, N.: Dealing with the task of imbalanced, multidimensional data classification using ensembles of exposers. In: Proceedings of Machine Learning Research (2017)
14. Kuncheva, L.: Fuzzy Classifier Design, vol. 49. Springer, Heidelberg (2000). https://doi.org/10.1007/978-3-7908-1850-5
15. Kuncheva, L.I., Bezdek, J.C., Duin, R.P.: Decision templates for multiple classifier fusion: an experimental comparison. Pattern Recogn. **34**(2), 299–314 (2001)
16. Kuncheva, L.I., Bezdek, J.C., Sutton, M.A.: On combining multiple classifiers by fuzzy templates. In: Annual Conference of the North American Fuzzy Information Processing Society - NAFIPS (1998)
17. Mitchell, T.M.: The Discipline of Machine Learning. Machine Learning (2006)

18. Moreo, A., Esuli, A., Sebastiani, F.: Distributional random oversampling for imbalanced text classification. In: SIGIR 2016 - Proceedings of the 39th International ACM SIGIR Conference on Research and Development in Information Retrieval (2016)
19. Ohsaki, M., Wang, P., Matsuda, K., Katagiri, S., Watanabe, H., Ralescu, A.: Confusion-matrix-based kernel logistic regression for imbalanced data classification. IEEE Trans. Knowl. Data Eng. **29**(9), 1806–1819 (2017)
20. Prusa, J., Khoshgoftaar, T.M., Dittman, D.J., Napolitano, A.: Using random undersampling to alleviate class imbalance on tweet sentiment data. In: Proceedings - 2015 IEEE 16th International Conference on Information Reuse and Integration, IRI 2015 (2015)
21. Rodriguez-Torres, F., Carrasco-Ochoa, J.A., Martínez-Trinidad, J.F.: Deterministic oversampling methods based on SMOTE. J. Intell. Fuzzy Syst. **36**(5), 4945–4955 (2019)
22. Wang, Q., Luo, Z.H., Huang, J.C., Feng, Y.H., Liu, Z.: A novel ensemble method for imbalanced data learning: bagging of extrapolation-SMOTE SVM. Comput. Intell. Neurosci. (2017)
23. Woźniak, M., Graña, M., Corchado, E.: A survey of multiple classifier systems as hybrid systems. Inf. Fusion **16**, 3–17 (2014)
24. Xu, Y., Yang, Z., Zhang, Y., Pan, X., Wang, L.: A maximum margin and minimum volume hyper-spheres machine with pinball loss for imbalanced data classification. Knowl.-Based Syst. **95**, 75–85 (2016)
25. Zhang, Y.: Deep generative model for multi-class imbalanced learning. ProQuest Dissertations and Theses (2018)

Employing One-Class SVM Classifier Ensemble for Imbalanced Data Stream Classification

Jakub Klikowski$^{(\boxtimes)}$ and Michał Woźniak

Wrocław University of Science and Technology, Wrocław, Poland
{jakub.klikowski,michal.wozniak}@pwr.edu.pl

Abstract. The classification of imbalanced data streams is gaining more and more interest. However, apart from the problem that one of the class is not well represented, there are problems typical for data stream classification, such as limited resources, lack of access to the true labels and the possibility of occurrence of the *concept drift*. Possibility of *concept drift* appearing enforces design in the method adaptation mechanism. In this article, we propose the OCEIS classifier (*One-Class support vector machine classifier Ensemble for Imbalanced data Stream*). The main idea is to supply the committee with one-class classifiers trained on clustered data for each class separately. The results obtained from experiments carried out on synthetic and real data show that the proposed method achieves results at a similar level as the state of the art methods compared with it.

Keywords: One-class classification · Imbalanced data · Data streams · Ensemble learning

1 Introduction

Currently, the classification of difficult data is a frequently selected topic of research. One of many examples of this type of data is data streams. Such data should be processed for a limited time, having appropriate memory restrictions and performing only one-time use of incoming data. Also, the classifiers are required to be adaptable. A common phenomenon accompanying streams is the *concept drift*, which causes a change in the incoming data distribution. These changes may occur indefinitely.

Another problem is the imbalance of data, when it is combined with streams, significantly increases the difficulty. Uneven distribution of the number of classes is a fairly common phenomenon occurring in real data sets. This is not a problem when the differences are small, but it becomes serious when the difference between the number of objects from minority and majority classes is significantly huge. One of the known ways to deal with these difficulties is data sampling methods. These methods are designed to reduce the number of objects in the dominant class or to generate artificial objects of the minority class [2].

© Springer Nature Switzerland AG 2020
V. V. Krzhizhanovskaya et al. (Eds.): ICCS 2020, LNCS 12140, pp. 117–127, 2020.
https://doi.org/10.1007/978-3-030-50423-6_9

Designing methods with mechanisms for adapting to this type of data is another approach. One of this kind of approach is Learn++CDS [6] method, which combines the Learn++NSE [7] for nonstationary streams and SMOTE [2] for oversampling data. The next method in this paper is Learn++NIE, which is similar to the previous one, but with little difference. The classification error is introduced and some variation of *bagging* is used for balancing data. Wang et al. [19] design a method that uses the k-Mean clustering algorithm for undersampling data by prototype generation from centroids. The REA method proposed by Chen and He [4]. It is extension of the SERA [3] and the MuSeRA [5]. This family of methods uses a strategy for estimating similarity between previous samples of minority classes and the current minority data from the chunk.

One of the demanding situations when classifying imbalanced data streams is the temporary disappearance of the minority class or their appearance only in later stages. This type of phenomenon can cause a significant decrease in quality or sometimes prevent the typical classifier from working. The solution that raises this type of problem is the use of one-class classifiers that can make decisions based only on objects from one class only. Krawczyk et al. [11] proposed to the form an ensemble of one-class classifiers. Clustered data within samples from each class is used to train new models and expand ensemble. J. Liu et al. [14] designed a modular committee of single-class classifiers based on data density analysis. This is a similar approach, where clusters are created as part of a single-class data set. Krawczyk and Woźniak [10] presented various metrics enabling the creation of effective one-class classifier committees.

This paper proposes an ensemble method for classifying imbalanced data streams. The purpose of this work is to conduct preliminary experiments and analyze the obtained results, which will confirm whether the designed method can deal with imbalanced data streams competing in tests with the methods of state of the art. The main contributions of this work are as follows:

- A proposal for an OCEIS method for classifying imbalanced data streams based on one-class SVM classifiers
- Introduction of an appropriate combination rule allowing full use of the potential of the one-class SVM classifier ensemble
- Designing the proper learning procedure for the proposed method using division of data into classes and *k-mean* clustering
- Experimental evaluation of the proposed OCEIS method using real and synthetic generated imbalanced data streams and a comparison with the state-of-the-art methods

2   Proposed Method

The proposed method **O**ne **C**lass support vector machine classifier **E**nsemble for **I**mbalanced data **S**tream (*OCEIS*) is a combination of different approaches to data classification. The main core of this idea is the use of one-class support

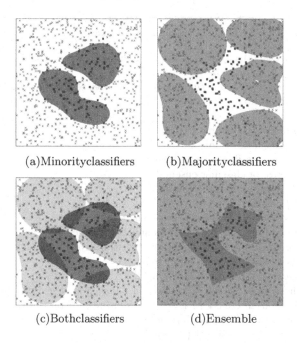

(a)Minorityclassifiers   (b)Majorityclassifiers

(c)Bothclassifiers   (d)Ensemble

Fig. 1. Decision regions visualisation on the paw dataset from the Keel.es repository [1]

vector machines ($OCSVM$) to classify imbalanced binary problems. This method is the chunk-based data stream method.

In the first step of the Algorithm 1, the chunk of training data is divided into a minority (D_{min}) and a majority set (D_{maj}). Then these sets of data are divided into clusters. Krawczyk et al. [11] indicate the importance of this idea. This decomposition of data over the feature space allows achieving less overlap of classifiers decision areas in the ensemble (Fig. 1). The k-$means$ algorithm [15] is used to create clusters. The key aspect is choosing the right number of clusters. Silhouette Value (SV) [18] comes with help, which allows calculating how similar an object is to its own cluster compared to other clusters. Kaufman et al. [9] introduced the Silhouette Coefficient (SC) for the maximum value of the mean SV over the entire dataset.

Minority and majority data is divided into clusters sets ($Cmin_{t,k}$, $Cmaj_{t,k}$) with a different number of centroids from 1 to K_{max}. The number of clusters with the highest value of SC is selected (K_{best}). This process is performed for minority and majority data. Then the formed clusters are used to fit new models ($h_{t,i}$, $h_{t,j}$) of OCSVM. These models are included in the pool of classifier committees (H_{min}, H_{maj}). The method is designed by default to operate on data streams. For this reason, a simple forgetting mechanism, also known as incremental learning, was implemented. This allows using models trained only on data with a certain time interval. When the algorithm reaches a set number (S) of chunks (t), in each iteration, the models built on the oldest chunk are removed from the ensemble.

Algorithm 1. OCEIS - Train

Input:

$D_t = \{(x_1^t, i_1^t), (x_2^t, i_2^t), (x_N^t, i_N^t)\}$ - training chunk of data stream

$x_k^t \in \mathcal{X}$, where $\mathcal{X}$ stands for the feature space

$i_k^t \in \mathcal{M} = \{minority, majority\}$, where $\mathcal{M}$ denotes set of the possible labels

t - current timestamp

N - chunk size

$Dmaj_t$ - majority data chunk

$Dmin_t$ - minority data chunk

$OCSVM$ - SVM classifier for one-class classification

S - maximum size of classifier ensemble

K_{max} - maximum number of clusters

k - number of clusters

$SilhouetteCoefficient$ - clusters consistency value [9]

K_{best} - number of clusters with best Silhouette Coefficient

$KMeanClustering$ - k-mean clustering algorithm [15]

$Cmaj_{t,k}$ - clusters of minority data $Dmaj_t$

$Cmin_{t,k}$ - clusters of minority data $Dmin_t$

$h_{t,j}$ - hypothesis from $OCSVM$ trained on $Cmaj_{t,j}$ cluster data

$h_{t,i}$ - hypothesis from $OCSVM$ trained on $Cmin_{t,i}$ cluster data

H_{maj} - majority hypothesis set (ensemble)

H_{min} - minority hypothesis set (ensemble)

1: **for** $t = 1, 2, ...$ **do**
2:     Split D_t into majority ($Dmaj_t$) and minority ($Dmin_t$) data
3:     **for** $k = 1, 2, ..., K_{max}$ **do**
4:         $Cmaj_{t,k} \leftarrow$ Call $KMeanClustering$ with k on $Dmaj_t$
5:         $Cmin_{t,k} \leftarrow$ Call $KMeanClustering$ with k on $Dmin_t$
6:     **end for**
7:     $K_{best} \leftarrow$ max Silhouette Coefficient on $Cmaj_{t,k}$
8:     **for** $i = 1, 2, ..., K_{best}$ **do**
9:         $h_{t,i} \leftarrow$ Call $OCSVM$ on $Cmaj_{t,i}$ cluster data
10:         Add $h_{t,i}$ to H_{maj}
11:     **end for**
12:     $K_{best} \leftarrow$ max Silhouette Coefficient on $Cmin_{t,k}$
13:     **for** $j = 1, 2, ..., K_{best}$ **do**
14:         $h_{t,j} \leftarrow$ Call $OCSVM$ on $Cmin_{t,i}$ cluster data
15:         Add $h_{t,j}$ to H_{min}
16:     **end for**
17:     **if** $t > S$ **then**
18:         Remove all $h_{t,i}$ where $t = t - S$ from H_{maj}
19:         Remove all $h_{t,j}$ where $t = t - S$ from H_{min}
20:     **end if**
21: **end for**

Algorithm 2. OCEIS - Prediction

Input:

$D_t = \{(x_1^t, i_1^t), (x_2^t, i_2^t), (x_N^t, i_N^t)\}$ - training chunk of data stream

$x_k^t \in \mathcal{X}$, where $\mathcal{X}$ stands for the feature space

$i_k^t \in \mathcal{M} = \{minority, majority\}$, where $\mathcal{M}$ denotes set of the possible labels

t - current timestamp

N - chunk size

DecisionFunction - Signed distance to the separating hyperplane.
Returns positive value inside and negative outside hyperplane.

$Dist_{i,m}$ - distance from h_i decision boundary to x_m

$Dist_{j,m}$ - distance from h_j decision boundary to x_m

D_{maj} - maximum value of distance from h_j decision boundary to x_m

D_{min} - maximum value of distance from h_i decision boundary to x_m

1: **for** $t = 1, 2, ...$ **do**
2:   **for each** h_j **in** H_{maj} **do**
3:     $Dist_{j,m} \leftarrow$ Compute *DecisionFunction* for h_j on each x_m in D_t
4:   **end for**
5:   **for each** h_i **in** H_{min} **do**
6:     $Dist_{i,m} \leftarrow$ Compute *DecisionFunction* for h_i on each x_m in D_t
7:   **end for**
8:   **for** $m = 1, 2, ..., N$ **do**
9:     $D_{maj} \leftarrow$ max value of $Dist_{j,m}$ for x_m
10:    $D_{min} \leftarrow$ max value of $Dist_{i,m}$ for x_m
11:    **if** $D_{maj} > D_{min}$ **then**
12:      Predict majority class for x_m
13:    **else**
14:      Predict minority class for x_m
15:    **end if**
16: **end for**
17: **end for**

A crucial component of any classifier ensemble is the combination rule, which makes decisions based on the predictions of the classifier ensemble. Designing a good decision rule is vital for proper operation and obtaining satisfactory classification quality. First of all, OCEIS uses one-class classifiers and class clustering technique, which changes the way how the ensemble works. Well-known decision making based on majority voting [20] does not allow this kind of committee to make correct decisions. The number of classifiers for individual classes may vary significantly depending on the number of clusters. In this situation, there is a considerable risk that the decision will mainly base on majority classifiers.

OCEIS uses the original combination rule (Algorithm 2) based on distance from the decision boundary of classifiers to predicted samples. In the first step, the distances ($Dist_{i,m}$, $Dist_{j,m}$) are calculated from all objects of the predicted data to the hypersphere of the models forming the minority and the majority committee. The *DecisionFunction* calculates these values. When the examined object is inside the checked hypersphere, it obtains a positive value, when it is outside, it receives a negative value. Then the highest value (D_{maj}, D_{min}) is determined from the majority and minority committees for each sample. When the best value (D_{maj}) for the model from the majority subensemble is greater than the best value (D_{min}) for the model from the minority subensemble, it means that this object belongs to the majority class. Similarly, when D_{min} is greater than D_{maj}, the object belongs to a minority class.

3   Experimental Evaluation

The main purpose of this experiment was to check how good the proposed method performed with comparison to the other methods for classifying imbalanced data streams. The following research hypothesis was formulated:

It is possible to design a method with a statistically better or equal classification quality of imbalanced data streams compared to the selected state of the art methods.

3.1   Experiment Setup

All tests were carried out using 24 generated streams and 30 real streams (Table 1). The generated data comes from stream-learn [12] generator. These generated data differ in the level of imbalance: 10%, 20%, 30%. Label noise: 0% or 10% and type of drift: incremental or sudden. All generated data streams have 10 features, two classes and consist of 100,000 objects each. The proposed method has been tested with the selected state of the art methods:

- L++CDS [6]   - REA [4]
- L++NIE [6]   - OUSE [8]
- KMC [19]   - MLPC [16] (as a baseline)

The SVM implementation from the *scikit-learn* framework [17] was used as the base classifier in all committees. OCEIS implementation and the experimental environment is available on public github repository.[1] Four metrics were used to measure the quality: Gmean, precision, recall and specificity. The results obtained in this way were compared using Wilcoxon statistical pair-tests. Each method was compared with OCEIS and these wins, lost and draw are shown in Fig. 2 and Fig. 3.

---

[1] https://github.com/w4k2/oceis-iccs2020.

Table 1. Overview of real datasets used in experimental evaluation (KEEL [1] and PROMISE Software Engineering Repository [13]), IR - Imbalance Ratio

Dataset	IR	SAMPLES	FEATURES
abalone-17_vs_7-8-9-10	39	2338	8
australian	1.2	690	14
elecNormNew	1.4	45312	8
glass-0-1-2-3_vs_4-5-6	3.2	214	9
glass0	2.1	214	9
glass1	1.8	214	9
heart	1.2	270	13
jm1	5.5	2109	21
kc1	5.5	2109	21
kc2	3.9	522	21
kr-vs-k-three_vs_eleven	35	2935	6
kr-vs-k-zero-one_vs_draw	27	2901	6
page-blocks0	8.8	5472	10
pima	1.9	768	8
segment0	6	2308	19
shuttle-1vs4	14	1829	9
shuttle-1vsA	3.7	57999	9
shuttle-4-5vsA	3.8	57999	9
shuttle-4vsA	5.5	57999	9
shuttle-5vsA	17	57999	9
vehicle0	3.3	846	18
vowel0	10	988	13
wisconsin	1.9	683	9
yeast-0-2-5-6_vs_3-7-8-9	9.1	1004	8
yeast-0-2-5-7-9_vs_3-6-8	9.1	1004	8
yeast-0-3-5-9_vs_7-8	9.1	506	8
yeast-0-5-6-7-9_vs_4	9.4	528	8
yeast-2_vs_4	9.1	514	8
yeast1	2.5	1484	8
yeast3	8.1	1484	8

3.2   Results Analysis

The obtained results of the Wilcoxon rank-sum pair statistical tests show that OCEIS can classify with the similar quality compared to the tested methods. For tested synthetic data streams (Fig. 2) there is a certain advantage of the L++CDS method over other methods. In second place can be put L++NIE and OCEIS. For the OUSE and L++NIE methods, there is a noticeable tendency to classify objects of the minority class, which is manifested by the higher results in the Recall (TPR) metric, but this causes a significant drop in Specifity (TNR). The worst in this test was the REA method, which shows a huge beat in the direction of the majority class. The results are more transparent for real data sets (Fig. 3). Despite many ties, the best performing method is OCEIS. The exceptions are Recall for OUSE and Specifity for REA.

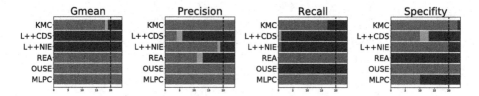

Fig. 2. Wilcoxon pair rank sum tests for synthetic data streams. Dashed vertical line is a critical value with a confidence level 0.05 (green - win, yellow - tie, red - lose) (Color figure online)

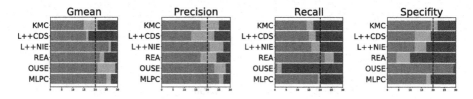

Fig. 3. Wilcoxon pair rank sum tests for real data streams. Dashed vertical line is a critical value with a confidence level 0.05 (green - win, yellow - tie, red - lose) (Color figure online)

Charts of Gmean score over the data chunks provide some useful information about obtained results. To get a much better readability, the data before plotting was processed using a Gaussian filter. This procedure smoothes the edges of the results, which allows getting much more information from the results. The first observation is that for an incremental drift stream (Fig. 4), OCEIS does not degrade quality over time. The negative effect of the concept drift can be seen on the KMC and REA methods, where the quality deteriorates significantly with the inflow of subsequent data chunks.

In sudden *concept drift* (Fig. 5), a certain decrease is noticeable, which is more or less reflected on every tested method. However, L++CDS, L++NIE and OCEIS can quickly rebuild this quality drop. This does not affect the overall quality of the classification significantly. Other methods perform a little bit randomly on sudden drifts. An example of the real-time shuttle-4vsA stream (Fig. 6) shows the clear advantage of the OCEIS method over the other tested methods. A similar observation can be seen in other figures for real streams.

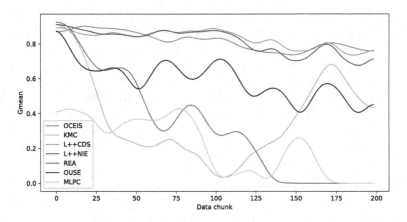

Fig. 4. Gmean score over the data chunks for synthetic data with incremental drift

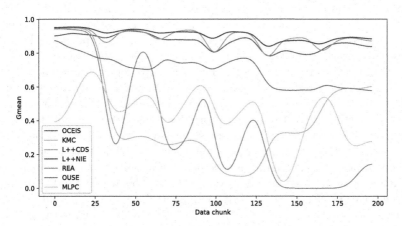

Fig. 5. Gmean score over the data chunks for synthetic data with sudden drift

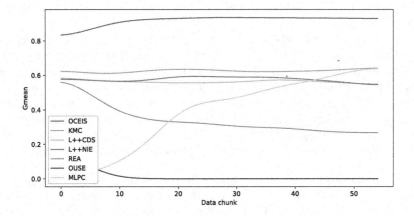

Fig. 6. Gmean score over the data chunks for real stream shuttle-4-5vsA

When analyzing the results, one should pay attention to the significant divergences in the performance of the proposed method for synthetic and real data streams. A large variety characterized real data streams, while artificial streams were generated using one type of generator (of course, for different settings). However, generated data streams are biased towards one type of data distribution, which probably was easy to analyze by some of the models, while the bias of the rest of them was not consistent with this type of data generator. Therefore, in the future, we are going to carry out the experimental research on the expanded pool of synthetic streams generated by other different generators.

4   Conclusions

We proposed an imbalanced data streams classification algorithm based on the one-class classifier ensemble. Based on the results obtained from reliable experiments, the formulated research hypothesis seems to be confirmed. OCEIS achieves results at a similar level to the compared methods, but it is worth noticing that it performs best on real stream data, which is its important advantage. Another advantage is that there is no tendency towards the excessive classification of objects from one of the classes. This was a problem in experiments carried out for the REA and OUSE methods. Such "stability" contributes significantly to improving the quality of classification and obtaining satisfactory results.

For synthetic data streams, the proposed algorithm is not the worst-performing one. However, one can see some dominance of the methods from the Learn++ family, because the decision made by OCEIS is built based on all classifiers as part of the committee. One possible way to change this would be to break down newly created models by data chunks. This would build subcommittees (the Learn++NIE method works similarly). Then decisions would be made for each subcommittee separately. Expanding this by the weighted voting decision may significantly improve predictive performance. Another modernization of the method that would allow for some improvement would be the introduction of a drift detector. This mechanism would enable the ensemble to clean up after detecting *concept drift.*

The conducted research indicates the potential hidden in the presented method. It is worth considering extending the research to streams with other types of concept drifts. It is also beneficial to increase the number of real streams to test to get a broader spectrum of knowledge about how this method works on real data. One of the ideas for further research that arose while working on this paper is to test the operation on streams where the imbalance ratio changes over time. A very interesting would be an experiment on imbalanced data streams where the minority class temporarily disappears or appears after some time.

Acknowledgment. This work was supported by the Polish National Science Centre under the grant No. 2017/27/B/ST6/01325 as well as by the statutory funds of the Department of Systems and Computer Networks, Faculty of Electronics, Wroclaw University of Science and Technology.

References

1. Alcalá-Fdez, J., et al.: Keel data-mining software tool: data set repository, integration of algorithms and experimental analysis framework. J. Multiple-Valued Logic Soft Comput. **17** (2011)
2. Chawla, N.V., Bowyer, K.W., Hall, L.O., Kegelmeyer, W.P.: Smote: synthetic minority over-sampling technique. J. Artif. Intell. Res. **16**, 321–357 (2002)
3. Chen, S., He, H.: Sera: selectively recursive approach towards nonstationary imbalanced stream data mining. In: 2009 International Joint Conference on Neural Networks, pp. 522–529. IEEE (2009)
4. Chen, S., He, H.: Towards incremental learning of nonstationary imbalanced data stream: a multiple selectively recursive approach. Evol. Syst. **2**(1), 35–50 (2011)
5. Chen, S., He, H., Li, K., Desai, S.: Musera: multiple selectively recursive approach towards imbalanced stream data mining. In: The 2010 International Joint Conference on Neural Networks (IJCNN), pp. 1–8. IEEE (2010)
6. Ditzler, G., Polikar, R.: Incremental learning of concept drift from streaming imbalanced data. IEEE Trans. Knowl. Data Eng. **25**(10), 2283–2301 (2012)
7. Elwell, R., Polikar, R.: Incremental learning of concept drift in nonstationary environments. IEEE Trans. Neural Netw. **22**(10), 1517–1531 (2011)
8. Gao, J., Ding, B., Fan, W., Han, J., Philip, S.Y.: Classifying data streams with skewed class distributions and concept drifts. IEEE Internet Comput. **12**(6), 37–49 (2008)
9. Kaufmann, L., Rousseeuw, P.J.: Finding Groups in Data: An Introduction to Cluster Analysis. Wiley, New York (1990)
10. Krawczyk, B., Woźniak, M.: Diversity measures for one-class classifier ensembles. Neurocomputing **126**, 36–44 (2014)
11. Krawczyk, B., Woźniak, M., Cyganek, B.: Clustering-based ensembles for one-class classification. Inf. Sci. **264**, 182–195 (2014)
12. Ksieniewicz, P., Zyblewski, P.: Stream-learn-open-source python library for difficult data stream batch analysis. arXiv preprint arXiv:2001.11077 (2020)
13. Lima, M., Valle, V., Costa, E., Lira, F., Gadelha, B.: Software engineering repositories: expanding the promise database. In: Proceedings of the XXXIII Brazilian Symposium on Software Engineering, pp. 427–436. ACM (2019)
14. Liu, J., Miao, Q., Sun, Y., Song, J., Quan, Y.: Modular ensembles for one-class classification based on density analysis. Neurocomputing **171**, 262–276 (2016)
15. MacQueen, J., et al.: Some methods for classification and analysis of multivariate observations. In: Proceedings of the Fifth Berkeley Symposium on Mathematical Statistics and Probability, Oakland, CA, USA, vol. 1, pp. 281–297 (1967)
16. Pal, S.K., Mitra, S.: Multilayer perceptron, fuzzy sets, classifiaction (1992)
17. Pedregosa, F., et al.: Scikit-learn: machine learning in Python. J. Mach. Learn. Res. **12**, 2825–2830 (2011)
18. Rousseeuw, P.J.: Silhouettes: a graphical aid to the interpretation and validation of cluster analysis. J. Comput. Appl. Math. **20**, 53–65 (1987)
19. Wang, Y., Zhang, Y., Wang, Y.: Mining data streams with skewed distribution by static classifier ensemble. In: Chien, B.C., Hong, T.P. (eds.) Opportunities and Challenges for Next-Generation Applied Intelligence, pp. 65–71. Springer, Heidelberg (2009). https://doi.org/10.1007/978-3-540-92814-0_11
20. Xu, L., Krzyzak, A., Suen, C.Y.: Methods of combining multiple classifiers and their applications to handwriting recognition. IEEE Trans. Syst. Man Cybern. **22**(3), 418–435 (1992)

Clustering and Weighted Scoring in Geometric Space Support Vector Machine Ensemble for Highly Imbalanced Data Classification

Paweł Ksieniewicz◉ and Robert Burduk(✉)◉

Department of Systems and Computer Networks,
Wroclaw University of Science and Technology, Wroclaw, Poland
{pawel.ksieniewicz,robert.burduk}@pwr.edu.pl

Abstract. Learning from imbalanced datasets is a challenging task for standard classification algorithms. In general, there are two main approaches to solve the problem of imbalanced data: algorithm-level and data-level solutions. This paper deals with the second approach. In particular, this paper shows a new proposition for calculating the weighted score function to use in the integration phase of the multiple classification system. The presented research includes experimental evaluation over multiple, open-source, highly imbalanced datasets, presenting the results of comparing the proposed algorithm with three other approaches in the context of six performance measures. Comprehensive experimental results show that the proposed algorithm has better performance measures than the other ensemble methods for highly imbalanced datasets.

Keywords: Imbalanced data · Ensemble of classifiers · Class imbalance · Decision boundary · Scoring function

1 Introduction

The goal of the supervised classification is to build a mathematical model of a real-life problem using a labeled dataset. This mathematical model is used to assign the class label to each new recognized object, which, in general, does not belong to the training set. The individual classification model is called a base classifier. Ensemble methods are a vastly used approach to improve the possibilities of base classifiers by building a more stable and accurate *ensemble of classifiers* (*EoC*) [23,28]. In general, the procedure for building *EoC* consists of three steps: generation, selection, and integration [18]. An imbalanced data problem occurs when the *prior* probability of classes in a given dataset is very diverse. There are many real-life problems in which we deal with imbalanced data [11,25], e.g., network intrusion detection [2,14], source code fault detection [8], or in general fraud detection [1].

There exist two main approaches to solve the problem of imbalanced data: a data-level [9,26] and an algorithm-level solution [29]. EoC is one of the

© Springer Nature Switzerland AG 2020
V. V. Krzhizhanovskaya et al. (Eds.): ICCS 2020, LNCS 12140, pp. 128–140, 2020.
https://doi.org/10.1007/978-3-030-50423-6_10

approaches to solve the imbalanced data classification problem which improve classification measure compared to single models and is highly competitive and robust to imbalanced data [10,13,16]. The use of not only voting in the *EoC* integration phase is one of the directions to solve a problem with imbalanced data [15]. Therefore, this article concerns about calculating the weighted scoring function to be applied in the weighted voting process.

In the process of *EoC* generation, we use the *K-Means* clustering algorithm [5] for each class label separately. The base linear classifier – *Support Vector Machine* [7] – is trained on cluster combination. The weighted scoring function takes into account the distance of a classified object from the decision boundary and cluster centroids used to learn the proper base classifier. Regardless of the number of learning objects in a given cluster, the cluster centroid is always determined. The proposed method for determining the scoring function is, therefore, insensitive to the number of objects defining the cluster. As shown in the article, the proposed approach is useful for imbalanced data.

The main objectives of this work are summarized as follows:

- A proposal of a new weighted scoring function that uses the location of the cluster centroids and distance to the decision boundary.
- The proposition of an algorithm that uses clustering and the proposed function.
- A new experimental setup to compare the proposed method with other algorithms on highly imbalanced datasets.

The paper is structured as follows: Sect. 2 introduces the base concept of *EoC* and presents the proposed algorithm. In Sect. 3, the experiments that were carried out are presented, while results and the discussion appear in Sect. 4. Finally, we conclude the paper in Sect. 5.

2 Clustering and Weighted Scoring

2.1 Basic Notation

From a probabilistic point of view the recognition algorithm Ψ maps the feature space $\mathcal{X} \subseteq \mathfrak{R}^d$ to the set of class labels $\Omega = \{\omega_1, \omega_2, \ldots, \omega_C\}$ according to the general formula:

$$\Psi : \mathcal{X} \to \Omega. \tag{1}$$

For the feature vector $x \in \mathcal{X}$, that represents the recognized object the Eq. (1) can be expressed as:

$$\Psi(x) = \omega_c. \tag{2}$$

Let us assume that L different base classifiers $\Psi_1, \Psi_2, \ldots, \Psi_L$, are employed to solve the classification task. This set of classifiers defines *EoC*. One of the most popular methods to integrate outputs of the base classifiers set L is the majority vote rule. In this method, each base model has the same impact on the final decision of *EoC*. This method allows counting base classifiers outputs as a vote for a class and assigns the input pattern to the class with the greatest count of votes. It is defined as follows:

$$\Psi_{MV}(x) = \arg\max_{\omega_c} \sum_{k=1}^{L} I(\Psi_k(x) = \omega_c), \qquad (3)$$

where $I(\cdot)$ is the indicator function.

In the weighted majority voting rule, the integration phase includes probability estimators or other factors of base models to the final decision of EoC [19], like in Eq. 4:

$$\Psi_{MV}(x) = \arg\max_{\omega_c} \sum_{k=1}^{L} w_k I(\Psi_k(x) = \omega_c), \qquad (4)$$

where w_k is the weight assigned to the classifier Ψ_k.

Over the last years, the issue of calculating the weights in the voting rule has been considered many times. The article [30] presents an approach in which the weights are combining with local confidence. The classifier trained on a subset of training data should be limited to the area it spans in an impact on the resulting classifier. The problem of generalization of majority voting was studied in [12]. The authors are using a probability estimate calculated as percentage of properly classified validation objects over geometric constraints. Separately, regions that are functionally independent are considered. A significant improvement in the classification quality was observed when using the proposed algorithm, although knowledge of the domain is needed to provide a proper division. The authors are using a retinal image and classify over anatomic regions.

The weights of the base classifier are also considered in the context of the interval-valued fuzzy sets [6]. The upper weight of base the classifier refers to the situation in which the definite base classifier was correct, while the other classifiers proved the correct prediction. The lower weight describes the situation in which the definite base classifier made errors, while the other classifiers didn't make any errors.

In the paper [22] weights are determined for each label separately over the entire validation dataset. This can lead to the improvement of the performance of the resulting integrated classifier.

The following article is a proposition of an algorithm assigning weights not to base classifiers, but recognized objects. The weight for each object depends on its location in the feature space. Therefore, the weight of an object is determined by the score function calculated in the geometric space.

2.2   Proposed Method

We propose that the score function of the object x depends on its position in the geometric space. In particular, the distance from the decision boundary of the base classifier Ψ_l from EoC and the cluster centroids used to learn this classifier are used to calculated the unweighted score function:

$$sf_l(x) = \|\psi_l(x)\| + \sum_{c=1}^{C} d_c, \tag{5}$$

where d_c is the distance from x to cluster centroid in *Manhattan* metric. This metric was chosen because of the lowest calculation cost among all the considered alternatives. The calculation of the distance occurs between the centroids of all clusters and all tested patterns, so the computational complexity of the prediction procedure is very much dependent on the chosen metric, which is the reason for minimizing its impact.

We propose the following scoring weighting method:

$$wsf_l(x) = 1 - \frac{sf_l(x)}{\sum_{l=1}^{L} sf_l(x)}, \tag{6}$$

which includes all scoring functions obtained for each classifier from EoC.

Figure 1 shows how to calculate the object's score function for a linear dichotomic classifier and two cluster centroids. A solid red line marks the decision boundary of the base linear classifier Ψ_l constructed for the selected class cluster combination – C^{ω_1} and C^{ω_2}. Blue points are cluster centroids. The sum of the dashed sections indicates the value of the score function for the tested object x. The red dashed line – the distance to decision boundary $\|\psi_l(x)\|$, the blue dashed line – the distance to cluster centroids determined by the *Manhattan* metric d_1 and d_2.

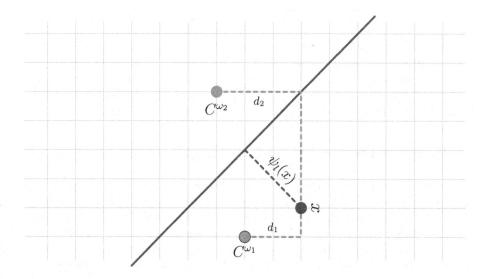

Fig. 1. Schema for calculating of the score function for the object x.

Algorithm 1: Clustering and Weighted Scoring in Geometric Space algorithm – for binary problem

---

Input: Learning set D, Number of clusters K – equal in each class label, object x

Output: The ensemble classifier decision

1 Cluster D into K clusters label using the K-means clustering procedure separately for each class. The final cluster combination equals $L = 2 * K$ for the binary problem.

2 Find the cluster centroids $C_1^{\omega_1}, \ldots, C_K^{\omega_1}, C_1^{\omega_2}, \ldots, C_K^{\omega_2}$ as the means of the points in the respective clusters.

3 Train base classifier $\Psi_1, \ldots, \Psi_L$ using each combination of clusters from different class labels.

4 Calculate weighted scoring functions for the object x:

$$ws f_l(x) = 1 - \frac{s f_l(x)}{\sum_{l=1}^{L} s f_l(x)},$$

where

$$s f_l(x) = \|\psi_l(x)\| + \sum_{c=1}^{2} d_c.$$

5 The ensemble classifier decision:

$$\Psi_{CWS}(x) = sign \left(\sum_{l=1}^{L} ws f_l(x) \Psi_l(x) \right),$$

where $\Psi(x)$ is the prediction returned by base classifier $\Psi(x) \in \{-1, 1\}$.

Algorithm 1 presents the pseudocode of the proposed approach to EoC with clustering and weighted scoring in the geometric space. In addition, Algorithm 1 concerns the dichotomous division of the learning set into class labels. These types of highly imbalanced datasets were used in the experimental research.

3   Experiments Set-Up

The experimental evaluation conducted for the needs of verification of the method proposed in the following work was based of 30 highly imbalanced datasets contained in the KEEL repository [3]. Datasets selected for the study are characterized by an *imbalance ratio* (the proportion between minority and majority class) ranging from 1:9 up to 1:40. Besides, due to the preliminary nature of the conducted research, the pool of datasets includes only binary classification problems.

The basis of the used division methodology was *Stratified K-Fold Crossvalidation* with $k = 5$, necessary to ensure the presence of minority class patterns in each of the analyzed training subsets. Statistical tests, for both pair and rank tests, were carried out using the *Wilcoxon test* with the significance level $\alpha = 0.05$ [4].

The analysis was conducted following the four classification approaches:

- (SVC) *Support Vector Machine*—the base experimental model with the scaled gamma and linear kernel [21].
- (CWS) *Clustering and Weighted Scoring*—*EoC* with the pool diversified by pairs of clusters and integrated geometrically by the rules introduced in Sect. 2.
- (CMV) *Clustering and Majority Vote*—*EoC* identical with CWS but integrated using the majority vote [24].
- (CSA) *Clustering and Support Accumulation*—*EoC* identical with CWS and CMV but integrated using the support accumulation rule [27].

In the construction of each *EoC*, in order to limit the number of the presented tables and readability of the analysis, each time we construct the ensemble by dividing classes into two clusters, thus building a pool of four members. In the case of data as strongly imbalanced as those from the selected databases, often only a few (literally four or five) minority class objects remain in a single fold so that a more substantial number would treat almost every minority object as a separate cluster.

The whole experimental evaluation was performed in Python, using the *scikit-learn* API [20] to implement the CWS method and is publicly available on the GIT repository[1]. As metrics for the conducted analysis, due to the imbalanced nature of the classification problem, three aggregate measures (*balanced accuracy score, F1-score*, and *G-mean*) and three base measures constituting their calculation (*precision, recall*, and *specificity*) were applied, using their implementation included in the *stream-learn* package [17].

4   Experimental Evaluation

For the readability of the analysis, the full results of the experiment, along with the presentation of the relation between the algorithms resulting from the conducted paired tests, are presented only for the *balanced accuracy score* (Table 1) and *recall* (Table 2) metrics.

As may be observed, for aggregate metrics (results are consistent for both *balanced accuracy* and *G-mean*, only in *F1-score* presenting a slightly smaller scale of differences) the use of majority voting (CMV) for *EoC* diversified with clustering, often leads to deterioration of the classification quality even concerning a single base classifier. Integration by support accumulation (CSA) performs slightly better, due to taking into consideration the certainty (*support*) of the decisions of each classifier but ignoring their areas of competence. The use of

---

[1] https://github.com/w4k2/geometric-integration.

Table 1. Results achieved by the analyzed method for the *balanced accuracy score* metric.

Dataset	IR	[1] SVC	[2] CWS	[3] CMV	[4] CSA
glass-0-4-vs-5	1:9	0.738 ± 0.160	0.938 ± 0.125	0.696 ± 0.247	0.856 ± 0.174
		−	[3]	−	[3]
ecoli-0-1-4-7-vs-5-6	1:12	0.867 ± 0.076	0.839 ± 0.060	0.713 ± 0.063	0.856 ± 0.025
		[3]	[3]	−	[3]
ecoli-0-6-7-vs-5	1:10	0.890 ± 0.103	0.915 ± 0.061	0.710 ± 0.056	0.882 ± 0.108
		−	[3]	−	−
ecoli-0-1-vs-2-3-5	1:9	0.880 ± 0.083	0.871 ± 0.115	0.692 ± 0.086	0.780 ± 0.142
		−	−	−	−
ecoli-0-3-4-6-vs-5	1:9	0.845 ± 0.118	0.890 ± 0.085	0.720 ± 0.081	0.879 ± 0.092
		−	[3]	−	−
yeast-0-3-5-9-vs-7-8	1:9	0.537 ± 0.081	0.597 ± 0.129	0.549 ± 0.078	0.539 ± 0.080
		−	−	−	−
ecoli4	1:16	0.570 ± 0.140	0.753 ± 0.088	0.619 ± 0.149	0.500 ± 0.000
		−	[1,4]	−	−
ecoli-0-1-4-7-vs-2-3-5-6	1:11	0.796 ± 0.120	0.790 ± 0.098	0.666 ± 0.130	0.756 ± 0.085
		−	−	−	−
ecoli-0-3-4-7-vs-5-6	1:9	0.766 ± 0.097	0.744 ± 0.114	0.829 ± 0.057	0.779 ± 0.139
		−	−	−	−
shuttle-c2-vs-c4	1:20	1.000 ± 0.000	1.000 ± 0.000	0.756 ± 0.026	1.000 ± 0.000
		[3]	[3]	−	[3]
yeast-2-vs-8	1:23	0.774 ± 0.120	0.774 ± 0.120	0.600 ± 0.093	0.650 ± 0.093
		−	−	−	−
ecoli-0-4-6-vs-5	1:9	0.839 ± 0.118	0.867 ± 0.070	0.611 ± 0.103	0.875 ± 0.093
		[3]	[3]	−	[3]
yeast-2-vs-4	1:9	0.667 ± 0.093	0.693 ± 0.070	0.634 ± 0.068	0.627 ± 0.039
		−	−	−	−
ecoli-0-6-7-vs-3-5	1:9	0.853 ± 0.077	0.835 ± 0.101	0.728 ± 0.185	0.807 ± 0.102
		−	−	−	−
ecoli-0-1-4-6-vs-5	1:13	0.829 ± 0.125	0.863 ± 0.077	0.631 ± 0.143	0.829 ± 0.063
		−	[3]	−	−
ecoli-0-2-3-4-vs-5	1:9	0.875 ± 0.75	0.859 ± 0.119	0.745 ± 0.100	0.804 ± 0.129
		[3]	−	−	−
glass-0-6-vs-5	1:11	0.639 ± 0.195	0.924 ± 0.103	0.744 ± 0.186	0.766 ± 0.113
		−	−	−	−
ecoli-0-2-6-7-vs-3-5	1:9	0.860 ± 0.115	0.833 ± 0.092	0.628 ± 0.079	0.785 ± 0.123
		[3]	[3]	−	−

(continued)

Table 1. (*continued*)

Dataset	IR	1 SVC	2 CWS	3 CMV	4 CSA
ecoli-0-3-4-vs-5	1:9	0.822 ± 0.123 [3]	0.886 ± 0.090 [3]	0.761 ± 0.143	0.911 ± 0.055 [3]
glass4	1:15	0.554 ± 0.093	0.914 ± 0.071 [all]	0.640 ± 0.124	0.568 ± 0.139
glass5	1:23	0.544 ± 0.087	0.882 ± 0.121	0.704 ± 0.188	0.737 ± 0.162
glass-0-1-5-vs-2	1:9	0.500 ± 0.000	0.500 ± 0.000	0.601 ± 0.159	0.507 ± 0.086
yeast-0-2-5-6-vs-3-7-8-9	1:9	0.509 ± 0.021	0.581 ± 0.101	0.532 ± 0.066	0.504 ± 0.010
yeast3	1:8	0.630 ± 0.035 [2]	0.500 ± 0.042	0.632 ± 0.050	0.701 ± 0.045 [all]
ecoli-0-1-vs-5	1:11	0.880 ± 0.093	0.932 ± 0.061	0.895 ± 0.123	0.864 ± 0.090
shuttle-c0-vs-c4	1:14	1.000 ± 0.000 [3]	1.000 ± 0.000 [3]	0.785 ± 0.030	0.992 ± 0.090 [3]
yeast6	1:41	0.500 ± 0.000	0.528 ± 0.055	0.500 ± 0.000	0.500 ± 0.000
yeast4	1:28	0.500 ± 0.000	0.510 ± 0.020	0.500 ± 0.000	0.500 ± 0.000
yeast-0-2-5-7-9-vs-3-6-8	1:9	0.704 ± 0.099 [4]	0.840 ± 0.158 [4]	0.669 ± 0.072 [4]	0.578 ± 0.067
vowel0	1:10	0.767 ± 0.119	0.719 ± 0.129	0.786 ± 0.079	0.787 ± 0.079

areas of competence present in the CWS method allows for substantial improvement in classification results, often leading to a statistically significant advantage. The primary source of advantage in results is a significant improvement in the *recall* metric in this approach.

These observations become even clearer when we look at the results of the ranking tests presented in Table 3. In the case of *balanced accuracy* and *G-mean*, the CWS method is statistically significantly better than in all other cases. For the *F1-score* metric, despite the numerical advantage, the statistical significance of the base method disappears, which is due to the symmetry of the *F1-score* metric relative to problem classes, which accepts the equal cost of a wrong decision to the minority and majority class.

Table 2. Results achieved by the analyzed method for the *recall* metric.

Dataset	IR	[1] SVC	[2] CWS	[3] CMV	[4] CSA
glass-0-4-vs-5	1:9	0.600 ± 0.374	1.000 ± 0.0	0.500 ± 0.447	0.800 ± 0.244
ecoli-0-1-4-7-vs-5-6	1:12	0.800 ± 0.178	0.760 ± 0.80	0.680 ± 0.097	0.800 ± 0.000
ecoli-0-6-7-vs-5	1:10	0.800 ± 0.187	0.900 ± 0.122	0.450 ± 0.100	0.800 ± 0.187
ecoli-0-1-vs-2-3-5	1:9	0.800 ± 0.178	0.760 ± 0.233 [3]	0.430 ± 0.218	0.620 ± 0.271
ecoli-0-3-4-6-vs-5	1:9	0.750 ± 0.273	0.850 ± 0.200	0.500 ± 0.158	0.850 ± 0.200
yeast-0-3-5-9-vs-7-8	1:9	0.080 ± 0.160	0.200 ± 0.252	0.100 ± 0.154	0.080 ± 0.160
ecoli4	1:16	0.150 ± 0.300	0.550 ± 0.244	0.250 ± 0.316	0.000 ± 0.000
ecoli-0-1-4-7-vs-2-3-5-6	1:11	0.687 ± 0.289	0.720 ± 0.231 [1,4]	0.540 ± 0.257	0.680 ± 0.263
ecoli-0-3-4-7-vs-5-6	1:9	0.640 ± 0.265	0.600 ± 0.219	0.840 ± 0.79	0.680 ± 0.271
shuttle-c2-vs-c4	1:20	1.000 ± 0.000	1.000 ± 0.000	1.000 ± 0.000	1.000 ± 0.000
yeast-2-vs-8	1:23	0.550 ± 0.244	0.550 ± 0.244	0.200 ± 0.187	0.300 ± 0.187
ecoli-0-4-6-vs-5	1:9	0.750 ± 0.273	0.800 ± 0.187	0.300 ± 0.187	0.850 ± 0.200
yeast-2-vs-4	1:9	0.335 ± 0.186	0.396 ± 0.146 [3]	0.276 ± 0.149	0.256 ± 0.83 [3]
ecoli-0-6-7-vs-3-5	1:9	0.740 ± 0.146	0.780 ± 0.203	0.490 ± 0.361	0.640 ± 0.185
ecoli-0-1-4-6-vs-5	1:13	0.700 ± 0.244	0.850 ± 0.200	0.350 ± 0.339	0.850 ± 0.200
ecoli-0-2-3-4-vs-5	1:9	0.800 ± 0.187	0.900 ± 0.122	0.600 ± 0.200	0.850 ± 0.122
glass-0-6-vs-5	1:11	0.300 ± 0.399 [3]	0.900 ± 0.200	0.600 ± 0.374	0.800 ± 0.244
ecoli-0-2-6-7-vs-3-5	1:9	0.760 ± 0.224	0.770 ± 0.203	0.310 ± 0.215	0.610 ± 0.261
ecoli-0-3-4-vs-5	1:9	0.800 ± 0.187 [3]	0.900 ± 0.122	0.650 ± 0.254	0.900 ± 0.122

(*continued*)

Table 2. (*continued*)

Dataset	IR	**1** **SVC**	**2** **CWS**	**3** **CMV**	**4** **CSA**
glass4	1:15	0.233 ± 0.200	0.933 ± 0.133	0.300 ± 0.266	0.200 ± 0.266
		–	all	–	–
glass5	1:23	0.200 ± 0.400	0.900 ± 0.200	0.500 ± 0.316	0.600 ± 0.374
		–	–	–	–
glass-0-1-5-vs-2	1:9	0.000 ± 0.000	0.000 ± 0.000	0.633 ± 0.335	0.633 ± 0.335
		–	–	–	–
yeast-0-2-5-6-vs-3-7-8-9	1:9	0.031 ± 0.40	0.197 ± 0.252	0.074 ± 0.147	0.011 ± 0.21
		–	–	–	–
yeast3	1:8	0.264 ± 0.69	0.190 ± 0.69	0.412 ± 0.104	0.430 ± 0.85
		–	–	1	1,2
ecoli-0-1-vs-5	1:11	0.800 ± 0.187	0.900 ± 0.122	0.800 ± 0.244	0.750 ± 0.158
		–	–	–	–
shuttle-c0-vs-c4	1:14	1.000 ± 0.000	1.000 ± 0.000	0.661 ± 0.179	0.984 ± 0.19
		–	–	–	–
yeast6	1:41	0.000 ± 0.000	0.057 ± 0.114	0.000 ± 0.000	0.000 ± 0.000
		–	–	–	–
yeast4	1:28	0.000 ± 0.000	0.020 ± 0.400	0.000 ± 0.000	0.000 ± 0.000
		–	–	–	–
yeast-0-2-5-7-9-vs-3-6-8	1:9	0.421 ± 0.189	0.711 ± 0.310	0.351 ± 0.135	0.160 ± 0.135
		4	4	4	–
vowel0	1:10	0.589 ± 0.284	0.489 ± 0.288	0.611 ± 0.185	0.600 ± 0.183
		–	–	–	–

Table 3. Results for mean ranks achieved by analyzed methods with all metrics included in the evaluation.

Metric	**1** **SVC**	**2** **CWS**	**3** **CMV**	**4** **CSA**
Balanced accuracy	2.500	3.250	1.867	2.383
	–	all	–	–
F1-score	2.733	3.217	1.767	2.283
	3	3,4	–	–
G-mean	2.467	3.283	1.900	2.350
	–	all	–	–
Precision	3.000	2.783	1.900	2.317
	3,4	3	–	–
Recall	2.333	3.350	1.917	2.400
	–	all	–	–
Specificity	2.817	2.017	2.533	2.633
	2	–	–	–

The design of the recognition algorithm dedicated to imbalanced data is almost always based on the calibration of factors measurable by the base classification metrics. As in the case of the CWS method, we try to increase the *recall* so that the inevitable reduction of *precision* or *specificity* will further give us a significant statistical advantage in the chosen aggregate metric, selected to define the cost of the incorrect classification that is relevant to us. In this case, the CWS method turns out to be much better than the other methods of *EoC* integration with a pool diversified by clusters and allows a statistically significant improvement of the base method in the case of highly imbalanced data.

5   Conclusions

This paper presented a new clustering and weighted scoring algorithm dedicated to constructing *EoC*. We proposed that the scoring function should take into account the distance from the decision boundary of each base classifier and the cluster centroids necessary to learn this classifier. In the proposed weighting scoring function the distance to the decision boundary and sum of the distances to the centroids have the same weight. The proposed approach applies to imbalanced datasets because each cluster centroid can be calculated regardless of the number of objects in this cluster.

Comprehensive experiments are presented on thirty examples of highly imbalanced datasets. The obtained results show that the proposed algorithm is better than other algorithms in the context of statistical tests and some performance measures. In particular, in the case of the *balanced accuracy* and *G-mean* classification measures, the proposed in this paper method is statistically significantly better than all others methods used in the experiments.

A possible future work is to be considered: other distance measures to calculate the distance to cluster centroids, the impact of the use of heterogeneous base classifiers in the proposed method of building *EoC* or another scoring weighting method. In particular, we suggest that the distance from the decision boundary can be scaled or weighting factors regarding the distance of the object from the decision boundary and the distance from the cluster centroids can be introduced. Additionally, we can assign weights for cluster centroids depending on the number of objects that were used to determine these centroids.

Acknowledgements. This work was supported by the Polish National Science Centre under the grant No. 2017/25/B/ST6/01750 as well as by the statutory funds of the Department of Systems and Computer Networks, Faculty of Electronics, Wroclaw University of Science and Technology.

References

1. Abdallah, A., Maarof, M.A., Zainal, A.: Fraud detection system: a survey. J. Netw. Comput. Appl. **68**, 90–113 (2016)
2. Abdulhammed, R., Faezipour, M., Abuzneid, A., AbuMallouh, A.: Deep and machine learning approaches for anomaly-based intrusion detection of imbalanced network traffic. IEEE Sens. Lett. **3**(1), 1–4 (2018)
3. Alcalá-Fdez, J., et al.: Kee data-mining sotware tool: dat set repository, integration of algrithms and experimental nalysis framewor. J. Multiple-Valued Logic Soft Comput. **17**, 255–287 (2011)
4. Alpaydin, E.: Introduction to Machine Learning. MIT Press, Cambridge (2014)
5. Basu, S., Banerjee, A., Mooney, R.: Semi-supervised clustering by seeding. In: Proceedings of 19th International Conference on Machine Learning, ICML 2002. Citeseer (2002)
6. Burduk, R.: Classifier fusion with interval-valued weights. Pattern Recogn. Lett. **34**(14), 1623–1629 (2013)
7. Cao, X., Wu, C., Yan, P., Li, X.: Linear SVM classification using boosting hog features for vehicle detection in low-altitude airborne videos. In: 2011 18th IEEE International Conference on Image Processing (ICIP), pp. 2421–2424. IEEE (2011)
8. Choraś, M., Pawlicki, M., Kozik, R.: Recognizing faults in software related difficult data. In: Rodrigues, J.M.F., et al. (eds.) ICCS 2019. LNCS, vol. 11538, pp. 263–272. Springer, Cham (2019). https://doi.org/10.1007/978-3-030-22744-9_20
9. Fotouhi, S., Asadi, S., Kattan, M.W.: A comprehensive data level analysis for cancer diagnosis on imbalanced data. J. Biomed. Inform. **90**, 103089 (2019)
10. Galar, M., Fernandez, A., Barrenechea, E., Bustince, H., Herrera, F.: A review on ensembles for the class imbalance problem: bagging-, boosting-, and hybrid-based approaches. IEEE Trans. Syst. Man Cybern. Part C (Appl. Rev.) **42**(4), 463–484 (2011)
11. Haixiang, G., Yijing, L., Shang, J., Mingyun, G., Yuanyue, H., Bing, G.: Learning from class-imbalanced data: review of methods and applications. Expert Syst. Appl. **73**, 220–239 (2017)
12. Hajdu, A., Hajdu, L., Jonas, A., Kovacs, L., Toman, H.: Generalizing the majority voting scheme to spatially constrained voting. IEEE Trans. Image Process. **22**(11), 4182–4194 (2013)
13. Klikowski, J., Ksieniewicz, P., Woźniak, M.: A genetic-based ensemble learning applied to imbalanced data classification. In: Yin, H., Camacho, D., Tino, P., Tallón-Ballesteros, A.J., Menezes, R., Allmendinger, R. (eds.) IDEAL 2019. LNCS, vol. 11872, pp. 340–352. Springer, Cham (2019). https://doi.org/10.1007/978-3-030-33617-2_35
14. Kozik, R., Choras, M., Keller, J.: Balanced efficient lifelong learning (B-ELLA) for cyber attack detection. J. UCS **25**(1), 2–15 (2019)
15. Krawczyk, B.: Learning from imbalanced data: open challenges and future directions. Prog. Artif. Intell. **5**(4), 221–232 (2016). https://doi.org/10.1007/s13748-016-0094-0
16. Krawczyk, B., Woźniak, M., Schaefer, G.: Cost-sensitive decision tree ensembles for effective imbalanced classification. Appl. Soft Comput. **14**, 554–562 (2014)
17. Ksieniewicz, P., Zyblewski, P.: Stream-learn-open-source python library for difficult data stream batch analysis. arXiv preprint arXiv:2001.11077 (2020)
18. Kuncheva, L.I.: Combining Pattern Classifiers: Methods and Algorithms. Wiley, Hoboken (2004)

19. Mao, S., Jiao, L., Xiong, L., Gou, S., Chen, B., Yeung, S.K.: Weighted classifier ensemble based on quadratic form. Pattern Recogn. **48**(5), 1688–1706 (2015)
20. Pedregosa, F., et al.: Scikit-learn: machine learning in Python. J. Mach. Learn. Res. **12**, 2825–2830 (2011)
21. Platt, J.C.: Probabilistic outputs for support vector machines and comparisons to regularized likelihood methods. In: Advances in Large Margin Classifiers, pp. 61–74. MIT Press (1999)
22. Rahman, A.F.R., Alam, H., Fairhurst, M.C.: Multiple classifier combination for character recognition: revisiting the majority voting system and its variations. In: Lopresti, D., Hu, J., Kashi, R. (eds.) DAS 2002. LNCS, vol. 2423, pp. 167–178. Springer, Heidelberg (2002). https://doi.org/10.1007/3-540-45869-7_21
23. Rokach, L.: Pattern Classification Using Ensemble Methodsd, vol. 75. World Scientific, Singapore (2010)
24. Ruta, D., Gabrys, B.: Classifier selection for majority voting. Inf. Fusion **6**(1), 63–81 (2005)
25. Sun, Y., Wong, A.K., Kamel, M.S.: Classification of imbalanced data: a review. Int. J. Pattern Recognit. Artif. Intell. **23**(04), 687–719 (2009)
26. Szeszko, P., Topczewska, M.: Empirical assessment of performance measures for preprocessing moments in imbalanced data classification problem. In: Saeed, K., Homenda, W. (eds.) CISIM 2016. LNCS, vol. 9842, pp. 183–194. Springer, Cham (2016). https://doi.org/10.1007/978-3-319-45378-1_17
27. Wozniak, M.: Hybrid Classifiers: Methods of Data, Knowledge, and Classifier Combination, vol. 519. Springer, Heidelberg (2013). https://doi.org/10.1007/978-3-642-40997-4
28. Woźniak, M., Graña, M., Corchado, E.: A survey of multiple classifier systems as hybrid systems. Inf. Fusion **16**, 3–17 (2014)
29. Zhang, C., et al.: Multi-imbalance: an open-source software for multi-class imbalance learning. Knowl.-Based Syst. **174**, 137–143 (2019)
30. Sultan Zia, M., Hussain, M., Arfan Jaffar, M.: A novel spontaneous facial expression recognition using dynamically weighted majority voting based ensemble classifier. Multimedia Tools Appl. **77**(19), 25537–25567 (2018). https://doi.org/10.1007/s11042-018-5806-y

Performance Analysis of Binarization Strategies for Multi-class Imbalanced Data Classification

Michał Żak$^{(\boxtimes)}$⬡ and Michał Woźniak⬡

Department of Systems and Computer Networks, Wrocław University of Science and Technology, Wyb. Wyspiańskiego 27, 50-370 Wrocław, Poland
{michal.zak,michal.wozniak}@pwr.edu.pl

Abstract. Multi-class imbalanced classification tasks are characterized by the skewed distribution of examples among the classes and, usually, strong overlapping between class regions in the feature space. Furthermore, frequently the goal of the final system is to obtain very high precision for each of the concepts. All of these factors contribute to the complexity of the task and increase the difficulty of building a quality data model by learning algorithms. One of the ways of addressing these challenges are so-called binarization strategies, which allow for decomposition of the multi-class problem into several binary tasks with lower complexity. Because of the different decomposition schemes used by each of those methods, some of them are considered to be better suited for handling imbalanced data than the others. In this study, we focus on the well-known binary approaches, namely One-Vs-All, One-Vs-One, and Error-Correcting Output Codes, and their effectiveness in multi-class imbalanced data classification, with respect to the base classifiers and various aggregation schemes for each of the strategies. We compare the performance of these approaches and try to boost the performance of seemingly weaker methods by sampling algorithms. The detailed comparative experimental study of the considered methods, supported by the statistical analysis, is presented. The results show the differences among various binarization strategies. We show how one can mitigate those differences using simple oversampling methods.

Keywords: Multi-class classification · Imbalanced data · Binarization strategies

1 Introduction

The goal of the supervised learning is to build a data model capable of mapping inputs x to outputs y with a good generalization ability, given a labeled set of input-output pairs $\mathcal{D} = (x_i, y_i)_{i=1}^{N}$, $\mathcal{D}$ being the training set and N being the number of training examples. Usually, each of the training inputs x_i is a d-dimensional vector of numbers and nominal values, the so-called features that

© Springer Nature Switzerland AG 2020
V. V. Krzhizhanovskaya et al. (Eds.): ICCS 2020, LNCS 12140, pp. 141–155, 2020.
https://doi.org/10.1007/978-3-030-50423-6_11

characterize a given example, but x_i might as well be a complex structured object like an image, a time series or an email message. Similarly, the type of the output variable can in principle be anything, but in most cases it is of a continuous type $y_i \in \mathbb{R}$ or a nominal type $y_i \in \mathbb{C}$, where, considering an m class problem, $\mathbb{C} = \{c_1, ..., c_m\}$. In the former case, it is a regression problem, while in the latter, it is a classification problem [10,22]. Classification problems are very common in a real-world scenario and machine learning is widely used to solve these types of problems in areas such as fraud detection [6,24], image recognition [17,26], cancer treatment [3] or classification of DNA microarrays [19].

In many cases, classification tasks involve more than two classes forming so-called multi-class problems. This characteristic often imposes some difficulties on the machine learning algorithm, as some of the solutions were designed strictly for binary-class problems and may not be applicable to those kinds of scenarios. What is more, problems, where multiple classes are present, are often characterized by greater complexity than binary tasks, as the decision boundaries between classes tend to overlap, which might lead to building a poor quality model by a given classifier. Usually, it is simply easier to build a model to distinguish only between two classes than to consider a multi-class problem. One approach to overcome those challenges is to use binarization strategies that reduce the task to multiple binary classification subproblems - in theory, with lower complexity - that can be solved separately by dedicated models, the so-called base learners [2,11,13,14]. The most commonly used binarization strategies are One-Vs-All (OVA) [25], One-Vs-One (OVO) [12,16] and Error-Correcting Output Codes (ECOC) [9], which is a general framework for the binary decomposition of multi-class problems.

In this paper, we focus on the performance of the aforementioned binarization strategies in the context of multi-class imbalanced problems. We aim to determine whether there are statistically significant differences among the performances of those methods, provided the most suitable aggregation scheme for a given problem. If so - whether or not one can nullify those differences by improving the quality of base learners within each binarization method with sampling algorithms. The main contributions of this work are:

- an exhaustive experimental study on the classification of multi-class imbalanced data with the use of OVA, OVO and ECOC binarization strategies.
- a comparative study of the aforementioned approaches with regard to a number of base classifier and aggregation schemes for each of the them.
- a study on the performance of the binarization strategies with the sampling algorithms used to boost the quality of their base classifiers.

The rest of this paper is organized as follows. In Sect. 2, an overview of binarization strategies used in the experiments is given. In Sect. 3 the experimental framework set-up is presented, including the classification and sampling algorithms, performance measures and datasets used in the study. The empirical analysis of obtained results has been carried out in Sect. 4. In Sect. 5 we make our concluding remarks.

2   Decomposition Strategies for Multi-classification

The underlying idea behind binarization strategies is to undertake the multi-class problems using binary classifiers with divide and conquer strategy [13]. A transformation like this is often performed with the expectation that the resulting binary subproblems will have lower complexity than the original multi-class problem. One of the drawbacks of such approach is the necessity to combine the individual responses of the base learners into the final output of the decision system. What is more, building a dedicated model for each of the binary subproblems significantly increases the cost of building a decision system in comparison to undertaking the same problem with a single classifier. However, the magnitude of this problem varies greatly depending on the chosen binarization strategy as well as the number of classes under consideration and the size of the training set itself. In this study, we focus on the most common binarization strategies: OVA, OVO, and ECOC.

2.1   One-Vs-All Strategy

OVA binarization strategy divides an m-class problem into m binary problems. In this strategy, m binary classifiers are trained, each responsible for distinguishing instances of a given class from the others. During the validation phase, the test pattern is presented to each of the binary models and the model that gives a positive output indicates the output class of the decision system. This approach can potentially result in ambiguously labeled regions of the input space. Usually, some tie-breaking techniques are required [13, 22].

While relatively simple, OVA binarization strategy is often preferred to more complex methods, provided that the best available binary classifiers are used as the base learners [25]. However, in this strategy, the whole training set is used to train each of the base learners. It dramatically increases the cost of building a decision system with respect to the single multi-class classifier. Another issue is that each of the binary subproblems is likely to suffer from the aforementioned class imbalance problem [13, 22].

2.2   One-Vs-One Strategy

OVA binarization strategy divides an m-class problem into $\frac{m \times (m-1)}{2}$ binary problems. In this strategy, each binary classifier is responsible for distinguishing instances of different pair of classes (c_i, c_j). The training set for each of the binary classifiers consists only of instances of the two classes forming a given pair, while the instances of the remaining classes are discarded. During the validation phase, the test pattern is presented to each of the binary models. The output of a model given by $r_{ij} \in [0, 1]$ is the confidence of the binary classifier discriminating classes i and j in favour of the former class. If the classifier does not provide it, the confidence for the latter class is computed by $r_{ji} = 1 - r_{ij}$ [12,13,22,29]. The class with the higher confidence value is considered as the output class of the decision system. Similarly to OVA strategy - this approach can also result in ambiguities [22].

Although the number of base learners in this strategy is of m^2 order, the growth in the number of learning tasks is compensated by the learning set reduction for each of the individual problems, as demonstrated in [12]. Also, one has to keep in mind that in this method, each of the base classifiers is trained using only the instances of two classes, deeming their output not significant for the instances of all the remaining classes. Usually, the assumption is that the base learner will make a correct prediction within its domain of expertise [13].

2.3   Error-Correcting Output Codes Strategy

ECOC binarization strategy is a general framework for the binary decomposition of multi-class problems. In this strategy, each class is assigned a unique binary string of length n, called *code word*. Next, n binary classifiers are trained, one for each bit in the string. During the training phase on an example from class i, the desired output of a given classifier is specified by the corresponding bit in the code word for this class. This process can be visualized by a $m \times n$ binary code matrix. As an example, Table 1 shows a 15-bit error-correcting output code for a five-class problem, constructed using exhaustive technique [9]. During the validation phase, the test pattern is presented to each of the binary models. Then the binary code word is formed from their responses. The class which code word was the nearest to the code word formed from the base learners' responses, according to the Hamming distance, indicates the output class of the decision system.

Table 1. A 15-bit error-correcting output code for a five class problem.

Class	Code word														
	1	2	3	4	5	6	7	8	9	10	11	12	13	14	15
1	1	1	1	1	1	1	1	1	1	1	1	1	1	1	1
2	0	0	0	0	0	0	0	0	1	1	1	1	1	1	1
3	0	0	0	0	1	1	1	1	0	0	0	0	1	1	1
4	0	0	1	1	0	0	1	1	0	0	1	1	0	0	1
5	0	1	0	1	0	1	0	1	0	1	0	1	0	1	0

In contrast to OVA and OVO strategies, ECOC method does not have a predefined number of binary models that will be used to solve a given multiclass problem. This number is determined purely by the algorithm one chooses to generate the ECOC code matrix. A measure of the quality of error-correcting code is the minimum Hamming distance between any pair of code words. If the minimum Hamming distance is l, then the code can correct at least $\frac{l-1}{2}$ single-bit errors.

2.4 Aggregation Schemes for Binarization Techniques

For the binarization techniques mentioned above, an aggregation method is necessary to combine the responses of an ensemble of base learners. In the case of ECOC binarization strategy, this aggregation method is embedded in it. An exhaustive comparison study has been carried out in [13], including various aggregation methods for both OVA and OVO binarization strategies. For our experimental study, the implementations of the following methods for OVA and OVO decomposition schemes have been used:

- OVA
    1. *Maximum Confidence Strategy*;
    2. *Dynamically Ordered One-Vs-All.*
- OVO
    1. *Voting Strategy*;
    2. *Weighted Voting Strategy*;
    3. *Learning Valued Preference for Classification*;
    4. *Decision Directed Acyclic Graph*

For ECOC strategy, the exhaustive codes were used to generate the code matrix if the number of classes m in the problem under consideration satisfied $3 \leq m \leq 7$. In other cases, the random codes were used as implemented in [23].

3 Experimental Framework

In this section, the set-up of the experimental framework used for the study is presented. The classification and sampling algorithms used to carry out the experiments are described in Sect. 3.1. Next, the performance measure used to evaluate the built models is presented in Sect. 3.2. Section 3.3 covers the statistical tests used to compare the obtained results. Finally, Sect. 3.4 describes the benchmark datasets used in the experiments.

3.1 Classification Used for the Study

One of the goals of the empirical study was to ensure the diversity of the classifiers used as base learners for binarization strategies. A brief description of the used algorithms is given in the remainder of this section.

- *Naïve Bayes* [22] is a simple model that assumes the features are conditionally independent given the class label. In practice, even if Naïve Bayes assumption is not true, it often performs fairly well.
- *k-Nearest Neighbors (k-NN)* [22] is a non-parametric classifier that simply uses chosen distance metric to find k points in the training set that are nearest to the test input x, and returns the most common class among those points as the estimate.

- *Classification and Regression Tree (CART)* [22] models are defined by recursively partitioning the input space, and defining a local model in each resulting region of input space.
- *Support Vector Machines (SVM)* [27] maps the original input space into a high-dimensional feature space via so-called kernel trick. In the new feature space, the optimal separating hyperplane with maximal margin is determined in order to minimize an upper bound of the expected risk instead of the empirical risk.
- *Logistic Regression* [22] is the generalization of the linear regression to the (binary) classification, so called Binomial Logistic Regression. Further generalization to Multi-Class Logistic Regression is often achieved via OVA approach.

During the building phase, for each of aforementioned base classifiers an exhaustive search over specified hyperparameter values was performed in attempt to build the best possible data model for a given problem - the values of hyperparameters used in the experiments are shown in Table 2. Furthermore, various sampling methods were used to boost the performance of base learners, namely SMOTE [7], Borderline SMOTE [15], SMOTEENN [4] and SMOTETomek [5]. All of the experiments were conducted using the Python programming language and libraries from the SciPy ecosystem (statistical tests and data manipulation) as well as scikit-learn (classifier implementations and feature engineering) and imbalanced-learn (sampling algorithms implementations) [18,23,28].

Table 2. Hyperparameter specification for the base learners used in the experiments.

Algorithm	Hyperparameters
Naive Bayes	—
k-Nearest Neighbors	$k \in \{1, 3, 5\}$
	Distance metric = Minkowski metric
CART	Split criterion $\in$ {Gini Impurity, Information Gain}
	Maximum depth = $(3, 11)$
	Minimum leaf samples $\in \{1, 3, 5\}$
SVM	Kernel type $\in$ {RBF, Linear}
	Regularization parameter $\in \{0.001, 0.01, 0.1, 1\}$
	Kernel coefficient $\in \{0.0001, 0.001, 0.01, 0.1, 1\}$
Logistic Regression	Regularization parameter $\in \{0.001, 0.01, 0.1, 1\}$
	Penalty $\in \{l1, l2\}$

3.2   Performance Measures

Model evaluation is a crucial part of an experimental study, even more so when dealing with imbalanced problems. In the presence of imbalance, evaluation metrics that focus on overall performance, such as overall accuracy, have a tendency to ignore minority classes because as a group they do not contribute much to the general performance of the system. To our knowledge, at the moment there is no consensus as to which metric should be used in imbalance data scenarios, although several solutions have been suggested [20,21]. Our goal was to pick a robust metric that ensures reliable evaluation of the decision system in the presence of strong class imbalance and at the same time is capable of handling multi-classification problems. Geometric Mean Score (G-Mean) is proven metric that meets both of these conditions - it focuses only on recall of each class and aggregates it multiplicatively across each class:

$$G - Mean = (\prod_{i=1}^{m} r_i)^{1/m}, \tag{1}$$

where r_i represents recall for $i - th$ class and m represents number of classes.

3.3   Statistical Tests

The non-parametric tests were used to provide statistical support for the analysis of the results, as suggested in [8]. Specifically, the Wilcoxon Signed-Ranks Test was applied as a non-parametric statistical procedure for pairwise comparisons. Furthermore, the Friedman Test was used to check for statistically significant differences between all of the binarization strategies, while the Nemenyi Test was used for posthoc comparisons and to obtain and visualize critical differences between models. The fixed significance level $\alpha = 0.05$ was used for all comparisons.

3.4   Datasets

The benchmark datasets used to conduct the research were obtained from the KEEL dataset repository [1]. The set of benchmark datasets was specially selected to ensure the robustness of the study and includes data with a varying numbers of instances, number and type of class attributes and the imbalance ratio of classes. The characteristics of the datasets used in the experiments are shown in Table 3 - for each dataset, it includes the number of instances

(#Inst.), the number of attributes (#Atts.), the number of real, integer and nominal attributes (respectively #Real., #Int., and #Nom.), the number of classes (#Cl.) and the distribution of classes (#Dc.). All numerical features were normalized, and categorical attributes were encoded using the so-called *one-hot encoding*.

4 Experimental Study

In this section, the results of the experimental study are presented. Table 4 shows the results for the best variant of each binarization strategy for the benchmark datasets without internal sampling. As we can see, in this case the OVO strategy outperformed the other two methods. Friedman Test returned $p - Value = p = 0.008$, pointing to a statistically significant difference between the results of those methods. However, Nemenyi Test revealed only the statistically significant difference between OVO and ECOC methods. Results obtained for each binarization strategy and critical differences for posthoc tests are visualized respectively in Fig. 1 and Fig. 2.

Table 3. Summary description of the datasets.

Dataset	#Inst.	#Atts.	#Real.	#Int.	#Nom.	#Cl.	#Dc.
Automobile	159	25	15	0	10	6	48/46/29/20/13/3
Balance	625	4	4	0	0	3	288/288/49
Car	1728	6	0	0	6	4	1210/384/69/65
Cleveland	297	13	13	0	0	5	160/54/35/35/13
Contraceptive	1473	9	6	0	3	3	629/511/333
Dermatology	358	34	0	34	0	6	111/71/60/48/48/20
Ecoli	336	7	7	0	0	8	143/77/52/35/20/5/2/2
Flare	1066	11	0	0	11	6	331/239/211/147/95/43
Glass	214	9	9	0	0	6	76/70/29/17/13/9
Hayes_roth	160	4	0	4	0	3	65/64/31
Led7digit	500	7	7	0	0	10	57/57/53/52/52/51/49/47/45/37
Lymphography	148	18	3	0	15	4	81/61/4/2
New_thyroid	215	5	4	1	0	3	150/35/30
Pageblocks	548	10	10	0	0	5	492/33/12/8/3
Thyroid	720	21	6	0	15	3	666/37/17
Vehicle	846	18	0	18	0	4	218/217/212/199
Wine	178	13	13	0	0	3	71/59/48
Winequality_red	1599	11	11	0	0	6	681/638/199/53/18/10
Yeast	1484	8	8	0	0	10	463/429/244/163/51/44/35/30/20/5
Zoo	101	16	0	0	16	7	41/20/13/10/8/5/4

Table 4. G-mean results for tested binarization strategies without sampling.

Dataset	OVA		OVO		ECOC	
	G-mean	Rank	G-mean	Rank	G-mean	Rank
Automobile	0.51 ± 0.17	3	0.57 ± 0.17	2	0.58 ± 0.17	1
Balance	0.91 ± 0.02	3	0.94 ± 0.02	2	0.95 ± 0.02	1
Car	0.92 ± 0.02	2	0.94 ± 0.02	1	0.81 ± 0.05	3
Cleveland	0.20 ± 0.06	1	0.17 ± 0.08	2	0.14 ± 0.06	3
Contraceptive	0.53 ± 0.02	1	0.50 ± 0.02	2	0.49 ± 0.01	3
Dermatology	0.97 ± 0.01	1.5	0.96 ± 0.01	3	0.97 ± 0.01	1.5
Ecoli	0.25 ± 0.01	2	0.25 ± 0.01	2	0.25 ± 0.01	2
Flare	0.47 ± 0.04	1	0.46 ± 0.08	2	0.41 ± 0.08	3
Glass	0.51 ± 0.15	2	0.55 ± 0.10	1	0.44 ± 0.11	3
Hayes_roth	0.83 ± 0.02	1.5	0.83 ± 0.03	1.5	0.74 ± 0.08	3
Led7digit	0.72 ± 0.02	2	0.75 ± 0.01	1	0.71 ± 0.02	3
Lymphography	0.67 ± 0.14	1	0.57 ± 0.26	2	0.38 ± 0.23	3
New_thyroid	0.94 ± 0.02	1.5	0.94 ± 0.02	1.5	0.90 ± 0.05	3
Pageblocks	0.50 ± 0.23	3	0.57 ± 0.21	1	0.54 ± 0.25	2
Thyroid	0.88 ± 0.07	3	0.90 ± 0.07	2	0.92 ± 0.05	1
Vehicle	0.80 ± 0.02	2	0.81 ± 0.03	1	0.77 ± 0.03	3
Wine	0.99 ± 0.01	1	0.98 ± 0.01	2	0.97 ± 0.01	3
Winequality_red	0.18 ± 0.06	1.5	0.18 ± 0.06	1.5	0.10 ± 0.03	3
Yeast	0.40 ± 0.04	1.5	0.40 ± 0.04	1.5	0.37 ± 0.05	3
Zoo	0.84 ± 0.13	1.5	0.84 ± 0.12	1.5	0.79 ± 0.19	3
Avg. rank	—	1.8	—	**1.675**	—	2.525

Table 5 shows results for binarization strategies after enhancing the performance of base learners with sampling methods. Although the results are visibly better than they were obtained using pure binarization schemes, the hierarchy seems to be preserved with OVO outperforming the other two techniques, which is confirmed by the Friedman Test returning $p - Value = p = 0.006$ pointing to statistically significant difference and Nemenyi Test revealing only statistically significant difference between OVO and ECOC strategies. Those results seem to be consistent with the study carried out in [11], which points out that OVO approach confronts a lower subset of instances and, therefore, is less likely to obtain a highly imbalanced training sets during binarization. Results obtained for each binarization strategy with the usage of internal sampling algorithms and critical differences for posthoc tests are visualized respectively in Fig. 3 and Fig. 4.

Wilcoxon Signed-Ranks Test was performed to determine whether or not there is a statistically significant difference between each strategy pure variant and variant enhanced with sampling algorithms. As shown in Table 6, in every case, the usage of sampling algorithms to internally enhance base models significantly improved the overall performance of the binarization strategy.

Table 5. G-mean results for tested binarization strategies with sampling.

Dataset	OVA*sampling*		OVO*sampling*		ECOC*sampling*	
	G-mean	Rank	G-mean	Rank	G-mean	Rank
Automobile	0.56 ± 0.22	3	0.61 ± 0.19	1	0.60 ± 0.20	2
Balance	0.88 ± 0.08	3	0.61 ± 0.19	1	0.92 ± 0.02	2
Car	0.91 ± 0.02	2	0.93 ± 0.02	1	0.82 ± 0.07	3
Cleveland	0.24 ± 0.06	2	0.25 ± 0.05	1	0.18 ± 0.06	3
Contraceptive	0.53 ± 0.02	1	0.52 ± 0.03	2	0.48 ± 0.02	3
Dermatology	0.96 ± 0.01	2.5	0.96 ± 0.01	2.5	0.97 ± 0.02	1
Ecoli	0.26 ± 0.01	2	0.26 ± 0.01	2	0.26 ± 0.01	2
Flare	0.56 ± 0.03	2	0.57 ± 0.03	1	0.52 ± 0.03	3
Glass	0.65 ± 0.07	1	0.62 ± 0.05	3	0.64 ± 0.06	2
Hayes_roth	0.84 ± 0.04	1	0.83 ± 0.04	2	0.68 ± 0.07	3
Led7digit	0.72 ± 0.02	2.5	0.74 ± 0.01	1	0.72 ± 0.02	2.5
Lymphography	0.66 ± 0.20	1	0.58 ± 0.26	2	0.55 ± 0.33	3
New_thyroid	0.94 ± 0.02	1.5	0.94 ± 0.04	1.5	0.92 ± 0.06	3
Pageblocks	0.57 ± 0.26	3	0.63 ± 0.20	1	0.59 ± 0.16	2
Thyroid	0.90 ± 0.06	3	0.92 ± 0.06	2	0.95 ± 0.05	1
Vehicle	0.81 ± 0.02	1.5	0.81 ± 0.02	1.5	0.80 ± 0.02	3
Wine	0.98 ± 0.01	1.5	0.98 ± 0.01	1.5	0.97 ± 0.01	3
Winequality_red	0.36 ± 0.08	1	0.33 ± 0.08	2	0.14 ± 0.05	3
Yeast	0.50 ± 0.03	2	0.51 ± 0.03	1	0.41 ± 0.05	3
Zoo	0.85 ± 0.13	1	0.84 ± 0.12	2	0.80 ± 0.16	3
Avg. rank	—	1.875	—	**1.6**	—	2.525

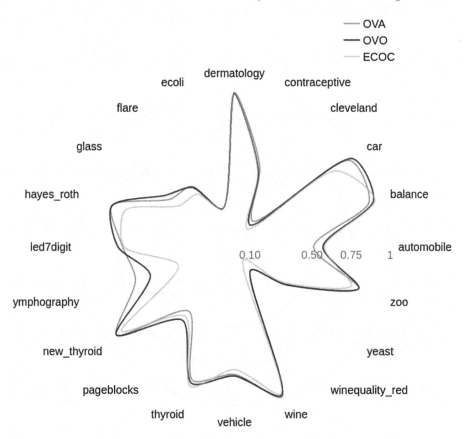

Fig. 1. G-mean results for tested binarization strategies without sampling.

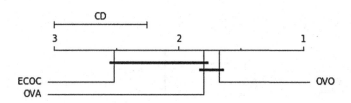

Fig. 2. Critical differences for Nemenyi Test for tested binarization strategies without sampling.

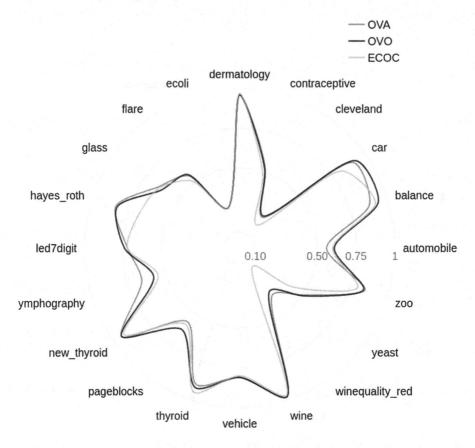

Fig. 3. G-mean results for tested binarization strategies with sampling.

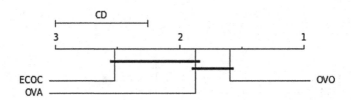

Fig. 4. Critical differences for Nemenyi Test for tested binarization strategies with sampling.

Table 6. Wilcoxon Signed-Ranks Test to compare binarization strategies variants with and without internal sampling. R^+ corresponds to the sum of the ranks for pure binarization strategy and R^- for variant with sampling.

Binarization strategy	R^+	R^-	Hypothesis ($\alpha = 0.05$)	p-value
OVA	43	161	Rejected for OVA *sampling*	0.02612
OVO	18	164	Rejected for OVO *sampling*	0.00518
ECOC	33	174	Rejected for ECOC *sampling*	0.00836

5 Concluding Remarks

In this paper, we carried out an extensive comparative experimental study of One-Vs-All, One-Vs-One, and Error-Correcting Output Codes binarization strategies in the context of imbalanced multi-classification problems. We have shown that one can reliably boost the performance of all of the binarization schemes with relatively simple sampling algorithms, which was then confirmed by a thorough statistical analysis. Another conclusion from this work is that the data preprocessing methods are able to partially mitigate the quality differences among different strategies, however the statistically significant difference among obtained results persists and OVO binarization seems to be the most robust of all three - this conclusion confirms the results of previous studies carried out in this field.

Acknowledgement. This work is supported by the Polish National Science Center under the Grant no. UMO-2015/19/B/ST6/01597 as well the statutory funds of the Department of Systems and Computer Networks, Faculty of Electronics, Wrocław University of Science and Technology.

References

1. Alcalá-Fdez, J., et al.: Keel data-mining software tool: data set repository, integration of algorithms and experimental analysis framework. J. Multiple-Valued Logic Soft Comput. **17**, 255–287 (2011)
2. Allwein, E., Schapire, R., Singer, Y.: Reducing multiclass to binary: a unifying approach for margin classifiers. J. Mach. Learn. Res. **1**, 113–141 (2000)
3. Anand, A., Suganthan, P.: Multiclass cancer classification by support vector machines with class-wise optimized genes and probability estimates. J. Theor. Biol. **259**(3), 533–540 (2009)
4. Batista, G., Bazzan, B., Monard, M.: Balancing training data for automated annotation of keywords: a case study. In: WOB, pp. 10–18 (2003)
5. Batista, G., Prati, R., Monard, M.: A study of the behavior of several methods for balancing machine learning training data. SIGKDD Explor. **6**(1), 20–29 (2004)

6. Chan, P., Stolfo, S.: Toward scalable learning with non-uniform class and cost distributions: a case study in credit card fraud detection. In: Proceedings of the Fourth International Conference on Knowledge Discovery and Data Mining, pp. 164–168 (1998)

7. Chawla, N.V., Bowyer, K.W., Hall, L.O., Kegelmeyer, W.P.: Smote: synthetic minority over-sampling technique. arXiv e-prints arXiv:1106.1813 (2011)

8. Demšar, J.: Statistical comparisons ofclassifiers over multiple data sets. J. Mach. Learn. Res. **7**, 1–30 (2006)

9. Dietterich, T., Bakiri, G.: Solving multiclass learning problems via error-correcting output codes. J. Artif. Intell. Res. **2**, 263–286 (1995)

10. Duda, R., Hart, P., Stork, D.: Pattern Classification, 2nd edn. Wiley-Interscience, Hoboken (2000)

11. Fernández, A., López, V., Galar, M., Jesus, M.D., Herrera, F.: Analysing the classification of imbalanced data-sets with multiple classes: binarization techniques and ad-hoc approaches. Knowl.-Based Syst. **42**, 97–110 (2013)

12. Fürnkranz, J.: Round robin classification. J. Mach. Learn. Res. **2**, 721–747 (2002)

13. Galar, M., Fernández, A., Barrenechea, E., Bustince, H., Herrera, F.: An overview of ensemble methods for binary classifiers in multi-class problems: experimental study on one-vs-one and one-vs-all schemes. Pattern Recogn. **44**(8), 1761–1776 (2011)

14. Galar, M., Fernández, A., Barrenechea, E., Herrera, F.: Empowering difficult classes with a similarity-based aggregation in multi-class classification problems. Inf. Sci. **264**, 135–157 (2014)

15. Han, H., Wang, W.-Y., Mao, B.-H.: Borderline-SMOTE: a new over-sampling method in imbalanced data sets learning. In: Huang, D.-S., Zhang, X.-P., Huang, G.-B. (eds.) ICIC 2005. LNCS, vol. 3644, pp. 878–887. Springer, Heidelberg (2005). https://doi.org/10.1007/11538059_91

16. Hastie, T., Tibshirani, R.: Classification by pairwise coupling. Ann. Stat. **26**(2), 451–471 (1998)

17. He, K., Zhang, X., Ren, S., Sun, J.: Deep residual learning for image recognition. arXiv e-prints arXiv:1512.03385 (2015)

18. Lemaître, G., Nogueira, F., Aridas, C.K.: Imbalanced-learn: a python toolbox to tackle the curse of imbalanced datasets in machine learning. J. Mach. Learn. Res. **18**(17), 1–5 (2017)

19. Liu, K., Xu, C.: A genetic programming-based approach to the classification of multiclass microarray datasets. Bioinformatics **25**(3), 331–337 (2009)

20. Luque, A., Carrasco, A., Martin, A., Heras, A.: The impact of class imbalance in classification performance metrics based on the binary confusion matrix. Pattern Recogn. **91**, 216–231 (2019)

21. Mosley, L.: A balanced approach to the multi-class imbalance problem. Graduate theses and dissertations, Iowa State University (2013)

22. Murphy, K.: Machine Learning: A Probabilistic Perspective. The MIT Press, Cambridge (2012)

23. Pedregosa, F., et al.: Scikit-learn: machine learning in Python. J. Mach. Learn. Res. **12**, 2825–2830 (2011)

24. Phua, C., Lee, V., Smith, K., Gayler, R.: A comprehensive survey of data mining-based fraud detection research. arXiv e-prints arXiv:1009.6119 (2010)

25. Rifkin, R., Klautau, A.: In defense of one-vs-all classification. J. Mach. Learn. Res. **5**, 101–141 (2004)
26. Simonyan, K., Zisserman, A.: Very deep convolutional networks for large-scale image recognition. arXiv e-prints arXiv:1409.1556 (2014)
27. Vapnik, V.: Statistical Learning Theory. Wiley-Interscience, Hoboken (1998)
28. Virtanen, P., et al.: SciPy 1.0: fundamental algorithms for scientific computing in Python. Nat. Methods **17**, 261–272 (2020). https://doi.org/10.1038/s41592-019-0686-2
29. Zhang, Z., Krawczyk, B., Garcìa, S., Rosales-Pérez, A., Herrera, F.: Empowering one-vs-one decomposition with ensemble learning for multi-class imbalanced data. Knowl.-Based Syst. **106**, 251–263 (2016)

Towards Network Anomaly Detection Using Graph Embedding

Qingsai Xiao[1,2], Jian Liu[1,2], Quiyun Wang[1], Zhengwei Jiang[1,2], Xuren Wang[1,3], and Yepeng Yao[1,2(✉)]

[1] Institute of Information Engineering, Chinese Academy of Sciences, Beijing, China
{xiaoqingsai,liujian6,wangqiuyun,jiangzhengwei,yaoyepeng}@iie.ac.cn
[2] School of Cyber Security, University of Chinese Academy of Sciences, Beijing, China
[3] College of Information Engineering, Capital Normal University, Beijing, China
wangxuren@cnu.edu.cn

Abstract. In the face of endless cyberattacks, many researchers have proposed machine learning-based network anomaly detection technologies. Traditional statistical features of network flows are manually extracted and rely heavily on expert knowledge, while classifiers based on statistical features have a high false-positive rate. The communications between different hosts forms graphs, which contain a large number of latent features. By combining statistical features with these latent features, we can train better machine learning classifiers. Therefore, we propose a novel network anomaly detection method that can use latent features in graphs and reduce the false positive rate of anomaly detection. We convert network traffic into first-order and second-order graph. The first-order graph learns the latent features from the perspective of a single host, and the second-order graph learns the latent features from a global perspective. This feature extraction process does not require manual participation or expert knowledge. We use these features to train machine learning algorithm classifiers for detecting network anomalies. We conducted experiments on two real-world datasets, and the results show that our approach allows for better learning of latent features and improved accuracy of anomaly detection. In addition, our method has the ability to detect unknown attacks.

Keywords: Network anomaly detection · Graph embedding · Feature engineering · Unknown attack discovery

1 Introduction

With the rapid development of the Internet and more network devices are connected to it, modern cyberattacks have emerged in recent years. Attackers aiming to exploit networked systems must take a number of relevant attack steps in order to achieve their goals. In order to deal with these attack steps, significant attention has been given to identifying correlated anomalous network flow

© Springer Nature Switzerland AG 2020
V. V. Krzhizhanovskaya et al. (Eds.): ICCS 2020, LNCS 12140, pp. 156–169, 2020.
https://doi.org/10.1007/978-3-030-50423-6_12

from the network traffic in recent academic research. In addition, for each of the attack steps that compose the entire attack scenario to be interconnected, looking at the network context of individual attack steps is helpful to understand the correlation between anomalous and benign streams and is important in a counterattack against network attacks.

Since modern network anomaly detection systems should not be highly dependent on human expert knowledge, they need to have the ability to autonomously adapt to the evolution of all network behaviors. Therefore, machine learning has become a very common method for detecting network anomalies, and feature engineering has become an important step in machine learning. However, the quality of the features directly determines the effectiveness of the machine learning algorithm. Researchers have proposed a number of ways to extract features from network traffic. Some feature extraction methods have also published as public tools such as CICFlowMeter [7]. Most of these features are statistical features. In that way, the existing feature extraction methods rely heavily on expert knowledge.

The communication between different hosts can be constructed into attributed graphs. These graphs contain huge number of latent features. For example, if the port usage of the two hosts is very similar, the distance of the two hosts in graphs should be close. These latent features cannot be extracted manually, but they are very helpful for machine learning-based algorithms.

In this paper, we propose a network anomaly detection method based on graph embedding. We convert the network traffic into first-order graph and second-order graph. The first-order graph learns the latent features from the perspective of a single host, and the second-order graph learns the latent features from a global perspective. This feature extraction process does not require manual participation or expert knowledge. We use these features to train machine learning algorithm classifiers for detecting network anomalies.

In general, our main contributions are as follows:

- We propose first-order and second-order graph of network traffic, and learn the low-dimensional vector representation of the nodes in the diagram.
- We design a network anomaly detection method based on graph embedding. By combining first-order and second-order low-dimensional vector representations of network traffic with several statistical features, it is possible to improve detection accuracy and detect unknown attacks.
- We built a prototype of the network anomaly detection framework and evaluate the detection accuracy and the ability to discover unknown attacks on real-world datasets.

The rest of the paper is organized as follows. We first introduce the threat model in Sect. 2. Section 3 is the graph embedding algorithm. Section 4 defines the proposed network anomaly detection framework. Experimental results are presented in Sect. 5. Finally, we discuss the related work in Sect. 6 and conclude in Sect. 7.

2   Threat Model

In order to better understand the method proposed in this paper, we will describe the threat model in this section. The attacker launches an attack on victims through the network. Here, we take remote control Trojan as an example. The attacker sends a control command to the Trojan through the C&C server through the phishing email of the victim, and accepts the stolen data. In many cases, the attacker will use encryption to communicate, but the victim's ability to know the content of the communication is limited. The victim needs to look for abnormal traffic in the network traffic and find attack information (Fig. 1).

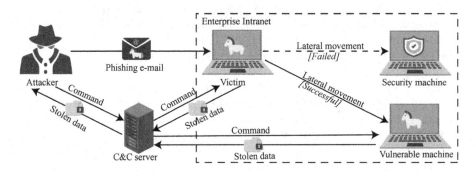

Fig. 1. A use case for the threat model

3   Graph Embedding Algorithm

In this section, we introduce the first-order graph and second-order graph of network traffic, then propose the graph embedding algorithm for these two graphs. At last, we also adopt two optimization methods to reduce the complexity of the proposed algorithm.

3.1   First-Order Graph Embedding

Definition 1 (*First-Order Graph*). *A bipartite graph consisting of IP addresses and ports (e.g. Fig. 2) is denoted as $G_1 = (V_{ip} \bigcup V_{port}, E)$, where V_{ip} is the set of IP addresses and V_{port} is the set of ports. E is the set of edges between IP addresses and ports. $\omega_{i,j}$ is the weight between v_{ip_i} and v_{port_j}. The weight can be the number of packets or the number of bits that flowing out from $PORT_j$ of IP_i. In the following experiments, we consider the weight as the number of packets that flowing out from $PORT_j$ of IP_i.*

The first-order graph learns the latent features from the perspective of a single host. The purpose of first-order graph embedding is that if two hosts have similar port usage distribution, their distance in the vector space should be close.

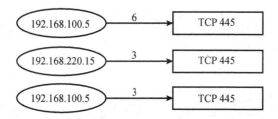

Fig. 2. Example of first-order graph

For $\forall\, v_{ip_i} \in V_{ip}, v_{port_j} \in V_{port}$, let $\overrightarrow{u_{ip_i}}$ and $\overrightarrow{u_{port_j}}$ are embedding vectors for v_{ip_i} and v_{port_j}. We define the following conditional probabilities:

$$p(v_{ip_i}|v_{port_j}) = \frac{\exp\left(\overrightarrow{u_{ip_i}}^T \cdot \overrightarrow{u_{port_j}}\right)}{\sum_{v_{ip_{i'}} \in V_{ip}} \exp\left(\overrightarrow{u_{ip_{i'}}}^T \cdot \overrightarrow{u_{port_j}}\right)} \tag{1}$$

$$p(v_{port_j}|v_{ip_i}) = \frac{\exp\left(\overrightarrow{u_{port_j}}^T \cdot \overrightarrow{u_{ip_i}}\right)}{\sum_{v_{port_{j'}} \in V_{port}} \exp\left(\overrightarrow{u_{port_{j'}}}^T \cdot \overrightarrow{u_{ip_i}}\right)} \tag{2}$$

According to Eq. 1 and Eq. 2, we can get the conditional distribution $p(\cdot|v_{port_j})$ and $p(\cdot|v_{ip_i})$.

To understand the latent features of IP addresses and ports, we minimize the distance between conditional distribution and empirical distribution. Therefore, we get the following objective function:

$$O_1 = \sum_j distance\left(\hat{p}(\cdot|v_{port_j}), p(\cdot|v_{port_j})\right) + \sum_i distance\left(\hat{p}(\cdot|v_{ip_i}), p(\cdot|v_{ip_i})\right) \tag{3}$$

where $\hat{p}(v_{ip_i}|v_{port_j}) = \frac{w_{i,j}}{\sum_{i'} w_{i',j}}$, $distance(\cdot, \cdot)$ is the distance between two distributions(e.g. Bhattacharyya distance, Jeffries–Matusita distance, KL-divergence distance). In the following experiments, we use the KL-divergence distance.

We use stochastic gradient descent for optimizing Eq. 3, then we are able to get the first-order low-dimensional vector representation $\{\overrightarrow{u_i^{ip}}\}_{i=1,2,3,...,|V_{ip}|}$ and $\{\overrightarrow{u_i^{port}}\}_{i=1,2,3,...,|V_{port}|}$.

3.2   Second-Order Graph Embedding

Definition 2 (Second-Order Graph). *A hypergraph consisting of source IP addresses, source ports, destination IP addresses, and destination ports (e.g. Fig. 3) is denoted as $G_2 = (V_{sip} \bigcup V_{sport} \bigcup V_{dip} \bigcup V_{dport}, E)$, where V_{sip} is a set of all source IP addresses; V_{sport} is a set of all source ports; V_{dip} is a set of all destination IP addresses; V_{dport} is a set of all destination ports; E is a set* of hyperedges consisting of four points: source IP address, source port, destination IP address and destination ports. $\omega_{i,j,k,l} \in E$ is the weight of hyperedge $< v_{sip_i}, v_{sport_j}, v_{dip_k}, v_{dport_l} >$.

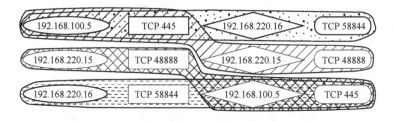

Fig. 3. Example of second-order graph

The second-order graph learns latent features from a global perspective.

For $\forall v \in V_{sip} \bigcup V_{sport} \bigcup V_{dip} \bigcup V_{dport}$, we define C is the context of v. For example, for a hyperedge $< v_i, v_j, v_k, v_l >$, (v_i, v_k, v_l) is the context of v_j. v_i, v_k and v_l are similar. Inspired by [6], we define the following conditional probabilities:

$$p(v|C) = \frac{\exp\left(Sim(v, C)\right)}{\sum_{v' \in V_C} \exp\left(Sim(v', C)\right)} \quad (4)$$

where V_C is a set of nodes which context is C. $Sim(v, C) = \frac{1}{|V_C|} \sum_{v' \in V_C} \vec{u_v}^T * u_{v'}$ is the similarity between v and C.

Similarity to Eq. 3, we minimize the distance between the conditional distribution and the empirical distribution. Therefore, we get the following objective function:

$$O_2 = \sum_{C \in \mathcal{P}} distance(\hat{p}(\cdot|C), p(\cdot|C)) \quad (5)$$

where $\mathcal{P}$ is a set of C, $\hat{p}(v|C) = \frac{\omega_{v,C}}{\sum_{v' \in V_C} \omega_{v',C}}$, $distance(\cdot, \cdot)$ is the distance between the two distributions. In the following experiments, we use the KL-divergence distance.

We use stochastic gradient descent for optimizing Eq. 5, then we could get the second-order low-dimensional vector representation $\{\overrightarrow{u_i^{sip}}\}_{i=1,2,\ldots,|V_{sip}|}$, $\{\overrightarrow{u_i^{sport}}\}_{i=1,2,\ldots,|V_{sport}|}$, $\{\overrightarrow{u_i^{dip}}\}_{i=1,2,\ldots,|V_{dip}|}$ and $\{\overrightarrow{u_i^{dport}}\}_{i=1,2,\ldots,|V_{dport}|}$.

3.3 Optimization

In order to speed up the calculation speed, we use the following two optimization methods.

Fast Sigmoid Algorithm: When optimizing objective function, $sigmoid(x)$ function needs to be calculated many times. In the computer, the $\exp(x)$ function is usually calculated by the series summation ($\exp(x) = \sum_{k=0}^{\infty} \frac{x^k}{k!} = 1 + x + \frac{x^2}{2} + \frac{x^3}{6} + \cdots$). A large number of multiplications and divisions need to be calculated, which takes a lot of time. So we use a sigmoid table instead of calculating $sigmoid(x)$ function. Firstly, we initialize this sigmoid table by calculating $sigmoid(-\frac{table\_size}{2})$, $sigmoid(-\frac{table\_size}{2}+1)$, $\cdots$, $sigmoid(\frac{table\_size}{2})$. When we need to calculate $sigmoid(x)$ in each iteration, we use the point closest to x in the sigmoid table as $sigmoid(x)$. The time complexity of calculating $sigmoid(x)$ is only $O(1)$, and the space complexity is $O(table\_size)$. The size of $table\_size$ is related to the precision of $sigmoid(x)$ function. In the following experiments, $table\_size$ takes 2000.

Alias Method for Sampling: The nodes need to be sampled during each iteration of the training. For nodes that appear frequently, we should have a greater probability of sampling. Let $\Omega = \{\omega_1, \omega_2, ..., \omega_{|V|}\}$ denotes the weight set of nodes, $\omega_{sum} = \sum_{i=0}^{|V|} \omega_i$ is the sum of all weights. A simple way is that the sampling probability of each node is $\{\frac{\omega_1}{\omega_{sum}}, \frac{\omega_2}{\omega_{sum}}, ..., \frac{\omega_{|V|}}{\omega_{sum}}\}$. There is a very naive sampling method that satisfies this sampling requirement: randomly select an integer r from $[1, \omega_{sum}]$, if $r \in [\sum_{i=1}^{j-1} \omega_i, \sum_{i=1}^{j} \omega_i)$, then select the j-th node. We can use prefix sum array and binary search algorithm to search for $\sum_{i=1}^{j-1} \omega_i \leq r < \sum_{i=1}^{j} \omega_i$. The time complexity of this method is $O(\log |V|)$. Another faster method is alias table method [8], which is also the method we used in our experiments. This method generates two auxiliary tables based on Ω at first. Then we can randomly select an integer of $[1, |V|]$ and a decimal of $[0, 1)$ to select a node. This method has a time complexity of $O(1)$ per sample, which saves a lot of time when the number of iterations is large.

4   Network Anomaly Detection Framework

Our framework is composed of five primary modules: network probe, embedding, training, detection and database as shown in Fig. 4.

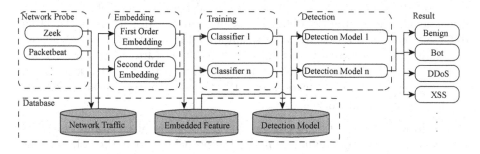

Fig. 4. Overview of the network anomaly detection framework

Network Probe. The network probe module captures traffic from network and generates traffic logs. The network probe module can be deployed on devices such as personal computers, firewalls, and servers. As mentioned in Sect. 3, our algorithm is to calculate the low latitude quantization representation of IP and port using IP, port, protocol, the number of packets and bytes (e.g. Table 1). So the generated log must contain these fields. There are many open source tools (e.g. Zeek [20], Packetbeat [4], and Joy [10]) that generate network traffic logs that meet our requirements. We employ these existing tools to generate and unify network traffic logs.

Table 1. An example of network traffic logs

Source IP address	Source port	Destination IP address	Destination port	Protocol	Packets	Bytes
192.168.100.5	445	192.168.220.16	58844	TCP	1	108
192.168.100.5	445	192.168.220.15	48888	TCP	1	108
192.168.220.15	48888	192.168.100.5	445	TCP	2	174
192.168.220.16	58844	192.168.100.5	445	TCP	2	174
192.168.100.5	445	192.168.220.15	48888	TCP	1	108
192.168.220.16	58844	192.168.100.5	445	TCP	2	174

Embedding. In the embedding module, we use the method in Sect. 3 to learn the first-order and second-order low-dimensional vector representation of the nodes in these graphs. For the initial training, we randomly initialize the embedding vector of each node. For each subsequent training, we use the embedding vector of the last training as the initial value of this training, and then use the new data for iterative training. At the end of each training, we store the results in the database for later use.

Training. In the training module, we train multiple machine learning classifiers using features learned in the embedding phase and some statistical features (e.g. flow duration, total forward packets, total backward packets). There are differences in the accuracy of different machine learning algorithms for detecting different types of attacks. In order to improve the accuracy of detection, we calculated the weights of different classifiers on different attack categories on the test set. $acc_{i,j}$ is the accuracy of the i-th classifier on the j-th attack category. $w_{i,j} = \frac{acc_{i,j}}{\sum_i acc_{i,j}}$ is the classification weight of the i-th machine learning classifier on the j-th attack category.

Detection. In the detection module, we use multiple machine learning classifiers to detect new network traffic log. $p_{i,j}$ is the probability that the i-th machine learning classifier considers this network traffic log belongs to the j-th attack category, where $\sum_j p_{i,j} = 1$. The final category of the output of the detection phase is $\arg \max_j \sum_i w_{i,j} * p_{i,j}$.

Database. Database module is mainly to store network traffic logs, embedding vector and detection models. There are many open source data storage system, including general relational database, distributed storage framework and so on.

In our experimental environment, we use MySQL for storage. When the network bandwidth is very large, a lot of network traffic logs will be generated. In this case, we should consider using a distributed storage framework.

5   Evaluation

In this section, we provide three thorough experiments of graph embeddings and network anomaly detection. The first experiment is a clustering of IP addresses. We verified on the CIDDS-001 dataset [13] whether the first-order and second-order low-dimensional vector representation can learn the latent features of IP addresses. The second experiment is network anomaly detection. We evaluate the precision, recall and F1 measure of our method on the CICIDS 2017 dataset [15]. The third experiment is an unknown threat discovery. We use a portion of the attack categories to train our model and then to evaluate if new attack types can be detected.

5.1   Experiment Setup

A proof-of-concept version of our graph embedding algorithms are implemented by C++ with compiler version g++ 4.8.5, which are running on CentOS 7.6 OS. The server is DELL R440 with 48 CPU cores and 128 GB of memory. We also implement machine learning algorithm for anomaly detection via Python3 and Scikit-learn [11].

5.2   IP Address Cluster

For an effective graph embedding algorithm, hosts of the same category in the network should be close in low-dimensional vector space. When running the clustering algorithm, hosts of the same category should be clustered into a cluster. So in this experiment, we use the DBScan clustering algorithm to cluster IP addresses.

We evaluate the effects of the clustering algorithm from three indicators: *accuracy*, *homogeneity*, and *completeness*. *Homogeneity* and *completeness* are independent of the labels.

The baseline method is *IP2Vec* [12]. *IP2Vec* obtains the low-dimensional vector representation of the IP addresses by training a neural network with single hidden layer. We experimented with the same dataset and experimental method as *IP2Vec*.

In the third week of the CIDDS-001 dataset [13], there are a total of 6,439,783 network traffic flows. We use all data during graph embedding, and only cluster the intranet IP addresses (192.168.100.0/24, 192.168.200.0/24, 192.168.210.0/24, 192.18.220.0/24). We divide the IP address into *server* and *client* according to the role that the host plays in the network. We treat *server* and *printer* as *server*; *windows client* and *linux client* as *client*. In the end, there are a total of 7 *servers* and 19 *clients*.

Table 2. Assignment matrix for the CIDDS-001 dataset

Method	Classes	Cluster 1	Cluster 2	Num. outliers
Our method	Server	7	0	0
	Client	0	19	0
IP2Vec	Server	6	·0	1
	Client	0	19	0

Table 2 shows the result of using DBScan to cluster IP addresses. As shown in the table, our method can distinguish between *server* and *client* precisely. But *IP2Vec* has a point that cannot be divided into any cluster. Table 3 shows the results of two methods on *accuracy, homogeneity* and *completeness*. For the *accuracy* and *completeness*, our method outperforms *IP2Vec*.

Table 3. Comparision of our method and *IP2Vec* for clustering the IP Address within the CIDDS-001 dataset

Similarity measure	Accuracy	Homogeneity	Completeness	Num. outliers
Our method	1.0	1.0	1.0	0
IP2Vec	0.9615	1.0	0.8406	1

5.3   Network Anomaly Detection

In this experiment, we used the CICIDS 2017 [15] dataset. The CICIDS 2017 dataset contains a total of 2,830,743 network flows. The attack types include DDoS, DoS, Brute Force, XSS, SQL Injection, Infiltration, Port Scan and Botnet. We remove the attack categories of less than 100 flows. For the remaining categories, we used 60% of each category of data as a training set and 40% as a test set.

There are 80 features in CICIDS 2017. In [15], the authors evaluated the importance of each feature on detecting attacks and selected eight most important features: flow duration, total forward packets, total backward packets, total length of forward packets, total length of backward packets, flow bytes per seconds, flow packets per seconds, down/up ratio. Most of the feature extraction tools can extract these eight features. Therefore, in this experiment we use first-order vector representation, second-order vector representation and these eight features training machine learning algorithms.

We did three times evaluations in this experiment. For the first time, we only use the raw features. For the second time, we only use embedding features. For the third time, we use both the raw features and the embedding features. We use the random forest algorithm to train and evaluate the results from three indicators: *Precision, Recall* and *F1 measure.*

Table 4. Performances of random-forest classifier over raw features, embedding features and all features

Attack scenarios	Raw features			Embedding features			All features		
	Precision	Recall	F1	Precision	Recall	F1	Precision	Recall	F1
Benign	0.9862	0.9860	0.9861	**0.9999**	**1.0000**	**1.0000**	0.9998	0.9999	0.9999
Bot	1.0000	0.0140	0.0276	**1.0000**	**0.9117**	**0.9538**	0.9502	0.8098	0.8744
DDoS	0.9969	0.9929	0.9949	0.9950	0.5596	0.7163	**0.9993**	**0.9995**	**0.9994**
DoS GoldenEye	0.9178	0.7402	0.8195	0.0000	0.0000	0.0000	**0.9945**	**0.9900**	**0.9922**
DoS Hulk	0.8806	0.9507	0.9143	0.7411	0.9987	0.8509	**0.9979**	**0.9998**	**0.9988**
DoS Slowhttptest	0.9812	0.8291	0.8987	0.0000	0.0000	0.0000	**0.9968**	**0.9998**	**0.9988**
DoS Slowloris	**1.0000**	0.4705	0.6399	0.0000	0.0000	0.0000	0.9857	**0.9885**	**0.9871**
FTP-Patator	**1.0000**	0.4921	0.6596	0.9730	0.9971	0.9849	0.9979	**1.0000**	**0.9990**
PortScan	0.9936	0.9912	0.9924	0.9998	0.9938	0.9968	0.9998	**0.9984**	**0.9991**
SSH-Patator	**1.0000**	0.4960	0.6631	0.9650	0.9977	0.9811	**1.0000**	**0.9989**	**0.9994**
Brute Force	**1.0000**	0.0381	0.0735	0.0000	0.0000	0.0000	0.7085	**0.9248**	**0.8023**
XSS	**1.0000**	0.0077	0.0152	0.0000	0.0000	0.0000	0.6400	**0.1641**	**0.2612**

Table 4 shows the experimental results. As can be seen from the table, using only raw features to train the classifier, the detection accuracy of the classifier is very high, but the recall is relatively low, especially bot, XSS and brute force. Therefore, the F1 measure of the classifier is low. In addition, the results of a classifier trained using only embedding features are also less effective. Although it has high detection accuracy and recall rate for several attack scenarios (e.g. Benign, Bot and Portscan), it has multiple attack scenarios that cannot be detected (e.g. DoS GoldenEye, DoS Slowhttptest and XSS). If both the raw feature and the embedding feature are used to train the classifier, the performance of the classifier can be greatly improved. It not only has a high detection accuracy rate, but also has a high recall rate.

Figure 5 shows the confusion matrix of the three times evaluations. As can be seen from the figure, a classifier trained using only raw features classifies a large amount of malicious traffic into benign traffic, which has low attack false positive rate. Although the classifier trained with embedding features cannot accurately detect the attack scenarios, it rarely categorizes malicious traffic as benign traffic, and the attack false negative rate is very low. A classifier trained using both raw features and embedding features has the advantages of both. It can not only identify malicious traffic, but also accurately detect different attack scenarios.

5.4   Unknown Threat Discovery

In this experiment, we use 60% of benign traffic and all Bot, PortScan, DoS Hulk, DoS slowloris, FTP-Patator, Brute Force attack traffic as training set, and the remaining 40% of benign traffic and other attack types as test set. The raw random forest algorithm is adopted to train a binary classifier, then to evaluate the ability of our algorithm to detect unknown attack traffic.

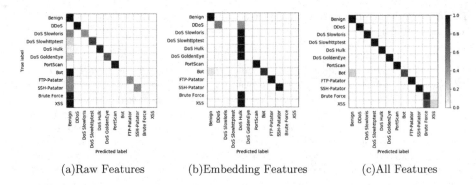

(a)Raw Features   (b)Embedding Features   (c)All Features

Fig. 5. Confusion matrix of random-forest classifier over different feature sets

Table 5. Results of unknown threat discovery

True classes	Benign	Anomaly	Total
Benign	906262	173	906435
DDoS	36	127991	128027
DoS Slowhttptest	0	5499	5499
DoS GoldenEye	0	10293	10293
Heartbleed	11	0	11
SSH-Patator	2	5895	5897
Web Attack-XSS	0	652	652
Web Attack-SQL Injection	0	21	21
Infiltration	36	0	36
Total	906347	150524	-

Table 5 shows the prediction result on the test set. 173 out of 906435 benign flows in the test set were predicted as anomaly flows, with an accuracy of 99.98%. 83 of the 150436 attack traffic in the test set were predicted as normal traffic, with an accuracy of 99.94%. Most of the attack traffic can be accurately detected, but Heartbleed and Infiltration is all predicted to be normal traffic. We analyzed the reason may be that the behaviours of these two attack categories are different from the categories of attacks in the training set and were not recognized correctly. Our approach can detect variants of known attacks, but is limited in its ability to detect brand-new attacks.

6   Related Work

The network anomaly detection technology has been divided into three categories according to [5]: statistical-based, knowledge-based and machine learning-based. Deep learning-based network anomaly detection has emerged in recent years due to the development of deep learning.

6.1   Statistical-Based Network Anomaly Detection

Caberera et al. [3] proposed a statistical flow modeling method that can detect unknown network attacks. The authors constructed a *Network Activity Models* and *Application Models* and evaluated it on the DARPA'98 dataset. The experimental results showed that the *Network Activity Models* can detect Denial-of-Service and Probing attacks; *Application Models* can distinguish between the normal telnet connections and attacks using telnet connections.

6.2   Knowledge-Based Network Anomaly Detection

By integrating specification-based with anomaly-based intrusion detection technology, Sekar et al. [14] proposed a intrusion detection approach by using a state-machine to begin with the specification of network protocol, then expanding the state of the state-machine using existing knowledge to detect the anomaly state.

6.3   Machine Learning-Based Network Anomaly Detection

Ariu et al. [2] proposed *HMMPayl*, an intrusion detection system using Hidden Markov Models (HMM). *HMMPayl* represented the HTTP payload as a sequence of bytes, and detected if it is an attack. *HMMPayl* has a detection rate of over 98% on the DARPA'99 dataset [9]. Syarif et al. [16] evaluated the anomaly detection accuracy of five different clustering algorithms: K-Means, improved K-Means, K-Medoids, EM clustering and distance-based outlier detection algorithms. Xu et al. [18] proposed an anomaly detection framework integrating artificial intelligence engine with fog computing, and used semi-supervised machine learning to train a robust detection model.

6.4   Deep Learning-Based Network Anomaly Detection

Zhu et al. [21] proposed a intrusion detection approach based on Attention-based Multi-Flow LSTM by using features of multiple flows rather than a single flow as input of LSTM, so that it can capture temporal feature between flows. They also add the attention mechanism to LSTM by achieving a high accuracy on the CICIDS 2017 dataset [15]. Wang et al. [17] proposed a CNN architecture similar to LeNet-5 to train raw binary traffic and classify it. Aldwairi et al. [1] used NetFlow to train the Restricted Boltzmann Machine and achieved a high accuracy on the CICIDS 2017 dataset. Yao et al. [19] proposed a framework for solving network anomaly detection tasks by integrating the graph kernel with deep neural networks, and evaluated different integrating methods.

7   Conclusion

In this paper, we propose a network anomaly detection framework based on graph embedding. We convert the network traffic data into first-order and second-order graph. First-order graphs learn the latent features from the perspective of a single host, and second-order graphs learn the latent features from a global perspective. This feature extraction process does not require human involvement at all. After training the machine learning classifiers using the first-order and second-order vector representation and some statistical features, we use these classifiers to detect new network flows. Three experiments on two real-world datasets showed that our first-order and second-order vector representation can learn latent features and can improve the efficiency of the anomaly detection framework. We also used some of the attack types traffic to train and predict new attack types in network traffic, then evaluated our framework's ability to discover unknown attack types.

Acknowledgement. This research is supported by Key Laboratory of Network Assessment Technology, Chinese Academy of Sciences and Beijing Key Laboratory of Network Security and Protection Technology. We thank the anonymous reviewers for their insightful comments on the paper.

References

1. Aldwairi, T., Perera, D., Novotny, M.A.: An evaluation of the performance of restricted Boltzmann machines as a model for anomaly network intrusion detection. Comput. Netw. **144**, 111–119 (2018)
2. Ariu, D., Tronci, R., Giacinto, G.: HMMPayl: an intrusion detection system based on hidden Markov models. Comput. Secur. **30**(4), 221–241 (2011)
3. Caberera, J., Ravichandran, B., Mehra, R.K.: Statistical traffic modeling for network intrusion detection. In: Proceedings 8th International Symposium on Modeling, Analysis and Simulation of Computer and Telecommunication Systems (Cat. No. PR00728), pp. 466–473. IEEE (2000)
4. Elasticsearch: Packetbeat: Network analytics using elasticsearch (2020). https://www.elastic.co/products/beats/packetbeat
5. Garcia-Teodoro, P., Diaz-Verdejo, J., Maciá-Fernández, G., Vázquez, E.: Anomaly-based network intrusion detection: techniques, systems and challenges. Comput. Secur. **28**(1–2), 18–28 (2009)
6. Gui, H., Liu, J., Tao, F., Jiang, M., Norick, B., Han, J.: Large-scale embedding learning in heterogeneous event data. In: 2016 IEEE 16th International Conference on Data Mining (ICDM), pp. 907–912. IEEE (2016)
7. Lashkari, A.H., Zang, Y., Owhuo, G.: Netflowmeter (2020). http://netflowmeter.ca/
8. Li, A.Q., Ahmed, A., Ravi, S., Smola, A.J.: Reducing the sampling complexity of topic models. In: Proceedings of the 20th ACM SIGKDD International Conference on Knowledge Discovery and Data Mining, pp. 891–900. ACM (2014)
9. Lippmann, R., Haines, J.W., Fried, D.J., Korba, J., Das, K.: The 1999 darpa offline intrusion detection evaluation. Comput. Netw. **34**(4), 579–595 (2000)
10. McGrew, D., Anderson, B.: Joy (2020). https://github.com/cisco/joy

11. Pedregosa, F., et al.: Scikit-learn: machine learning in Python. J. Mach. Learn. Res. **12**(Oct), 2825–2830 (2011)
12. Ring, M., Dallmann, A., Landes, D., Hotho, A.: IP2Vec: learning similarities between IP addresses. In: 2017 IEEE International Conference on Data Mining Workshops (ICDMW), pp. 657–666. IEEE (2017)
13. Ring, M., Wunderlich, S., Grüdl, D., Landes, D., Hotho, A.: Flow-based benchmark data sets for intrusion detection. In: Proceedings of the 16th European Conference on Cyber Warfare and Security. ACPI, pp. 361–369 (2017)
14. Sekar, R., et al.: Specification-based anomaly detection: a new approach for detecting network intrusions. In: Proceedings of the 9th ACM Conference on Computer and Communications Security, pp. 265–274. ACM (2002)
15. Sharafaldin, I., Lashkari, A.H., Ghorbani, A.A.: Toward generating a new intrusion detection dataset and intrusion traffic characterization. In: ICISSP, pp. 108–116 (2018)
16. Syarif, I., Prugel-Bennett, A., Wills, G.: Unsupervised clustering approach for network anomaly detection. In: Benlamri, R. (ed.) NDT 2012. CCIS, vol. 293, pp. 135–145. Springer, Heidelberg (2012). https://doi.org/10.1007/978-3-642-30507-8_13
17. Wang, W., Zhu, M., Zeng, X., Ye, X., Sheng, Y.: Malware traffic classification using convolutional neural network for representation learning. In: 2017 International Conference on Information Networking (ICOIN), pp. 712–717. IEEE (2017)
18. Xu, S., Qian, Y., Hu, R.Q.: A semi-supervised learning approach for network anomaly detection in fog computing. In: ICC 2019–2019 IEEE International Conference on Communications (ICC), pp. 1–6. IEEE (2019)
19. Yao, Y., Su, L., Zhang, C., Lu, Z., Liu, B.: Marrying graph kernel with deep neural network: a case study for network anomaly detection. In: Rodrigues, J.M.F., et al. (eds.) ICCS 2019. LNCS, vol. 11537, pp. 102–115. Springer, Cham (2019). https://doi.org/10.1007/978-3-030-22741-8_8
20. Zeek.org: The Zeek network security monitor (2020). https://www.zeek.org/
21. Zhu, M., Ye, K., Wang, Y., Xu, C.-Z.: A deep learning approach for network anomaly detection based on AMF-LSTM. In: Zhang, F., Zhai, J., Snir, M., Jin, H., Kasahara, H., Valero, M. (eds.) NPC 2018. LNCS, vol. 11276, pp. 137–141. Springer, Cham (2018). https://doi.org/10.1007/978-3-030-05677-3_13

Maintenance and Security System for PLC Railway LED Sign Communication Infrastructure

Tomasz Andrysiak$^{(\boxtimes)}$ and Łukasz Saganowski

Institute of Telecommunications and Computer Science,
Faculty of Telecommunications, Information Technology and Electrical
Engineering, UTP University of Science and Technology,
Al. Prof. Kaliskiego 7, 85-796 Bydgoszcz, Poland
{andrys,luksag}@utp.edu.pl

Abstract. LED marking systems are currently becoming key elements of every Smart Transport System. Ensuring proper level of security, protection and continuity of failure-free operation seems to be not a completely solved issue. In the article, a system is present allowing to detect different types of anomalies and failures/damage in critical infrastructure of railway transport realized by means of Power Line Communication. There is also described the structure of the examined LED Sign Communications Network. Other discussed topics include significant security problems and maintenance of LED sign system which have direct impact on correct operation of critical communication infrastructure. A two-stage method of anomaly/damage detection is proposed. In the first step, all the outlying observations are detected and eliminated from the analysed network traffic parameters by means of the Cook's distance. So prepared data is used in stage two to create models on the basis of autoregressive neural network describing variability of the analysed LED Sign Communications Network parameters. Next, relations between the expected network traffic and its real variability are examined in order to detect abnormal behaviour which could indicate an attempt of an attack or failure/damage. There is also proposed a procedure of recurrent learning of the exploited neural networks in case there emerge significant fluctuations in the real PLC traffic. A number of scientific research was realized, which fully confirmed efficiency of the proposed solution and accuracy of autoregressive type of neural network for prediction of the analysed time series.

Keywords: Anomaly and fault detection · Time series analysis · Outliers detection · Network traffic prediction · Autoregressive neural networks · Critical infrastructure · LED sign communications network

1 Introduction

Intelligent Transport Systems (ITS) are different types of solutions which are an answer to increasing demand for goods and human mobility. By their means we can create vast, fully-functional and efficient systems of managing transport in real time. To achieve these aims, there are used diverse information and telecommunication technologies (e.g. Internet of Things (IoT), Wireless Sensor Network (WSN) or Power Line

© Springer Nature Switzerland AG 2020
V. V. Krzhizhanovskaya et al. (Eds.): ICCS 2020, LNCS 12140, pp. 170–183, 2020.
https://doi.org/10.1007/978-3-030-50423-6_13

Communication (PLC)) but also automation solutions for mobile objects including infrastructure, vehicles and their users. The main objective of ITS is boosting the capacity of transport systems and enhancement of quality of their services [1]. A special solution of ITS are variable LED sign systems, which visualize traditional signs used in the utility transport system. These signs are switched on by the operator, or they can be connected to the smart steering system. Their typical function is showing appropriate marking and/or defined information [2]. An important aspect of every Intelligent Transport System is proper solving of potential problems connected with protection and security of its own infrastructure. One of possible solutions is anomaly/damage detection systems allowing to recognize occurring threats, i.e. noticing variation from normal behavior or situation. Identification of abnormal incidents which are an aftermath of an attack, abuse or harm is crucial from the system's security point of view, because it may lead to critical states in the protected infrastructure and may require instant remedial activities [3].

The article presents possible solutions allowing to detect different type of anomalies and failures/damage for critical infrastructure of railway signs. There is proposed and described the structure of constructed LED Sign Communications Network realized with the use of PLC technology. There are presented key security problems which have direct impact on proper operation of the critical communication infrastructure. Furthermore, numerous experiments were conducted which fully confirmed efficiency and efficacy of the suggested solution.

The article is organized as follows. After the Introduction, Sect. 2 presents related work on existing security and maintenance solutions for PLC Railway LED Sign Communication Infrastructure. Next, Sect. 3 presents the structure and operation of the proposed solution. In Sect. 4, experimental results are presented and discussed. Finally, Sect. 5 concludes our work.

2 Related Work

Internet of things and solutions in industrial and consumer applications has been spreading rapidly in recent years. As an example, we can mention solutions connected to Smart Grids like metering systems [4, 5], smart lights [6, 7], intelligent transportation and city systems [8, 9]. Anomalies in communication systems may be caused by various factors, i.e. deliberate or undeliberate human activities, damage of elements of communication infrastructure or any possible sorts of abuse. When analyzing related literature, one can notice many works concentrating on solutions of anomaly detection in computer network as well as in Smart Grid systems [4] including the last mile of communication network [10]. There exist various solutions for detecting anomalies in wireless sensor networks, intelligent measure networks or PLC infrastructure [11, 12]. They usually concentrate on protecting communication in last mile network.

The problems of anomaly detection are also noticeable in different smart-solutions concerning environment protection. In [13] authors propose a completely data-driven and machine-learning-based approach for water demand forecasting and online anomaly detection (such as smart meter faults, frauds or possible cyber-physical attacks) through the comparison between forecast and actual values. Another solution to the problem are methods for finding anomalies in gas consumption that can identify causes of wasting energy, presented in [14].

Power Line Communication technology so far was used mainly in smart lights and smart metering systems [5, 7] where the requirements regarding safety and speed of operation are much smaller than in railway critical infrastructure application [15]. In the article we propose original solution for control, maintenance and security of railway signs critical infrastructure.

So far, new LED based signalization devices usually replace old signalization without additional functionalities for control and maintenance. Railway automation systems control only level of current consumption for such devices and detect only on/off and failure states of a signalization device. Such solutions are provided today by main suppliers of railway automation systems, like Bombardier Transportation [16]. Even in computer based railway, automation systems' state of railway signs is controlled by means of current level measurement by microcontroller cards dedicated for a given railway sign circuit.

That is why we have noticed the need for proposing and implementing solution of LED railway signs with new control and maintenance functionalities. We can control the state of the railway sign and transmit by means of Power Line Communication technology packets between LED sign and LED sign controller interface maintenance information to railway automation system without additional investments in cable infrastructure.

In the article we propose solution for improving safety and maintenance functionalities in the network consisting of the proposed railway PLC LED signs.

3 Maintenance and Security System: The Proposed Solution

Traffic from railway LED signs is transmitted by means of PLC point to point links between a LED sign controller and a LED sign. We collected cumulative PLC traffic from point to point links with the use of LED signs controllers interfaces (e.g. RS232, RS485, Ethernet depending on installation). There are two main steps in the proposed method (see Fig. 1).

In the first step we calculated railway PLC signs traffic models for cumulative traffic of railway signs. At the beginning we select and calculate traffic in a form of univariate time series of PLC traffic features presented in Table 1. Next, all the outlying observations are detected and eliminated from the analyzed network traffic parameters by means of the Cook's distance (see Sect. 3.1). Subsequently, traffic features' time series are used for neural network autoregression learning (see Sect. 3.2). Based on neural network prediction intervals and Bollinger bands, we achieve models of variability for every railway PLC sign traffic feature (see Sect. 3.3).

Second branch of the proposed method consist of real time steps for railway LED signs anomaly/attack detection method. First, we select and calculate traffic features from cumulative traffic of railway PLC signs. Next, we check if every value of univariate time series representing traffic feature does not exceed boundaries represented by calculated models in the first step of the proposed method. If values are outside boundaries set for a given traffic feature, we generate detection report. The proposed methodology has also possibility of traffic model recalculation in case of significant changes in traffic characteristic of the examined network. Condition of models recalculation/update is presented in Sect. 3.4.

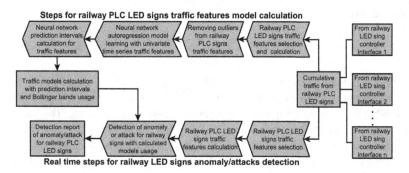

Fig. 1. Block representation of main steps in the proposed algorithm for anomaly/attack detection in railway PLC signs network.

3.1 Outliers' Detection and Elimination Based On Cook's Distance

The Cook's Distance [17] was chosen to recognize outliers in the examined PLC traffic parameters. By means of this approach we calculate the distance stating the level of data corresponding for two models: *(i)* a full model with all observations from the learning set, and *(ii)* a model lacking one observation *i* from its data set

$$D_i = \frac{\sum_{j=1}^{n} \left(\hat{Y}_j - \hat{Y}_{j(i)} \right)^2}{m \cdot MSE}, \tag{1}$$

where $\hat{Y}_j$ is the forecasted value of x variable for observations number j in the full model, i.e. built on the complete learning set; $\hat{Y}_{j(i)}$ is the predicted value of x variable for observations number j in the model built on the set in which the i - number observation was temporarily deactivated, MSE is the mean-model error, and m is the number of parameters used in the analyzed model.

For the Cook's distance D_i threshold value, over which the given observation should be understood as an outlier, according to criterion (1), 1 is accepted, or alternatively $4/(n - m - 2)$, where n is the number of observations in the learning set. The above rules are performed in order to detect and eliminate outliers from the PLC network traffic parameters. So prepared data is ready for stage of creating models.

3.2 The PLC Traffic Features Forecasting Using Neural Networks

The nonlinear autoregressive model of order p, $NAR(p)$, defined as

$$z_t = h\left(z_{t-1}, \ldots, z_{t-p} \right) + \in_t \tag{2}$$

is a direct generalization of linear AR model, where $h(\cdot)$ is a nonlinear known function [18]. It is presumed that $\{\in_t\}$ is a sequence of random independent variables identically distributed with zero mean and finite variance σ^2. The autoregressive neural network (NNAR) is a feedforward network and constitutes a nonlinear approximation $h(\cdot)$, which is defined as

$$\widehat{z}_t = \widehat{h}(z_{t-1}, \ldots, z_{t-p}), \; \widehat{z}_t = \beta_0 + \sum_{i=1}^{I} \beta_i f\left(\alpha_i + \sum_{j=1}^{P} \omega_{ij} z_{t-j}\right), \tag{3}$$

where $f(\cdot)$ is the activation function, and $\Theta = (\beta_0, \ldots, \beta_I, \alpha_1, \ldots, \alpha_I, \omega_{11}, \ldots, \omega_{IP})$ is the parameters vector, p denotes the number of neurons in the hidden layers [18].

The NNAR model is a parametric non-linear model of forecasting. The process of forecasting is conducted in two steps. In the first stage, we determine the auto-regression order for the examined time series. It indicates the number of former values on which the current values of time series depend. In the second stage, we train the NN by means of the set previously prepared with order of auto-regression. Next, we determine the total of input nodes in the auto-regression order, the inputs to the NN being the former, lagged observations in forecasting of univariate time series. Finally, the forecasted values constitute the NN model's output. There are two possibilities to check for hidden nodes, namely, trial-and-error and experimentation, as there is no constituted theoretical ground for their selection. It is crucial though that the number of iterations is correct not to meet the issue of over-fitting [19].

3.3 Estimation of the Forecast Variability Based on Bollinger Bands

Bollinger's Bands is a tool of technical analysis invented at the beginning of 1980-ties [20]. The main idea of this tool is the condition that when variability of data is low (their standard variation is decreasing) then the bands are shrinking. On the other hand, in case of increase of data changeability, the bands are expanding. Therefore, this tool presents dynamics of data variation in a given time window. In the presented solution, we used the Bollinger's Bands to estimate changeability of forecasts of the used models. As the middle band (not presented in the pictures) we accepted the calculated values of used models' forecasts, and with upper and lower bands we tied their double standard variation [21].

3.4 The Condition of Neural Network Model's Update

It is highly likely that the character and nature of the examined parameters of the railway LED Sign Communication Network imply possibility of appearance of significant data variabilities in the analyzed time series. The reasons of such phenomenon are to be found in possible changes in the communication infrastructure (ageing of devices, exchange into new/different models, or extension/modification of already existing infrastructure). Therefore, the following statistical condition can be formulated, fulfilling of which should cause launching of the recurrent learning procedure of the neural network

$$x_i \notin (\mu - 3\sigma, \mu + 3\sigma), \ i = 1, 2, \ldots, n, \tag{4}$$

where $\{x_1, x_2, \ldots, x_n\}$ is time series limited by n elements' analysis window, μ is mean estimated from forecasts of the neural network in the analysis window, and σ is standard deviation of elements of the examined time series in reference to such mean.

4 Experimental Results

The proposed new solution of railway LED Sign Communicating through PLC link is implemented to work with existing solutions of rail automation systems [16]. A PLC sign controller can work with classic analog systems and computer based systems. The novelty of the proposed solutions comes from the fact of existence of digital transmission with actual signaling cables to signs used for railway traffic control. In existing analog or computer based rail automation system the interface is analog (the state of device is controlled by level of current consumption [16]. In this article we put emphasis on security and maintenance issues of the proposed solution. In Fig. 2 we presented placing of our PLC controlled LED signs in typical part of rail automation system responsible for control and maintenance of railway signs. Every sign controller is connected to a dedicated interface responsible for a given LED sign.

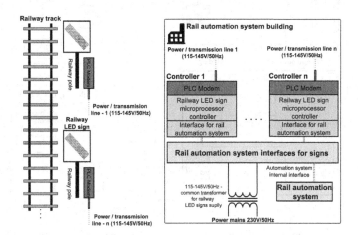

Fig. 2. Place of PLC controlled railway LED signs in typical rail automation system responsible for signs control and maintenance.

The railway sign is mounted on a railway pole on the one side of railway track. Communication between the sign controller and the LED sign is performed through standard signalization cable so the proposed solution can be implemented without big investments in new cable infrastructure. In Fig. 3 we can see internal block scheme of railway sign controller, LED signs and transmission links between signs and controllers. Every pair of a sign and a controller is connected by a PLC communication link where typical distance is approximately 1 km. Every sign controller is supplied from common power source (transformer). Between a sign controller and a LED sign, the packets are transmitted through point to point link. Every point-to-point link is separated by proposed by authors PLC transmission separation filter. Such a filter isolates transmission of PLC packets in common medium (signaling cable) for a given point-to-point link and avoids to reach packets from one point-to-point link to another point-to-point links connected to the same common medium (see Fig. 3). Separation of transmission is necessary to ensure safety and reliability in signs critical infrastructure. Sign's controller may be equipped with different communication interfaces (e.g. relay, RS232, RS485, CAN etc.) depending on railway automation system type. A sign controller and LED sign is constructed in order to meet highest Safety Integrity Level 4 (SIL4) standard [15].

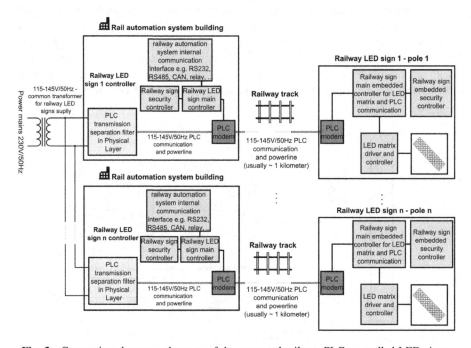

Fig. 3. Connections between elements of the proposed railway PLC controlled LED signs.

For analysis of PLC railway signs traffic's anomaly and attack detection we captured traffic features that are connected to network features (data link and network layers) from Table 1 and for maintenance purposes (see Table 2). Traffic features are processed into form of univariate time series where every sample arrives in constant period of time. After traffic features selection and calculation time series are used for neural network auto-regression models' learning.

Fig. 4. Part of experimental testbed used for measurements and methodology verification representing connections between main elements of railway PLC signs together with power and transmission line (1 km length).

Part of experimental test bed used for the proposed methodology verification is presented in Fig. 4. We can see there an example set of devices communicating through signaling cable where we can set point-to-point link from 1 to 5 km.

Table 1. Railway PLC signs network features.

Traffic feature	Railway PLC sign feature description
SF1P	RSSIP: received signal strength indication for PLC transmission in [dBm] for a given railway LED sign point to point link
SF2P	SNRP: signal-to-noise ratio in [dBu] for a given railway LED sign point to point link
SF3P	NPRP: number of packet retransmissions per time interval for a given railway LED sign point to point link
SF4P	PERP: packet error rate per time interval in [%] for a given railway LED sign point to point link
SF5P	CNPTP: Cumulative number of packets per time interval for railway signs point to point links
SF6P	ACKPRP: Number of acknowledgements of proper packet receiving and configuration sent by LED controller for a given point to point sign link
SF7P	SPSP: Number of status packet sent by LED sign for a given railway LED sign point to point link

We gathered traffic features connected with physical PLC signal parameters like SF1P (RSSIP: received signal strength indication for PLC), SF2P (SNRP: signal-to-noise ratio) and features related to transmission protocol e.g. SF3P (NPRP: number of packet retransmissions) or SF6P (ACKPRP: Number of acknowledgements of proper packet receiving and configuration).

Table 2. Railway PLC signs maintenance features.

Traffic feature	Railway PLC sign feature description
MF1P	LEDP: Health status of railway sign LED matrix for a given railway LED sign point-to-point link
MF2P	RSLLP: Railway LED sign luminosity level in [%] (changes from 0–100%) for a given railway LED sign
MF3P	MTP: Temperatures of PLC modems communicating through point-to-point link
MF4P	SCCP: Number of safety circuit activation for a given LED sign
MF5P	TOT: Railway LED sign total operation time for a given railway LED sign

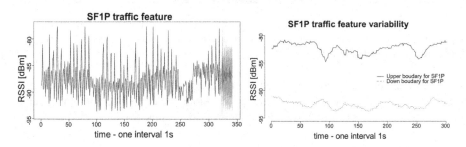

Fig. 5. Railway PLC sign traffic feature SF1P (RSSI [dBm]) neural network prediction intervals (20 samples horizon) with narrower (80%) and wider (90%) prediction intervals (on the left) and SF1P traffic feature variability used for model calculation (on the right).

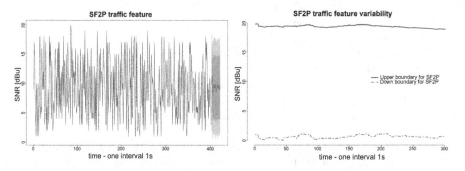

Fig. 6. Railway PLC sign traffic feature SF2P (SNR [dBu]) neural network prediction intervals (20 samples horizon) with narrower (80%) and wider (90%) prediction intervals (on the left) and SF2P traffic feature variability used for model calculation (on the right).

Maintenance traffic features from Table 2 are mainly used by railway automation system staff to assess railway LED signs condition and to plan mandatory technical inspections. Maintenance features are connected to status information sent by LED sign in packet payload to a sign controller. As an example, we can mention MF1P

(LEDP: Health status of railway sign LED matrix for a given railway LED sign) where information about a broken LED is transmitted (number of shortened LED and number of opened LED).

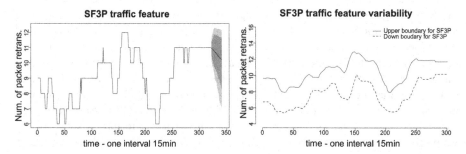

Fig. 7. Railway PLC sign traffic feature SF3P (number of packet retransmissions) neural network prediction intervals (20 samples horizon) with narrower (80%) and wider (90%) prediction intervals (on the left) and SF3P traffic feature variability used for model calculation (on the right).

Graphics representations of exemplary traffic features are presented in Fig. 5, 6 and Fig. 7. An instance of SF1P-SF3P traffic features captured from examined railway signs network are presented in a form of univariate time series (see left sides of Fig. 5, 6 and 7). We can also see there examples of 20 samples prediction intervals calculated by neural network auto-regression model in order to assess how this type of neural network manages with examined signals (time series). Two intervals represent 80% (narrower) and 90% (wider) prediction intervals. Proposed maintenance and security system is complementary to functions available inherently in railway automation system. We have to mention here that our solution is an advisory element for maintenance staff but for physical on/off operations is responsible railway automation system. First of all our additional system can't be responsible for increasing number of railway traffic stops and in consequence substantial economic losses caused by alarms from our systems. That's why prediction intervals from our traffic model have to be wide enough even in unusual situations caused by testing scenarios TS1–TS3. In standard work condition of our PLC signs network our anomaly detection system won't trigger an alarm. For every traffic feature we used these prediction intervals in order to achieve variability of traffic feature for a given traffic feature. In order to evaluate neural network prediction accuracy we presented in Table 3 Root Mean Square Error

Table 3. RMSE and MAE calculated for 10 sample prediction intervals for SF1P, SF2P and SF3P traffic features.

Neural Network	RMSE SF1P	MAE SF1P	RMSE SF2P	MAE SF2P	RMSE SF3P	MAE SF3P
NNAR	4.28	33.53	4.81	37.04	2.32	21.98

(RMSE) and Mean Absolute Error (MAE) parameters calculated with the use of 10 sample prediction intervals. Values were calculated for SF1P–SF3P traffic features. In the next step we calculate Bollinger bands (see Sect. 3.3) in order to achieve upper and down boundary values for a given traffic feature.

Variability for SF1P – SF3P traffic features are presented on the right sides of Fig. 5, 6 and 7. The final PLC railway signs traffic model can be generated based on one on more traffic feature variability time series depending on observation period of time. When more than one set of Bollinger bands is calculated for given period of time for a given traffic feature then final boundaries are calculated as a time series representing maximum or minimum values for higher or down boundary respectively (see Sect. 3.3). Variabilities of traffic features represent models of traffic feature behavior in our case. Online steps of our algorithm start with railway traffic features selection and calculation. Selected values of time series for a given traffic feature is subsequently compared to calculated models representing traffic features variability. If value of traffic feature exceeds boundaries set by calculated model then we generate detection report for a given railway sign and traffic feature. We also propose condition for models recalculation (see Sect. 3.4). Models recalculation is necessary in case of significant change of railway signs PLC traffic behavior or changes in physical structure of the examined network. Without models recalculation false positive values would rise to unacceptable levels. In order to evaluate anomaly/attack detection solution we proposed subsequent Testing Scenarios TS1–TS3:

- **TS1**

  First testing methodology requires generation of disturbance signals by means of equipment used for Electromagnetic Compatibility (EMC) conformance tests. Tests were performed by generating for example Electrical Fast Transient (EFT)/Burst disturbance signal according to IEC 61000-4-4 or Radio Frequency disturbances by current injection clamp according to IEC 61000-4-6 standard. Simpler attack may be performed also by connecting capacitor close to PLC modems. These methods are used for attacking Physical layer of PLC signs communication link. As a result of the proposed attacks, we are degrading physical parameters of PLC transmission line. Week parameters of PLC transmission signal has a big impact on communication reliability also in higher layers of PLC communication protocol stack.

- **TS2**

  Next testing scenario is based on connecting additional PLC communication device to railway sign communication link. Fake transmission node with PLC modem generates and transmits random packets. In different mode untrusted device capture arriving PLC packets, change/disturb them and retransmit to railway PLC sign modems. These packets disturb communication process between PLC sign controller and LED sign. Influence of this type of attack can be observed especially for traffic features connected to data link and network layers.

- **TS3**

  Subsequent testing scenario requires adding devices that create untrusted communication tunnel with the use of the same carrier frequency that is used by railway sign modems. One of the fake PLC node captures arriving packets and transmit them to other untrusted device. Another fake communication tunnel has an impact on reliability of communication between PLC sign controller and LED sign. Another way of attack is a replay attack where untrusted device copy the received PLC packets that arrive to its PLC modem and transmit copy of this packet to legitimate PLC modems with certain delay. Abuses described in this scenario have the biggest impact especially on traffic features connected to data link and network layer and may have indirect influence on some maintenance features.

Taking into consideration all simulated attack or anomalies described in Testing Scenarios TS1–TS3, we achieve cumulative results presented in Table 4. Detection rate (DR) changes from 98.22%–90.14%, while false positive (FP) 5.83%–2.82%. The best results were achieved for SF4P (PERP: packet error rate per time interval) and SF2P (SNRP: signal-to-noise ratio in [dBu]). Based on literature analysis and solutions proposed by railway industry [16] we couldn't make straight comparison to similar solution for railway LED signs controlling. Present used interfaces for railway LED signs are based on analog interface to digital railway automation systems [16]. Our solution based on PLC transmission for railway LED signs is our novel proposition to digital control of railway LED signs signalization by existing railway infrastructure (classic signalization cable). From these reasons we can only indirectly compare our solution to anomaly/attack detections systems which utilize PLC transmission as a communication. For anomaly detection class systems (which also utilize PLC transmission) where we try to recognize abuses with unknown behavior signature, false positive values about 5% are treated as acceptable [10, 22]. We have to mention that anomaly detection class systems try to recognize unknown traffic behavior (so called 0 day attacks) on the contrary to the Intrusion Detection Systems (IDS) where patterns of malicious activity are already known. That's why false positive indications from anomaly detection system can be higher than in case of IDS systems.

Table 4. Detection rate and false positive for railway PLC signs network features.

Traffic feature	DR [%]	FP [%]
SF1P	96.20	3.20
SF2P	97.40	2.82
SF3P	95.63	4.35
SF4P	98.22	3.27
SF5P	90.14	5.83
SF6P	95.45	4.65
SF7P	90.27	5.64

There can also be observed some correlations between different types of testing scenarios. For example generation of electromagnetic disturbances or hardware modifications have an impact on testing scenarios TS2 and TS3 by disturbing packet exchange process in data link and network layers by sign controller and LED sign. The same type of coincidences can also be observed between different traffic features. For example, when values of SF2P (SNRP: signal-to-noise ratio in [dBu]) decrease in consequence SF3P (NPRP: number of packet retransmissions per time interval) and SF4P (PERP: packet error rate per time interval) increases. Less obvious and indirect coincidences can be observed for maintenance features from Table 2. Maintenance features are usually used by railway automatic system engineers to assess technical condition of railway signs by analyzing features like MF1P (LEDP: Health status of railway sign LED matrix), MF2P (RSLLP: Railway LED sign luminosity level) or MF4P (SCCP: Number of safety circuit activation for a given LED sign) in order to plan service schedule for given signs. For example, coincidence may be observed when SF5P (CNPTP: Cumulative number of packets per time interval) increases then MF3P (MTP: Temperatures of PLC modems communicating through point to point link) also rises.

5 Conclusions

Continuous monitoring of resources and systems of critical infrastructures in order to ensure proper level of security and protection is currently a field of intense research. It is apparent that due to their nature, rail marking systems, especially those based on PLC technology, are susceptible to a great number of threats originating both inside and outside their own infrastructure. Significant problems connected to their safety are caused by attacks with increasingly great range and complexity level, as well as failures and damage of communication infrastructure elements. Most often implemented solutions which are supposed to ensure adequate level of security and protection are methods of detection and classification which allow to identify untypical behaviors reflected in the analyzed network traffic. In the present work, there were provided proposals of a system allowing to detect different types of anomalies and failures/damage in critical infrastructure of rail transport realized with the use of PLC technology. The structure and features of the examined LED Sign Communication Network were described. Furthermore, key aspects of security and system maintenance were analyzed, which influence correct operation of the critical communication infrastructure. There were also performed numerous experiments which confirmed effectiveness and efficiency of the proposed solution. We evaluated proposed solution by means of real world railway LED signs test bed. We analyzed 7 network features and 5 maintenance features in order to detect anomaly or attack in network of LED signs. Achieved DR changes from 98.22%–90.14%, while FP 5.83%–2.82%.

References

1. An, S., Lee, B., Shin, D.: A survey of intelligent transportation systems. In: Proceedings of the 3rd International Conference on Computational Intelligence, Communication Systems and Networks, pp. 332–337 (2011)
2. Qureshi, K., Abdullah, A.: A survey on intelligent transportation systems. Middle East J. Sci. Res. 15(5), 629–642 (2013)
3. Fadlil, J., Pao, H.K., Lee, Y.J.: Anomaly detection on ITS data via view association. In: Proceedings of the ACM SIGKDD Workshop on Outlier Detection and Description, pp. 22–30 (2013)

4. Rossi, B., Chren, S., Buhnova, B., Pitner, T.: Anomaly detection in smart grid data: an experience report. In: Proceedings of the 2016 IEEE International Conference on Systems, Man, and Cybernetics, pp. 9–12 (2016)
5. Lloret, J., Tomas, J., Canovas, A.: An integrated IoT architecture for smart metering. IEEE Commun. Mag. 54(12), 50–57 (2016)
6. Daely, L., Red, P.T., Satrya, H.T., Kim, J.W., Shin, S.Y.: Design of smart LED streetlight system for smart city with web-based management system. IEEE Sens. J. 17(18), 6100–6110 (2017)
7. Mahoor, M., Salmasi, F.R., Najafabadi, T.A.: A hierarchical smart street lighting system with brute-force energy optimization. IEEE Sens. J. 17(9), 2871–2879 (2017)
8. Garcia-Font, V., Garrigues, C., Rifà-Pous, H.: Attack classification schema for smart city WSNs. Sensors 17(4), 1–24 (2017)
9. Leccese, F., Cagnetti, M., Trinca, D.: A smart city application: a fully controlled street lighting isle based on Raspberry-Pi card, a ZigBee sensor network and WiMAX. Sensors 14(12), 24408–24424 (2014)
10. Xie, M., Han, S., Tian, B., Parvin, S.: Anomaly detection in wireless sensor networks: a survey. J. Netw. Comput. Appl. 34(4), 1302–1325 (2011)
11. Rajasegarar, S., Leckie, C., Palaniswami, M.: Anomaly detection in wireless sensor networks. IEEE Wirel. Commun. Mag. 15(4), 34–40 (2008)
12. Yau, K., Chow, K.P., Yiu, S.M., Chan, C.F.: Detecting anomalous behavior of PLC using semi-supervised machine learning. In: Proceedings of the 2017 IEEE Conference on Communications and Network Security, pp. 580–585 (2017)
13. Candelieri, A.: Clustering and support vector regression for water demand forecasting and anomaly detection. Sensors 9(3), 1–19 (2017)
14. De Nadai, M., Someren, M.: Short-term anomaly detection in gas consumption through ARIMA and artificial neural network forecast. In: Proceedings of the 2015 IEEE Workshop on Environmental, Energy and Structural Monitoring Systems, pp. 250–255 (2015)
15. IEC 61508: Functional safety of electrical/electronic/programmable electronic safety-related systems (2010). https://www.iec.ch/functionalsafety/. Accessed 10 Jan 2020
16. Bombardier Transportation. https://www.bombardier.com/en/transportation.html. Accessed 12 Jan 2020
17. Cook, R.D.: Detection of influential observations in linear regression. Technometrics 19(1), 15–18 (1977)
18. Cogollo, M.R., Velasquez, J.D.: Are neural networks able to forecast nonlinear time series with moving average components? IEEE Lat. Am. Trans. 13(7), 2292–2300 (2015)
19. Zhang, G.P., Patuwo, B.E., Hu, M.Y.: A simulation study of artificial neural networks for nonlinear time series forecasting. Comput. Oper. Res. 28, 381–396 (2001)
20. Bollinger, J.: Bollinger on Bollinger Bands. McGraw Hill (2002)
21. Vervoort, S.: Smoothing the Bollinger bands. Tech. Anal. Stocks Commod. 28(6), 40–44 (2010)
22. Garcia-Font, V., Garrigues, C., Rifà-Pous, H.: A comparative study of anomaly detection techniques for smart city wireless sensor networks. Sensors 16(6), 868 (2016)

Fingerprinting of URL Logs: Continuous User Authentication from Behavioural Patterns

Jakub Nowak[1], Taras Holotyak[2], Marcin Korytkowski[1],
Rafał Scherer[1]([⊠]), and Slava Voloshynovskiy[2]

[1] Częstochowa University of Technology, Al. Armii Krajowej 36,
42-200 Częstochowa, Poland
{jakub.nowak,marcin.korytkowski,rafal.scherer}@pcz.pl
[2] Department of Computer Science, University of Geneva, Geneva, Switzerland
svolos@unige.ch

Abstract. Security of computer systems is now a critical and evolving issue. Current trends try to use behavioural biometrics for continuous authorization. Our work is intended to strengthen network user authentication by a software interaction analysis. In our research, we use HTTP request (URLs) logs that network administrators collect. We use a set of full-convolutional autoencoders and one authentication (one-class) convolutional neural network. The proposed method copes with extensive data from many users and allows to add new users in the future. Moreover, the system works in a real-time manner, and the proposed deep learning framework can use other user behaviour- and software interaction-related features.

Keywords: URL logs · Computer networks · Continuous authentication · Behavioural biometrics · Software interaction · Autoencoder · Convolutional

1 Introduction

For the past twenty years, the Internet and its utilisation have grown at an explosive rate. Moreover, for several years computer network users have been using various devices, not only personal computers. We also have to manage with many appliances being constantly online and small Internet of Things devices. Efficient computer network intrusion detection and user profiling are substantial for providing computer system security. Along with the proliferation of online devices, we witness more sophisticated security threats. It is possible to enumerate many ways to harm networks, starting from password weakness. Malicious software can be illicitly installed on devices inside the network to cause harm, steal information or to perform large tasks. Another source of weakness can be Bring Your Own Device schemes, where such devices can be infected outside

V. V. Krzhizhanovskaya et al. (Eds.): ICCS 2020, LNCS 12140, pp. 184–195, 2020.
https://doi.org/10.1007/978-3-030-50423-6_14

the infrastructure. At last, social engineering can be used to acquire access to corporate resources and data.

Each network user leaves traces, some of them are generated directly by the user, e.g. on social networks, others are closely related to the computer network mechanisms. Thanks to network traffic-filtering devices, network administrators nowadays have an enormous amount of data related to network traffic at their disposal. Authorising users based on their behaviour can be done in many ways, depending on the available data and methods. Identification can be based on facial features [15,17] or based on spoken instructions [2]. In [3,11] data from smartphone sensors were used to analyse user behaviour. Similarly, the way how the user unlocks the smartphone [4] can be explored, and based on the collected data, they show the uniqueness of using the phone. This is related to certain user preferences, habits as well as to the physical conditions of individual users, i.e. the way the phone is held. Another option is to authenticate the user with the signature [16]; in this solution, a signature is not only analysed as an image but also the dynamics of the signature creation using a haptic sensor.

In our solution, we test whether the logged-in user has access to a given resource and does not impersonate someone else by breaking initial security measures based on, e.g., a password. Our research can be used in new generation firewall devices working in layer 7 of the OSI model however, in our case, the security rules will be based on the analysis of the pages visited. Our method provides a continuous authentication based on software interaction patterns.

Last years brought learned semantic hashes to information retrieval. Semantic hashing [13] aims at generating compact vectors which values reflect semantic content of the objects. Thus, to retrieve similar objects, we can search for similar hashes which is much faster and takes much less memory than operating directly on the objects. The term was coined in [13]. They used for the first time a multilayer neural network to generate hashes and showed that semantic hashing obtains much better results than Latent Semantic Analysis (LSA) [7] used earlier. A similar method for generating hashes can be using the HashGAN network [5]; this solution is based on generative adversarial networks [8].

In the presented solution, we use autoencoders to create semantic compact hashes for the behaviour of computer network users from their URL request sequences. Our approach is sparked by the aforementioned studies that use hashes to analyze data, especially in NLP. After training the autoencoders, we use the encoder parts to generate hashes and fed them to the input of a one-class convolutional network that performs the final user authentication (Architecture 2). Schematic diagram of the system located in the computer network infrastructure is presented in Fig. 1. We also propose two smaller systems (Architecture 1 and 3) with worse accuracy. Through this research, we highlight the following features and contributions of the proposed system.

- We present three different approaches to URL-based computer network user software interaction behavioural continuous authentication. Up to now, the network traffic was usually analysed by some hand-crafted statistics.

– Our work provides new insights, showing that the system of autoencoders and convolutional neural network can be trained efficiently for one-class authentication for nearly any number of users.
– The method can use nearly any kind of data as a features.
– The proposed system is fast and can be used in real-time in various IT scenarios.

The remainder of the paper is organised as follows. In Sect. 2, we discuss the problem of behavioural authenticating users in computer networks. The proposed data representation and three Architectures are presented in Sect. 3. Experiments on real-world data from a large local municipal network, showing accuracy and a comparison of three presented Architectures, are shown in Sect. 4. Finally, conclusions and discussions of the paper are presented in Sect. 5.

2   Problem Formulation

The aim is to create an additional security layer to verify users in IT systems using data collected by computer network administrative infrastructure. The additional authorisation is carried out without the user's knowledge. The proposed system constantly monitors HTTP request patterns for every computer in the network. The requests come from browsing websites or applications sending queries with URL addresses. In other words, the method provides a software interaction-based behavioural biometrics. It should be remembered here that the addresses stored in the firewall logs apply to both pages opened in WWW browsers and programs running in the background, such as anti-virus or operating system updates. This article is based on data collected from a LAN network infrastructure, which is used by residents of four districts in Poland, as well as network users who are employees of the local government offices and their organisational units, e.g. schools, hospitals, etc. Internet access to the analysed network is done with the help of two CISCO ASR edge routers that route packets using RIP version 2. The data for neural network training was collected between June 2017 and January 2018, and for testing in February 2018. The size of raw logs was approximately 460 GB, and 0.9 GB after preprocessing (selecting time, date, user ID and URL).

   Based on the previously collected data, we examine whether an anomaly occurred, which is supposed to indicate a possible attack or use of the account by another user. To solve the problem, we use autoencoders and convolutional networks using various methods of data representation. We divided the task into several stages. The first is to create a session based on registered URL logs from the database. In our case, a session means a set of consecutive, non-repeating URLs for a given user. Each URL address was truncated to 45 characters. This size comes from the average length of addresses and observation of their construction. We have assumed that the most important information is at the beginning of the address. We primarily care about the domain of the website visited, the protocol that was used, and the basic parameters of the GET method. When creating a session, it happened that a query consisting of several URLs was sent

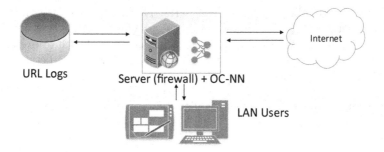

Fig. 1. System location in the computer network infrastructure.

at the same time. In this case, certain sets of addresses were associated in a very specific way; between two identical addresses, another one different from the others was added. To remove duplicate addresses, an additional sorting by address name was used. An example of a set requested at the same time is "address1 address2 address1 address3 address1". We tend rather to have the set "address1 address2 address3". This is an exceptional case; however, omitting increases the authenticating CNN error. The session to be analyzed consists of a minimum of 20 different URLs with a maximum length of 45 characters included in the dictionary. Another limitation was the maximum size of the session to be analyzed. We used up to 200 different URLs for fast neural network operation. The interval between the recorded addresses in one session cannot be longer than 30 min. If this time has been exceeded, successive addresses form a new session.

3   Neural Network Architecture

We scrutinized three neural configurations: Architecture 1 with a convolutional network (CNN) with two-dimensional filters used for text classification, Architecture 2 consisting of one-class CNN with unique autoencoder for every user, and Architecture 3 with one-class CNN network and one autoencoder for all the users.

3.1   Data Representation

A URL is an address that allows locating a website on the Internet. The user encounters it mainly when using a web browser. However, URLs can be requested by applications running in the background such as antiviruses, system updates, etc. Each user's computer uses different applications and at a different time, which allows to even better distinguish them. Very often text is represented by some dictionaries. The construction of the URL has been repeatedly addressed in various articles [1,20]. The majority of the previous works created some handcrafted features based on URL statistics. URLs do not consist regular words; thus using word dictionaries is not viable. In our experiment, we encode the entire address without dividing its subsequent parts. The condition is that the characters that constitute the URL should be in the previously defined set containing 64 unique characters. We were inspired here by Zhang et al. [19] to present text data in the form of a one-hot vector at the character level. The dictionary consisted of the following characters:

```
abcdefghijklmnopqrstuvwxvyz_0123456789
-;.!?:/\| @#$%^& * ~'+=<>()[]
```

If a character was not from the above alphabet it was removed. At the input of the neural network, in addition to the session of addresses used for classification, we also provide user identification data. In our case, we concatenate the user ID data with the URL input session. Therefore, in one input column, we have two values equal to 1, the rest of the rows of one column are filled with zeros. The first one is placed on the position in the range <1, 64> which defines the letter from the dictionary, the second one on the position in the range <65, 119> denoting the user ID, because each user has its unique number. The input size to the CNN network in Architecture 1 was 119 × 4096, where 4096 means the maximum length of the session we can provide at the network input. The coding scheme is presented in Fig. 2.

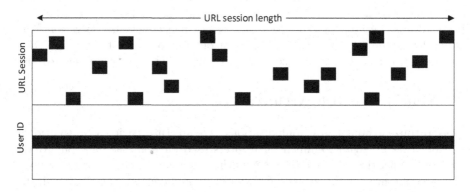

Fig. 2. URL text coding scheme for convolutional networks in Architectures 1–3. The upper part is one-hot character-level text encoding, and the lower part is one-hot user ID encoding.

3.2 Architecture 1

Our first attempt was to use a convolutional neural network with two-dimensional filtering presented in Fig. 3, similar to [10]. The network architecture is as follows:

```
ConvLayer 128 FMs, filter 7x119, stride 3, ReLU
MaxPoolingLayer    3
ConvLayer 256 FMs, filter 5x128, stride 2, ReLU
MaxPoolingLayer    3
ConvLayer 256 FMs, filter 3x256, stride 1, ReLU
MaxPoolingLayer    3
Fully connected 512 + Dropout, ReLU
Fully connected 256 + Dropout, ReLU
Output   2 softmax
```

This method proved to be a weak solution to the problem because the single convolutional neural network had to cope with a highly complex problem and demonstrated a significant classification error. The accuracy of the anomaly recognition exceeded barely 63%, which is slightly higher than the random response and not viable in real-world computer network infrastructure.

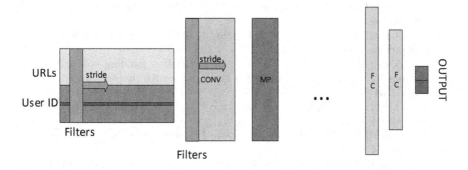

Fig. 3. Convolutional network used in Architecture 1. Detailed meta-parameters are provided in the text. All the filters are large enough to use a one-directional stride. CNNs in Architecture 2 and 3 are similar with different meta-parameters, and the autoencoder latent space instead of the URL session.

3.3   Architecture 2

Architecture 2 was inspired by the one-class neural network [6], and here each user (class) has a different autoencoder. The task of the network is to detect an anomaly in a given class. In our case, we add the user ID to "ask" the network whether given network traffic belongs to this particular user. Initially, we tried to use modified convolutional networks of the U-Net [12] structure without connections between feature maps of the same size (skip connections), what transpired to have an unacceptable training error. In the decoder part of the autoencoder, we implemented a sub-pixel convolutional layer in one dimension inspired by [14]. It changes the size of the input to the convolutional layer by increasing the width of the channels at the expense of their number.

Training data for the autoencoder is created similarly to one sentence in NLP consisting of URLs instead of words. We do not use a separator between URLs; addresses are given as words in a sentence. The structure of the autoencoder for text is different from the structure of the autoencoder for images; here we were inspired somehow by [18]. In our case, the latent space (bottleneck) layer in the autoencoder is 128×64. In the adopted architecture with pooling, each addition of a layer reduces the size of the smallest, latent space layer in the autoencoder. The size of the latent space is a trade-off determined experimentally between the accuracy and the input size to the one-class CNN. The autoencoder structure used in the article is presented in Fig. 4a), and the detailed meta-parameters are as follows:

– Encoding part
  1. ConvLayer1 64 FMs, filters 3×64, stride 1, ReLU + padding + MaxPooling
     Output 64×2048
  2. ConvLayer1 64 FMs, filters 3×64, stride 1, ReLU + padding + MaxPooling
     Output 64×1024
  3. ConvLayer1 64 FMs, filters 3×64, stride 1, ReLU + padding + MaxPooling
     Output 64×512
  4. ConvLayer1 64 FMs, filters 3×64, stride 1, ReLU + padding + MaxPooling
     Output 64×256
  5. ConvLayer1 64 FMs, filters 3×64, stride 1, ReLU + padding + MaxPooling
     Output 64×128 (after training, it is the input for the discriminator)
– Decoding part
  1. ConvLayer1 64 FMs, filters 3×32, stride 1, ReLU + padding + UpPooling
     Output 64×512
  2. ConvLayer1 64 FMs, filters 3×32, stride 1, ReLU + padding + UpPooling
     Output 64×512
  3. ConvLayer1 64 FMs, filters 3×32, stride 1, ReLU + padding + UpPooling
     Output 64×1024
  4. ConvLayer1 64 FMs, filters 3×32, stride 1, ReLU + padding + UpPooling
     Output 64×2048
  5. ConvLayer1 64 FMs, filters 3×32, stride 1, ReLU + padding + UpPooling
  6. ConvLayer1 64 FMs, filters 3×32, stride 1, Sigmoid activation
     Output 64×4096

Our solution also utilizes a discriminator as a convolution network. The problem posed by us was to check whether the recorded session belongs to the user and whether a given set of URLs could be generated by a specific user. It was, therefore, necessary to create a suitable discriminator for the autoencoder. The idea of the discriminator is similar to the GAN network [8]; we only care about assessing the mapping of data in the autoencoder and whether the given set could be created by a specific user.

The user identifier has been moved in relation to Architecture 1 from the system input to the discriminator input, i.e. the last coding (latent space) layer in the autoencoder, the user coding method is identical as in the case of Architecture 1. The autoencoder, in this case, is unique for each user (unlike in Architecture 3). The discriminator input in Architecture 2 and 3 was 183×64, where 183 is made up from 128 (column size with hash from the autoencoder latent space) and 55 (user ID added in the same way in Architecture 1). The value 64 comes from the number of feature maps from the autoencoder. Detailed meta-parameters of the discriminator in Architecture 2 and 3 are as follows:

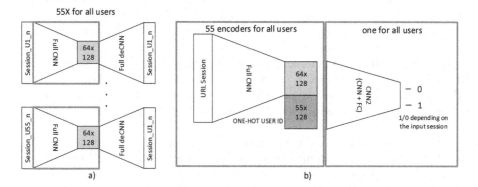

Fig. 4. Architecture 2 consists of many autoencoders and one convolutional network (discriminator). The left part a) is a set of as many autoencoders as the users in the network. The right part b) is the authenticating infrastructure using encoders from a) and the convolutional network with fully connected authentication layer.

- Input from autoencoder + User ID 183 × 64
- ConvLayer 32 FM, filters 5 × 183, stride 2, batch normalization, ReLU
- MaxPoolingLayer 2
- ConvLayer 64 FM, filters 3 × 32, stride 1, batch normalization, ReLU
- MaxPoolingLayer 2
- ConvLayer 128 FM, filters 3 × 64, stride 1, batch normalization, ReLU
- MaxPoolingLayer 2
- Fully connected 512 + dropout 0.5, ReLU
- Fully connected 128 + dropout 0.5, ReLU
- Output 2 softmax

3.4 Architecture 3

We combined the two previous frameworks to create something in between in terms of size and complexity. Creating separate autoencoders for each user is somehow problematic in terms of logistics, where it is easy to make a mistake in processing sessions for the user. Here we use one autoencoder as a uniform way of representing URLs for all users. The size of the autoencoder is the same as in the previous Architecture 2. This solution improved the results of the CNN from Architecture 1; however, it was worse than Architecture 2 with user-wise unique autoencoders.

4 Experiments

We performed experiments on a database with logs of visited URLs for 55 users. In the case of our database, the most active user had 6,137 training sessions, and the least active user had 440 URL sessions, which was about 14 times less (details are presented in Table 3. In training all Architectures 1–3, we had to generate illegitimate user sessions for the training purposes. We did not decide to create synthetic user sessions because this is a challenging issue, and it could result in generating data different from the existing distribution. In our research, the data that the discriminator has to evaluate negatively was created based on existing sessions. However, we provide them to the network as if they belonged to another user. These sessions are randomly selected from among the entire dataset.

The cross-entropy with softmax loss function from the CNTK package was used to train CNNs (discriminators). The training coefficient for CNNs was taken on as follows: 0.0008 for 5 epochs, 0.0002 for 10, 0.0001 for 10, 0.00005 for 10 epochs, and 0.00001 from then on to a maximum of 300 learning epochs.

We used the binary cross-entropy loss function for training all the autoencoders. The best universal effects for each user were obtained when the learning coefficient was 0.0001 for the first two epochs then 0.00001 for 200 epochs. We trained all the architectures with the momentum stochastic gradient descent (SGD) algorithm with momentum set at 0.9 for both autoencoder and CNN.

To assess the accuracy of the autoencoder (Table 2), the Sørensen similarity coefficient $QS = \frac{2C}{A+B}$ was used, where A and B are compared elements, in our case the output and input to the autoencoder and C is the number of elements common for both layers. After each convolutional layer we used Batch Normalization [9] with RELU activation function. The results are summarized in Table 1.

Table 1. Overall accuracy of the three presented Architectures

Architecture	Accuracy
1. CNN	61.00 %
2. OC and 1 autoencoder	68.00 %
3. OC and 55 autoencoders	84.24 %

The best solution turned out to be Architecture 2 using dedicated autoencoders and one discriminator. The detailed results for each user are presented in Table 3. The number of training sessions for a given user does not affect accuracy.

During the implementation of neural networks, the only limitation turned out to be the available GPU memory. We used four Nvidia GTX 1080 Ti graphics cards with 11 GB of memory. In the case of training autoencoders, we could divide each task into four graphics cards without the need for special techniques enabling multiprocessing on multiple graphics cards. Autoencoder networks can

Table 2. Accuracy of the encoders trained on user URL sessions. After training, the encoder parts are fed to the one-class CNN (discriminator).

User ID	1	2	3	4	5	6	7	8	9	10
Accuracy	87.4%	83.4%	87.5%	83.3%	83.5%	89.8%	85.9%	79.0%	87.3%	87.3%
User ID	11	12	13	14	15	16	17	18	19	20
Accuracy	87.1%	88.1%	89.0%	77.5%	85.0%	87.0%	86.2%	83.5%	85.1%	86.7%
User ID	21	22	23	24	25	26	27	28	29	30
Accuracy	83.4%	87.5%	85.9%	83.2%	75.9%	82.9%	82.5%	84.1%	86.3%	80.5%
User ID	31	32	33	34	35	36	37	38	39	40
Accuracy	77.1%	79.8%	80.4%	77.9%	80.9%	90.8%	72.8%	87.4%	79.9%	88.9%
User ID	41	42	43	44	45	46	47	48	49	50
Accuracy	86.2%	84.7%	78.4%	86.8%	81.9%	87.3%	91.6%	77.8%	79.2%	80.5%
User ID	51	52	53	54	55					
Accuracy	73.9%	74.9%	82.0%	88.5%	85.4%					

Table 3. Number of training and testing sessions, and accuracy of detecting anomalies by the global, one-class CNN for every computer network user.

User ID	1	2	3	4	5	6	7	8	9	10
Training	6137	5199	3184	2892	2434	2368	2120	1928	1841	1674
Testing	1534	1299	795	723	608	591	530	481	460	418
Accuracy [%]	93.82	85.84	82.92	70.59	80.13	93.06	78.05	84.98	83.93	89.24
User ID	11	12	13	14	15	16	17	18	19	20
Training	1663	1650	1266	1236	1206	1181	1168	1091	1047	987
Testing	415	412	316	308	301	295	292	272	261	246
Accuracy [%]	73.67	90.46	84.22	83.08	90.92	80.85	83.44	90.97	87.54	88.80
User ID	21	22	23	24	25	26	27	28	29	30
Training	984	969	948	920	920	840	835	817	810	731
Testing	246	242	237	229	230	209	208	204	202	182
Accuracy [%]	82.96	83.26	81.97	71.10	80.37	79.10	80.62	87.08	87.60	79.19
User ID	31	32	33	34	35	36	37	38	39	40
Training	715	705	705	703	686	670	664	643	1303	604
Testing	178	176	176	175	171	167	166	160	325	151
Accuracy [%]	79.04	88.11	74.16	94.25	76.32	86.47	92.65	84.94	72.86	87.57
User ID	41	42	43	44	45	46	47	48	49	50
Training	588	579	573	567	544	500	479	474	470	446
Testing	147	144	143	141	136	125	119	118	117	111
Accuracy [%]	86.79	80.73	86.93	84.86	86.09	91.97	92.07	86.35	88.92	85.18
User ID	51	52	53	54	55					
Training	440	621	637	885	1267					
Testing	110	155	159	221	316					
Accuracy [%]	81.53	76.91	80.47	87.48	85.23					

be trained independently of each other. To speed up the learning of the discriminator, we use data from the autoencoder training instead of computing the encoder output again. Otherwise, only 40 users could be trained on the aforementioned equipment without using this technique. To train more users, a machine with more GPU memory is required. This limitation, however, was not valid after training and if we had a machine with more memory only for training, it would be possible to use the trained system on the equipment we had at our disposal. Another indicator is the number of URL sessions processed by the system in a specific time. Using the above-mentioned GPU, we are able to process a set (mini-batch) of 200 sessions in 36.8 ms per package, which shows that the system can be used in real-time scenarios. This is the number of sessions that can be loaded at once on a GPU with 11 GB memory.

5   Conclusion

Our system based on autoencoders and one-class CNN, is a new approach to system security and anomaly detection in user behaviour. It provides continuous authentication of computer network users by software interaction analysis. We used real text data from the network traffic instead of hand-crafted traffic statistics as in the case of the previous approaches. Moreover, the proposed framework is a universal anomaly detection system applied in the paper for user authentication. We proposed three architectures that differ in size and complexity. The best architecture presented in the paper allows to add and remove any user without having to retrain the whole system. Thanks to this, we can save both time and computational resources. The presented solution can be used for other behavioural security solutions to create user profiles utilizing other available data. Future research would involve alternate methods to create autoencoders to improve accuracy. In the presented article, we used the same way of training the autoencoder for each user. To obtain better results, it would be beneficial to create a dedicated autoencoder architecture for each user, but this would involve a change in the implementation of the discriminator.

Acknowledgement. The project financed under the program of the Polish Minister of Science and Higher Education under the name "Regional Initiative of Excellence" in the years 2019–2022 project number 020/RID/2018/19, the amount of financing 12,000,000.00 PLN.

References

1. Blum, A., Wardman, B., Solorio, T., Warner, G.: Lexical feature based phishing URL detection using online learning. In: Proceedings of the 3rd ACM Workshop on Artificial Intelligence and Security, pp. 54–60. ACM (2010)
2. Boles, A., Rad, P.: Voice biometrics: deep learning-based voiceprint authentication system. In: 12th System of Systems Engineering Conference (SoSE), pp. 1–6, June 2017

3. Buriro, A., Crispo, B., Delfrari, F., Wrona, K.: Hold and sign: a novel behavioral biometrics for smartphone user authentication. In: IEEE Security and Privacy Workshops (SPW), pp. 276–285, May 2016
4. Buriro, A., Crispo, B., Conti, M.: AnswerAuth: a bimodal behavioral biometric-based user authentication scheme for smartphones. J. Inf. Secur. Appl. **44**, 89–103 (2019)
5. Cao, Y., Liu, B., Long, M., Wang, J.: HashGan: deep learning to hash with pair conditional Wasserstein GAN. In: Proceedings of the IEEE Conference on Computer Vision and Pattern Recognition, pp. 1287–1296 (2018)
6. Chalapathy, R., Menon, A.K., Chawla, S.: Anomaly detection using one-class neural networks. arXiv preprint arXiv:1802.06360 (2018)
7. Deerwester, S., Dumais, S.T., Furnas, G.W., Landauer, T.K., Harshman, R.: Indexing by latent semantic analysis. J. Am. Soc. Inform. Sci. **41**(6), 391–407 (1990)
8. Goodfellow, I., et al.: Generative adversarial nets. In: Advances in Neural Information Processing Systems, pp. 2672–2680 (2014)
9. Ioffe, S., Szegedy, C.: Batch normalization: accelerating deep network training by reducing internal covariate shift. arXiv preprint arXiv:1502.03167 (2015)
10. Kwon, D., Natarajan, K., Suh, S.C., Kim, H., Kim, J.: An empirical study on network anomaly detection using convolutional neural networks. In: IEEE 38th International Conference on Distributed Computing Systems (ICDCS), pp. 1595–1598. IEEE (2018)
11. Mahfouz, A., Mahmoud, T.M., Eldin, A.S.: A survey on behavioral biometric authentication on smartphones. J. Inf. Secur. Appl. **37**, 28–37 (2017)
12. Ronneberger, O., Fischer, P., Brox, T.: U-Net: convolutional networks for biomedical image segmentation. In: Navab, N., Hornegger, J., Wells, W.M., Frangi, A.F. (eds.) MICCAI 2015. LNCS, vol. 9351, pp. 234–241. Springer, Cham (2015). https://doi.org/10.1007/978-3-319-24574-4_28
13. Salakhutdinov, R., Hinton, G.: Semantic hashing. Int. J. Approximate Reasoning **50**(7), 969–978 (2009). Special Section on Graphical Models and Information Retrieval
14. Shi, W., et al.: Real-time single image and video super-resolution using an efficient sub-pixel convolutional neural network. In: Proceedings of the IEEE Conference on Computer Vision and Pattern Recognition, pp. 1874–1883 (2016)
15. Sun, Y., Chen, Y., Wang, X., Tang, X.: Deep learning face representation by joint identification-verification. In: Advances in Neural Information Processing Systems, pp. 1988–1996 (2014)
16. Xiao, G., Milanova, M., Xie, M.: Secure behavioral biometric authentication with leap motion. In: 4th International Symposium on Digital Forensic and Security (ISDFS), pp. 112–118, April 2016
17. Zhang, P., You, X., Ou, W., Chen, C.P., Cheung, Y.M.: Sparse discriminative multi-manifold embedding for one-sample face identification. Pattern Recognit. **52**, 249–259 (2016)
18. Zhang, X., LeCun, Y.: Byte-level recursive convolutional auto-encoder for text. arXiv preprint arXiv:1802.01817 (2018)
19. Zhang, X., Zhao, J., LeCun, Y.: Character-level convolutional networks for text classification. In: Proceedings of the 28th International Conference on Neural Information Processing Systems - Volume 1. NIPS 2015, pp. 649–657. MIT Press, Cambridge (2015)
20. Zouina, M., Outtaj, B.: A novel lightweight URL phishing detection system using SVM and similarity index. Hum. Cent. Comput. Inf. Sci. **7**(1), 17 (2017)

On the Impact of Network Data Balancing in Cybersecurity Applications

Marek Pawlicki[1,2](✉), Michał Choraś[1,2], Rafał Kozik[1,2],
and Witold Hołubowicz[2]

[1] ITTI Sp. z o.o., Poznań, Poland
[2] UTP University of Science and Technology, Bydgoszcz, Poland
{Marek.Pawlicki,chorasm}@utp.edu.pl

Abstract. Machine learning methods are now widely used to detect a wide range of cyberattacks. Nevertheless, the commonly used algorithms come with challenges of their own - one of them lies in network dataset characteristics. The dataset should be well-balanced in terms of the number of malicious data samples vs. benign traffic samples to achieve adequate results. When the data is not balanced, numerous machine learning approaches show a tendency to classify minority class samples as majority class samples. Since usually in network traffic data there are significantly fewer malicious samples than benign samples, in this work the problem of learning from imbalanced network traffic data in the cybersecurity domain is addressed. A number of balancing approaches is evaluated along with their impact on different machine learning algorithms.

Keywords: Data imbalance · Machine learning · Classifiers · Cybersecurity

1 Introduction

The importance of cybersecurity rises with every passing year, along with the number of connected individuals and the growing number of devices utilising the Internet for various purposes [1,2]. The antagonistic forces, be it hackers, crackers, state-sponsored cyberforces or a range of other malicious actors employ a variety of methods to cause harm to common users and critical infrastructure alike [3,4]. The massive loads of data transmitted every single second exceeded the human capacity to deal with them long time ago. Thus, a myriad of machine learning (ML) methods were successfully implemented in the domain [5–7]. As rewarding as they are, AI-related approaches come with their own set of problems. One of them is the susceptibility to data imbalance.

The data imbalance problem refers to a situation in which one or multiple classes have significantly more learning samples as compared to the remaining classes. This often results in misclassification of the minority samples by a substantial number of classifiers, a predicament especially pronounced if the minority classes are the ones that bear the greatest importance - like malignant cancer

© Springer Nature Switzerland AG 2020
V. V. Krzhizhanovskaya et al. (Eds.): ICCS 2020, LNCS 12140, pp. 196–210, 2020.
https://doi.org/10.1007/978-3-030-50423-6_15

samples, fraud events, or, as in the case of this work, network intrusions. Additionally, the deterioration of a given model might go unnoticed if the method is only evaluated on the basis of accuracy.

With the significance of the above-mentioned difficulty in plenty of high-stake practical settings, various methods to counter that issue have been proposed. These fall roughly into three categories: undersampling, oversampling and cost-sensitive methods. In this work numerous approaches to dataset balancing are examined, the influence each method has on a number of ML classifiers is highlighted and in conclusion the best experimentally found approach in the case of network intrusion detection is chosen.

The major contribution and the unique value presented in this work comes in the form of highlighting the notion that the impact dataset balancing methods have on the behaviour of ML classifiers is not always a straightforward and intuitive one. A number of balancing approaches is thoroughly evaluated and their impact on both the dataset and the behaviour of classifiers is showcased. All of this in the context of a practical, vital domain that is network intrusion detection.

In the era of big data, undersampling approaches need to be thoroughly researched as their reduced computational cost could become a major benefit in contrast to oversampling methods.

The paper is structured as follows: in Sect. 2 the pipeline of network intrusion detection is illustrated and described, and the ML algorithms utilised are succinctly introduced, in Sect. 3 the chosen balancing methods are characterised.

Table 1. Encoded labels and number of instances in intrusion detection evaluation dataset used in this work (see Sect. 4)

No of training instances	Class label	Encoded label
1459377	BENIGN	0
207112	DoS Hulk	4
142924	PortScan	9
115222	DDoS	2
9264	DoS GoldenEye	3
7141	FTP-Patator	7
5307	SSH-Patator	10
5216	DoS slowloris	6
4949	DoS Slowhttptest	5
2713	Web Attack Brute Force	11
1760	Bot	1
1174	Web Attack XSS	13
38	Web Attack SQL Injection	12
10	Heartbleed	8

Section 4 expresses the experiments undertaken and finally Sect. 5 showcases the obtained results.

2   Machine Learning Approach Enhanced with Data Balancer

The focus of this research lies on the impact the balance of the instance numbers among classes in a dataset has on the performance of ML-based classification methods. In general, the step-by-step process of ML-based Intrusion Detection System (IDS) can be succinctly summarised as follows: a batch of annotated data is used to train a classifier. The algorithm 'fits' to the training data, creating a model. This is followed by testing the performance of the acquired model on the testing set - a batch of unforeseen data. In order to alleviate the data balancing problem present in the utilised IDS dataset an additional step is undertaken before the algorithm is trained (as seen in Fig. 1).

Fig. 1. IDS training pipeline with dataset balancing

The ML-based classifier block of Fig. 1 can be realised by an abundance of different machine learning methods. In fact, recent research showcases numerous novel approaches including deep learning [7,8], ensemble learning [9,10], various augmentations to classical ML algorithms [11] etc. In this work three basic models were chosen to put emphasis on the data balancing part. These are:

– Artificial Neural Network [12,13]
– Random Forest [14]
– Naive Bayes [15]

These represent three significantly different approaches to machine learning and were selected to cover possibly the widest range of effects dataset balancing could have on the effectiveness of ML.

The ANN in use is set up as follows: two hidden layers of 40 neurons, with the Rectified Linear Unit as activation function, and the ADAM optimizer, batch size of 100 and 35 epochs. The setup emerged experimentally.

3   Balancing Methods

In the cases suffering from the data imbalance problem the number of training samples belonging to some classes is larger in contrast to other classes (Table 2).

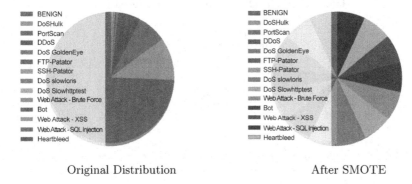

Original Distribution   After SMOTE

Fig. 2. Class distribution in CICIDS 2017 - original unbalanced distribution and after SMOTE

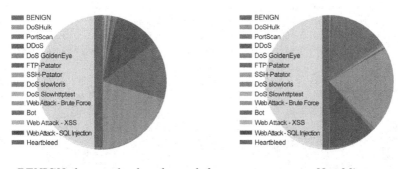

BENIGN class randomly subsampled   NearMiss

Fig. 3. Class distribution in CICIDS 2017 - after performing random undersampling and NearMiss

The conundrum of data imbalance has recently been deeply studied in the area of machine learning and data mining. In numerous cases, this predicament impacts the machine learning algorithms and in result deteriorates the effectiveness of the classifier [16]. Typically in such cases, classifiers will achieve higher predictive accuracy over the majority class, but poorer predictive accuracy over the minority class. In general, solutions to this problem can be categorised as (i) data-related, and (ii) algorithm-related.

In the following paragraphs, these two categories of balancing methods will be briefly introduced. The focus of the analysis was on the practical cybersecurity-related application that faces the data imbalance problem.

3.1 Data-Related Balancing Methods

Two techniques, belonging to this category, that are commonly used to cope with imbalanced data use the principle of acquiring a new dataset out of the

Table 2. CICIDS2017 (full set)/Unbalanced

	ANN ACC: 0.9833			RandomForest ACC: 0.9987			NaiveBayes ACC: 0.2905			Support
	precision	recall	f1-score	precision	recall	f1-score	precision	recall	f1-score	
0	0.99	0.99	0.99	1.00	1.00	1.00	1.00	0.10	0.18	162154
1	0.97	0.35	0.52	0.88	0.68	0.77	0.01	0.65	0.01	196
2	1.00	1.00	1.00	1.00	1.00	1.00	0.94	0.95	0.94	12803
3	0.99	0.97	0.98	1.00	0.99	1.00	0.09	0.93	0.16	1029
4	0.95	0.94	0.94	1.00	1.00	1.00	0.74	0.70	0.72	23012
5	0.89	0.98	0.93	0.96	0.98	0.97	0.00	0.67	0.01	550
6	0.99	0.98	0.99	1.00	0.99	0.99	0.05	0.52	0.09	580
7	0.99	0.98	0.99	1.00	1.00	1.00	0.10	0.99	0.18	794
8	1.00	1.00	1.00	1.00	1.00	1.00	1.00	1.00	1.00	1
9	1.00	1.00	1.00	1.00	1.00	1.00	0.99	0.99	0.99	15880
10	1.00	0.49	0.66	1.00	1.00	1.00	0.08	0.99	0.15	590
11	0.85	0.10	0.17	0.86	0.99	0.92	0.00	0.07	0.00	301
12	0.00	0.00	0.00	1.00	1.00	1.00	0.01	1.00	0.02	4
13	1.00	0.02	0.05	0.95	0.61	0.74	0.08	0.93	0.14	130
Macro avg	0.90	0.70	0.73	0.97	0.95	0.96	0.36	0.75	0.33	218024
Weighted avg	0.98	0.98	0.98	1.00	1.00	1.00	0.95	0.29	0.34	218024

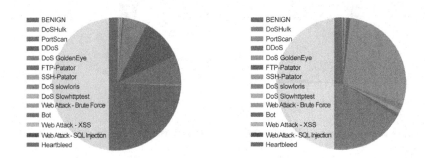

After cleaning the Tomek-links   Cluster-Centers undersampling

Fig. 4. Class distribution in CICIDS 2017 - after cleaning the Tomek-Links and performing ClusterCenters undersampling

existing one. This is realised with data sampling approaches. There are two widely recognised approaches called data over-sampling and under-sampling.

Under-sampling balances the dataset by decreasing the size of the majority class. This method is adopted when the number of elements belonging to the majority class is rather high. In that way, one can keep all the samples belonging to the minority class and randomly (or not) select the same number of elements representing the majority class. In our experiments we tried a number of under-sampling approaches, one of those was **Random Sub-sampling**. The effect random subsampling has on the dataset is illustrated in Fig. 3. The results the method has in conjunction with the selected ML algorithms is showcased in Table 3.

Table 3. CICIDS2017 (full set)/Random subsampling

	ANN ACC: 0.9812			RandomForest ACC: 0.9980			NaiveBayes ACC: 0.2911			Support
	precision	recall	f1-score	precision	recall	f1-score	precision	recall	f1-score	
0	1.00	0.98	0.99	1.00	1.00	1.00	1.00	0.10	0.18	162154
1	0.50	0.63	0.56	0.91	0.92	0.91	0.01	0.65	0.01	196
2	1.00	1.00	1.00	1.00	1.00	1.00	0.94	0.95	0.94	12803
3	0.98	0.98	0.98	1.00	1.00	1.00	0.09	0.93	0.16	1029
4	0.90	0.99	0.95	1.00	1.00	1.00	0.74	0.70	0.72	23012
5	0.90	0.99	0.94	0.98	0.99	0.99	0.00	0.67	0.01	550
6	0.97	0.98	0.97	0.99	0.99	0.99	0.05	0.52	0.09	580
7	0.99	0.98	0.98	1.00	1.00	1.00	0.10	0.99	0.19	794
8	1.00	1.00	1.00	1.00	1.00	1.00	1.00	1.00	1.00	1
9	1.00	1.00	1.00	1.00	1.00	1.00	0.99	0.99	0.99	15880
10	0.97	0.49	0.65	1.00	0.99	1.00	0.08	0.99	0.15	590
11	0.59	0.23	0.33	0.80	0.97	0.88	0.00	0.07	0.00	301
12	0.00	0.00	0.00	1.00	0.80	0.89	0.01	1.00	0.02	4
13	0.80	0.03	0.06	0.96	0.40	0.57	0.08	0.93	0.15	130
Macro avg	0.83	0.73	0.74	0.97	0.93	0.94	0.36	0.75	0.33	218024
Weighted avg	0.99	0.98	0.98	1.00	1.00	1.00	0.95	0.29	0.34	218024

There are also approaches that introduce some heuristics to the process of sampling selection. The algorithm called **NearMiss** [17] is one of them. This approach engages algorithm for nearest neighbours analysis (e.g. k-nearest neighbour) in order to select the dataset instances to be under-sampled. The NearMiss algorithm chooses these samples for which the average distance to the closest samples of the opposite class is the smallest. The effect the algorithm has on the dataset is illustrated in Fig. 3, the results obtained are found in Table 4

Another example of algorithms falling into the undersampling category is called **TomekLinks** [18]. The method performs under-sampling by removing Tomek's links. Tomek's link exists if the two samples are the nearest neighbours of each other. More precisely, A Tomek's link between two samples of different class x and y is defined as $d(x,y) < d(x,z)$ and $d(x,y) < d(y,z)$ for any sample z. The effect removing Tomek-links has on the dataset is illustrated in Fig. 4, the effect it has on ML models is found in Table 5.

A different approach to under-sampling involves centroids obtained from a clustering method. In that type of algorithms the samples belonging to majority class are first clustered (e.g. using k-means algorithm) and replaced with the cluster centroids. In the experiments this approach is indicated as **Cluster Centroids**. The results of the clustering procedure are illustrated in Fig. 4 and in Table 6.

On the other hand, the oversampling method is to be adopted when the size of the original dataset is relatively small. In that approach, one takes the minority class and increases its cardinality in order to achieve the balance among classes. This can be done by using a technique like bootstrapping. In that case, the minority class is sampled with repetitions. Another solution is to use **SMOTE**

Table 4. CICIDS2017 (full set)/NearMiss

	ANN ACC: 0.7725			RandomForest ACC: 0.7116			NaiveBayes ACC: 0.3744			Support
	precision	recall	f1-score	precision	recall	f1-score	precision	recall	f1-score	
0	1.00	0.70	0.82	1.00	0.61	0.76	1.00	0.21	0.35	162154
1	0.02	0.71	0.04	0.03	0.81	0.06	0.01	1.00	0.02	196
2	0.90	1.00	0.95	0.52	1.00	0.68	0.91	0.96	0.93	12803
3	0.99	0.97	0.98	0.97	0.99	0.98	0.22	0.93	0.35	1029
4	0.66	1.00	0.80	0.51	1.00	0.68	0.65	0.70	0.68	23012
5	0.58	0.99	0.73	0.57	0.98	0.72	0.00	0.64	0.01	550
6	0.27	0.98	0.43	0.07	0.99	0.13	0.07	0.82	0.13	580
7	0.19	1.00	0.32	0.25	1.00	0.40	0.10	1.00	0.18	794
8	1.00	1.00	1.00	1.00	1.00	1.00	1.00	1.00	1.00	1
9	0.45	1.00	0.62	0.89	1.00	0.94	1.00	0.99	0.99	15880
10	0.12	0.99	0.21	0.07	1.00	0.13	0.11	0.99	0.20	590
11	0.35	0.56	0.43	0.09	0.99	0.16	0.00	0.08	0.01	301
12	0.00	0.00	0.00	0.05	1.00	0.10	0.01	1.00	0.02	4
13	0.01	0.02	0.02	0.06	0.49	0.11	0.17	0.92	0.29	130
Macro avg	0.47	0.78	0.52	0.43	0.92	0.49	0.38	0.80	0.37	218024
Weighted avg	0.91	0.77	0.81	0.90	0.71	0.75	0.94	0.37	0.46	218024

Table 5. CICIDS2017 (full set)/Tomek Links

	ANN ACC: 0.9836			RandomForest ACC: 0.9986			NaiveBayes ACC: 0.5263			Support
	precision	recall	f1-score	precision	recall	f1-score	precision	recall	f1-score	
0	0.99	0.99	0.99	1.00	1.00	1.00	1.00	0.10	0.18	162154
1	0.92	0.37	0.53	0.81	0.78	0.80	0.01	0.65	0.01	196
2	1.00	1.00	1.00	1.00	1.00	1.00	0.94	0.95	0.94	12803
3	0.99	0.97	0.98	1.00	0.99	0.99	0.09	0.93	0.16	1029
4	0.94	0.95	0.95	1.00	1.00	1.00	0.74	0.70	0.72	23012
5	0.90	0.99	0.94	0.97	0.98	0.98	0.00	0.67	0.01	550
6	0.99	0.98	0.98	0.99	0.99	0.99	0.05	0.52	0.09	580
7	0.99	0.98	0.99	1.00	1.00	1.00	0.10	0.99	0.18	794
8	1.00	1.00	1.00	1.00	1.00	1.00	1.00	1.00	1.00	1
9	1.00	1.00	1.00	1.00	1.00	1.00	0.99	0.99	0.99	15880
10	1.00	0.49	0.66	1.00	0.99	1.00	0.08	0.99	0.15	590
11	0.85	0.07	0.13	0.84	0.97	0.90	0.00	0.07	0.00	301
12	0.00	0.00	0.00	1.00	0.75	0.86	0.01	1.00	0.02	4
13	1.00	0.02	0.05	0.91	0.55	0.68	0.08	0.93	0.14	130
Macro avg	0.90	0.70	0.73	0.97	0.93	0.94	0.36	0.75	0.33	218024
Weighted avg	0.98	0.98	0.98	1.00	1.00	1.00	0.95	0.29	0.34	218024

(Synthetic Minority Over-Sampling Technique) [19]. There are various modification to the original SMOTE algorithm. The one evaluated in this paper is named **Borderline SMOTE**. In this approach the samples representing the minority class are first categorised into three groups: danger, safe, and noise. The sample x is considered to belong to category *noise* if all nearest-neighbours of x are from a different class than the analysed sample, *danger* when only a half belongs to

Table 6. CICIDS2017 (full set)/ClusterCentroids

	ANN ACC: 0.4569			RandomForest ACC: 0.2560			NaiveBayes ACC: 0.2832			Support
	precision	recall	f1-score	precision	recall	f1-score	precision	recall	f1-score	
0	1.00	0.47	0.64	1.00	0.00	0.00	1.00	0.09	0.16	162154
1	0.01	0.28	0.02	0.03	1.00	0.07	0.01	0.65	0.01	196
2	0.90	0.63	0.74	0.74	1.00	0.85	0.93	0.95	0.94	12803
3	0.77	0.68	0.72	0.75	1.00	0.85	0.08	0.93	0.16	1029
4	0.86	0.62	0.72	0.81	1.00	0.90	0.67	0.70	0.69	23012
5	0.15	0.69	0.25	0.82	0.99	0.89	0.00	0.67	0.01	550
6	0.35	0.22	0.27	0.25	0.99	0.40	0.05	0.52	0.09	580
7	0.06	0.47	0.11	0.71	1.00	0.83	0.10	0.99	0.18	794
8	0.01	1.00	0.01	0.50	1.00	0.67	1.00	1.00	1.00	1
9	0.57	0.00	0.00	1.00	1.00	1.00	1.00	0.99	0.99	15880
10	0.00	0.00	0.00	0.10	1.00	0.18	0.08	0.99	0.15	590
11	0.00	0.00	0.00	0.17	0.98	0.29	0.00	0.07	0.00	301
12	0.00	0.00	0.00	0.18	1.00	0.31	0.01	1.00	0.02	4
13	0.00	0.03	0.00	0.05	0.65	0.09	0.09	0.93	0.16	130
Macro avg	0.34	0.36	0.25	0.51	0.90	0.52	0.36	0.75	0.33	218024
Weighted avg	0.93	0.46	0.60	0.95	0.26	0.23	0.94	0.28	0.32	218024

different class, and *safe* when all nearest-neighbours are from the same class. In Borderline SMOTE algorithm, only the *safe* data instances are over-sampled [19]. The effect of this procedure on the dataset is expressed in Fig. 2. The results are placed in Table 7.

A final note concluding this section would be the observation that there is no silver bullet putting one sampling method over another. In fact, their application depends on the use case scenarios and the dataset itself. For the sake of clear illustration the original dataset's class distribution is depicted in Fig. 2, the results the ML algorithms have achieved are found in Table 1.

3.2 Algorithm-Related Balancing Methods

Utilizing unsuitable evaluation metrics for the classifier trained with the imbalanced data can lead to wrong conclusions about the classifier's effectiveness. As the majority of machine learning algorithms do not operate very well with imbalanced datasets, the commonly observed scenario would be the classifier totally ignoring the minority class. This happens because the classifier is not sufficiently penalized for the misclassification of the data samples belonging to the minority class. This is why the algorithm-related methods have been introduced as a part of the modification to the training procedures. One technique is to use other performance metrics. The alternative evaluation metrics that are suitable for imbalanced data are:

- precision - indicating the percentage of relevant data samples that have been collected by the classifier

Table 7. CICIDS2017 (full set)/BORDERLINE SMOTE

	ANN ACC: 0.9753			RandomForest ACC: 0.9920			NaiveBayes ACC: 0.5263			Support
	precision	recall	f1-score	precision	recall	f1-score	precision	recall	f1-score	
0	1.00	0.97	0.99	1.00	0.99	1.00	1.00	0.52	0.68	162154
1	0.17	0.94	0.29	0.17	0.98	0.30	0.00	0.65	0.01	196
2	0.99	1.00	1.00	1.00	1.00	1.00	0.85	0.95	0.90	12803
3	0.94	0.99	0.96	1.00	1.00	1.00	0.05	0.87	0.10	1029
4	0.93	0.99	0.96	0.99	1.00	1.00	0.68	0.70	0.69	23012
5	0.64	0.96	0.77	0.94	0.98	0.96	0.01	0.20	0.02	550
6	0.78	0.51	0.62	1.00	0.97	0.98	0.01	0.03	0.01	580
7	0.87	0.98	0.92	0.99	1.00	1.00	0.02	0.47	0.05	794
8	0.50	1.00	0.67	1.00	1.00	1.00	1.00	1.00	1.00	1
9	0.99	1.00	0.99	1.00	1.00	1.00	0.01	0.00	0.00	15880
10	0.64	0.53	0.58	1.00	0.89	0.94	0.07	0.50	0.12	590
11	0.20	0.26	0.22	0.85	0.84	0.84	0.02	0.89	0.05	301
12	0.01	0.75	0.01	1.00	1.00	1.00	0.01	1.00	0.03	4
13	0.16	0.77	0.27	0.67	0.90	0.77	0.00	0.00	0.00	130
Macro avg	0.63	0.83	0.66	0.90	0.97	0.91	0.27	0.56	0.26	218024
Weighted avg	0.98	0.98	0.98	1.00	0.99	0.99	0.87	0.53	0.64	218024

- recall (or sensitivity)- indicating the total percentage of all relevant instances that have been detected.
- f1-score - computed as the harmonic mean of precision and recall.

Another technique that is successfully used in the field is a cost-sensitive classification. Recently this learning procedure has been reported to be an effective solution to class-imbalance in the large-scale settings. Without losing the generality, let us define the cost-sensitive training process as the following optimisation formula:

$$\hat{\theta} = \min_{\theta} \left\{ \frac{1}{2}||\theta||^2 + \frac{1}{2} \sum_{i=1}^{N} C_i ||e_i||^2 \right\} \tag{1}$$

where θ indicates the classifier parameters, e_i the error in the classifier response for the i-th (out of N) data samples, and C_i the importance of the i-th data sample.

In cost-sensitive learning, the idea is to give a higher importance C_i to the minority class, so that the bias towards the majority class is reduced. In other words, we are producing a cost function that is penalizing the incorrect classification of the minority class more than incorrect classifications of the majority class.

In this paper we have focused on **Cost-Sensitise Random Forest** as an example of cost-sensitive meta-learning. This is mainly due to the fact the Random Forest classifier in that configuration yields the most promising results. These can be found in Table 10.

4   Experiments and Results

Dataset Description - Intrusion Detection Evaluation Dataset - CICIDS2017

CICIDS2017 [20] is an effort to create a dependable and recent cybersecurity dataset. The Intrusion Detection datasets are notoriously hard to come by, and the ones available display at least one of frustrating concerns, like the lack of traffic diversity, attack variety, insufficient features etc. The authors of CICIDS2017 offer a dataset with realistic benign traffic, created as an interpolation of the behaviour of 25 users using multiple protocols. The dataset is a labelled capture of 5 days of work, with 4 days putting the framework under siege by a plethora of attacks, including malware, DoS attacks, web attacks and others. This work relies on the captures from Tuesday, Wednesday, Thursday and Friday. CICIDS2017 constitutes one of the newest datasets available to researchers, featuring over 80 network flow characteristics. The Imbalance Ratio of the Majority Class to the sum of all the numbers of samples of the rest of the classes was calculated to be 2.902. The sample counts for particular classes in the training set are showcased in Table 1.

Results and Perspectives

CICIDS 2017 dataset consists of 13 classes - 12 attacks and 1 benign class. As depicted in Fig. 2, there is a wide discrepancy among the classes in terms of the number of instances, especially the benign class as compared to the attack classes. The number of instances in the respective classes in the training set is displayed in Table 1.

During the tests the initial hypothesis was that balancing the classes would improve the overall results. Random Subsampling (Table 3) along a slew of other subsampling methods were used to observe the influence dataset balancing has on the performance of 3 reference ML algorithms - an Artificial Neural Network (ANN), a RandomForest algorithm and a Naive Bayes classifier. Finally, Borderline SMOTE was conducted as a reference oversampling method. The results of those tests are to be witnessed in Table 4, 7, 5 and 6. It is immediately apparent from inspecting the recall in the unbalanced dataset (Table 1) that some instances of the minority classes are not recognised properly (class 1 and 13). Balancing the benign class to match the number of samples of all the attacks combined changed both the precision and the recall achieved by the algorithm. It also became apparent that none of the subsampling approaches outperformed simple random subsampling in the case of CICIDS2017. The tests revealed an interesting connection among the precision, recall and the imbalance ratio of the dataset. Essentially, there seems to exist a tradeoff between precision and recall that can be controlled by the number of the instances of classes in the training dataset. To evaluate that assertion further tests were conducted. Random Forest algorithm was trained on the Unbalanced dataset and then all the classes were subsampled to match the number of samples in one of the minority classes (Table 9 - 1174 instances per class and Table 8 - 7141 instances per class).

The tests proved that changing the balance ratio undersampling the majority classes improves the recall of the minority classes, but degrades the precision of

Table 8. CICIDS2017/Random subsampling down to 7141 instances per class/RandomForest

	precision	recall	f1-score	Support
0	1.00	0.98	0.99	162154
1	0.13	0.99	0.23	196
2	1.00	1.00	1.00	12803
3	0.92	1.00	0.96	1029
4	0.98	1.00	0.99	23012
5	0.85	0.99	0.92	550
6	0.93	0.99	0.96	580
7	0.93	1.00	0.96	794
8	0.17	1.00	0.29	1
9	1.00	1.00	1.00	15880
10	0.73	1.00	0.85	590
11	0.63	0.98	0.77	301
12	0.07	1.00	0.14	4
13	0.32	0.48	0.39	130
Accuracy			0.9872	218024
Macro avg	0.69	0.96	0.74	218024
Weighted avg	0.99	0.99	0.99	218024

the classifier on those classes. This basically means that dataset balancing causes the ML algorithms to misclassify the (previously) majority classes as instances of the minority classes, thus boosting the false positives.

Finally, a cost-sensitive random forest algorithm was tested. After trying different weight setups results exceeding any previous undersampling or oversampling methods were attained (Table 10). It is noteworthy that the achieved recall for class 13 is higher while still retaining a relatively high precision. A relationship between class 11 and class 13 was also discovered, where setting a higher weight for class 13 would result in misclassification of class 11 samples as class 13 samples and the other way round.

Statistical Analysis of Results

To provide further insight into the effects of dataset balancing statistical analysis was performed with regards to balanced accuracy [21]. The tests revealed that: the cost-sensitive random forest has better results than simple random subsampling, with the t-value at 2.07484 and the p-value at 0.026308. The result is significant at $p < 0.05$. The random forest classifier over the dataset randomly subsampled down to 7141 samples in each majority class performed better

Table 9. CICIDS2017/Random Subsampling down to 1174 instances per class/RandomForest

	precision	recall	f1-score	Support
0	1.00	0.96	0.98	162154
1	0.07	1.00	0.13	196
2	0.99	1.00	1.00	12803
3	0.69	1.00	0.82	1029
4	0.94	0.99	0.97	23012
5	0.76	0.99	0.86	550
6	0.86	0.99	0.92	580
7	0.81	1.00	0.89	794
8	0.17	1.00	0.29	1
9	1.00	1.00	1.00	15880
10	0.44	1.00	0.61	590
11	0.23	0.65	0.34	301
12	0.07	1.00	0.13	4
13	0.13	0.95	0.23	130
Accuracy			0.9657	218024
Macro avg	0.58	0.97	0.65	218024
Weighted avg	0.99	0.97	0.97	218024

than when just the 'benign' class was randomly subsampled with the t-value at 2.96206 and the p-value is 0.004173. The result is significant at $p < 0.05$. The cost-sensitive random forest was not significantly better than the random forest trained on the randomly subsampled dataset in the 7141 variant (the t-value is 1.23569; the p-value is 0.11623. The result is not significant at $p < 0.05$). Cutting the Tomek-links did not prove as good a method as random subsampling in the 7141 variant, with the t-value at 3.69827, the p-value at 0.000823. The result is significant at $p < 0.05$. Removing the Tomek-links wasn't significantly better than just using the imbalanced dataset, with the t-value at 0.10572. The p-value at 0.458486. Both the 7141 variant of random subsampling and the cost-sensitive random forest were better options over just using the imbalanced dataset, with the t-value at 2.96206. The p-value at 0.004173 for random subsampling and the t-value at 2.65093 and the p-value at 0.008129 for the cost-sensitive classifier.

Table 10. CICIDS2017/Cost-Sensitive RandomForest

	precision	recall	f1-score	Support
0	1.00	1.00	1.00	162154
1	0.34	0.91	0.50	196
2	1.00	1.00	1.00	12803
3	1.00	0.99	0.99	1029
4	1.00	1.00	1.00	23012
5	0.97	0.98	0.97	550
6	1.00	0.99	0.99	580
7	1.00	1.00	1.00	794
8	1.00	1.00	1.00	1
9	1.00	1.00	1.00	15880
10	1.00	1.00	1.00	590
11	0.98	0.85	0.91	301
12	1.00	1.00	1.00	4
13	0.72	0.96	0.83	130
Accuracy			0.9973	218024
Macro avg	0.93	0.98	0.94	218024
Weighted avg	1.00	1.00	1.00	218024

5   Conclusions

In this paper the evaluation of a number of dataset balancing methods for the ML algorithms in the cybersecurity doman was presented. The conducted experiments revealed a number of interesting details about those methods. Firstly, in the case of the CICIDS2017 dataset, random subsampling was just as good or better than other undersampling methods and the results were on par with Borderline SMOTE. Secondly, the final proportions of the dataset can bear just as much an impact on the results of ML classification as the choice of the balancing procedure itself. Thirdly, there is a relationship among the size of the majority classes, the precision and the recall achieved, which is simply expressed by the number of majority samples falsely classified as minority samples.

Acknowledgement. This work is funded under the SPARTA project, which has received funding from the European Union's Horizon 2020 research and innovation programme under grant agreement No. 830892.

References

1. Parekh, G., et al.: Identifying core concepts of cybersecurity: results of two Delphi processes. IEEE Trans. Educ. **61**(1), 11–20 (2018)
2. Tabasum, A., Safi, Z., AlKhater, W., Shikfa, A.: Cybersecurity issues in implanted medical devices. In: International Conference on Computer and Applications (ICCA), pp. 1–9, August 2018
3. Bastos, D., Shackleton, M., El-Moussa, F.: Internet of things: a survey of technologies and security risks in smart home and city environments. In: Living in the Internet of Things: Cybersecurity of the IoT - 2018, pp. 1–7 (2018)
4. Kozik, R., Choraś, M., Ficco, M., Palmieri, F.: A scalable distributed machine learning approach for attack detection in edge computing environments. J. Parallel Distrib. Comput. **119**, 18–26 (2018)
5. Sewak, M., Sahay, S.K., Rathore, H.: Comparison of deep learning and the classical machine learning algorithm for the malware detection. In: 19th IEEE/ACIS International Conference on Software Engineering, Artificial Intelligence, Networking and Parallel/Distributed Computing (SNPD), pp. 293–296, June 2018
6. Choraś, M., Kozik, R.: Machine learning techniques applied to detect cyber attacks on web applications. Logic J. IGPL **23**(1), 45–56 (2015)
7. Özkan, K., Işı, Ş., Kartal, Y.: Evaluation of convolutional neural network features for malware detection. In: 6th International Symposium on Digital Forensic and Security (ISDFS), pp. 1–5, March 2018
8. Nguyen, K.D.T., Tuan, T.M., Le, S.H., Viet, A.P., Ogawa, M., Minh, N.L.: Comparison of three deep learning-based approaches for IoT malware detection. In: 10th International Conference on Knowledge and Systems Engineering (KSE), pp. 382–388, November 2018
9. Wang, Y., Shen, Y., Zhang, G.: Research on intrusion detection model using ensemble learning methods. In: 7th IEEE International Conference on Software Engineering and Service Science (ICSESS), pp. 422–425, August 2016
10. Gautam, R.K.S., Doegar, E.A.: An ensemble approach for intrusion detection system using machine learning algorithms. In: 8th International Conference on Cloud Computing, Data Science Engineering (Confluence), pp. 14–15, January 2018
11. Kunal, Dua, M.: Machine learning approach to IDS: a comprehensive review. In: 3rd International conference on Electronics, Communication and Aerospace Technology (ICECA), pp. 117–121, June 2019
12. Skansi, S.: Introduction to Deep Learning. UTCS. Springer, Cham (2018). https://doi.org/10.1007/978-3-319-73004-2
13. Sonawane, H.A., Pattewar, T.M.: A comparative performance evaluation of intrusion detection based on neural network and PCA. In: International Conference on Communications and Signal Processing (ICCSP), pp. 0841–0845, April 2015
14. Breiman, L.: Random forests. Mach. Learn. **45**(1), 5–32 (2001)
15. Maimon, O., Rokach, L.: Data Mining and Knowledge Discovery Handbook, 2nd edn. Springer, Boston (2010). https://doi.org/10.1007/978-0-387-09823-4
16. Kozik, R., Choraś, M.: Solution to data imbalance problem in application layer anomaly detection systems. In: Martínez-Álvarez, F., Troncoso, A., Quintián, H., Corchado, E. (eds.) HAIS 2016. LNCS (LNAI), vol. 9648, pp. 441–450. Springer, Cham (2016). https://doi.org/10.1007/978-3-319-32034-2_37
17. Zhang, J., Mani, I.: KNN approach to unbalanced data distributions: a case study involving information extraction. In: Proceedings of the ICML 2003 Workshop on Learning from Imbalanced Datasets (2003)

18. Tomek, I.: Two modifications of CNN. IEEE Trans. Syst. Man Cybern. SMC 6(11), 769–772 (1976)
19. Han, H., Wang, W.-Y., Mao, B.-H.: Borderline-SMOTE: a new over-sampling method in imbalanced data sets learning. In: Huang, D.-S., Zhang, X.-P., Huang, G.-B. (eds.) ICIC 2005. LNCS, vol. 3644, pp. 878–887. Springer, Heidelberg (2005). https://doi.org/10.1007/11538059_91
20. Sharafaldin, I., Lashkari, A.H., Ghorbani, A.A.: Toward generating a new intrusion detection dataset and intrusion traffic characterization. In: Proceedings of the 4th International Conference on Information Systems Security and Privacy - Volume 1: ICISSP, pp. 108–116. INSTICC, SciTePress (2018)
21. Brodersen, K.H., Ong, C.S., Stephan, K.E., Buhmann, J.M.: The balanced accuracy and its posterior distribution. In: 20th International Conference on Pattern Recognition, pp. 3121–3124 (2010)

Pattern Recognition Model to Aid the Optimization of *Dynamic Spectrally-Spatially Flexible Optical Networks*

Paweł Ksieniewicz[1]([⊠])[iD], Róża Goścień[1][iD], Mirosław Klinkowski[2][iD], and Krzysztof Walkowiak[1][iD]

[1] Wroclaw University of Science and Technology, Wyb. Wyspianskiego 27, 50-370 Wroclaw, Poland
{pawel.ksieniewicz,roza.goscien,krzysztof.walkowiak}@pwr.edu.pl
[2] National Institute of Telecommunications, Szachowa 1, 04-894 Warszawa, Poland
m.klinkowski@itl.waw.pl

Abstract. The following paper considers pattern recognition-aided optimization of complex and relevant problem related to optical networks. For that problem, we propose a four-step dedicated optimization approach that makes use, among others, of a regression method. The main focus of that study is put on the construction of efficient regression model and its application for the initial optimization problem. We therefore perform extensive experiments using realistic network assumptions and then draw conclusions regarding efficient approach configuration. According to the results, the approach performs best using *multi-layer perceptron* regressor, whose prediction ability was the highest among all tested methods.

Keywords: Spectrally-spatially flexible optical networks · Routing · Space and spectrum optimization · Network optimization · Ensemble learning · Pattern regression

1 Introduction

According to *Cisco* forecasts, the global consumer traffic in the Internet will grow on average with annual compound growth rate (CAGR) of 26% in years 2017–2022 [3]. The increase in the network traffic is a result of two main trends. Firstly, the number of devices connected to the internet is growing due to the increasing popularity of new services including *Internet of Things* (*IoT*). The second important trend influencing the traffic in the internet is popularity of bandwidth demanding services such as video streaming (e.g., *Netflix*) and cloud computing. The Internet consists of many single networks connected together, however, the backbone connecting these various networks are optical networks based on fiber connections. Currently, the most popular technology in optical

© Springer Nature Switzerland AG 2020
V. V. Krzhizhanovskaya et al. (Eds.): ICCS 2020, LNCS 12140, pp. 211–224, 2020.
https://doi.org/10.1007/978-3-030-50423-6_16

networks is WDM (*Wavelength Division Multiplexing*), which is expected to be not efficient enough to support increasing traffic in the nearest future. In last few years, a new concept for optical networks has been deployed, i.e., architecture of *Elastic Optical Networks* (EONs). However, in the perspective on the next decade some new approaches must be developed to overcome the predicted "capacity crunch" of the Internet.

One of the most promising proposals is *Spectrally-Spatially Flexible Optical Network* (SS-FON) that combines *Space Division Multiplexing* (SDM) technology [14], enabling parallel transmission of co-propagating spatial modes in suitably designed optical fibers such as multi-core fibers (MCFs) [1], with flexible-grid EONs [4] that enable better utilization of the optical spectrum and distance-adaptive transmissions [15]. In MCF-based SS-FONs, a challenging issue is the inter-core crosstalk (XT) effect that impairs the quality of transmission (QoT) of optical signals and has a negative impact on overall network performance. In more detail, MCFs are susceptible to signal degradation as a result of the XT that happens between adjacent cores whenever optical signals are transmitted in an overlapping spectrum segment. Addressing the XT constraints significantly complicates the optimization of SS-FONs [8].

Besides numerous advantages, new network technologies bring also challenging optimization problems, which require efficient solution methods. Since the technologies and related problems are new, there are no benchmark solution methods to be directly applied and hence many studies propose some dedicated optimization approaches. However, due to the problems high complexity, their performance still needs a lot of effort to be put [6,8]. We therefore observe a trend to use *artificial intelligence* techniques (with the high emphasis on *pattern recognition* tools) in the field of optimization of communication networks. According to the literature surveys in this field [2,10,11,13], the researchers mostly focus on discrete labelled *supervised* and *unsupervised learning* problems, such as traffic classification. Regression methods, which are in the scope of that paper, are mostly applied for traffic prediction and estimation of quality of transmission (QoT) parameters such as delay or bit error rate.

This paper extends our study initiated in [7]. We make use of pattern recognition models to aid optimization of dynamic MCF-based SS-FONs in order to improve performance of the network in terms of minimizing bandwidth blocking probability (BBP), or in other words to maximize the amount of traffic that can be allocated in the network. In particular, an important topic in the considered optimization problem is selection of a modulation format (MF) for a particular demand, due to the fact that each MF provides a different tradeoff between required spectrum width and transmission distance. To solve that problem, we define applicable distances for each MF (i.e., minimum and maximum length of a routing path that is supported by each MF). To find values of these distances, which provide best allocation results, we construct a regression model and then combine it with *Monte Carlo* search. It is worth noting that this work does not address dynamic problems in the context of changing the concept over time, as is often the case with processing large sets, and assumes static distribution of the concept [9].

The main novelty and contribution of the following work is an in-depth analysis of the basic regression methods stabilized by the structure of the estimator ensemble [16] and assessment of their usefulness in the task of predicting the objective function for optimization purposes. In one of the previous works [7], we confirmed the effectiveness of this type of solution using a regression algorithm of the nearest weighted neighbors, focusing, however, much more on the network aspect of the problem being analyzed. In the present work, the main emphasis is on the construction of the prediction model. Its main purpose is:

- A proposal to interpret the optimization problem in the context of pattern recognition tasks.
- Construction of the prediction model based on the introduced interpretation.
- Extensive experimental verification of the *state-of-art* regression algorithms efficiency in a problems defined by datasets being the result of computer simulations performed for the purposes of network optimization.
- Proposition and verification of the use of such models in optimization of the problem.

The rest of the paper is organized as follows. In Sect. 2, we introduce studied network optimization problem. In Sect. 3, we discuss out optimization approach for that problem. Next, in Sect. 4 we evaluate efficiency of the proposed approach. Eventually, Sect. 5 concludes the work.

2 Optimization of Spectrally-Spatially Flexible Optical Network

The optimization problem is known in the literature as dynamic *Routing, Space and Spectrum Allocation* (RSSA) in SS-FONs [5]. We are given with an SS-FON topology realized using MCFs. The topology consists of nodes and physical link. Each physical link comprises of a number of spatial cores. The spectrum width available on each core is divided into arrow and same-sized segments called *slices*.

The network is in its operational state – we observe it in a particular time perspective given by a number of iterations. In each iteration (i.e., a time point), a set of demands arrives. Each demand is given by a source node, destination node, duration (measured in the number of iterations) and bitrate (in *Gbps*). To realize a demand, it is required to assign it with a light-path and reserve its resources for the time of the demand duration. When a demand expires, its resources are released. A light-path consists of a routing path (a set of links connecting demand source and destination nodes) and a channel (a set of adjacent slices selected on one core) allocated on the path links. The channel width (number of slices) required for a particular demand on a particular routing path depends on the demand bitrate, path length (in kilometres) and selected modulation format. Each incoming demand has to be realized unless there is not enough free resources when it arrives. In such a case, a demand is rejected. Please note that the selected light-paths in i-th iteration affect network state and allocation possibilities in the next iterations. The objective function is defined here as bandwidth blocking probability (BBP) calculated as a summed bitrate of all rejected demands divided by the summed bitrate of all offered demands. Since we aim to support as much traffic as it is possible, the objective criterion should be minimized [5,8].

The light-paths' allocation process has to satisfy three basic RSSA constraints. First, each channel has to consists of adjacent slices. Second, the same channel (i.e., the same slices and the same core) has to be allocated on each link included in a light-path. Third, in each time point each slice on a particular physical link and a particular core can be used by at most one demand [8].

There are four modulation formats available for transmissions—8-QAM, 16-QAM, QPSK and BPSK. Each format is described by its spectral efficiency, which determines number of slices required to realize a particular bitrate using that modulation. However, each modulation format is also characterized by the maximum transmission distance (MTD) which provides acceptable value of optical signal to noise ratio (OSNR) at the receiver side. More spectrally-efficient formats consume less spectrum, however, at the cost of shorter MTDs. Moreover, more spectrally-efficient formats are also vulnerable to XT effects which can additionally degrade QoT and lead to demands' rejection [7,8]. Therefore, the selection of the modulation format for each demand is a compromise between spectrum efficiency and QoT.

To answer that problem, we use the procedure introduced in [7] to select a modulation format for a particular demand and routing path [7]. Let $m = 1, 2, 3, 4$ denote modulation formats ordered in increasing MTDs (and in decreasing spectral efficiency at the same time). It means that $m = 1$ denotes 8-QAM and $m = 4$ denotes BPSK. Let $MTD = [mtd_1, mtd_2, mtd3, mtd_4]$ be a vector of MTDs for modulations 8-QAM, 16-QAM, QPSK, BPSK respectively. Moreover, let $ATD = [atd_1, atd_2, atd3, atd_4]$ (where $atd_i <= mtd_i, i = 1, 2, 3, 4$) be the vector of applicable transmission distances. For a particular demand and a routing path we select most spectrally-efficient modulation format i for which atd_i is grater of equal to the selected path length and the XT effect is on an acceptable level. For each candidate modulation format, we asses the XT level based on the adjacent resources' (i.e., slices and cores) availability using procedure proposed in [7]. It is important to note that we do not indicate atd_4 (for BPSK) since we assume that this modulation is able to support transmission on all candidate routing paths regardless of their length. Please also note that when XT level is too high for all modulation formats, the demand is rejected regardless of the light-paths' availability.

3   Pattern Recognition Model to Aid Optimization of SS-FONs

In Sect. 2 we have studied RSSA problem and emphasised the importance of efficient modulation selection task. For that task we have proposed solution method whose efficiency strongly depends on the applied ATD vector. Therefore, we aim to find ATD* vector that provides best results. The vector elements have to be positive and have upper bounds given by vector MTD. Moreover, the following condition have to be satisfied: $atd_i < atd_{i+1}, i = 1, 2$. Since solving RSSA instances is a time consuming process, it is impossible to evaluate all possible ATD vectors in a reasonable time. We therefore make use of regression methods and propose a scheme to find ATD* depicted in Fig. 1.

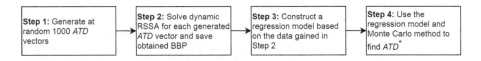

Fig. 1. Optimization procedure for dynamic SS-FON.

A representative set of 1000 different ATD vectors is generated. Then, for each of them we simulate allocation of demands in SS-FON (i.e., we solve dynamic RSSA). For the purpose of demands allocation (i.e., selection of light-paths), we use a dedicated algorithm proposed in [7]. For each considered ATD vector we save obtained BBP. Based on that data, we construct a regression model, which predicts BBP based on an ATD vector. Having that model, we use *Monte Carlo* method to find ATD* vector, which is recommended for further experiments.

3.1 Solving RSSA

To solve an RSSA instance for a particular ATD vector, we use heuristic algorithm proposed in [7]. We work under the assumption that there are 30 candidate routing paths for each traffic demand (generated using Dijkstra algorithm). Since the paths are generated in advance and their lengths are known, we can use an ATD vector and preselect for these paths modulation formats based on the procedure discussed in Sect. 2. Therefore, RSSA is reduced to the selection of one of the candidate routing paths and a communication channel with respect to the resource availability and assessed XT levels.

3.2 Construction and Validation of the Model

From the perspective of pattern recognition methods, the abstraction of the problem is not the key element of processing. The main focus here is the representation available to construct a proper decision model. For the purposes of considerations, we assume that both input parameters and the objective function take only quantitative and not qualitative values, so we may use probabilistic pattern recognition models to process them. If we interpret the optimization task as searching for the extreme function of many input parameters, each simulation performed for their combination may also be described as a label for the training set of *supervised learning* model.

In this case, the set of parameters considered in a single simulation becomes a vector of object features (x_n), and the value of the objective function acquired around it may be interpreted as a continuous object label (y_n). Repeated simulation for randomly generated parameters allows to generate a data set (X) supplemented with a label vector (y). A supervised machine learning algorithm can therefore gain, based on such a set, a generalization abilities that allows for precise estimation of the simulation result based on its earlier runs on the random input values.

A typical pattern recognition experiment is based on the appropriate division of the dataset into training and testing sets, in a way that guarantees their separability (most often using cross-validation), avoiding the problem of data peeking and a sufficient number of repetitions of the validation process to allow proper statistical testing of mutual model dependencies hypotheses. For the needs of the proposal contained in this paper, the usual 5-fold cross validation was adopted, which calculates the value of the r^2 metric for each loop of the experiment.

3.3   Finding ATD^*

Having constructed regression model, we are able to predict BBP value for a sample ATD vector. Please note that the time required for a single prediction is significantly shorter that the time required to simulate a dynamic RSSA. The last step of our optimization procedure is to find ATD*—vector providing lowest estimated BBP values. To this end, we use *Monte Carlo* method with a number of guesses provided by the user.

4   Experimental Evaluation

4.1   Experiments Set-Up

The RSSA problem was solved for two network topologies—DT12 (12 nodes, 36 links) and *Euro*28 (28 nodes, 82 links). They model *Deutsche Telecom* (German national network) and European network, respectively. Each network physical link comprised of 7 cores wherein each of the cores offers 320 frequency slices of 12.5 GHz width. We use the same network physical assumptions and XT levels and assessments as in [7]. Traffic demands have randomly generated end nodes and birates uniformly distributed between 50 Gbps and 1 Tbps, with granularity of 50 Gbps. Their arrival follow *Poisson* process with an average arrival rate λ demands per time unit. The demand duration is generated according to a negative exponential distribution with an average of $1/\mu$. The traffic load offered is λ/μ normalized traffic units (NTUs). For each testing scenario, we simulate arrival of 10^6 demands. Four modulations are available (8-QAM, 16-QAM, QPSK, BPSK) wherein we use the same modulation parameters as in [7].

For each topology we have generated 9 different datasets, each consists of 1000 samples of ATD vector and corresponding BBP. The datasets differ with the XT coefficient ($\mu = 1 \cdot 10^{-9}$ indicated as "XT1", $\mu = 2 \cdot 10^{-9}$ indicated as "XT2", for more details we refer to [7]) and network links scaling factor (the multiplier used to scale lengths of links in order to evaluate if different lengths of routing paths influence performance of the proposed approach). For DT12 we use following scaling factors: 0.4, 0.6, 0.8, ..., 2.0. For *Euro28* the values are as follows: 0.104, 0.156, 0.208, 0.260, 0.312, 0.364, 0.416, 0.468, 0.520. We indicate them as "*Sx.xxx*" where *x.xxx* refers to the scaling factor value. Using these datasets we can evaluate whether XT coefficient (i.e., level of the vulnerability to XT effects) and/or average link length influence optimization approach performance.

The experimental environment for the construction of predictive models, including the implementation of the proposed processing method, was implemented in Python, following the guidelines of the *state-of-art* programming interface of the *scikit-learn* library [12]. Statistical dependency assessment metrics for paired tests were calculated according to the Wilcoxon test, according to the implementation contained in *scipy* module. Each of the individual experiments was evaluated by r^2 score – a typical quality assessment metric for regression problems. The full source code, supplemented with employed datasets is publicly available in a GIT repository[1].

Five simple recognition models were selected as the base experimental estimators:

- KNR—*k-Nearest Neighbors* regressor with five neighbors, leaf size of 30 and euclidean metric approximated by Minkowski distance,
- dKNR—KNR regressor weighted by distance from closest patterns,
- MLP—a *Multilayer Perceptron* with one hidden layer of one hundred neurons, with the ReLU activation function and *adam* optimizer,
- DTR—CART tree with MSE split criterion,
- LIN—*Linear Regression* algorithm.

5 Experimental Evaluation

In this section we evaluate performance of the proposed optimization approach. To this end, we conduct three experiments. Experiment 1 focuses on the number of patterns required to construct a reliable prediction model. Experiment 2 assesses the statistical dependence of built models. Eventually, experiment 3 verifies efficiency of the proposed approach as a function of number of guesses in the *Monte Carlo* search.

5.1 Experiment 1

The first experiment carried out as part of the approach evaluation is designed to verify how many patterns – and thus how many repetitions of simulations – must be passed to individual regression algorithms to allow the construction of a reliable prediction model. The tests were carried out on all five considered regressors in two stages. First, the range from 10 to 100 patterns was analyzed, and in the second, from 100 to 1000 patterns per processing. It is important to

---

[1] https://github.com/w4k2/regression-aided-optimization.

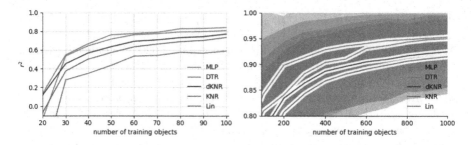

Fig. 2. Dependency between number of training objects and quality of predictions of analyzed regression models according to r^2 metric for DT12 topology.

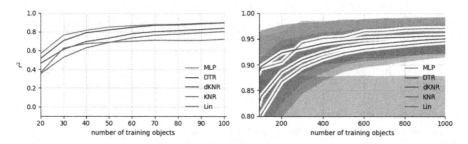

Fig. 3. Dependency between number of training objects and quality of predictions of analyzed regression models according to r^2 metric for $Euro$28 topology.

note that due to the chosen approach to cross-validation, in each case the model is built on 80% of available objects.

The analysis was carried out independently on all available data sets, and due to the non-deterministic nature of sampling of available patterns, its results were additionally stabilized by repeating a choice of the objects subset five times.

In order to allow proper observations, the results were averaged for both topologies. Plots for the range from 100 to 1000 patterns were additionally supplemented by marking ranges of standard deviation of r^2 metric acquired within the topology and presented in the range from the .8 value.

The results achieved for averaging individual topologies are presented in Figs. 2 and 3. For DT12 topology, MLP and DTR algorithms are competitively the best models, both in terms of the dynamics of the relationship between the number of patterns and the overall regression quality. The *Linear Regression* clearly stands out from the rate. A clear observation is also the *saturation* of the models, understood by approaching the maximum predictive ability, as soon as around 100 patterns in the data set. The best algorithms already achieve quality within .8, and with 600 patterns they stabilize around .95. The relationship between each of the recognition algorithms and the number of patterns takes the form of a logarithmic curve in which, after fast initial growth, each subsequent object gives less and less potential for improving the quality of prediction. This suggests that it is not necessary to carry out further simulations to extend the training set, because it will not significantly affect the predictive quality of the developed model.

Very similar observations may be made for *Euro*28 topology, however, noting that it seems to be a simpler problem, allowing faster achievement of the maximum model predictive capacity. It is also worth noting here the fact that the *standard deviation* of results obtained by MLP is smaller, which may be equated with the potentially greater stability of the model achieved by such a solution.

5.2 Experiment 2

The second experiment extends the research contained in Experiment 1 by assessing the statistical dependence of models built on a full datasets consisting of a thousand samples for each case. The results achieved are summarized in Tables 1a and b.

Table 1. Mean r^2 scores achived by partricular regression algorithms for all cases of both analyzed topologies.

(a) DT12 topology.

DATASET	KNR	dKNR	MLP	DTR	LIN
XT1-S0.4	0.621	0.631	0.725	0.789	0.000
	5	1, 5	1, 2, 5	all	—
XT1-S0.6	0.902	0.928	0.927	0.928	0.585
	5	1, 5	1, 5	5	—
XT1-S0.8	0.934	0.946	0.967	0.964	0.516
	5	1, 5	1, 2, 5	1, 2, 5	—
XT1-S1.0	0.928	0.941	0.959	0.962	0.654
	5	1, 5	1, 5	1, 2, 5	—
XT1-S1.2	0.833	0.852	0.891	0.915	0.582
	5	1, 5	1, 2, 5	all	—
XT1-S1.4	0.883	0.903	0.962	0.954	0.479
	5	1, 5	1, 2, 5	1, 5	—
XT1-S1.6	0.896	0.912	0.976	0.975	0.481
	5	1, 5	1, 2, 5	1, 2, 5	—
XT1-S1.8	0.961	0.968	0.987	0.988	0.623
	5	1, 5	1, 2, 5	1, 2, 5	—
XT1-S2.0	0.964	0.969	0.989	0.992	0.725
	5	1, 5	1, 2, 5	1, 2, 5	—
XT2-S0.4	0.966	0.969	0.954	0.987	0.776
	3, 5	1, 3, 5	5	all	—
XT2-S0.6	0.952	0.958	0.963	0.978	0.681
	5	1, 5	1, 5	all	—
XT2-S0.8	0.965	0.966	0.970	0.968	0.752
	5	1, 5	5	5	—
XT2-S1.0	0.951	0.953	0.953	0.944	0.857
	4, 5	1, 4, 5	4, 5	5	—
XT2-S1.2	0.888	0.901	0.908	0.907	0.744
	5	1, 5	5	5	—
XT2-S1.4	0.879	0.897	0.965	0.941	0.500
	5	1, 5	all	1, 2, 5	—
XT2-S1.6	0.933	0.945	0.983	0.971	0.539
	5	1, 5	all	1, 2, 5	—
XT2-S1.8	0.967	0.971	0.989	0.980	0.700
	5	1, 5	all	1, 2, 5	—
XT2-S2.0	0.976	0.979	0.989	0.984	0.809
	5	1, 5	all	1, 2, 5	—

(b) *Euro*28 topology.

DATASET	KNR	dKNR	MLP	DTR	LIN
XT1-S0.104	0.971	0.974	0.977	0.967	0.875
	5	1, 5	1, 5	5	—
XT1-S0.156	0.966	0.972	0.990	0.991	0.746
	5	1, 5	1, 2, 5	all	—
XT1-S0.208	0.969	0.970	0.969	0.986	0.751
	5	5	5	all	—
XT1-S0.260	0.852	0.858	0.889	0.830	0.534
	4, 5	1, 4, 5	all	5	—
XT1-S0.312	0.960	0.963	0.971	0.963	0.752
	5	5	5	5	—
XT1-S0.364	0.957	0.962	0.967	0.951	0.861
	5	1, 4, 5	1, 4, 5	5	—
XT1-S0.416	0.907	0.918	0.970	0.945	0.677
	5	1, 5	all	1, 2, 5	—
XT1-S0.468	0.933	0.942	0.983	0.971	0.607
	5	1, 5	all	1, 2, 5	—
XT1-S0.520	0.960	0.965	0.988	0.986	0.644
	5	1, 5	1, 2, 5	1, 2, 5	—
XT2-S0.104	0.956	0.959	0.972	0.947	0.810
	5	1, 5	all	5	—
XT2-S0.156	0.975	0.978	0.993	0.992	0.777
	5	1, 5	all	1, 2, 5	—
XT2-S0.208	0.950	0.951	0.956	0.959	0.893
	5	5	5	5	—
XT2-S0.260	0.952	0.958	0.965	0.964	0.696
	5	1, 5	1, 5	5	—
XT2-S0.312	0.957	0.962	0.978	0.963	0.670
	5	1, 5	1, 2, 5	5	—
XT2-S0.364	0.911	0.921	0.951	0.911	0.698
	5	1, 5	all	5	—
XT2-S0.616	0.897	0.911	0.972	0.961	0.483
	5	1, 5	1, 2, 5	1, 2, 5	—
XT2-S0.468	0.938	0.949	0.986	0.979	0.551
	5	1, 5	all	1, 2, 5	—
XT2-S0.520	0.955	0.962	0.990	0.989	0.660
	5	1, 5	1, 2, 5	1, 2, 5	—

As may be seen, for the DT12 topology, the LIN algorithm clearly deviates negatively from the other methods, in absolutely every case being a worse solution than any of the others, which leads to the conclusion that we should completely reject it from considering as a base for a stable recognition model. Algorithms based on neighborhood (KNR and dKNR) are in the middle of the rate, in most cases statistically giving way to MLP and DTR, which would also suggest departing from them in the construction of the final model. The statistically best solutions, almost equally, in this case are MLP and DTR.

For *Euro*28 topology, the results are similar when it comes to LIN, KNR and dKNR approaches. A significant difference, however, may be seen for the achievements of DTR, which in one case turns out to be the worst in the rate, and in many is significantly worse than MLP. These observations suggest that in the final model for the purposes of optimization lean towards the application of *neural networks*.

What is important, the highest quality prediction does not exactly mean the best optimization. It is one of the very important factors, but not the only one. It is also necessary to be aware of the shape of the decision function. For this purpose, the research was supplemented with visualizations contained in Fig. 4.

Algorithms based on neighborhood (KNN, dKNN) and decision trees (DTR) are characterized by a discrete decision boundary, which in the case of visualization resembles a picture with a low level of quantization. In the case of an ensemble model, stabilized by cross-validation, actions are taken to reduce this property in order to develop as continuous a border as possible. As may be seen in the illustrations, compensation occurs, although in the case of KNN and dKNN leads to some disturbances in the decision boundary (interpreted as thresholding the predicted label value), and for the DTR case, despite the general correctness of the performed decisions, it generates image artifacts. Such a model may still retain high predictive ability, but it has too much tendency to overfit and leads to insufficient continuity of the optimized function to perform effective optimization.

Clear decision boundaries are implemented by both the LIN and MLP approaches. However, it is necessary to reject LIN from processing due to the linear nature of the prediction, which (i) in each optimization will lead to the selection of the extreme value of the analyzed range and (ii) is not compatible with the distribution of the explained variable and must have the largest error in each of the optimas.

Summing up the observations of Experiments 1 and 2, the MLP algorithm was chosen as the base model for the optimization task. It is characterized by (i) statistically best predictive ability among the methods analyzed and (ii) the clearest decision function from the perspective of the optimization task.

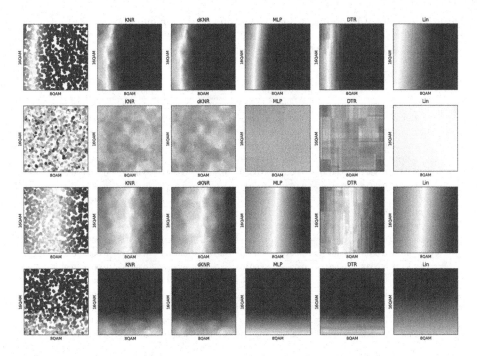

Fig. 4. Exemplary subspace visualization of decision function modeled by different regression algorithms supplemented by scatter plot of original objects inside subspace.

5.3   Experiment 3

The last experiment focuses on the finding of best ATD vector based on the constructed regression model. To this end, we use *Monte Carlo* method with different number of guesses. Tables 2 and 3 present the obtained results as a function of number of guesses, which changes from 10^1 up to 10^9. The results quality increases with the number of guesses up to some threshold value. Then, the results do not change at all or change only a little bit. According to the presented values, *Monte Carlo* method applied with 10^3 guesses provides satisfactory results. We therefore recommend that value for further experiments.

Table 2. Results of Experiment 3 for DT12 (*min*Y-R is minimum observed BBP (in the dataset), *min*Y-E is minimum estimated BBP, Y-MC*(g)* is BBP estimation for ATD found using regression model and *Monte Carlo* with *g* guesses.

DATASET	MIN Y-R	MIN Y-E	Y-MC(G=...)						r^2
			10	100	1 K	10 K	100 K	1 M	
XT1-S0.4	0.0062	0.0075	0.0085	0.0076	0.0075	0.0074	0.0074	0.0074	0.731
XT1-S0.6	0.0001	0.0003	0.0003	0.0001	0.0001	0.0001	0.0001	0.0001	0.933
XT1-S0.8	0.0000	0.0001	0.0001	−0.0000	−0.0002	−0.0002	−0.0002	−0.0002	0.971
XT1-S1.0	0.0011	0.0010	0.0024	0.0012	0.0008	0.0008	0.0007	0.0007	0.965
XT1-S1.2	0.0030	0.0039	0.0042	0.0038	0.0035	0.0034	0.0034	0.0034	0.899
XT1-S1.4	0.0042	0.0050	0.0053	0.0050	0.0048	0.0047	0.0047	0.0047	0.966
XT1-S1.6	0.0024	0.0026	0.0036	0.0025	0.0024	0.0024	0.0024	0.0024	0.976
XT1-S1.8	0.0034	0.0044	0.0046	0.0029	0.0027	0.0027	0.0026	0.0026	0.987
XT1-S2.0	0.0035	0.0047	0.0049	0.0040	0.0033	0.0032	0.0032	0.0032	0.989
XT2-S0.4	0.0000	−0.0002	−0.0000	−0.0002	−0.0003	−0.0003	−0.0003	−0.0004	0.951
XT2-S0.6	0.0000	−0.0001	−0.0002	−0.0003	−0.0003	−0.0004	−0.0004	−0.0004	0.963
XT2-S0.8	0.0006	0.0009	0.0010	0.0009	0.0008	0.0008	0.0008	0.0008	0.971
XT2-S1.0	0.0037	0.0041	0.0068	0.0046	0.0039	0.0038	0.0038	0.0038	0.954
XT2-S1.2	0.0047	0.0050	0.0078	0.0063	0.0042	0.0042	0.0042	0.0041	0.922
XT2-S1.4	0.0048	0.0059	0.0083	0.0068	0.0055	0.0054	0.0054	0.0054	0.965
XT2-S1.6	0.0059	0.0065	0.0091	0.0074	0.0063	0.0061	0.0061	0.0061	0.982
XT2-S1.8	0.0073	0.0080	0.0102	0.0085	0.0074	0.0072	0.0071	0.0071	0.989
XT2-S2.0	0.0060	0.0066	0.0086	0.0066	0.0061	0.0057	0.0057	0.0056	0.988

Table 3. Results of experiment 3 for *Euro*28 (*min*Y-R is minimum observed BBP (in the dataset), *min*Y-E is minimum estimated BBP, Y-MC*(g)* is BBP estimation for ATD found using regression model and *Monte Carlo* with *g* guesses.

DATASET	MIN Y-R	MIN Y-E	Y-MC(G=...)						r^2
			10	100	1K	10 K	100 K	1 M	
XT1-S0.104	0.0023	0.0030	0.0056	0.0032	0.0024	0.0023	0.0022	0.0022	0.979
XT1-S0.156	0.0055	0.0068	0.0076	0.0068	0.0066	0.0065	0.0064	0.0064	0.989
XT1-S0.208	0.0075	0.0091	0.0093	0.0090	0.0088	0.0088	0.0087	0.0087	0.970
XT1-S0.260	0.0033	0.0048	0.0048	0.0043	0.0043	0.0042	0.0042	0.0042	0.891
XT1-S0.312	0.0016	0.0027	0.0027	0.0024	0.0024	0.0023	0.0023	0.0023	0.971
XT1-S0.364	0.0052	0.0063	0.0100	0.0063	0.0055	0.0053	0.0053	0.0053	0.968
XT1-S0.416	0.0053	0.0057	0.0094	0.0063	0.0054	0.0054	0.0053	0.0053	0.969
XT1-S0.468	0.0055	0.0068	0.0095	0.0070	0.0064	0.0061	0.0061	0.0061	0.983
XT1-S0.520	0.0050	0.0058	0.0080	0.0060	0.0057	0.0056	0.0056	0.0056	0.988
XT2-S0.104	0.0015	0.0023	0.0033	0.0023	0.0016	0.0015	0.0015	0.0014	0.971
XT2-S0.156	0.0078	0.0092	0.0106	0.0090	0.0090	0.0084	0.0084	0.0084	0.993
XT2-S0.208	0.0006	0.0013	0.0005	0.0003	0.0004	0.0003	0.0002	0.0002	0.951
XT2-S0.260	0.0000	0.0000	0.0001	−0.0000	−0.0001	−0.0001	−0.0001	−0.0001	0.964
XT2-S0.312	0.0002	0.0005	0.0005	0.0003	0.0003	0.0003	0.0003	0.0003	0.977
XT2-S0.364	0.0030	0.0034	0.0054	0.0041	0.0033	0.0033	0.0033	0.0033	0.957
XT2-S0.416	0.0038	0.0043	0.0060	0.0048	0.0040	0.0040	0.0039	0.0039	0.975
XT2-S0.468	0.0050	0.0052	0.0078	0.0063	0.0053	0.0052	0.0051	0.0051	0.986
XT2-S0.520	0.0064	0.0079	0.0087	0.0073	0.0069	0.0064	0.0063	0.0063	0.990

6   Conclusions

The following work has considered the topic of employing pattern recognition methods to support SS-FON optimization process. For a wide pool of generated cases, analyzing two real network topologies, the effectiveness of solutions implemented by five different, typical regression methods was analyzed, starting from *Logistic Regression* and ending with *neural networks*.

Conducted experimental analysis shows, with high probability obtained by conducting proper statistical validation, that MLP is characterized by the greatest potential in this type of solutions. Even with a relatively small pool of input simulations, constructing a data set for learning purpouses, interpretable in both the space of optimization and machine learning problems, simple networks of this type achieve both high quality prediction measured by the r^2 metric, and continuous decision space creating the potential for conducting optimization.

Basing the model on the stabilization realized by using ensemble of estimators additionally allows to reduce the influence of noise on optimization, which – in a *state-of-art* optimization methods – could show a tendency to select invalid optimas, burdened by the nondeterministic character of the simulator. Further research, developing ideas presented in this article, will focus on the generalization of the presented model for a wider pool of network optimization problems.

Acknowledgements. The work of Paweł Ksieniewicz, Róża Goścień, and K. Walkowiak was supported by National Science Centre, Poland under Grant 2017/27/B/ST7/00888. The work of Mirosław Klinkowski was supported by National Science Centre, Poland under Grant No. 2018/31/B/ST7/03456. The work of Róża Goścień was also partially supported by the Foundation for Polish Science (FNP).

References

1. Awaji, Y., et al.: High-capacity transmission over multi-core fibers. Opt. Fiber Technol. **35**, 100–107 (2017)
2. Boutaba, R., et al.: A comprehensive survey on machine learning for networking: evolution, applications and research opportunities. J. Internet Serv. Appl. **9**(1), 1–99 (2018). https://doi.org/10.1186/s13174-018-0087-2
3. CISCO: Cisco Visual Networking Index: Forecast and Trends, 2017–2022. Technical report (2019)
4. Gerstel, O., Jinno, M., Lord, A., Yoo, S.J.B.: Elastic optical networking: a new dawn for the optical layer? IEEE Comm. Mag. **50**(2), 12–20 (2012)
5. Goścień, R.: On the efficient dynamic routing in spectrally-spatially flexible optical networks. In: 2019 11th International Workshop on Resilient Networks Design and Modeling (RNDM), pp. 1–8, October 2019
6. Goścień, R., Lechowicz, P.: On the complexity of RSSA of any cast demands in spectrally-spatially flexible optical networks. In: 9th International Network Optimization Conference Avignon, France, 12–14 June 2019, p. 7 (2019)
7. Klinkowski, M., Ksieniewicz, P., Jaworski, M., Zalewski, G., Walkowiak, K.: Machine learning assisted optimization of dynamic crosstalk-aware spectrally-spatially flexible optical networks. J. Lightwave Technol. 1 (2020). https://doi.org/10.1109/JLT.2020.2967087

8. Klinkowski, M., Lechowicz, P., Walkowiak, K.: Survey of resource allocation schemes and algorithms in spectrally-spatially flexible optical networking. Opt. Switch. and Netw. **27**, 58–78 (2018)
9. Ksieniewicz, P., Woźniak, M., Cyganek, B., Kasprzak, A., Walkowiak, K.: Data stream classification using active learned neural networks. Neurocomputing **353**, 74–82 (2019)
10. Mata, J., et al.: Artificial intelligence (AI) methods in optical networks: a comprehensive survey. Opt. Switch. and Netw. **28**, 43–57 (2018). https://doi.org/10.1016/j.osn.2017.12.006
11. Musumeci, F., et al.: An overview on application of machine learning techniques in optical networks. IEEE Commun. Surv. Tutorials **21**(2), 1383–1408 (2018)
12. Pedregosa, F., et al.: Scikit-learn: machine learning in python. J. Mach. Learn. Res. **12**, 2825–2830 (2011)
13. Rafique, D., Velasco, L.: Machine learning for network automation: overview, architecture, and applications [invited tutorial]. IEEE/OSA J. Opt. Commun. Netw. **10**(10), D126–D143 (2018). https://doi.org/10.1364/JOCN.10.00D126
14. Saridis, G.M., Alexandropoulos, D., Zervas, G., Simeonidou, D.: Survey and evaluation of space division multiplexing: from technologies to optical networks. IEEE Commun. Surv. Tutorials **17**(4), 2136–2156 (2015)
15. Walkowiak, K.: Modeling and Optimization of Cloud-Ready and Content-Oriented Networks. SSDC, vol. 56. Springer, Cham (2016). https://doi.org/10.1007/978-3-319-30309-3
16. Zyblewski, P., Ksieniewicz, P., Woźniak, M.: Classifier selection for highly imbalanced data streams with *Minority Driven Ensemble*. In: Rutkowski, L., Scherer, R., Korytkowski, M., Pedrycz, W., Tadeusiewicz, R., Zurada, J.M. (eds.) ICAISC 2019. LNCS (LNAI), vol. 11508, pp. 626–635. Springer, Cham (2019). https://doi.org/10.1007/978-3-030-20912-4_57

Missing Features Reconstruction Using a Wasserstein Generative Adversarial Imputation Network

Magda Friedjungová[✉][ID], Daniel Vašata[ID], Maksym Balatsko[ID],
and Marcel Jiřina

Faculty of Information Technology, Czech Technical University in Prague,
Prague, Czech Republic
{magda.friedjungova,daniel.vasata,balatmak,marcel.jirina}@fit.cvut.cz

Abstract. Missing data is one of the most common preprocessing problems. In this paper, we experimentally research the use of generative and non-generative models for feature reconstruction. Variational Autoencoder with Arbitrary Conditioning (VAEAC) and Generative Adversarial Imputation Network (GAIN) were researched as representatives of generative models, while the denoising autoencoder (DAE) represented non-generative models. Performance of the models is compared to traditional methods k-nearest neighbors (k-NN) and Multiple Imputation by Chained Equations (MICE). Moreover, we introduce WGAIN as the Wasserstein modification of GAIN, which turns out to be the best imputation model when the degree of missingness is less than or equal to 30%. Experiments were performed on real-world and artificial datasets with continuous features where different percentages of features, varying from 10% to 50%, were missing. Evaluation of algorithms was done by measuring the accuracy of the classification model previously trained on the uncorrupted dataset. The results show that GAIN and especially WGAIN are the best imputers regardless of the conditions. In general, they outperform or are comparative to MICE, k-NN, DAE, and VAEAC.

Keywords: Imputation methods · Feature reconstruction · Missing data · Generative models · Autoencoders · Wasserstein GAN

1 Introduction

When working with real-world datasets one of the standard problems that needs solving as part of the data preprocessing phase is dealing with missing data. The missingness can be represented by either individual missing data randomly located in instances or by the absence of entire features.

To our best knowledge, not much attention is paid to the second scenario where entire features are missing, i.e., there are no clear answers to questions such as how to face the situation, how the standard imputation method will perform or if there is a need to approach this challenge in a different way.

© Springer Nature Switzerland AG 2020
V. V. Krzhizhanovskaya et al. (Eds.): ICCS 2020, LNCS 12140, pp. 225–239, 2020.
https://doi.org/10.1007/978-3-030-50423-6_17

The aim of our work is to study these issues by experimentally comparing several state-of-the art imputation methods in real-world scenarios where one needs to impute (i.e., reconstruct) entire features. This work follows up on our previous work presented in paper [12], where we focus on the comparison of traditional (k-NN, linear regression, MICE) and modern (multi-layer perceptron, extreme gradient boosted trees) imputation methods.

In the current paper, we research more universal imputers represented by autoencoders and generative neural network models. These models have a common advantage in that one does not need to know which features are missing in advance. On the contrary, regular imputation methods need to be trained for each combination of missing features separately. A typical example where a universal imputer is needed is the prediction of a classification model from sensor data, where a sensor breakdown leads to missing data in one or more features. Usually, the prediction model itself is not able to handle this situation without a significant decrease in its performance. Furthermore, one typically does not know in advance which sensor is going to be broken. The best approach would be to retrain the model using data without missing features. However, in a production setting model retraining is impossible as the existing model needs to respond to corrupted data immediately.

We consider a situation where the prediction model is trained on a complete preprocessed dataset with numeric features, and we study its accuracy changes on new unseen data with imputed missing features. The amount of missing data (i.e. features) varies between 10% and 50%. Experiments are performed on ten real and two artificial datasets. The impact of imputation is measured as the classification accuracy change of the best performing from six commonly used classification models: logistic regression, multi-layer perceptron, k-NN, naive Bayes, extreme gradient boosted trees [7], and random forest. Besides accuracy we also use root mean squared error (RMSE) (which was also used in [6,17,35]) as a measure of the quality of the imputation.

We compare the denoising autoencoder (DAE) [33], Generative Adversarial Imputation Network (GAIN) [35], and Variational Autoencoder with Arbitrary Conditioning (VAEAC) [17] with k-NN and MICE [4], which are considered to be successful traditional imputation methods. Moreover, we introduce Wasserstein Generative Adversarial Imputation Network (WGAIN), a Wasserstein based modification of GAIN, see [2]. WGAIN is a generative imputation model and generally outperforms other presented models on the tested datasets. The Earth-Mover distance and the corresponding discriminator's critic of the Wasserstein approach do not suffer from vanishing gradients in the way that a vanilla GAN would. This enables the model to capture the desired distribution better.

The paper is organized as follows. In Sect. 2, we briefly review related work in this field. In Sect. 3 the WGAIN model is introduced. Section 4 is devoted to the description of experiments performed, including the evaluation of their results. Finally, the paper is concluded in Sect. 5.

2   Related Work

There are many traditional imputation methods, such as e.g., [11,24,32]. Some of the most common and successful are k-nearest neighbors imputation (k-NN) [18] and multivariate imputation by chained equations (MICE) [29,32].

Approaches based on deep learning have been under active development for the last few years. They use many variants of neural networks starting from multi-layer perceptron, e.g., in [3,30]. A more advanced approach is based on the autoencoder as a specific kind of neural network aiming to reconstruct inputs on its outputs. Here, one of the most commonly used models is the denoising autoencoder (DAE) [33], e.g., [5,8,10,15,34]. Typically, they are used in a discriminative way (see [15] for difference), meaning they impute a single value, which is deterministic once the network is trained.

On the other hand, the most recent research focuses on generative models which enables one to sample from the distribution conditioned on the observed features and thus get information about the uncertainty in imputed values. There are two groups of deep learning generative models. First, there are models based on the variational autoencoder (VAE) [19] and its conditional alternations, see [25,26,31,36]. In this group, some of the most successful imputation models are VAEAC [17] and HI-VAE [27].

The second group contains models based on the Generative Adversarial Network (GAN) [16]. Notably, one can encounter them in image reconstruction tasks (i.e., image inpainting), see [20,22,28]. One of the most prominent methods based on GAN is the GAIN [35], which uses the generator discriminator mechanism to achieve learning of the desired distribution. The generator observes some components of a real data vector, imputes the missing components conditioned on what is observed, and outputs a completed vector. The discriminator then takes a completed vector and attempts to determine which components were observed and which were imputed. The GAIN forms the base for our modification of the imputation method based on Wasserstein GAN [2], which is introduced in the next section. Only recently, GAIN was outperformed by the previously mentioned VAEAC and HI-VAE. However, for numeric variables, HI-VAE achieves a comparable error to the rest of the methods [27]. Therefore we have chosen only VAEAC for the experimental comparison.

3   Wasserstein Generative Adversarial Imputation Network

In this section, the WGAIN model is introduced as GAIN adapting the discriminative approach from Wasserstein GAN.

Let us denote $\mathcal{X} = \mathbb{R}^d$ the d-dimensional numeric data domain and let $X = (X_1, \ldots, X_d)$ be a random vector with values in $\mathcal{X}$ whose distribution is denoted by $P(X)$. Let the mask be a random binary vector M, i.e., random vector with values in $\{0,1\}^d$. The mask corresponds to unobserved values of X so that the value 0 of its jth component means that the jth feature of X_j is missing and the value 1 means that the jth feature of X_j is not missing. The distribution of M corresponds to the distribution of missingness in the data. Let us further denote by $\tilde{X}$ the vector X having zeros in place of missing values given by

$$\tilde{X} = X \odot M,$$

where $\odot$ denotes element-wise multiplication. Our aim is to impute missing values in $\tilde{X}$ based on information from non-missing features of $\tilde{X}$ and M. It is done in a generative way and it means that we want to learn the conditional distribution $\mathrm{P}(X|\tilde{X} = \tilde{x}, M = m)$ of X given $\tilde{X} = \tilde{x}$ and $M = m$. To do this let Z be a random vector with identically distributed independent components having normal distribution $\mathrm{N}(0, \sigma^2)$ with variance σ^2 and define

$$\tilde{X}_Z = Z \odot (1 - M) + X \odot M,$$

i.e. $\tilde{X}_Z$ is $\tilde{X}$ with missing components replaced by normal random variables.

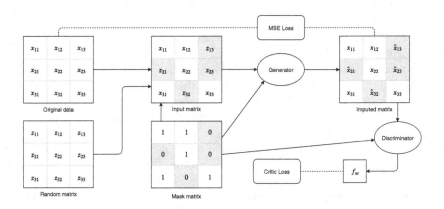

Fig. 1. WGAIN structure and mini-batch data flow.

The WGAIN model consists of two parts, the generator g and the critic f, both represented by deep neural networks. The generator g is constructed as a mapping $g : \mathcal{X} \times \{0,1\}^d \to \mathcal{X}$ so that

$$\hat{X}_Z = g(\tilde{x}_Z, m) \odot (1 - m) + \tilde{x} \odot m$$

is a random vector whose conditional distribution $\mathrm{P}(\hat{X}_Z|\tilde{X} = \tilde{x}, M = m)$, determined by the distribution $\mathrm{P}(Z)$ of Z, should be close to the conditional distribution $\mathrm{P}(X|\tilde{X} = \tilde{x}, M = m)$. Note that $g(\tilde{x}_Z, m)$ is a random vector corresponding to $\tilde{x}$ with all missing components imputed.

In order to train it, we employ the standard squared loss function

$$L_{\mathrm{MSE}}(\hat{x}_z, x) = \|\hat{x}_z - x\|^2,$$

forcing the output $\hat{\boldsymbol{X}}_Z$ to be close to the original data $\boldsymbol{X}$. However, it turns out that this condition alone is not sufficient for learning the proper conditional distribution. To improve the performance of the generator, one may introduce a discriminator trying to find out which components of $\hat{\boldsymbol{X}}_Z$ were imputed and use the discriminator for adversarial training. This approach was introduced in [35] and is the base of WGAIN.

In this paper we present a similar way how to improve the conditional distribution of the generator's output. It is based on the Earth-Mover (EM) distance between two probability distributions $\mathrm{P}(X), \mathrm{P}(Y)$ defined by

$$W\big(\mathrm{P}(X),\mathrm{P}(Y)\big) = \inf_{\gamma \in \boldsymbol{\Pi}(\mathrm{P}(X),\mathrm{P}(Y))} \mathrm{E}_{(X,Y)\sim\gamma}\|X - Y\|,$$

where $\boldsymbol{\Pi}(\mathrm{P}(X),\mathrm{P}(Y))$ denotes the set of all joint distributions (X,Y) whose marginals are respectively $\mathrm{P}(X)$ and $\mathrm{P}(Y)$. The term $\mathrm{E}_{(X,Y)\sim\gamma}\|X - Y\|$ might be understood as a measure of how much probability mass has to be transported in order to transform the distributions $\mathrm{P}(X)$ into the distribution $\mathrm{P}(Y)$ when the joint distribution is γ. The EM distance can thus be seen as the cost of the optimal transport plan, see [2] and references therein for more details. The EM distance is usually expressed using the Kantorovich-Rubinstein duality as

$$W\big(\mathrm{P}(X),\mathrm{P}(Y)\big) = \sup_{\|f\|_L \leq 1} \mathrm{E}_{X\sim\mathrm{P}(X)} f(X) - \mathrm{E}_{Y\sim\mathrm{P}(Y)} f(Y), \tag{1}$$

where $\|f\|_L$ means that f is Lipschitz continuous with Lipschitz constant 1 which might be changed to any constant K since it just multiplies $W\big(\mathrm{P}(X),\mathrm{P}(Y)\big)$ by the same constant.

In Wasserstein GAN one approximates (1) by training the neural network f_w parametrized with weights $\boldsymbol{w}$ in some compact space $\mathcal{W}$, thus enforcing the Lipschitz continuity. The function f_w is called the *critic* and is trained to maximize the expectations difference in (1). For a single dimensional generator g trying to transform random variable Z so that it has the distribution $\mathrm{P}(X)$ one maximizes

$$\max_{\boldsymbol{w}\in\mathcal{W}} \mathrm{E}_{X\sim\mathrm{P}(X)} f_w(X) - \mathrm{E}_{Z\sim\mathrm{P}(Z)} f_w(g(Z)).$$

In our case we want to minimize the EM distance between $\mathrm{P}(\hat{\boldsymbol{X}}_Z|\tilde{\boldsymbol{X}} = \tilde{\boldsymbol{x}}, \boldsymbol{M} = \boldsymbol{m})$ and $\mathrm{P}(\boldsymbol{X}|\tilde{\boldsymbol{X}} = \tilde{\boldsymbol{x}}, \boldsymbol{M} = \boldsymbol{m})$. Hence, we take the mask $\boldsymbol{M}$ as the second argument of the critic as additional information to the first argument given by $\boldsymbol{X}$ with correct features behind the mask $\boldsymbol{M}$. The critic is therefore a mapping $f_w : \mathcal{X} \times \{0,1\}^d \to \mathbb{R}$ trained to maximize

$$\max_{\boldsymbol{w}\in\mathcal{W}} \mathrm{E}_{\boldsymbol{X}\sim\mathrm{P}(X)} f_w(\boldsymbol{X},\boldsymbol{M}) - \mathrm{E}_{Z\sim\mathrm{P}(Z)} f_w(\hat{\boldsymbol{X}}_Z,\boldsymbol{M}),$$

which is usually estimated by sample means from mini-batches. The overall structure of WGAIN is depicted in Fig. 1.

Algorithm 1: WGAIN training pseudo-code

Input: α - the learning rate; $w_{\max}$ - maximal norm used in clipping; m - the mini-batch size

Draw m samples from the dataset $\{x_j\}_{j=1}^m$;

Draw m samples from the mask distribution $\{m_j\}_{j=1}^m$;

Draw m samples from the normal distribution of Z, $\{z_j\}_{j=1}^m$;

while *not converged* **do**

$\quad \tilde{x}_{z_j} \leftarrow z_j \odot (1 - m_j) + x_j \odot m_j$;

$\quad \hat{x}_{z_j} \leftarrow g(\tilde{x}_{z_j}, m_j) \odot (1 - m_j) + x_j \odot m_j$;

$\quad$ Update weights w of f_w using RMSProp with learning rate α and gradient
$\quad \nabla J(f_w) = \lambda_{f_w} \nabla \left[\frac{1}{m} \sum_{i=1}^m f_w \left(\hat{x}_{z_j}, m_j \right) - \frac{1}{m} \sum_{i=1}^m f_w \left(x_j, m_j \right) \right]$;

$\quad$ Clip the norm of w by $w_{\max}$;

$\quad$ Update weights of g using RMSProp with learning rate α and gradient
$\quad \nabla J(g) = \nabla \left[-\lambda_g \frac{1}{m} \sum_{i=1}^m f_w \left(\hat{x}_{z_j}, m_j \right) + \lambda_{\mathrm{MSE}} \frac{1}{m} \sum_{i=1}^m \| \hat{x}_{z_j} - x_j \|^2 \right]$;

end

3.1   Training

The critic f_w is used in adversarial training of both the generator g and the critic itself. There the generator and the critic play an iterative two-player minimax game when the critic wants to recognize the imputed values from the real ones and the goal of the generator is to trick the critic so it cannot recognize them. Moreover, the generator's output is tighten to the correct output by the squared loss function L_{MSE}.

Putting it all together, we have two objective functions to minimize. The first corresponds to training of the discriminator given by

$$J(f_w) = \lambda_{f_w} \left(\mathrm{E}_{Z \sim \mathrm{P}(Z)} f_w(\hat{X}_Z, M) - \mathrm{E}_{X \sim \mathrm{P}(X)} f_w(X, M) \right),$$

where the weight λ enables one to increase or decrease the influence of the corresponding gradient. Second is the objective for the generator,

$$J(g) = -\lambda_g \mathrm{E}_{Z \sim \mathrm{P}(Z)} f_w(\hat{X}_Z, M) + \lambda_{\mathrm{MSE}} \mathrm{E}_{X \sim \mathrm{P}(X), Z \sim \mathrm{P}(Z)} L_{\mathrm{MSE}}(\hat{X}_Z, X),$$

where the first term λ_g and λ_{MSE} are weights enabling one to strengthen or weaken the influence of squared loss function. The optimization is done via alternating gradient descent, where the first step is updating the critic f_w and the second step is updating the generator g. Hence, when perfectly trained, the discriminator gives negative values to cases with imputed features and positive values for cases with true features. On the other hand, the generator entering the critic will be pushed to obtain large positive values of the critic as it gives to real values.

The pseudo-code of the WGAIN training is given in Algorithm 1.

4   Experiments

An experimental validation of WGAIN using ten real and two artificial publicly available datasets is presented below. These datasets contain numeric data only and are devoted to the classification task. Their overview, together with the corresponding best performing classification models, is given in Table 2.

During the experiments, all datasets were divided as follows: 70% of data was used to train all classification and imputation models and 30% was used as a test set to evaluate imputation performance. The imputation models were trained to impute in scenarios where randomly selected combinations of multiple features are missing. The amount of missingness varies from 10% to 50% of missing features. Finally, evaluation of the accuracy of the classification model combined with all imputation methods is performed on the test dataset.

4.1   Imputation Models and Their Parameters

Let us start with the presented WGAIN model. The generator and the critic architectures were the same for all datasets and are described in Table 1. During the training, the following settings were used:

- The original data X are sampled in mini-batches of size $m = 128$.
- The missingness is introduced using the mask M with the following distribution: for each training point, the portion of missingness is sampled from a uniform distribution between 0 and maximum missing rate, which was chosen to be 0.3. Then the binary elements of M were independently sampled with this portion of missingness, i.e., its item is 0 with a probability which was previously sampled.
- The components of random vector Z are i.i.d. with normal distribution having 0 mean and standard deviation 0.01.
- The weights of the objectives functions $J(f_w)$ and $J(g)$ are $\lambda_{f_w} = 10$, $\lambda_g = 2$, and $\lambda_{\mathrm{MSE}} = 1$.
- Maximal norm used in clipping of the critic weights is $w_{\max} = 1$.
- We use RMSProp with learning rate $\alpha = 0.0001$ as optimizers.
- The number of training epochs is 8000.

The GAIN implementation follows the original paper [35] and is analogous to the described WGAIN with the following differences:

- The generator architecture differs only in the sizes of layers, which are all equal to the input dimension.
- The discriminator architecture is analogous to the generator architecture except for the sigmoid activation function on the last layer.
- The binary elements of M are independently sampled with the common portion of missingness, which is 0.2.
- The hint rate used for the hint matrix is 0.9.
- As an optimizer, we use Adam with learning rate of 0.0001.
- The number of training epochs is 7000.

Table 1. Architecture details of the WGAIN. Abbreviation: FC = fully connected layer.

Layer	Generator
	Concatenate data and mask
1	FC-(1.5 input dimension), ReLU
2	FC-(1.25 input dimension), ReLU
3	FC-(input dimension), Linear
Layer	Critic
	Concatenate data and mask
1	FC-(1.5 input dimension), ReLU
2	FC-(1.25 input dimension), ReLU
3	FC-(1), Linear

In the case of DAE, we follow the structure presented in [15]. For the hyperparameters search, the hyperband [21] algorithm was used. The typical best setup is the following: ELU as an activation function, three layers in both the encoder and decoder parts, the size of the code is twice the input dimension, and no regularization is used.

DAE, GAIN, and WGAIN models were implemented in the `TensorFlow` library[1].

The implementation of VAEAC was based on the repository[2] corresponding to the original paper [17]. All hyper-parameters stayed in the default settings.

For the MICE method (*mice*), we used the `IterativeImputer` class from the `scikit-learn` library[3]. In the default settings, the implementation uses Bayesian ridge regression as the internal imputation model and multiple imputations are pooled by the mean.

The k-NN imputation (*knn*) was implemented using the `fancyimpute` library[4]. A missing value is imputed by sampling the mean of the values of its neighbors weighted proportionally to their inverse distances. In the case where multiple features are missing, we impute all missing values at once (per row). For the hyper-parameter k values $11, 13, 15, 17, 19, 21, 23, 25$ were tested. The best k was chosen based on the RMSE value.

4.2 Evaluation

The impact of imputation is evaluated using the classification accuracy changes of the best performing classification model chosen from the six commonly used

---

[1] `TensorFlow` platform: https://www.tensorflow.org.
[2] `VAEAC` implementation: https://github.com/tigvarts/vaeac.
[3] `Scikit-learn` library: https://scikit-learn.org.
[4] `Fancyimpute` repository: https://github.com/iskandr/fancyimpute.

ones: logistic regression (LR), multi-layer perceptron (MLP), k-nearest neighbors (k-NN), naive Bayes (NB), extreme gradient boosted trees (XGBT) (for details see [7]), and random forest (RF). The best hyperparameters for each model were found using randomized search algorithm. The accuracy of the best performing model for each dataset is shown in Table 2. Furthermore, the root mean squared error (RMSE) between the original and the imputed data is also used for evaluation, e.g., [6,17,35].

Table 2. Details of datasets with the corresponding best performing classification model and its accuracy on the test set. The number of features (# f.) does not include the target label. The # r. stands for the number of records.

Name	Type	# f.	# r.	Model name	Accuracy
Cancer [23]	Real	9	683	RF	0.975
EEG [23]	Real	14	14980	k-NN	0.952
MAGIC [23]	Real	10	19020	XGBT	0.868
Ozone-1 [23]	Real	72	1846	k-NN	0.977
Ozone-8 [23]	Real	72	1848	LR	0.941
QSAR [23]	Real	41	1055	MLP	0.868
Shuttle [23]	Real	9	57998	RF	0.999
Spambase [23]	Real	57	4597	MLP	0.940
Waveform [23]	Real	21	5000	LR	0.869
Yeast [23]	Real	8	1484	XGBT	0.578
Ringnorm [1]	Art	20	7400	NB	0.979
Twonorm [1]	Art	20	7400	MLP	0.979

After all classification models were trained, and the most accurate one for each dataset was chosen, they were combined with imputation methods. Then, the accuracies of classification models on the imputed test dataset were measured.

Since it is not sound to compare accuracies for different datasets, we use a rank comparison. To do so, the algorithms are ranked for each dataset separately, the best performing algorithm getting the rank of 1, the second-best rank 2, etc. An example of accuracies and corresponding ranks for 10% of missingness is presented in Tables 4 and 5. Even in cases when WGAIN is not the best, its performance is always comparable to the best performers. The only exception is the EEG dataset, where k-NN imputation performs the best and the WGAIN is in second place with a difference of almost two percent.

The algorithms can be compared, taking the mean over the datasets. The results can be seen in Table 9. When the degree of missingness varies from 10% to 30% the WGAIN performs the best. When the degree of missingness is upwards of 30% the GAIN outperforms the WGAIN.

Table 3. Mean ranks of the RMSE for different degrees of missingness.

Method	Degree of missingness				
	10%	20%	30%	40%	50%
k-NN	2.67	2.67	2.67	3.08	2.75
MICE	3.17	3.50	3.33	3.00	2.91
DAE	5.08	5.08	5.17	5.33	4.91
VAEAC	3.25	3.33	3.42	3.17	3.50
GAIN	**2.17**	**2.08**	**2.17**	**2.08**	**2.50**
WGAIN	4.67	4.33	4.25	4.33	4.42

Table 4. Mean of the accuracies for 10% of missing features.

	k-NN	MICE	DAE	VAEAC	GAIN	WGAIN
Cancer	0.9700	0.9744	0.9744	0.9749	0.9739	**0.9755**
EEG	**0.9226**	0.9046	0.8994	0.6374	0.9028	0.9052
MAGIC	**0.8562**	0.8465	0.8459	0.8527	0.8522	0.8511
Ozone-1	0.9754	0.9763	**0.9768**	0.9762	0.9759	0.9763
Ozone-8	0.9404	**0.9407**	0.9405	0.9405	0.9406	0.9406
QSAR	0.8608	0.8619	0.8615	0.8619	0.8609	**0.8626**
Shuttle	0.9995	**0.9996**	0.9945	0.9994	0.9992	0.9995
Spambase	**0.9363**	0.9278	0.9307	0.9303	0.9339	0.9296
Waveform	0.8603	0.8604	0.8585	0.8596	**0.8605**	0.8593
Yeast	0.5516	0.5507	0.5533	0.5496	0.5541	**0.5558**
Ringnorm	0.9668	0.9671	0.9672	0.9673	0.9674	**0.9680**
Twonorm	0.9711	0.9716	0.9716	0.9716	0.9719	**0.9723**

The results of the ranking evaluation can be statistically evaluated using the Friedman test [13,14] and the corresponding posthoc tests. For more details, see [9]. P-values of Friedman χ_F^2 and F_F tests are shown in Table 8. One can see that from 20% to 40% of missing data the null-hypothesis, that all methods perform the same, can be rejected at a 10% significance level. However, when the Bonferroni-Dunn post-hoc test is applied the performance of WGAIN is significantly better than DAE only and just for 20% and 30% of missing data.

Table 5. Ranks of accuracies of the imputation methods for 10% of missing features.

	k-NN	MICE	DAE	VAEAC	GAIN	WGAIN
Cancer	6	3.5	3.5	2	5	1
EEG	1	3	5	6	4	2
MAGIC	1	5	6	2	3	4
Ozone-1	6	2	1	4	5	3
Ozone-8	6	1	5	4	2.5	2.5
QSAR	6	2.5	4	2.5	5	1
Shuttle	2	1	6	3.5	5	3.5
Spambase	1	6	3	4	2	5
Waveform	3	2	6	4	1	5
Yeast	4	5	3	6	2	1
Ringnorm	6	5	4	3	2	1
Twonorm	6	4	4	4	2	1

Table 6. Mean of the RMSE for 10% of missing features.

	k-NN	MICE	DAE	VAEAC	GAIN	WGAIN
Cancer	**0.1905**	0.1960	0.2219	0.1943	0.1959	0.2087
EEG	**16.4752**	27.8197	29.1700	293.9315	21.8986	34.2722
MAGIC	**0.1821**	0.2067	0.2072	0.1866	0.1844	0.1928
Ozone-1	0.1364	**0.0826**	0.1204	0.1047	0.1038	0.1051
Ozone-8	0.1549	**0.0972**	0.1473	0.1233	0.1230	0.1206
QSAR	**0.2356**	0.3115	0.2505	0.2445	0.2376	0.2492
Shuttle	**0.0954**	0.1022	0.1316	0.1085	0.1053	0.1097
Spambase	**0.2404**	0.2723	0.2692	0.2659	0.2587	0.2731
Waveform	0.2312	0.2304	0.2690	0.2301	**0.2278**	0.2429
Yeast	**0.3542**	0.3610	0.3666	0.3585	0.3560	0.3631
Ringnorm	0.3222	**0.3184**	0.3187	0.3191	0.3190	0.3282
Twonorm	0.2967	0.2948	0.3081	0.2935	**0.2918**	0.2975

The same ranking process is repeated for RMSE with results in Table 3. An example of RMSE and corresponding ranks for 10% of missingness is presented in Tables 6 and 7. Interestingly, the WGAIN performance is one of the worst, whereas the GAIN performs the best. This is in contrary to the fact that the WGAIN imputes the best from the accuracy point of view. Hence, we can see that low RMSE, which is usually taken as a measure of imputation quality may not lead to the desired performance on the target task. On the other hand, the RMSE differences are relatively small as can be seen in Table 6.

Table 7. Ranks of RMSE of the imputation methods for 10% of missings.

	k-NN	MICE	DAE	VAEAC	GAIN	WGAIN
Cancer	1	4	6	2	3	5
EEG	1	3	4	6	2	5
MAGIC	1	5	6	3	2	4
Ozone-1	6	1	5	3	2	4
Ozone-8	6	1	5	4	3	2
QSAR	1	6	5	3	2	4
Shuttle	1	2	6	4	3	5
Spambase	1	5	4	3	2	6
Waveform	4	3	6	2	1	5
Yeast	1	4	6	3	2	5
Ringnorm	5	1	2	4	3	6
Twonorm	4	3	6	2	1	5

Table 8. P-values of Friedman χ_F^2 and F_F test.

	Degree of missingness				
	10%	20%	30%	40%	50%
χ_F^2 test	0.252	0.049	0.106	0.020	0.477
F_F test	0.253	0.041	0.099	0.014	0.490

Table 9. Mean ranks of the accuracy changes for different degrees of missingness.

	Degree of missingness				
Method	10%	20%	30%	40%	50%
k-NN	4.00	3.54	3.63	3.17	3.25
MICE	3.33	4.04	3.75	4.21	3.71
DAE	4.21	4.79	4.67	4.71	4.33
VAEAC	3.75	3.13	3.50	3.59	3.54
GAIN	3.21	2.83	2.83	**2.21**	**2.79**
WGAIN	**2.50**	**2.67**	**2.63**	3.12	3.38

5 Conclusion

We propose a Wasserstein Generative Adversarial Imputation Network as a new deep learning imputation model. It is inspired by the GAIN. However, the discriminator is replaced by the Wasserstein critic. It is known that the Wasserstein approach does not suffer from vanishing gradients in the way that a vanilla GAN does. This enables the model to capture the desired distribution better. One may assume such benefits in WGAIN as well. We experimentally showed that in the imputation performance measured by classification accuracy, the WGAIN outperforms the other methods when the degree of missingness is lower than or equal to 30%. In other cases, it is competitive. In future work, we would like to focus on the use of WGAIN in image inpainting tasks.

Acknowledgements. This research has been supported by SGS grant No. SGS20/213/OHK3/3T/18 and by GACR grant No. GA18-18080S.

References

1. Alcalá-Fdez, J., et al.: Keel data-mining software tool: data set repository, integration of algorithms and experimental analysis framework. J. Mult. Valued Logic Soft Comput. **17**, 255–287 (2011)
2. Arjovsky, M., Chintala, S., Bottou, L.: Wasserstein gan (2017)
3. Arroyo, Á., Herrero, Á., Tricio, V., Corchado, E., Woźniak, M.: Neural models for imputation of missing ozone data in air-quality datasets. Complexity **2018**, 14 (2018)
4. Azur, M.J., Stuart, E.A., Frangakis, C., Leaf, P.J.: Multiple imputation by chained equations: what is it and how does it work? Int. J. Methods Psychiatric Res. **20**(1), 40–49 (2011)
5. Beaulieu-Jones, B.K., Moore, J.H.: Missing data imputation in the electronic health record using deeply learned autoencoders. In: Pacific Symposium on Biocomputing 2017, pp. 207–218. World Scientific (2017)
6. Camino, R.D., Hammerschmidt, C.A., State, R.: Improving missing data imputation with deep generative models. CoRR, abs/1902.10666 (2019)
7. Chen, T., Guestrin, C.: Xgboost: a scalable tree boosting system. In: Proceedings of the 22Nd ACM SIGKDD International Conference on Knowledge Discovery and Data Mining, KDD 2016, pp. 785–794. ACM, New York (2016)
8. Costa, A.F., Santos, M.S., Soares, J.P., Abreu, P.H.: Missing Data Imputation via Denoising Autoencoders: The Untold Story. In: Duivesteijn, W., Siebes, A., Ukkonen, A. (eds.) IDA 2018. LNCS, vol. 11191, pp. 87–98. Springer, Cham (2018). https://doi.org/10.1007/978-3-030-01768-2_8
9. Demšar, J.: Statistical comparisons of classifiers over multiple data sets. J. Mach. Learn. Res. **7**, 1–30 (2006)
10. Duan, Y., Lv, Y., Liu, J.-L., Wang, F.-Y.: An efficient realization of deep learning for traffic data imputation. Transp. Res. Part C Emerg. Technol. **72**, 168–181 (2016)
11. Farhangfar, A., Kurgan, L.A., Dy, J.G.: Impact of imputation of missing values on classification error for discrete data. Pattern Recogn. **41**, 3692–3705 (2008)
12. Friedjungová, M., Jiřina, M., Vašata, D.: Missing features reconstruction and its impact on classification accuracy. In: Rodrigues, J.M.F., et al. (eds.) ICCS 2019. LNCS, vol. 11538, pp. 207–220. Springer, Cham (2019). https://doi.org/10.1007/978-3-030-22744-9_16
13. Friedman, M.: The use of ranks to avoid the assumption of normality implicit in the analysis of variance. J. Am. Statist. Assoc. **32**(200), 675–701 (1937)

14. Friedman, M.: A comparison of alternative tests of significance for the problem of m rankings. Ann. Math. Statist. **11**(1), 86–92 (1940)
15. Gondara, L., Wang, K.: MIDA: Multiple imputation using denoising autoencoders. In: Phung, D., Tseng, V.S., Webb, G.I., Ho, B., Ganji, M., Rashidi, L. (eds.) PAKDD 2018. LNCS (LNAI), vol. 10939, pp. 260–272. Springer, Cham (2018). https://doi.org/10.1007/978-3-319-93040-4_21
16. Goodfellow, I.J.,et al.: Generative adversarial nets. In: Advances in Neural Information Processing Systems 27: Annual Conference on Neural Information Processing Systems 2014, 8–13 December 2014, Montreal, Quebec, Canada, pp. 2672–2680 (2014)
17. Ivanov, O., Figurnov, M., Vetrov, D.: Variational autoencoder with arbitrary conditioning. In: International Conference on Learning Representations (2019)
18. Jonsson, P., Wohlin, C.: An evaluation of k-nearest neighbour imputation using likert data. In: 10th International Symposium on Software Metrics, 2004. Proceedings, pp. 108–118. IEEE (2004)
19. Kingma, D.P., Welling, M.: Auto-encoding variational bayes. In: 2nd International Conference on Learning Representations, ICLR 2014, Banff, AB, Canada, 14–16 April 2014, Conference Track Proceedings (2014)
20. Lee, D., Kim, J., Moon, W.-J., Ye, J.C.: Collagan: Collaborative GAN for missing image data imputation. CoRR, abs/1901.09764 (2019)
21. Li, L., Jamieson, K., DeSalvo, G., Rostamizadeh, A., Talwalkar, A.: Hyperband: a novel bandit-based approach to hyperparameter optimization. J. Mach. Learn. Res. **18**(1), 6715–6816 (2017)
22. Li, S.C.-X., Jiang, B., Marlin, B.M.: Misgan: learning from incomplete data with generative adversarial networks. CoRR, abs/1902.09599 (2019)
23. Lichman, M.: UCI machine learning repository (2013). http://archive.ics.uci.edu/ml
24. Little, R.J.A., Rubin, D.B.: Statistical Analysis with Missing Data, vol. 333. Wiley, Hoboken (2014)
25. Lopez-Martin, M., Carro, B., Sanchez-Esguevillas, A., Lloret, J.: Conditional variational autoencoder for prediction and feature recovery applied to intrusion detection in IoT. Sensors **17**(9) (2017)
26. McCoy, J.T., Kroon, S., Auret, L.: Variational autoencoders for missing data imputation with application to a simulated milling circuit. IFAC-PapersOnLine **51**(21), 141–146 (2018)
27. Nazábal, A., Olmos, P.M., Ghahramani, Z., Valera, I.: Handling incomplete heterogeneous data using vaes. CoRR, abs/1807.03653 (2018)
28. Pathak, D., Krahenbuhl, P., Donahue, J., Darrell, T., Efros, A.A.: Context encoders: feature learning by inpainting. In: Proceedings of the IEEE Conference on Computer Vision and Pattern Recognition, pp. 2536–2544 (2016)
29. Schafer, J.L.: Analysis of Incomplete Multivariate Data. Chapman and Hall, London (1997)
30. Silva-Ramírez, E.-L., Pino-Mejías, R., López-Coello, M.: Single imputation with multilayer perceptron and multiple imputationcombining multilayer perceptron and k-nearest neighbours for monotonepatterns. Appl. Soft Comput. **29**, 65–74 (2015)
31. Sohn, K., Lee, H., Yan, X.: Learning structured output representation using deep conditional generative models. In: Cortes, C., Lawrence, N.D., Lee, D.D., Sugiyama, M., Garnett, R. (eds.) Advances in Neural Information Processing Systems 28, pp. 3483–3491. Curran Associates Inc. (2015)
32. Van Buuren, S.: Flexible Imputation of Missing Data. Chapman and Hall/CRC, Boca Raton (2018)

33. Vincent, P., Larochelle, H., Bengio, Y., Manzagol, P.-A.: Extracting and composing robust features with denoising autoencoders. In: Proceedings of the 25th International Conference on Machine Learning ACM (2008)
34. Wong, L.Z., Chen, H., Lin, S., Chen, D.C.: Imputing missing values in sensor networks using sparse data representations. In: Proceedings of the 17th ACM International Conference on Modeling, Analysis and Simulation of Wireless and Mobile Systems, MSWiM 2014, pp. 227–230. ACM, New York (2014)
35. Yoon, J., Jordon, J., van der Schaar, M.: GAIN: missing data imputation using generative adversarial nets. In: Jennifer Dy and Andreas Krause, editors, Proceedings of the 35th International Conference on Machine Learning, volume 80 of Proceedings of Machine Learning Research, pp. 5689–5698. PMLR, Stockholmsmässan, Stockholm Sweden, 10–15 Jul 2018
36. Zadeh, A., Lim, Y.C., Liang, P.P., Morency, L.-P.: Variational auto-decoder. CoRR, abs/1903.00840 (2019)

Complex Social Systems Through the Lens of Computational Science

Cooperation for Public Goods Under Uncertainty

Jeroen Bruggeman[1]([envelope])[ORCID] and Rudolf Sprik[2]([envelope])[ORCID]

[1] Department of Sociology, University of Amsterdam, Amsterdam, The Netherlands
j.p.bruggeman@uva.nl
[2] van der Waals-Zeeman Institute, University of Amsterdam,
Amsterdam, The Netherlands
r.sprik@uva.nl

Abstract. Everyone wants clean air, peace and other public goods but is tempted to freeride on others' efforts. The usual way out of this dilemma is to impose norms, maintain reputations and incentivize individuals to contribute. In situations of high uncertainty, however, such as confrontations of protesters with a dictatorial regime, the usual measures are not feasible, but cooperation can be achieved nevertheless. We use an Ising model with asymmetric spins that represent cooperation and defection to show numerically how public goods can be realized. Under uncertainty, people use the heuristic of conformity. The turmoil of a confrontation causes some individuals to cooperate accidentally, and at a critical level of turmoil, they entail a cascade of cooperation. This critical level is much lower in small networks.

Keywords: Public goods · Cooperation · Uncertainty · Ising model

1 Introduction

A main benefit for people living in groups, ranging from families to empires, is that collectively, they can achieve more than the sum of individuals can independently, in particular collective goods [10]. Cases in point are infrastructure, health care, defense and education for a country's population. People first have to agree on the public good and the way to realize it, but even if they arrive at a consensus, the provision of these goods is non-obvious because individuals are tempted to freeride on others' efforts. Cooperation is a dilemma [18] that usually requires time to solve, during which people develop norms and form a network to transmit information (gossip) about one another, on the basis of which reputations are established that in their turn are used to reward cooperators and punish defectors [8,17].

This scenario is well-covered by the literature, but there are important cases where there is no time to develop it. Examples are disasters where victims need urgent help, protests against dictatorial regimes that prevent critics to organize themselves, and unplanned violent confrontations between groups. These

© Springer Nature Switzerland AG 2020
V. V. Krzhizhanovskaya et al. (Eds.): ICCS 2020, LNCS 12140, pp. 243–251, 2020.
https://doi.org/10.1007/978-3-030-50423-6_18

cases have high uncertainty in common. People then realize that cooperation might yield a valuable outcome but they can't assess their payoffs, among others because they don't know if they will get hurt. Our questions are how cooperation under uncertainty is self-organized, and how this is influenced by participants' network.

The first question has been addressed by critical mass theory [11,14], which does not rely on norms, reputations and selective incentives (i.e. rewards and punishments). It argues that if a critical number of actors starts cooperating, the others are won over to join in, thereby making cooperation self-reinforcing. Individuals are assumed to know at each moment how many cooperators there are in total, and to have fairly accurate expectations of their marginal payoff. On the basis thereof they can rationally decide to contribute to the public good or to freeload. In highly uncertain situations, however, these rationality assumptions are unlikely to hold true. Moreover, the theory does not show the effect of network topology, and does not explain the cooperation of the initiative takers before the critical point is reached. We, in contrast, explicate the effect of topology, explain the initiative takers endogenously, and instead of rationality assume one simple heuristic: under high uncertainty, people become conformists to the majority of their network neighbors [20]. This may turn out bad for them in a specific situation but may serve them on average over many occasions [2]. Conformism can be observed as behavioral synchronization [15]. When people conform to their initially defecting network neighbors, they might eventually cooperate when many others do, but who would start?

Pending (or starting) violence and disaster are characterized by increasing turmoil, for example in terms of opponents' threats, insults and violence. Turmoil can be measured by participant's heart rates that indicate their arousal [13] or, in an information theoretic manner, by accelerating situational updates to the same effect [12]. Arousal causes "trembling hands" [6] as game theorists say, denoting a chance that some individuals accidentally cooperate, which in turn might entail a cascade of cooperation. An example of turmoil is the mutual provocation of soccer fans of opposing camps to a boiling point when fighting breaks out; their public good is the humiliation of their opponents. To show that turmoil can drive cooperation, we use an Ising model [1,4], for which we make the behavioral options (spins) asymmetric, reflecting the asymmetry of defection versus cooperation. This provides a parsimonious explanation of cooperation without the usual mechanisms developed over longer time, and without rationality assumptions.

2   Model

At the beginning there is a group of n individuals who may not know one another yet, in an uncertain situation. Each of them has at least visual contact with some others, and identifies with, and is therefore inclined to conform to, others who share an interest in a given public good. The network of visual or verbal ties A_{ij} denoting that i pays attention to j is represented by a row-normalized adjacency matrix with cells $a_{ij} = A_{ij}/\sum_j A_{ij}$, as in many models of social influence [9], hence $\sum_j a_{ij} = 1$. Ties are bi-directional but asymmetric, and the network may have multiple disconnected components. An individual can cooperate (C) or defect (D) with $C > D > 0$, and everybody defects at the beginning. The behavioral variable S_i can take the value $S_i = C$ or $S_i = -D$. Behavior and network are expressed in the Hamiltonian of the Ising model

$$H = -\sum_{i \neq j}^{n} a_{ij} S_i S_j. \tag{1}$$

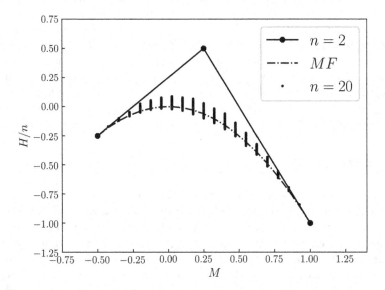

Fig. 1. Ising model with $S = \{1, -1/2\}$. Mean dissatisfaction H/n at different levels of cooperation M in a network with $n = 2$, and in a random net with $n = 20$ and density $= 0.8$. In the latter, vertical line segments are composed of dots, each representing a micro state. The dash-dotted line is drawn with the mean field approach (MF).

The mean degree of cooperation is described by an order parameter $M = 1/n \sum_{i=1}^{n} S_i$. A cooperator's payoff is a share of the public good at a cost $P_C = r(M + C/n) - C$, with a synergy parameter $r > 1$; a defector's payoff is $P_D = r(M - D/n)$ [19]. In conventional game theory, $C = 1$ and $D = 0$ [19], but since payoffs are used only comparatively, it does not matter if they are negative here. If people believe that $C \leq D$, this payoff function does not apply and people will not cooperate anyway. If they believe that $C > D$, they may cooperate provided that enough others do. Because under high uncertainty, participants can't assess the synergy of their collective effort, don't know the costs and benefits, and in large groups can't see all cooperators, they can't calculate their payoff. The conformism heuristic then works as a collective lever that minimizes H and can maximize the mean payoff indirectly, even without individuals knowing the payoff function. Individuals' heuristic decision making is implemented computationally by the Metropolis algorithm [1]. Over a large number of Monte Carlo steps, a node i is chosen randomly, its behavior S_i is flipped from D to C (or the other way around), and H is compared to H' that has the flip implemented and is otherwise identical to H. The flip is accepted if for a random number $0 \leq c_r \leq 1$,

$$c_r < exp\left(\frac{-(H'-H)}{T}\right), \qquad (2)$$

where T is the level of turmoil, or temperature in the original model [1]. H/n can be regarded as mean dissatisfaction with respect to the conformity heuristic, biased towards cooperation for the public good. The relation between H/n and M is illustrated in Fig. 1 for $S = \{1, -1/2\}$ on a network with $n = 2$, for a random net with $n = 20$ including all micro states (configurations of C's and D's) therein, and in a mean field approach (MF) for comparison. The Metropolis minimization of H starts at the defective state (at the left) and evolves by random steps over all micro states until the lowest value of H is reached at the cooperative state (at the right).

3   Results

The relation between cooperation, M, and turmoil, T, turns out to be qualitatively the same for all networks. At low turmoil, there is no cooperation but it emerges at a critical level T_c, illustrated in Fig. 2. Then the turmoil overcomes the "energy" barrier (the mountain in Fig. 1), and drives the transition to the cooperative state. For as long as $T < T_c$, the chance that this happens in a finite number of Monte Carlo steps is zero. If T keeps increasing beyond T_c, the effect of turmoil becomes progressively more dominant until individuals randomly alternate cooperation and defection. Here, turmoil trumps conformism. If not by this chaos, cooperation will end when the public good is realized, others intervene, the participants get exhausted, or if they get to understand the situation (i.e. their uncertainty lowers) and start behaving strategically (i.e. freeride).

Social networks, if large, are clustered and sparse with a skewed degree distribution [16]. We therefore examine the effects of clustering, density, degree distribution and size on the tipping point T_c, and subsequently zoom in on local variations of S. As a characteristic example of clustering, we start with ten cliques (fully connected networks) with $n = 10$ each (Fig. 2). When randomly rewiring ties with a probability p, the network becomes more random at higher p and thereby less clustered, as in small world networks [16]. Decreasing clustering results in increasing T_c. In Fig. 3 this is demonstrated by starting with a strongly clustered network, and by rewiring the links, which will often form random cross-cluster connections. From the most strongly clustered network onward, the largest effect on T_c is at small numbers of rewired ties. The effect of network size on the tipping point is larger than of clustering, though,

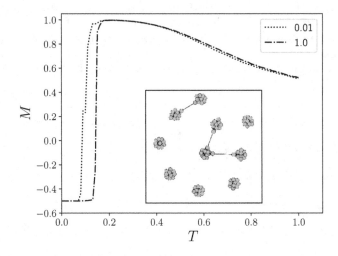

Fig. 2. Ising model with $S = \{1, -1/2\}$, showing M with increasing T for a network with $n = 100$ and 10 clusters, $n = 10$ each. There is a chance p for ties to be rewired. If $p = 0$, the clusters are fully connected cliques, mutually disconnected. The network at $p = 0.01$ is shown at the inset and with a dotted line in the $M - T$ plot. If $p = 1$ the network becomes essentially a random graph, with a higher T_c.

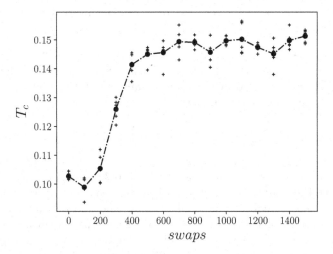

Fig. 3. Effect of increasing rewiring on the tipping point T_c in a $n = 100$ graph with 10 clusters, $n = 10$ each, as in Fig. 2. Again, $S = \{1, -1/2\}$. The horizontal axis indicates the number of swaps. A swap is generated by randomly choosing three nodes with 1 and 2 connected but not 1 and 3. Then 1's tie to 2 is relayed to 3. The simulation was repeated five times (plus signs) and resulted in an average value indicated by the dots connected by the dash-dotted line.

in particular at the smallest network sizes (Fig. 4). T_c is also lower in sparser networks but the effect of density is much weaker than of size and more variable across simulation runs (not shown). When varying the degree distribution between Poisson and power law, T_c does not change at all. In sum, changes of T_c are largest at small network size, high clustering, and low density.

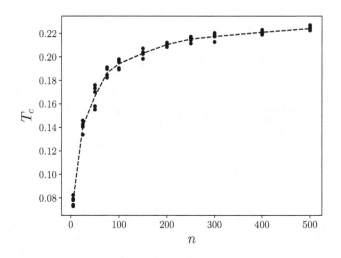

Fig. 4. Effect of increasing network size n on the tipping point T_c in random networks with density = 0.8 and $S = \{1, -1/2\}$. The dots show small variation across simulation runs.

To investigate the effect of local variations of S, we generalize the values of $S = \{1, -1/2\}$ that we used in the above examples. To this end we rewrite the asymmetric Ising model in a symmetric form by the mapping

$$S = \{C, -D\} \rightarrow \{S_0 + \Delta, S_0 - \Delta\}, \tag{3}$$

with a bias $S_0 = (C - D)/2$ and an increment $\Delta = (C + D)/2$. Accordingly, the values chosen in our examples imply $S_0 = 0.25$ and $\Delta = 0.75$. Elsewhere we show that the Hamiltonian, once written in a symmetric form, becomes a conventional Ising model plus a global external field and a locally varying field that depends on the network [3]. We use this generalized form in a mean field analysis, from which we infer that for given Δ, higher S_0 (that corresponds to a stronger interest in the public good) results in lower T_c [3].

We now use different S_0 in a network with two clusters, keeping $\Delta = 0.75$. When $p = 0$ the two groups are disconnected, but within-group links can be randomly rewired with a probability p and thereby connect the two groups. In Fig. 5, two-cluster networks ($n = 20$) with $p = 0$ and $p = 0.1$, respectively, are illustrated at the top and bottom row; $T = 0.09$ for both. In each network, all members of one cluster (top) have $S_0 = 0.50$ and all members of the other cluster

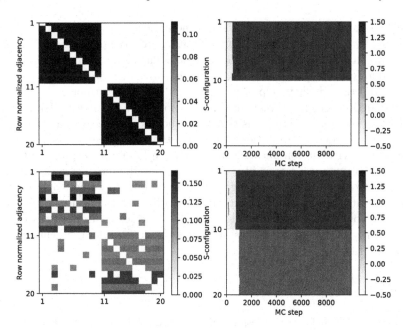

Fig. 5. Networks with two groups at $T = 0.09$. Top row left: adjacency matrix of two disconnected groups ($p = 0, n = 10$), with $S_0 = 0.50$ (nodes 1–10) and $S_0 = 0.25$ (nodes 11–20), respectively. The top row right panel has Monte Carlo steps on the horizontal axis. The group with higher S_0 (nodes 1–10) cooperates after few Monte Carlo steps whereas the group with lower S_0 (11–20) does not cooperate. Bottom row left: for the same two groups, ties are randomly rewired with a chance $p = 0.1$, resulting in connections between the groups that are visible in the adjacency matrix. Bottom row right: the group with higher S_0 (1–10) wins over the other group (11–20) to cooperate after few Monte Carlo steps. The different gray tones for the two cooperating subgroups are due to their different S_0 values.

(bottom) have $S_0 = 0.25$. For the network with disconnected groups, the group with lower S_0 does not reach its tipping point. When the groups are connected ($p = 0.1$), the group with higher S_0 helps the other group to cooperate at lower T_c than it would on its own, whereas the former is somewhat held back by the latter. This result generalizes to small numbers of initiative takers or leaders with higher S_0 embedded in a network of many others with lower S_0, where the initiative takers reduce T_c for the majority (not shown).

4   Conclusions

The model points out that under high uncertainty, the dilemma of cooperation can in principle be solved by situational turmoil among conformists. This solution holds for a broad range of networks, although the critical level of turmoil is markedly lower in small ones, which implies that turmoil-driven cooperation is more likely in small than in large groups. This result is consistent with empirical

250 J. Bruggeman and R. Sprik

findings on violent group confrontations, where most of the physical violence is committed by small subgroups [5]. Small fighting groups increase overall turmoil, though, which may agitate larger groups to join. The model also shows that there is no need for especially zealous initiative takers (with high S_0) to get cooperation started, even though they help to get it started at lower turmoil.

Our contributions are threefold. First, we made an asymmetric Ising model, which has a different phase transition than the widely used symmetric model. Second, we provided computational evidence that the dilemma of cooperation can be solved without complicated and costly mechanisms such as reputations, selective incentives and rationality. Third, we showed which networks are most conductive to cooperation under high uncertainty. Future studies could confront the model with empirical data, perhaps even of different species, for example buffalo herd bulls who chase a lion together [7]. One could also investigate more systematically the numbers of necessary Monte Carlo steps for different networks and distributions of S_0.

References

1. Barrat, A., Barthelemy, M., Vespignani, A.: Dynamical Processes on Complex Networks. Cambridge University Press, New York (2008)
2. Van den Berg, P., Wenseleers, T.: Uncertainty about social interactions leads to the evolution of social heuristics. Nat. Commun. **9**, 2151 (2018)
3. Bruggeman, J.B., Sprik, R., Quax, R.: Spontaneous cooperation for public goods (2020). arXiv:1812.05367
4. Castellano, C., Fortunato, S., Loreto, V.: Statistical physics of social dynamics. Rev. Mod. Phys. **81**, 591 (2009)
5. Collins, R.: Violence: A Micro-Sociological Theory. Princeton University Press, Princeton (2008)
6. Dion, D., Axelrod, R.: The further evolution of cooperation. Science **242**, 1385–1390 (1988)
7. Estes, R.: The Behavior Guide to African Mammals. University of California Press Berkeley, London (1991)
8. Fehr, E., Fischbacher, U.: The nature of human altruism. Nature **425**, 785 (2003)
9. Friedkin, N.E., Johnsen, E.C.: Social Influence Network Theory: A Sociological Examination of Small Group Dynamics. Cambridge University Press, Cambridge (2011)
10. Gavrilets, S.: Collective action problem in heterogeneous groups. Philos. Trans. Roy. Soc. B **370**, 20150016 (2015)
11. Granovetter, M.: Threshold models of collective behavior. Am. J. Sociol. **83**, 1420–1443 (1978)
12. Johnson, N.F., et al.: New online ecology of adversarial aggregates: isis and beyond. Science **352**, 1459–1463 (2016)
13. Konvalinka, I., et al.: Synchronized arousal between performers and related spectators in a fire-walking ritual. Proc. Natl. Acad. Sci. **108**, 8514–8519 (2011)
14. Marwell, G., Oliver, P.: The Critical Mass in Collective Action. Cambridge University Press, Cambridge, MA (1993)
15. McNeill, W.H.: Keeping Together in Time: Dance and Drill in Human History. Harvard University Press, Cambridge, MA (1995)

16. Newman, M.: Networks, 2nd edn. Oxford University Press, Oxford (2018)
17. Nowak, M.A., Sigmund, K.: Evolution of indirect reciprocity. Nature **437**, 1291–1298 (2005)
18. Olson, M.: The Logic of Collective Action: Public Goods and the Theory of Groups. Harvard University Press, Harvard (1965)
19. Perc, M., Jordan, J.J., Rand, D.G., Wang, Z., Boccaletti, S., Szolnoki, A.: Statistical physics of human cooperation. Phys. Rep. **687**, 1–51 (2017)
20. Wu, J.J., Li, C., Zhang, B.Y., Cressman, R., Tao, Y.: The role of institutional incentives and the exemplar in promoting cooperation. Sci. Rep. **4**, 6421 (2014)

An Information-Theoretic and Dissipative Systems Approach to the Study of Knowledge Diffusion and Emerging Complexity in Innovation Systems

Guillem Achermann[1] , Gabriele De Luca[1($\boxtimes$)] ,
and Michele Simoni[2]

[1] RUDN University, Moscow, Russia
gabriele.deluca@mail.ru
[2] University of Naples Parthenope, Naples, Italy

Abstract. The paper applies information theory and the theory of dissipative systems to discuss the emergence of complexity in an innovation system, as a result of its adaptation to an uneven distribution of the cognitive distance between its members. By modelling, on one hand, cognitive distance as noise, and, on the other hand, the inefficiencies linked to a bad flow of information as costs, we propose a model of the dynamics by which a horizontal network evolves into a hierarchical network, with some members emerging as intermediaries in the transfer of knowledge between seekers and problem-solvers. Our theoretical model contributes to the understanding of the evolution of an innovation system by explaining how the increased complexity of the system can be thermodynamically justified by purely internal factors. Complementing previous studies, we demonstrate mathematically that the complexity of an innovation system can increase not only to address the complexity of the problems that the system has to solve, but also to improve the performance of the system in transferring the knowledge needed to find a solution.

Keywords: Knowledge diffusion · Innovation system · Hierarchical networks · Dissipative systems

1 A Network and Systems Approach to Problem-Solving

1.1 Networks of Innovation as Information and Knowledge Processing Systems

The analysis of networks through sociometry has encouraged social scientists since the beginning of the twentieth century [1] to calculate degrees of strength or density of connections between different organisations located in a network. At the same time, efforts to understand the innovation process have led researchers to abandon the idea of a linear process of innovation, and to propose instead evolutionary models where the

---

Authors are listed in alphabetical order.

© Springer Nature Switzerland AG 2020
V. V. Krzhizhanovskaya et al. (Eds.): ICCS 2020, LNCS 12140, pp. 252–265, 2020.
https://doi.org/10.1007/978-3-030-50423-6_19

formalization and organization of a network becomes strategic to accelerate the flow of information and knowledge and the emergence of innovation [2]. Several forms of reticular organization (hierarchical, heterarchical, according to the centrality of elements, according to the transitivity of element, etc.) can be conceptualized within that context. Evolutionary economics and technology studies highlight (neo-Schumpeterian) models to understand the plurality of evolution cases, depending on the initial forms of organization, but also on the ability of a system to adapt to systemic crises.

In this work we study, from an information-theoretical perspective, the relationship between the structure of an innovation network, the noise in its communication channels and the energy costs associated with the network's maintenance. An innovation network is here considered to encompass a variety of organisations who, through their interactions and the resulting relationships, build a system conducive to the emergence of innovation. This system is identified by the literature [3] with different terms, such as innovation ecosystem, [4] problem-solving network, [5] or innovation environment [6]. In this system, the information channels transfer a multitude of information and knowledge which, depending on the structural holes, [7, 8] but also on the absence of predetermined receivers, are subject to information "noise" [9]. The more the information is distorted in the network, the more energy is needed to transfer accurate information, in order to keep performance of the innovation network high. The idea we propose is that the structure of an innovation system evolves to address the heterogeneity in the quality of communication that takes place between its members. In particular, we argue that the noise in a network increases the complexity of the network structure required for the accurate transfer of information and knowledge, and thus leads to the emergence of hierarchical structures. These structures, thanks to their fractal configuration, make it possible to combine high levels of efficiency in the transmission of information, with low network maintenance costs. This idea complements previous studies that have analysed the relationship between the structure of an innovation network, on one hand, and the complexity of the problem to be solved and the resulting innovation process, on the other, [10] by focusing on communication noise and cost of network structure maintenance. To the existing understanding of this phenomenon we contribute by identifying a thermodynamically efficient process which the network follows as it decreases in entropy while simultaneously cutting down its costs.

This model is based on the analysis of a network composed of two classes or categories of organisations, which operate within the same innovation system [11]. These classes are represented by a central organisation called *seeker*, which poses a research question to a group of other organisations, called problem-*solvers*, and from which in turn receives a solution. It has been suggested [12] that one of the problems that the innovation system has to solve, and for which it self-organises, is the problem of effective diffusion of knowledge between problem-*solvers* and solution-*seekers*, as this can be considered as a problem *sui generis*. The theory on the diffusion of knowledge in an innovation system suggests that this problem is solved through the evolution of modular structures in the innovation network, which implies the emergence of organisations that act as intermediary conduits of knowledge between hyperspecialised organisations in the same innovation environment [13]. A modular structure is, in network theory, connected to the idea of a hierarchical or fractal structure of the network, [14] and is also characterised by scale-invariance; [15] the latter is a particularly important property, because if innovation systems have it as an emergent property of their behaviour, this allows them to be considered as complex adaptive systems [16]. It has been suggested that scale-invariance property of an innovation system might emerge as the result of horizontal cooperation between its elements, [17] which try to reach the level of complexity required to solve a complex problem; but it is not yet clear how does a complex structure emerge when the complexity of the problem does not vary, which is a phenomenon observed empirically [18, 19]. In this paper we show how complexity can also vary as a result of a non-uniform distribution of the cognitive distance between organisations of the network, and of the adaptation required to solve the problem of knowledge diffusion among them. Our contribution to the theoretical understanding on the self-organising properties of innovation systems is that, by framing the problem of heterogeneous cognitive distance between organisations under the theory of dissipative systems, we can explain in thermodynamically efficient terms the reduction in entropy of an innovation system, as an emergent adaptation aimed at reducing costs of maintenance of the system's structure.

1.2   Self-organisation and Complexity in Dissipative Innovation Systems

The theoretical framework which we use for this paper is comprised by four parts. First, we will frame the innovation system as a thermodynamically-open system, which is a property that derives from the fact that social systems also are [20]. Second, we will see under what conditions a system can undertake self-organisation and evolution. This will allow us to consider an innovation system as a complex adaptive system, should it be found that there are emergent properties of its behaviour which lead to an increase in complexity. Third, we will frame the innovation system as a dissipative system, which is a property also shared by social systems [21]. Dissipative systems are characterised by the fact that a variation in the level of their entropy tends to happen as a consequence of their changed ability to process inputs, and we will see how this applies for innovation systems. Lastly, we will study cognitive distance as it applies to a network of innovation, in order to show how a spontaneous reduction in it leads to an increase in complexity of the network.

Thermodynamically-Open Innovation Systems. An open thermodynamic system is defined as a system which exchanges matter and energy with its surrounding environment, [22] and among them are found all social systems, which are open systems due to their exchanging of energy with the surrounding environment [23]. Social systems are also dynamical systems, because their structure changes over time through a process of dynamical evolution [24]. Innovation systems are some special classes of social systems, [25] which can thus also be considered as open systems [26]. In addition to this, like all social systems, innovation systems are also capable of self-organisation, [27] which is a property that they inherit from social systems [28]. There is however a property which distinguishes innovation systems from the generic social system: that is, the capacity of the former to act as problem-solving environments [11]. An innovation system possesses the peculiar function of developing knowledge, [29] which is not necessarily possessed by the general social system [30]. It has been theorised that developing and distributing knowledge [31] is the method by which the innovation system implements the function of solving problems, [32, 33] and we will be working within this theoretical assumption. The innovation system, for this paper, is therefore framed as a thermodynamically-open social system which solves problems through the development and diffusion of knowledge.

Evolution and Self-organisation. Like all other social systems, [34] an innovation system undertakes evolution [35] and changes in complexity over time [36]. The change in complexity of a system, in absence of any central planning or authority, is called in the literature *self-organisation* [37]. Self-organisation in a system implies that the system's components do not have access to global information, but only to information which is available in their immediate neighbourhood, and that upon that information they then act [28].

Innovation systems evolve, with a process that may concern either their members, [38] their relationships and interactions, [39] the technological channels of communication, [40] the policies pursued in them, [41] or all of these factors simultaneously [42]. For the purpose of this work we will limit ourselves to consider as evolution of an innovation system the modification of the existing relationships between its members, and the functions which they perform in their system. This process of evolution of the innovation system is characterised by self-organisation, [43] and it occurs along the lines of both information [44] and knowledge flows within the system [45]. The self-organisation of an innovation system is also the result of evolutionary pressures, [46] and we will here argue that one form of such pressures is cognitive distance between organisations within a network of innovation, whose attempt at reduction may lead to modifications in the relationships within the system and to the emergence of complex structures. While it has also been suggested that variations in the complexity of an innovation system might be the consequence of intrinsic complexity of the problems to be solved, [47] it has also been suggested that problems related to the transfer of knowledge within the elements of the system can, by themselves, generate the emergence of complex network structures, through a process which is thermodynamically advantageous.

Dissipative Innovation Systems. As the innovation system acquires a more complex structure, its entropy decreases. If one assumes that the decrease in entropy follows the expenditure of some kind of energy by the system, without which its evolution towards a lower-entropy state is not possible, then it follows that the innovation system can be framed as a dissipative system. This is a consequence of the theory which, in more general terms, suggests that all social systems can be considered as dissipative systems; [48] and, among them, innovation systems can thus also be considered as dissipative systems [49].

The application of the theory of dissipative structures [50] to the study of social systems has already been done in the past, [51, 52] and it has also been applied to the study of innovation systems specifically, [53, 54] to understand the process by which new structures evolve in old organisational networks [55].

By framing the problem in this manner the emergence of a hierarchical structure in a dissipative innovation system can be considered as a process through which the innovation system reaches a different level of entropy in its structure, [56] by means of a series of steps which imply sequential minimal variations in the level of entropy of the system, and lead to the emergence of complexity [57].

Cognitive Distance as Noise. The process of transferring knowledge between organisations presumes the existence of knowledge assets that are transferred [58]. Companies embedded in an innovation system are therefore characterised by an intellectual or knowledge capital, [59] which is the sum of the knowledge assets which they possess, [60] and which in turn are the result of the individual organisation's path of development, [61] and of the knowledge possessed by the human and technological components of the organisation [62]. Any two organisations do not generally share the same intellectual capital, and therefore there are differences in the knowledge assets which they possess, and in the understanding and representation which they create about the world. This difference is called "cognitive distance" in the literature on knowledge management, and it refers to the difficulty in transferring knowledge between any two organisations [63].

The theory suggests that an innovation network has to perform a trade-off between increasing cognitive distance between organisations, which means higher novelty value, and increasing mutual understanding between them, which gives higher transfer of knowledge at the expenses of novelty [64]. It has been argued that if alliances (that is, network connections) are successfully formed between organisations with high cognitive distance between their members, this leads to a higher production of innovation by that alliance, [65] as a consequence of the relationship between cognitive distance and novelty, as described above. It has also been argued that the measure of centrality of a organisation in an innovation network is a consequence of the organisation's impact on the whole knowledge governance process, with organisations contributing more to it located more centrally in the network [66]. We propose that this known mechanism might play a role in the dynamic evolution of an innovation system, in a manner analogous to that of noise in an information system. The idea is that an organisation generally possessing a lower cognitive distance between multiple components of a network might spontaneously become a preferential intermediary for the transfer of knowledge within the innovation system, and as a consequence of this a hierarchical network structure emerges out of a lower-ordered structure.

2   The Structure of the Network and Its Evolution

2.1   The Structure of the Network

The modeling of the process of evolution of a network of innovation is conducted as follows. First, we imagine that there are two different structures of the ego-network of an innovation seeker company that are the subject of our analysis. The first is a horizontal network, in which a seeker organisation is positioned in a network of solvers, which are all directly connected with the seeker organisation in question. The second is a hierarchical or fractal network, in which a structure exists that represents the presence of intermediaries in the transfer of knowledge between the seeker organisation and the solving organisations in the same network.

All nodes besides the seeker organisation being studied in the first scenario, and all nodes at the periphery of the hierarchical structure of the second scenario, are from here on called *solvers*. There are N nodes in the ego-network of an innovation seeker company. The N nodes in the horizontal network are all solver nodes, while the N nodes in the hierarchical network are divided into two classes of nodes: the *intermediaries* comprised of M nodes, and the *solvers*, comprised of M^2 nodes (Fig. 1).

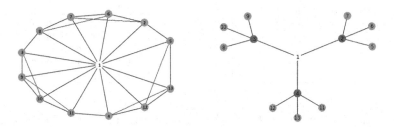

Fig. 1. In a horizontal network (to the left), all nodes in the ego-network of a seeker organisation are solver nodes (in green). In a hierarchical network (to the right), all nodes in the ego-network are either solver nodes (in green), or intermediaries (in grey). (Color figure online)

In order to make the two network structures comparable we impose the additional condition that the total number of nodes in the two networks is the same, which is satisfied for $N = M^2 + M$. We also impose the additional condition that each of the N solver nodes in the periphery of the horizontal network has at least M link neighbours belonging to N, as this allows us to describe a dynamical process which leads from the horizontal network to the hierarchical network without the creation of new links.

2.2   The Entropy of the Network

The hierarchical network always possesses a lower entropy than the horizontal network comprised of the same number of nodes. This can be demonstrated by using as a measure of entropy Shannon's definition, [67] which calculates it as the amount of information required to describe the current status of a system, accordingly to the formula below:

$$H(X) = -\sum\nolimits_{i=1}^{n} p(x_i)\log_2 p(x_i) \tag{1}$$

This measure of entropy can be applied to a social network by assigning the random variable X to the flattened adjacency matrix of the edges of the network, as done by others [68]. The adjacency matrices of the two classes of networks in relation to the size $N + 1$ of the same network are indicated in the tables below, for the specific values $M = 2 \rightarrow N = 6$ (Table 1).

Table 1. Adjacency matrices of two network structures for $M = 2$.

Horizontal Network									Hierarchical Network							
node	1	2	3	4	5	6	7		node	1	2	3	4	5	6	7
1	0	1	1	1	1	1	1		1	0	1	1	0	0	0	0
2	1	0	0	1	1	0	0		2	1	0	0	1	1	0	0
3	1	0	0	0	0	1	1		3	1	0	0	0	0	1	1
4	1	1	0	0	0	0	1		4	0	1	0	0	0	0	0
5	1	1	0	0	0	1	0		5	0	1	0	0	0	0	0
6	1	0	1	0	1	0	0		6	0	0	1	0	0	0	0
7	1	0	1	1	0	0	0		7	0	0	1	0	0	0	0

In general, for any value $M \geq 2$, if a horizontal network has N *solver* nodes, one *seeker* node is connected to all other N nodes, and all *solver* nodes are additionally connected to M solver nodes each, where $M + M^2 = N$. In a hierarchical network with M *intermediary* nodes and M^2 *solver* nodes, the *seeker* node is connected to M *intermediary* nodes, and each of the *intermediary* nodes is connected to M *solver* nodes. The general formulation of the adjacency matrix is indicated below, in relation to the value of M (Table 2).

Table 2. General structure of the adjacency matrices

Horizontal network		Hierarchical network	
Potential links	Existing links	Potential links	Existing links
$(N+1)^2$	$2N + NM$	$(N+1)^2$	$2M + 2M^2$

The adjacency matrices can be flattened by either chaining all rows or all columns together, in order to obtain a vector X which univocally corresponds to a given matrix. This vector has a dimensionality of $(N+1)^2$, having been derived from an $N+1$ by $N+1$ matrix. The vector X which derives from flattening can then be treated as the probability distribution over a random binary variable, and Shannon's measure of entropy can be computed on it. For the horizontal network, the vector $X_{horizontal}$ has value 1 two times for each of the peripheral nodes because of their connection to the centre, and then again twice for each of the peripheral nodes. This means that the vector $X_{horizontal}$ corresponds to the probability distribution (2).

$$X_{horizontal} = \left\{ \begin{array}{c} p(x_1) = \frac{2N+NM}{(N+1)^2} = \frac{(M+M^2)(M+2)}{(M+M^2+1)^2}; \\ p(x_0) = 1 - p(x_1) \end{array} \right\} \tag{2}$$

For the hierarchical network, the vector $X_{hierarchical}$ has value 1 two times for each of the M intermediary nodes, and then 2 times for each of the M^2 nodes. The probability distribution associated with the vector $X_{hierarchical}$ is therefore (3)

$$X_{hierarchical} = \left\{ \begin{array}{l} p(x_1) = \frac{2M + 2M^2}{(M + M^2 + 1)^2}; \\ p(x_0) = 1 - p(x_1) \end{array} \right\} \tag{3}$$

The hierarchical network systematically possesses a lower level of entropy than a horizontal network with the same number of nodes, as shown in the graph below (Fig. 2).

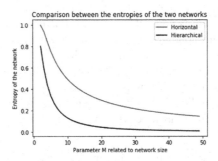

Fig. 2. Comparison between the levels of entropy of the two types of network structures.

Since we consider the network as a dissipative system, the lower level of entropy implies an expected higher energetic cost of maintenance for the lower-entropy structure. It follows from this theoretical premise that the hierarchical network should either allow the system to receive a higher input, or emit a lower output, or both simultaneously, lest its structure would decay to a higher entropy form, the horizontal one.

2.3   The Evolutionary Dynamics of the Network

An innovation system which starts evolving from a horizontal structure would tend to develop a hierarchical structure as a solution to the problem of transfer of knowledge in a network where cognitive distance is not uniformly distributed, as we will see in this paragraph. This can be shown by considering the hierarchical network as an attractor for the dynamical evolution of a horizontal network, under condition that the cognitive distance between pairs of nodes is distributed non-uniformly.

Stationary States. For the context of this paper, as we model a finite-state network which operates on discrete time, which models the dynamics of a dissipation systems which evolves over time [69]. These functions have the form depicted below, with $x(k)$ being the state of the system at time k, $u(k)$ being the input to the system at k, and $y(k)$ being the output of the system.

$$x(k+1) = f(x(k), u(k)) \tag{7}$$

$$y(k) = h(x(k), u(k)) \tag{8}$$

If the system does not undertake change in its internal structure, having already reached a stationary state, then $x(k+1) = x(k)$. As we want to study whether the system spontaneously evolves from a horizontal to a hierarchical structure, we can assume that $x(k+1) = f_{hierarchical}(x(k), u(k)) = x(k)$ which can only be true if either the input $u(k)$ is 0, which is not the case if the system is active, or if $u(k+1) = u(k)$. For the innovation system this condition is valid if minor variations in the structure of the network associated with it do not lead to a significant variation of the input to the system, which means that no advantages in the receipt by the *seeker* of solutions found by the *solver* should be found. If this is true, and if the hierarchical structure is an attractor for the corresponding horizontal network, then we expect the input of the horizontal network to increase as it acquires a modular structure and develops into a hierarchical network.

Input of the System. The input function of the system depends on the receipt by the *seeker* organisation of a solution to a problem found by one of the peripheral *solver* organisations, as described above. Let us imagine that at each timestep the *solver* organisations do indeed find a solution, and that thus the input $u(k)$ depends on the number of *solver* nodes, and for each of them on the probability of correct transmission of knowledge from them to the *seeker* organisation, which increases as the cognitive distance between two communicating nodes decreases. If this is true, then the input to the horizontal network is a function of the form $u_{horizontal}(N_k, p_k)$, where N is the number of *solver* nodes, and p is the cognitive distance in the knowledge transmission channel. Similarly, the input to the hierarchical network is a function of the form $u_{hierarchical}(M_k^2, q_k)$ which depends on the M^2 *solver* nodes in the hierarchical network, and on the parameter q which describes the cognitive distance. N and M are such that as they increase so do, respectively, $u_{horizontal}$ and $u_{hierarchical}$; while p and q are such that, as they decrease, so do respectively $u_{horizontal}$ and $u_{hierarchical}$ increase. It can be also noted that $M^2 < N \rightarrow \forall (M, N) \in \mathbb{N}$, under condition $N = M^2 + M$ defined above. It can then be argued that if $p \leq q$ then $u_{horizontal} > u_{hierarchical}$, which means that the system would not evolve into a hierarchical network. It can also be noted that, if N and M^2 are sufficiently large, then $\lim_{N,M \to +\infty} (N/M^2) = 1$ and therefore any difference between the number of solvers in the two network structures would not play a role in the input to the innovation system. From this follows that $u_{hierarchical} > u_{horizontal} \rightarrow q < p$; that is, that the input to the innovation system with a hierarchical structure is higher than the input to the innovation system with a horizontal structure, if the cognitive distance between the members of the former is lower than the cognitive distance between the members of the latter.

Output of the System. As per the output of the system, we can imagine that there is a cost to be paid for the maintenance of the communication channels from which the *seeker* receives solutions from the *solvers*. If the system is in a stationary state, the condition $y(k+1) = y(k)$ must be valid, as it follows from the considerations that $u(k+1) = u(k)$. If the system is not in a stationary state, as the input to the system increases, so should the output, under the hypothesis of dissipative system described above. A graphical representation of the evolution of the system from higher to lower entropy state is thus presented below (Fig. 3).

Horizontal network   Evolution   Hierarchical network

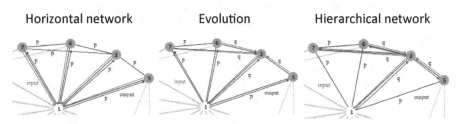

Fig. 3. Evolution of a branch of the innovation network from a higher to a lower entropy structure, from left to right. The letters p and q define respectively a high and a low cognitive distance between peers.

The *seeker* organisation would at each step receive a solution transferred by one of its link neighbours, with the indication of the full path through which the communication has reached it. The *seeker* would then pay a certain cost, an *output* with the terminology of dissipative systems, for the maintenance of the channel through which the solution has been transferred to it successfully. Such channels increase in intensity or weight, and are more likely to be used in subsequent iterations. On the contrary, channels through which a solution has not been received in a given iteration are decreased in intensity or weight, and are less likely to be used in the future. A process such as the one described would eventually, if enough iterations are performed, lead to the withering links between nodes with higher cognitive distance, and to the preservation of links between nodes with a lower cognitive distance. New connections are not formed, because cognitive distance is considered to be an exogenous parameter in this model, which does not vary once the innovation system starts evolving. Real-world phenomena are not characterised by this restriction, which should be considered when analysing real-world systems under this model.

3   Conclusions and Future Work

The originality of this paper consists in the framing of an innovation system under different theoretical approaches, such as that of thermodynamically-open systems, self-organisation and evolution, dissipative systems, and cognitive distance, which, when combined, highlight another way of understanding the overall operation and the evolution of innovation systems. From this perspective, the process which we here describe accounts for an emergent complexity of the innovation system, which can occur without central planning and on the basis of information locally available by its members. This seems to confirm the theory according to which innovation systems can self-organise to solve, among others, the problem of transfer of knowledge among their members. This seems also to suggest that, if the only form of proximity which matters is cognitive, and not geographical, organisational, or other, it might be possible to infer cognitive distance between the members of an innovation system on the basis of the way in which their relationships change over time. The theoretical prediction which this model allows to make is that, should a connection between members of an innovation system be preserved while others are dropped, this means that the cognitive distance between pairs of nodes with surviving connections is lower than that of other nodes in their ego-networks. The modelling of the evolution of an innovation system that we propose also shows that, if an innovation system starts its evolution with a centrally, highly-connected organisation in a largely horizontal network of solver, where the cognitive distance between each pair of nodes is not uniformly distributed, then the system would evolve towards a lower-entropy hierarchical structure, in order to solve the problem of transfer of knowledge from the organisations at the periphery of the innovation system to the central organisation. Our finding is consistent with the theory on modularity as an emergent property of complex adaptive innovation systems. Subsequent research might apply the mathematical model described in this paper to a longitudinal study of the evolution of real-world innovation networks, in order to test whether the theory related to the spontaneous emergence of a hierarchical structure of innovation networks can be empirically supported. On the theoretical plane, further research could expand the understanding of the evolution of an innovation network by adding considerations related to the role which geographical and organisational proximity have in the development of the network, and add these factors to the model proposed. Issues related to perturbation of the network, limit cycle of its evolution, and self-organised criticality in connection to our model may also be explored in subsequent works.

References

1. Simmel, G.: Sociologie et épistémologie, Paris, PUF, collection «Sociologie», 1981 (1917)
2. Taalbi, J.: Evolution and structure of technological systems – an innovation output network (2018). https://arxiv.org/pdf/1811.06772.pdf
3. Russell, M.G., Smorodinskaya, N.V.: Leveraging complexity for ecosystemic innovation. Technol. Forecast. Soc. Chang. 136, 114–131 (2018)
4. Jackson, D.J.: What is an innovation ecosystem. National Science Foundation, 1 (2011)
5. Parker, L.A.: Networks for innovation and problem solving and their use for improving education: a comparative overview (1977)
6. Lubik, S., Garnsey, E., Minshall, T., Platts, K.: Value creation from the innovation environment: partnership strategies in university spin-outs. R&D Management 43(2), 136–150 (2013)
7. Burt, R.S.: Structural Holes: The Social Structure of Competition. Harvard university press, Cambridge (2009)

8. Roos, G.: Knowledge management, intellectual capital, structural holes, economic complexity and national prosperity. J. Intell. Capital **18**(4), 745–770 (2017)
9. Leydesdorff, L., Rotolo, D., de Nooy, W.: Innovation as a nonlinear process, the scientometric perspective, and the specification of an 'innovation opportunities explorer'. Technol. Anal. Strateg. Manag. **25**(6), 641–653 (2013)
10. Sáenz-Royo, C., Gracia-Lázaro, C., Moreno, Y.: The role of the organization structure in the diffusion of innovations. PLoS ONE **10**(5), e0126076 (2015)
11. Terwiesch, C., Xu, Y.: Innovation contests, open innovation, and multiagent problem solving. Manage. Sci. **54**(9), 1529–1543 (2008). https://doi.org/10.1287/mnsc.1080.0884
12. Cowan, R., Jonard, N.: Network structure and the diffusion of knowledge. J. Econ. Dyn. Control **28**(8), 1557–1575 (2004)
13. Brusoni, S.: The limits to specialization: problem solving and coordination in "modular networks". Organ. Stud. **26**(12), 1885–1907 (2005). https://doi.org/10.1177/0170840605059161
14. Ravasz, E., Barabási, A.-L.: Hierarchical organization in complex networks. Phys. Rev. E, **67**(2) (2003) https://doi.org/10.1103/physreve.67.026112
15. Barabási, A.L., Dezső, Z., Ravasz, E., Yook, S.H., Oltvai, Z.: Scale-free and hierarchical structures in complex networks. In: AIP Conference Proceedings, vol. 661, no. 1, pp. 1–16. AIP, April 2003
16. Katz, J.S.: What is a complex innovation system? PLoS ONE **11**(6), e0156150 (2016)
17. Katz, J.S., Ronda-Pupo, G.A.: Cooperation, scale-invariance and complex innovation systems: a generalization. Scientometrics (2019). https://doi.org/10.1007/s11192-019-03215-8
18. Zhang, J.: Growing Silicon Valley on a landscape: an agent-based approach to high-tech industrial clusters. In: Industry And Labor Dynamics: The Agent-Based Computational Economics Approach, pp. 259–283 (2004)
19. Spielman, D.J., Ekboir, J., Davis, K.: Developing the art and science of innovation systems enquiry: alternative tools and methods, and applications to sub-Saharan African agriculture. In: Innovation Africa, pp. 98–112. Routledge (2012)
20. Brent, S.B.: Prigogine's model for self-organization in nonequilibrium systems. Hum. Dev. **21**(5–6), 374–387 (1978)
21. Harvey, D.L., Reed, M.H.: The evolution of dissipative social systems. J. Soc. Evol. Syst. **17**(4), 371–411 (1994)
22. Chick, V., Dow, S.: The meaning of open systems. J. Econ. Methodol. **12**(3), 363–381 (2005)
23. Luhmann, N.: Social Systems. Stanford University Press, Stanford (1995)
24. Fischer, M.M., Fröhlich, J. (Eds.) Knowledge, Complexity and Innovation Systems. Advances in Spatial Science (2001). https://doi.org/10.1007/978-3-662-04546-6
25. Amable, B.: Institutional complementarity and diversity of social systems of innovation and production. Rev. Int. Polit. Econ. **7**(4), 645–687 (2000)
26. Gómez-Uranga, M., Etxebarria, G.: Thermodynamic properties in the evolution of firms and innovation systems (2015). SSRN 2697747
27. Saviotti, P.P.: Networks, national innovation systems and self-organisation. In: Fischer, M.M., Fröhlich, J. (eds.) Knowledge, Complexity and Innovation Systems. Advances in Spatial Science, pp. 21–45. Springer, Heidelberg (2001). https://doi.org/10.1007/978-3-662-04546-6_2
28. Hemelrijk, C. (ed.): Self-organisation and Evolution of Biological and Social Systems. Cambridge University Press, Cambridge (2005)

29. Hekkert, M.P., Suurs, R.A.A., Negro, S.O., Kuhlmann, S., Smits, R.E.H.M.: Functions of innovation systems: a new approach for analysing technological change. Technol. Forecast. Soc. Chang. **74**(4), 413–432 (2007). https://doi.org/10.1016/j.techfore.2006.03.002

30. Collins, R.: On the sociology of intellectual stagnation: the late twentieth century in perspective. Theory Cult. Soc. **9**(1), 73–96 (1992)

31. Yan, E.: Disciplinary knowledge production and diffusion in science. J. Assoc. Inf. Sci. Technol. **67**(9), 2223–2245 (2016)

32. Nickerson, J.A., Zenger, T.R.: A knowledge-based theory of the organisation—the problem-solving perspective. Organ. Sci. **15**(6), 617–632 (2004)

33. Eisner, H.: Thinking: A Guide to Systems Engineering Problem-solving. CRC Press, Boca Raton (2019)

34. Kappeler, P.M., Clutton-Brock, T., Shultz, S., Lukas, D.: Social complexity: patterns, processes, and evolution (2019)

35. Pyka, A., Foster, J. (eds.): The Evolution of Economic and Innovation Systems. ECE. Springer, Cham (2015). https://doi.org/10.1007/978-3-319-13299-0

36. Frenken, K.: A complexity-theoretic perspective on innovation policy. Complex. Innov. Policy **3**(1), 35–47 (2017)

37. De Wolf, T., Holvoet, T.: Emergence versus self-organisation: different concepts but promising when combined. In: Brueckner, Sven A., Di Marzo Serugendo, G., Karageorgos, A., Nagpal, R. (eds.) ESOA 2004. LNCS (LNAI), vol. 3464, pp. 1–15. Springer, Heidelberg (2005). https://doi.org/10.1007/11494676_1

38. Kastelle, T., Potts, J., Dodgson, M.: The evolution of innovation systems. In: DRUID Summer Conference, June 2009

39. Feldman, M.P., Feller, I., Bercovitz, J.L., Burton, R.: Understanding evolving university-industry relationships. In: Feldman, M.P., Link, A.N. (eds.) Innovation policy in the knowledge-based economy. Economics of Science, Technology and Innovation, 23rd edn, pp. 171–188. Springer, Boston (2001). https://doi.org/10.1007/978-1-4615-1689-7_8

40. Murray, F.: Innovation as co-evolution of scientific and technological networks: exploring tissue engineering. Res. Policy **31**(8–9), 1389–1403 (2002)

41. Cooke, P.: From technopoles to regional innovation systems: the evolution of localised technology development policy. Can. J. Region. Sci. **24**(1), 21–40 (2001)

42. Trippl, M., Grillitsch, M., Isaksen, A., Sinozic, T.: Perspectives on cluster evolution: critical review and future research issues. Eur. Plan. Stud. **23**(10), 2028–2044 (2015)

43. Silverberg, G., Dosi, G., Orsenigo, L.: Innovation, diversity and diffusion: a self-organisation model. Econ. J. **98**(393), 1032–1054 (1988)

44. Fuchs, C. (2002). Social information and self-organisation

45. Fuchs, C., Hofkirchner, W.: Self-organization, knowledge and responsibility. Kybernetes **34** (1/2), 241–260 (2005)

46. Etzkowitz, H., Leydesdorff, L.: The dynamics of innovation: from National Systems and "Mode 2" to a Triple Helix of university–industry–government relations. Res. Policy **29**(2), 109–123 (2000). https://doi.org/10.1016/s0048-7333(99)00055-4

47. Brusoni, S., Marengo, L., Prencipe, A., Valente, M.: The value and costs of modularity: a problem-solving perspective. Eur. Manag. Rev. **4**(2), 121–132 (2007). https://doi.org/10.1057/palgrave.emr.1500079

48. Xiong, F., Liu, Y., Zhu, J., Zhang, Z.J., Zhang, Y.C., Zhang, Y.: A dissipative network model with neighboring activation. Eur. Phys. J. B **84**(1), 115–120 (2011). https://doi.org/10.1140/epjb/e2011-20286-7

49. Diao, Z.F., Zhang, F.S.: The Analysis of dissipative structure in the technological innovation system of enterprises. J. Jiangnan Univ. (Human. Soc. Sci. Edn.) **2** (2009)

50. Prigogine, I., Kondepudi, D.: Modern Thermodynamics: From Heat Engines to Dissipative Structures, 2nd edn. Wiley, New York (1998)
51. Kiel, L.D.: Lessons from the nonlinear paradigm: applications of the theory of dissipative structures in the social sciences. Soc. Sci. Q. (1991)
52. Schieve, W.C., Allen, P.M. (eds.): Self-organization and Dissipative Structures: Applications in the Physical and Social Sciences. University of Texas Press, Austin (1982)
53. Leifer, R.: Understanding organizational transformation using a dissipative structure model. Hum. Relat. 42(10), 899–916 (1989)
54. Jenner, R.A.: Technological paradigms, innovative behavior and the formation of dissipative enterprises. Small Bus. Econ. 3(4), 297–305 (1991)
55. Gemmill, G., Smith, C.: A dissipative structure model of organization transformation. Hum. Relat. 38(8), 751–766 (1985). https://doi.org/10.1177/001872678503800804
56. Li, Z., Jiang, J.: Entropy model of dissipative structure on corporate social responsibility. In: IOP Conference Series: Earth and Environmental Science, vol. 69, no. 1, p. 012126). IOP Publishing, June 2017
57. Basile, G., Kaufmann, H.R., Savastano, M.: Revisiting complexity theory to achieve strategic intelligence. Int. J. Foresight Innov. Policy 13(1–2), 57–70 (2018)
58. Girard, J., Girard, J.: Defining knowledge management: toward an applied compendium. Online J. Appl. Knowl. Manag. 3(1), 1–20 (2015)
59. Laperche, B.: Enterprise Knowledge Capital. Wiley, Hoboken (2017)
60. Marr, B., Schiuma, G., Neely, A.: Intellectual capital–defining key performance indicators for organizational knowledge assets. Bus. Process Manag. J. 10(5), 551–569 (2004)
61. Moustaghfir, K.: The dynamics of knowledge assets and their link with firm performance. Measuring Bus. Excell. 12(2), 10–24 (2008)
62. Wu, W.W., Kan, H.L., Liu, Y.X., Kim, Y.: Management mechanisms, technological knowledge assets and firm market performance. Stud. Sci. Sci. 5, 14 (2017)
63. Nooteboom, B.: Problems and solutions in knowledge transfer (2001)
64. Wuyts, S., Colombo, M.G., Dutta, S., Nooteboom, B.: Empirical tests of optimal cognitive distance. J. Econ. Behav. Organ. 58(2), 277–302 (2005). https://doi.org/10.1016/j.jebo.2004.03.019
65. Filiou, D., Massini, S.: Industry cognitive distance in alliances and firm innovation performance. R&D Manag. 48(4), 422–437 (2018)
66. Zhao, J., Xi, X., Guo, T.: The impact of focal firm's centrality and knowledge governance on innovation performance. Knowl. Manag. Res. Pract. 16(2), 196–207 (2018). https://doi.org/10.1080/14778238.2018.1457004
67. Shannon, C.E., Weaver, W.: The mathematical theory of communication. University of Illinois Press, Urbana (1949)
68. De Domenico, M., Granell, C., Porter, M.A., Arenas, A.: The physics of spreading processes in multilayer networks. Nat. Phys. 12(10), 901–906 (2016)
69. Tan, Z., Soh, Y.C., Xie, L.: Dissipative control for linear discrete-time systems. Automatica 35(9), 1557–1564 (1999). https://doi.org/10.1016/s0005-1098(99)00069-2

Mapping the Port Influence Diffusion Patterns: A Case Study of Rotterdam, Antwerp and Singapore

Peng Peng and Feng Lu[✉]

State Key Laboratory of Resources and Environmental Information System,
Institute of Geographic Sciences and Natural Resources Research,
Chinese Academy of Sciences, Beijing 100101, China
{pengp,luf}@lreis.ac.cn

Abstract. Ports play a vital role in global oil trade and those with significant influence implicitly have better control over global oil transportation. To provide a better understanding of port influence, it is necessary to analyze the development of the mechanisms underlying port influence. In this study, we adopt a port influence diffusion model to modelling diffusion patterns using vessel trajectory data from 2009 to 2016. The results of the case study of Rotterdam, Antwerp and Singapore ports shows: 1) ports with a strong direct influence control their neighboring ports, thereby building a direct influence area; 2) directly influenced ports show path-dependent characteristics, reflecting the importance of geographical distance; 3) the indirect influence of the initial diffusion port creates hierarchical diffusion, with directly influenced ports affected by previous diffusion-influenced ports. 4) a port's indirect influence and efficiency can be increased via an increase in the number of significant ports it influences directly or by increasing its influence on significant ports in an earlier diffusion stage.

Keywords: Global oil transport patterns · Vessel trajectory · Ports · Influence diffusion · Direct influence · Indirect influence

1 Introduction

The global oil trade continues to show tremendous growth [1], and maritime transportation is considered the most important trade mode for global oil. By combining locational advantages with long-term high-quality operational processes, some ports have achieved greater competitiveness and influence. Ships and oil companies will prioritize these ports for selection when designing routes. Hence, the growing transport network has been concentrating around these hub ports over time, and traffic distribution has shown place-dependent characteristics as well [2]. As a result, the emergence of patterns of port influence has produced the Matthew effect [3]; that is, the influence of these significant ports has progressively grown. Studying the formation and evolution of port influence is necessary to help optimize port trade relations, and provide theoretical support for port development.

© Springer Nature Switzerland AG 2020
V. V. Krzhizhanovskaya et al. (Eds.): ICCS 2020, LNCS 12140, pp. 266–276, 2020.
https://doi.org/10.1007/978-3-030-50423-6_20

Recently, scholars have built a series of index systems [4–6] based on the characteristics of a port's natural conditions, locational advantages, and operational efficiencies to design port evaluation methods [7–9] and study the competitiveness of different ports, reflecting port influence from new perspectives.

Networks are also a vital medium for spreading information on port competitiveness. They are the conduits by which the innovative efforts of oil companies and port technology are disseminated. Information externalities and spillover from the concentration of oil and tanker companies in a port city can influence the port network. In turn, the spread of a port's influence can reinforce its reputation and competitiveness, which attracts more oil traffic. Common network models reveal port influence via certain network indicators [10–13] such as degree and betweenness centrality.

Although several scholars have studied port competitiveness and network indicators to reflect port influence, the methods used to date have some limitations. Among these, port-related data and statistics are not always completely available and updated. Thus, it is often difficult to obtain complete infrastructure data for all ports involved in the world oil trade, making it problematic to verify a large-scale port competitiveness assessment. In addition, Peng et al. found that there were several transshipment characteristics in oil transportation [14]; specifically, port influence had multiple propagation and cumulative effects, which could not be measured by network centrality indicators alone. In a quantitative study of the influence diffusion of ports involved in oil transportation, Peng et al. found that port influence had multiple diffusion characteristics [15].

Against this backdrop, the aim of this study is to fill the gaps in the literature by adopting a port influence diffusion model to explore the evolution of the influence of different oil ports, considering direct and indirect influence. We do this through a case study of Rotterdam, Antwerp and Singapore using worldwide vessel trajectory data from 2009 to 2016.

2 Data and Model

2.1 Data for Global Oil Transport Network

In this study, we used vessel trajectories generated from AIS data from 2009 to 2016 to construct global oil transport networks. We adopted a weighted directed graph $G(N_y, E_y, W_y)$ to represent the global oil transport network, N_y represents all ports involved in oil trade in year y (namely, 2009,…,2016); E represents all edges (namely, routes between port pairs) linking pairs of nodes in N_y; and W_y the weights for all routes, expressed by the annual total freight volume on the route in each year y. Due to the cargo volume of each voyage cannot be accurately calculated by the data. The deadweight of each vessel was assumed to reflect the vessel's transport capacity, and it is all offloaded at a calling port.

2.2   Port Influence Diffusion Model

To study the evolution of port influence, our approach was to model the diffusion paths of the ports and the scale and geographical distribution of the ports affected at each diffusion stage. Figure 1 shows a schematic diagram of the port influence diffusion model.

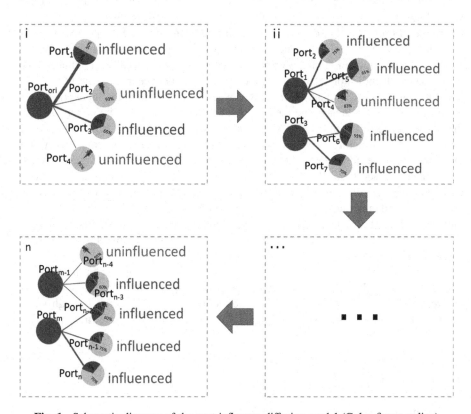

Fig. 1. Schematic diagram of the port influence diffusion model (Color figure online)

Figure 1(i) shows the first diffusion stage of the port influence mechanism. Here, $Port_{ori}$ refers to the port with the initial influence diffusion; the blue line refers to the diffusion path, where the thickness captures the freight volume from the original port to the target port, and the green sector captures the proportion of the freight volume from $Port_{ori}$ to the affected port out of all the import freight volume to the affected port. Therefore, the influence coefficient b_{uv}^i of port u on port v in the i th diffusion, $i = (1, 2, \ldots)$ is given as follows:

$$b_{uv}^i = \frac{FreightVolume_u^{out}(v)}{\sum\limits_{w \in N(v)^i} FreightVolume_v^{in}(w)} \tag{1}$$

where $FreightVolume_u^{out}(v)$ is the total freight volume from port u to port v (using formula (2)). $FreightVolume_v^{in}(w)$ denotes the total freight volume from port w to port v (using formula (3)), where $N(v)^i$ is the group of all ports that transport cargo directly from other ports in the network to port v in the i th diffusion, $i = (1, 2, \ldots)$.

$$FreightVolume_u^{out}(v) = \sum VesselFreightVolume_u^{out}(v) \tag{2}$$

$$FreightVolume_v^{in}(w) = \sum VesselFreightVolume_v^{in}(w) \tag{3}$$

where $VesselFreightVolume_u^{out}(v)$ denotes the freight volume of a vessel departing from port u to port v. $VesselFreightVolume_v^{in}(w)$ is the freight volume of a vessel from port w to port v, expressed in terms of the vessel's deadweight.

At this stage, the cumulative influenced value θ_v^i of port v at the i th diffusion stage is expressed as:

$$\theta_v^i = \sum_{u \in N(v)^i} b_{uv}^i \tag{4}$$

As an influenced port's influence spreads over the network, other uninfluenced ports in the network may become "influenced" (labeled "influenced" in Fig. 1) or may remain "uninfluenced" (labeled "uninfluenced"). The model is expressed in formula (5).

$$f_{influence_i}(v) = \begin{cases} influenced & \theta_v^i \geq \theta_v \\ uninfluenced & \theta_v^i < \theta_v \end{cases} \tag{5}$$

where $f_{influence_i}(v)$ indicates whether port v is influenced (labeled influenced around the node) in the i th diffusion, $i = (1, 2, \ldots)$. θ_v is the threshold of being influenced by adjacent nodes. When the accumulated influence of all neighboring influential ports of port v is greater than the threshold θ_v, port v changes to influenced status at the ith $+ 1$ diffusion stage, and remains with an influenced status until the diffusion ends. Otherwise, the diffusion fails to occur, but the influence coefficient is accumulated until the next stage. Without loss of generality, we set the threshold for all ports at 0.2. This meant that the node was influenced when its cumulative impact (freight volume) reached at least 20% of its total impact (freight volume). This is substantial enough to represent a significant influence of the original port on a target port.

At the first diffusion stage, port$_1$ and port$_3$ become influenced, while other ports do not change their status. The influence in the first diffusion of the port also denotes the direct influence of the initial diffusion port.

Figure 1(ii) shows the second diffusion stage. Ports that become influenced at the first diffusion stage are considered diffusion ports at the second diffusion stage. Moreover, the influence coefficient at the second diffusion stage can also be calculated by formula (1), and influence accumulated using the influence coefficient in the first diffusion stage (formula (4)) to calculate the cumulative influence coefficients of all ports at the second diffusion stage.

For port v, the accumulated influence weight of all ports is bounded by the unit value and expressed as follows:

$$\sum_{u \in N(v)^i} b_{uv}^i \leq 1 \tag{6}$$

It should be noted that at this stage, some ports are activated by the influence of stage two only, such as $port_5$ and $port_7$; while some ports are activated by the influence of both stage one and stage two, such as $port_2$.

According to the above process, we can calculate successive influence diffusion stages of the ports iteratively. Figure 1(n) shows the n_{th} stage influence diffusion, that is, the last effective influence diffusion to obtain all the ports affected by the initial diffusion port. The ports affected at the first diffusion stage are all directly influenced ports, while after the first diffusion stage are all indirectly influenced ports. Moreover, the higher the diffusion stage, the lower efficiency of the port influence diffusion.

3   Case Study

We use the above-mentioned method to calculate the influence of Rotterdam, Antwerp, and Singapore ports from 2009 to 2016. We set the influence of the port's first stage based on years, using 2009, 2013, and 2016 as representative years to analyze the development of port influence. Visualizing the path of influence diffusion at different stages and we focus on the first diffusion stage in our model as well, that is, the direct influence of the ports.

Figure 2(a)-(c) shows the first influence diffusion stage of Rotterdam. From the perspective of the spatial distribution of the influenced ports, Rotterdam's oil trade developed early as its facilities were relatively complete. In 2009, it was important in the global oil trade, and its influence area was far greater than other ports. However, due to the relatively regional scale of its oil trade, its influenced areas were concentrated in Northwest Europe, the Mediterranean region, and North America (Fig. 2(a)). At the first diffusion stage, there were 84 ports influenced by Rotterdam. This includes six ports of influence that only influenced one other port.

With the rapid development of navigation technology and the increase in the size of ships, long-distance transportation has increased, which has accelerated the spread of port influence to a certain extent. By 2013, Rotterdam's direct influence on other ports showed expansion geographically, influencing 313 ports in 71 countries, and even 12 East Asia ports including Shenzhen Port (Fig. 2(b)). Rotterdam's *direct influence ratio* reached 77.28% (out-degree of 405, *direct influence ratio* $= \frac{number\ of\ directly\ influenced\ ports}{out-degree\ of\ the\ port}$), indicating that most of the ports it has trade relations with were under its direct influence. In 2013, it directly influenced 96 ports in the first diffusion stage compared with 2009, but the influence among these ports was small, including only one port that could influence more than 10 ports (Kiel). In 2016, the number of directly influenced ports increased rapidly to 391 in 89 countries, and the proportion of directly affected ports reached 82.84% (out-degree of 472), indicating that the scale of directly influenced ports grew. Rotterdam had increased its influence over adjacent ports, such as the number of ports influenced in the UK and the US, which increased from 34 and 17 in

2013, to 42 and 28 in 2016, respectively. However, there was a decrease in the number of influenced ports farther away geographically from Rotterdam. For example, there was only one influenced port in East Asia (Fig. 2(c)). Among the ports directly influenced by Rotterdam in 2016, 209 were the same as those directly influenced by it in 2013, which indicates a path dependence among the ports directly influenced by Rotterdam. In addition, the number of ports of influence increased slightly, reaching 112. However, at that time, the number included some ports with significant influence including Kiel (influencing 32 ports), Montreal (influencing 18 ports), Quintero (influencing 10 ports), and Colon (influencing 7 ports).

With the rapid increase in the direct influence of a given port, its indirect influence also showed obvious growth. Not only did the number of ports influenced through fewer diffusion stages show rapid growth, but the number of ports with more influence also reflected significant growth.

In Fig. 3, we show the number of ports of influence affected by Rotterdam, Antwerp, and Singapore, at different diffusion stages (1, 2, ..., 8 denote the diffusion stages). As shown in the figure, there was only one port of influence that was influenced by Rotterdam during the second diffusion stage in 2009. By 2013, there were 74 and 68 ports of influence affected by Rotterdam at diffusion stages two and three, respectively. Notably, the top 10 most influential ports in the world, except Ichihara (located in Japan, influenced in the third diffusion stage), were all influenced in the second diffusion stage. In 2016, 83 and 98 ports of influence had been influenced by Rotterdam in the second and third diffusion stages, respectively. Moreover, most of the ports with significant influence were affected in the second stage including Antwerp and Istanbul. However, the ports with significant influence in East Asia, which are relatively farther away from Rotterdam, were all influenced in the third diffusion stage, including Ichihara and Yokohama in Japan, Yeosu in South Korea, and Shanghai in China.

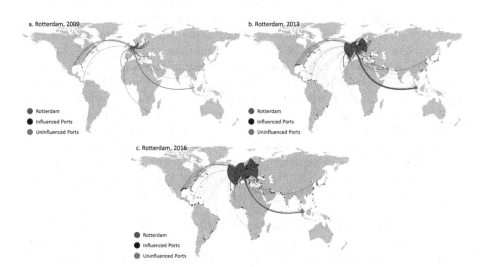

Fig. 2. Influence diffusion at first stage: Rotterdam (Color figure online)

Note: The orange node represents the initial diffusion port, while the blue line represents the real route from the initial diffusion port to other ports worldwide. The red node connected with the blue line represents a node that had been influenced at the first stage, namely, the direct influenced ports. The green node represents a node not yet influenced.

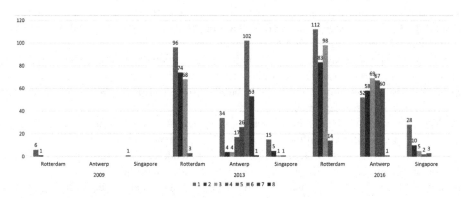

Fig. 3. Number of influenced ports of influence at different diffusion stages

Figure 4(a)-(c) shows Antwerp's first influence diffusion stage. The initial influence of Antwerp Port was relatively small. It only influenced 12 ports in 2009, and these ports were limited to Northwest Europe (Fig. 4(a)). However, the direct influence of Antwerp Port shows strong growth over time. By 2013, it directly influenced 160 ports in 49 countries (out-degree of 325), while its *directinfluenceratio* reached 49.23%. Its direct influence on the geographical distribution of the ports also reflected a relative expanding trend, included Northwest Europe, the Mediterranean region, and the Americas (Fig. 4(b)). The directly influenced ports included 34 ports all with less influence (Fig. 3). For example, the most influential port was Kristiansund, which only influenced five other ports. By 2016, the direct influence of the ports had expanded further with Antwerp affecting 228 ports in 66 countries (out-degree of 373), with a *directinfluenceratio* reaching 61.13%, and the geospatial scope of the influenced ports expanded further. In addition to the previous regions, the expansion included ports in East Africa and South Asia (Fig. 4(c)). Among the directly influenced ports, there were 52 ports of influence, although these ports had relatively little influence. Of them, Kiel was the most influential port, followed by San Vicente (affecting seven ports). The number of ports directly influenced by Antwerp, along with the number of ports of influence, and their geographical distribution were all less than that of Rotterdam. Therefore, ultimately, its direct influence was less than that of Rotterdam.

In terms of indirect influence, Antwerp reflected significant growth there as well, although still much smaller than Rotterdam. Although in 2009, Antwerp had no indirect influence, by 2013, its indirect influence showed rapid growth, which could influence almost all ports through the seven diffusion stages. Yet, the number of ports of influence it affected were all influenced in the later diffusion stages. Only four ports of influence were affected in the second and third diffusion stages. Except for Kiel, the

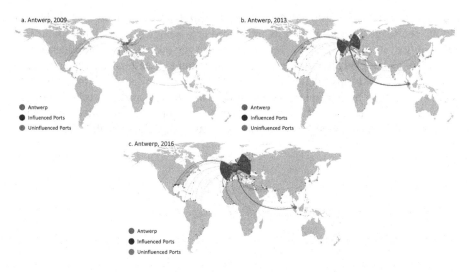

Fig. 4. Influence diffusion at first stage: Antwerp

top 10 ports with significant influence were affected in the fifth or sixth diffusion stages, while most were influenced at the sixth diffusion stage. The conclusion here is that the majority of the ports were influenced in the sixth and seventh diffusion stages, indicating that although the number of ports influenced was relatively large, the indirect influence diffusion efficiency was relatively low. However, in 2016, Antwerp's indirect influence increased significantly. Specifically, among the top 10 significant ports, Rotterdam, Istanbul, Amsterdam, and Balboa were all influenced during the third diffusion stage and most of the ports of influence were affected before the fifth diffusion stage; in the second and third stages the numbers increased to 58 and 69 ports, respectively (Fig. 3). However, Rotterdam's diffusion efficiency was greater than Antwerp's. For example, Istanbul, which ranked fourth, was affected during the second diffusion stage, and Antwerp was influenced in the third diffusion stage. With an increase in direct influence and an increase in the number of ports of influence at lower diffusion stages, diffusion efficiency of port influence improves. The most direct embodiment of this is the number of ports influenced by Antwerp's first five diffusion stages, which accounts for 91.80% of all ports influenced compared with 2013.

Figure 5(a)-(c) presents Singapore's first influence diffusion stage. Singapore's direct influence is relatively small compared with both Rotterdam's and Antwerp's. In 2019, it only influenced 14 ports, and the geographical distribution of these ports was relatively limited to the neighboring Middle East region and Southeast Asia (Fig. 5(a)). Furthermore, its first diffusion stage only influenced one port of influence, namely, Piraeus in Greece. In 2013, the number of ports influenced by Singapore's first diffusion stage increased to 115, and its *directinfluenceratio* increased to 32.86% (outdegree of 350). These directly influenced ports began to spread rapidly to East Asia, the Americas, and the Mediterranean region (Fig. 5(b)). During the first diffusion stage, Singapore was able to influence 15 ports of influence, but the influence of these ports was small (a maximum of two ports were influenced). In 2016, Singapore influenced

138 ports at the first diffusion stage, and its *directinfluenceratio* increased slightly to 34.67% (out-degree of 398), with its influenced area somewhat more concentrated in the vicinity of its port (Fig. 5(c)). The affected ports included 28 ports of influence, but these ports had a relatively small influence. For example, San Francisco had the most influence but could only influence five ports. The relatively small number of ports directly influenced by Singapore, the relatively concentrated spatial distribution of the influenced ports, and the relatively small number of ports with significant influence resulted in the direct influence of Singapore ports being much smaller than that of Rotterdam and Antwerp.

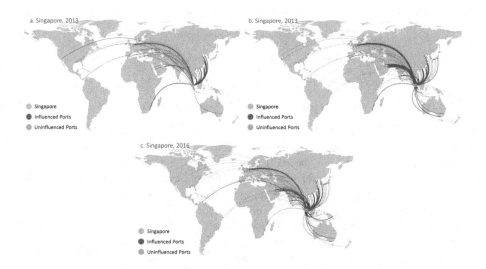

Fig. 5. Influence diffusion at first stage: Singapore

Singapore's indirect influence did, however, show an upward trend. However, the number of influential ports affected through its multiple diffusion stages remained relatively small (Fig. 3), and the number of ports affected by these ports of influence were also small, which, to some extent, limited the further expansion of its indirect influence. In 2009, Singapore had two diffusion stages, but the second diffusion stage did not affect any port of influence. In 2013 and 2016, the number of influential ports affected by the second diffusion stage increased slightly to 5 and 10, respectively, while the numbers of influential ports affected by other diffusion stages were lower. Among these ports, Sharjah in the United Arab Emirates (influenced at the third diffusion stage in 2013) was the most influential port, but still only influenced three ports. The number of indirectly influenced ports was also relatively small, and the spatial scope of the influenced ports relatively concentrated, resulting in Singapore's indirect influence being far less than that of Rotterdam or Antwerp.

4 Conclusion

Our study developed a port influence diffusion model to comprehensively analyze the evolution of the influence of Rotterdam, Antwerp and Singapore ports. The methodology considers two aspects: direct and indirect port influence based on global oil vessel trajectory data. The model identifies the influence of each port via their diffusion patterns and recursively the number of ports influenced in the network. We can get the following conclusion. The geospatial range of the ports directly influenced by these three ports continues to expand in scope to neighboring areas, and reflects rapid expansion globally. As ports with greater influence continuously strengthen their direct control, their direct influence continues to expand and grow rapidly, and the directly influenced ports become path dependent. In addition, the strong direct influence of the port improves its indirect influence to a certain extent (i.e., the number of ports influenced after two stages of diffusion) and the diffusion efficiency of the port (i.e., the number of ports influenced through fewer diffusion stages increases), forming a hierarchical diffusion pattern. Moreover, by increasing the number of ports of influence via direct influence while expanding geographical scope, indirect influence is also increased further. Thus, port diffusion efficiency can be improved by strengthening port influence on ports of influence at a lower diffusion stage. It should be noted that there is no foreshadowed relationship between the number of routes and direct port influence.

Acknowledgement. This research is supported by the China Postdoctoral Science Foundation (2019M660774).

References

1. Dudley, B.: BP statistical review of world energy (2018)
2. Ducruet, C.: Multilayer dynamics of complex spatial networks: the case of global maritime flows (1977–2008). J. Transp. Geogr. **60**, 47–58 (2017)
3. Perc, M.: The Matthew effect in empirical data. J. R. Soc. Interface **11**(98), 20140378 (2014)
4. Yeo, G.-T., Roe, M., Dinwoodie, J.: Evaluating the competitiveness of container ports in Korea and China. Transp. Res. Part A Policy Pract. **42**(6), 910–921 (2008)
5. van Dyck, G., Ismael, H.M.: Multi-criteria evaluation of port competitiveness in West Africa using analytic hierarchy process (AHP). Am. J. Ind. Bus. Manage. **5**(06), 432 (2015)
6. Peng, P., Yang, Y., Lu, F., Cheng, S., Mou, N., Yang, R.: Modelling the competitiveness of the ports along the Maritime Silk Road with big data. Transp. Res. Part A Policy Pract. **118**, 852–867 (2018)
7. Ren, J., Dong, L., Sun, L.: Competitiveness prioritisation of container ports in Asia under the background of China's belt and road initiative. Transp. Rev. **38**(1), 1–21 (2018)
8. Chen, L., Zhang, D., Ma, X., Wang, L., Li, S., Wu, Z., et al.: Container port performance measurement and comparison leveraging ship GPS traces and maritime open data. IEEE Trans. Intell. Transp. Syst. **17**(5), 1227–1242 (2016)
9. Min, H., Park, B.: A two-dimensional approach to assessing the impact of port selection factors on port competitiveness using the Kano model. Marit. Econ. Logist. 1–30 (2019). https://doi.org/10.1057/s41278-019-00117-7

10. Kaluza, P., Kölzsch, A., Gastner, M., Blasius, B.: The complex network of global cargo ship movements. J. R. Soc. Interface **7**(48), 1093–1103 (2010)
11. Ducruet, C., Notteboom, T.: The worldwide maritime network of container shipping: spatial structure and regional dynamics. Glob. Netw. J. Transnatl. Aff. **12**(3), 395–423 (2012)
12. Li, Z., Xu, M., Shi, Y.: Centrality in global shipping network basing on worldwide shipping areas. GeoJournal **80**(1), 47–60 (2014). https://doi.org/10.1007/s10708-014-9524-3
13. Peng, P., et al.: A fine-grained perspective on the robustness of global cargo ship transportation networks. J. Geog. Sci. **28**(7), 881–889 (2018). https://doi.org/10.1007/s11442-018-1511-z
14. Peng, P., Yang, Y., Cheng, S., Lu, F., Yuan, Z.: Hub-and-spoke structure: characterizing the global crude oil transport network with mass vessel trajectories. Energy **168**, 966–974 (2019)
15. Peng, P., Poon, J., Yang, Y., Lu, F., Cheng, S.: Global oil traffic network and diffusion of influence among ports using real time data. Energy **172**, 333–342 (2019)

Entropy-Based Measure for Influence Maximization in Temporal Networks

Radosław Michalski[1,2]([✉])(ID), Jarosław Jankowski[2](ID), and Patryk Pazura[2](ID)

[1] Faculty of Computer Science and Management, Department of Computational Intelligence, Wrocław University of Science and Technology, Wrocław, Poland
radoslaw.michalski@pwr.edu.pl
[2] Department of Computer Science and Information Technology, West Pomeranian University of Technology, Szczecin, Poland
{jjankowski,ppazura}@wi.zut.edu.pl

Abstract. The challenge of influence maximization in social networks is tackled in many settings and scenarios. However, the most explored variant is looking at how to choose a seed set of a given size, that maximizes the number of activated nodes for selected model of social influence. This has been studied mostly in the area of static networks, yet other kinds of networks, such as multilayer or temporal ones, are also in the scope of recent research. In this work we propose and evaluate the measure based on entropy, that investigates how the neighbourhood of nodes varies over time, and based on that and their activity ranks, the nodes as possible candidates for seeds are selected. This measure applied for temporal networks intends to favor nodes that vary their neighbourhood highly and, thanks to that, are good spreaders for certain influence models. The results demonstrate that for the Independent Cascade Model of social influence the introduced entropy-based metric outperforms typical seed selection heuristics for temporal networks. Moreover, compared to some other heuristics, it is fast to compute, thus can be used for fast-varying temporal networks.

Keywords: Social networks · Influence maximization · Entropy · Seed selection · Temporal networks

1 Introduction

Social influence maximization is a research topic that has been posed in 2003 by Kempe et al. [16]. Assuming a static social network and a given influence model, typical social influence maximization task is to choose a set of k nodes that will result with the highest influence (number of activations) across all methods. As presented in Related Work, this area has been investigated at the beginning for the case of static networks. However, recently researchers extended their interest to other kinds of networks, including multilayer [18] or temporal networks [11,23]. This is caused by the fact, that for some applications these

© Springer Nature Switzerland AG 2020
V. V. Krzhizhanovskaya et al. (Eds.): ICCS 2020, LNCS 12140, pp. 277–290, 2020.
https://doi.org/10.1007/978-3-030-50423-6_21

networks are more suitable for representing several types of processes, especially related to information diffusion. Here, an aggregated graph is not capable to represent the ordering of events, crucial for studying these processes. This is the reason why temporal networks are the model chosen in the case of modelling diffusion processes. Following this direction, the researchers exploring influence maximization area also started to investigate temporal networks [39].

In this work, we propose a method based on entropy for influence maximization in temporal networks. Entropy as a measure of diversity is capable of providing information about diversity - in this case how the neighbourhood of the nodes changes over time in a temporal network. This feature can be considered important in the area of information spreading, since a node - when activated - by changing its neighbourhood is increasing the chances of activating other nodes. Otherwise, if the neighbourhood of a node will be the same, this node usually would not contribute to the spreading process anymore after contacting its neighbours at the beginning.

The purpose of this work is to compare introduced entropy-based seed selection method against other heuristics commonly used in the area of social influence maximization. To do so, we studied the performance of the proposed method of seed selection for Independent Cascade Model of social influence using real-world datasets representing temporal networks of different kind and dynamics.

The work is organized as follows. In the next section we shortly describe the related work in the area covered by this research. Next, in Sect. 3, we present the experimental setting. Section 4 presents and discusses the results, while in Sect. 5 summary and future work directions are presented.

2   Related Work

Information spreading processes within social networks attract attention of researchers from various fields. It resulted with new sub-disciplines and research areas like network science [1]. In the background, they use theoretical models, network evolution mechanisms, multilayer and dynamic structures, methods for community detection, modeling and analysis of ongoing processes. While information spreading processes within complex networks are observed in various areas, many studies focus on their modeling and analysis.

One of the key topics is related to influence maximisation and selection of starting nodes for initialisation of spreading processes. It was defined with main goal to increase the coverage within the network with properly selected seed nodes [16]. Due to the difficulty of finding the exact solution, several heuristics, with the most effective greedy approach [16] and its extensions with adjustable computational performance [9] were proposed trough the course of last years. Apart from that, other seed selection methods were explored, including heuristics based on centrality measures [32], community seeding [38], k-shell decomposition [17], genetic algorithms [34,35] and other solutions [10]. One of their goals is avoiding overlapping the regions of influence and increase the distance between seeds. It was a key concept behind Vote-Rank method [36] and the studies with its extensions.

Initial works in the area of information diffusion were related to static networks, and later evolved towards temporal networks. They focused on seed selection in dynamic networks and comparison of different approaches [20]. They also focused on links, changing topologies and nodes availability [14]. Recent studies focused on topological features of temporal networks figured out temporal versions of static centrality measures [26,30,31] and they can be used for seed selection. As a generalization of the closeness centrality for static networks, Pan and Saramäki define the temporal closeness centrality [25]. Takaguchi proposed method to represent the importance of a temporal vertex defined as temporal coverage centrality [29]. Beside the topological features, recent studies highlighted the dynamics of temporal networks as significant factor of information diffusion [12]. They showed that dynamics-sensitive centrality in temporal networks performed seeds selection much better than topological centrality. In [2] temporal sequence of retweets in Twitter cascades was used to detect a set of influential users by identifying anchor nodes. It is also worth mentioning that in parallel to temporal networks, recently information cascades in multilayer networks are also studied [4,13,22].

At this point, it is also important to mention that social influence is a process that is difficult to observe and measure. Individual decisions of the members of a social networks are distributed over time and rarely it is possible to observe them, as they are often internal and not directly bound with actions. This is why we often models are often used to represent the process and their parameters are often derived either based on data [8] or a number of small scale experiments conducted in real world [5,6].

The approach presented in this study is based on Shannon's entropy [27] that measures the amount of information. As such, the application of entropy is not a new concept in the area of complex network analysis. Most of the studies, however, focused on the predictive capabilities or global quantification of the dynamics of the network. For instance, Takaguchi et al. evaluated the predictability of the partner sequence for individuals [28]. In [37] authors proposed entropy-based measure to quantify how many typical configurations of social interactions can be expected at any given time, taking into account the history of the network dynamic processes. Shannon entropy has been also used in order to show how Twitter users are focusing on topics comparing to the entire system [33]. Another work is using measures based on entropy for analyzing the human communication dynamics and demonstrating how the complex systems stabilise over time [19]. The results presented in the last work are important to understand that the implications of stabilisation are significantly affecting the diffusion. In the case when our social circles do not change, information has less possibilities to reach different areas of the network. As a consequence, if information has limited chances to appear in some parts of the network, the same would apply to social influence. This observation was the inspiration of our work in which we wanted to find the nodes that have the most-varying neighbourhood, thus the highest chances to spread information to others. This resulted with a entropy-based measure that looks at the variability of neighbourhood, but also takes into account the overall number of different neighbours the nodes has contacted over time.

3   Entropy-Based Measure of Variability

The proposed measure for quantifying the variability of neighbourhood is using the entropy to calculate the diversity of neighbours of a given node v_i across neighbouring time windows in a temporal network. Introduced measure is focusing on finding the nodes that have the highest exchange in the neighbourhood over consecutive time windows and the values of it can be considered as a diversity of a node. However, to not to fall into specific cases, some additional factors have to be considered as well - these are discussed after introducing the definition of this measure.

The proposed entropy-based measure is expressed in the following way:

$$Var(v_i) = -\sqrt{|e_{ij}|} \sum_{n=1}^{N-1} p(NB_{v_i,n,n+1}) \ln(p(NB_{v_i,n,n+1})), \tag{1}$$

where e_{ij} is the number of all unique edges originating at node v_i and $NB_{v_i,n,n+1}$ is the neighbourhood of a node in windows that are next to each other. Defined this way, the measure promotes the nodes that vary the neighbourhood over subsequent windows and incorporates also the size of the neighbourhood itself.

For directed networks, the neighbourhood is expressed as outgoing links, for undirected ones as all links. The last factor is promoting the variability of neighbours in such a way that the set of neighbours should be the widest, since without this element only varying the neighbours across windows next to each other would be enough to maximize the measure.

After calculating this measure for all nodes for given network settings, it has been used as the utility score for choosing seeds - top k percent of nodes have been selected as seeds and activated during window $T_{N/2+1}$.

It is worth underlining that the measure is computed at the node-level and only looks at the neighbourhood of nodes, similarly to the degree-based measures. As a consequence, it has similar computational complexity that is much lower than for betweenness centrality. This makes the measure still applicable for large graphs.

4   Experimental Setting

4.1   Datasets

The experiments have been conducted using two datasets: manufacturing company email communication [24] consisting from emails sent between employees of a manufacturing company over a course of nine months and a Haggle dataset representing contacts between people measured by carried wireless devices [3]. The statistics of the datasets presented in Table 1 indicate that albeit similar in size, the datasets differ in a number of factors contributing to information diffusion, e.g. in an average degree or power law exponent. These datasets have been then converted into a temporal social network according to the model presented in the next subsection. Resulting temporal networks also significantly differ in terms of how nodes and events distribute over windows showing different nature of datasets.

Table 1. Properties of evaluated datasets - manufacturing company emails and Haggle. The statistics have been computed to the aggregated networks.

Property	Manufacturing	Haggle
Individual type	Employee	Person
Event type	Email	Contact
Format	Directed	Undirected
Edge weights	Multiple unweighted	Multiple unweighted
Size	167 vertices	274 vertices
Volume	82,927 edges	28,244 edges
Average degree (overall)	993.14 edges/vertex	206.16 edges/vertex
Maximum degree	9,053 edges	2,092 edges
Largest connected component	167 vertices	274 vertices
Power law exponent (estimated)	4.6110 ($d_{min} = 53$)	1.5010 ($d_{min} = 1$)
Gini coefficient	61.9%	84.2%
Clustering coefficient	5.412664×10^{-1}	5.66×10^{-1}
Diameter	5 edges	4 edges
90-percentile effective diameter	2.50 edges	2.79 edges
Mean shortest path length	1.96 edges	2.42 edges
Preferential att. exponent	1.2284 ($\epsilon = 2.4955$)	1.1942 ($\epsilon = 7.8302$)

4.2   Temporal Social Network

The temporal social network is based on time windows. For all the evaluated datasets, the periods they cover have been split into n windows of equal time [11]. Then, all the events within a particular window have been a source for building a temporal social network snapshot $T_n = (V, E)$, $n \in 1 \ldots N$ consisting from the set of nodes V and the set of directed edges E. The edge weights are defined according to the following formula:

$$w_{ij} = \frac{\sum e_{ij}}{\sum e_i}, \tag{2}$$

where $\sum e_{ij}$ is the total number of contacts originated at node i to node j and $\sum e_i$ is the total number of contacts originated at node i. All self loops were removed. Each snapshot T_n can be then considered as a static graph where events do collapse into same time frame, but since the snapshots are time-ordered, temporal aspects are preserved at a certain level of granularity. The reasons why this temporal network model was used is that it enables using established seed selection heuristic and diffusion model known for static networks. An exemplary temporal network consisting from five time windows has been shown in Fig. 1. The evaluated values of N have been the following: $8, 16, 32, 64$. Half of the windows has been used for training purposes - building rankings of nodes based on evaluated methods - and half for evaluation. In case of this experiment, training purposes mean building the seed set based on the information about behaviour of nodes in the training windows and evaluating their performance is interpreted as a total number of nodes activated at the end of the influence process.

In order to verify whether the datasets chosen for evaluation are different, we investigated the number of unique nodes and number of unique events - Fig. 2 shows how these values evolve over subsequent time windows. It can be observed that whilst for the manufacturing temporal network the number of nodes in each window is relatively stable, this is not the case for Haggle network, since this value undergoes bigger changes. This also impacts the number of events.

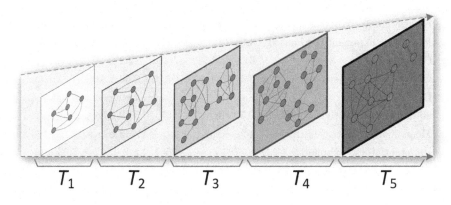

Fig. 1. A visualisation of a temporal network model based on windows used for experiments.

It should be noted that the introduced measure in this form gives more importance to the nodes that are present in consecutive time windows. Otherwise the $NB_{v_i,n,n+1}$ of Eq. 1 will be contributing to decreasing the value of the measure for nodes that are not present on subsequent time windows. The reason why it is important to look at the presence of nodes in time windows that are next to each other is depicted in Fig. 3. Here one can see that with the increasing number of windows the number of nodes that exist in all the windows starts to drop significantly after reaching certain critical value.

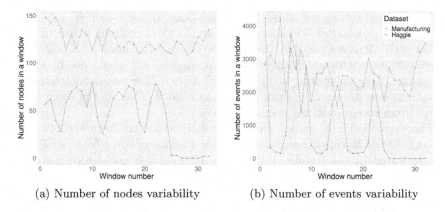

(a) Number of nodes variability   (b) Number of events variability

Fig. 2. Number of nodes and number of events for each time window for evaluated datasets: manufacturing company and Haggle for 32 windows.

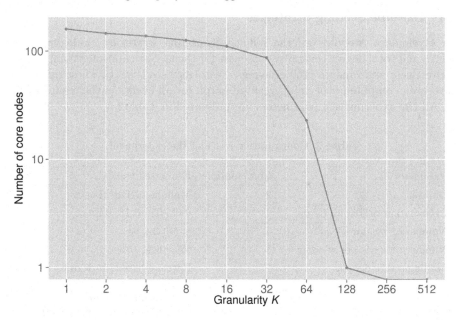

Fig. 3. The number of nodes existing in the all time windows dependent of the number of windows (the granularity of the split) for the manufacturing dataset.

4.3   Social Influence Model

The model chosen for social influence is *Independent Cascade Model* (IC) [7]. This model assumes that a node has a single chance of activating its neighbour expressed as a probability p. If the node will succeed, this neighbour will become activated and will be attempting to activate its neighbours in the following iterations. As the basic version of this model was proposed for static networks, in the temporal setting we added to modifications to the model: (i) *exhaustion of spreading capabilities* - in every snapshot T_n the iterations follow until no further activations are possible, (ii) *single attempt of activation* - if a node failed to activate its neighbour, in the subsequent snapshots it would not be able to try again. These extensions to the base IC model allow to adequately spread activations over the span of evaluation time windows, but at the same time - restrict from reaching all the nodes too early.

The evaluated independent cascade probabilities have been the following: $0.05, 0.1, 0.15, 0.2$, whilst the seed set was the fraction of $5\%, 10\%, 15\%$ of nodes that appeared in the training set. As the IC model is not deterministic, for each parameter combination we run a 1,000 simulations of the diffusion process. Moreover, in order to make results comparable, we followed the coordinated execution procedure introduced in [15] - the results of drawings have been the same for all the runs.

4.4 Baseline Seeding Strategies

As a reference, we did use the following baseline strategies: in-degree, out-degree and total degree centrality, closeness for the largest connected component, betweenness and random (100 drawings - averaged) seed set [1]. These measures have been computed over an aggregated graph for all events in the training set.

All the experimental parameters are presented in Table 2.

Table 2. Configuration space of the experiment

Parameter	Combinations	Evaluated values
Dataset - d	2	Manufacturing, Haggle
Propagation probability - p	4	0.05, 0.1, 0.15, 0.2
Number of windows - k	4	8, 16, 32, 64
Number of seeds per window - j	3	5%, 10%, 15%
Seeding strategies - h	7	Entropy-based, in-degree, out-degree, total degree, closeness, betweenness, random (100 runs, averaged)

5 Results

In total, we did evaluate 672 combination of all parameters - 96 per measure, as shown in Table 2. Results presented in Fig. 4 show how the measures performed relatively to the entropy-based measure being a reference. As a measure of performance we compare the number of nodes activated at the end of the seeding process - this is a typical approach for comparing different heuristics for social influence.

What is observed, the proposed method outperforms the others and the out-degree approach is the second leading one. The second position of out-degree is linked to two factors. Firstly, to the structure of typical social networks. Usually they follow the preferential attachment, so in the case of high degree nodes, these are linked to other nodes with high degree and so on. Secondly, the Independent Cascade Model, as name suggests, compared to some other models like linear threshold does not require any fraction of nodes for activating the neighbours, so the cascades can spark independently of each other. That is why choosing nodes with high outdegree maximizes the chances of starting in the central part of the typical social network.

Regarding the variability measure proposed in this work, it also contains the factor promoting the nodes with high number of neighbours (see Eq. 3), however also requires these nodes to vary its neighbours over windows. This, in turn,

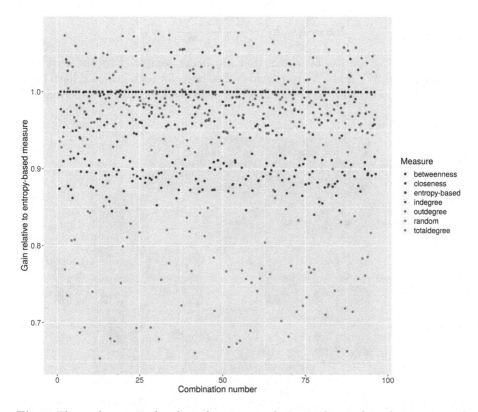

Fig. 4. The performance of evaluated measures relative to the number of activations of the entropy-based measure. Overall, the entropy-based measure has been outperforming others in 41.5% of cases, the second best-performing heuristic was out-degree with 16.5%. Next: betweenness (13.7%), closeness (12.05%), in-degree (7.7%), total degree (6.7%) and random (2.2%).

increases to the possibility of reaching other areas of the directed network that could not have been potentially activated by other seeding strategies.

The differences in terms of number of activated nodes between particular seeding strategies, three seeding strategies: entropy-based, out-degree and betweenness have been performing on average 5–7% better then other degree-based methods. However, among these three, the differences were smaller, on average ranging from 2.5–7% in most cases. This part of experimental analysis shows that the entropy-based measure is capable of generating well-performing seed sets.

One of the research questions we wanted to find answer on is how the seed sets differ. This would indicate whether are there similarities against the seed sets meaning that the nodes also share similar properties in terms of measures. To do so, we compared all the seed sets built by each evaluated heuristics by using the Jaccard index defined as follows:

$$Jaccard(h_i, h_j) = \frac{|V^{h_i} \cap V^{h_j}|}{|V^{h_i} \cup V^{h_j}|}, i \neq j, \tag{3}$$

where h_i and h_j are the heuristics that are being compared by the seed sets they generated, V^{h_i} and V^{h_j}, respectively. Seed sets for each heuristic have been compared pairwise with seed sets of other heuristics for all the experimental

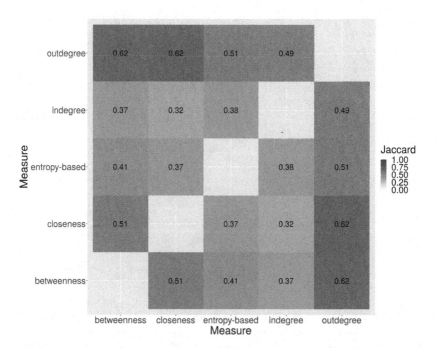

Fig. 5. The similarity of seed sets generated by different heuristics evaluated in this work, including proposed entropy-based one. The metric used for computing the similarity of seed sets is the Jaccard index.

parameters (see Table 2) and then the results have been averaged pairwise. Since the random heuristic did not provide coherent results for each run, it has not been compared against others. The results demonstrating the similarity of seed sets generated by different heuristics are presented in Fig. 5. This analysis indicates that there is only a partial similarity between seed sets generated by different heuristics and the introduced entropy-based measure is the most similar to outdegree (0.51) and betweenness (0.41). However, in general, it is observed that the similarity of seed sets is only partial and different nodes are selected for initial activation.

6 Conclusions and Future Work

In this work we proposed a method based on entropy that can be used for seed selection in temporal social networks. The method is basing on the variability of neighbourhood that leads to increasing the spread of influence in the social network. The evaluation of the method demonstrates that in many cases the results for the introduced method outperform other heuristics. Moreover, comparing to some other heuristics that require computing shortest paths in a graph, such as betweenness or closeness, introduced entropy-based measure is simple to compute.

However, it must be noted that this method of influence maximization is suited mostly for models such as Independent Cascade that do not require a committed neighbourhood for activations. Nevertheless, many real-life diffusion cascades are actually following the independent cascade schema, since members of many social networks decide upon adoption of an idea based shortly afterwards observing activities of others. This is often observed in social media where people decide whether to share a content just after being exposed to it. On the other hand, in the case of more complex decisions, these decisions could follow other models. This is why it is a necessity to understand which models apply for certain situations before deciding on the social influence seed selection method.

In the case of temporal networks, one needs to remember that there are new factors that substantially impact the process compared to static network scenario [21]. One of the most important ones relates to the seed set. Contrary to static networks, not all nodes selected for initial activation can appear in the subsequent time windows. This means that a part of the budget can be wasted if some nodes would not appear. Another aspect is the fact that some of the time windows could potentially contain a limited number of edges or, in the worst case, no edges at all. This will impact the dynamics of the spreading process. This is why when considering developing heuristics for social influence in networks, one needs to consider these factors by appropriately looking at the historical behaviour of nodes and the temporal network itself.

Regarding the future work directions, it is planned to follow a number of them. Firstly, we would like to investigate in detail how the nodes selected by the entropy-based method penetrate different areas of the network compared to other heuristics. Next, as results indicate, the seed sets provided by different heuristics only partially overlap, yet some heuristics still produced well-performing seed sets. The idea is to propose a combined mixture-based measure that will take advantage of different properties of nodes in order to be even more successful in activating others. The third direction requires investigating how the proposed measure performs for other models of social influence, e.g. linear threshold.

Acknowledgments. This work was supported by the National Science Centre, Poland, grant no. 2016/21/B/HS4/01562.

References

1. Barabási, A.L., et al.: Network Science. Cambridge University Press, Cambridge (2016)
2. Bhowmick, A.K., Gueuning, M., Delvenne, J.C., Lambiotte, R., Mitra, B.: Temporal sequence of retweets help to detect influential nodes in social networks. IEEE Trans. Comput. Soc. Syst. **6**(3), 441–455 (2019)
3. Chaintreau, A., Hui, P., Crowcroft, J., Diot, C., Gass, R., Scott, J.: Impact of human mobility on opportunistic forwarding algorithms. IEEE Trans. Mob. Comput. **6**(6), 606–620 (2007)
4. Erlandsson, F., Bródka, P., Borg, A.: Seed selection for information cascade in multilayer networks. In: Cherifi, C., Cherifi, H., Karsai, M., Musolesi, M. (eds.) COMPLEX NETWORKS 2017 2017. SCI, vol. 689, pp. 426–436. Springer, Cham (2018). https://doi.org/10.1007/978-3-319-72150-7_35
5. Friedkin, N.E., Cook, K.S.: Peer group influence. Soc. Methods Res. **19**(1), 122–143 (1990)
6. Friedkin, N.E., Johnsen, E.C.: Social Influence Network Theory: A sociological Examination of Small Group Dynamics, vol. 33. Cambridge University Press, Cambridge (2011)
7. Goldenberg, J., Libai, B., Muller, E.: Talk of the network: a complex systems look at the underlying process of word-of-mouth. Mark. Lett. **12**(3), 211–223 (2001). https://doi.org/10.1023/A:1011122126881
8. Goyal, A., Bonchi, F., Lakshmanan, L.V.: Learning influence probabilities in social networks. In: Proceedings of the third ACM international conference on Web search and data mining, pp. 241–250 (2010)
9. Goyal, A., Lu, W., Lakshmanan, L.V.: Celf++: optimizing the greedy algorithm for influence maximization in social networks. In: Proceedings of the 20th international conference companion on World wide web, pp. 47–48. ACM (2011)
10. Hinz, O., Skiera, B., Barrot, C., Becker, J.U.: Seeding strategies for viral marketing: an empirical comparison. J. Mark. **75**(6), 55–71 (2011)
11. Holme, P., Saramäki, J.: Temporal networks. Phys. Rep. **519**(3), 97–125 (2012)
12. Huang, D.W., Yu, Z.G.: Dynamic-sensitive centrality of nodes in temporal networks. Sci. Rep. **7**, 41454 (2017)
13. Jankowski, J., Michalski, R., Bródka, P.: A multilayer network dataset of interaction and influence spreading in a virtual world. Sci. Data **4**, 170144 (2017)
14. Jankowski, J., Michalski, R., Kazienko, P.: Compensatory seeding in networks with varying avaliability of nodes. In: 2013 IEEE/ACM International Conference on Advances in Social Networks Analysis and Mining, ASONAM 2013, pp. 1242–1249. IEEE (2013)

15. Jankowski, J., et al.: Probing limits of information spread with sequential seeding. Sci. Rep. **8**(1), 13996 (2018)
16. Kempe, D., Kleinberg, J., Tardos, É.: Maximizing the spread of influence through a social network. In: Proceedings of the ninth ACM SIGKDD international conference on Knowledge discovery and data mining, pp. 137–146 (2003)
17. Kitsak, M., et al.: Identification of influential spreaders in complex networks. Nat. Phys. **6**(11), 888 (2010)
18. Kivelä, M., Arenas, A., Barthelemy, M., Gleeson, J.P., Moreno, Y., Porter, M.A.: Multilayer networks. J. Complex Netw. **2**(3), 203–271 (2014)
19. Kulisiewicz, M., Kazienko, P., Szymanski, B.K., Michalski, R.: Entropy measures of human communication dynamics. Sci. Rep. **8**(1), 1–8 (2018)
20. Michalski, R., Kajdanowicz, T., Bródka, P., Kazienko, P.: Seed selection for spread of influence in social networks: temporal vs static approach. New Gener. Comput. **32**(3–4), 213–235 (2014)
21. Michalski, R., Kazienko, P.: Maximizing social influence in real-world networks—the state of the art and current challenges. In: Król, D., Fay, D., Gabryś, B. (eds.) Propagation Phenomena in Real World Networks. ISRL, vol. 85, pp. 329–359. Springer, Cham (2015). https://doi.org/10.1007/978-3-319-15916-4_14
22. Michalski, R., Kazienko, P., Jankowski, J.: Convince a dozen more and succeed-the influence in multi-layered social networks. In: 2013 International Conference on Signal-Image Technology & Internet-Based Systems, pp. 499–505. IEEE (2013)
23. Michalski, R., Palus, S., Bródka, P., Kazienko, P., Juszczyszyn, K.: Modelling social network evolution. In: Datta, A., Shulman, S., Zheng, B., Lin, S.-D., Sun, A., Lim, E.-P. (eds.) SocInfo 2011. LNCS, vol. 6984, pp. 283–286. Springer, Heidelberg (2011). https://doi.org/10.1007/978-3-642-24704-0_30
24. Nurek, M., Michalski, R.: Combining machine learning and social network analysis to reveal the organizational structures. Appl. Sci. **10**(5), 1699 (2020)
25. Pan, R.K., Saramäki, J.: Path lengths, correlations, and centrality in temporal networks. Phys. Rev. E **84**(1), 016105 (2011)
26. Pfitzner, R., Scholtes, I., Garas, A., Tessone, C.J., Schweitzer, F.: Betweenness preference: quantifying correlations in the topological dynamics of temporal networks. Phys. Rev. Lett. **110**(19), 198701 (2013)
27. Shannon, C.E.: A mathematical theory of communication. Bell Syst. Tech. J. **27**(3), 379–423 (1948)
28. Takaguchi, T., Nakamura, M., Sato, N., Yano, K., Masuda, N.: Predictability of conversation partners. Phys. Rev. X **1**(1), 011008 (2011)
29. Takaguchi, T., Yano, Y., Yoshida, Y.: Coverage centralities for temporal networks. Eur. Phys. J. B **89**(2), 1–11 (2016). https://doi.org/10.1140/epjb/e2016-60498-7
30. Tang, J., Scellato, S., Musolesi, M., Mascolo, C., Latora, V.: Small-world behavior in time-varying graphs. Phys. Rev. E **81**(5), 055101 (2010)
31. Taylor, D., Myers, S.A., Clauset, A., Porter, M.A., Mucha, P.J.: Eigenvector-based centrality measures for temporal networks. Multiscale Model. Simul. **15**(1), 537–574 (2017)
32. Wang, X., Zhang, X., Zhao, C., Yi, D.: Maximizing the spread of influence via generalized degree discount. PLoS ONE **11**(10), e0164393 (2016)
33. Weng, L., Flammini, A., Vespignani, A., Menczer, F.: Competition among memes in a world with limited attention. Sci. Rep. **2**, 335 (2012)
34. Weskida, M., Michalski, R.: Evolutionary algorithm for seed selection in social influence process. In: 2016 IEEE/ACM International Conference on Advances in Social Networks Analysis and Mining, ASONAM, pp. 1189–1196. IEEE (2016)

35. Weskida, M., Michalski, R.: Finding influentials in social networks using evolutionary algorithm. J. Comput. Sci. **31**, 77–85 (2019)
36. Zhang, J.X., Chen, D.B., Dong, Q., Zhao, Z.D.: Identifying a set of influential spreaders in complex networks. Sci. Rep. **6**, 27823 (2016)
37. Zhao, K., Karsai, M., Bianconi, G.: Models, entropy and information of temporal social networks. In: Holme, P., Saramäki, J. (eds.) Temporal Networks. Understanding Complex Systems, pp. 95–117. Springer, Heidelberg (2013)
38. Zhao, Y., Li, S., Jin, F.: Identification of influential nodes in social networks with community structure based on label propagation. Neurocomputing **210**, 34–44 (2016)
39. Zhuang, H., Sun, Y., Tang, J., Zhang, J., Sun, X.: Influence maximization in dynamic social networks. In: 2013 IEEE 13th International Conference on Data Mining, pp. 1313–1318. IEEE (2013)

Evaluation of the Costs of Delayed Campaigns for Limiting the Spread of Negative Content, Panic and Rumours in Complex Networks

Jaroslaw Jankowski[(✉)], Piotr Bartkow, Patryk Pazura, and Kamil Bortko

Faculty of Computer Science and Information Technology,
West Pomeranian University of Technology, Szczecin, Poland
jjankowski@wi.zut.edu.pl

Abstract. Increasing the performance of information spreading processes and influence maximisation is important from the perspective of marketing and other activities within social networks. Another direction is suppressing spreading processes for limiting the coverage of misleading information, spreading information helping to avoid epidemics or decreasing the role of competitors on the market. Suppressing action can take a form of spreading competing content and it's performance is related to timing and campaign intensity. Presented in this paper study showed how the delay in launching suppressing process can be compensated by properly chosen parameters and the action still can be successful.

Keywords: Information spreading · Competing processes · Limiting the spread · Viral marketing · Misinformation · Rumours · Social networks

1 Introduction

Information spreading processes within the networks are usually analysed from the perspective of the performance with main goal to maximize their coverage. It can be achieved by a proper selection of initial nodes starting propagation among their neighbours. The problem defined as influence maximisation was presented together with greedy solution for finding set of nodes delivering results close to optimum [17]. Apart from greedy approach other possibilities use heuristics based on selection of nodes with high centrality measures [20]. Increasing the spread is central problem for viral marketing, diffusion of products and innovations [12].

The purpose of information spreading can be suppression of other processes [5] in case of the spread of misleading or harmful information, rumours and by marketing companies to compete with other products. One of directions is studying the factors affecting the dominance over other processes [3]. It can be used for spreading competing products, opinion, ideas or digital content in a form of memes, videos [29] or gifts [7]. Information spreading can be also used for stopping real epidemics, with the use of educational information, awareness [10] and related to the rumors

© Springer Nature Switzerland AG 2020
V. V. Krzhizhanovskaya et al. (Eds.): ICCS 2020, LNCS 12140, pp. 291–304, 2020.
https://doi.org/10.1007/978-3-030-50423-6_22

and panic [25]. For planning and launching competing processes, the habituation effect should also be taken into account, because of dropping response to multiple viral marketing messages received in a short time [6].

The performance of the suppressing process is related to various factors including proper timing and campaign parameters. Presented in this paper study shows relation between delays of suppressing actions and the costs required to make them successful. It is assumed that delayed action still can be successful, if the strength of the process is increased, when compared to the harmful process. However, it is related to increased costs of the action, for example better, more appealing content, the number of seeds, more effective way of selection seeds or incentives used.

The remainder of the paper is organised as follows: Sect. 2 includes the related work, the assumptions are presented within the Sect. 3 and are followed by empirical results within the Sect. 4. Obtained results are concluded in the last section.

2   Related Work

Online platforms based on social networks created infrastructures for information spreading processes [2]. Initially activated network members can spread information to their neighbours, in the next step they can spread content to their neighbours and so on. The process continues till saturation point, when new transmissions are not possible. At the end, the process reaches some fraction of nodes or whole network. In most cases the main goal is the increasing the spread of the content to achieve a high number of network nodes influenced [27].

Spread of information can be modelled with various models. Some of approaches use models derived from epidemiology like SIS and SIR with their further extensions [16] while other use branching processes [14]. From the perspective of network structures the most often and well studied models include Independent Cascade Model (ICM) and Linear Threshold Model (LT) [17]. Both approaches were used for various applications and further extensions taking into account time factors or network dynamics [15].

Apart from increasing the process dynamics, quite opposite goals can be taken, to use techniques to block spreading information [26]. To stop epidemics, spreading information about pathogens can be used [10]. For example, several studies were carried on to study spreading information in communication networks to increase awareness [4].

Not only pathogens can be treated as harmful with the need to block spreading. Recently more and more attention is put on spreading the information and digital content which can be potentially harmful or even dangerous for target users. Processes of this type can be based on misleading information and fake news, false medical information, panic and rumours which can negatively influence audiences or promote bad behaviours. It is important to spread alternative information to suppress the negative content and prevent harmful rumours spreading [13]. Similar situation takes place on the market where companies are trying to launch viral campaign or spread information competing with other campaigns [30].

Factors affecting competition were analysed in terms of network structures for multi-layer networks [4]. Another studies take into account immunization strategies based on vaccination of nodes [28]. Another aspects are related to intervals between marketing messages because the ability to process information is limited [19] and habituation effect takes place [24]. As a result high intensity of viral marketing messages received in a short time can be treated as a SPAM [6].

Together with the need of modeling multiple processes within the networks extension of models were proposed. For example Linear Threshold model was extended towards Competitive Linear Threshold Model (CLTM) [11]. The influence blocking maximization problem (IBM) was defined. Authors assumed that positive $(+)$ and negative $(-)$ information spreads within the network. Network nodes can be in three states inactive, +active, and -active. Second main spreading model, the Independent Cascade Model, was extended towards Multi-Campaign Independent Cascade Model (MCICM) [5]. It assumes two campaigns spreading simultaneously within the network with competition mechanism. One of processes is treated as the primary campaign and the secondary, treated as limiting campaign, is decreasing the dynamics and the coverage of the first one. Like in the ICM model activated node had the single chance to activate it's neighbours. The objective of the study was to protect nodes from activation by the first process by activation with the second one.

3   Research Assumptions and Propagation Model

In this paper we analyse the role of timing for suppressing campaign and the relation between the costs of delayed campaigns and their performance. In our research we assume that costs are related to propagation probability and the number of seeds. Propagation probability can be related to incentives, samples quality or others ways to motivate users to propagate the content. Number of seeds is related to fraction of target audience selected as seed and is directly related to campaign costs. The goal is to analyse the costs required to launch successful campaign (treated as a Positive Process) even the delay in the relation to Negative Process takes place. It is assumed that Positive Process is the reaction, that's why the Negative Process starts first.

Positive and negative propagation processes considered in this paper are modelled within network $N(V, E)$ based on vertex set $V = v_1, v_2, ..., v_m$ and edges set $E = e_1, e_2, ..., e_n$. According to the used Independent Cascade model [17] node $u \in V$ is contacting all neighbours, nodes with relation represented by edge $(u, v) \in E$, within the network N and has only one chance to activate node $v \in V$, in the step $t + 1$ with propagation probability $PP(u, v)$ under condition that node v was activated at time t. For our case, Independent Cascade Model is adapted to two concurrent cascades. Probability $PP_{NP}(u, v)$ denotes the probability that node u activates node v one step after node u is activated by a Negative Process. Probability $PP_{PP}(u, v)$ denotes the probability that node u activates node v one step after u is activated by a Positive Process. Two separate seed sets are used to initialise Negative Process and Positive Process. Seed set denoted by $S_{NP} \subseteq V$ is used to initialise the Negative Process. The ranking method R_{NP} is used to select number of seeds according to seeding fraction SF_{NP}. Seed set for Positive Process denoted by $S_{PP} \subseteq V$ such as $S_{NP} \cup S_{PP} = \varnothing$ is used to initialise the negative process. The ranking method R_{PP} is used to select a number of seeds according to seeding fraction SF_{PP}. Every seed node $s_{NP} \subseteq S_{PN}$ is activated in time $t_{NP} = t1$, Every seed node $s_{PP} \subseteq S_{PP}$ is activated in time $t_{PP} \subseteq T$, $T = \{t_1, t_2, ..., t_n\}$. Lets denote by A_{NP}, t the set of active nodes $A_{NP,t} \in V$ possessing the negative information at time t, activated in time point $t - 1$ by a Negative Process, and by A_{PP}, t the set of active nodes possessing the positive information $A_{PP,t} \in V$ at time t, activated by a Positive Process in time $t - 1$. Let's denote by set of not active nodes $A_{\varnothing,t} \in \{V - A_{NP,t} - A_{PP,t}\}$. Selection of nodes $a_{NP,t}$ newly activated by a Negative Process among all neighbours n_i such as $(n_i, v_i) \in E$ takes place according to the formula:

$$\bigvee_{v_i \in A_{NP,t}} a_{NP} \in \{n \in N(v_i) | n \in A_{NP,t}, n \in (A_{\varnothing,t} + A_{PP,t})\} \qquad (1)$$

with probability PP_{NP}. Selection of candidates for activations with positive process a_{PP} takes place among all neighbours n_i such as $(n_i, v_i) \in E$ not active or activated by a Negative Process:

$$\bigvee_{v_i \in A_{PP,t}} a_{PP} \in \{n \in N(v_i) | n \in A_{PP,t}, n \in (A_{\varnothing,t} + A_{NP,t})\} \qquad (2)$$

with probability PP_{PP}. The sequence of steps (1) and (2) is taken randomly to deliver equal chances to positive and negative process. All newly activated nodes are included in active nodes sets for the next time point $t + 1$, respectively for Negative and Positive Process as $A_{NP,t+1} = a_{NP}$ and $A_{PP,t+1} = a_{PP}$. Process is continued until no more new activations are observed. Final results are represented by coverage $C_{\varnothing}$ with nodes not activated by any process, C_{NP} with nodes activated by a Negative Process and C_{PP} with nodes activated by a Positive Process such as $V = C_{\varnothing} \bigcup C_{NP} \bigcup C_{PP}$.

3.1   Illustrative Example

To clarify the approach, the toy example presents three scenarios. In Fig. 1 a Positive Process (green) is started in the same step like Negative Process (red) and has high chance to suppress it. In the second scenario presented in Fig. 2 Positive Process was started too late and parameters where not enough to dominate Negative Process. Third scenario in Fig. 3 shows that delayed process can survive but parameters like propagation probability should be properly adjusted. It is using a simplified graph with weights assigned to edges. Information flows only if weight on the edge is lower or equal to propagation probability $PP1$ for

Negative Process and $PP2$ for Process. Both $PP1$ and $PP2$ are the same to all nodes according to Independent Cascade Model. Graph has 10 nodes and 17 edges. Both processes are competing. Also the seeds selected for the spreading are given, for Negative Process it is node number 1 and for Positive Process node number 10. Last parameter is delay Di that is responsible for start suppression process (green). For this example process consists of simulation steps. Each step is divided into two stages: stage for Negative Process, and stage for Positive Process. Figure 1 - Figure 3 show how delay can affect the results of reaction depending om process parameters.

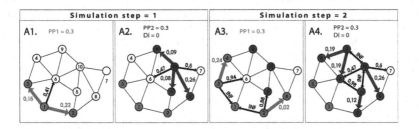

Fig. 1. Example for competing processes with same parameters. In $SIMULATION\ STEP\ =\ 1$ **(A1)** infected node 1 begins and try to infect all his neighbors $(2,3,6)$. Propagation probability $(PP1 = 0,3)$ allows to infect node 2 $(0,22 < 0,3)$ and 3 $(0,15 < 0,3)$ but prevent to infect node 6 $(0,41 > 0,3)$. **(A2)** node 10 starts suppressing campaign and tries to infect his neighbors $(5, 6, 7, 8, 9)$. $PP2 = 0,3$ allows to infect nodes 5, 8 and 9. In $SIMULATION\ STEP\ =\ 2$ cycle repeats for every node with S1 **(A3)** and then with S2 **(A4)**. In this case process 2 begins to defeat process 1. In the next step suppressing process $(S2)$ finishes with more nodes activated than process $(S1)$. (Color figure online)

3.2 Assumptions for Experimental Study

Information spreading for both Negative Process (NP) and Positive Process (PP) are divided into simulation steps. In each step, each process has a chance to increase coverage with activated nodes contacting their neighbours. Simulation starts with choosing seed nodes according to seeds selection strategy. The spreading process starts with selection of seeds according to their ranking, R_{NP} for Negative Process and R_{PP} for Positive Process. Negative Process can be initialized by choosing random nodes (like in the real world, a disease itself cannot choose node while the carrier does), otherwise marketing strategies are usually based on strictly selected nodes. To try to suppress these two ways of contamination with competing process we choose three seeding strategies based on tree rankings: Random, Degree based and third Effective Degree. We choose selecting nodes by the most common centrality measure, the node degree, treated as a one of main and relatively effective heuristics for seed selection as well as a reference method [12]. Additionally we use effective version of degree, computed before launching Positive Process. It is not based on the total number of the neighbours, but on the number of nodes infected by Negative Process.

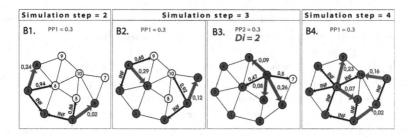

Fig. 2. Competing processes with the same Propagation probability but process no 2 starts with given $Di = 2$. **(B1)** In $SIMULATION\ STEP = 2$ (step $= 1$ for process no 1 is the same as in Fig. 1 **(A1)**) infected nodes 2 and 3 infect nodes 4 and 8. **(B2)** In $SIMULATION\ STEP = 3$ nodes infected in last step infect next nodes (6 and 7). **(B3)** $Di = 2$ allows process no 2 to start in $SIMULATION\ STEP = 3$ Propagation probability ($PP2 = 0, 3$) allows to infect node (5, 8 and 9). **(B4)** shows that process no 1 will win in $SIMULATION\ STEP = 3$. Comparison with Fig. 1. shows that same Propagation probability do not guarantee success if we delay our reaction (Color figure online)

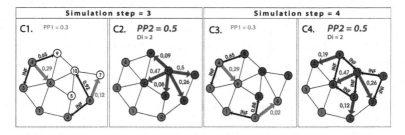

Fig. 3. According to examples above Fig. 3. shows that if we wait with reaction, we need to increase the cost of our campaign to success. **(C1)** In $SIMULATION\ STEP = 3$ (step 1 same as in Fig. 1. **(A1)** and 2 as in Fig. 2. **(B1)**) nodes 4 and 6 infect. In **(C2)** process no 2 starts with same Di as in Fig. 2. but with increased $PP1$ (from 0.3 to 0.5). This affects on possibility to infect nodes 6 ($0, 47 < 0, 5$) and 7 ($0, 5 = 0, 5$) which wasn't possible earlier with $PP2 = 0.3$. $SIMULATION\ STEP = 4$ repeats competing process **(C3 and C4)** which results in victory of a process 2 in next step of a simulation. (Color figure online)

The experiment assumes five different sizes of seed set selected in each used network. Experiments verifies the efficiency of increasing number of seeds. Number of seeds, seeding fraction (SF_{NP}, SF_{PP}) is equal to 1%, 2%, 3%, 4% or 5% and represents the percentage of nodes selected as seeds. Suppressing process (Positive Process) will start with the given delay Di. It can start at the same step or, later (delay: 0–8), to test consequences of late reaction. And finally for both of the competing strategies we give propagation probability (PP_{PP}, (PP_{NP})) equal 0.1, 0.2, 0.3, 0.4 or 0.5. Propagation probability is responsible for the chance to infect node and represents propagation probability according to Independent Cascade Model [17]. During the process for each edge possible for transmission random value is dynamically generated. if it takes value lower or equal to propagation probability (different for Positive Process and Negative Process) activation of contacted node takes place. All values used for all parameters are presented in Table 1.

Table 1. Networks and diffusion parameters used in simulations for Positive Process (PP) and Negative Process (NP)

Symbol	Parameter	Values	Variants
R_{NP}	Ranking type for NP	2	Random, Degree
R_{PP}	Ranking type for PP	3	Random, Degree, Effective Degree
PP_{NP}	NP Propagation Probability	5	0.1, 0.2, 0.3, 0.4, 0.5
PP_{PP}	PP Propagation Probability	5	0.1, 0.2, 0.3, 0.4, 0.5
SF_{NP}	Seeds Fraction for NP	5	1%, 2%, 3%, 4%, 5%
SF_{PP}	Seeds Fraction for PP	5	1%, 2%, 3%, 4%, 5%
Di	Delay in PP initialisation	9	0, 1, 2, 3, 4, 5, 6, 7, 8
N	Network	5	Real networks

Performance of Positive Process can be measured with the use of several metrics. Performance Factor (PF) is represented by a total number of nodes activated by a positive process by the number of nodes activated by negative process for the same configuration parameters. Another metrics, Success Rate (SR), represents percentage of spreading processes with winning Positive Process.

4   Results from Empirical Study

4.1   Experimental Setup

Simulations were performed on five real networks $N1$–$N5$ UoCalifornia [23], Political blogs [1], Net science [21], Hamsterster friendships [18] and UC Irvine forum [22] available from public repositories, having from 899 to 1899 nodes and from 2742 to 59835 edges. We obtained total $R_{NP} \times R_{PP} \times PP_{NP} \times PP_{PP} \times SF_{NP} \times SF_{PP} \times Di \times N$ with the total number 168,750 of simulation configurations. For each of them ten runs were repeated and averaged. The main goal is to investigate the influence of increasing the efficiency of nodes in contaminating their environment. In order to gather necessary knowledge each combination of parameters to find the most successful way to suppress spreading potentially dangerous process as compared. The loop searches through nodes activated by Positive Process, for each infected node the script is looking through its neighbours and tries to infect every neighbour who is not infected with the same disease. Nodes that are infected in this step of contamination cannot spread the disease in the same step. If $Di = 0$, seeds for Negative Process are selected from 'healthy' nodes and the loop repeats spreading but searches through the nodes activated by Positive Process and tries to infect with positive content every node activated by negative process or neutral. If $Di > 0$, Positive Process starts after $Di + 1$ cycles of Positive Process spreading. In this case the simulation step ends when Positive Process step ends, until the suppressing process is activated. The competing lasts until one of the strategies defeat the competitor and spread all over the network or network states stabilize.

4.2   Overall Results

In this section, results from agent-based simulations are presented. During analysis we estimated costs of making effective delayed process. The main goal was answer the question to as far we need to increase propagation probability (PP) to obtain certain success rate (SR) of suppressing process under varying delay steps. Figure 4 shows significance of suppression process. As we can notice, cases with no delay ($Di = 0$) provides the best performance of Positive Processes. Overall, for cases with no delay suppression campaign achieved 31% coverage. Subsequently differences between Positive Process and Negative Process are grown. Negative Process reached the best performance when the Positive Process was most delayed. It was analysed for eight steps of delay ($Di = 8$), and for this case overall negative campaign performance is 62.4%.

For a more detailed evaluation of the diffusion of Positive Process we figured out three factors, presented in the Fig. 5. Used propagation probability (PP), seeds fraction (SF) and networks (N) were analysed. Propagation probability causes the biggest increment of coverage performance. Along with propagation probability, coverage performance is increasing as following: 2.00%, 7.88%, 12.10%, 15.98%, 21.44%. A similar relationships can be seen for seeds fraction (SF) values. However there the growth of performance isn't so drastic. The following results we obtained: 9.72%, 11.10%, 12.13%, 12.86%, 13.59%. In terms of coverage performance $N3$ achieves the highest performance 18.43%. The worst outcome was obtained within $N5$, it 7.39%. Therefore, it is evidence that effect are with relation to network topology.

4.3   Sensitivity Analysis

For sensitivity analysis to determine the key parameters affecting coverage of the positive process the meta-modeling based on the Treed Gaussian Process (TGP) was used. Briefly, TGP is one of the significant machine learning methods, developed by Gramacy [8]. Gramacy et al. further extended TGP to be suitable for sensitivity analysis [9]. Since then, after constructing TGP models, sensitivity analysis can be used to identify key variables in the models by using the variance-based method. We used here two sensitivity indexes: first order and total effects. The first order index represents the main effect and the contribution of each input variable to the variance of the output. The total effects include both main effects and interactions between input variables.

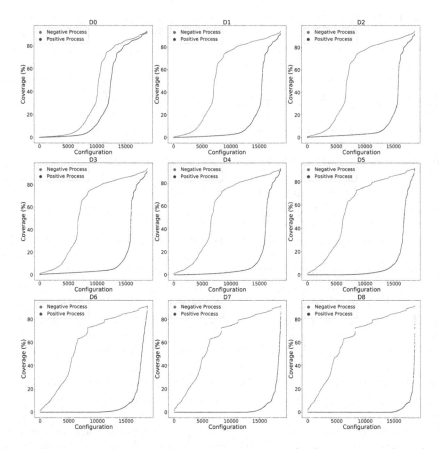

Fig. 4. Comparison of two spreading processes, negative (red) and positive (green) for each combination of configuration parameters. Figure presents relation between both processes, from delay equal to 0 up to delay equal to 8 steps. Along with steps of delay the significance of suppression decrease and distance between both processes grows. Together with increased delay of Positive Process, the Negative Process changes from s-shaped towards increased dynamics. (Color figure online)

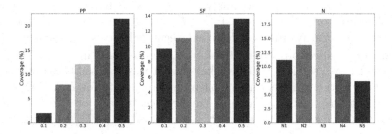

Fig. 5. Average coverage performance of spreading positive process figured for individual values of PP, SF and N respectively. (Color figure online)

The Fig. 6A shows the slopes of the various parameters used in simulations. It provides information on whether the output, performance of the Positive Process, is an increasing or decreasing function of the respective input data. Solid lines are mean values that are within the 95% confidence interval. It was observed that the changes of four parameters SF_{NP}, R_{NP}, R_{PP}, SF_{PP} had only small influence on the output. In addition, with the increase of SF_{PP} and R_{PP}, the main effects caused by the increase of these variables slightly increased, suggesting that the impact of individual differences between SF_{PP} and R_{PP} on changes in the examined networks was slightly improved in such conditions. However, a clear improvement can be seen with the increase of PP_{PP}. Here we see a clearly noticeable increase, which indicates a significant impact of the variable under the existing conditions. Impact of N shows that results were highly dependent on used networks. The increase in the R_{NP} and SF_{NP} variables indicates that the effects caused by their increase have worsened to a lesser extent. It can be clearly seen that increases in delay Di, and PP_{NP} negatively affect the coverage of positive process This means that the variables have a negative impact and their significance deteriorates under the circumstances.

In Fig. 6B first order sensitivity indicators quantify changes in output variables suitably caused by individual input variables while in Fig. 6C sensitivity indicators reflect the interactive effects of all input variables on the output variable. Figure 6B clearly shows that PP_{NP} is the main contributor to the network coverage of PP. Di and PP_{PP} are classified as second and third factors respectively contributing to network coverage of positive process. This differs to some extent from the individual effects shown in Fig. 6A, which can be explained by the combined effects of PP_{NP}, Di and PP_{PP}. The role of remaining variables is approximately the same, sharing small values of the network response coverage of positive process. The cumulative effects Fig. 6C increases when we consider the interactions between all variables, especially for PP_{NP}, to a slightly lesser extent for Di. The sensitivity indicators are not sum to one and it indicates that interactive effects between two or more variables are important for the individual assessment.

4.4   Evaluation the Costs of Delayed Suppressing Process

Another step of analysis includes analysis of Success Rate for different propagation probabilities of both processes. Figure 7 shows Success Rates for each pair of probabilities for Positive Process (PP) and Negative Process (NP). Success Rate (SR) for propagation probability 0.1 for both processes is marked with RED within the Delay 0 section. If delay takes place it was impossible to obtain same Success Rate without increasing Propagation Probability for PP. For example for Delay 4 it was possible to achieve similar Success Rate for $PP_{PP} = 0.2$ and for Delay 5 with $PP_{PP} = 0.5$ (reference cells marked with RED). Success Rate

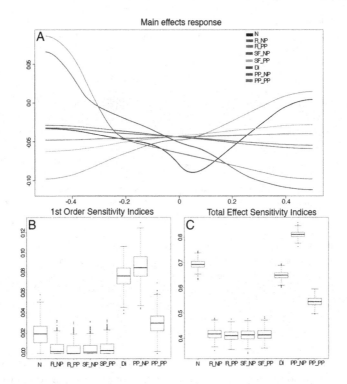

Fig. 6. Sensitivity analysis results for used parameters. (**A**) Main effects response, (**B**) 1st Order Sensitivity Indices and (**C**) Total Effect Sensitivity Indices. (Color figure online)

values marked with $\boxed{\text{GREEN}}$ are related to processes competing with Negative Process spreading with $PP_{NP} = 0.2$. In similar way reference values are marked for $PP_{NP} = 0.3, 0.4, 0.5$ with colors $\boxed{\text{BLUE}}$, $\boxed{\text{YELLOW}}$ and $\boxed{\text{VIOLET}}$ respectively. If Positive Process is delayed one, two or three steps ($Di = 1$, $Di = 2$ and $Di = 3$) properly increased propagation probability makes possible obtaining high Success Rate above 80%. It Positive Process starts with delay four or five steps ($Di = 3$, $Di = 4$) even with high probability only in few cases Success Rate exceeded 50%. Results show that further delay with Positive Process is resulted dropping Success Rate to low ranges. If delay is longer than five steps it was impossible to obtain Success Rate higher than 15% even if probability of negative process was at the level 0.1 and the positive process was launched with probability 0.5%.

Delay 0	0.1	0.2	0.3	0.4	0.5
0.1	33,93%	87,73%	97,73%	99,60%	99,87%
0.2	2,67%	33,73%	61,33%	81,73%	93,73%
0.3	0,53%	8,53%	32,80%	50,53%	68,93%
0.4	0,00%	0,93%	15,33%	32,00%	44,80%
0.5	0,00%	0,00%	8,60%	19,20%	31,73%

Delay 1	0.1	0.2	0.3	0.4	0.5
0.1	12,93%	68,40%	90,00%	96,93%	98,27%
0.2	1,73%	8,40%	20,93%	45,07%	68,00%
0.3	0,13%	1,73%	6,67%	13,33%	28,93%
0.4	0,00%	0,13%	1,33%	4,13%	10,40%
0.5	0,00%	0,00%	0,13%	1,33%	4,00%

Delay 2	0.1	0.2	0.3	0.4	0.5
0.1	4,13%	53,60%	81,60%	88,93%	88,93%
0.2	1,73%	6,67%	14,00%	35,47%	61,87%
0.3	0,13%	1,33%	4,27%	11,20%	24,93%
0.4	0,00%	0,00%	0,80%	2,40%	8,53%
0.5	0,00%	0,00%	0,13%	0,53%	2,00%

Delay 3	0.1	0.2	0.3	0.4	0.5
0.1	0,53%	45,07%	72,40%	78,40%	80,27%
0.2	0,67%	4,80%	11,87%	30,27%	58,53%
0.3	0,00%	0,40%	3,07%	9,73%	18,53%
0.4	0,00%	0,00%	0,80%	1,87%	6,00%
0.5	0,00%	0,00%	0,00%	0,40%	0,53%

Delay 4	0.1	0.2	0.3	0.4	0.5
0.1	0,00%	40,67%	57,47%	62,93%	62,00%
0.2	0,13%	2,80%	5,47%	21,73%	37,20%
0.3	0,00%	0,53%	1,87%	7,73%	12,93%
0.4	0,00%	0,00%	0,13%	0,93%	4,53%
0.5	0,00%	0,00%	0,00%	0,00%	0,80%

Delay 5	0.1	0.2	0.3	0.4	0.5
0.1	0,00%	28,40%	36,00%	36,93%	37,07%
0.2	0,00%	0,67%	1,87%	9,87%	12,53%
0.3	0,00%	0,00%	0,67%	5,60%	8,53%
0.4	0,00%	0,00%	0,13%	0,80%	3,07%
0.5	0,00%	0,00%	0,00%	0,00%	0,13%

Delay 6	0.1	0.2	0.3	0.4	0.5
0.1	0,00%	11,47%	12,93%	13,73%	14,53%
0.2	0,00%	0,00%	0,00%	0,67%	1,47%
0.3	0,00%	0,00%	1,07%	2,67%	5,60%
0.4	0,00%	0,00%	0,13%	0,53%	2,27%
0.5	0,00%	0,00%	0,00%	0,00%	0,00%

Delay 7	0.1	0.2	0.3	0.4	0.5
0.1	0,00%	2,27%	2,27%	3,20%	2,67%
0.2	0,00%	0,00%	0,00%	0,00%	0,00%
0.3	0,00%	0,00%	0,13%	0,40%	2,13%
0.4	0,00%	0,00%	0,00%	0,13%	1,60%
0.5	0,00%	0,00%	0,00%	0,00%	0,00%

Delay 8	0.1	0.2	0.3	0.4	0.5
0.1	0,00%	0,00%	0,00%	0,27%	0,00%
0.2	0,00%	0,00%	0,00%	0,00%	0,00%
0.3	0,00%	0,00%	0,00%	0,13%	0,13%
0.4	0,00%	0,00%	0,00%	0,13%	0,53%
0.5	0,00%	0,00%	0,00%	0,00%	0,13%

Fig. 7. Success Rate (SR) represented by a percentage of winning Positive Processes with corresponding Negative Processes for each pair of propagation probabilities (PP_{NP}, PP_{PP}) and the delay from $Di = 0$ to $Di = 8$. Rows are denoted with PP_{NP} and columns are denoted with (PP_{PP}). Colors for Delay 0 denote selected cases for propagation probability, and same colors in tables with other delays show when was possible to obtain same or higher SR with given PP_{NP} and what required PP_{PP} for delayed Positive Process. (Color figure online)

5   Conclusions

Information transmitted with the use of electronic media spreads with high dynamics. Apart from neutral or positive content online social networks can be used for information or content potentially harmful. Misleading information, rumour or textual information may cause panic and lead to bad behaviours. From that perspective suppressing information spreading processes is challenging and important direction in the area of network science, what was confirmed by earlier studies. Possible actions taken against negative content can be based on launching competing campaigns with the main goal for limiting the dynamics and coverage of negative processes. Later the action is taken it can be less effecting and stopping negative content can be problematic.

Presented study showed how delays in launching positive process is influencing its performance and ability to reduce negative process. It can be achieved by proper adjusting the parameters of limiting process when compared to negative process. Increasing propagation probabilities increases the ability to cope with the negative process, even if limiting action is taken with delay at the moment when large fraction of network is covered by negative content.

Future directions include the role of intervals between messages on campaign performance. Another possible areas include experiments within temporal networks and investigation of the role of changing network topology on performance of limiting actions.

Acknowledgments. This work was supported by the National Science Centre, Poland, grant no. 2017/27/B/HS4/01216.

References

1. Adamic, L.A., Glance, N.: The political blogosphere and the 2004 US election: divided they blog. In: Proceedings of the 3rd International Workshop on Link Discovery, pp. 36–43 (2005)
2. Bakshy, E., Rosenn, I., Marlow, C., Adamic, L.: The role of social networks in information diffusion. In: Proceedings of the 21st International Conference on World Wide Web, pp. 519–528 (2012)
3. Bharathi, S., Kempe, D., Salek, M.: Competitive influence maximization in social networks. In: Deng, X., Graham, F.C. (eds.) WINE 2007. LNCS, vol. 4858, pp. 306–311. Springer, Heidelberg (2007). https://doi.org/10.1007/978-3-540-77105-0_31
4. Bródka, P., Musial, K., Jankowski, J.: Interacting spreading processes in multilayer networks: a systematic review. IEEE Access **8**, 10316–10341 (2020)
5. Budak, C., Agrawal, D., El Abbadi, A.: Limiting the spread of misinformation in social networks. In: Proceedings of the 20th International Conference on World Wide Web, pp. 665–674 (2011)
6. Datta, S., Majumder, A., Shrivastava, N.: Viral marketing for multiple products. In: 2010 IEEE International Conference on Data Mining, pp. 118–127. IEEE (2010)
7. Goode, S., Shailer, G., Wilson, M., Jankowski, J.: Gifting and status in virtual worlds. Journal of Management Information Systems **31**(2), 171–210 (2014)
8. Gramacy, R.B., Taddy, M.A.: Tgp: Bayesian treed gaussian process models. R Package Version, 2–1 (2008)
9. Gramacy, R.B., Taddy, M.: Categorical inputs, sensitivity analysis, optimization and importance tempering with TGP version 2, an R package for treed gaussian process models. University of Cambridge Statistical Laboratory Technical report (2009)
10. Granell, C., Gómez, S., Arenas, A.: Competing spreading processes on multiplex networks: awareness and epidemics. Phys. Rev. E **90**(1), 012808 (2014)
11. He, X., Song, G., Chen, W., Jiang, Q.: Influence blocking maximization in social networks under the competitive linear threshold model. In: Proceedings of the 2012 SIAM International Conference on Data Mining, pp. 463–474. SIAM (2012)
12. Hinz, O., Skiera, B., Barrot, C., Becker, J.U.: Seeding strategies for viral marketing: an empirical comparison. J. Mark. **75**(6), 55–71 (2011)
13. Huang, J., Jin, X.: Preventing rumor spreading on small-world networks. J. Syst. Sci. Complexity **24**(3), 449–456 (2011)
14. Jankowski, J., Michalski, R., Kazienko, P.: The multidimensional study of viral campaigns as branching processes. In: Aberer, K., Flache, A., Jager, W., Liu, L., Tang, J., Guéret, C. (eds.) SocInfo 2012. LNCS, vol. 7710, pp. 462–474. Springer, Heidelberg (2012). https://doi.org/10.1007/978-3-642-35386-4_34

15. Jankowski, J., Michalski, R., Kazienko, P.: Compensatory seeding in networks with varying avaliability of nodes. In: 2013 IEEE/ACM International Conference on Advances in Social Networks Analysis and Mining, ASONAM 2013, pp. 1242–1249. IEEE (2013)

16. Kandhway, K., Kuri, J.: How to run a campaign: optimal control of sis and sir information epidemics. Appl. Math. Comput. **231**, 79–92 (2014)

17. Kempe, D., Kleinberg, J., Tardos, É.: Maximizing the spread of influence through a social network. In: Proceedings of the ninth ACM SIGKDD international conference on Knowledge discovery and data mining, pp. 137–146 (2003)

18. KONECT: Hamsterster friendships network dataset (2016)

19. Lang, A.: The limited capacity model of mediated message processing. J. Commun. **50**(1), 46–70 (2000)

20. Mochalova, A., Nanopoulos, A.: On the role of centrality in information diffusion in social networks. In: ECIS, p. 101 (2013)

21. Newman, M.E.: Finding community structure in networks using the eigenvectors of matrices. Phys. Rev. E **74**(3), 036104 (2006)

22. Opsahl, T.: Triadic closure in two-mode networks: redefining the global and local clustering coefficients. Soc. Netw. **35**(2), 159–167 (2013)

23. Opsahl, T., Panzarasa, P.: Clustering in weighted networks. Soc. Netw. **31**(2), 155–163 (2009)

24. Vance, A., Kirwan, B., Bjornn, D., Jenkins, J., Anderson, B.B.: What do we really know about how habituation to warnings occurs over time? a longitudinal FMRI study of habituation and polymorphic warnings. In: Proceedings of the 2017 CHI Conference on Human Factors in Computing Systems, pp. 2215–2227 (2017)

25. Wang, J., Zhao, L., Huang, R.: Siraru rumor spreading model in complex networks. Phys. A **398**, 43–55 (2014)

26. Wang, S., Zhao, X., Chen, Y., Li, Z., Zhang, K., Xia, J.: Negative influence minimizing by blocking nodes in social networks. In: Workshops at the Twenty-Seventh AAAI Conference on Artificial Intelligence (2013)

27. Wang, Z., Chen, E., Liu, Q., Yang, Y., Ge, Y., Chang, B.: Maximizing the coverage of information propagation in social networks. In: Twenty-Fourth International Joint Conference on Artificial Intelligence (2015)

28. Wang, Z., Zhao, D.W., Wang, L., Sun, G.Q., Jin, Z.: Immunity of multiplex networks via acquaintance vaccination. EPL (Europhys. Lett.) **112**(4), 48002 (2015)

29. Wei, X., Valler, N.C., Prakash, B.A., Neamtiu, I., Faloutsos, M., Faloutsos, C.: Competing memes propagation on networks: a network science perspective. IEEE J. Sel. Areas Commun. **31**(6), 1049–1060 (2013)

30. Wu, H., Liu, W., Yue, K., Huang, W., Yang, K.: Maximizing the spread of competitive influence in a social network oriented to viral marketing. In: Dong, X.L., Yu, X., Li, J., Sun, Y. (eds.) WAIM 2015. LNCS, vol. 9098, pp. 516–519. Springer, Cham (2015). https://doi.org/10.1007/978-3-319-21042-1_53

From Generality to Specificity: On Matter of Scale in Social Media Topic Communities

Danila Vaganov[✉], Mariia Bardina, and Valentina Guleva[✉]

ITMO University, 49 Kronverksky Pr., St. Petersburg 197101, Russia
{vaganov,guleva}@itmo.ru

Abstract. Research question stated in current paper concerns measuring significance of interest topic to a person on the base of digital footprints, observed in on-line social media. Interests are represented by online social groups in VK social network, which were marked by topics. Topic significance to a person is supposed to be related to the fraction of representative groups in user's subscription list. We imply that for each topic, depending on its popularity, relation to geographical region, and social acceptability, there is a value of group size which is significant. In addition, we suppose, that professional clusters of groups demonstrate relatively higher inner density and unify common groups. Therefore, following groups from more specific clusters indicate higher personal involvement to a topic – in this way, representative topical groups are marked. We build social group similarity graph, which is based on the number of common followers, extract subgraphs related to a single topic, and analyse bins of groups, build with increase of group sizes. Results show topics of general interests have higher density at larger groups in contrast to specific interests, which is in correspondence with initial hypothesis.

Keywords: Topic communities · On-line social media · Similarity group network · Personal involvement · General interest · Specific interest · Scaling phenomena

1 Introduction

Interests play a big role in people's lives [22]. They are influenced by our life as well as they are causing influence on it. In psychological studies, researchers usually explore development of interest in relation to academic field and career path. They examine how interest can affect motivation in order to understand how to increase engagement in studying. Rotgans and Schmidt [26] propose a model of interest formation that focuses on how prior relationship to the topic

This research is financially supported by The Russian Science Foundation, Agreement #17-71-30029 with co-financing of Bank Saint Petersburg.

affects formation of interest and discuss how interest can emerge when person is confronted with a problem. Harackiewicz and Knogler emphasize how future choices and career path can be influenced by interests in all stages of development [12]. The advertisement industry is also interested in finding what is behind an interest. They aim at determining current personal interests and patterns of their possible change. Modern day involvement of people in social media is high, almost 3.5 billion people actively use social media around the world and over 70 million people in Russia[1], which makes data, collected from social media, a great asset in Social studies.

Personal interest analysis, performed by means of social media data, requires estimation of group importance as a marker of topic interest and involvement. Big groups of general interest attract people of different preferences, education, and occupation. Groups presenting specific context are less related to our daily life. In this way, they can imply entry threshold, restricting maximal number of people involved [2]. Therefore, topic popularity affects the upper threshold of group size.

In addition to the effects of participation costs, there exist an influence of topic consistency cues. This is promoted by attraction-selection-attrition (ASA) theory [2], which concerns factors, attracting people to communities, in the model of community evolution. In this way, they concern both, group size related to its age and ability of people to get content and be satisfied.

Other aspect, characterising personal involvement is the number of groups related to a topic and containing similar sets of members. Let think about critical case of such a phenomena as a professional community. In this way, we can formalise "specificity" of groups union if they are intersected enough and related to similar topics. Then, relation to a topic with strong intersections and number of groups from the cluster is implied to be an indicator of personal involvement.

This assumption is reinforced by Cinelli et al. [5], showing experts being active in several groups, while majority of people are satisfied by following outstanding one.

In this study we are aimed at exploration the groups of different topics and sizes to obtain patterns, being able to characterise personal involvement to a topic on the base of subscriptions to social groups. We suppose that there are values of group sizes, which are significant to a certain topic. For this purpose we consider group similarity graph with edges, weighted as group intersection, divide groups between bins of different sizes, and explore density of a topic cluster relatively to the rest of the topics. Similarly, we fix a topic and explore densities related to interscale densities. This allows for distinguishing the most connected clusters and corresponding group size, and mitigate effects of topic popularity. In this way, we conclude that the smallest groups related to the most dense topic clusters indicate higher personal involvement to the topic.

---

[1] https://datareportal.com/reports/digital-2019-russian-federation.

2   Related Literature

Categorisation of interests is mainly related to personal energy associated with a topic, an active personal concern for a certain object or activity [7]. Interests can be triggered by an event (situational) or remain stable over considered amount of time (individual) [25], like emotions and feelings. Interests development from situational to individual can be considered as four phase process [13,24], depending on interest "stability".

The evolution of interests is driven by perceived values [14], which is related to subject utility, and can be reasoned by both, social [18] and personal values. Possibility of communication with other people triggers interest development [30], as well as sense of social belonging, showing increase in interest and involvement of people working in groups in contrast to people working alone [3]. In addition, acceptability of interest may result in interest formation [15].

Interest acceptability is strongly related to personal background. Social status, gender, age, and other features, affecting personal perception, sensibility, and ability to be focused form initial base and background for further scope of interests. Women are less engaged in science, engineering, and math [4], pupil of different age and socioeconomic status were shown to differ in music interests [19], age and gender influence interest in learning about cooking [35], factors like school location (urban or rural) and racial composition affect differences in gender gap and arts consumption patterns [27].

Applied studies are related to the exploration of existing structures in friendship networks, e.g. user segmentation approaches [6], and their relation to personal interests [16]. In this way, correlation between interest similarity and demographic factors (gender, age, and location) is studied [11]. To measure similarity of interests, the relative number of common interests can be taken. Authors take how rare common interests are among the general population [11]. Guleva et al. concluded, that topology of friendships in a social group significantly depends on the topic, particularly in terms of degree assortativity and clustering coefficient [10]. Backstrom et al. [1] show that the transitivity of friendship (clustering coefficient), being in common community, affect the decision to join them; not only absolute number of friends there. Correlation between structure of interactions and interests topic category [33] show, that topics can be ordered according to their social importance.

Other class of applied studies are related to categories of interest prediction [36], usage of recommending systems [9,17], and topic hierarchy trees [17], like Wikipedia graphs [8,23]. Wikipedia graphs are also used to build user interest profiles. Modelling of short- and long-term interests uses neural networks on the base of click-streams [29].

3   Data Set and Preprocessing

We gather profiles of 28,520 groups from social media VK.com, selecting groups which subscribers are self-affiliated with Saint-Petersburg. For each group we collected captions, brief description (up to 200 words), status, list of followers and up to 20 textual posts. During the preprocessing, we combined group captions, statuses, descriptions, and collected posts into a single document. Then we remove all special symbols, numbers, non-cyrillic words, and stop-words.

For each group, we assigned the number of followers by measuring length of each collected follower list. To perform separation into the levels, we used the logarithmic binning and divided the whole data-set of groups into $p = 10$ levels (number of bins), where, for bin of order i, the size s_i is defined as follows:

$$s_i = \min(L) \cdot \frac{\max(L)}{\min(L)}^{\frac{i-1}{p}}, \tag{1}$$

where L is the set of sizes of all considered groups in the data set. It is important to emphasize, that further in the paper we refer to a certain bin of group size as the term "level".

Figure 1 reflects the distribution of group sizes and corresponding binning, i.e. separation on levels. One can observe that this fits power-law distribution, and in this work we consider the groups with sizes from several hundreds to tens of millions subscribers. The highest fraction of groups is placed on the 5th level, where group size varies from 54 K to 160 K. Levels 3 and 6 are also prevalent.

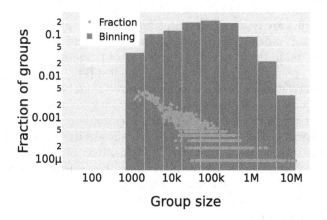

Fig. 1. Distribution of the group size and obtained log bins (levels)

4   Method

In this study, we consider a case of scaling phenomena observed in on-line social groups, also called subscriptions, which represent internet communities related to a certain topic of interest. In this way, for each user, a group list reflects their involvement into different topics. We assume that the scale, corresponding to group size reflects a lot of characteristics such as group popularity or specificity of particular interest. To explore these scales separately we apply logarithm binning and extract several levels of groups. For the extracted levels we build similarity network, demonstrating their collaborative relation, i.e. how strong is the tendency of users to prefer each pair of groups, and do topic modeling to build a connection with characteristics of extracted levels.

4.1 Building Similarity Network of Groups

Similarity between two groups is taken as normalised intersection between their subscribers. To compute similarity θ between groups $\gamma_i, \gamma_j \in \Gamma$, we consider the cosine measure of the corresponding subscriber sets $V(\gamma_i)$, $V(\gamma_j)$:

$$\theta(\gamma_i, \gamma_j) = \frac{|V(\gamma_i) \cap V(\gamma_j)|}{\sqrt{|V(\gamma_i)| \cdot |V(\gamma_j)|}} \tag{2}$$

After the calculation of similarity between all groups, one obtain a weighted complete graph, where nodes correspond to groups and edge weights reflect similarity between them. In order to separate it into the clusters of closest groups, one should find the edge between degree of separation and connectivity of clusters as described in [32]. We vary a threshold of the lowest possible similarity between groups and look for the best intercluster separation.

Then, for each group we describe a prevailing topic, using topic modelling techniques, aimed at comparison of similarity networks and topics at different scales. The final goal is to build interconnection between scales and to describe semantic differences.

4.2 Topic Modelling

Topic modelling is performed to extract key words, describing group topic. Posts published by a group are collected in a document. After that, lemmatization on the set of collected documents is performed by means of a morphological analyzer Mystem [28]. For topic modeling we use Additive Regularization of Topic Models (ARTM) [34], model implemented in BigARTM library[2]. Key feature of this model is the ability to assign combination of regularizers (a criterion to be maximized) for better model tuning. To train a model we use combination of two regularizers, SmoothSparseThetaRegularizer and DecorrelatorPhiRegularizer. First regulazier is responsible for smoothing or sparsing topics. The second one provides decorrelation of topics, which is needed to make the learned topics more interpretable. Both regulaziers are controlled by the coefficients of regularization (τ_1 and τ_2, respectively). The optimal number of topics and values of regularizers are chosen based on perplexity and coherence measure. The perplexity measure indicates the level and speed of convergence of the model. It is defined as

$$\mathcal{P} = exp\left(-\frac{1}{n}\sum_{d \in D}\sum_{w \in d} n_{dw}\, ln\, p(w|d)\right), \tag{3}$$

---

[2] BigARTM open source project: *bigartm.org*.

where D is set of documents, n_{dw} is frequency of word w in document d, $p(w|d)$ is the probability of a term w to occur in a document d. The coherence measure is well correlated with human evaluation of interpretability [20] and is defined as

$$C_t = \frac{2}{k(k-1)} \sum_{i=1}^{k-1} \sum_{j=i+1}^{k} value(w_i, w_j), \tag{4}$$

where k is a number of the most representative tokens for the topic and $value$ is pairwise information about tokens, for example, as used in [20], the pointwise mutual information (PMI) is:

$$value(w_i, w_j) = PMI(w_i, w_j) = \left[log \frac{p(w_i, w_j)}{p(w_i)p(w_j)} \right], \tag{5}$$

which is used to measure the similarity of words w_i and w_j based on co-occurrence statistics.

4.3   Topic Density on a Certain Level

The main idea of our study is to measure how strong group connections are inside a certain topic in relation to the groups outside the topic on a certain level of group sizes, an example illustrated on (Fig. 2).

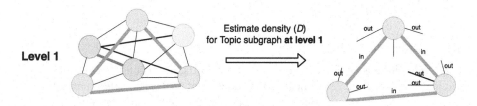

Fig. 2. Illustration of a subgraph extraction for evaluation of topic relative density at level 1.

Formally the relative density for a given subgraph is described by equation:

$$D = \frac{\sum_{e \in E_T} w(e)/(N_T \cdot (N_T - 1))}{\sum_{e \in E \backslash E_T \backslash E_{\bar{T}}} w(e)/(N_T \cdot N_{\bar{T}})}, \tag{6}$$

where $w(e)$ is weight of an edge e, showing similarity between groups; N_T is the number of nodes related to topic T; numerator is weighted network density for groups, related to topic T, and denominator reflects weighted sum of edges between T and all other group topics, related to all possible edge weights between them. Maximal edge weight is supposed to be 1 (due to Eq. (2)). For this measure we consider each level separately.

5 Results

5.1 Similarity Graph

Following Sect. 4.1, we use groups, having at least 10 posts, to construct a similarity graph. A calculated optimal weight threshold is $\theta \leqslant 0.066$, which guarantees that all groups are in the same connected component. As a result, the number of nodes is 12,092 and number of edges is 917 K.

5.2 Topic Modelling

After lemmatization, we removed all documents consisting of less than 5 lemmas. To determine the optimal number of topics, we trained ARTM model without regularizers with number of passes set to 20. As a result, the number of topics varied from 20 to 110 with an increment of 5 as presented in Fig. 3. Based on coherence measure, the optimal number of topics was set to 80. For training with regulaziers, which were described in Sect. 4.2, we conducted an experiment where τ_1 varied from -0.05 to -0.4 with step 0.05 and τ_2 varies from $2 \cdot 10^4$ to $16 \cdot 10^4$ with step $2 \cdot 10^4$. The best result, obtained during training with different variables of regularizers, was achieved at $\tau_1 = -0.2$ and $\tau_2 = 10^5$.

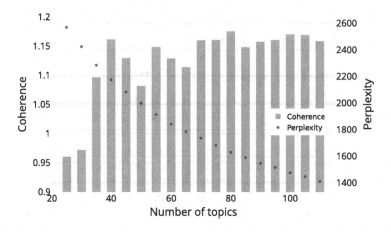

Fig. 3. Results of the experiment for finding right number of topics

After that, we gathered 80 topics represented by top-5 most probable words in each topic. On the base of that representation, each topic was manually assigned by keywords. Then, the topics marked as "noise" were removed. After that, we assigned the most probable topics to each group. To leave only representative groups, we need to choose a threshold for the value of probability, according to which topics were assigned to the groups. Each group, having the probability of being attributed to the topic less than the value of threshold, was excluded.

We looked into the distribution of the number of groups in a topic and set a value of threshold to 0.4. The chosen distribution and its relationship to the original set of groups is shown in Fig. 4.

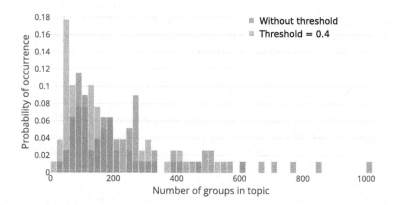

Fig. 4. Distribution of the number of groups in topic with and without threshold set up

After filtering of (non-representative) topics, showing less than 20 related groups, we obtain 28 topics, containing more than 7K groups. In the rest of the study we consider similarity subgraph for the chosen groups.

5.3   Relation Between Topics and Group Preference

In this section we begin with the exploration of the interrelation between the obtained topics and the graph of groups similarity, and then we investigate an influence of group sizes.

First, consider an illustration of giant component of obtained graph, where the color of node corresponds to group topic (Fig. 5). For better representation, we prune weak ties by the weight, with threshold of similarity $\theta \leqslant 0.1$, and remove some groups, which are not in giant component. One can observe, that groups of the same topics form natural clusters based on the similarity of users' subscriptions. Some topics form independent weakly connected modules, for instance, a blue cluster in the right part of the figure corresponds to the "furniture" topic, while huge pink module is "handcraft" topic.

To estimate a strength of relation between topics and overlapping group affiliation numerically, we calculate, for each topic $t_i \in T$, the maximum intersection between groups from each topic ($g_t \in t_i$) and groups in graph clusters ($g_c \in c_j$) as:

$$I_{max_i} = max(\frac{|t_i \cup c_j|}{|t_i|}), c_j \in C \tag{7}$$

where clusters ($c_j \in C$) were obtained by means of the Leiden algorithm [31] and an optimal value of modularity (defined by Newman et al. [21]) is 0.54,

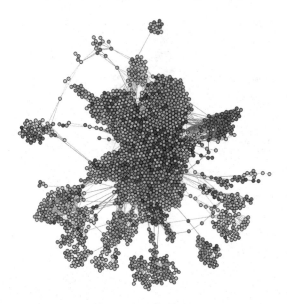

Fig. 5. The obtained graph of groups similarity (nodes colored by topics).

which means that considered graph is well separated (i.e. groups forms a certain blocks by interests). As a result, we obtain 38 clusters. In Fig. 6 one can observe the values of maximum intersections I_{max} obtained for some topics. As we have seen before, the topic cluster "furniture" gets the highest intersection with one of the modularity clusters and close to the maximum. In contrast, "activity in Saint-Petersburg" reaches the minimal value of intersection: the possible reason is that different activities can be associated with different topics, but not in terms of dominant words. The same situation is possible with groups related to the topic of "motivation". A median value of maximum intersection between clusters and topics is 0.489, which suggests there is a strong interrelation between the obtained topics and group similarity in terms of users preference. That gives us confidence, that we are able to combine similarity graph and topic modelling in order to obtain interpretable picture of group preference in a scale of the whole social network.

5.4   Topics and Group Scale

At this stage, we perform an analysis of the dependency between the relative density of topic communities on each level (see details in Sect. 4.3). To present obtained results in an appropriate way, we divide topics according to levels, at which they reach peaks of relative density. As a result, we obtain 4 dominant levels (Fig. 7). Despite observing groups at level 1 and 10 in the distribution, the relative density equals to 0 for both of them because of the weak ties between groups in similar topics. Possibly it is an issue of data collection process, as we collect relatively large groups. From the other side, level 1 is poorly represented

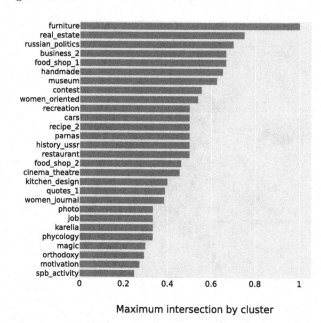

Fig. 6. Maximum intersection with cluster for each topic.

due to scale-free effect (instances are rare), level 10 is also poor. It is important to emphasize, that maximum possible value of relative density decreases with growing of dominant level, that should signify a tendency of small communities to be more specific, i.e. if one is interested in a certain topic, they subscribe for a different groups with higher probability.

Topics with the highest relative density at levels 2 and 3 (Fig. 7a and 7b) show similar patterns of having the highest density on rather small membership scale (up to 6k and 18k respectively), followed by decrease on the next levels, with a small possible rise afterwards. This could indicate that for the subset of topics, users tend to express more interest to less crowded groups in relation to bigger groups of the same topic. At the level 2 one can observe, that topic related to "Job" demonstrates smooth decrease with increasing level. Levels 3 and 4 have also strong preference, moreover, on level 6 there is another significant peak. Such results can be interpreted in the following way: small groups (level 2–3) represent jobs in a certain field, and users tend to subscribe them in order to find a job; for level 6 we have a peak possibly because there is some big groups-aggregators with a wide range of professions. Similar trends are observed for "Photo" and "Parnas" (district of Saint-Petersburg) at level 2 and for "Cinema and theatre", "Recipe" at level 3. There is another interesting pattern: a topic starts with slightly low relative density and scatter relatively equally among other levels. This possibly means, that such kinds of topics do not tend to be specific and are relatively general at all levels: "Restaurant", "Saint-Petersburg activity", "Motivation", or "Women journals" (except level 3 with high specificity).

In plots *c* and *d* of Fig. 7 the dominant topics at level 4 and 5 show the same patterns: relative density starts from the maximum value and then decreases among the next levels ("Kitchen design", "Furniture", "Karelia" at level 4 and "Orthodoxy", "Magic" and "Real-estate" on 5) or equally scattered over the next levels ("Quotes", "Psychology" on 4, "Women oriented" on 5). However, these two dominant levels bring us to a new pattern: some topics tend to show growth, the presence of one peak followed by a decrease. Appearance of such picture, especially if value of relative density on previous or next level is close, means, that such topics are popular among the social network, as a border between scale levels does not have a significant difference and there is a lot of groups with similar behaviour. In this case, the existence of a single peak become significant, as it is able to describe the tendency of a topic to be more general or specific. In this way, for instance, for "Business" topic at level 4, density of the "growth" to the peak is prevalent, which suggests, that people tends to follow smaller groups more, because they are related to a more specific business. The same trend is observed for "Russian politics" at level 5. In contrast, for "Handmade" topic at level 5, the decrease in density over levels is prevalent, which means users tends subscribe to general groups, aggregating wider range of interesting things to make by own hands (which is more attractable for users).

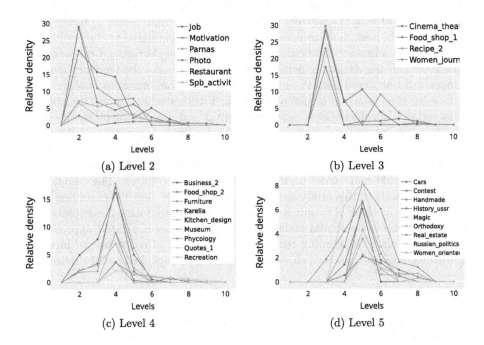

Fig. 7. Dominant topics on different levels

6   Conclusion and Future Work

Current paper concerns measuring the degree significance of interest topic in terms of their generality or specificity. To address this question, we divide all collected groups into 10 levels and measure the similarity between them on the based of the number of mutual followers. Then we perform topic modelling based on post texts in such groups. To obtain an interpretable picture of groups preference by users, we combine similarity graph with topic modelling and estimate their conformity: a median value of maximum intersection between clusters of similarity graph and topics is 0.489, which suggests a strong interrelation between the groups semantics and their coincidence in terms of users preference. Finally, we analyse the density inside the topic in relation to all other topics, at each considered level of group size.

Based on the analysis of the dependency between levels (in terms of group size) and relative density of topics, we conclude, that in general, relative density decreases with increasing of group size, which means small communities are more specific. Moreover, we uncover three patterns: 1) a topic appears at a certain level with a maximum relative density and followed by downfall on the next levels, i.e. this topic is clearly specific on the smallest level where it appears; 2) a topic starts with slightly low relative density and scattered equally among other levels: that a topic may be of general interest, but is divided in multiple subtopics, which become observable by a relatively high density on multiple levels; 3) topics tend to show growth, peak, and decrease of density, which is related to the popularity of the topic as a border between scale levels become neglected, but one can still distinguish the degree of specificity for them by analysis of the tendency density changes. All uncovered patterns interplay well with the semantics of topics.

Results give us a lot of perspectives for future studies, as we are able to characterise a specificity of a particular group and also characterise the level of topic involvement for a user or a local group. There is a great possibility to model interscale dependency between individuals, local groups of individuals, and the whole topics of interests. Moreover, obtained results give a good perspective to study and model topic interrelation and evolution. The complexity of other methods in this study is not higher than $O(N \cdot \log N)$, excluding only the proposed method of graph construction. It's computational complexity is $O(N^2)$, therefore one should elaborate on the more sophisticated approaches, for instance, machine learning techniques in the field of collaborative filtering algorithms, which allows for processing millions of groups, instead of thousands.

References

1. Backstrom, L., Huttenlocher, D., Kleinberg, J., Lan, X.: Group formation in large social networks: membership, growth, and evolution. In: Proceedings of the 12th ACM SIGKDD International Conference on Knowledge Discovery and Data Mining, pp. 44–54 (2006)
2. Butler, B.S., Bateman, P.J., Gray, P.H., Diamant, E.I.: An attraction-selection-attrition theory of online community size and resilience. MIS Q. **38**(3), 699–729 (2014)

3. Carr, P.B., Walton, G.M.: Cues of working together fuel intrinsic motivation. J. Exp. Soc. Psychol. **53**, 169–184 (2014)
4. Cheryan, S., Plaut, V.C.: Explaining underrepresentation: a theory of precluded interest. Sex Roles **63**(7–8), 475–488 (2010)
5. Cinelli, M., Brugnoli, E., Schmidt, A.L., Zollo, F., Quattrociocchi, W., Scala, A.: Selective exposure shapes the Facebook news diet. arXiv preprint arXiv:1903.00699 (2019)
6. Derevitskii, I., Severiukhina, O., Bochenina, K.: Clustering interest graphs for customer segmentation problems. In: 2019 Sixth International Conference on Social Networks Analysis, Management and Security (SNAMS), pp. 321–327. IEEE (2019)
7. Dewey, J.: Interest and Effort in Education. Houghton Mifflin, Boston (1913)
8. Faralli, S., Stilo, G., Velardi, P.: Automatic acquisition of a taxonomy of microblogs users' interests. J. Web Semant. **45**, 23–40 (2017)
9. Gao, L., Wu, J., Zhou, C., Hu, Y.: Collaborative dynamic sparse topic regression with user profile evolution for item recommendation. In: Thirty-First AAAI Conference on Artificial Intelligence (2017)
10. Guleva, V., Vaganov, D., Voloshin, D., Bochenina, K.: Topology of thematic communities in online social networks: a comparative study. In: Shi, Y., et al. (eds.) ICCS 2018. LNCS, vol. 10860, pp. 260–273. Springer, Cham (2018). https://doi.org/10.1007/978-3-319-93698-7_20
11. Han, X., Wang, L., Crespi, N., Park, S., Cuevas, Á.: Alike people, alike interests? Inferring interest similarity in online social networks. Decis. Support Syst. **69**, 92–106 (2015)
12. Harackiewicz, J.M., Knogler, M.: Theory and application. In: Handbook of Competence and Motivation: Theory and Application, p. 334 (2017)
13. Hidi, S., Renninger, K.A.: The four-phase model of interest development. Educ. Psychol. **41**(2), 111–127 (2006)
14. Hulleman, C.S., Thoman, D.B., Dicke, A.-L., Harackiewicz, J.M.: The promotion and development of interest: the importance of perceived values. In: O'Keefe, P.A., Harackiewicz, J.M. (eds.) The Science of Interest, pp. 189–208. Springer, Cham (2017). https://doi.org/10.1007/978-3-319-55509-6_10
15. Hulleman, C., Dicke, A., Kosovich, J., Thoman, D.: The role of perceived social norms and parents' value in the development of interest in biology. In: Poster Presented at the Annual Meeting of the Society for Personality and Social Psychology, San Diego, CA (2016)
16. Ji, L., Liu, J.G., Hou, L., Guo, Q.: Identifying the role of common interests in online user trust formation. PloS One **10**(7), e0121105 (2015). https://doi.org/10.1371/journal.pone.0121105
17. Kapanipathi, P., Jain, P., Venkataramani, C., Sheth, A.: User interests identification on twitter using a hierarchical knowledge base. In: Presutti, V., d'Amato, C., Gandon, F., d'Aquin, M., Staab, S., Tordai, A. (eds.) ESWC 2014. LNCS, vol. 8465, pp. 99–113. Springer, Cham (2014). https://doi.org/10.1007/978-3-319-07443-6_8
18. McCaslin, M.: Co-regulation of student motivation and emergent identity. Educ. Psychol. **44**(2), 137–146 (2009)
19. McPherson, G.E., Osborne, M.S., Barrett, M.S., Davidson, J.W., Faulkner, R.: Motivation to study music in Australian schools: the impact of music learning, gender, and socio-economic status. Res. Stud. Music Educ. **37**(2), 141–160 (2015)

20. Newman, D., Lau, J.H., Grieser, K., Baldwin, T.: Automatic evaluation of topic coherence. In: Human Language Technologies: The 2010 Annual Conference of the North American Chapter of the Association for Computational Linguistics, pp. 100–108. Association for Computational Linguistics, Los Angeles (2010)
21. Newman, M.E.: Modularity and community structure in networks. Proc. Natl. Acad. Sci. **103**(23), 8577–8582 (2006)
22. Philippe, F.L., Vallerand, R.J., Lavigne, G.L.: Passion does make a difference in people's lives: a look at well-being in passionate and non-passionate individuals. Appl. Psychol.: Health Well-Being **1**(1), 3–22 (2009)
23. Piao, G., Breslin, J.G.: Inferring user interests for passive users on Twitter by leveraging followee biographies. In: Jose, J.M., et al. (eds.) ECIR 2017. LNCS, vol. 10193, pp. 122–133. Springer, Cham (2017). https://doi.org/10.1007/978-3-319-56608-5_10
24. Renninger, K.A., Hidi, S.: The Power of Interest for Motivation and Engagement. Routledge, Abingdon (2015)
25. Renninger, K.A., Hidi, S., Krapp, A., Renninger, A.: The Role of Interest in Learning and Development. Psychology Press (2014)
26. Rotgans, J.I., Schmidt, H.G.: The role of interest in learning: knowledge acquisition at the intersection of situational and individual interest. In: O'Keefe, P.A., Harackiewicz, J.M. (eds.) The Science of Interest, pp. 69–93. Springer, Cham (2017). https://doi.org/10.1007/978-3-319-55509-6_4
27. Schmutz, V., Stearns, E., Glennie, E.J.: Cultural capital formation in adolescence: high schools and the gender gap in arts activity participation. Poetics **57**, 27–39 (2016)
28. Segalovich, I.: A fast morphological algorithm with unknown word guessing induced by a dictionary for a web search engine. In: MLMTA, pp. 273–280. Citeseer (2003)
29. Song, Y., Elkahky, A.M., He, X.: Multi-rate deep learning for temporal recommendation. In: Proceedings of the 39th International ACM SIGIR Conference on Research and Development in Information Retrieval, pp. 909–912 (2016)
30. Thoman, D.B., Sansone, C., Geerling, D.: The dynamic nature of interest: embedding interest within self-regulation. In: O'Keefe, P.A., Harackiewicz, J.M. (eds.) The Science of Interest, pp. 27–47. Springer, Cham (2017). https://doi.org/10.1007/978-3-319-55509-6_2
31. Traag, V.A., Waltman, L., van Eck, N.J.: From Louvain to Leiden: guaranteeing well-connected communities. Sci. Rep. **9**(1), 1–12 (2019)
32. Vaganov, D., Sheina, E., Bochenina, K.: A comparative study of social data similarity measures related to financial behavior. Proc. Comput. Sci. **136**, 274–283 (2018)
33. Vaganov, D.A., Guleva, V.Y., Bochenina, K.O.: Social media group structure and its goals: building an order. In: Aiello, L.M., Cherifi, C., Cherifi, H., Lambiotte, R., Lió, P., Rocha, L.M. (eds.) COMPLEX NETWORKS 2018. SCI, vol. 813, pp. 473–483. Springer, Cham (2019). https://doi.org/10.1007/978-3-030-05414-4_38
34. Vorontsov, K.V.: Additive regularization for topic models of text collections. Dokl. Math. **89**(3), 301–304 (2014). https://doi.org/10.1134/S1064562414020185
35. Worsley, A., Wang, W., Ismail, S., Ridley, S.: Consumers' interest in learning about cooking: the influence of age, gender and education. Int. J. Consum. Stud. **38**(3), 258–264 (2014)
36. Zarrinkalam, F., Kahani, M., Bagheri, E.: User interest prediction over future unobserved topics on social networks. Inf. Retrieval J. **22**(1–2), 93–128 (2019)

Computational Health

Hybrid Text Feature Modeling for Disease Group Prediction Using Unstructured Physician Notes

Gokul S. Krishnan$^{(\boxtimes)}$ (iD) and S. Sowmya Kamath (iD)

Healthcare Analytics and Language Engineering (HALE) Lab,
Department of Information Technology, National Institute of Technology Karnataka,
Surathkal, Mangalore 575025, India
gsk1692@gmail.com, sowmyakamath@nitk.edu.in

Abstract. Existing Clinical Decision Support Systems (CDSSs) largely depend on the availability of structured patient data and Electronic Health Records (EHRs) to aid caregivers. However, in case of hospitals in developing countries, structured patient data formats are not widely adopted, where medical professionals still rely on clinical notes in the form of unstructured text. Such unstructured clinical notes recorded by medical personnel can also be a potential source of rich patient-specific information which can be leveraged to build CDSSs, even for hospitals in developing countries. If such unstructured clinical text can be used, the manual and time-consuming process of EHR generation will no longer be required, with huge person-hours and cost savings. In this article, we propose a generic ICD9 disease group prediction CDSS built on unstructured physician notes modeled using hybrid word embeddings. These word embeddings are used to train a deep neural network for effectively predicting ICD9 disease groups. Experimental evaluation showed that the proposed approach outperformed the state-of-the-art disease group prediction model built on structured EHRs by 15% in terms of AUROC and 40% in terms of AUPRC, thus proving our hypothesis and eliminating dependency on availability of structured patient data.

Keywords: Topic modeling · Word embedding · Natural language processing · Machine learning · Healthcare informatics · Mortality prediction

1 Introduction

Electronic Health Records (EHRs) and predictive analytics have paved the way for Clinical Decision Support Systems (CDSSs) development towards realizing intelligent healthcare systems. The huge amount of patient data generated in hospitals in the form of discharge summaries, clinical notes, diagnostic scans

---

This work is funded by the Government of India's DST-SERB Early Career Research Grant (ECR/2017/001056) to the second author.

© Springer Nature Switzerland AG 2020
V. V. Krzhizhanovskaya et al. (Eds.): ICCS 2020, LNCS 12140, pp. 321–333, 2020.
https://doi.org/10.1007/978-3-030-50423-6_24

etc., present a wide array of opportunities to researchers and data scientists for developing effective methods to improve patient care. CDSSs built on techniques like data mining and machine learning (ML) have been prevalent in health-care systems, aiding medical personnel to make informed, rational decisions and execute timely interventions for managing critical patients. Some existing ML based CDSSs include – mortality prediction [5,10,11,16,21,23], hospital read-mission prediction [12,14,22], length of stay prediction [1,11,24], generic disease or ICD9[1] group prediction [6,19,21] and so on. However, it is significant to note that, most existing CDSSs depend on structured patient data in the form of Electronic Health Records (EHRs), which favors developed countries due to large scale adoption of EHRs. However, healthcare personnel in developing countries continue to rely on clinical notes in the form of free and unstructured text, as structured EHR adoption rate is extremely low in these countries. Moreover, the conversion of patient information recorded in the form of text/clinical reports to a standard structured EHR is a labor-intensive, manual process that can result in loss of critical patient-specific information that might be present in the clinical notes. For this reason, there is a need for alternative approaches of developing CDSSs that do not rely on structured EHRs.

ICD9 disease coding is an important task in a hospital, during which a trained medical coder with domain knowledge assigns disease-specific, standard-ized codes called ICD9 codes to a patient's admission record. As hospital billing and insurance claims are based on the assigned ICD9 codes, the coding task requires high precision, but is often prone to human error, which has resulted in an annual spending of $25 billion in the US to improve coding efficacy [7,25]. Hence, automated disease coding approaches have been developed as a solution to this problem. Currently, this is an area of active research [2–4,17,25,26], but performance that has been recorded so far is below par indicating huge scope for improvement, thus making automated ICD9 coding an open research prob-lem. Existing ICD9 coding methods make use of only discharge summaries for coding, similar to the process adopted by human medical coders. However, as the records are digitized, other textual records under the same admission iden-tifier might be available, which may provide additional patient-specific insights pertaining to the diagnosed diseases. Further, for ICD9 disease coding to be effective, the correct determination of generic disease categories or groups prior to specific coding is very crucial as information regarding generic disease groups can be informative for specific ICD9 coding. With this in mind, we explore other textual sources of patient data, utilizing physician notes to predict the generic ICD9 disease groups for the patient, following which actual ICD9 disease code prediction can be achieved. In this article, we focus on effective prediction of ICD9 disease groups and the code prediction will be considered in future work.

While most existing works aim to predict ICD9 codes rather than ICD9 disease groups, Purushotham et al. [21] performed benchmarking of three CDSS tasks - ICU mortality prediction, ICD9 group prediction and hospital

---

[1] ICD-9-CM: International Classification of Diseases, Ninth Revision, Clinical Modi-fication.

readmission prediction on ICU patients data using Super Learner and Deep Learning techniques. The ICD9 group prediction task discussed in their work is a generic disease group prediction task that is a step prior to ICD9 coding, which can also be a viable disease risk prediction CDSS for caregivers. Their work uses feature subsets extracted from structured patient data, where the best performing model uses multitude of numerical features such as input events (volume of input fluids through IV), output events (quantity of output events like urinary and rectal), prescriptions and other ICU events like chart events (readings recorded in the ICU) and lab events (lab test values). As deep learning models can effectively learn from numerous raw feature values quite well, their approach achieved good results. However, in a real world scenario, to measure all these values and then using them as inputs to a prediction model for an outcome, will cause an inevitable and significant time delay, during which the patient condition can worsen. CDSS prediction models that use unstructured clinical text data can predict desired outcomes with lower latency and also help improve the prediction performance as these raw text sources contain more patient-specific information.

Some recent works [8,9,15] focussed on investigating the effectiveness of modeling various unstructured text reports such as radiology and nursing notes of ICU patients for disease group prediction. With the objective of continuing such analyses on other kinds of unstructrured clinical reports, we intend to focus on physician notes in this work. Additionally, we also analyze how they can be used in predicting ICD9 disease groups by effectively modeling these physician notes. In this article, a hybrid feature modeling approach that uses hybrid clinical word embeddings to generate quality features which are used to train and build a deep neural network model to predict ICD9 disease groups is presented. We show the results of a benchmarking study of our proposed model (modeled on unstructured physician notes) against the current state-of-the-art model for disease group prediction (built on structured clinical data). The rest of this article is structured as follows: Sect. 2 describes the proposed approach in detail, followed by experimental results and discussion in Sect. 3. We conclude the article with prospective future work and research directions.

2   Materials and Methods

The proposed approach for disease group prediction is depicted in Fig. 1. We used physicians' clinical notes in unstructured text form from the MIMIC-III [13] dataset for our experiments. MIMIC-III contains data relating to 63,000 ICU admissions of more than 46,000 patients admitted in Beth Israel Hospital, New York, USA, during 2001–2012. We extracted only the physician notes from the 'NOTEEVENTS' table, which resulted in a total of 141,624 physician notes generated during 8,983 admissions. As per MIMIC documentation, physicians have reported some identified errors in notes present in the 'NOTEEVENTS' table. As these notes can affect the training negatively, records with physician identified errors were removed from the cohort. Additionally, those records with

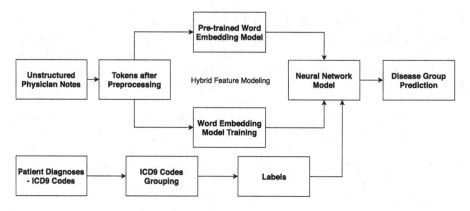

Fig. 1. Proposed ICD9 disease group prediction process

less than 15 words are removed and finally, the remaining 141,209 records are considered for the study. Some characteristic features of the physician notes corpus are tabulated in Table 1.

We observed that there were 832 kinds of physician notes available in the MIMIC-III dataset such as Physician Resident Progress note, Intensivist note, etc. The frequency statistics of the top ten kinds of physician notes are tabulated in Table 2. A particular patient may suffer from multiple diagnoses during a particular admission and hence, it is necessary for the prediction to be a multi-label prediction task. Therefore, for each physician note, all the diagnosed disease groups during that particular admission were considered as labels and given binary values - 0 (if the disease was not diagnosed) and 1 (if the disease was diagnosed).

2.1 Preprocessing and Feature Modeling

The physician notes corpus is first preprocessed using basic Natural Language Processing (NLP) techniques such as tokenization and stop word removal. Using

Table 1. Physician notes corpus characteristics

Feature	Total records
Physician notes	141,209
Unique words	635,531
Words in longest note	3,443
Words in shortest note	16
Average word length of notes	858
Unique diseases	4,208
Disease groups	20

Table 2. Top 10 physician notes types

Note type	Occurrences
Physician Resident Progress	62,550
Intensivist	26,028
Physician Attending Progress	20,997
Physician Resident Admission	10,611
ICU Note - CVI	4,481
Physician Attending Admission (MICU)	3,307
Physician Resident/Attending Progress (MICU)	1,519
Physician Surgical Admission	1,102
Physician Fellow/Attending Progress (MICU)	970
Physician Attending Admission	873

tokenization, the clinical text corpus is broken down into basic units called tokens and by the stopping process, unimportant words are filtered out. The preprocessed tokens are then fed into a pre-trained word embedding model to generate the word embedding vector representation of the corpus that can be considered as textual features. The pre-trained word embedding model used in this study is an openly available model that is trained on biomedical articles in PubMed and PMC along with texts extracted from an English Wikipedia dump [20], hence capturing relevant terms and concepts in the biomedical domain, which helps generate quality feature representation of the underlying corpus. The pre-trained model generates word embedding vectors of size 1×200 for each word and these vectors were averaged to generate a representation such that each preprocessed physician note is represented as a 1×200 vector. The preprocessed tokens are then used to train a Word2Vec model [18], a neural network based word representation model that generates word embeddings based on co-occurrence of words. The Skipgram model of Word2Vec was used for training the physician notes tokens with a dimension size of 200 (same as the pre-trained model) with an initial learning rate of 0.025. The averaged word vector representation for each report tokens are extracted and then fed into the neural network model along with the vector representation extracted from the pre-trained model and the ICD9 disease group labels.

2.2 ICD9 Disease Code Grouping

ICU Patients' diagnoses in terms of ICD9 disease codes from the 'DIAG-NOSES_ICD' table of MIMIC-III dataset were grouped as per standards[2] (similar to the approach adopted by Purushotham et al. [21]). A total of 4,208 unique ICD9 disease codes thus obtained were grouped into 20 ICD9 disease groups,

---

[2] Available online http://tdrdata.com/ipd/ipd_SearchForICD9CodesAndDescriptions. aspx.

i.e., potential labels. As the ICD9 group prediction task is considered as a binary classification of multiple labels, 20 labels (disease groups) were considered with binary values: 0 (negative diagnosis of the disease) and 1 (positive diagnosis of the disease). The physician notes modeled into two feature matrices of shape 141209×200 each, along with 20 ICD9 disease groups (labels) is now used to train the neural network model.

2.3   Neural Network Model

The proposed Deep Neural Network Prediction Model is illustrated in Fig. 2. The neural network architecture is divided into 2 parts – the first for determining the weights for the hybrid combination of features dynamically and the next for multi-label classification of ICD9 disease codes. The process of dynamically modeling the weightage to be assigned for the combination of pre-trained word embeddings and the word embeddings generated using the physicians notes is

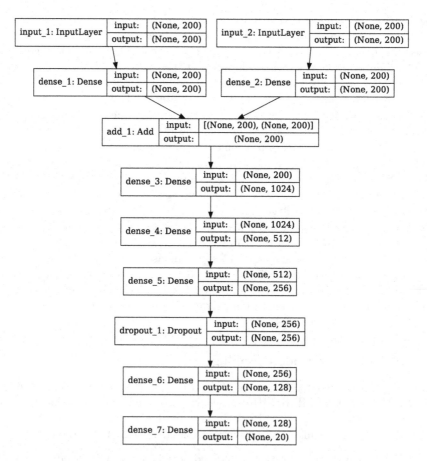

Fig. 2. Overall neural network model for ICD9 group prediction

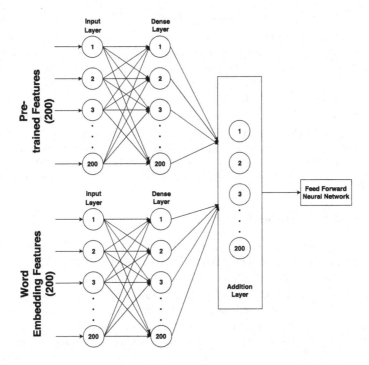

Fig. 3. Hybrid features using dynamic weighted addition of feature representations

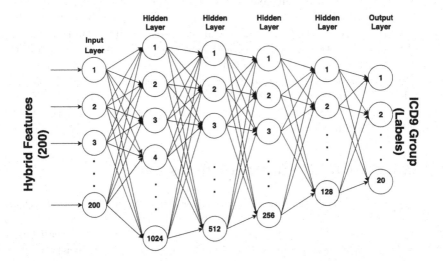

Fig. 4. Feed-forward neural network - disease group prediction model

performed as shown in Fig. 3. The two feature set are fed as inputs into the neural network model, where both the input layers consists of 200 neurons, equal to the number of features generated from both models. The addition layer merges the two sets of input features in a weighted combination which is dynamically determined through backpropagation of the overall neural network architecture thereby ensuring the optimal and effective combination that offers the best classification performance possible. This architecture also ensures that the weights for the hybrid combination of features is always determined dynamically and hence can be used for any clinical text corpus. The combined set of features, i.e., the hybrid features, are then fed on to the dense feed forward neural network model which performs the training for multi-label classification of ICD9 disease code groups.

2.4   Disease Prediction Model

The dynamically weighted feature matrix consisting of the hybrid word embedding features, along with ICD9 group labels are next used for training a Feed Forward Neural Network (FFNN) used as the prediction model (depicted in Fig. 4). The input layer consists of 1024 neurons with input dimension as 200 (number of input features); followed by three hidden layers with 512, 256 and 128 neurons respectively and finally an output layer with 20 neurons, each representing an ICD-9 disease group. To prevent overfitting, a dropout layer, with a dropout rate of 20% was also added to the FFNN model (see Fig. 2). As this is a binary classification for multiple labels, binary cross entropy was used as a loss function, while Stochastic Gradient Descent (SGD) was used as the optimizer with a learning rate of 0.01. Rectified Linear Unit $ReLU$ activation function was used as the input and hidden layer activation functions as the feature matrix values are standardized to the range -1 and 1. The major hyperparameters for the FFNN model – the optimizer, learning rate of the optimizer and the activation function, were tuned empirically over several experiments using the GridSearchCV function in Python sklearn library. Finally, the output layer activation function is a sigmoid, again as the classification is two-class for each of the 20 labels. Training was performed for 50 epochs and then the model was applied to the validation set to predict disease groups after which the results were observed and analyzed.

3   Experimental Results and Discussion

To evaluate the proposed approach, we performed several experiments using standard metrics like accuracy, precision, recall, F-score, Area under Receiver Operating Characteristic curve (AUROC), Area under Precision Recall Curve (AUPRC) and Matthew's Correlation Coefficient (MCC). We measured these metrics on a sample-wise basis, i.e., for each report, the predicted and actual disease groups were compared and analyzed. It can be observed from the Table 3 that, the proposed model achieved promising results: AUPRC of 0.85 and

AUROC of 0.89. The accuracy, precision, MCC and F-score of 0.79, 0.82, 0.57 and 0.79 respectively also indicate a good performance.

For evaluating the proposed hybrid text feature modeling approach, we designed experiments to compare its performance against that of baseline feature modeling approaches – TF-IDF based bag-of-words approach, trained word embedding approach (only Word2Vec model) and a pre-trained word embedding approach (using word embedding model trained on PubMed, PMC and Wikipedia English articles). The bag-of-words approach is based on calculated term Frequency and inverse document frequency ($tfXidf$) scores determined from the physician notes corpus. The Sklearn English stopword list was used to filter the stopwords and n-gram ($n = 1, 2, 3$) features were considered. Finally, the top 1000 features were extracted from the corpus and then fed into the neural network model for training. The other two baselines were kept the same as explained in Sect. 2. It is to be noted that in the neural network configuration, only one input is present, i.e., there is no hybrid weighted addition layer. The results of comparison are tabulated in Table 3. It can be observed that the proposed approach that involves a hybrid weighted combination of pre-trained and trained word embeddings is able to perform comparatively better in terms of all metrics.

Next, a comparative benchmarking against the state-of-the-art ICD9 disease group prediction model developed by Purushotham et al. [21] was performed. For each hospital admission ID (HADM_ID) considered by Purushottam et al. [21], the physician notes (if present) were extracted from 'NOTEVENTS' table. Although the number of records under consideration for both the studies are different, it is to be noted that the labels are distributed similarly (statistics shown in Fig. 5) and is therefore considered a fair comparison. We consider this comparison in order to study the effect of the disease group prediction models built on structured patient data (state-of-the-art approach [21]) and unstructured patient data (physician notes in this case). The results of the benchmarking are tabulated in Table 4, which clearly shows the proposed approach outperformed Purushotham et al.'s model Purushotham et al. [21] by 15% in terms of AUROC and 40% in terms of AUPRC. This shows that the predictive power of a model built on unstructured patient data exceeds that of those built on structured data, where some relevant information may be lost during the coding process. To encourage comparative studies that use physician notes in MIMIC-III dataset, certain additional metrics were also considered. The Recall & F-Score performance as well as the MCC values of the proposed model over our easily reproducible patient cohort data subset are also observed and provided.

Discussion. From our experiments, we observed a significant potential in developing prediction based CDSS using unstructured text reports directly, eliminating the dependency on the availability of structured patient data and EHRs. The proposed approach that involves a textual feature modeling and a neural network based prediction model was successful in capturing the rich and latent clinical information available in unstructured physician notes, and using it to effectively learn disease group characteristics for prediction. The Word2Vec

Table 3. Experimental results – baseline comparison

Parameter	Proposed approach	Bag-of-words (TF-IDF)	Only Word2Vec	Only pre-trained
AUROC	0.89	0.87	0.88	0.88
AUPRC	0.85	0.81	0.84	0.82
Accuracy	0.79	0.78	0.79	0.79
Precision	0.82	0.80	0.82	0.80
Recall	0.79	0.78	0.79	0.78
F-Score	0.79	0.78	0.79	0.79
MCC	0.58	0.53	0.57	0.56

Table 4. Experimental results – comparison with the state-of-the-art

Parameter	Our approach	Purushotham et al. [21]
Type of data	Unstructured text	Structured data
AUROC	0.89	0.77
AUPRC	0.85	0.60
Accuracy	0.79	*
Precision	0.82	*
Recall	0.79	*
F-Score	0.79	*
MCC	0.58	*

* Results not reported in the study

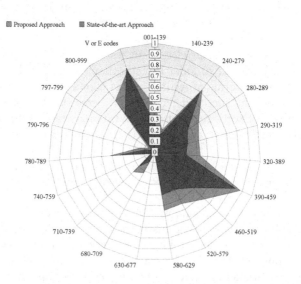

Fig. 5. ICD9 group labels statistics comparison

model was trained on the physician notes corpus with optimized parameter configuration generates effective word embedding features to be fed into the neural network model. The hybrid combination of these with the embedding features of the corpus generated by the pre-trained model trained on PubMed and PMC articles further enhanced and enriched the semantics of the textual features with biomedical domain knowledge. This is clear from the baseline comparison shown in Table 3 and it is this combination that has further enabled the FFNN to generalize better and learn the textual feature representation, effectively improving prediction performance when compared to the state-of-the-art model built on structured data.

It is interesting to note that the patient data was modeled using only textual features, without any EHRs, structured data or other processed information. The high AUPRC and AUROC values obtained in comparison to the state-of-the-art's (based on structured data) performance is an indication that the unstructured text clinical notes (physician notes in this case) contain abundant patient-specific information that was beneficial for predictive analytics applications. Moreover, the conversion process from unstructured patient text reports to structured data can be eliminated, thereby saving huge man hours, cost and other resources. The proposed approach also eliminates any dependency on structured EHRs, thus making it suitable for deployment in developing countries.

We found that the data preparation pipeline adopted for this study could be improved, as it sometimes resulted in some conflicting cases during training. This is because of the structure of the MIMIC-III dataset, in which, the physician notes do not have a direct assignment to ICD9 disease codes. To overcome this problem, we designed an assignment method in which we extracted ICD9 codes from the DIAGNOSES_ICD table and assigned them to all patients with the same SUBJECT_ID and HADM_ID in the physician notes data subset. A side-effect of this method is that, at times, the ICD9 disease codes/groups assigned to the physician notes may not be related to that particular disease as a patient under same admission can have multiple physician notes. However, the model achieved promising results and the good performance indicates that prediction model was able to capture disease-specific features using the information present in the notes during the admission.

4 Conclusion and Future Work

In this article, a deep neural network based model for predicting ICD9 disease groups from physician notes in the form of unstructured text is discussed. The proposed approach is built on a hybrid feature set consisting of word embedding features generated from a Word2Vec Skipgram model and also from a pre-trained word embedding model trained on biomedical articles from PubMed and PMC. The ICD9 disease codes were categorized into 20 standard groups and then used to train a binary classifier for multi-label prediction. The two sets of word embedding features were input into a neural network model as inputs and the weights of hybrid addition was determined dynamically using backpropagation.

The combined features were further fed into a FFNN architecture to train the classifier and the prediction model was validated and benchmarked against state-of-the-art ICD9 disease group prediction model. The experiments highlighted the promising results achieved by the proposed model, outperforming the state-of-the-art model by 15% in terms of AUROC and 40% in terms of AUPRC.

The promising results achieved by the proposed model underscores its usefulness as an alternative to current CDSS approaches, as it eliminates the dependency on structured clinical data, ensuring that hospitals in developing countries with low structured EHR adoption rate can also make use of effective CDSS in their functioning. As part of future work, we plan to address the issues observed in the current data preparation strategy, and enhance it by sorting out the disease group assignment problems. We also intend to explore other techniques to further optimize topic and feature modeling of textual representations to study their effect on disease prediction. Additionally, we intend to evaluate and compare other disease group prediction approaches on the patient cohort used in this study. Furthermore, we also aim to perform a detailed efficiency analysis and comparison with respect to training times of the proposed approach and various baselines. Finally, extensive benchmarking of approaches for modeling unstructured clinical notes on the physician notes, for further insights into the adapatability of the proposed approach in real-world scenarios will also be explored.

Acknowledgement. We gratefully acknowledge the use of the facilities at the Department of Information Technology, NITK Surathkal, funded by Govt. of India's DST-SERB Early Career Research Grant (ECR/2017/001056) to the second author.

References

1. Appelros, P.: Prediction of length of stay for stroke patients. Acta Neurol. Scand. **116**(1), 15–19 (2007)
2. Ayyar, S., Don, O., Iv, W.: Tagging patient notes with ICD-9 codes. In: Proceedings of the 29th Conference on Neural Information Processing Systems (2016)
3. Baumel, T., Nassour-Kassis, J., Cohen, R., Elhadad, M., Elhadad, N.: Multi-label classification of patient notes: case study on ICD code assignment. In: Workshops at the Thirty-Second AAAI Conference on Artificial Intelligence (2018)
4. Berndorfer, S., Henriksson, A.: Automated diagnosis coding with combined text representations. Stud. Health Technol. Inf. **235**, 201 (2017)
5. Calvert, J., et al.: Using electronic health record collected clinical variables to predict medical intensive care unit mortality. Ann. Med. Surg. **11**, 52–57 (2016)
6. Choi, E., Bahadori, M.T., Schuetz, A., Stewart, W.F., Sun, J.: Doctor AI: predicting clinical events via recurrent neural networks. In: Machine Learning for Healthcare Conference, pp. 301–318 (2016)
7. Farkas, R., Szarvas, G.: Automatic construction of rule-based ICD-9-CM coding systems. In: BMC Bioinformatics, vol. 9, p. S10. BioMed Central (2008)
8. Gangavarapu, T., Krishnan, G., Kamath, S., Jeganathan, J.: FarSight: long-term disease prediction using unstructured clinical nursing notes. IEEE Trans. Emerg. Top. Comput. **01**, 1 (2020)

9. Gangavarapu, T., Jayasimha, A., Krishnan, G.S., Kamath, S.: Predicting ICD-9 code groups with fuzzy similarity based supervised multi-label classification of unstructured clinical nursing notes. Knowl.-Based Syst. **190**, 105321 (2020)
10. Ge, W., Huh, J.W., Park, Y.R., Lee, J.H., Kim, Y.H., Turchin, A.: An interpretable ICU mortality prediction model based on logistic regression and recurrent neural networks with LSTM units. In: AMIA Annual Symposium Proceedings, vol. 2018, p. 460. American Medical Informatics Association (2018)
11. Harutyunyan, H., Khachatrian, H., Kale, D.C., Galstyan, A.: Multitask learning and benchmarking with clinical time series data. arXiv preprint arXiv:1703.07771 (2017)
12. Jiang, S., Chin, K.S., Qu, G., Tsui, K.L.: An integrated machine learning framework for hospital readmission prediction. Knowl.-Based Syst. **146**, 73–90 (2018)
13. Johnson, A.E., et al.: MIMIC-III, a freely accessible critical care database. Sci. Data **3**, 160035 (2016)
14. Kansagara, D., et al.: Risk prediction models for hospital readmission: a systematic review. JAMA **306**(15), 1688–1698 (2011)
15. Krishnan, G.S., Kamath, S.S.: Ontology-driven text feature modeling for disease prediction using unstructured radiological notes. Comput. Sistemas **23**(3), 915–922 (2019)
16. Krishnan, G.S., Kamath, S.S.: A novel GA-ELM model for patient-specific mortality prediction over large-scale lab event data. Appl. Soft Comput. **80**, 525–533 (2019). http://www.sciencedirect.com/science/article/pii/S1568494619302108
17. Li, M., et al.: Automated ICD-9 coding via a deep learning approach. IEEE/ACM Trans. Comput. Biol. Bioinf. **16**, 1193–1202 (2018)
18. Mikolov, T., Chen, K., Corrado, G., Dean, J.: Efficient estimation of word representations in vector space. arXiv preprint arXiv:1301.3781 (2013)
19. Miotto, R., Li, L., Kidd, B.A., Dudley, J.T.: Deep patient: an unsupervised representation to predict the future of patients from the electronic health records. Sci. Rep. **6**, 26094 (2016)
20. Nédellec, C., et al.: Overview of BioNLP shared task 2013. In: Proceedings of the BioNLP Shared Task 2013 Workshop, pp. 1–7 (2013)
21. Purushotham, S., Meng, C., Che, Z., Liu, Y.: Benchmarking deep learning models on large healthcare datasets. J. Biomed. Inf. **83**, 112–134 (2018)
22. Reddy, B.K., Delen, D.: Predicting hospital readmission for lupus patients: an RNN-LSTM-based deep-learning methodology. Comput. Biol. Med. **101**, 199–209 (2018)
23. Shickel, B., Loftus, T.J., Adhikari, L., Ozrazgat-Baslanti, T., Bihorac, A., Rashidi, P.: DeepSOFA: a continuous acuity score for critically ill patients using clinically interpretable deep learning. Sci. Rep. **9**(1), 1879 (2019)
24. Van Houdenhoven, M., et al.: Optimizing intensive care capacity using individual length-of-stay prediction models. Crit. Care **11**(2), R42 (2007)
25. Xie, P., Xing, E.: A neural architecture for automated ICD coding. In: Proceedings of the 56th Annual Meeting of the Association for Computational Linguistics (Volume 1: Long Papers), vol. 1, pp. 1066–1076 (2018)
26. Zeng, M., Li, M., Fei, Z., Yu, Y., Pan, Y., Wang, J.: Automatic ICD-9 coding via deep transfer learning. Neurocomputing **324**, 43–50 (2019)

Early Signs of Critical Slowing Down in Heart Surface Electrograms of Ventricular Fibrillation Victims

Berend Nannes[1]($\boxtimes$), Rick Quax[1,2], Hiroshi Ashikaga[3], Mélèze Hocini[4,5,6], Remi Dubois[5,6], Olivier Bernus[5,6], and Michel Haïssaguerre[4,5,6]

[1] Computational Science Lab, University of Amsterdam,
Amsterdam, The Netherlands
berend.nannes@quicknet.nl
[2] Institute of Advanced Studies, University of Amsterdam,
Amsterdam, The Netherlands
[3] Cardiac Arrhythmia Service, Johns Hopkins University School of Medicine,
Baltimore, MD, USA
[4] Electrophysiology and Cardiac Stimulation, Bordeaux University Hospital,
Bordeaux, France
[5] IHU LIRYC, Electrophysiology and Heart Modeling Institute, Bordeaux, France
[6] University of Bordeaux, Bordeaux, France

Abstract. Ventricular fibrillation (VF) is a dangerous type of cardiac arrhythmia which, without intervention, almost always results in sudden death. Implantable automatic defibrillators are among the most successful devices to prevent sudden death by automatically applying a shock to the heart when fibrillation occurs. However, the electric shock is very painful and could lead to dangerous situations when a patient is, for example, driving or biking. An early warning signal for VF could reduce the risk in such situations or, in the future, reduce the need for defibrillation altogether. Here, we test for the presence of critical slowing down (CSD), which has proven to be an early warning indicator for critical transitions in a range of different systems. CSD is characterized by a buildup of autocorrelation; we therefore study the residuals of heart surface electrocardiograms (ECGs) of patients that suffered VF to investigate if we can measure positive trends in autocorrelation. We consider several methods to extract these residuals from the original signals. For three out of four VF victims, we find a significant amount of positive autocorrelation trends in the residuals, which might be explained by CSD. We show that these positive trends may not be measurable from the original body surface ECGs, but only from certain areas around the heart surface. We argue that additional experimental studies involving heart surface ECG data of subjects that did not suffer VF are required to quantify the prediction accuracy of the promising results we get from the data of VF victims.

Keywords: Critical slowing down · Ventricular fibrillation · Critical transition · Early warning signal

© Springer Nature Switzerland AG 2020
V. V. Krzhizhanovskaya et al. (Eds.): ICCS 2020, LNCS 12140, pp. 334–347, 2020.
https://doi.org/10.1007/978-3-030-50423-6_25

1   Introduction

Ventricular Fibrillation (VF) is a dangerous type of cardiac arrhythmia where the ventricles, instead of beating normally, tremble uncontrollably. As a result, the heart is unable to regulate the blood circulation around the body, resulting almost always in sudden death. The most important contributions to the treatment of VF to date are automated defibrillators. For example, an implantable cardioverter-defibrillator (ICD) can be used for patients at an increased risk of suffering dangerous arrhythmia. The ICD detects cardiac arrhythmia and automatically intervenes by applying an electric shock. For these devices it is crucial to correctly detect the onset of VF in real time to be able to prevent permanent damage or death of the patient. In earlier research, several algorithms have been proposed to this end [1]. The ICD does not prevent the arrhythmia itself, but provides treatment by shock after the occurence. Intervention with ICD is not without drawbacks: the electric shock is experienced by patients as very painful [2] and can in some cases cause psychological problems [3]. Moreover, the shock could lead to dangerous situations if the patient is, for example, driving or biking. In the latter case, an early warning signal seconds before the shock could greatly reduce risk of accidents. Furthermore, early warning signals in combination with future methods to prevent VF from happening would reduce the need for defibrillation. Many possible indicators have been proposed in earlier work [4] from which Heart Rate Variability (HRV) is the most promising. However, contrary evidence has also been found, as HRV is influenced by characteristics - for example, medical conditions - of the individual patients [5]. Evidenty, a generic early warning signal for VF that is independent of patient characteristics is highly desirable.

If we look at VF from a general perspective, we can argue that the shift from a state where the ventricles are pumping normally to a state in which they quiver is a sudden transition between two different dynamical regimes for which the manifestation of VF is the tipping point. Such abrupt changes in dynamical behavior are seen in many real complex systems in nature and are generally called "critical transitions". In critical transitions, once the tipping point is exceeded, it is not easy to return to the previous state. A real-life example of this is desertification: once a patch of land reaches a barren state, it is hard for vegetation to reappear. This "irreversible" character has made possible predictors of these critical transitions much sought-after. Earlier research [6] has shown that there exists a domain-free early-warning signal for critical transitions in complex dynamical systems in different fields of research: critical slowing down (CSD). The theory behind CSD is based on the fact that, mathematically, some critical transitions in real systems can be interpreted as catastrophic bifurcations. It has been shown [7] that systems that approach such bifurcations experience a decrease in resilience; the system needs more time to recover from perturbations when a critical event is close. This decrease in resilience can be measured by increasing autocorrelation and standard deviation in the corresponding time series data (as we will explain in the Method section). These symptoms have indeed been identified in a wide range of real complex systems. For example:

the ending of multiple different glacial periods by abrupt climate changes are preceded by the building up of autocorrelation in deuterium measurements [8], and brain activity shows increasing amounts of variance when close to an epileptic seizure [9]. Here, the same underlying principle applies, independent of the scientific field. If we assume that the onset of VF can be viewed as such a critical transition, we expect that we can detect the same early warning signals that are found in a variety of other real systems.

Historically, there has been controversy about the mechanism that drives VF [10], with some research suggesting that fibrillation represents a chaotic system [11], and others stating that it is rather similar to a nonchaotic random signal [12]. Nowadays it is commonly believed that the onset of VF is a transition to spatiotemporal chaos [13,16] and thus a shift to a different, chaotic attractor. This transition is initiated by a wavebreak that arises when a wavefront and waveback of the cardiac excitation meet [15]. Under normal circumstances this never occurs, but in certain conditions the propagating impulse does not die out but returns to reexite the heart (called reentry [14]). When reentry triggers a wavebreak, it can in turn produce daughter waves, causing new wavebreaks etc., quickly degenerating into spatiotemporal chaos and VF. The transition from normal heart rhythm to abnormal, chaotic, heart rhythm (the process from reentry to VF) takes place through a series of bifurcations [16]. There exist different theories about how we can exactly describe this route to chaos [17]. In the context of nonlinear dynamics, however, we can consider the shift from a normal heart rhythm to VF as a change in topology of electrical wave dynamics or a transition between two states with different basins of attraction [18]. This drastic change in dynamical regime may therefore bear resemblance to critical transitions in other complex systems and may possibly be signaled by CSD.

In this thesis, we investigate if CSD can be observed in the residuals of heart surface electrocardiogram (ECG) recordings from patients that suffered VF. We analyze data sets from four patients provided by IHU LIRYC (Electrophysiology and Heart Modeling Institute) in Bordeaux, France. All patients suffered sudden cardiac arrest due to documented VF resulting from ischemic heart disease (n = 2), early repolarization syndrome (n = 1), or idiopathic VF (n = 1). All patients were male and the age range was between 15 and 74 years old. Each set contains around 1400 ECG signals (leads) over the heart surface, which are estimated using body surface ECG measurements by solving an ill-posed, inverse problem [19]. The original body surface potential maps consist of around 250 leads. Each lead in the set is a 20-s ECG recording; 10 s of normal heart rhythm followed by 10 s of arrhythmia, with a sampling rate of 1 kHz. We examine the 10 s of the signal that precede the tipping point looking for CSD. Specifically, we look for a significant increase in autocorrelation in the residual.

The main challenge of our research is the actual extraction of the residuals; the short-term fluctuations relative to the main ECG waves. To expose the residual, we have to filter out the wave components that are typical to the ECG. The difficulty resides in the high frequency characteristic of some of these typical waves, which impedes the use of a simple frequency filter. In this thesis, we go over several alternatives to correctly obtain the residuals, which is essential for credible autocorrelation measurements.

Electrocardiography is traditionally prone to different types of noise like power line interference, electrode contact noise and motion artifacts. We have to keep in mind that some of this noise (in particular, high frequency noise) we can not distinguish from fluctuations in the real signal and will end up in the residual. The ill-posed nature of the inverse solution may also induce errors in the residual. Therefore, to quantify the prediction accuracy of autocorrelation measurements for ECG data from VF victims, we have to conduct the same analysis on heart surface ECG estimations of subjects that did not suffer VF. The main goal of this thesis to show the necessity of such data.

2   Method

Critical slowing down (CSD) is the increase of recovery time needed by the system when it is perturbed from the stable state. In our case, this stable state is described by the well-known features of an ECG signal. Therefore, to be able to measure a possible CSD effect, we aim to find the fluctuations relative to the typical ECG curve: the residual. In this section we discuss several methods we considered to extract the residuals from the original signals and argue which method is the most preferable. To look for CSD, we have to analyze the time evolution of the autocorrelation of the extracted residuals. Because autocorrelation measures the likeness of a signal with a lagged version of itself, a slowly varying signal has a higher autocorrelation than a rapidly fluctuating signal. For that reason it is expected that, if residuals show a slowing down effect, we measure a significant positive trend in autocorrelation.

2.1   Measuring the Autocorrelation

The most straightforward autocorrelation measure for equispaced data is the lag-1 autocorrelation, where the state of the signal at time t is directly compared to its state in the previous time unit $t - 1$. The lag-1 autocorrelation can be estimated by treating the signal as a first-order autoregressive (AR(1)) process and calculating the corresponding lag-1 autoregression coefficient. We estimate this parameter using Yule-Walker equations with Python library *statsmodels.tsa.stattools*. To capture the time evolution of the autocorrelation, the autoregression coefficient is calculated over a moving window. The window length is chosen to be exactly half the signal length, so that there is a reasonable trade-off between a sufficiently long window to compute the autocorrelation, and a long enough sequence of autocorrelation values to be able to study its time-evolution. (The influence of different window lengths on our result is shown in Appendix A.10)

2.2   Extracting Fluctuations

The typical ECG signal is composed of a set of main wave components (PQRST) formed by electrical currents produced by the depolarization and re-polarization of different heart chambers. Depolarization is responsible for the contraction of cardiac muscle, while with re-polarization, cardiac muscle relaxes. A schematic representation is provided in Fig. 1. The P-wave is produced by the depolarization and contraction of both atria. The QRS-complex is composed of the electrical signals from both the depolarization of the ventricles and the re-polarization of the atria. Finally, the re-polarization of the ventricles produces the T-wave. The orientation of the waves is dependent on the polarity (positive or negative) of the electrode.

Normally, an effective technique to extract short term fluctuations from a signal would be to filter out the lower frequency components using by applying a high-pass filter. However, the QRS-complex has a significantly higher frequency range than the P- and T-waves. A high-pass filter that removes all characteristic waves requires a high cutoff frequency, and thus imposes risk of filtering out short-term fluctuations that possibly show a CSD effect. The same problem arises for methods using wavelet decomposition: by removing high frequency sub-bands from the signal to remove the QRS complex we might accidentally remove fluctuations we want to analyze. For this reason, we have to consider other methods to extract the residuals: by utilizing the knowledge of the recurring wave components we create a model for the signal containing all characteristic waves and subtract it from the original signal.

In the following subsections we go over techniques we considered to extract the residues of the ECG leads.

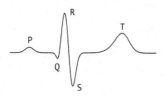

Fig. 1. The main wave components of an ECG; the P-wave, QRS-complex and T-wave.

Pre-processing. We aim to remove characteristic components (Fig. 1) from the signal to extract the fluctuations. Baseline wandering in the signal makes it hard to distinguish wave components from the zero-volt level. Moreover, the height of characteristic waves may vary between heartbeats. In all methods discussed in this section, the baseline trend is first removed from the signal. Some methods require the signal to be cut into segments of one heartbeat cycle. These pre-processing steps are described in detail in Appendix A.1.

Fitting Gaussian Curves. To isolate the fluctuations in the ECG signal, we can directly subtract the characteristic ECG waves for every cardiac cycle. By fitting a mixture of Gaussian curves to the original ECG that approximate the PQRST-waves we can construct an characteristic version (as shown in Fig. 1) of the signal, which can in turn be subtracted to reveal the residual. The implementation of this method is explained in more detail in Appendix A.2.

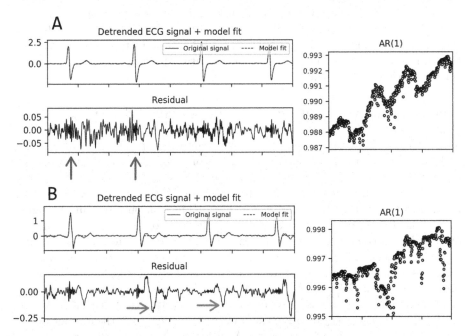

Fig. 2. Segments of detrended ECG signals, their corresponding model fit and the residuals after subtracting the model. **A:** The model seemingly fits the signal. However, we observe high frequency errors in the residual (red arrows) resulting in peaks and dips in the AR(1) measurements (right) occurring at the same frequency as the QRS-complexes. **B:** The signal, while we can identify the PQRST-waves, slightly deviates from the characteristic shape portrayed in Fig. 1. The fitting method fails to reproduce the characteristic waves resulting in periodically recurring errors in the residual (green arrows) (Color figure online).

Techniques modeling ECG waveforms using Gaussian curves have been around for some time [20] and are generally used to extract clinical features like the location, height and width of the characteristic waves. When we apply our method to the ECG data we observe that indeed, for signals that resemble the PQRST-composition portrayed in Fig. 1 like the example given in Fig. 2A, the Gaussian fitting method seemingly provides a good model. However, when we subtract the model fit, the parts of the residual at the location of the QRS-complex clearly have higher frequency compared to other parts of the residual, which indicates that they are error artifacts induced by the fitting method. The effect of these errors is clearly visible in the AR(1) measurements: rapidly fluctuating parts of the signal have relatively low autocorrelation; combined with the moving window this results in peaks and dips in AR(1) values. These peaks and dips recur with the same frequency as the QRS-complexes. Clearly, while this modeling technique may be suitable for clinical feature extraction, it is not ideal to extract correct residuals of the signals. Moreover, to obtain a correct model, the signal has to resemble the characteristic shape from Fig. 1. When the signal deviates from this shape the fitting technique becomes infeasible, resulting in errors in the residual. An example of this is given in Fig. 2B. We conclude that this method is ineffective for the extraction of fluctuations from the signal since the majority of the signals in the provided data set do not have the required characteristic PQRST-composition.

Computing an Average Beat. The method described above to extract fluctuations from the signal turns out to be ineffective as it requires each signal to have a certain characteristic shape. In reality, characteristics may be different for each signal. To capture characteristic features of a signal, one can also construct an average beat using the individual beat cycles. Characteristic features recur periodically and are thus automatically present in the average curve. Everything relative to the average beat can be classified as fluctuations. We cut the signal into segments containing one beat cycle. The fluctuations can be extracted beat by beat by subtracting the average beat from every cycle. However, the height as well as the overall shape of the characteristic waves is not constant over the whole signal. Therefore, the average curve must be adjusted for each beat to provide a good model for the signal. A detailed description of the average beat fit and these adjustments is given in Appendix A.3. It turns out that in almost all signals one or more QRS-complexes deviate too much from the average curve, even after the adjustments are made. As a result, the average curve is unable to fit the signal and errors occur in the residual (Fig. 3). For this reason, we conclude that the average beat method is unusable for credible autocorrelation measurements.

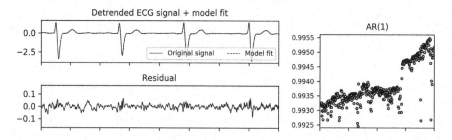

Fig. 3. Segment of an ECG signal and the average beat fit. The average beat is not able to fit all QRS-complexes, leading to jumps in the AR(1) measurements similar to Fig. 2A. This is typical for almost all signals in the data sets.

Excluding QRS-Complexes. In the methods discussed above, we encounter the problem that we can not (correctly) extract the fluctuations of parts of the signal around the QRS-complex. This inability is mainly caused by the fact that during the QRS-complex, a lot of change in y-axis value takes place in a limited amount of time points. Here, small fitting errors on the time axis can lead to big errors in y-axis value in the residual, and this heavily influences autocorrelation calculations. Therefore, to get more reliable results, it might be preferable to remove the QRS-complexes, or, generally, all parts with high first-order differences, from the signal altogether. This has a clear disadvantage; we discard information by cutting parts of the signal. However, if the autocorrelation in the residual is gradually building up, this effect should still be observable, even if the signal is not complete. We implement a sequence of steps to remove unwanted parts of the ECG. This procedure accounts for the fact that, when cutting out parts of the signal, the difference in y-value at the edges might induce sudden "jumps" in the signal which affect any autocorrelation measurements. The cutting process is illustrated in Appendix A.4. Besides the fact that this procedure solves the problems we encountered in the methods mentioned above, it also opens up the possibility to use a frequency filter to extract the residual. As mentioned in the introduction of Sect. 2.2, this filtering technique was not possible before due to the high-frequency characteristic of the QRS-complexes. Now that we are excluding these parts of the signal, we are able to extract the fluctuations using a simple low-pass filter. This is shown in Fig. 4: after removing the QRS-complexes the resulting signal is filtered using a 10 Hz cutoff frequency, which is sufficient to filter out any recurring ECG features. With this method, we are able to extract the residues of almost all signals without the major errors we encountered using other methods. We keep in mind that while the cutting procedure avoids major jumps in the resulting residual for most leads, it might still cause unwanted disruptions for some (for example, very noisy) signals. However, we have no reason to assume that this would result in more positive than negative trends in autocorrelation.

3   Results

We measure the trend of the lag-1 autocorrelation in the residuals calculated using the filtering method in Sect. 2.2. The parameter setting can be found in Appendix A.6. The trend is obtained by fitting a linear function to the calculated $AR(1)$-coefficients using a least-squares method and taking the slope. Evidently, applying this method to signals that exhibit CSD should result in significant positive slopes.

To determine the statistical significance of the slope of the $AR(1)$ coefficients, we generate a distribution of $AR(1)$-slopes from 1000 surrogate time series. The surrogate time series are created by taking the Fourier transform of the residual, multiplying the computed coefficients by random phases and transforming back. In the transformation, linear properties (amplitudes) are preserved and nonlinear properties (phase angles) are randomized. This way, the power spectrum is

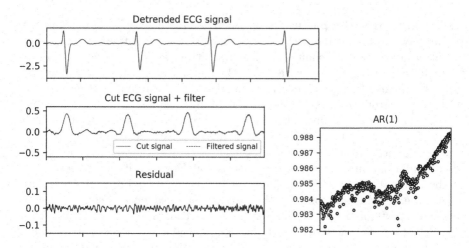

Fig. 4. ECG signal from Fig. 3 before (top) and after (middle) applying the cutting procedure. The residual (bottom) is calculated by subtracting a filtered version of the signal (dotted line). A low-pass filter with a cutoff frequency of 10 Hz is used. The relative change in width of the plots represents the portion of the signal that is cut. The resulting AR(1) measurements form a smooth curve compared to the results from methods discussed above.

preserved and the surrogate time series have the same overall autocorrelation as the original residual but are random otherwise [21]. The AR(1)-slopes of the surrogates are normally distributed. A slope is considered significant if it does not fall within the two-sided 95% confidence interval of the obtained distribution. An example of this significance test is given in Appendix A.5.

Our null hypothesis is a situation where the heart is not close to VF, and no CSD is found in the corresponding ECG data. In that case, given the significance testing method described above, we should find an equal amount of positive and negative significant trends. We represent this by letting significant positive and negative slopes be drawn from a binomial distribution with success probability 0.5. $H_0 : p = 0.5$. We reject this null hypothesis if $p > 0.5$ with a significance level of 5% (i.e. if the probability of $p = 0.5$ is lower than 5%). For cases where the null hypothesis is rejected the alternative hypothesis $H_a : p > 0.5$ is accepted and are considered to have a substantial amount of significant positive trends that may possibly be explained by CSD.

We evaluate the trend of the autocorrelation in the ECG data of four patients that suffered VF. Each data set contains around 1400 leads. In Fig. 5 the slope of the AR(1) coefficients is plotted against the power (root mean square) of the residual. Each scatter plot represents results from a different patient. Significant positive or negative AR(1)-slopes are highlighted in red. The figure shows that in three of the four cases there is a substantial majority of leads that show a significant positive trend in the lag-1 autocorrelation, compared to the number of significant negative trends. For these cases, H_0 is rejected and CSD might be at play. The corresponding scatter plots seem to have a similar shape, with the number of positive trends increasing as the power of the residual decreases. A possible explanation for this could be that residuals with high power contain a bigger proportion of measurement noise of the ECG recording, distorting the component of the residual that could contain CSD.

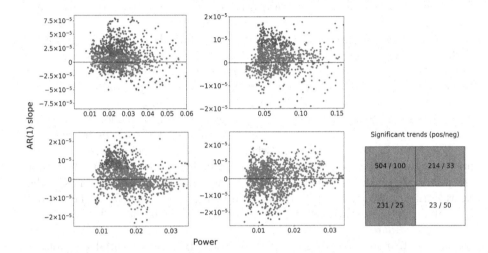

Fig. 5. Trend of the lag-1 autocorrelation in ECG residuals of four different VF events, plotted against the power of the residual. The trend is measured by the slope. Each dot represents one lead. Significant slopes are colored red. The table on the right shows the amount of significant positive/negative trends for the corresponding plot. For three out of the four patients H_0 is rejected (red cells) (Note that some points may be out of bounds for the sake of better visualization) (Color figure online)

Before we draw the conclusion that the number of significant positive slopes is in fact the result of a CSD effect we examine sets of test data that consist of ECG signals from hearts that are not close to a VF event. Because we do not have access to similar heart surface electrograms for this category we use open source ECG data from PhysioNet [22] for this purpose. We analyze 9 sets of test data consisting of 10-s samples from 24-h, 250 Hz Holter ECG recordings. As expected, we find no substantial majority of significant positive trends in any of these test sets. The results of the analysis of the test data are shown in Appendix A.7.

We also perform our analysis on the data set containing the original body surface ECGs that are used to compute the inverse solution. Strikingly, we do not observe the same substantial majority of positive trends we see for three of the four data sets of estimated heart surface ECGs. If our measurements are in fact the result of CSD, it seems this is not measurable with the original data and that the transformation to the inverse solution does provide extra information necessary to observe this effect.

An overview of all AR(1)-trend measurements is given in Appendix A.11. Given our significance testing method, under the null hypothesis (no CSD) we expect that around 5% of the autocorrelation trends we observe will be significant. Remarkably, for most data sets where H_0 is valid, we find that more than 5% of the trends are significant. It is likely that a portion of these significant trends is the result of residuals that are corrupted by noise, either directly (the noise directly influences the autocorrelation), or indirectly (the noise forces errors in the extraction of the residual, which in turn influences the autocorrelation). We can see that significant trends are relatively common for residuals with high power. These results indicate that, while artifacts can cause more significant trends than expected, they occur both in positive or negative form (see Appendix A.7), and are therefore not likely to lead to a rejection of H_0.

The sets of heart surface ECGs include triangulation coordinates, mapping each ECG signal to a point in 3D-space that corresponds to the location around the heart for which the inverse solution is calculated. We use this information to reproduce the plots from Fig. 5 where every point is color-coded based on its coordinates. These plots are shown in Appendix A.9. The results show that the points are clustered, which means that the significant positive trends in the heart surface data (red in Fig. 5) can be measured from specific angles, rather than all around the heart surface. We do not have enough information to couple the 3D-coordinates to a physical location of the heart surface; this would not be difficult to realize in future data acquisition. If the substantial amounts of significant trends are the result of CSD this can be valuable information, since one would know exactly where to look for possible early warning signals.

If, in further research, we can prove the presence of CSD in heart surface electrograms of VF victims, it may be possible for implantable devices such as ICDs to detect this effect and provide an early warning signal for an oncoming arrhythmia. If, additionally, we can isolate from which area around the heart surface it can be measured, this may even be done by using just a single lead rather than the full potential map of the heart surface we used for this experiment.

4   Conclusion and Disussion

CSD has been used as a generic early warning signal for critical transition in a wide range of systems ranging from finance to climate. We reason that the heart as a complex system may bear similarities to such systems, since, in the context of dynamical systems theory, the transition from a normal heart rhythm to VF can be understood as a shift between two states with a different attractor. We hypothesized that when the heart is close to VF (i.e., close to the basin of attraction of the chaotic attractor) it may show decreasing resilience to perturbations, which can be measured as CSD. To test this hypothesis we investigate heart surface ECG signals right before the onset of VF.

In our results we indeed find signs of CSD: for three out of four VF victims, we find a substantial majority of significant positive autocorrelation trends in the residuals of heart surface ECG signals compared to the amount of significant negative trends. The heart surface ECGs are estimated by solving an inverse problem using body surface ECGs. If we perform the same analysis on the original body surface data we do not find such a majority, suggesting that, if CSD is in fact present, we have to compute the inverse solution to observe this effect. Furthermore, triangulation coordinates of the heart surface ECGs suggest that the possible CSD effect can only be measured from specific, yet unspecified angles surrounding the heart. We compare the results of the heart surface ECGs of VF victims to results from Holter recordings of subjects that are not close to a VF event. For the latter (no VF) data, we find that none of the nine recordings we analyze contain a substantial amount of significant positive autocorrelation trends and thus exhibit no CSD.

It has become clear, however, that this test data does not serve as an appropriate comparison the VF data for a number of reasons. Firstly, the test signals have a lower sampling rate of 250 Hz compared to the 1000 Hz of the VF data. It is possible that fluctuations in the residuals that show CSD can be captured by a sampling rate of 1000 Hz but not by a sampling rate of 250 Hz. If this is in fact the case, using a sampling rate 250 Hz can not prove the absence of CSD in the test data. Secondly, with Holter recordings, the electrodes are placed on the chest of the patient. If we assume that the substantial amount of significant positive trends in the heart surface data is caused by CSD, our analysis of the data used to compute the inverse solution already seems to indicate that this effect is not as easily measured on the body surface. It would therefore only be logical if the test also does not show CSD. Lastly, the triangulation coordinates of the heart surface ECGs suggest that, if the data shows CSD, it can probably only be measured from specific angles surrounding the heart surface. Since the test data consists Holter recordings with only one lead, and thus only one angle of incidence, it is already unlikely to measure CSD for these signals.

The test data we use serves more as a validation of our method, rather than a validation of our result. While we do not expect CSD, the test signals are prone to the same types of noise as the VF data, which could influence autocorrelation measurements. Our analysis has shown, however, that this does not lead to a substantial majority of significant positive trends in any of the test data sets. This could indicate that the signs of CSD we find for the VF data are no artifacts of measurement noise. On the other hand, we can argue that some types of measurement noise (for example, noise caused by the movement of a patient) could affect multiple leads at once for the heart surface data since each lead covers that same ten-second time span. For this reason, the test data is still not sufficient to rule out the possibility that measurement noise influences our results.

To prove that the predominant amount of significant positive autocorrelation trends we find in the residuals of heart surface ECGs of patients that suffered VF is in fact the result of CSD, we have to directly compare it to data of subjects that did not suffer VF, recorded in a similar manner. We therefore advocate further data acquisition. An important first step would be to obtain the 1400-lead heart surface ECG of subjects that did not suffer VF by solving the inverse problem using the body surface potentials. If the analysis of such data does not result in a majority of significant positive trends, this would be a strong indication that our measurements are showing CSD. Additional data from patients that suffered VF would also enable better statistical grounding. For future ECG recordings it is important to be able to map each signal to an exact physical location so that, if we can indeed prove CSD, we can also pinpoint angles from which this effect is measurable. So far we have checked for CSD by looking for a buildup in autocorrelation in the 10 s prior to the VF event. However, it is possible that this buildup initiates at an earlier time. If we can indeed use CSD as an early warning signal in this setting, it would be valuable to detect this as early as possible. For future data sets, it might therefore be useful to record ECG even longer before the onset of VF, where possible.

We concluded that for the extraction of the residuals from the signals it is best practice to remove parts of the signal with high first-order differences, which are otherwise hard to filter out. Evidently, for future research it is desirable to develop a method that can correctly extract residuals for the full signal, for example by improving the average beat method proposed in Sect. 2.2 or developing more advanced modeling techniques than the Gaussian fitting method in Sect. 2.2.

Acknowledgements. This work was partly supported by the Fondation Leducq Transatlantic Network of Excellence 16CVD02. RQ thanks EU Horizon 2020 project TO_AITION (848146) for support.

A   Appendix

The full version of the paper including the appendix is available online at: https://tinyurl.com/ukjfwuc.

References

1. Amann, A., Tratnig, K., Unterkofler, K.: Reliability of old and new ventricular fibrillation detection algorithms for automated external defibrillators. Biomed. Tech./Biomed. Eng. **48**, 216–217 (2003)
2. Ahmad, M., Bloomstein, L.Z., Roelke, M., Bernstein, A.D., Parsonnet, V.: Patients' attitudes toward implanted defibrillator shocks. Pacing Clin. Electrophysiol. **23**(6), 934–938 (2000)
3. Sears, S.F., Conti, J.B.: Quality of life and psychological functioning of ICD patients. Heart **87**(5), 488–493 (2002)
4. Lane, R.: Prediction and prevention of sudden cardiac death in heart failure. Heart **91**(5), 674–680 (2005)
5. Reed, M., Robertson, C., Addison, P.: Heart rate variability measurements and the prediction of ventricular arrhythmias. QJM **98**(2), 87–95 (2005)
6. Scheffer, M., et al.: Early-warning signals for critical transitions. Nature **461**(7260), 53–59 (2009)
7. Wissel, C.: A universal law of the characteristic return time near thresholds. Oecologia **65**(1), 101–107 (1984)

8. Scheffer, M., Dakos, V., Nes, E., Brovkin, V., Petoukhov, V., Held, H.: Slowing down as an early warning signal for abrupt climate change. Proc. Natl. Acad. Sci. **105**(38), 14308–14312 (2008)
9. McSharry, P., Smith, L., Tarassenko, L.: Prediction of epileptic seizures: are nonlinear methods relevant? Nat. Med. **9**(3), 241–242 (2003)
10. Goldberger, A., Bhargava, V., West, B., Mandell, A.: Some observations on the question: is ventricular fibrillation "chaos"? Phys. D **19**(2), 282–289 (1986)
11. Surawicz, B.: Ventricular fibrillation. Am. J. Cardiol. **28**(3), 268–287 (1971)
12. Kaplan, D., Cohen, R.: Is fibrillation chaos? Circ. Res. **67**(4), 886–892 (1990)
13. Garfinkel, A., et al.: Quasiperiodicity and chaos in cardiac fibrillation. J. Clin. Invest. **99**(2), 305–314 (1997)
14. Wit, A., Cranefield, P.: Reentrant excitation as a cause of cardiac arrhythmias. Am. J. Physiol. Heart Circ. Physiol. **235**(1), H1–H17 (1978)
15. Weiss, J., et al.: The dynamics of cardiac fibrillation. Circulation **112**(8), 1232–1240 (2005)
16. Weiss, J., Garfinkel, A., Karagueuzian, H., Qu, Z., Chen, P.: Chaos and the transition to ventricular fibrillation. Circulation **99**(21), 2819–2826 (1999)
17. Qu, Z.: Chaos in the genesis and maintenance of cardiac arrhythmias. Prog. Biophys. Mol. Biol. **105**(3), 247–257 (2011)
18. Qu, Z., Hu, G., Garfinkel, A., Weiss, J.: Nonlinear and stochastic dynamics in the heart. Phys. Rep. **543**(2), 61–162 (2014)
19. Gulrajani, R.: The forward and inverse problems of electrocardiography. IEEE Eng. Med. Biol. Mag. **17**(5), 84–101, 122 (1998)
20. Suppappola, S., Sun, Y., Chiaramida, S.: Gaussian pulse decomposition: an intuitive model of electrocardiogram waveforms. Ann. Biomed. Eng. **25**(2), 252–260 (1997)
21. Schreiber, T., Schmitz, A.: Surrogate time series. Phys. D **142**(3–4), 346–382 (2000)
22. Goldberger, A., et al.: PhysioBank, PhysioToolkit, and PhysioNet. Circulation **101**(23), e215–e220 (2000)

A Comparison of Generalized Stochastic Milevsky-Promislov Mortality Models with Continuous Non-Gaussian Filters

Piotr Śliwka[(⊠)] and Leslaw Socha

Department of Mathematics and Natural Sciences, Cardinal Stefan Wyszynski University, Woycickiego 1/3, 01-938 Warsaw, Poland
{p.sliwka,l.socha}@uksw.edu.pl

Abstract. The ability to precisely model mortality rates $\mu_{x,t}$ plays an important role from the economic point of view in healthcare. The aim of this article is to propose a comparison of the estimation of the mortality rates based on a class of stochastic Milevsky-Promislov mortality models. We assume that excitations are modeled by second, fourth and sixth order polynomials of outputs from a linear non-Gaussian filter. To estimate the model parameters we use the first and second moments of $\mu_{x,t}$. The theoretical values obtained in both cases were compared with theoretical $\widehat{\mu_{x,t}}$ based on a classical Lee-Carter model. The obtained results confirm the usefulness of the switched model based on the continuous non-Gaussian processes used for modeling $\mu_{x,t}$.

Keywords: Forecasting of mortality rates · Hybrid mortality models · Switched models · Lee-Carter model · Ito stochastic differential equations

1 Introduction

The determination of the mortality models is one of the basic problems not only in the field of life insurance but recently particularly in the economics of healthcare [1,3–5]. Currently, the most frequently used model is Lee-Carter [9,11, 12,16,17] not only the change in mortality associated with age x and calendar year t but also takes into account the influence of belonging to a particular generation (cohort effect) and takes the form: $ln(\mu_{x,t}) = \alpha_x + \beta_x k_t + \varepsilon_{x,t}$. The assumption that the estimated a_x and b_x are fixed at time t causes a wave of criticism, especially from the point of view of forecasting. Therefore, there is a need to look for other methods predicting mortality rates that take into account the variability of parameters over time. One of these propositions may be the approach recently proposed in Rossa and Socha [18], Rossa et al. [19], Sliwka and Socha [20], Sliwka [21,22], and based on the Milevsky-Promislov family of models [15] with extensions [2,7,13]. Methods of modeling $\mu_{x,t}$ taking into account the causes of death have been characterized in article [23].

© Springer Nature Switzerland AG 2020
V. V. Krzhizhanovskaya et al. (Eds.): ICCS 2020, LNCS 12140, pp. 348–362, 2020.
https://doi.org/10.1007/978-3-030-50423-6_26

Works on switchings model, which consist of several subsystems with the same structure and different parameters and which can switch over time according to an unknown switching rule, have been taken, among others in Sliwka and Socha [20]. In the mentioned paper it was shown that modeling of empirical mortality coefficients $\mu_{x,t}$ using the non-Gaussian linear scalar filters model second order with switchings (nGLSFo2s) allows a more precise estimate of $\mu_{x,t}$ than using the Gaussian linear scalar filters model with switchings (GLSFs) and Lee-Carter with switchings (LCs) model for some fixed ages x.

In this paper we first propose three extended Milevsky and Promislov models with continuous non-Gaussian filters. We assume that excitations are modeled not only by the second but also by the fourth (nGLSFo4) and the sixth order polynomials (nGLSFo6) of outputs from a linear nGLSF. To estimate the model parameters we use the first and second moments of mortality rates. We show that in considered models some of the parameters can be estimated. Next, we use these models to create hybrid models, where submodels have the same structure and possible different parameters. To estimate the model parameters we use the first and second moments of mortality rates. According to our knowledge, the mortality models proposed above, their hybrid versions and methods for estimating their parameters and switching are new in the field of life insurance.

The paper is organized as follows. In Sect. 2 basic notations and definitions of stochastic hybrid systems are entered. Three new basic models represented by even-order polynomials of outputs from linear Gaussian filter are introduced and the non–stationary solutions of corresponding moment equations are presented in Sect. 3. The derivation of these non–stationary solutions are derived in Appendix. In Sect. 4 the procedure of the parameters estimation and determination of switching points is presented. Based on the adapted numerical algorithm of a nonlinear minimization problem, parameter estimation is performed. In Sect. 5 we have compared empirical mortality rates with theoretical ones obtained from proposed models as well as from standard LC model in two versions with switchings and without switchings. The last Section summarizes the obtained results.

2   Mathematical Preliminaries

Throughout this paper we use the following notation. Let $|\cdot|$ and $<\cdot>$ be the Euclidean norm and the inner product in $\mathbb{R}^n$, respectively. We mark $\mathbb{R}_+ = [0, \infty)$, $\mathbb{T} = [t_0, \infty)$, $t_0 \geq 0$. Let $\Xi = (\Omega, \mathcal{F}, \{\mathcal{F}_t\}_{t \geq 0}, \mathbb{P})$ be a complete probability space with a filtration $\{\mathcal{F}_t\}_{t \geq 0}$ satisfying usual conditions. Let $\sigma(t) : \mathbb{R}_+ \to \mathbb{S}$ be the switching rule, where $\mathbb{S} = \{1, \ldots, N\}$ is the set of states. We denote switching times as $\tau_1, \tau_2, \ldots$ and assume that there is a finite number of switches on every finite time interval. Let $W_k(t)$ be the independent Brownian motions. We assume that processes $W_k(t)$ and $\sigma(t)$ are both $\{\mathcal{F}_t\}_{t \geq 0}$ adapted.

By the stochastic hybrid system we call the vector Itô stochastic differential equations with a switching rule described by

$$dx(t) = \mathbf{f}(\mathbf{x}(t), t, \sigma(t))dt + \mathbf{g}(\mathbf{x}(t), t, \sigma(t))dW(t), \ (\sigma(t_0), x(t_0)) = (\sigma_0, x_0), \quad (1)$$

where $x \in R^n$ is the state vector, (σ_0, x_0) is an initial condition, $t \in T$ and M is a number of Brownian motions. $f(x(t), t, \sigma(t))$ and $g(x(t), t, \sigma(t))$ are defined by sets of $f(x(t), t, l)$ and $g(x(t), t, l)$, respectively i.e.

$$f(x(t), t, \sigma(t)) = f(x(t), t, l), \quad g(x(t), t, \sigma(t)) = g(x(t), t, l) \quad \text{for } \sigma(t) = l.$$

Functions $f : R^n \times T \times S \to R^n$ and $g : R^n \times T \times S \to R^n$ are locally Lipschitz and such that $\forall l \in S, t \in T, f(\mathbf{0}, t, l) = g(\mathbf{0}, t, l) = \mathbf{0}, k = 1, \dots, M$. These conditions together with these enforced on the switching rule σ ensure that there exists a unique solution to the hybrid system (1).

Hence it follows that Eq. (1) can be treated as a family (set) of subsystems defined by

$$d\mathbf{x}(t, l) = \mathbf{f}(\mathbf{x}(t), t, l)dt + \sum_{k=1}^{M} \mathbf{g}_k(\mathbf{x}(t), t, l)dW_k(t), \quad l \in S \qquad (2)$$

where $\mathbf{x}(t, l) \in \mathbb{R}^n$ is the state vector of l- subsystem.

We assume additionally that the trajectories of the hybrid system are continuous. It means, when the stochastic system is switched from l_1 subsystem to l_2 subsystem in the moment τ_j, then

$$\mathbf{x}(\tau_j, l_1) = \mathbf{x}(\tau_j, l_2), \quad l_1, l_2 \in S. \qquad (3)$$

3 Models with Continuous Non-Gaussian Linear Scalar Filters

We consider a family of mortality models with a continuous nGLSF described by

$$\mu_x(t, l) = \mu_{x0}^l \exp\{\alpha_x^l t + \sum_{i=1}^{m} q_{x_i}^l y^i(t, l)\}, \qquad (4)$$

$$dy(t, l) = -\beta_{x_1}^l y(t, l)dt + \gamma_{x_1}^l dW(t), \qquad (5)$$

where $\mu_x(t, l)$ is a stochastic process representing a mortality rate for a person aged x ($x \in X = 0, 1, \dots, \omega$) at time t; $\alpha_x^l, \beta_{x_1}^l, q_{x_i}^l, i = 1, \dots, m, \mu_{x_0}^l, \gamma_{x_1}^l$ are constant parameters, $l \in S$; $W(t)$ is a standard Wiener process.

We will show that the proposed model (4), (5) can be transformed to the formula (2) for all $l \in S$.

Introducing new variables $y_1(t, l) = y(t, l)$, $y_i(t, l) = y^i(t, l)$, $i = 1, \dots m$, $l \in S$ and applying Ito formula we obtain

$$dy_1(t, l) = -\beta_{x_1}^l y_1(t, l)dt + \gamma_{x_1}^l dW(t), \qquad (6)$$

$$dy_2(t,l) = [-2\beta_{x_1}^l y_2(t,l) + (\gamma_{x_1}^l)^2]dt + 2\gamma_{x_1}^l y_1(t,l)dW(t), \tag{7}$$

$$\vdots$$

$$dy_m(t,l) =$$
$$\left[-m\beta_{x_1}^l y_m(t,l) + \frac{m(m-1)}{2}(\gamma_{x_1}^l)^2 y_{m-2}(t,l)\right] dt + m\gamma_{x_1}^l y_{m-1}(t,l)dW(t).$$

Taking natural logarithm of both sides of Eq. (4) and applying Ito formula for all $l \in \mathbb{S}$ we find

$$d\ln\mu_x(t,l) = \alpha_x^l + \sum_{i=1}^{m} q_{x_i}^l dy_i(t,l) = \alpha_x^l - \sum_{i=1}^{m}\left[i\beta_{x_1}^l q_{x_i}^l y_i(t,l)\right] \tag{8}$$

$$+ \frac{i(i-1)}{2}q_{x_i}^l(\gamma_{x_1}^l)^2 y_i(t,l)^{i-2}\right] dt + \sum_{i=1}^{m} i\gamma_{x_1}^l q_{x_i}^l y_{i-1}(t,l)dW(t) \tag{9}$$

Now we consider in details three cases of model (4) and (6)–(7), namely for $m = 2$, 4 and 6.

3.1   Model with Six Order Output of a Scalar Linear Filter

Equations (4) and (6)–(7) for $m = 6$ take the form

$$\mu_x(t,l) = \mu_{x0}^l \exp\{\alpha_x^l t + \sum_{i=1}^{6} q_{x_i}^l y^i(t,l)\}, \tag{10}$$

$$dy_1(t,l) = -\beta_{x_1}^l y_1(t,l)dt + \gamma_{x_1}^l dW(t), \tag{11}$$

$$dy_2(t,l) = [-2\beta_{x_1}^l y_2(t,l) + (\gamma_{x_1}^l)^2]dt + 2\gamma_{x_1}^l y_1(t,l)dW(t), \tag{12}$$

$$dy_3(t,l) = [-3\beta_{x_1}^l y_3(t,l) + 3(\gamma_{x_1}^l)^2 y_1(t,l)]dt + 3\gamma_{x_1}^l y_2(t,l)dW(t), \tag{13}$$

$$dy_4(t,l) = [-4\beta_{x_1}^l y_4(t,l) + 6(\gamma_{x_1}^l)^2 y_2(t,l)]dt + 4\gamma_{x_1}^l y_3(t,l)dW(t), \tag{14}$$

$$dy_5(t,l) = [-5\beta_{x_1}^l y_5(t,l) + 10(\gamma_{x_1}^l)^2 y_3(t,l)]dt + 5\gamma_{x_1}^l y_4(t,l)dW(t), \tag{15}$$

$$dy_6(t,l) = [-6\beta_{x_1}^l y_6(t,l) + 15(\gamma_{x_1}^l)^2 y_4(t,l)]dt + 6\gamma_{x_1}^l y_5(t,l)dW(t). \tag{16}$$

Introducing a new vector state

$$\mathbf{z}_x(t,l) = [z_{x_1}(t,l), z_{x_2}(t,l), \cdots, z_{x_7}(t,l)]^T = \tag{17}$$

$$[\ln\mu_x(t,l), y_1(t,l), y_2(t,l), y_3(t,l), y_4(t,l), y_5(t,l), y_6(t,l)]^T, \tag{18}$$

Equations (10)–(16) one can rewrite in a vector form

$$d\mathbf{z}_x(t,l) = \left[\mathbf{A}_x^6(l)\mathbf{z}_x(t,l) + \mathbf{b}_x^6(l)\right] dt + \left[\mathbf{C}_x^6(l)\mathbf{z}_x(t,l) + \mathbf{g}_x^6(l)\right] dW(t) \tag{19}$$

where

$$\mathbf{A}_x^6(l) = [a_{ij}^l], \mathbf{b}_x^6(l) = [b_i^l], \mathbf{C}_x^6(l) = [c_{ij}]^l, \mathbf{g}_x^6(l) = [g_i^l]. \tag{20}$$

The elements of the matrices $\mathbf{A}_x^6(l), \mathbf{C}_x^6(l)$ and vectors $\mathbf{b}_x^6(l), \mathbf{g}_x^6(l)$ are defined by:

$$a_{12}^l = -\beta_{x_1}^l q_{x_1}^l + 3q_{x_3}^l (\gamma_{x_1}^l)^2, a_{13}^l = -2\beta_{x_1}^l q_{x_2}^l + 6q_{x_4}^l (\gamma_{x_1}^l)^2,$$

$$a_{14}^l = -3\beta_{x_1}^l q_{x_3}^l + 10q_{x_5}^l (\gamma_{x_1}^l)^2, a_{15}^l = -4\beta_{x_1}^l q_{x_4}^l + 15q_{x_6}^l (\gamma_{x_1}^l)^2,$$

$$a_{16}^l = -5\beta_{x_1}^l q_{x_5}^l, a_{17}^l = -6\beta_{x_1}^l q_{x_6}^l, a_{22}^l = -\beta_{x_1}^l, a_{33}^l = -2\beta_{x_1}^l, a_{42}^l = 3(\gamma_{x_1}^l)^2,$$

$$a_{44}^l = -3\beta_{x_1}^l, a_{53}^l = 6(\gamma_{x_1}^l)^2, a_{55}^l = -4\beta_{x_1}^l, a_{64}^l = 10(\gamma_{x_1}^l)^2, a_{66}^l = -5\beta_{x_1}^l,$$

$$a_{75}^l = 15(\gamma_{x_1}^l)^2, a_{77}^l = -6\beta_{x_1}^l, b_1^l = \alpha_x^l + q_{x_2}^l (\gamma_{x_1}^l)^2, b_3^l = (\gamma_{x_1}^l)^2,$$

$$c_{12}^l = 2q_{x_2}^l \gamma_{x_1}^l, c_{14}^l = 4q_{x_4}^l \gamma_{x_1}^l, c_{15}^l = 5q_{x_5}^l \gamma_{x_1}^l, c_{16}^l = 6q_{x_6}^l \gamma_{x_1}^l, c_{32}^l = 2\gamma_{x_1}^l,$$

$$c_{43}^l = 3\gamma_{x_1}^l, c_{54}^l = 4\gamma_{x_1}^l, c_{65}^l = 5\gamma_{x_1}^l, c_{76}^l = 6\gamma_{x_1}^l,$$

$$g_1^l = q_{x_1}^l \gamma_{x_1}^l, g_2^l = \gamma_{x_1}^l.$$

We note that similarly to Eq. (2) we may treat Eq. (19) as a family (set) of subsystems. It means, we have obtained new mortality hybrid model. The unknown parameters in family of Eq. (19) are

$$\ln \mu_{x_0}^l (= \alpha_{0_x}^l), \alpha_x^l, \beta_{x_1}^l, q_{x_1}^l, q_{x_2}^l, q_{x_3}^l, q_{x_4}^l, q_{x_5}^l, q_{x_6}^l, \gamma_{x_1}^l. \tag{21}$$

3.2   Nonstationary Solutions

Using linear vector stochastic differential Eq. (19) and Ito formula we derive differential equations for the first order moments $E[z_{x_i}(l)]$ and second order moments $E[z_{x_i}(l)z_{x_j}(l)]$, $i, j = 1, ..., 7$. Next, we find the nonstationary solutions of the first moment of the processes $z_{x_i}(t, l)$ for nGLSF of all order models, i.e. (nGLSFo1), (nGLSFo2), ... (nGLSFo6) models

$$E[z_{x_1}(t,l)] = \alpha_x^l t + \alpha_{0_x}^l, \quad l \in \mathbb{S} \tag{22}$$

In the case of second moment of the processes $z_{x_i}(t, l)$ we find first the nonstationary solutions for nGLSF even order models. In the case of sixth order model it has the form

$$E[z_{x_1}^2(t,l)] = (\alpha_x^l)^2 t^2 - 2\alpha_x^l \left[-\alpha_{0_x}^l + q_{x_2}^l \frac{(\gamma_{x_1}^l)^2}{2\beta_{x_1}^l} \right. \tag{23}$$

$$\left. +3q_{x_4}^l \left(\frac{(\gamma_{x_1}^l)^2}{2\beta_{x_1}^l} \right)^2 + 15q_{x_6}^l \left(\frac{(\gamma_{x_1}^l)^2}{2\beta_{x_1}^l} \right)^3 \right] t + c_{0_x}^l \tag{24}$$

where $l \in \mathbb{S}$, $q_{x_2}^l = q_{x_4}^l = q_{x_6}^l = 1$, $q_{x_2}^l = q_{x_4}^l = q_{x_6}^l = 1$, and $\alpha_{0_x}^l, c_{0_x}^l$ are constants of integration (see Sect. A).

To obtain the moment equations for nGLSF second and fourth order models and the corresponding stationary and nonstationary solutions we assume that:

- in the case of second order model the parameters $q_{x_2}^l = 1$, and $q_{x_4}^l = q_{x_6}^l = 0$,
- in the case of fourth order model the parameters $q_{x_2}^l = q_{x_4}^l = 1$, and $q_{x_6}^l = 0$.

The corresponding nonstationary solution for the second moment of the process $z_{x_1}(t, l)$ takes the form:

$$E[z_{x_1}^2(t,l)] = (\alpha_x^l)^2 t^2 + 2\alpha_x^l \alpha_{0_x}^l t - 2\alpha_x^l q_{x_2}^l pt + c_{0_x}^l \tag{25}$$

for nGLSF second order model, where $c_{0_x}^l$ is an integration constant, and

$$E[z_{x_1}^2(t,l)] = (\alpha_x^l)^2 t^2 - 2\alpha_x^l \left[-\alpha_{0_x}^l + q_{x_2}^l p + 3q_{x_4}^l p^2 \right] t + c_{0_x}^l \tag{26}$$

for nGLSF fourth order model, where $c_{0_x}^l$ is an integration constant and $p = \frac{(\gamma_{x_1}^l)^2}{2\beta_{x_1}^l}$.

It can be proved that in the case odd order models the nonstationary solutions have similar forms, i.e. in the case of the first order (nGLSFo1) model

$$E[z_{x_1}^2(t,l)] = (\alpha_x^l)^2 t^2 + 2\alpha_x^l \alpha_{0_x}^l t + c_{0_x}^l \tag{27}$$

and in the case of other odd order models, i.e. (nGLSFo3), (nGLSFo5), (nGLSFo7) models the nonstationary solutions are the same as the nonstationary solutions for nGLSF even order models, i.e. (nGLSFo2), (nGLSFo4), (nGLSFo6) models, respectively.

4 The Procedure of the Parameters Estimation and Determination of Submodels (based on Switching Points)

4.1 The Procedure

Simultaneous estimation of parameters: $\alpha_0^l, \alpha_x^l, \beta_x^l, \gamma_x^l, c_{0_x}^l, q_x^l$ (where $l \in \mathbb{S}$) nGLSF models of 2, 4 or 6 order given by formulas (22)–(26) using traditional methods does not provide unambiguous results (this problem has already been considered in [20] part 4.1.1, in particular by considering the analytical formula for estimating parameters of GLSFo2 model). Therefore, in this case, a two-step procedure was used to estimate the parameters. In the first step, the α_{0_x} and α_x of the first moment $E[z_{x_1}(t)]$ of the process $z_{x_1}(t)$ were estimated. In the second step, c_{0_x} and p_x of the second moment $E[z_{x_1}^2(t)]$ were estimated based on the already known $\widehat{\alpha_{0_x}^l}, \widehat{\alpha_x^l}$, where p_x was defined as follows: $p_x = \frac{\gamma_{x_1}^2}{2\beta_{x_1}}$. The applied procedure allows to obtain unambiguous estimates of all parameters assuming that $q_{x_i} = 1, \forall_{i=1,\dots,6}$.

One of the fundamental problems in the field of switching models is to find the set of switching points. This problem is closely related to the problem of segmentation of a time series discussed in many papers (see for instance [10, 14]).

In our considerations we propose a procedure which is a combination of a statistical test (based on [6]) and so called Top-Down algorithm. It has the following form.

First we introduce some notations. We assume that an extracted time series (Input) consists of n empirical values $y_{emp_1}, y_{emp_2}, \ldots, y_{emp_n}$ defined in time points $t_1, t_2, \ldots, t_n$, respectively. By $<t_1, t_2>$ we denote an interval that begins at t_1 and ends in t_2. We define three sets

$\mathcal{P}$ - the set of non-verified intervals,
$\mathcal{R}$ - the set of intervals without switching points,
$\mathcal{T}$ - the set of switching points.

Then the initial conditions have the form

$$<t_1, t_n> \in \mathcal{P}, \quad \mathcal{R} = \phi, \quad \mathcal{T} = \phi.$$

Step 1

We calculate the values of function $L(*)$ given by formula (28)

$$L(\alpha^l_{0_x}, \alpha^l_x, \sigma^2, \tau) = -\frac{\tau}{2} ln(2\pi) - \frac{\tau}{2} ln(\sigma^2) - \frac{1}{2\sigma^2} \sum_{i=1}^{\tau} (y_{emp_i} - E[z_{x_1}(i,l)])^2$$

$$-\frac{n-\tau}{2} ln(2\pi) - \frac{n-\tau}{2} ln(\sigma^2) - \frac{1}{2\sigma^2} \sum_{i=\tau+1}^{n} (y_{emp_i} - E[z_{x_1}(i,l)])^2 \quad (28)$$

for all points from an interval $<t_1, t_n>$ and assuming the random component $\epsilon_t \sim N(\mu, \sigma^2)$.

If $\tilde{L}(\tau_1) = maxL(*)$ is found at the beginning or at the end of the considered interval, then there is not a switching point in this interval. Then we receive

$$\mathcal{P} = \phi, \quad <t_1, t_n> \in \mathcal{R}, \quad \mathcal{T} = \phi,$$

If $\tilde{L}(\tau_1) = maxL(*)$ is found inside the interval for $\tau_1 = t_k$, then

$$<t_1, t_k>, <t_{k+1}, t_n> \in \mathcal{P}, \quad \mathcal{R} = \phi, \quad \tau_1 \in \mathcal{T}.$$

Step 2

Choose an interval from the set $\mathcal{P}$ and check if its length is greater than 2.

Step 3

If "no", then transfer this interval from the set $\mathcal{P}$ to the set $\mathcal{R}$ and go back to **Step 2**, if "yes", go back to **Step 1**.

Step 4

The procedure is ended when
$\mathcal{P} = \phi$,
$\mathcal{R}$ consists only with subintervals without switching points,
$\mathcal{T}$ consists of all switching points that can be sorted from the smallest to the greatest one.

4.2   The Determination of Submodels

In Subsect. 4.1 we have established the switching points set, which allow to define submodels. From (21) and further considerations we find that unknown parameters in family of (19) are

$$\alpha_{0_x}^l, \alpha_x^l, p^l, c_{0_x}^l \tag{29}$$

where $p^l = \frac{(\gamma_{x_1}^l)^2}{2\beta_{x_1}^l}$, and parameters $q_{x_2}, q_{x_4}, q_{x_6}$ are equal 0 or 1.

Based on the numerical algorithm of nonlinear minimization with additional conditions of $\alpha_{0_x}^l$ ($\forall x \; \alpha_{0_x}^l < 0$) parameters (29) given in the formula (23)–(26) were assessed. The algorithm works by generating a population of random starting points and next uses a local optimization method from each of the starting points to converge to a local minimum. As the solution, the best local minimum was chosen.

For a fixed sex, fixed age x, and knowing the switching points (designated in accordance with the procedure described above) two sets of time series of $\widehat{\mu_{x,t}}$ values were created. In the first case, the estimation of $\widehat{\mu_{x,t}}$ was based on empirical data from 1958–2010 (using the next 6 years for ex-post error evaluation). Similar estimation based on the years 1958–2016 was done in the second case. In both cases the choice of the theoretical value $\widehat{\mu_{x,t}}$ at a fixed moment t from the theoretical values of the models (nGLSFo2), (nGLSFo4) and (nGLSFo6) was based on minimization of the absolute error (AE), i.e.

$$\min_{i=2,4,6} |\widehat{\mu_{x,t}}^{nGLSFo_i} - \mu_{x,t}|.$$

In addition, point forecasts for the period 2017–2025 have been determined. The parameters for the Lee-Carter model with switchings were estimated based on the formulas given in the literature [11] and using the same set of switches as in the case of the nGLSF model.

We note that the hybrid model (19) is continuous. However, the moment equations of the first and second-order defined by (22)–(27) are not continuous in switching points because the empirical data of mortality rates we have used were discrete, and these moments are determined separately for every submodel.

5   Results

Selected results for a 45-year old and a 60-year old woman and man presented in Figs. 1 and 2 (source of empirical data: [8]). In Figs. 1 and 2, blue circular points indicate empirical data, red, black and green solid lines indicate the theoretical values of the models: Lee-Carter (LCs), nGLSF order 2 (nGso2) and nGLSF mixed order 2,4, and 6 (nGs) with switchings respectively, while the solid purple line indicates the forecast of the nG (nGsf) for the next five years.

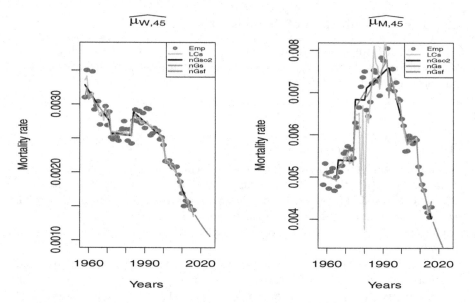

Fig. 1. Mortality rates for women (left side) and men (right side) aged 45 and empirical, theoretical values based on the following models: LCs, nGs, and forecasts (Color figure online)

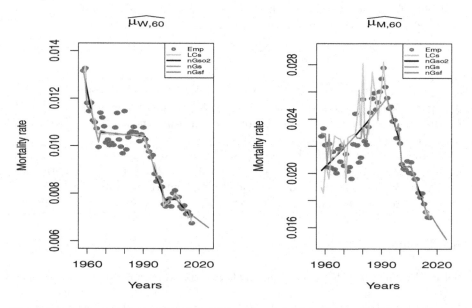

Fig. 2. Mortality rates for women (left side) and men (right side) aged 60 and empirical, theoretical values based on the following models: LCs, nGs, and forecasts (Color figure online)

To verify the goodness of fit of the proposed nGs models with switchings to the empirical mortality rates and compared with Lee-Carter model the mean squared errors (MSE) between empirical mortality $\mu_{x,t}$ and theoretical values $\widehat{\mu_{x,t}}$ in the years 1958–2010 ('10) and 1958–2016 ('16) as well as the 95% confidence interval for MSE has been calculated. Selected results (45 and 60-year old female and male) are presented in Table 1 (where: CI_L-lower -, CI_U-upper confidence interval, $\{W, M\}_{X,MSE}$ - MSE value for {female, male} aged X). The results in column 5th illustrate the model (nGso2) considered in [20].

Table 1. Goodness of fit measures (woman-W, man-M) based on MSE.

Sex & Age	EMP-LCs$^{'10}$	EMP-nGs$^{'10}$	EMP-LCs$^{'16}$	EMP-nGso2$^{'16}$	EMP-nGs$^{'16}$
W_{45,CI_L}	7.145E−09	8.373E−09	9.468E−09	9.296E−09	7.380E−09
$W_{45,MSE}$	1.010E−08	1.190E−08	1.320E−08	1.290E−08	1.030E−08
W_{45,CI_U}	1.541E−08	1.806E−08	1.960E−08	1.925E−08	1.528E−08
W_{60,CI_L}	8.639E−08	8.983E−08	8.351E−08	9.293E−08	8.190E−08
$W_{60,MSE}$	1.220E−07	1.270E−07	1.160E−07	1.290E−07	1.140E−07
W_{60,CI_U}	1.863E−07	1.937E−07	1.729E−07	1.924E−07	1.696E−07
M_{45,CI_L}	3.176E−07	5.247E−08	2.882E−07	7.714E−08	4.546E−08
$M_{45,MSE}$	4.490E−07	7.430E−08	4.010E−07	1.070E−07	6.330E−08
M_{45,CI_U}	6.851E−07	1.132E−07	5.967E−07	1.597E−07	9.412E−08
M_{60,CI_L}	1.995E−06	8.113E−07	1.827E−06	9.966E−07	6.578E−07
$M_{60,MSE}$	2.820E−06	1.148E−06	2.540E−06	1.390E−06	9.160E−07
M_{60,CI_U}	4.303E−06	1.750E−06	3.784E−06	2.063E−06	1.362E−06

MSE values calculated on the basis of empirical and theoretical data from 1958–2016 and included in Table 1 and Figs. 1 and 2 provide the following conclusions:

- the theoretical values of the mortality rate $\widehat{\mu_{x,t}^{nGs}}$ based on the non-Gaussian linear scalar filters with switching provide closer estimates to empirical values than $\widehat{\mu_{x,t}}^{LCs}$ based on LC model and $\widehat{\mu_{x,t}}^{nGso2}$ with switching for both a 45-year-old and a 60-year-old woman and man,
- the range confidence interval is the smallest for the nGs model compared to all other models given in Table 1, which means greater precision of the proposed nGs for forecasting than the other models presented here,
- the empirical mortality rates for women are more accurately fitted using the proposed nGs model than for men (lower MSE value),
- based on graphical results (Fig. 1–Fig. 2), it can be seen that the proposed method of modeling $\mu_{x,t}$ using nGs more precisely adapts to empirical data, especially for data with a large variance than the LC model (e.g. see empirical data from 1980–1990 for a 60-year-old man on Fig. 2, right side).

Moreover, taking into account all results for people aged $x = 0, \ldots, 100$ years (also partly included in Table 1) it can be seen that the proposed nGs model fits more accurately to the empirical data for younger than older (lower MSE for 45 years old than for 60 years old man and woman).

6   Conclusions

In this paper, three extended Milevsky and Promislov models with excitations modeled by the second, the fourth and the sixth order polynomials of outputs from a linear non-Gaussian filter are proposed and adopted to Polish mortality data. To obtain hybrid models the procedures of parameters estimation and the determination of switching points were proposed. Based on the theoretical values obtained from these three models, one series of theoretical values based on the AE criterion was constructed and compared with the theoretical mortality rates based on classical the Lee–Carter model. In addition, a point forecast was computed. The obtained results confirm the usefulness of the switched model based on the continuous non-Gaussian process for modeling mortality rates.

A natural extension of the research contained in this article is the Markov chain application (homogeneous or heterogeneous), which will be used to describe the space of states built on extended Milevsky and Promislov models with excitations modeled by the second, the fourth and the sixth order polynomials. The issues discussed above will be examined in the next article.

A   Appendix

The derivation of stationary and nonstationary solutions of moment equations in nGLSF six order model
Using linear vector stochastic differential equation (19) and Ito formula we derive differential equations for the first order moments $E[z_{x_i}(l)]$ and second order moments $E[z_{x_i}(l)z_{x_j}(l)]$, $i, j = 1, \ldots, 7$.

Next we find the stationary solutions for the first order moments $E[z_{x_i}(l)]$, $i = 2, 3, \ldots, 7$ and for second order moments $E[z_{x_i}(l)z_{x_j}(l)]$, $i, j = 1, \ldots, 7$, $(i, j) \neq (1, 1)$ equating to zero the corresponding time derivatives, i.e.

$$\frac{dE[z_{x_i}(t,l)]}{dt} = 0, \quad i = 2, 3, \ldots, 7 \tag{30}$$

$$\frac{dE[z_{x_i}(t,l)z_{x_j}(t,l)]}{dt} = 0, \quad i, j = 1, \ldots, 7, (i, j) \neq (1, 1) \tag{31}$$

Then we obtain

$$E[z_{x_2}(t, l)] = E[z_{x_4}(t, l)] = E[z_{x_6}(t, l)] = 0, E[z_{x_3}(t, l)] = \frac{(\gamma_{x_1}^l)^2}{2\beta_{x_1}^l}, \tag{32}$$

$$E[z_{x_3}(t,l)] = \frac{\gamma_{x_1}^2}{2\beta_{x_1}}, E[z_{x_5}(t,l)] = 3\left(\frac{\gamma_{x_1}^2}{2\beta_{x_1}}\right)^2, E[z_{x_7}(t,l)] = 15\left(\frac{\gamma_{x_1}^2}{2\beta_{x_1}}\right)^3 \tag{33}$$

Hence, from conditions (32)–(33) and equality

$$\frac{E[z_{x_1}(t,l)]}{dt} = \alpha_x^l \tag{34}$$

we find the nonstationary solution for the first moment of the process $z_{x_1}(t,l)$

$$E[z_{x_1}(t,l)] = \alpha_x^l t + \alpha_0^l, \tag{35}$$

where α_0^l is an integration constant.

Next, taking into account conditions (30)–(31), (32)–(33) and (35) we obtain

$$E[z_{x_2}^2(l)] = \frac{(\gamma_{x_1}^l)^2}{2\beta_{x_1}^l}, E[z_{x_3}^2(l)] = 3\left(\frac{(\gamma_{x_1}^l)^2}{2\beta_{x_1}^l}\right)^2, E[z_{x_2}z_{x_3}(l)] = 0, \tag{36}$$

$$E[z_{x_2}z_{x_4}(l)] = 3\left(\frac{(\gamma_{x_1}^l)^2}{2\beta_{x_1}^l}\right)^2, E[z_{x_4}^2(l)] = 15\left(\frac{(\gamma_{x_1}^l)^2}{2\beta_{x_1}^l}\right)^3, E[z_{x_4}z_{x_5}(l)] = 0, \tag{37}$$

$$E[z_{x_3}z_{x_4}(l)] = 0, E[z_{x_2}(l)z_{x_5}(l)] = 0, \quad E[z_{x_3}z_{x_5}(l)] = 15\left(\frac{(\gamma_{x_1}^l)^2}{2\beta_{x_1}^l}\right)^3, \tag{38}$$

$$E[z_{x_5}^2(l)] = 105\left(\frac{(\gamma_{x_1}^l)^2}{2\beta_{x_1}^l}\right)^4, E[z_{x_2}z_{x_6}(l)] = 15\left(\frac{(\gamma_{x_1}^l)^2}{2\beta_{x_1}^l}\right)^3, E[z_{x_3}z_{x_6}(l)] = 0 \tag{39}$$

$$E[z_{x_4}z_{x_6}(l)] = 105\left(\frac{(\gamma_{x_1}^l)^2}{2\beta_{x_1}^l}\right)^4, E[z_{x_5}z_{x_6}(l)] = 0, E[z_{x_6}^2(l)] = 945\left(\frac{(\gamma_{x_1}^l)^2}{2\beta_{x_1}^l}\right)^5 \tag{40}$$

$$E[z_{x_2}(l)z_{x_7}(l)] = 0, \quad E[z_{x_3}z_{x_7}(l)] = 105\left(\frac{(\gamma_{x_1}^l)^2}{2\beta_{x_1}^l}\right)^4, E[z_{x_4}z_{x_7}(l)] = 0, \tag{41}$$

$$E[z_{x_5}(l)z_{x_7}(l)] = 945\left(\frac{(\gamma_{x_1}^l)^2}{2\beta_{x_1}^l}\right)^5, E[z_{x_6}z_{x_7}(l)] = 0, E[z_{x_7}^2(l)] = 10395\left(\frac{(\gamma_{x_1}^l)^2}{2\beta_{x_1}^l}\right)^6 \tag{42}$$

$$E[z_{x_1}(l)z_{x_2}(l)] = q_{x_1}^l \frac{(\gamma_{x_1}^l)^2}{2\beta_{x_1}^l} + 3q_{x_3}^l\left(\frac{(\gamma_{x_1}^l)^2}{2\beta_{x_1}^l}\right)^2 + 15q_{x_5}^l\left(\frac{(\gamma_{x_1}^l)^2}{2\beta_{x_1}^l}\right)^3, \tag{43}$$

$$E[z_{x_1}(l)z_{x_3}(l)] = \frac{1}{2\beta_{x_1}^l}\left[(\gamma_{x_1}^l)^2 E[z_{x_1}(l)] + \alpha_x^l \frac{(\gamma_{x_1}^l)^2}{2\beta_{x_1}^l} + 4\beta_{x_1}^l q_{x_2}^l\left(\frac{(\gamma_{x_1}^l)^2}{2\beta_{x_1}^l}\right)^2 \right.$$

$$\left. + 24\beta_{x_1}^l q_{x_4}^l\left(\frac{(\gamma_{x_1}^l)^2}{2\beta_{x_1}^l}\right)^3 + 180\beta_{x_1}^l q_{x_6}^l\left(\frac{(\gamma_{x_1}^l)^2}{2\beta_{x_1}^l}\right)^4\right], \tag{44}$$

$$E[z_{x_1}(l)z_{x_4}(l)] = 3q_{x_1}^l \left(\frac{(\gamma_{x_1}^l)^2}{2\beta_{x_1}^l}\right)^2 + 15q_{x_3}^l \left(\frac{(\gamma_{x_1}^l)^2}{2\beta_{x_1}^l}\right)^3 + 105q_{x_5}^l \left(\frac{(\gamma_{x_1}^l)^2}{2\beta_{x_1}^l}\right)^4 \quad (45)$$

$$E[z_{x_1}(l)z_{x_5}(l)] = \frac{1}{4\beta_{x_1}^l}\left[6(\gamma_{x_1}^l)^2 \frac{(\gamma_{x_1}^l)^2}{2\beta_{x_1}^l}E[z_{x_1}(l)] + 9\alpha_x^l \left(\frac{(\gamma_{x_1}^l)^2}{2\beta_{x_1}^l}\right)^2 + \right.$$
$$\left. + 384\beta_{x_1}^l q_{x_4}^l \left(\frac{(\gamma_{x_1}^l)^2}{2\beta_{x_1}^l}\right)^4 + 3600\beta_{x_1}^l q_{x_6}^l \left(\frac{(\gamma_{x_1}^l)^2}{2\beta_{x_1}^l}\right)^5\right], \quad (46)$$

$$E[z_{x_1}z_{x_6}(l)] = 15q_{x_1}^l \left(\frac{(\gamma_{x_1}^l)^2}{2\beta_{x_1}^l}\right)^3 + 105q_{x_3}^l \left(\frac{(\gamma_{x_1}^l)^2}{2\beta_{x_1}^l}\right)^4 + 945q_{x_5}^l \left(\frac{(\gamma_{x_1}^l)^2}{2\beta_{x_1}^l}\right)^5, \quad (47)$$

$$E[z_{x_1}(l)z_{x_7}(l)] = 15(\frac{(\gamma_{x_1}^l)^2}{2\beta_{x_1}^l})^3 E[z_{x_1}(l)] + \frac{55}{4\beta_{x_1}^l}\alpha_{x_1}^l \left(\frac{(\gamma_{x_1}^l)^2}{2\beta_{x_1}^l}\right)^3 + 90q_{x_2}^l \left(\frac{(\gamma_{x_1}^l)^2}{2\beta_{x_1}^l}\right)^4$$
$$+ 900q_{x_4}^l \left(\frac{(\gamma_{x_1}^l)^2}{2\beta_{x_1}^l}\right)^5 + 10170q_{x_6}^l \left(\frac{(\gamma_{x_1}^l)^2}{2\beta_{x_1}^l}\right)^6.$$
$$(48)$$

Substituting quantities (36)–(48) to equation for the derivative $\frac{dE[z_{x_1}^2(tl)]}{dt}$ we obtain

$$\frac{dE[z_{x_1}^2(t,l)]}{dt} = 2\alpha_x^l E[z_{x_1}(t,l)]$$
$$- 2\alpha_x^l \left[q_{x_2}^l \frac{(\gamma_{x_1}^l)^2}{2\beta_{x_1}^l} + 3q_{x_4}^l (\frac{(\gamma_{x_1}^l)^2}{2\beta_{x_1}^l})^2 + 15q_{x_6}^l (\frac{(\gamma_{x_1}^l)^2}{2\beta_{x_1}^l})^3\right] \quad (49)$$

Hence, from Eq. (49) and equality (35) we find the nonstationary solution for the second moment of the process $z_{x_1}(t,l)$

$$E[z_{x_1}^2(t,l)] = (\alpha_x^l)^2 t^2 - 2\alpha_x^l \left[-\alpha_{0_x}^l + q_{x_2}^l \frac{(\gamma_{x_1}^l)^2}{2\beta_{x_1}^l}\right. \quad (50)$$

$$\left. + 3q_{x_4}^l \left(\frac{(\gamma_{x_1}^l)^2}{2\beta_{x_1}^l}\right)^2 + 15q_{x_6}^l \left(\frac{(\gamma_{x_1}^l)^2}{2\beta_{x_1}^l}\right)^3\right]t + c_{0_x}^l \quad (51)$$

where $c_{0_x}^l$ is an integration constant.

References

1. Booth, H., Tickle, L.: Mortality modelling and forecasting: a review of methods. ADSRI Working Paper 3 (2008)
2. Boukas, E.K.: Stochastic Hybrid Systems: Analysis and Design. Birkhauser, Boston (2005)
3. Cairns, A.J.G., et al.: Modelling and management of mortality risk: a review. Scand. Actuar. J. **2–3**, 79–113 (2008)
4. Cairns, A.J.G., et al.: A quantitative comparison of stochastic mortality models using data from England and Wales and the United States. North Am. Actuar. J. **13**, 1–35 (2009)
5. Cao, H., Wang, J., et al.: Trend analysis of mortality rates and causes of death in children under 5 years old in Beijing, China from 1992 to 2015 and forecast of mortality into the future: an entire population-based epidemiological study. BMJ Open **7** (2017). https://doi.org/10.1136/bmjopen-2017-015941
6. Chow, G.C.: Tests of equality between sets of coefficients in two linear regressions. Econometrica **28**, 591–605 (1960)
7. Giacometti, R., et al.: A stochastic model for mortality rate on Italian Data. J. Optim. Theory Appl. **149**, 216–228 (2011)
8. GUS: Central Statistical Office of Poland (2015). http://demografia.stat.gov.pl/bazademografia/TrwanieZycia.aspx
9. Jahangiri, K., et al.: Trend forecasting of main groups of causes-of-death in Iran using the Lee-Carter model. Med. J. Islam. Repub. Iran **32**(1), 124 (2018)
10. Keogh, E., et al.: Segmenting time series: a survey and novel approach. In: Last, M., Bunke, H., Kandel, A. (eds.) Data Mining in Time Series Databases, vol. 83, pp. 1–22. World Scientific, Singapore (2004)
11. Lee, R.D., Carter, L.: Modeling and forecasting the time series of U.S. mortality. J. Am. Stat. Assoc. **87**, 659–671 (1992)
12. Lee, R.D., Miller, T.: Evaluating the performance of the Lee-Carter method for forecasting mortality. Demography **38**, 537–549 (2001)
13. Liberzon, D.: Switching in Systems and Control, Boston, Basel. Birkhauser, Berlin (2003)
14. Lovrić, M., et al.: Algorithmic methods for segmentation of time series: an overview. J. Contemp. Econ. Bus. Iss. **1**, 31–53 (2014)
15. Milevsky, M.A., Promislov, S.D.: Mortality derivatives and the option to annuitise. Insur. Math. Econ. **29**(3), 299–318 (2001)
16. Renshaw, A., Haberman, S.: Lee-Carter mortality forecasting with age-specific enhancement. Insur. Math. Econ. **33**(2), 255–272 (2003)
17. Renshaw, A., Haberman, S.: A cohort-based extension to the Lee-Carter model for mortality reduction factor. Insur. Math. Econ. **38**(3), 556–570 (2006)
18. Rossa, A., Socha, L.: Proposition of a hybrid stochastic Lee-Carter mortality model. Metodol. zvezki (Adv. Methodol. Stat.) **10**, 1–17 (2013)
19. Rossa, A., Socha, L., Szymanski, A.: Hybrid Dynamic and Fuzzy Models of Mortality, 1st edn. WUL, Lodz (2018)
20. Sliwka, P., Socha, L.: A proposition of generalized stochastic Milevsky-Promislov mortality models. Scand. Actuar. J. **2018**(8), 706–726 (2018). https://doi.org/10.1080/03461238.2018.1431805
21. Sliwka, P.: Proposed methods for modeling the mortgage and reverse mortgage installment. In: Recent Trends in the Real Estate Market and Its Analysis, pp. 189–206. SGH, Warszawa (2018)

22. Śliwka, P.: Application of the Markov chains in the prediction of the mortality rates in the generalized stochastic Milevsky–Promislov model. In: Mondaini, R.P. (ed.) Trends in Biomathematics: Mathematical Modeling for Health, Harvesting, and Population Dynamics, pp. 191–208. Springer, Cham (2019). https://doi.org/10.1007/978-3-030-23433-1_14

23. Sliwka, P.: Application of the model with a non-Gaussian linear scalar filters to determine life expectancy, taking into account the cause of death. In: Rodrigues, J.M.F., et al. (eds.) ICCS 2019. LNCS, vol. 11538, pp. 435–449. Springer, Cham (2019). https://doi.org/10.1007/978-3-030-22744-9_34

Ontology-Based Inference for Supporting Clinical Decisions in Mental Health

Diego Bettiol Yamada[1(✉)], Filipe Andrade Bernardi[2],
Newton Shydeo Brandão Miyoshi[1], Inácia Bezerra de Lima[3],
André Luiz Teixeira Vinci[1], Vinicius Tohoru Yoshiura[1],
and Domingos Alves[1]

[1] Ribeirao Preto Medical School, University of Sao Paulo, Ribeirao Preto, Brazil
diego.yamada@usp.br
[2] Bioengineering Postgraduate Program, University of Sao Paulo,
Sao Carlos, Brazil
[3] School of Nursing of Ribeirão Preto, University of Sao Paulo,
Ribeirao Preto, Brazil

Abstract. According to the World Health Organization (WHO), mental and behavioral disorders are increasingly common and currently affect on average 1/4 of the world's population at some point in their lives, economically impacting communities and generating a high social cost that involves human and technological resources. Among these problems, in Brazil, the lack of a transparent, formal and standardized mental health information model stands out, thus hindering the generation of knowledge, which directly influences the quality of the mental healthcare services provided to the population. Therefore, in this paper, we propose a computational ontology to serve as a common knowledge base among those involved in this domain, to make inferences about treatments, symptoms, diagnosis and prevention methods, helping health professionals in clinical decisions. To do this, we initially carried out a literature review involving scientific papers and the most current WHO guidelines on mental health, later we transferred this knowledge to a formal computational model, building the proposed ontology. Also, the Hermit Reasoner inference engine was used to deduce facts and legitimize the consistency of the logic rules assigned to the model. Hence, it was possible to develop a semantic computational artifact for storage and generate knowledge to assist mental health professionals in clinical decisions.

Keywords: Ontology-based inference · Knowledge representation · Mental health

1 Introduction

Mental and behavioral disorders represent a large part of public health problems worldwide, it is estimated that by 2030 depression will be the largest cause of disability on the planet [1]. In countries such as the United States and Canada, it is estimated that this type of disorder is already the leading cause of disability for people aged 15–44 [2]. In Brazil, the situation is even more complicated, as the country lacks a formal and

© Springer Nature Switzerland AG 2020
V. V. Krzhizhanovskaya et al. (Eds.): ICCS 2020, LNCS 12140, pp. 363–375, 2020.
https://doi.org/10.1007/978-3-030-50423-6_27

standardized model of mental health information, which makes it difficult for health professionals involved in this area to generate and use knowledge [3]. In Brazil, the Unified Health System ('SUS'; Portuguese: 'Sistema Único de Saúde') offers integral treatment at all levels of care, and these treatments often involve multi-professional care and procedures performed in different health services. Thus, it is essential to have a standardized model that facilitates communication between the various nodes of the health network, to help the units to communicate efficiently [4].

In this scenario, interoperability between Health Information Systems (HIS) presents itself as a viable solution to promote information sharing, exchange, and reuse [5]. Interoperability can be classified into four levels: fundamental, structural, semantic and organizational. Fundamental interoperability occurs through the exchange of data between HIS without the receiver having the ability to interpret the data. Its main applications range from direct database connections and service-oriented architectures using, for example, web services. Structural interoperability is an intermediate level that defines the syntax of data exchange to ensure that data exchanged between HIS can be interpreted and its meaning preserved and unchanged. Its main feature is based on the concept of corporate service buses using standards for message formats, for example, Digital Imaging and Communications in Medicine (DICOM) and Health Level Seven (HL7) [6].

The highest level of interoperability is achieved by the semantic level when two or more systems or elements can exchange information [7]. To achieve this, initially it is necessary to establish a standard structure of vocabularies and terminologies, expressive relations and description of processes for knowledge representation. One way to provide this type of entity standardization and taxonomy for HIS integration is by building and implementing computational ontologies [8]. By definition, an ontology is a set of explicit formal specifications of terminologies (classes or entities) and relationships (properties) between these elements in a given domain of knowledge [9]. Ontologies have contributed to facilitate the processing of information with added semantic value, mediating the exchange of information between machines and humans through computer systems and serving as a schema for knowledge bases, smart applications, health observatories, and inference knowledge models [7, 8].

Finally, we can also consider organizational interoperability, which is concerned with how different organizations collaborate to achieve their objectives by maintaining different internal structures and varying business processes. Even with the standardization of concepts and terminologies, organizations have different operating models or work processes. Thus, standardization of work processes, where two organizations need to have the same vision, is called organizational interoperability [3, 7].

In addition to the interoperability problems involving the mental health domain, it has been found that clinical decisions are often made without consulting a consolidated knowledge base [4, 10]. Therefore, semantic applications, such as ontologies, are relevant in health because besides having originally integrated knowledge, this tool can be constantly developed with the dynamic participation of the community that involves this domain of knowledge.

Decision support tools are essential to guide the practice of health care and support the decisions that will directly influence the quality of care provided to the population. Making a reference data set available as a Decision Support System (DSS), provided with consistent ontologies, for integrating, analyzing, comparing and viewing health data through the integration of heterogeneous and dispersed databases is the main aspect to be considered [11].

In this way, we merge the best practices of ontology development with the knowledge collected on mental health to build an expressive semantic repository through upper-ontology [8, 12]. The World Wide Web Consortium (W3C) best practices for building semantic applications involve 3 basic blocks: a standard data model; a query protocol; and a set of reference vocabulary and terminologies. Resource Description Framework (RDF), Simple Protocol and RDF Query Language (SPARQL), and ontologies developed in Ontology Web Language (OWL) refer to these basic blocks respectively [13]. Therefore, we consider that mental health processes will be more efficient if based on a formal structure that will be proposed through the use of the mentioned set of standards and technologies.

The main purpose of this paper is to develop a computational ontology capable of representing the reality of the mental health domain and making evidence-based inferences, assisting health professionals in clinical decisions and promoting semantic interoperability between mental health information systems. The rest of the paper is organized as follows. In Sect. 2, related studies on the use of ontologies to support clinical decisions will be discussed. In Sect. 3, the proposed methodology will be presented in detail. Experimental results will be presented in Sect. 4. Finally, the conclusion will be described in Sect. 5.

2 Literature Review

The W3C has been committed to projects to improve integration, standardization, and data sharing in the World Wide Web. Such initiatives involve the use of the Semantic Web, defined as a set of technologies that, not only links documents, but is also recognize the meaning of the data from these documents and, through ontologies, promote inferences that help management and decision making. Thus, ontologies have been used on a large scale in the medical field, as this domain presents a complex and dispersed data in HIS with low standardization and integration [5].

Many studies have identified that human errors are frequent in clinical settings. The three most common types of errors in these environments relate to non-compliance with guidelines, hasty decision-making, and lack of awareness of the responsibilities and roles of each type of practitioner [14]. These are problems that can easily be avoided by an ontology-based decision support system [15]. Cases of development of this type of application can be observed in various areas of health.

In midwifery, for example, the detection of characteristics that frame women in a risky pregnancy is essential for proper and personalized medical follow-up. However, in many countries, the lack of human resources in the medical field means that these diagnoses are performed without proper scrutiny, leading many women to death due to complications in pregnancy. Thus, in Pakistan, an ontology-based clinical decision support system has been developed to assist in the diagnosis of high-risk pregnant women and refer them to qualified physicians for timely treatment [15].

In the field of Alzheimer's disease, the early detection of pathology is a complex task for medical staff, usually, the approach involves medical image processing, psychological testing, and neurological testing, and such exams produce data that can generate knowledge. In the United States, one of the ways found to mitigate this complexity was by developing a tool to support clinical decision making, where ontologies and semantic reasoning play a key role in making inferences about diagnoses through the collected patient data [16].

In recent years the population has grown older and the number of patients with chronic respiratory-related diseases has increased [17]. In Taiwan, an intelligent ontology-based tool has been built to assist medical staff in recognizing changes in clinical examinations for the detection of chronic respiratory diseases. This system recognizes patterns that usually lead patients to develop respiratory problems and points out these characteristics to the medical team, who in possession of this information can more accurately indicate the degree of complexity of pathologies related to a patient's respiratory system [18].

Breast cancer is the most common cancer among women in many countries. In Canada, a Semantic Web-based application has been implemented to support physicians in tracking patients with this disease. The approach involves computerization and implementation of guidelines for monitoring women undergoing breast cancer treatment. For this, a domain ontology was built that models the knowledge inherent in these guidelines and practices and serves as a source of knowledge to determine specific recommendations for each patient [19].

Also, many other semantic applications already exist in healthcare and many more are under construction at this time. The justification for the development of these applications is the accessible implementation of these technologies in new and legacy systems, the low cost, and the encouragement of the use by various international entities to disseminate, integrate and reuse the knowledge generated [8, 15]. For the efficient use of a semantic tool, the information inserted in it must come from concrete and reliable sources. In this paper, we use studies and knowledge from the WHO's Mental Health Action Plan 2013–2020 [20] and the Diagnostic and Statistical Manual of Mental Disorders (DSM-5) [21].

3   Methods

In this paper, we propose a framework for inferring mental health knowledge through the use of a domain ontology. This framework consists of three main modules: 1) Structural Ontology Model, 2) Fact Inference Engine and 3) RDF Database.

3.1   Structural Ontology Model

An ontology is an explicit and formal specification of a set of concepts in a specific field of interest [9]. In this proposal, an ontology for the mental health domain was developed to assist in the management of the mental health care network in Brazil, to standardize concepts and to assist the integration and interoperability between different health units. Besides, this tool can infer knowledge from the inputted data in ontology to assist mental health professionals in clinical decisions.

The building of an ontology usually comprises a series of stages, therefore, for proper coordination of all these steps, the Methontology methodology was adopted. Developed at the Artificial Intelligence Department of the Polytechnic of Madrid (http://www.dia.fi.upm.es/), this ontology construction methodology stands out for presenting a rigid detail to explore and structure the processes of knowledge acquisition [22]. The knowledge acquisition phase for this project involved interviews with mental health professionals and a literature review that considered scientific articles, official documents, WHO guidelines and the DSM-5 standards and classifications.

Subsequently, we used the Protégé software (https://protege.stanford.edu/) to transfer the collected knowledge to the computational environment, thus building the desired ontology, with all its entities and properties developed in OWL, following the data model RDF. Protégé software has been selected for being open source, reliable and robust, and has wide acceptance and compatibility with other international projects [23].

Among the created entities, we store information about the types of mental disorders, symptoms, prevention methods, types of medications, side effects of these drugs, types of health units, management processes of psychiatric hospitals, documents that involve these processes, among other information from the knowledge acquisition phase. Information on mental disorder classes as well as their subtypes was imported from the 10th revision of the International Classification of Diseases (ICD-10) [24].

We then use the Jena Ontology Application Programming Interface (API) (https://jena.apache.org/documentation/ontology/) to perform fact deduction and legitimation of the consistency of the logical rules assigned to the semantic repository. This programming toolkit uses the Java language to provide the developer with standardized structures that connect the modules of a semantic platform [25], as detailed in Sect. 3.2.

3.2   Fact Inference Engine

After the ontology was built, we uploaded the proposed schema on the web via the BioPortal platform (https://bioportal.bioontology.org/) and we imported this model into the Jena Fuseki Server (https://jena.apache.org/). Using the API provided by this software and the Hermit Reasoner inference engine (http://www.hermit-reasoner.com/) it was possible to validate the entities that make up the ontology, as well as all its connective elements, aiming to generate consistent facts from the information inserted in the ontology [26]. Figure 1 shows the scheme used in this step, arrows mean the connected modules consume information from each other.

The elements of ontology had their structure verified by the aforementioned inference engine, this step avoids future data inconsistency problems. Posteriorly, the information was entered into the RDF triples that make up the structure of the ontology. The next section details the operation of this type of data repository.

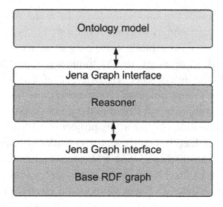

Fig. 1. Ontology-based application structure

3.3   RDF Database

Through Protégé software, we insert mental health information into the semantic repository schema created, then use the Jena Fuseki Server to create an endpoint to perform custom queries to the database using the SPARQL language. This query language has a syntax that encompasses commands capable of providing broad effectiveness in extracting content from a semantic base. Also, the results can be presented in various ways, according to the user's needs. Figure 2 shows how the search for information stored in triple RDF occurs.

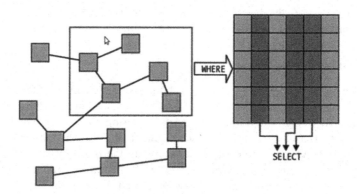

Fig. 2. Operation of a SPARQL query

The SELECT command is responsible for specifying the conditions for selecting the data relation under which the query will be executed, and the WHERE clause is responsible for selecting which data of this relation to display. This basic structure added to a range of other expressive commands allows the elucidation of facts and the asking of questions about knowledge stored in the triple form [27].

Thus, queries were held to test the use of the structure to prove its functionality to clarify relevant facts about the context of mental health. According to WHO, the use of contextual information about types of disorders, symptoms, treatments, preventions and drug side effects are essential topics for the continuous improvement of health services [1, 20]. The results of this process are in the later section.

4 Results and Discussion

A series of tests were conducted to demonstrate the effective functioning of the proposal. In this section, we will present the main results of the work.

The experiments focused on elucidating facts about mental disorders and their symptoms, treatments, prevention recommendations and medications. According to WHO, this information is of great importance for clinical decisions within the mental health field [20]. To exemplify these features, in this paper, we chose depression as the target disorder. This disorder was chosen due to its high tendency of social and economic impact [1]. Besides, according to the Mental Health Information System ('SISAM'; Portuguese: 'Sistema de Informação de Saúde Mental'), which coordinates mental health in 26 Brazilian municipalities, mood disorders are the second leading cause of psychiatric hospitalizations in the region, accounting for 31% of all cases of psychiatric hospitalizations [3]. In the United States, depression is also one of the most common psychiatric disorders, with an estimated lifetime prevalence rate of 16.2% [28]. Figure 3 shows the location of the concept of "Mental Depression" in the structure of the ontology.

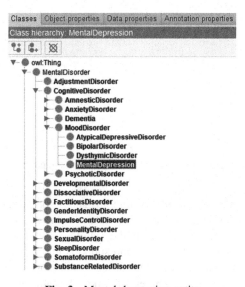

Fig. 3. Mental depression entity

Therefore, we relate this disorder with the knowledge about it collected. Through the Property Assertions function of Protégé software, we use the relationships created to bind the classes with their respective instances, so finally we could perform SPARQL queries to extract knowledge.

Figure 4 shows how semantic inference is made through a SPARQL query that looks for the main symptoms of depression, using a standardized terminology according to the guidelines of WHO and DSM-5. We can see that the set of results provides relevant and helpful information for clinical decision making, as the structured knowledge of the most frequent symptoms presented by depressive patients provides the psychologist or psychiatrist with a consistent basis for better identification of the disorder [29].

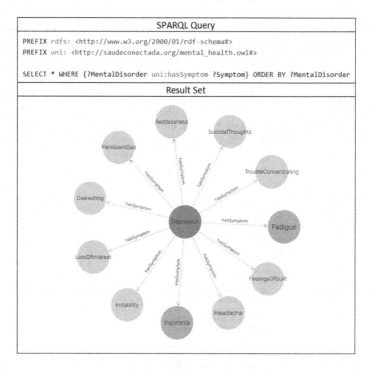

Fig. 4. Inference about depression symptoms

To diagnose depression, 5 of these symptoms should be present almost every day for 2 weeks, and one of them must be persistent sadness or loss of interest [21]. Therefore, the knowledge structured in the ontology is characterized as a solid repository of evidence-based information and supports the fast and accurate diagnosis identification, so the patient can then be referred to the most appropriate treatment.

Knowledge about the most recommended types of treatments for a disorder is also of great importance to aid a medical decision [20]. The entities referring to the different types of treatments for mental disorders are also structured in our ontology. Figure 5 shows some classes of treatments associated with depression [21]. The standardization of this type of information avoids inconsistencies in data sharing between different health services [11, 15].

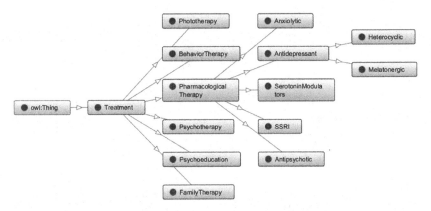

Fig. 5. Structure of treatments for depression

Each type of treatment has its characteristics, but knowledge about the consequences of recommending psychoactive drugs should also be highlighted in the model. Since on average 38% of patients who have already taken antidepressants have experienced at least one of the side effects of this drug [30] and these effects are more detrimental to long-term health [29], the healthcare professional must have at its disposal a standardized digital base with the possible side effects of each class of medicine. Figure 6 shows an inference about the main side effects related to the use of antidepressant drugs.

Other side effects, treatments, symptoms, drugs, disorders, or any entities may be inserted or removed from the platform as discoveries are made through scientific research, it is important to remember that this type of tool is available on the web for future updates and maintenance as needed. An ontology must represent the reality of a domain, whereas reality is changeable the ontology must also provide scope for alteration to correctly represent the desired domain [7, 9].

Considering the many side effects caused by the continued use of psychoactive drugs [29], and the currently increasing demand for mental health treatments and services [4], the public health policies are essential and must be focused on preventing these disorders, minimizing social problems and resource expenditures from more complex health services [30]. Therefore, our model also can infer prevention recommendations for mental disorders. Figure 7 shows the results of this type of query for depression.

It is important to emphasize that an ontology must minimize any kind of ambiguity or subjectivity in the representation of knowledge [6, 12]. This way, many additional resources can be used to complement information stored in this type of tool. We can use, for example, semantic annotations of the RDF-Schema (RDF-S) data model, an extension of the RDF, to relate concepts of our vocabulary with other relevant information on the web.

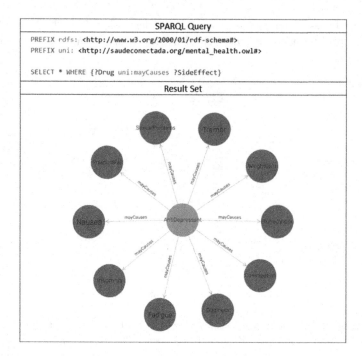

Fig. 6. Inference about antidepressant side effects

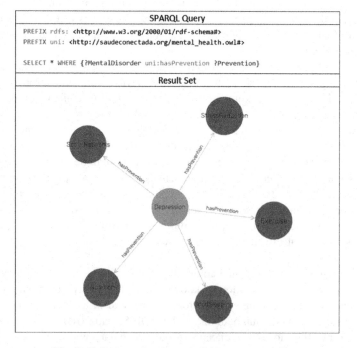

Fig. 7. Inference about prevention recommendations

For example, the "GoodSleeping" concept, presented as a preventive aspect in the Result set of Fig. 7, can bring subjectivity about how much sleep would be ideal for each individual. Therefore, we can create semantic annotations for this concept, such annotations may refer to concepts from other web pages or other ontologies available on the web [9]. In this case, we use the National Sleep Foundation website (https:// www.sleepfoundation.org/) [31] to couple with this concept, thus highlighting the recommendations of sleep hours for each age group.

In this paper, specifically, we show results of depression-related inferences, but many other mental disorders can be explored. There are no limits to the knowledge stored in this platform, the semantic repository is dynamic and is available on the Web to be constantly fed with new knowledge from professionals and researchers involved in this area.

The developed ontology comprises 361 classes, 37 relationships among these classes, 72 individuals, and the maximum depth is 9. This proposal not only addresses the biomedical aspects of mental health but also relates them to the day-to-day operations of health services, allowing information systems to be aligned with the strategies of the mental health network, which is its most innovative point. Even in a clinical setting supported by the use of traditional information systems and electronic medical records, the rate of medical errors can still reach 24.4% [32]. In this way, computational ontologies are presented as an extension of these traditional systems, offering additional ways to structure information, infer and make available evidence-based knowledge.

However, ontology still needs the insertion of a wider range of knowledge and the establishment of relationships with other health areas, thus increasing its completeness. For example, drug entities may be related to other pharmaceutical ontologies, providing the exact description of each type of medication and their characteristics. Nutrition-related entities can be connected with other nutritionism ontologies, thus providing highly expressive and informative structures to users.

Aiming at this kind of integration, we make available the developed ontology (Mental Health Management Ontology - MHMO) in the international repository of BioPortal biomedical ontologies so that it can be used in other health projects around the world. The BioPortal is an open database that provides access to biomedical ontologies via Web services, facilitating community participation in ontology assessment and evolution and providing additional resources for terminology mapping and criteria review [33].

5   Conclusion

In conclusion, this computational artifact contains stored knowledge that can be extracted through personalized consultations to assist mental health professionals in clinical decisions and can be used as a common basis for knowledge sharing between humans and machines, promoting improvements in interoperability issues of HIS. It can also be used as a layer for building other decision-support information systems, health observatories, and varied smart applications, being useful in areas such as machine learning and artificial intelligence.

This study is part of a larger project. As future work, we aim to use the ontology to develop a Mental Health Observatory for a Brazilian Public Health Network, to improve the monitoring, analysis, and visualization of mental health issues, providing information to support evidence-based decisions, health policies elaboration, public planning, and data sharing. The ontology can be constantly fed with new knowledge to continuously assist the mental health services provided to the population and can be used to optimizing processes, reducing resource consumption and bringing benefits to health professionals, managers, and patients.

References

1. World Health Organization - The global burden of disease. http://www.who.int/entity/healthinfo/global_burden_disease/GBD_report_2004update_full.pdf. Accessed 04 Dec 2019
2. National Advisory Mental Health Council Workgroup - From discovery to cure: Accelerating the development of new and personalized interventions for mental illness. https://www.nimh.nih.gov/about/advisory-boards-and-groups/namhc/reports/fromdiscoverytocure_103739.pdf. Accessed 04 Dec 2019
3. Miyoshi, N.S.B., De Azevedo-Marques, J.M., Alves, D., De Azevedo-Marques, P.M.: An eHealth platform for the support of a Brazilian regional network of mental health care (eHealth-Interop): development of an interoperability platform for mental care integration. JMIR Ment. Health 5(4), e10129 (2018)
4. Castro, S.A.D.: Adesão ao tratamento psiquiátrico, após alta hospitalar: acompanhamento na rede de serviços de saúde (Doctoral dissertation, Universidade de São Paulo) (2015)
5. Yoshiura, V.T., et al.: Towards a health observatory conceptual model based on the semantic web. Proc. Comput. Sci. 138, 131–136 (2018)
6. Jabbar, S., Ullah, F., Khalid, S., Khan, M., Han, K.: Semantic interoperability in heterogeneous IoT infrastructure for healthcare. Wireless Commun. Mob. Comput. 2017 (2017)
7. Abhishek, K., Singh, M.P.: An ontology based decision support for tuberculosis management and control in India. Int. J. Eng. Technol. 8(6), 2860–2877 (2016)
8. Yamada, D.B., et al.: Proposal of an ontology for mental health management in Brazil. Proc. Comput. Sci. 138, 137–142 (2018)
9. Shaaban, A.M., Gruber, T., Schmittner, C.: Ontology-based security tool for critical cyber-physical systems. In: Proceedings of the 23rd International Systems and Software Product Line Conference, vol. B, p. 91. ACM, September 2019
10. Vinci, A.L.T., Rijo, R.P.C.L., de Azevedo Marques, J.M., Alves, D.: Proposal of an evaluation model for mental health care networks using information technologies for its management. Proc. Comput. Sci. 100, 826–831 (2016)
11. Delfini, M. ., Miyoshi, N.S.B., Alves, D.: Minimum data consensus: essential information to continuing healthcare. In: 2015 IEEE 28th International Symposium on Computer-Based Medical Systems, pp. 205–207. IEEE, June 2015
12. Willner, A., et al.: The open-multinet upper ontology towards the semantic-based management of federated infrastructures. EAI Endor. Trans. Scalable Inf. Syst. 2(7) (2015)
13. W3C Semantic Web Standards. https://www.w3.org/standards/semanticweb/. Accessed 11 Dec 2019
14. Sanchez, E., et al.: A knowledge-based clinical decision support system for the diagnosis of Alzheimer disease. In: 2011 IEEE 13th International Conference on e-Health Networking, Applications and Services, pp. 351–357. IEEE, June 2011

15. Manzoor, U., Usman, M., Balubaid, M.A., Mueen, A.: Ontology-based clinical decision support system for predicting high-risk pregnant woman. System **6**(12), 203–208 (2015)
16. Adnan, M., Warren, J., Orr, M.: Ontology based semantic recommendations for discharge summary medication information for patients. In: 2010 IEEE 23rd International Symposium on Computer-Based Medical Systems (CBMS), pp. 456–461. IEEE, October 2010
17. Zwar, N., et al.: A systematic review of chronic disease management (2017)
18. Lee, C.-S., Wang, M.-H.: An ontology-based intelligent agent for respiratory waveform classification. In: Ali, M., Dapoigny, R. (eds.) IEA/AIE 2006. LNCS (LNAI), vol. 4031, pp. 1240–1248. Springer, Heidelberg (2006). https://doi.org/10.1007/11779568_131
19. Abidi, S.R., Hussain, S., Shepherd, M., Abidi, S.S.R.: Ontology-based modeling of clinical practice guidelines: a clinical decision support system for breast cancer follow-up interventions at primary care settings. In: Medinfo 2007: Proceedings of the 12th World Congress on Health (Medical) Informatics; Building Sustainable Health Systems, p. 845. IOS Press (2007)
20. World Health Organization: Mental health action plan 2013–2020 (2013)
21. American Psychiatric Association. Diagnostic and statistical manual of mental disorders (DSM-5). American Psychiatric Pub (2013)
22. González, M.A.V.: Building ontologies with methontology as a technical resource for specialized translation: ontoUAV project, a multilingual web ontology (EN/FR/ES) on Unmanned Aircraft Vehicles (UAV) for specialized translation. In: Temas actuales de terminología y estudios sobre el léxico, pp. 163–192. Comares (2017)
23. Musen, M.A.: The protégé project: a look back and a look forward. AI Matters **1**(4), 4 (2015)
24. WHO ICD 10 online version. https://www.who.int/classifications/icd/icdonlineversions/en/. Accessed 11 Dec 2019
25. Selvaraj, S., Choi, E.: A study on traditional medicine ontology. In: Proceedings of the 2nd International Conference on Software Engineering and Information Management, pp. 235–239. ACM, January 2019
26. Shearer, R., Motik, B., Horrocks, I.: HermiT: a highly-efficient OWL reasoner. In: Owled, vol. 432, p. 91, October 2008
27. Allemang, D., Hendler, J.: Semantic Web for the Working Ontologist: Effective Modeling in RDFS and OWL. Elsevier, Amsterdam (2011)
28. Kessler, R.C., et al.: The epidemiology of major depressive disorder: results from the National Comorbidity Survey Replication (NCS-R). JAMA **289**(23), 3095–3105 (2003)
29. Karyotaki, E., et al.: Combining pharmacotherapy and psychotherapy or monotherapy for major depression? A meta-analysis on the long-term effects. J. Affect. Disord. **194**, 144–152 (2016)
30. Cascade, E., Kalali, A.H., Kennedy, S.H.: Real-world data on SSRI antidepressant side effects. Psychiatry (Edgmont) **6**(2), 16 (2009)
31. National Sleep Foundation Recommends New Sleep Times. https://www.sleepfoundation. org/press-release/national-sleep-foundation-recommends-new-sleep-times. Accessed 12 Dec 2019
32. Shojania, K.G., Burton, E.C., McDonald, K.M., Goldman, L.: Changes in rates of autopsy-detected diagnostic errors over time: a systematic review. JAMA **289**(21), 2849–2856 (2003)
33. Salvadores, M., Alexander, P.R., Musen, M.A., Noy, N.F.: BioPortal as a dataset of linked biomedical ontologies and terminologies in RDF. Semant. Web **4**(3), 277–284 (2013)

Towards Prediction of Heart Arrhythmia Onset Using Machine Learning

Agnieszka Kitlas Golińska$^{1(\boxtimes)}$ ⓘ, Wojciech Lesiński^{1} ⓘ, Andrzej Przybylski^{2} ⓘ, and Witold R. Rudnicki^{1} ⓘ

^{1} Institute of Informatics, University of Białystok,
ul. Konstantego Ciołkowskiego 1M, 15-245 Białystok, Poland
`akitlas@ii.uwb.edu.pl`
^{2} Faculty of Medicine, University of Rzeszów, Rzeszów, Poland

Abstract. Current study aims at prediction of the onset of malignant cardiac arrhythmia in patients with Implantable Cardioverter-Defibrillators (ICDs) using Machine Learning algorithms. The input data consisted of 184 signals of RR-intervals from 29 patients with ICD, recorded both during normal heartbeat and arrhythmia. For every signal we generated 47 descriptors with different signal analysis methods. Then, we performed feature selection using several methods and used selected feature for building predictive models with the help of Random Forest algorithm. Entire modelling procedure was performed within 5-fold cross-validation procedure that was repeated 10 times. Results were stable and repeatable. The results obtained (AUC = 0.82, MCC = 0.45) are statistically significant and show that RR intervals carry information about arrhythmia onset. The sample size used in this study was too small to build useful medical predictive models, hence large data sets should be explored to construct models of sufficient quality to be of direct utility in medical practice.

Keywords: Arrhythmia · Implantable Cardioverter-Defibrillators · Artificial intelligence · Machine Learning · Random Forest

1 Introduction

Some types of cardiac arrhythmia, such as VF (ventricular fibrillation) or VT (ventricular tachycardia), are life-threatening. Therefore, prediction, detection, and classification of arrhythmia are very important issues in clinical cardiology, both for diagnosis and treatment. Recently research has concentrated on the two latter problems, namely detection and classification of arrhythmia which is a mature field [1]. These algorithms are implemented in Implantable Cardioverter-Defibrillators (ICD) [2], which are used routinely to treat cardiac arrhythmia [3]. However, the related problem of prediction of arrhythmia events still remains challenging.

© Springer Nature Switzerland AG 2020
V. V. Krzhizhanovskaya et al. (Eds.): ICCS 2020, LNCS 12140, pp. 376–389, 2020.
https://doi.org/10.1007/978-3-030-50423-6_28

In recent years we have observed an increased interest in application of Machine Learning (ML) and artificial intelligence methods in analysis of biomedical data in hope of introducing new diagnostic or predictive tools. Recently an article by Shakibfar et al. [4] describes prediction results regarding electrical storm (i.e. arrhythmic syndrome) with the help of Random Forest using daily summaries of ICDs monitoring. Authors then generated 37 predictive variables using daily ICD summaries from 19935 patients and applied ML algorithms, for construction of predictive models. They concluded that the use of Machine Learning methods can predict the short-term risk of electrical storm, but the models should be combined with clinical data to improve their accuracy.

In the current study ML algorithms are used for prediction of the onset of malignant cardiac arrhythmia using RR intervals. This is important problem, since the standard methods of prediction aim at stratification of patients into high- and low-risk groups using various sources of clinical data [3]. Then, the patients from the high-risk group undergo surgical implantation of ICD [3], which monitors the heart rate. The algorithms for identification of arrhythmia events implemented in these devices recognise the event and apply the electric signal that restarts proper functioning of the heart. Despite technological progress, inappropriate ICD interventions are still a very serious side-effect of this kind of therapy. About 10–30% of therapies delivered by ICD have been estimated as inappropriate [2]. These are usually caused by supraventricular tachyarrhythmias, T-wave oversensing, noise or non-sustained ventricular arrhythmias.

The goal of the current study is to examine whether one may predict an incoming arrhythmia event using only the signal available for these devices. If such predictions are possible with high enough accuracy, they might be communicated by ICD's to warn patients of incoming event, helping to minimise adverse effects or even possibly avoid them completely. One of the first studies considering this problem was a Master of Science thesis by P. Iranitalab [5]. In that study the author used time and frequency domain analysis of QRS-complex as well as R-R interval variability analysis for only 18 patients, but he concluded that none of these methods proved to be an effective predictor that could be applied to a large patient population successfully. This analysis was performed on normal (sinus) and pre-arrhythmia EGM (ventricular electrogram) data. The newest article considering prediction of ventricular tachycardia (VT) and ventricular fibrillation (VF) was published in September 2019 by Taye et al. [6]. Authors extracted features from HRV and ECG signals and used artificial neural network (ANN) classifiers to predict the VF onset 30 s before its occurrence. The prediction accuracy estimated using HRV features was 72% and using QRS complex shape features from ECG signals – 98.6%, but only 27 recordings were used for this study.

Other studies, which seem to be related [7–9] in fact consider different issues. In [7] authors investigate a high risk patients of an ICD and evaluate QT dispersion, which may be a significant predictor of cardiovascular mortality. They claim that QT dispersion at rest didn't predict the occurrence and/or reoccurrence of ventricular arrhythmias. In [8] authors proposed a new atrial fibrillation (AF) prediction algorithm to explore the prelude of AF by classifying ECG before AF into normal and abnormal states. ECG was transformed into spectrogram using short-time Fourier transform and then trained. In paper [9], it seems like it's more about detection or classification than prediction of onset of arrhythmia. Authors used a clustering approach and regression methodology to predict type of cardiac arrhythmia.

Machine Learning algorithms are powerful tools, but should be used with caution. Loring et al. mention in their paper [10] the possible difficulties in application of methods of this kind (e.g. critical evaluation of methodology, errors in methodology difficult to detect, challenging clinical interpretation). We have planned our research taking this into account.

2   Materials and Methods

The data used in the study, in the form of RR intervals, comes from patients with implanted ICD's; the details are described below. The raw RR intervals were transformed into descriptive variables using several alternative methods. Then, the informative variables were identified with the help of several alternative feature selection methods. Finally, predictive models for arrhythmia events were constructed using Machine Learning algorithms (Fig. 1). The details of this protocol are described in the following sections.

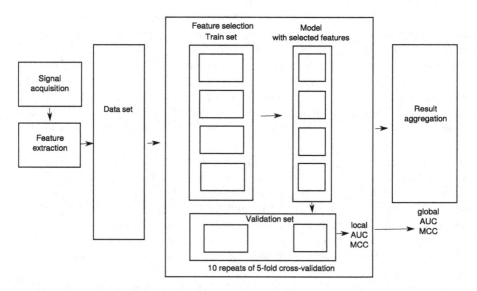

Fig. 1. Block diagram of data processing: data acquisition, feature extraction and modelling in cross-validation loop. AUC – area under receiver operator curve, MCC – Matthews correlation coefficient.

2.1 Data Set

The input data consisted of 184 tachograms (signals of RR-intervals i.e. beat-to-beat intervals, observed in ECG) from 29 patients with single chamber ICD implanted in the years 1995–2000 due to previous myocardial infarction. Only data from patients with devices compatible with the PDM 2000 (Biotronik) and STDWIN (Medtronic) programs were analysed in the study. Patients who had a predominantly paced rhythm were excluded from the study. The VF zone was active in all patients with the lower threshold from 277 ms to 300 ms. The VT zone was switched on in all patients. Antitachycardia pacing (ATP) was the first therapy in the VT zone. Ventricular pacing rate was 40–60 beats/min (bpm).

Samples were recorded both during normal heartbeat (121 events) and onset of arrhythmia – ventricular fibrillation (VF – 12 events) or ventricular tachycardia (VT – 51 events). Both types of arrhythmia were considered as a single class. The length of these signals varied from 1000 to 9000 RR intervals. The signals have been collected from patients from The Cardinal Wyszyński Institute of Cardiology in Warsaw [11]. Patient characteristics are presented in Table 1.

Table 1. Patients clinical characteristics ($n = 29$), ACEI – Angiotensin Converting Enzyme Inhibitors, ARB – Angiotensin Converting Enzyme Inhibitor, CABG – Coronary Artery Bypass Grafting, PCI – Percutaneous Transluminal Intervention, SCD – Sudden Cardiac Death.

	Age (years)	56.1 ± 9.8
	Male gender (%)	26 (90)
	Left ventricular ejection fraction %	32 ± 12.4
	PCI n (%)	8 (27.6)
	CABG n (%)	3 (10.3)
Indications for ICD implantation n (%)	Primary prophylaxis of SCD	9 (31)
	Secondary prophylaxis of SCD	20 (69)
Pharmacological treatment n (%)	Amiodarone	18 (62.5)
	Sotalol	2 (6.8)
	β-blockers	28 (96.5)
	ACEI and (or) ARB	29 (100)
	Statins	23 (82.4)
	Antiplatelet drugs	27 (93.1)
	Diuretics	10 (34.5)
	Aldosterone blockers	12 (41.3)
ICD manufacturer n (%)	Biotronik	13 (44.8)
	Medtronic	16 (54.2)

Typical signals of RR intervals from a patient with ICD during normal rhythm and during arrhythmia (VF) are shown in Fig. 2.

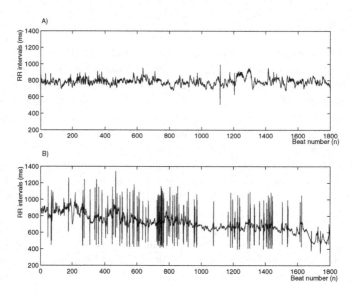

Fig. 2. RR intervals from patient with ICD: A) during normal rhythm, B) during arrhythmia (VF).

2.2 Data Preprocessing

Data preprocessing was performed with the help of the RHRV package for analysis of heart rate variability of ECG records [12] implemented in R [13]. We followed the basic procedure proposed by the authors of this package. First, the heart beat positions were used to build an instantaneous heart rate series. Then, the basic filter was applied in order to eliminate spurious data points. Finally, the interpolated version of data series with equally spaced values was generated and used in frequency analysis. The default parameters were used for the analysis, with the exception of the width of the window for further analysis, as described later. For every signal we generated descriptors – performed basic analysis in time domain, frequency domain and also we calculated parameters related to selected nonlinear methods.

2.3 Descriptors

The preprocessed data series was then used to generate 47 descriptors using following approaches: statistical analysis in time domain, analysis in frequency (Fourier analysis) and time-frequency (wavelet analysis) domains, nonlinear analysis (Poincaré maps, the detrended fluctuation analysis, and the recurrence quantification analysis). The detailed description of the parameters is presented below.

Statistical Parameters in Time Domain. Statistical parameters [12] calculated in time domain are:

- SDNN—standard deviation of the RR interval,
- SDANN—standard deviation of the average RR intervals calculated over short periods (50 s),
- SDNNIDX—mean of the standard deviation calculated over the windowed RR intervals,
- pNN50—proportion of successive RR intervals greater than 50 ms,
- SDSD—standard deviation of successive differences,
- r-MSSD—root mean square of successive differences,
- IRRR—length of the interval determined by the first and third quantile of the ΔRR time series,
- MADRR—median of the absolute values of the ΔRR time series,
- TINN—triangular interpolation of RR interval histogram,
- HRV index—St. George's index.

Parameters in Frequency Domain and Time-Frequency Domain. In frequency domain and time-frequency domain we performed Fourier transform and wavelet transform, obtaining a power spectrum for frequency bands.

Spectral analysis is based on the application of Fourier transform in order to decompose signals into sinusoidal components with fixed frequencies [14]. The power spectrum yields the information about frequencies occurring in signals. In particular we used RHRV package and we applied STFT (short time Fourier transform) with Hamming window (in our computations with parameters size = 50 and shift = 5, which, after interpolation, gives 262–376 windows, depending on the signal).

Wavelet analysis allows to simultaneously analyse time and frequency contents of signals [15]. It is achieved by fixing a function called mother wavelet and decomposing the signal into shifted and scaled versions of this function. It allows to precisely distinguish local characteristics of signals. By computing wavelet power spectrum one can obtain the information about frequencies occurring in the signal as well as when these frequencies occur. In this study we used Daubechies wavelets.

We obtained mean values and standard deviations for power spectrum (using Fourier and wavelet transform) for 4 frequency bands: ULF—ultra low frequency component 0–0.003 Hz, VLF—very low frequency component 0.003–0.03 Hz, LF—low frequency component 0.03–0.15 Hz, HF—high frequency component 0.15–0.4 Hz. We have also computed mean values and standard deviations of LF/HF ratio, using Fourier and wavelet transform.

Parameters from Nonlinear Methods

Poincaré Maps. We used standard parameters derived from Poincaré maps, They are return maps, in which each result of measurement is plotted as a function of the previous one. A shape of the plot describes the evolution of the system and allows us to visualise the variability of time series (here RR-intervals). There are standard descriptors used in quantifying Poincaré plot geometry, namely SD1 and SD2 [16,17], that are obtained by fitting an ellipse to the Poincaré map. We also computed $SD1/SD2$ ratio.

DFA Method. Detrended Fluctuation Analysis (DFA) quantifies fractal-like autocorrelation properties of the signals [18,19]. This method is a modified RMS (root mean square) for the random walk. Mean square distance of the signal from the local trend line is analysed as a function of scale parameter. There is usually a power-law dependence and an interesting parameter is the exponent. We obtained 2 parameters: short-range scaling exponent (fast parameter f.DFA) and long-range scaling exponent (slow parameter s.DFA) for time scales.

RQA Method. We computed several parameters from Recurrence Quantification Analysis (RQA) which allow to quantify the number and duration of the recurrences in the phase space [20]. Parameters obtained by RQA method [12] are:

- REC – recurrence, percentage of recurrence points in a recurrence plot,
- DET – determinism, percentage of recurrence points that form diagonal lines,
- RATIO – ratio between DET and REC, the density of recurrence points in a recurrence plot,
- Lmax – length of the longest diagonal line,
- DIV – inverse of Lmax,
- Lmean – mean length of the diagonal lines; Lmean takes into account the main diagonal,
- LmeanWithoutMain – mean length of the diagonal lines; the main diagonal is not taken into account,
- ENTR – Shannon entropy of the diagonal line lengths distribution,
- TREND – trend of the number of recurrent points depending on the distance to the main diagonal,
- LAM – percentage of recurrent points that form vertical lines,
- Vmax – longest vertical line,
- Vmean – average length of the vertical lines.

2.4 Identification of Informative Variables

We have used several methods to identify the descriptors generated from the signal that are related to the occurrence of arrhythmia, namely the straightforward t-test, importance measure from the Random Forest [21], relevant variables returned by Boruta algorithm for all-relevant feature selection [22], as well as relevant variables returned by the MDFS (Multi-Dimensional Feature Selection) algorithm [23,24]. Boruta is a wrapper on the Random Forest algorithm, whereas MDFS is a filter that relies on the multi-dimensional information entropy and therefore can take into account non-linear relationships and synergistic interactions between multiple descriptors and decision variable. We have applied MDFS in one and two-dimensional mode, using default parameters. All computations were performed in R [13], using R packages.

2.5   Predictive Models

Predictive models were built using Random Forest algorithm [21] and SVM (Support Vector Machine) [25]. The Random Forest model achieved better accuracy than the SVM model, which is consistent with the results presented by Fernández-Delgado et al. [26]. Hence, we focused on the Random Forest model exclusively, a method that can deal with complex, nonlinear relationships between descriptors and decision variable. It is routinely used as "out of the box" classifier in very diverse application areas. In a recent comprehensive test of 179 classification algorithms from 17 families, Random Forest was ranked as best algorithm overall [26]. It is an ensemble of decision tree classifiers, where each tree in the forest has been trained using a bootstrap sample of individuals from the data, and each split attribute in the tree is chosen from among a random subset of attributes. Classification of individuals is based upon aggregate voting over all trees in the forest. While there are numerous variants of Random Forest general scheme, we chose to use the classic algorithm proposed by Breiman implemented in the randomForest package in R [27]. Each tree in the Random Forest is built as follows:

- let the number of training objects be N, and the number of features in features vector be M,
- training set for each tree is built by choosing N times with replacement from all N available training objects,
- number $m << M$ is an amount of features on which to base the decision at that node. These features are randomly chosen for each node,
- each tree is built to the largest extent possible. There is no pruning.

Repetition of this algorithm yields a forest of trees, which all have been trained on bootstrap samples from training set. Thus, for a given tree, certain elements of training set will have been left out during training. The randomForest function was called with default parameters, with one modification – 1000 trees were used instead of 500.

Measuring Quality of Models and Validation of Modelling Procedure.
Three metrics were used to assess the quality of models: AUC (area under ROC curve) and MCC (Matthews Correlation Coefficient) [28] in addition to ordinary error level. Two former functions are more robust, in particular for imbalanced data sets.

It is well-known that variable selection can introduce significant over-fitting, especially when parameters selected within cross-validation are not highly informative [29]. To deal with the problem and to estimate the robustness of the models we applied the entire modelling was performed in five-fold cross-validation scheme. Then the procedure was repeated ten times and results are averaged to remove dependence on the particular split of data set into folds. This protocol is very demanding computationally, since entire modelling procedure is performed 50 times. In particular also the most time-consuming part of protocol, namely identification of informative variables, is performed 50 times. Nevertheless, these computations are essential for robust estimate of performance of the machine learning models.

3   Results and Discussion

3.1   Feature Selection

Feature selection was performed with the help of five algorithms using t-test, MDFS in 1 dimension (MDFS 1d) and 2 dimensions (MDFS 2d), Random Forest (RF) feature importance, and Boruta algorithm. Table 2 displays the number of times when variable was deemed relevant in fifty runs of each algorithm. The best results are presented according to the results obtained by Boruta.

Table 2. Number of occurrences of parameters in cross-validation loop for different feature selection methods.

Variable	Feature selection method				
	t-test	MDFS 1d	MDFS 2d	RF the best 10	Boruta
SD1/SD2	50	50	50	50	50
SD2	50	50	43	50	50
s.DFA	50	40	46	21	50
HRVi	47	49	49	49	45
mean.fULF	10	9	9	–	45
r-MSSD	50	49	34	49	31
SDNNIDX	19	29	16	44	31
IRRR	4	10	19	–	31
TINN	3	49	49	50	24
MADRR	2	11	13	46	23
SDSD	–	48	40	47	16
sd.fHF	41	17	–	–	13
sd.wHF	–	18	–	1	11
pNN50	50	47	34	–	9

For Random Forest feature importance one can see results for the best 10 features. The most frequently appearing parameters SD1/SD2 and SD2 are obtained from the Poincaré maps. The s.DFA arises from the Detrended Fluctuation Analysis. The HRVi, IRRR, r-MSSD, SDNNIDX, TINN, MADRR, SDSD and pNN50 variables are the statistical parameters in the time domain. The mean.fULF, sd.fHF and sd.wHF arise in the wavelet analysis. Interestingly, all methods agree on that variables arising from nonlinear analysis are most important. Then the relative importance of variables diverges among methods.

Most methods agree that statistical variables in time domain are important, but there are significant differences between methods with respect to which of them are most relevant. The largest disagreement concerns variables arising from spectral analysis, which are generally considered irrelevant by most methods, but some variables are considered very important by some methods.

3.2   Predictive Models

First, we tested whether predicting arrhythmia is even possible. The results of five point summary (Minimum, Maximum, Median, 1st Quartile, 3rd Quartile) statistics on a set of observations are presented in Table 3.

Table 3. Testing of arrhythmia prediction's possibility (*true* labels versus *random* labels).

Labels	Min	1st Quartile	Median	Mean	3rd Quartile	Max
true	0.14	0.25	0.31	0.30	0.33	0.58
random	0.22	0.42	0.50	0.50	0.58	0.83

We focused on the Random Forest model. The evaluation of prediction was done by 5-fold cross validation. Tests were carried out in two ways. First we performed 1000 iterations with true labels (Table 3 row labelled *true*). The result was poor: error median and mean were about 0.3. Nevertheless, it shows that it is possible to perform prediction. Then, we did the same procedure, but with random labels. Before each iteration a new set of labels was randomised. The next step was to perform the prediction using Random Forest in 5-fold cross validation. The results are in Table 3 (row labelled *random*). Mean and median of prediction error were 0.5. The comparison of the results described in Table 3 shows that there is a significant difference in prediction based on real and random labels.

The prediction results in cross-validation loop for different feature selection algorithms measured by AUC and MCC are presented in Table 4. The best results

Table 4. Results of prediction on different feature sets with selected parameters (mean value ± standard deviation of the mean).

Parameters set	Error	MCC	AUC
t-test	0.264 ± 0.008	0.432 ± 0.022	0.813 ± 0.009
MDFS 1d	0.274 ± 0.009	0.422 ± 0.022	0.807 ± 0.009
MDFS 2d	0.289 ± 0.010	0.404 ± 0.025	0.796 ± 0.011
RF the best 10	0.253 ± 0.008	0.450 ± 0.023	0.823 ± 0.010
Boruta	0.287 ± 0.008	0.396 ± 0.019	0.796 ± 0.010

were obtained for classifier that used 10 most relevant variables from the Random Forest. Results are stable and repeatable.

Each model was built using all variables that were deemed relevant by a feature selection algorithm in a given iteration of the cross-validation. Usually the number of relevant variables was close to 10—depending on applied feature selection method.

The prediction results in cross-validation loop for different feature selection algorithms measured by AUC and MCC are presented in Fig. 3. One can observe outlier points in MCC results of RF the best 10 features. Results are stable and repeatable.

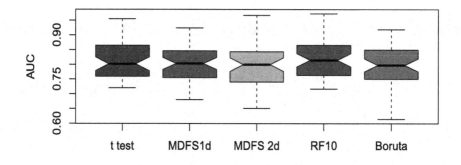

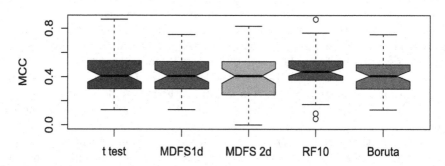

Fig. 3. Prediction results in cross-validation loop for different feature selection algorithms measured by AUC (top) and MCC (bottom).

In Fig. 4 we present AUC (area under ROC curve) for different feature selection algorithms.

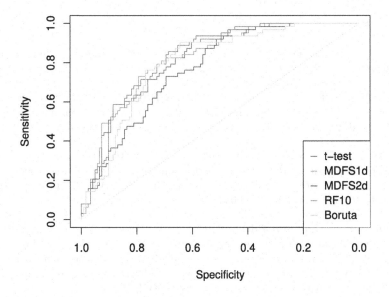

Fig. 4. AUC for different feature selection algorithms.

4   Conclusion

Based on obtained results we concluded that it's possible to find information about arrhythmia in RR intervals, but it's too weak to build useful medical predictive models using currently available methods. The subject requires further research to find algorithms better suited to the problem. In particular, a substantial increase of the size of the experimental sample, for instance by two or three orders of magnitude, should improve the quality of the models, as has been shown in numerous cases in applications of Machine Learning tools to different problems [30]. Additionally, it is likely that building individual models for each patient could yield better results.

Acknowledgements. This work was supported by the Polish Ministry of Science and Higher Education under subsidy for maintaining the research potential of the Institute of Informatics, University of Białystok (grant BST-144).

Conflict of Interest. A. P. was a consultant for Biotronik and receives lectures fees from Medtronic, Biotronik and Abbott. He receives also a proctoring contract from Medtronic.

References

1. Luz, E.J.S., Schwartz, W.R., et al.: ECG-based heartbeat classification for arrhythmia detection: a survey. Comput. Methods Program. Biomed. **127**, 144–164 (2016). https://doi.org/10.1016/j.cmpb.2015.12.008
2. Wilkoff, B.L., Fauchier, L., et al.: 2015 HRS/EHRA/APHRS/SOLAECE expert consensus statement on optimal implantable cardioverter-defibrillator programming and testing. EP Eur. **18**(2), 159–183 (2015). https://doi.org/10.1093/europace/euv411
3. Al-Khatib, S.M., Stevenson, W.G., et al.: 2017 AHA/ACC/HRS guideline for management of patients with ventricular arrhythmias and the prevention of sudden cardiac death: executive summary. Circulation **138**(13), e210–e271 (2018). https://doi.org/10.1161/CIR.0000000000000548
4. Shakibfar, S., Krause, O., et al.: Predicting electrical storms by remote monitoring of implantable cardioverter-defibrillator patients using machine learning. EP Eur. **21**, 268–274 (2019). https://doi.org/10.1093/europace/euy257
5. Iranitalab, I.: Prediction of arrythmia through analysis of the ventricular electrogram. A thesis presented to The Faculty of the Department of Chemical and Materials Engineering. San Jose State University (2009)
6. Taye, G.T., Shim, E.B., Hwang, H.-J., et al.: Machine learning approach to predict ventricular fibrillation based on QRS complex shape. Front. Physiol. **10**, 1193 (2019)
7. Blužaitė, I., Rickli, H., et al.: Assessment of QT dispersion in prediction of life-threatening ventricular arrythmias in recipients of implantable cardioverter defibrillator. Elek. Elektrotech. **75**(3), 73–76 (2007)
8. Cho, J., Kim, Y., Lee, M.: Prediction to atrial fibrillation using deep convolutional neural networks. In: Rekik, I., Unal, G., Adeli, E., Park, S.H. (eds.) PRIME 2018. LNCS, vol. 11121, pp. 164–171. Springer, Cham (2018). https://doi.org/10.1007/978-3-030-00320-3_20
9. Cp, P., Suresh, A., Suresh, G.: Prediction of cardiac arrhythmia type using clustering and regression approach (P-CA-CRA). In: 2017 International Conference on Advances in Computing, Communications and Informatics (ICACCI), pp. 51–54. IEEE (2017)
10. Loring, Z., Mehrotra, S., Piccini, J.P.: Machine learning in 'big data': handle with care. EP Eur. **21**(9), 1284–1285 (2019). https://doi.org/10.1093/europace/euz130
11. Przybylski, A., Baranowski, R., et al.: Verification of implantable cardioverter defibrillator (ICD) interventions by nonlinear analysis of heart rate variability - preliminary results. Eur. Eur. Pacing Arrhythm. Card. Electrophysiol. J. Work. Groups Card. Pacing Arrhythm. Card. Cell. Electrophysiol. Eur. Soc. Cardiol. **6**, 617–624 (2004). https://doi.org/10.1016/j.eupc.2004.08.001
12. Martínez, C.A.G., Quintana, A.O., et al.: Heart Rate Variability Analysis with the R Package RHRV. Springer, Heidelberg (2017). https://doi.org/10.1007/978-3-319-65355-6. https://www.springer.com/gp/book/9783319653549. Accessed 6 Sept 2019
13. R Development Core Team, R: A language and environment for statistical computing. R Foundation for Statistical Computing, Vienna, Austria (2008)
14. Challis, R.E., Kitney, R.I.: Biomedical signal processing (in four parts). Part 2. The frequency transforms and their inter-relationships. Med. Biol. Eng. Comput. **29**, 1–17 (1991)

15. Mallat, S.: A Wavelet Tour of Signal Processing: The Sparse Way. Academic Press, Cambridge (2008)
16. Brennan, M., Palaniswami, M., Kamen, P., et al.: Do existing measures of Poincaré plot geometry reflect nonlinear features of heart rate variability? IEEE Trans. Biomed. Eng. **48**, 1342–1347 (2001). https://doi.org/10.1109/10.959330
17. Tulppo, M.P., Mäkikallio, T.H., et al.: Quantitative beat-to-beat analysis of heart rate dynamics during exercise. Am. J. Physiol. **271**, H244–H252 (1996). https://doi.org/10.1152/ajpheart.1996.271.1.H244
18. Rodriguez, E., Echeverria, J.C., Alvarez-Ramirez, J.: Detrended fluctuation analysis of heart intrabeat dynamics. Phys. A: Stat. Mech. Appl. **384**, 429–438 (2007). https://doi.org/10.1016/j.physa.2007.05.022
19. Peng, C.K., Havlin, S., et al.: Quantification of scaling exponents and crossover phenomena in nonstationary heartbeat time series. Chaos **5**(1), 82–87 (1995). https://doi.org/10.1063/1.166141
20. Zbilut, J.P., Thomasson, N., Webber, C.L.: Recurrence quantification analysis as a tool for nonlinear exploration of nonstationary cardiac signals. Med. Eng. Phys. **24**, 53–60 (2002). https://doi.org/10.1016/S1350-4533(01)00112-6
21. Breiman, L.: Random Forests. Mach. Learn. **45**, 5–32 (2001). https://doi.org/10.1023/A:1010933404324
22. Kursa, M.B., Jankowski, A., Rudnicki, W.R.: Boruta - a system for feature selection. Fundam. Inf. **101**, 271–285 (2010)
23. Piliszek, R., Mnich, K., et al.: MDFS - Multidimensional feature selection in R. R J. **11**, 198–210 (2019)
24. Mnich, K., Rudnicki, W.R.: All-relevant feature selection using multidimensional filters with exhaustive search. Inf. Sci. (2020, in Press). https://doi.org/10.1016/j.ins.2020.03.024
25. Cortes, C., Vapnik, V.: Support-vector networks. Mach. Learn. **20**, 273–297 (1995). https://doi.org/10.1023/A:1022627411411
26. Fernández-Delgado, M., Cernadas, E., et al.: Do we need hundreds of classifiers to solve real world classification problems? J. Mach. Learn. Res. **15**, 3133–3181 (2014)
27. Liaw, A., Wiener, M.: Classification and Regression by randomForest. R News. **2**, 18–22 (2002)
28. Matthews, B.W.: Comparison of the predicted and observed secondary structure of T4 phage lysozyme. Biochim. Biophys. Acta (BBA) - Protein Struct. **405**, 442–451 (1975). https://doi.org/10.1016/0005-2795(75)90109-9
29. Cawley, G.C., Talbot, N.L.C.: On over-fitting in model selection and subsequent selection bias in performance evaluation. J. Mach. Learn. Res. **11**, 2079–2107 (2010)
30. Halevy, A., Norvig, P., Pereira, F.: The unreasonable effectiveness of data. IEEE Intell. Syst. **24**, 8–12 (2009). https://doi.org/10.1109/MIS.2009.36

Stroke ICU Patient Mortality Day Prediction

Oleg Metsker[1](✉), Vozniuk Igor[2], Georgy Kopanitsa[1] (iD),
Elena Morozova[2], and Prohorova Maria[2]

[1] ITMO University, Saint Petersburg, Russia
olegmetsker@gmail.com, georgy.kopanitsa@gmail.com
[2] The Saint Petersburg Research Institute of Emergency Medicine. I.I.
Dzhanelidze, Saint Petersburg, Russia
voznjouk@yandex.ru, novaj44@mail.ru, airty@mail.ru

Abstract. This article presents a study on development of methods for analysis of data reflecting the process of treatment of stroke inpatients to predict clinical outcomes at the emergency care unit. The aim of this work is to develop models for the creation of validated risk scales for early intravenous stroke with minimum number of parameters with maximum prognostic accuracy and possibility to calculate the time of "expected intravenous stroke mortality". The study of experience in the development and use of medical information systems allows us to state the insufficient ability of existing models for adequate data analysis, weak formalization and lack of system approach in the collection of diagnostic data, insufficient personalization of diagnostic data on the factors determining early intravenous stroke mortality.

In our study we divided patients into 3 subgroups according to the time of death - up to 1 day, 1 to 3 days, and 4 to 10 days. Early mortality in each subgroup was associated with a number of demographic, clinical, and instrumental-laboratory characteristics based on the interpretation of the results of calculating the significance of predictors of binary classification models by machine learning methods from the Scikit-Learn library. The target classes in training were "mortality rate of 1 day", "mortality rate of 1–3 days", "mortality rate from 4 days". AUC ROC of trained models reached 91% for the method of random forest. The results of interpretation of decision trees and calculation of significance of predictors of built-in methods of random forest coincide that can prove to correctness of calculations.

Keywords: ICU · Stroke · Mortality · Machine learning · Mortality prediction

1 Introduction

Stroke is the second most deadly cause of death worldwide. In Russia, brain stroke is the second leading cause of death after myocardial infarction. Every year around 450000 people suffer from stroke, in fact it is the population of a big city [1]. The mortality rate in Russia is 4 times higher than in the USA and Canada [2]. Among European countries, the mortality rate from cerebrovascular diseases is the highest in Russia. According to the All-Russian Center for Preventive Medicine, in our country 25% of men and 39% of women die from cerebrovascular diseases. In the largest cities

© Springer Nature Switzerland AG 2020
V. V. Krzhizhanovskaya et al. (Eds.): ICCS 2020, LNCS 12140, pp. 390–405, 2020.
https://doi.org/10.1007/978-3-030-50423-6_29

of the country the situation with this type of pathology is extremely unfavorable. In St. Petersburg, for example, the frequency of stroke is about 528 cases per 100,000 residents, while the mortality rate for ischemic stroke is 39%. It is necessary to emphasize the catastrophic consequences of ischemic stroke - up to 84–87% of patients die or remain disabled and only 16–13% of patients fully recover [2]. According to the findings of a large-scale study of recent years, some modern epidemiological trends have been identified [3]: In general, global statistics show a decline in stroke mortality over the past two decades due to the introduction of new treatments (thrombolysis, thrombectrosis), but the absolute number of people who have stroke is only increasing every year [4]. This nosology is still the strong second leading cause of death from cardiovascular disease (CVD), remaining the undisputed leader among all nosologies leading to severe disability. Hospital mortality remains one the most important quality indicator, which can be used to identify problems associated with the optimization of prehospital and hospital treatment process. It can be used to assess the effectiveness of primary and secondary care, routing, and the degree of implementation of modern diagnostic and treatment algorithms, including the quality of interaction between different levels of care [5]. It is important to note that regional characteristics of the populations may significantly differ from the global ones, and the development of specialized care programs for patients with a stroke has its national and institutional characteristics. Understanding the factors that contribute to the reduction of hospital mortality will allow us to develop a targeted strategy for the development of services providing care to patients with a stroke in Russia and in the world. Thus, development of personalized models and algorithms for planning of individual treatment tactics for the stroke patients can reduce mortality and increase the standard of life. The development of such models and algorithms will ensure better continuity and efficiency of medical care and help reducing the number of complications. The basis for such models can be the scales of calculation of patients' mortality risks in emergency units, which are also absent in Russia at present.

2   Related Works

Most statistics are accumulated in national stroke registries or national databases: China National Stroke Registry II (CNSR II) [6], the Nationwide Hospital Discharge Database (NHDD), Berlin Stroke Register (BSR), German Stroke Register, the Registry of the Canadian Stroke Network (RCSN), National Acute Stroke Israeli (NASIS) registry, FLENI Stroke Data Bank, Australian Stroke Clinical Registry (AuSCR), National Stroke Register of Ireland, the Austrian Stroke Registy. The analysis of available literature revealed rather heterogeneous values of the share of hospital mortality of patients with stroke in different countries. At the same time, direct indicators of the share of hospital mortality had significant differences from 1,4% in China [7] to 22,7% In Ethiopia [8]. Significant differences in data can be explained both by the quality of care and by the nature of statistical data collection. In particular, most of the reports took into account only the ischemic type of stroke [9–14], different exclusion criteria were applied in a number of observations - daily mortality and stay exceeding 180 days [9], inhospital stroke [15], patients in need of admission to the general intensive care unit [16] or a general department. It should also be noted that samples are heteroge-neous in terms of the number of patients: from 110 [8] to 12 million patients [17]. Hospital mortality rates vary considerably between facilities within the same country. For example, the average hospital mortality rate in Germany in 2011 was 4.6% when 26 Stroke units were evaluated. [13], at the same time as in the German study of 2015 on this parameter was 8.2% [18]. In Australia, hospital mortality also varies signifi-cantly (from 7% to 23%) depending on the level of the hospital. [15], in Germany, there is a dependence on the size of the hospital - from 0% to 25% in small hospitals and from 0.4% to 9.3% in large hospitals. [11]. Only 9 studies out of 22 provide data that allow tracking the dynamics of changes in the indicator of intra-hospital mortality. The average rate of decline in this indicator was 0.36% per year. Rapid changes in this parameter are more typical of the ischemic type of stroke, and mainly the faster rate of decline was associated with the introduction and expansion of the vascular center network for stroke (with mandatory Stroke Unit). The most significant example of Canada - where vascular center system was introduced, which led to the rate of change in the provinces was 0.28% per year, while in the provinces without the introduction of the vascular center system, the rate changed only by 0.11% per year [19]. The avail-ability of prognostic models and scales that are understandable to clinical staff and easy to operate, reduces hospital mortality and allows for a more targeted and individualized approach to therapy. Such models should take into account locally established prac-tices. Models should be available that can predict a fatal scenario for the disease, considering all relevant factors. To date, the international medical community has made repeated attempts to create such a prognostic scale. In the 2002 review, C. Counsell and M. Dennis analyzed 83 models with a total of 150 prognostic factors, and the assessment resulted in only 4 models meeting quality criteria [8]. The databases have a huge number of parameters including various tests and indicators. In some cases, the use of a large number of features leads to lower rates of learning and forecasting, reduces the predictive accuracy of the model, and prevents the model from being interpreted, which is an important requirement for models used in medicine. Thus, finding the best set of features in the context of our task is one of the key factors ensuring high quality of the predictive model.

On the basis of the analysis of 12 modern prognostic models from 10 countries we can identify some of the most stable (main) predictors for the causes of intra-hospital mortality: age [16, 20–24]; type of stroke [25]; lesion location [25]; level of con-sciousness [11, 20, 23, 25, 26] upon admission; NIHSS stroke severity [10, 21, 22, 24]; comorbidity [22, 27], Charlson comorbidity index [23], Atrial fibrillation [11, 22], case history Transitor ischemic attack (TIA) [31]; hospital complications (high intracranial pressure) [16], pneumonia, seizures, anxiety/depression, infections, limb pains and constipation [22, 27].

Among the predictors related to the organization of care, the time of admission to hospital can be noted - in a Japanese study, the 7-day mortality rate increased if the patient was admitted on weekends or holidays. [23], hospital delivery method had a predictive value as well [16], Both these parameters are included in the GWTG-Stroke program [14]. In order to identify priority areas for improving the outcome of the disease it is necessary to divide the selected factors (predictors) into modifiable and unmodifiable, respectively. Modifiable mortality predictors can be referred to: time and method of hospital delivery; qualifications of medical personnel; stroke care model; history of stroke or TIA, atrial fibrillation, diabetes mellitus, comorbidity index - parameters to which primary prevention should be directed; intra-hospital complications (high intracranial pressure pneumonia, seizures, anxiety/ depression, infection, extremity pain and constipation). A special form of complications in the form of extracerebral pathology - polyorgan failure syndrome - is distinguished separately. Special attention should be paid to the prevention of this syndrome. The unmodifiable factors of stroke mortality include: gender, age, type of stroke, localization of lesion. As for the assessment of the impact of comorbid diseases, it is important to consider not only the presence of individual pathologies, but also their combination. In particular, the following groups can be distinguished: Arterial hypertension + atrial fibrillation, Arterial hypertension + atrial fibrillation + Coronary heart disease, atrial fibrillation + postinfarction cardiosclerosis, and, Arterial hypertension + postinfarction cardiosclerosis + atrial fibrillation MA и + Diabetes mellitus. Only two studies presented clear prognostic scales containing a scoring system for rapid assessment of the risk (probability) of in hospital mortality [14, 21]. The PREMISE scale is simple, quick to calculate at > 85% of strokes and uses only variables that are available shortly after the onset of ischemic stroke when admitted to the Stroke Unit. It should be noted that the practical application of any analyzed scale above in different countries requires corrections to take into account regional peculiarities - social, geographical and medical and economic factors [26]. The creation of such scales and models in Russia would provide a tool to assess the efficiency of care. The goal of this work is to identify features for the creation of validated risk scales for early hospital mortality.

3 Methods

3.1 Cohort Description

The study includes data about 36450 episodes (17158 outpatient 5565-inpatient patients 200-lethal patients 5565 patients who has international criteria for diagnosis ICD i60 to i69.8) and were treated in the Almazov national research center from 2011 to 2019. Among the causes of admission: ischemic stroke, hemorrhagic stroke, embolic stroke, transitor attacks. As the initial data describing the condition the patient's examination data at the intake and use of clinical scales (NIHSS, mRS), conclusion of magnetic resonance imaging (MRI), conclusion of ultrasound investigation, data from laboratory tests, data on treatment events from the medical information system. A separate more detailed analysis of the group of only deceased patients from 100 people was carried out to identify differences and mortality factors in different time periods (1 day, 2–3 days, 4–15 days) on the basis of data from the The Saint Petersburg Research Institute of Emergency Medicine n.a. I.I. Dzhanelidze[1].

The data of the medical information system of the operating specialized center of MRI, ultrasound and other characteristics of the volume of cerebral and vascular stroke examination were compared with the data on the duration and outcomes and time of

---

[1] http://www.emergency.spb.ru/.

death. Information about hospital mortality was included in the study, if they met the following criteria: the fact of clinically confirmed diagnosis of acute cerebral circulation disorder (ischemic or hemorrhagic), with the presence of focal, general cerebral neurological syndromes, which lasted more than 24 h from the beginning of the disease; hospitalization in connection with stroke in the first day of the disease; the entire period of hospitalization in connection with acute case of the patient spent in one institution; lethal outcome was associated with an acute period of stroke. Information confirming lesions of the brain substance has been obtained from data from the CT scan and/or MRI of the brain, which have been repeated if necessary. The extent of precerebral and cerebral artery lesions was assessed using ultrasound duplex scanning, CT scan, MRI or cerebral angiography.

3.2   Analysis and Machine Learning Methods

To obtain the optimal set of features a combination of classical methods based on different correlation coefficients of features (Pearson correlation coefficient and Spearman correlation coefficient) were used. Ensemble algorithms, including ensembles built on the basis of models with the use of decision trees, and random forest are used as prognostic models. A Scikit-Learn library was used to implement machine learning methods. In the process of definition of hyperparameters of machine learning models, cross-validation by k blocks was applied. Precision and Recall (accuracy and completeness), as well as their harmonic mean (F-score) were used as metrics at this stage. Construction of the confusion matrix of multiclass classification allowed to analyze errors, improve data sampling used for model training and initialize the next iteration of model training. The data on treatment of real patients from the Almazov Center were used for validation of the final resulting models. The data of patients who did not participate in any stages of model training and adjustment of hyper-parameters were used. AUC ROC - the area under the error curve - was used as the result metrics. P-value was calculated using two methods. The essence of the first method is that for each sample of dead (<1 day, 1–3 days, 4–10 days) we have calculated P-value for every feature of the corresponding test. Chi-square criterion was used for categorical features, Kolmogorov-Smirnov's test was used for continuous features. The essence of the second method of calculating P-value by one attribute (mortality period) for three groups of patients according to the severity and type of stroke (group 1: ICH+PVH, IS +ICH; group 2: IS+Bilat atr, Is-foc 16-25hu; group 3: IS-1/3<16HU, Sub Tent ICV).

4   Results

The analysis obtained a general model of mortality for all patients with stroke AUC ROC-93% demonstrated random forest learned on the dataset with more than 60 laboratory and personal patient observation features. Three separate models have also been developed for patients with different lethality periods (up to 1 day, from 1 to 3 days, and from 4 to 15 days) using decision trees that showed an AUC ROC of 85 to 91%. For these models, the dataset consisted of more than 70 specialized features, including a score on neurologic scales, brain examination data, assessment of the patient's consciousness and somatic state. The importance of features for different duration of lethality was also compared. Moreover, a clinical interpretation of the comparison results is given below.

4.1 General Machine Learning Model

The models were trained on the dataset describing 5565 patients who were treated as a binary classification models by machine learning methods from the Scikit-Learn library. The following parameters were used as features: patients age, male, pressure, area of brain damage, the size of the hematoma. Moreover, the following laboratory tests were used as features: MCHC-red blood cell index, endothelin, interleukin-10, interleukin-8, interleukin-6, interleukin-4, interleukin-1-beta, INR, fibrogen, vitamin D, paratohormone, urine Nitrites, urine bilirubin, urine, bld urine, LEU urine, urine NIT, urine KET, urine glucose, urine PRO, urine Ph, urine color, D-dimmer, albumin, lipids, triglyceride, total cholesterol, prothrombin index, fibrinogen by Klaus, K+ (Vienna), neutrophils, monocytes, lymphocytes, MPV average, platelet volume, PDW Width of platelet distribution by volume, RDW Width of red blood cell distribution by volume, MCHC the average concentration of hemoglobin in eritr, MCH is the average hemoglobin content in 1 erythrocyte average volume of red blood cells, reactive protein, erythrocyte sedimentation rate, troponin, ALT, AST, HGB Hemoglobin, WBC white blood cells, RBC red blood cells, PLT platelets, creatinine, bilirubin, HCT Hematocrit, glucose level.

Random forest demonstrated the best AUC ROC-93%. The nine most importantly lethality features of the stroke patient further: systolic pressure (0.06), RBC red blood cells (0.05), interleukin-8 (0.05), HCT hematocrit (0.04), diastolic pressure (0.04), age (0.03), MCHC - red blood cell index(0.03), ventricular damage(0.03), hematoma volume (0.03).

The following conclusions emerge from the general analysis of the overall data: 1. *Terms of mortality.* All cases of death of patients, which were distributed within 14 days, were estimated, with the greatest number of lethal outcomes occurring within 5 days. 2. *Patients age* 60 to 90 years (at least 65% out of 200 dead) were most frequently encountered in the group of the deceased, the maximum frequency (25%) falls on the age of 60 to 70 years, in the same age group there is the maximum morbidity of stroke with their share is almost 35% of the number of diseased. 3. *Among the deceased*, men prevailed (by more than 25%). 4. *The proportion of deaths* in the hemorrhagic stroke cohort was twice as high compared with the proportion of deaths in ischemic stroke patients.

From the general analysis, several interlinked signs are evident indicating the likelihood of lethal outcomes in patients with cerebrovascular disease at an early stage: hemorrhagic type of stroke is most likely to be lethal in patients with acute cerebrovascular disease; stroke incidence and mortality are highest in patients aged 60 to 70 years; stroke with lethal outcomes are more likely in men; regardless of the type of stroke, lethal outcomes are most likely in patients aged 60 to 90 years.

4.2   Mortality Day Prediction Models

All patients were divided into 3 subgroups according to the time of death - up to 1 day, 1 to 3 days, and 4 to 10 days. Early mortality in each subgroup was associated with a number of demographic, clinical, and instrumental-laboratory characteristics based on the interpretation of the results of calculating the significance of predictors of binary classification models by machine learning methods from the Scikit-Learn library[2]. The target classes in training were "mortality rate of 1 day", "mortality rate of 1–3 days", "mortality rate from 4 days". AUC ROC of trained models reached 91% for the method of random forest. The results of interpretation of decision trees and calculation of significance of predictors of built-in methods of random forest coincide that can testify to correctness of calculations. As a result of the decision trees, the following conclusions were drawn regarding the time frame of death:

1. Factors that cause patients to be lethal on the first day: Patient's age over 67 years; male sex; significant volume of brain lesions (more than 1/3 of the middle cerebral artery pool) hemispheric ischemic (or hemorrhagic with impregnation of the ischemic focus) stroke or patients with intracerebral hematoma (both less than 50 ml and 50 to 100 ml) with a breakthrough into the ventricular system of the brain; more important was the combination of ischemic or hemorrhagic lesions with displacement of the medial structures due to perifocal edema; right hemispheric cerebral lesion; severe condition at entry (with severe neurological deficit, up to 23 NIHSS points); unstable systemic hemodynamics, expressed by fluctuations in blood pressure, appearance of tachycardia and tachyarrhythmia, i.e., in the ventricular system.h. with sharp rise (>200 mm Hg) or sharp decrease (<65 mm Hg).) systolic and diastolic blood pressure and heart rhythm disorders (tachycardia and tachyarrhythmia); manifestations of decompensated hypersympathicotonia accompanied by hyperthermia and polyuria (densephalic syndrome, irritation of the densephalic region of the brain) and hemoconcentration (hypercoagulation); high degree of comorbidity (presence of significant number of concomitant diseases at the decompensation stage, comorbidity index > 6.5).

2. Mortality in the group from 1 to 3 days is caused by the following factors: age over 55 years old; male gender; consciousness impairment not lower than stun; presence of extensive hemispheric ischemic (more than 1/3 of the middle cerebral artery basin) or large intracerebral hematoma against the background of pronounced brain atrophy, in some cases with hemorrhagic saturation of the ischemic focus; the greatest importance was given to the combination of ischemic or hemorrhagic lesions with displacement of the medial structures due to perifocal edema; lesion of the right hemisphere; instability of system hemodynamics - with indicators of sharp decrease (<65 mm Hg.st.) of systolic blood pressure, heart rate - with indicators of sharp decrease.st.) systolic blood pressure, heart rhythm disorders (bradiarrhythmia and tachycardia); or with a high degree of comorbidity (presence of a significant number of concomitant diseases at the decompensation stage, comorbidity index > 6.5); vivid manifestations of vegetative regulation decompensation (hypersympathicotonia),

---

[2] https://scikit-learn.org/.

accompanied by hyperthermia and polyuria (diencephal syndrome, irritation of the diencephalic region of the brain) and hemoconcentration (hypercoagulation); phenomena of systemic inflammatory reaction and presence of signs of hemoconcentration in blood tests;

3. The largest contribution to the patients' mortality from 4 to 10 days was made by the following factors: age from 30 to 90 years (the largest group of patients aged 60–70 years); female gender; extensive hemispheric ischemic (more than 1/3 of the pool of the middle cerebral artery) in combination with severe hemispheric atrophy, or the presence of intracerebral hematoma (much more often less than 50 ml), a breakthrough into the ventricles of the brain, the most important was the presence of dislocation, a combination of ischemic or hemorrhagic lesions with the displacement of medial structures due to general edema; conscious disturbance (stun, coma) or condition that required sedation (to provide prosthetics for breathing function); unstable systemic hemodynamics - with sharp rise (>200 mm Hg) or (<65 mm Hg) of systolic blood pressure; phenomena of moderate hemoconcentration and moderate systemic inflammatory response in blood tests; high degree of comorbidity (presence of a significant number of concomitant diseases at the decompensation stage, comorbidity index > 5). At the same time, it should be noted that in contrast to patients with 1–3 day mortality, in this case the side of the brain lesion did not matter.

4.3   Analysis of Mortality Factors in Subgroups Based on P-Value Assessment

Three groups of patients were compared by the terms of mortality (mortality in the first day, mortality from 1 to 3, lethality from 4 to 10 days) among themselves by means of standard t-test (non-parametric criterion Chi) with thirty one parameter. The value $P < 0.0005$ was considered significant. The results of the interpretation of the obtained test are presented in Tables 1, 2 and 3.

Table 1. Interpretation of the individual features

Feature	<1 day P	1–3 days P	4–10 days P	Interpretation
Age	0,4190	0,7048	0,6351	Age was not specific (significant) for the development of mortality in the groups under consideration, since the most typical for patients of all subgroups was the age from 60 to 90 years

(*continued*)

Table 1. (*continued*)

Feature	<1 day P	1–3 days P	4–10 days P	Interpretation
Gender	0,0003	0,0004	0,0048	Gender showed the significance of differences between all subgroups, with the groups with mortality of 4-10 days dominated by women, and between the subgroups of mortality up to 1 day and mortality of 1–3 days, with a general prevalence of incidence of men among deceased patients, the frequency of occurrence in the subgroups also significantly differed
Period of admission less than 4.5 h	0,0003	0,0003	0,7530	The difference between subgroups of up to 1 day and 1–3 days is insignificant, i.e. the fact of early hospitalization did not affect the earlier fatal outcome. The differences in subgroups 1–3 and 4–10 are significant
Period of admission more than 4.5 h	0,0003	0,0002	0,5291	The difference between subgroups of up to 1 day and 1–3 days is insignificant, i.e. the fact of later hospitalization did not affect earlier mortality. The differences in subgroups 1–3 and 4–10 are significant, for lighter patients (with lower comorbidity index or with severe atrophy)
Period of admission over 24 h	0,0002	0,0002	0,6332	The difference between subgroups of up to 1 day and 1–3 days is insignificant, i.e. the fact of later hospitalization did not affect earlier mortality. The differences in subgroups 1–3 and 4–10 are significant, for lighter patients (with lower comorbidity index or with severe atrophy)

The following calculation results have been obtained P-value using method 2: for the groups 1 and 2 P-Value = 0.0052; for the groups 2 and 3 P-Value = 0.0042; from the groups 1 and 3 P-Value = 0. On the basis of the analysis calculations it is possible to draw a conclusion about a significant difference between the 1st and the 3rd group, where group 1: ICH (Intracerebral hemorrhage) +PVH (periventricular hyperintensity), IS (ischemic stroke) +ICH (intracranial hemorrhage); group 2: IS+Bilat atr, IS-foc 16–25 hu; group 3: IS-1/3<16HU, Sub Tent ICV (Intracerebroventricular).

Table 2. Interpretation of the Clinical features of the acute period

Feature	<1 day P	1–3 days P	4–10 days P	Interpretation
NIHSS	0,6833	0,0057	0,0401	The difference between subgroups of up to 1 day and 1–3 days is significant, i.e. the fact of more expressed. The differences in subgroups of up to 1 day 4–10 days and in subgroups of 1–3 days and 4–10 days were insignificant
				Clinical features of the acute period
Charlson comorbidity index	0,0007	0,0007	1,0000	The indicator had the significance of differences when comparing the frequency of mortality in the group 4–10 days with the groups of mortality up to 1 day and mortality 1–3 days, because high values of comorbidity index (>5 and >6.5) played a role in the development of gross and life-compatible disorders in the first 3 days of stroke, which predetermined the lethal outcome
Consciousness disruption to coma on admission (any quantitative disturbances of consciousness) as well as respiratory failure	0,0001	0,000004	0,3451	The difference between subgroups of up to 1 day and 1–3 days is insignificant, i.e. the presence of a coma did not affect earlier mortality. The differences in subgroups 1–3 and 4–10 are significant, for patients entering a coma the treatment program provides for immediate prosthetics of vital functions
Patients admitted in stun and remaining conscious for up to 3 days	0,0002	0,0001	0,3451	The difference between subgroups of up to 1 day and 1–3 days is insignificant, i.e. the presence of a coma did not affect earlier mortality. Differences in subgroups 1–3 and 4–10 are significant

(*continued*)

Table 2. (*continued*)

Feature	<1 day P	1–3 days P	4–10 days P	Interpretation
Patients in need of sedation up to 3 days on admission, as well as medical ventilation	<0,0001	<0,0001	1,0000	The difference between subgroups of up to 1 day and 1–3 days is insignificant, i.e. the necessity to apply sedation is always an unfavorable sign accompanying hypersympathetic reactions, development of dencephalic syndrome or psychomotor excitation, which are associated with high probability of mortality. The differences in subgroups 1–3 and 4–10 are significant, since the use of sedation is aimed at "smoothing" the manifestation of hypersympathicotonia
Instability of system hemodynamics (including bradycardia)	<0,0001	<0,0001	1,0000	The difference between subgroups of up to 1 day and 1–3 days is insignificant, i.e. the presence of instability of system hemodynamics is always a factor associated with early mortality. The differences in subgroups of 1–3 days and 4–10 days are significant, so the appearance of bradycardia and instability of blood pressure are signs of dislocation (contraction) of the central structures of the brain (middle brain, trunk, mediobasal sections of the temporal lobe) - mortality is earlier in these manifestations
Presence of systemic inflammatory reaction (SIR) or hemoconcentration	0,0003	0,0002	0,5261	The difference between subgroups of up to 1 day and 1–3 days is insignificant. The differences in the subgroups of 1–3 days and 4–10 days are significant, since the development of SIR takes some time, these reactions are signs of a complicated course often accompanied by clinically proven pneumonia, urinary tract infection and/or polyorganic failure syndrome

Table 3. Interpretation of morphology

Hemorrhagic stroke	0,0001	0,000016	0,0772	The indicator predetermined a large share of lethal outcomes in the first 3 days, and significantly affected the death in the period up to 1 day and from 1 to 3 days. The indicator was significant when comparing the subgroup 4–10 mortality with the other two
Affection side of hemispheric stroke (as well as aphasia)	0,0003	0,0002	0,7245	The influence of the defeat side on the probability of a lethal outcome is equally significant only when comparing subgroups of 1–3 days and 4–10 days, these differences are more significant for the left hemisphere; differences in the frequency of mortality in subgroups of up to 1 day and 1–3 days and in the left and right hemispheres are insignificant
Edema	0,0002	0,0002	1,0000	The difference between the subgroups of up to 1 day and 1–3 days is insignificant, i.e. the fact of edema affected earlier mortality. the differences in subgroups 1–3 and 4–10 are significant due to the fact that edema developed later as a factor affecting mortality or did not determine the mortality (e.g. in patients with severe atrophy, small foci)
Dislocation	0,0003	0,0003	1,0000	The difference between the subgroups of up to 1 day and 1–3 days is insignificant, i.e. the fact of edema influenced earlier mortality. the differences in subgroups 1–3 and 4–10 are significant due to the fact that the brain substance dislocation developed later as a factor influencing mortality or also did not determine the mortality (e.g. in patients with severe atrophy, small focus, cortical-subcortical focus, without affecting the central structures of the brain)

(*continued*)

Table 3. (*continued*)

Hematoma volume < 50 ml	0,0003	<0,0001	0,000013	The difference between the subgroups is insignificant, i.e. in each case, the fact of the hematoma of a small volume did not in itself cause death, did not have a direct impact on mortality
Hemorrhagic transformation	0,0003	0,0003	0,3307	The difference between subgroups of up to 1 day and 1–3 days is insignificant, i.e. in each case the fact of hemorrhagic impregnation of the zone of brain matter ischemia affected mortality. In subgroups of 1–3 days and 4–10 days this difference is significant due to availability of reserve spaces due to brain atrophy and less probability of dislocation of brain substance
Amount of ischemia > 1/3 of the middle cerebral artery (MCA)	0,0002	0,0002	0,4306	The difference between subgroups of up to 1 day and 1–3 days is insignificant, i.e. in each case the fact of extensive ischemic lesion had an impact on mortality in earlier periods. In subgroups of 1–3 days and 4–10 days this difference is significant due to availability of reserve spaces in connection with brain atrophy and less probability of threatening dislocation (constriction) of brain substance even in presence of a large focal point of ischemia and consequently edema and tissue swelling
Expressed atrophic changes in brain matter	0,0003	0,0003	0,6586	The difference between subgroups of up to 1 day and 1–3 days is insignificant, In the subgroups of 1–3 days and 4–10 days this difference is significant as the availability of reserve spaces due to brain atrophy reduces the probability of dislocation of brain matter even in the presence of a large focal point of ischemia or hemorrhage (large hematoma)

5   Conclusion and Future Work

A detailed study of electronic medical records data and combinations of clinical and laboratory characteristics of patients made it possible to reveal dependencies and develop descriptive models between the degree of lesion and the time of intra-hospital lethality of patients. Further, based on a large array of correlated data, models were developed to identify major favorable and unfavorable patterns of early mortality of patients for control and correction of the treatment plan. Decision-making models for predicting the outcome and duration of treatment of stroke patients have been developed using systems analysis, statistical analysis, mathematical modeling and machine learning methods. As a result, clinical and morphological predictors of early hospital stroke mortality have been identified. Similar models can also be used to validate existing scales, to study the causes of mortality at the emergency stages and to develop clinical guidelines, including for the prevention, diagnosis and treatment of stroke.

As a result of this study, descriptive and prognostic models of mortality in stroke patients have been developed. The significance of predictors was ranked using statistical and machine learning methods. Clinical interpretation of the obtained results was made in the form of clear conclusions that can be used in the organization of continuity care for acute stroke patients, as well as the calculation of personal risks. Provided that all standards of specialized medical care for patients with stroke are complied with, first of all, monitoring and intra-hospital logistics, completeness of the diagnostic scope, it is possible to make a prognostic assessment to identify predictors of early hospital lethality. A number of clinical, pathomorphological and instrumental parameters may indicate a high probability of early lethality, namely: Charlson comorbidity index with a value greater than 3.0; six subtypes of stroke; for 3 subtypes, combination with an extended intracellular CMA clot, with an age greater than 64 years and Ind. Ch-4.5 b. For 1 and 2 subtypes the severity of lesion volume and presence of dislocation complications determine the high risk of mortality. For 4 subtypes the greatest risk is associated with the combination of an acute focus in the deep parts of the temporal lobe with moderate perifocal ischemic oedema, compression of medial structures, with the age over 64 years old and high and Ind. Ch-4.5 b. For subtype 6, a significant contribution is made by global (diffuse atrophy or the presence of a fresh acute focus in the deep regions of the temporal lobe on the side of the opposite marked atrophy (including post-stroke).

Acknowledgements. This work was financially supported by the Government of the Russian Federation through the ITMO fellowship and professorship program. This work is financially supported by National Center for Cognitive Research of ITMO University.

References

1. Johnson, C.O., et al.: Global, regional, and national burden of stroke, 1990–2016: a systematic analysis for the Global Burden of Disease Study 2016. Lancet Neurol. **18**, 439–458 (2019)
2. Jdanov, D., et al.: Human mortality database. In: Gu, D., Dupre, M. (eds.) Encyclopedia of Gerontology and Population Aging. Springer, Cham (2019). https://doi.org/10.1007/978-3-319-69892-2
3. Feigin, V.L.: Global burden of stroke and risk factors in 188 countries, during 1990–2013: a systematic analysis for the Global Burden of Disease Study 2013. Lancet Neurol. **15**(9), 913–924 (2016)
4. Feigin, V., et al.: Global and regional burden of stroke during 1990–2010: findings from the Global Burden of Disease Study 2010. Lancet Neurol. **383**(9913), 245–255 (2014)
5. Lee, M., et al.: Impact of microalbuminuria on incident stroke: a meta-analysis. Stroke **41**(11), 2625–2631 (2010)
6. Wang, Y., et al.: The China National Stroke Registry for patients with acute cerebrovascular events: design, rationale, and baseline patient characteristics. Int. J. Stroke **6**(4), 355–361 (2011)
7. Sun, S., et al.: GWTG risk model for all stroke types predicts in-hospital and 3-month mortality in Chinese patients with acute stroke. J. Stroke Cerebrovasc. Dis. **28**(3), 800–806 (2019)
8. Fekadu, G., Chelkeba, L., Kebede, A.: Burden, clinical outcomes and predictors of time to in hospital mortality among adult patients admitted to stroke unit of Jimma university medical center: a prospective cohort study. BMC Neurol. **19**(1), 213 (2019). https://doi.org/10.1186/s12883-019-1439-7
9. Chen, H., et al.: Analysis on geographic variations in hospital deaths and endovascular therapy in ischaemic stroke patients: an observational cross-sectional study in China (2019). bmjopen.bmj.com
10. Alemayehu Gebreyohannes, E., et al.: In-hospital mortality among ischemic stroke patients in Gondar University hospital: a retrospective cohort study. Stroke Res. Treat. (2019)
11. Heuschmann, P.U., et al.: Predictors of in-hospital mortality and attributable risks of death after ischemic stroke the German Stroke Registers Study Group. Arch. Internal Med. **164**(16), 1761–1768 (2004)
12. Koennecke, H., et al.: Factors influencing in-hospital mortality and morbidity in patients treated on a stroke unit. Neurology **77**(10), 965–972 (2011)
13. Minnerup, J., et al.: Explaining the decrease of in-hospital mortality from ischemic stroke. PLoS ONE **10**(7) (2015). Public Library of Science
14. Smith, E.E., et al.: Risk score for in-hospital ischemic stroke mortality derived and validated within the get with the guidelines-stroke program. Circulation **122**(15), 1496–1504 (2010)
15. Cadilhac, D.A., et al.: Risk-adjusted hospital mortality rates for stroke: evidence from the Australian Stroke Clinical Registry (AuSCR). Med. J. Aust. **206**(8), 345–350 (2017)
16. Steiner, M.M., Brainin, M.: The quality of acute stroke units on a nation-wide level: the Austrian Stroke Registry for acute stroke units. Eur. J. Neurol. **10**(4), 353–360 (2003)
17. Stepanova, M., et al.: Recent trends in inpatient mortality and resource utilization for patients with stroke in the United States: 2005–2009. J. Stroke Cerebrovas. Dis. **22**(4), 491–499 (2013)
18. Keller, K., et al.: Impact of atrial fibrillation on in-hospital mortality of ischemic stroke patients and identification of promoting factors of atrial thrombi–results from the German **98**(4) (2019). ncbi.nlm.nih.gov

19. Ganesh, A., et al.: Integrated systems of stroke care and reduction in 30-day mortality: a retrospective analysis. Neurology **86**(10), 898–904 (2016)
20. Counsell, C., et al.: Predicting outcome after acute and subacute stroke: development and validation of new prognostic models. Stroke **33**(4), 1041–1047 (2002)
21. Gattringer, T., et al.: Predicting early mortality of acute ischemic stroke: score-based approach. Stroke **50**(2), 349–356 (2019)
22. Ho, W.-M., et al.: Prediction of in-hospital stroke mortality in critical care unit. SpringerPlus **5**(1), 1–9 (2016). https://doi.org/10.1186/s40064-016-2687-2
23. Lee, J., et al.: Derivation and validation of in-hospital mortality prediction models in ischaemic stroke patients using administrative data. Cerebrovas. Dis. 2013(1), 73–80 (2013). karger.com
24. Weimar, C., et al.: Age and National Institutes of Health Stroke Scale Score within 6 hours after onset are accurate predictors of outcome after cerebral ischemia: development and external validation of prognostic models. Stroke **35**(1), 158–162 (2004)
25. Baptista, M., et al.: Prediction of in-hospital mortality after first-ever stroke: the Lausanne Stroke Registry. J. Neurol. Sci. **166**(2), 107–114 (1999)
26. Wang, Y., et al.: A prognostic index for 30-day mortality after stroke. J. Clin. Epidemiol. **54**(8), 766–773 (2001)
27. Tanne, D., et al.: Trends in management and outcome of hospitalized patients with acute stroke and transient ischemic attack: the national acute stroke Israeli (nasis) registry. Stroke **43**(8), 2136–2141 (2012)

Universal Measure for Medical Image Quality Evaluation Based on Gradient Approach

Marzena Bielecka[1](✉), Andrzej Bielecki[2], Rafał Obuchowicz[3], and Adam Piórkowski[4]

[1] Faculty of Geology, Geophysics and Environmental Protection,
Chair of Geoinformatics and Applied Computer Science, AGH University of Science and Technology, Mickiewicza Ave. 30, 30-059 Cracow, Poland
bielecka@agh.edu.pl
[2] Faculty of Electrical Engineering, Automation, Computer Science and Biomedical Engineering, Chair of Applied Computer Science, AGH University of Science and Technology, Mickiewicza Ave. 30, 30-059 Cracow, Poland
bielecki@agh.edu.pl, azbielecki@gmail.com
[3] Department of Diagnostic Imaging, Jagiellonian University Medical College, Kopernika Street 19, 31-501 Cracow, Poland
r.obuchowicz@gmail.com
[4] Faculty of Electrical Engineering, Automation, Computer Science and Biomedical Engineering, Chair of Biocybernetics and Biomedical Engineering, AGH University of Science and Technology, Mickiewicza Ave. 30, 30-059 Cracow, Poland
pioro@agh.edu.pl

Abstract. In this paper, a new universal measure of medical images quality is proposed. The measure is based on the analysis of the image by using gradient methods. The number of isolated peaks in the examined image, as a function of the threshold value, is the basis of the assessment of the image quality. It turns out that for higher quality images the curvature of the graph of the said function has a higher value for lower threshold values. On the basis of the observed property, a new method of no-reference image quality assessment has been created. The experimental verification confirmed the method efficiency. The correlation between the arrangement depending on the image quality done by an expert and by using the proposed method is equal to 0.74. This means that the proposed method gives a correlation of higher than the best methods described in the literature. The proposed measure is useful to maximize the image quality while minimizing the time of medical examination.

Keywords: Medical imaging · Imaging quality assessment · Gradient methods

1 Introduction

Computer imaging is one of the crucial problems in contemporary computer science. Medical imaging is strictly connected to this topic. Medical images obtained

© Springer Nature Switzerland AG 2020
V. V. Krzhizhanovskaya et al. (Eds.): ICCS 2020, LNCS 12140, pp. 406–417, 2020.
https://doi.org/10.1007/978-3-030-50423-6_30

by using various techniques - see Sect. 3 - are encoded in digital form. Since the images are analyzed by experts for diagnostic purposes, good image quality is necessary to detect all pathological changes, especially in the context of finding lesions at an early stage. On the other hand, a great number of medical images is the reason a great demand of their computer analysis. Various methods of artificial intelligence (see [12]) are widely used for this purpose. Good image quality is essential for the analyzing algorithm to work effectively. It is particularly important when syntactic methods are used, which are characterized by high sensitivity to any disturbances occurring in the analyzed image [13]. Syntactic methods are often used to analyze medical images [1–5,25–27]. Therefore it is important to get high quality images. However, in order to obtain a good quality image, a sufficiently long examination time is necessary, which in some cases, e.g. X-ray images, involves the patient's exposure to harmful radiation. Therefore, it is important to develop methods that will maximize image quality while minimizing time of medical examination. Obtaining a precise measure of image quality, which is the subject of this paper, is one of the issues in this undertaking.

2   Medical Imaging Quality in Computer Science

Image quality assessment is an important field not only in the IT or entertainment industry, but also in medicine. Based on the control of image parameters, the radiation dose or acquisition time is reduced. Unfortunately, standard parameters such as Signal-to-Noise Ratio (SNR) or Contrast-to-Noise Ratio (CNR) do not work in this area. Examples can be given where, despite the high parameters, the images have difficult or doubtful diagnostic value. For this reason, the search for new assessment methods is still a challenge [6].

Image quality assessment methods can be divided into methods based on reference images (Partial or Full Reference-Image Quality Assessment) and methods without reference (No Reference - Image Quality Assurance, NR-IQA) [7,33]. In the first case these are well-known measures such as SNR [10], PSNR (Peak Signal-to-Noise-Ratio) [17], Contrast-to-NoiseRatio (CNR) [15], and Structural Similarity Index (SSIM) [16]. These measurements require a reference to human-pointed images as reference. The second group (NR-IQA) are methods that specify a blind measure for images without referring to the indicated quality sample [8,28,30,31]. In this case, it is easier to compare the quality of images for different modalities, although these measures do not correlate so well with the subjective assessments of people.

No-reference measures have a different construction. For example, a measure, based on structural MRI and two types of analysis of variance, was proposed in [34]. An attempt, based on applying multidirectional filters to MR images and then examining the feature statistics, is described in [19]. In [29] the authors applied Bayes theory to the relationship between entropy and image quality attributes. Although the subject of pulse noise removal in magnetic resonance imaging is well known [21], no one has ever tried to evaluate the quality of this imaging by quantifying this noise. This issue is the subject of this work.

3 Medical Imaging Quality in Clinical Practice

To obtain best imaging quality in the techniques where X-rays are used both in conventional radiography (CR) and the computed tomography (CT) but also where RF beam is used - as in magnetic resonance (MR) - one has to meet two main criteria. First is to produce possibly best - hence diagnostic image, and second - to provide an examination in a way most safe and comfortable for the patient as it is possible. In techniques where X-rays are used the criterion of image quality is met where the contrast of structures with different absorption of X-rays represented on the image is sufficiently different and imaged structures can be recognized as separate elements. Reduction of the tube voltage (kV) and, as a consequence, the reduction of X-rays energy provides good differentiation of the imaged objects. On the other hand, higher energy of X-rays enables to reduce dose absorbed by the human body. Therefore images with higher kV are preferred. Image quality in radiography is also strongly affected by the noise (mottle) which may decrease quality of the imaged object borders and hence limits its proper recognition [18]. Both in CR and CT the mottle is usually caused by a decrease of photons quantity used to create the image and typically associated with a reduction of X-ray tube current - amperage (mAs). Although its increase will improve image contrast, one has to remember that the increase of the amperage will result in a proportional increase in the radiation dose. Therefore, good quality of the radiographic pictures comes as an effect of the interplay between radiation dose and diagnostic performance of the image, built on proper contrast to noise ratio. Choice of appropriate parameters for sufficient contrast of the image and acceptable noise should be influenced by the clinical situation (body region and structures needed to be visualized). Beam filtration medium is another important physical determinant of X-ray beam quality usually preselected to the specific imaging tasks [11]. Similar principles of the image creation are actual for the CT due to helical movement of the X-ray tube around the patient body. One must remember that scan time, proportional to radiation dose, is longer. Slice thickness also affects radiation dose entering each detector but also patients body. Currently most commonly used CTDI (CT dose index) is a parameter that describes the dose output of a scanner [20]. It should be stressed that spacing between slices is often equal to the slice thickness - there is a close relationship between consecutive slices that are practically continuous. Therefore one describes cumulative dose from slice series as multiple-slice average dose (MSAD) and describes 1.25 to 1,4 higher dose than in single slice dose. Cone beam CT (CBCT) is free of the aforementioned phenomena, as there is no helical movement of the X-ray tube. Therefore, cone shape X-ray beam is used and average dose can to ten times (in comparison to the helical scan of the head) but information according to soft tissue cannot be represented in diagnostically usable form what is the major drawback of the CBCT. As CBCT allows to a substantial reduction of the dose still mean radiation dose in CBCT exceeds over three times that in CR technique [32]. In magnetic resonance technique radiation is not used and technique is reported as harmless to the human body. The only consideration is scanning time - substantially longer than in CT technique what combined with high susceptibility to the motion artefacts makes MR useless for unconscious patients or persons unable to withstand longer in the still position due to pain or altered posture. Therefore in MR quality of an image is a tradeoff time and picture characteristics. There are a plethora of parameters that might influence image quality. From geometric features of the MR image, the increase of matrix size will decrease signal although image produced will be sharper due to the decrease of pixel volume. To overcome signal drop mostly averages increase is used however this is in the expense of a substantial increase of scanning time. At the opposite, one can put the field of view (FOV) changes - if the increased amount of signal is also increased what comes with pixel size multiplication (with the image sharpness reduction). On similar principle works another geometrical parameter of MR image - a slice size - can be changed. Amount of signal will be increased by slice size as more signal is captured in the bigger voxel. Clinical usefulness of such image, however, is not always provided. Sequence parameters as TR and TE have to be manipulated with caution as those parameters are strictly valued for appropriate sequences [14]. Currently, image quality in MR is estimated parametrically by comparison of the image noise to the background noise what is expressed as SNR and this value is displayed on the operator console. This parameter, however, not always expresses subjective assumption of the image and it is useful for the mathematical estimation of the presumptive image signal loss [22]. This assumption is particularly useful during sequence planning, where time reduction is desired because of the clinical situation. Awareness of the amount of signal perceived from the planned sequence is needed especially where useful tools as parallel imaging are used. This technique called Grappa by Simens vendor allows for substantial reduction of exam time on the expanse of undersampled signal [9]. Amount of the signal and time of examination in the case of MR but also dose quantity and image quality in CR and CT is a topic of continuous discussion and research. It is also a field of technological improvements. Therefore, objective and repetitive measures of the image quality are awaited by the medical community as tools indispensable to provide patient safety and comfort.

4   Impulse Noise Quantification

Image quality assessment is often subjective. The authors noticed that there is some correlation between inferior image quality ratings and the impulse noise present on it. Therefore, they decided to try to quantify this relationship.

The proposed method is based on pixel counting, which predominates brightness over its surroundings (4-connected neighborhood) by a given threshold t or more - see Formula (1). Two directions are considered: the selected points are lighter than their surroundings or darker than their surroundings. For a given image, it is possible to assess how this number changes with the increase of the threshold. To keep the reference level, all images are scaled to a range of 8 bits (0–255).

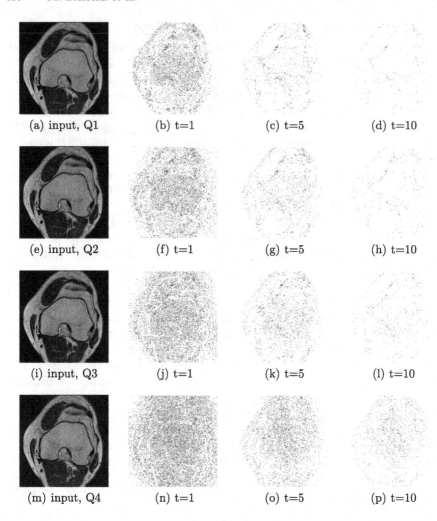

(a) input, Q1   (b) t=1   (c) t=5   (d) t=10

(e) input, Q2   (f) t=1   (g) t=5   (h) t=10

(i) input, Q3   (j) t=1   (k) t=5   (l) t=10

(m) input, Q4   (n) t=1   (o) t=5   (p) t=10

Fig. 1. Images with gradual reduction of acquisition time and decreasing quality.

Projections of the same patient, acquired by using magnetic resonance imaging, are shown in Fig. 1. Four successively reduced acquisition times, which resulted in a gradual decrease in quality, are presented - see also [24]. Pixels that dominated or that were dominated by the surroundings by level at least of 1, 5 and 10 were determined for these images, and they are presented in the following columns in the figure. It can be seen that as the set threshold increases, their number decreases, which is precisely presented in the plot - see Fig. 2 and Fig. 3. It should be noted that for better quality images, the number of indicated points for the same threshold is smaller than for images of lower quality - this process is gradual. The analysis of the presented relationship is the basis for determining the quality of the image.

Let us define the following logical conditions that describe whether the grey level $P(x, y)$ of the pixel (x, y) in the pixel matrix $M \times N$ differs from the grey level of neighboring pixels vertically and horizontally, respectively, by more than the threshold value t :

$$\phi_{ver}^+(x, y) : \ (P(x, y) > P(x, y + 1) + t) \wedge (P(x, y) > P(x, y - 1) + t),$$

$$\phi_{hor}^+(x, y) : \ (P(x, y) > P(x + 1, y) + t) \wedge (P(x, y) > P(x - 1, y) + t),$$

$$\phi_{ver}^-(x, y) : \ (P(x, y) > P(x, y + 1) - t) \wedge (P(x, y) > P(x, y - 1) - t),$$

$$\phi_{hor}^-(x, y) : \ (P(x, y) > P(x + 1, y) - t) \wedge (P(x, y) > P(x - 1, y) - t),$$

where $x \in \{1, ..., M-1\}$, $y \in \{1, ..., N-1\}$. The below logical condition described by the propositional function $\phi(x, y)$ is satisfied if the grey level of the pixel (x, y) differs from the grey level of neighboring pixels by more than the threshold value t :

$$\phi(x, y) = \left(\phi_y^+(x, y) \wedge \phi_x^+(x, y)\right) \vee \left(\phi_y^-(x, y) \wedge \phi_x^-(x, y)\right) \tag{1}$$

Let $A(t)$ be the set of the pixels that satisfy the specified condition (1), i.e. $A(t) = \{(x, y) : \ \phi(x, y)\}$ and let $I(t)$ denotes the number of such pixels, i.e. $I(t) = \#A(t)$. The study of some geometrical properties of the graph of the function $I(t)$ is the basis of the proposed method - see the next section.

5   Description of the Method

Gradient methods allow the researcher to analyze the local properties of the functions $F : \mathbb{R}^n \to \mathbb{R}$ and, as a consequence, the local properties of the graphs of the functions, i.e. the sets of the following form $\{(x_1, ..., x_n, f(x_1, ..., x_n))\}$. If $n = 2$, then the graph of the function has natural interpretation in the context of computer graphics. In particular, let us assume that $\mathbb{R}^2$ is discretized and $f(x_1, x_2)$ is defined on a finite rectangle - the pixel matrix - that is a subset of the discretized $\mathbb{R}^2$. If, furthermore, the values of the function f are values of gray level of individual pixels, then gradient methods can be used to analysis and processing of the digital image. In this paper such approach is applied to medical images.

In this sort of images, individual areas in the image correspond to individual tissues. Therefore, in good quality images, in the mentioned areas, the gray level should be more or less constant, whereas the boundaries between areas should be clear - see Fig. 2. As a consequence, the number of isolated peaks $I(t)$ (see Sect. 4), as a function of the threshold value t, in good quality images should decrease rapidly and then flatten - see Fig. 3. Thus, the value of the maximum curvature of the graph of this function and the point in which it is reached provide important information about the quality of the image.

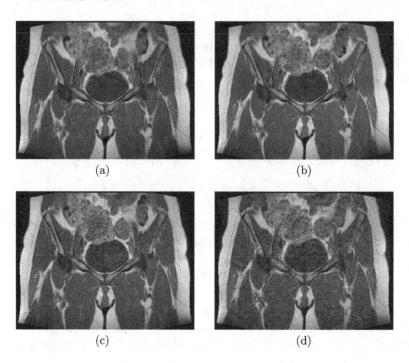

(a)   (b)

(c)   (d)

Fig. 2. Four images of the pelvis of the same patient P01. The images are sorted according to decreasing quality. Thus, image (a) is the best one, whereas image (d) is the worst one.

In the proposed method, the curve was fitted to the points obtained in this way. A few fitting functions were tested and it turned out that for the following formula

$$f(x) = a \cdot x^b, \tag{2}$$

where a and b are the fitted parameters, the goodness of fit R^2 is close to 1. The curvature κ at the point (x_0, y_0) of the curve described by a function $y = f(x)$ is given by the formula

$$\kappa(x_0, y_0) = \frac{|y_0''|}{\sqrt{(1 + y_0')^3}} \tag{3}$$

For each curve its maximum curvature has been calculated numerically. The point, in which this maximum is realized, has been detected as well.

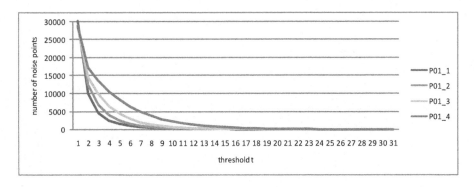

Fig. 3. The obtained curves for one patient P01. Each curve represents one image with different quality. The better quality, the higher maximum curvature. The blue, red, green and violet curves correspond to image (a), (b), (c) and (d) from Fig. 2, respectively. Thus, the blue curve corresponds to the best image, whereas the violet curve corresponds to the worst one. The vertical axis represents $I(t)$, whereas the horizontal one represents the threshold value t. (Color figure online)

6   Results

In order to verify the described non-reference method, the quality of MR images has been assessed. The calculations were based on images taken for 38 patients. The total number of images was 2708. The number of taken images varied from patient to patient and ranged from 52 to 120 pictures. For each patient, the images were ordered by an expert according to the decreasing quality. The result obtained for one patient is a series of images made with four different exposure times what caused various quality of the images. In addition, imaging was performed on various parts of the body such as the hip joint, spine, knee etc. For a given patient images of only one part of the body were taken, all in the same position. Data used in this paper were described in detail in the article [24].

As it has been noticed in the previous section, the number of pixels, identified as impulse noise for an image, with changing threshold value can be used to assess the quality of a given image. Therefore, each image is represented by a curve that describes the change in the number of peaks isolated with a changing threshold. Next, the function was fitted to each curve and its parameters a and b were estimated. The obtained determination coefficient R^2 in each case was above 0.95. Based on the estimated parameters of the fitted curves, the maximum curvature κ was calculated for each of them. An example of a sequence of values of κ for two patients is shown in Fig. 4.

The images were ordered according to decreasing quality. The better the image quality, the greater value of the maximum curvature κ. The relationship, however, is statistical. The adjusted trend for each κ coefficient sequence is a decreasing function. Also, the mean of the curvature coefficients for each of the four groups of images obtained for one patient indicates a decreasing image quality - see Fig. 5.

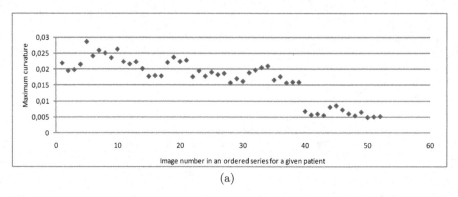

(a)

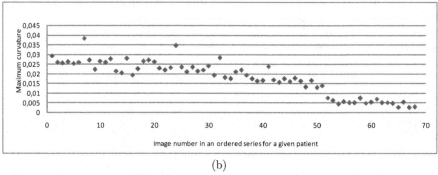

(b)

Fig. 4. Curvature coefficient values for series of images obtained for two patients.

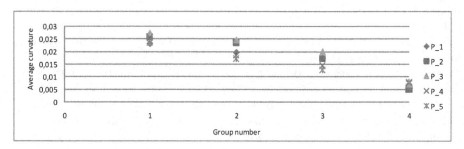

Fig. 5. The mean curvature coefficient values for each of the four highlighted groups for the five selected patients P_1, P_2, P_3, P_4, P_5

The statistical nature of the relationship between the maximum curvature of the obtained curve and the image quality means that outliers may appear in the neighboring groups. As a result of observation of the analyzed curves, it was noticed that for images of higher quality, the maximum curvature κ is higher and occurs for a lower threshold value than for images of lower quality - see Fig. 6. The correlation ratio between the maximum value of κ coefficient for a given patient and the image quality according to the arrangement done by an expert

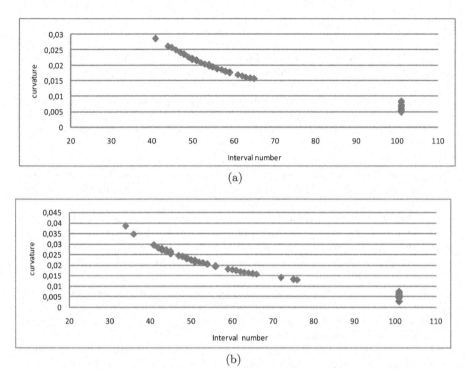

Fig. 6. Curvature coefficient values depending on the adopted threshold for the patients from Fig. 4

is $r = 0.74$. It should be stressed that the correlation is greater than for the method based on entropy, for witch $r = 0.67$ and for other methods - see [23].

7   Concluding Remarks

The obtained results confirm the strong relationship between image quality and the maximum curvature of the curve obtained on the basis of the quantification of impulse noise in the analyzed image. For an ordered sequence of images due to their quality, the obtained κ coefficient values form an ordered descending sequence of values concordant statistically with the sequence of images ordered by an expert. As it has been specified, the correlation ratio is equal to 74%. This means that for higher quality images, the κ ratio has, statistically, a higher value for lower threshold values. Thus, the method of automatic assessment of the image quality based on the proposed approach is more effective than the methods described in literature - see [23] and references given there. Therefore, the proposed method is a significant step towards automatic choice of the radiation dose during investigation. Another problem is how the proposed method will evaluate the image after filtering. Among other things, it is planned to be the subject of further studies.

References

1. Bielecka, M.: Syntactic-geometric-fuzzy hierarchical classifier of contours with application to analysis of bone contours in X-ray images. Appl. Soft Comput. **69**, 368–380 (2018)
2. Bielecka, M., et al.: Modified jakubowski shape transducer for detecting osteophytes and erosions in finger joints. In: Dobnikar, A., Lotrič, U., Šter, B. (eds.) ICANNGA 2011. LNCS, vol. 6594, pp. 147–155. Springer, Heidelberg (2011). https://doi.org/10.1007/978-3-642-20267-4_16
3. Bielecka, M., et al.: Application of shape description methodology to hand radiographs interpretation. In: Bolc, L., Tadeusiewicz, R., Chmielewski, L.J., Wojciechowski, K. (eds.) ICCVG 2010. LNCS, vol. 6374, pp. 11–18. Springer, Heidelberg (2010). https://doi.org/10.1007/978-3-642-15910-7_2
4. Bielecka, M., Korkosz, M.: Generalized shape language application to detection of a specific type of bone erosion in X-ray images. LNAI **9692**, 531–540 (2016)
5. Bielecka, M., Obuchowicz, R., Korkosz, M.: The shape language in application to the diagnosis of cervical vertebrae pathology. PLoS ONE **13**(10), 17 (2018). Article number e0204546
6. Chandler, D.M.: Seven challenges in image quality assessment: past, present, and future research. ISRN Sig. Process. 7 (2013). Article ID 356291
7. Chow, L.S., Paramesran, R.: Review of medical image quality assessment. Biomed. Signal Process. Control **27**, 145–154 (2016)
8. Chow, L.S., Rajagopal, H.: Modified-BRISQUE as no reference image quality assessment for structural MR images. Magn. Reson. Imaging **43**, 74–87 (2017)
9. Deshmane, A., Gulani, V., Griswold, M.A., Seiberlich, N.: Parallel MR imaging. J. Magn. Reson. Imaging **36**, 55–72 (2012)
10. Dietrich, O., Raya, J.G., Reeder, S.B., Reiser, M.F., Schoenberg, S.O.: Measurement of signal-to-noise ratios in MR images: influence of multichannel coils, parallel imaging, and reconstruction filters. J. Magn. Reson. Imaging **26**(2), 375–385 (2007)
11. Elojeimy, S., Tipnis, S., Huda, W.: Relationship between radiographic techniques (kilovolt and milliampere-second) and CTDIVOL. Radiat. Prot. Dosim. **141**(1), 43–49 (2010)
12. Flasiński, M.: Introduction to Artificial Intelligence. Springer, Cham (2016). https://doi.org/10.1007/978-3-319-40022-8
13. Flasiński, M.: Syntactic Pattern Recognition. World Scientific, Singapore (2019)
14. Gedamu, E.L., Collins, D., Arnold, D.L.: Automated quality control of brain MR images. J. Magn. Reson. Imaging **28**(2), 308–319 (2008)
15. Geissler, A., Gartus, A., Foki, T., Tahamtan, A.R., Beisteiner, R., Barth, M.: Contrast-to-noise ratio (CNR) as a quality parameter in fMRI. J. Magn. Reson. Imaging **25**(6), 1263–1270 (2007)
16. Hore, A., Ziou, D.: Image quality metrics: PSNR vs. SSIM. In: 20th International Conference on Pattern Recognition, pp. 2366–2369. IEEE (2010)
17. Huynh-Thu, Q., Ghanbari, M.: Scope of validity of PSNR in image/video quality assessment. Electron. Lett. **44**(13), 800–801 (2008)
18. Huda, W., Abrahams, R.B.: Radiographic techniques, contrast, and noise in X-ray imaging. Am. J. Roentgenol. **204**(2), 126–131 (2015)
19. Jang, J., Bang, K., Jang, H., Hwang, D.: Alzheimer's disease neuroimaging initiative quality evaluation of no-reference MR images using multidirectional filters and image statistics. Magn. Reson. Med. **80**(3), 914–924 (2018)

20. Ludlow, J.B., Ivanovic, M.: Comparative dosimetry of dental CBCT devices and 64-slice CT for oral and maxillofacial radiology. Oral Surg. Oral Med. Oral Pathol. Oral Radiol. Endod. **106**, 106–114 (2008)
21. Mafi, M., Martin, H., Adjouadi, M.: High impulse noise intensity removal in MRI images. In: IEEE Signal Processing in Medicine and Biology Symposium (SPMB), pp. 1–6 (2017)
22. Miao, J., Huang, F., Narayan, S., Wilson, D.L.: A new perceptual difference model for diagnostically relevant quantitative image quality evaluation: a preliminary study. Magn. Reson. Imaging **31**, 596–603 (2013)
23. Obuchowicz, R., Oszust, M., Bielecka, M., Bielecki, A., Piórkowski, A.: Magnetic resonance image quality assessment by using non-maximum suppression and entropy analysis. Entropy **22**(2) (2020). Article number e22020220
24. Obuchowicz, R., Piórkowski, A., Urbanik, A., Strzelecki, M.: Influence of acquisition time on MR image quality estimated with nonparametric measures based on texture features. Biomed Res. Int. 10 (2019). Article ID 3706581
25. Ogiela, M.: Languages of shape feature description and syntactic methods for recognition of morphological changes of organs in analysis of selected X-ray images. In: Proceedings of Medical Imaging 1998, vol. 3338, pp. 1295–1305 (1998)
26. Ogiela, M., Tadeusiewicz, R.: Syntactic pattern recognition for X-ray diagnosis of pancreatic cancer-algorithms for analysing the morphologic shape of pancreatic ducts for early diagnosis of changes in the pancreas. IEEE Eng. Med. Biol. Mag. **19**, 94–105 (2000)
27. Ogiela, M., Tadeusiewicz, R., Ogiela, L.: Image languages in intelligent radiological palm diagnostics. Pattern Recogn. **39**, 2157–2165 (2006)
28. Okarma, K., Fastowicz, J.: No-reference quality assessment of 3D prints based on the GLCM analysis. In: 21st International Conference on Methods and Models in Automation and Robotics (MMAR), pp. 788–793. IEEE (2016)
29. Osadebey, M., Pedersen, M., Arnold, D., Wendel-Mitoraj, K.: Bayesian framework inspired no-reference region-of-interest quality measure for brain MRI images. J. Med. Imaging **4**(2), 502–504 (2017)
30. Oszust, M.: No-reference image quality assessment using image statistics and robust feature descriptors. IEEE Signal Process. Lett. **11**(24), 1656–1660 (2017)
31. Oszust, M.: No-reference image quality assessment with local features and high-order derivatives. J. Vis. Commun. Image Represent. **56**, 15–26 (2018)
32. Schulze, D., Heiland, M., Thurmann, H., Adam, G.: Radiation exposure during midfacial imaging using 4- and 16-slice computed tomography, cone beam computed tomography systems and conventional radiography. Dentomaxillofac. Radiol. **33**(2), 83–86 (2004)
33. Sinha, N., Ramakrishnan, A.G.: Quality assessment in magnetic resonance images. Crit. Rev. Biomed. Eng. **38**(2), 127–141 (2010)
34. Woodard, J.P., Carley-Spencer, M.P.: No-reference image quality metrics for structural MRI. Neuroinformatics **4**(3), 243–262 (2006)

Constructing Holistic Patient Flow Simulation Using System Approach

Tesfamariam M. Abuhay[1,3](✉), Oleg G. Metsker[1],
Aleksey N. Yakovlev[1,2], and Sergey V. Kovalchuk[1]

[1] ITMO University, Saint Petersburg, Russia
tesfamariam.m.abuhay@gmail.com,
olegmetsker@gmail.com,
sergey.v.kovalchuk@gmail.com
[2] Almazov National Research Center, Saint Petersburg, Russia
ya-kovlev_an@almazovcentere.ru
[3] University of Gondar, Gondar, Ethiopia

Abstract. Patient flow often described as a systemic issue requiring a systemic approach because hospital is a collection of highly dynamic, interconnected, complex, ad hoc and multi-disciplinary sub-processes. However, studies on holistic patient flow simulation following system approach are limited and/or poorly understood. Several researchers have been investigating single departments such as ambulatory care unit, Intensive Care Unit (ICU), emergency department, surgery department or patients' interaction with limited resources such as doctor, endoscopy or bed, independently. Hence, this article demonstrates *how to achieve system approach in constructing holistic patient flow simulation, while maintaining the balance between the complexity and the simplicity of the model.* To this end, system approach, network analysis and discrete event simulation (DES) were employed. The most important departments in the diagnosis and treatment process are identified by analyzing network of hospital departments. Holistic patient flow simulation is constructed using DES following system approach. Case studies are conducted and the results illustrate that healthcare systems must be modeled and investigated as a complex and interconnected system so that the real impact of changes on the entire system or parts of the system could be observed at strategic as well as operational levels.

Keywords: Patient flow simulation · Network analysis · System approach · Discrete event simulation

1 Introduction

Healthcare systems all over the world are under pressure due to large share of aging population, pandemic (e.g., COVID-19), scarcity of resources and poor healthcare planning, organization and management. As a well-coordinated and collaborative care improves patient outcomes and decreases medical costs [1], there is a need for effective organization of healthcare processes.

© Springer Nature Switzerland AG 2020
V. V. Krzhizhanovskaya et al. (Eds.): ICCS 2020, LNCS 12140, pp. 418–429, 2020.
https://doi.org/10.1007/978-3-030-50423-6_31

Modeling and analyzing healthcare processes based on patient flow to and in a hospital is essential because patient flow demonstrates organizational structure, resource demand and utilization patterns, clinical and operational pathways, bottlenecks, prospect activities and "what if" scenarios [2, 3]. Patient flow can be investigated from *clinical or operational* perspectives [3]. From operational perspective, analysis of patient flow in a single department such as ambulatory care unit [4], Intensive Care Unit (ICU) [5], emergency department [6–8] or surgery department [9, 10] was investigated in detail. On the other hand, patients' interaction with limited resources such as doctor [11], endoscopy [12] or bed [13] was also studied.

However, hospital is a combination of highly dynamic, interconnected, complex, ad hoc and multi-disciplinary sub-processes [14–16]. In other words, the organizational behavior and result of a hospital are shaped by the interaction of its discrete components. For this reason, hospital systems cannot be fully understood by analyzing their individual components in separation [17, 18]. This indicates that constructing holistic patient flow simulation following system approach is essential because patient flow is often described as a systemic issue requiring a systemic approach [19]. So that true impact of changes on the whole and/or parts of a hospital system can be investigated at macro as well as micro levels.

Nevertheless, studies on holistic patient flow simulation are limited [20] and/or poorly understood [19]. For instance, Djanatliev [17] proposed theoretical approach which considered reciprocal influences between processes and higher level entities using hybrid simulation. Abuhay et al. [21] and Kovalchuk et al. [22] have proposed construction of patient flow in multiple departments of a hospital. Suhaimi et al. [23] built holistic simulation model that represents multiple clinics from different locations.

Since it is impossible to model all departments that exist in a hospital and include them in the patient flow simulation due to complexity, time and cost, there is a need to analyze departments and identify the most important ones in the diagnosis and treatment process. Gunal [16] mentioned that choosing services/departments of a hospital to be modelled is the modeler's task. However, the aforementioned authors did not discuss how and why departments/clinics/units were selected and included in their model. This prompts the authors of this article to ask the following question: *how to achieve system approach in constructing holistic patient flow simulation, while keeping the balance between the complexity and the simplicity of the model?*

Analyzing data-driven network of hospital departments based on patient transfer may provide an answer for the aforementioned question. Network or graph can be defined as a set of social entities such as people, groups, and organizations, with some relationships between them [24, 25]. Network analysis allows to investigate topological properties of a network, discover patterns of relations and identify the roles of nodes and sub-groups within a network [24, 26]. However, to the best of our knowledge, no one has investigated the network or collaboration of hospital departments and construct holistic patient flow simulation using system approach. Hence, *this article aims at demonstrating construction of holistic patient flow simulation using system approach, network analysis and discrete event simulation.*

The rest of the paper is organized as follows: Sect. 2 outlines model construction; Sect. 3 discusses case studies and Sect. 4 presents conclusion.

2   Model Construction

Automation of administrative operations of healthcare using Electronic Health Record (EHR) presents an opportunity of constructing data-driven decision support tools that facilitate modeling, analyzing, forecasting and managing operational processes of healthcare. Figure 1 depicts conceptual, methodological and architectural foundations of the proposed model. This study was conducted in collaboration with the Almazov National Medical Research Centre[1]. Different kinds of data such as data about length of stay, cost of treatment, inter-arrival rate of patients to a hospital, characteristics of patients, event log about movement of patients (transition matrix), laboratory test results and load of doctors can be extracted from the EHR and used as an input to build components/submodels of the proposed model.

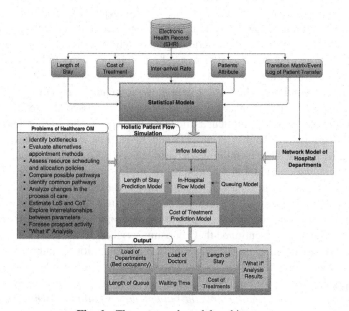

Fig. 1. The proposed model architecture.

Seven years, from 2010 to 2016, empirical data of 24902 Acute Coronary Syndrome (ACS) patients was collected from the aforementioned hospital. The event log data describes movement of patients from department to department with associated timestamp, Length of Stay (LoS), and Cost of Treatment (CoT). All departments visited by ACS patients from 2010 to 2016 are included in this study.

Network analysis [24] using Gephi 0.9.2 [27] is used to investigate network of hospital departments and Discrete Event Simulation (DES) method [16] is employed to construct a holistic patient flow simulation. Kernel Density Estimation (KDE) is used to

---

[1] http://www.almazovcentre.ru/?lang=en.

model LoS and CoT. Poisson distribution [28] is implemented to model patient inflow to a hospital. Movement of patients through departments is governed by probability law constructed as transition matrix [29] and access to limited resources is managed by First In First Out (FIFO) queuing method.

2.1 Analyzing Network of Hospital Departments

The objective of this section is to identify the most important departments in the diagnosis and treatment process of Acute Coronary Syndrome (ACS) patients. The result will be used as an input to construct holistic patient flow simulation. Each patient's event log data was sorted based on event date and network of departments was constructed based on the chronological transfer of ACS patients.

The network of departments is both directed (shows from where to where a patient was transferred) and weighted, representing the number of patients transferred among departments. To reduce potential noises, departments with less than 10 interaction in seven years are excluded.

The network contains 227 nodes that represent departments and 4305 edges that represent transfer of patients. Both degree and weighted degree distributions are positively skewed with a large majority of departments having a low degree and a small number of departments having a high degree.

The average degree and weighted degree account for 19 and 5800, in that order. Even though the network of departments is sparse with density equals to 0.1, the average path length is short which accounts for 2.3.

Table 1. Departments with high degree, weighted degree, betweeness and closeness centrality.

Departments	Degree	In degree	Out degree	Weighted Degree	Weighted Outdegree	Weighted Indegree	Betweeness centrality	Closeness centrality
Laboratory	240	124	116	277469	139210	138259	5089	0.68
Functional diagnostics	219	105	114	171371	85407	85964	2987	0.68
Cardiology2	182	90	92	96148	46623	49525	3879	0.60
Cardiology1	182	93	89	101915	49534	52381	3820	0.61
Admission	169	91	78	62683	37096	25587	1330	0.61
ICU1	144	73	71	71910	36055	35855	1174	0.59
Surgey2	134	68	66	31671	15702	15969	1053	0.59
Surgey1	128	62	66	34465	17096	17369	1521	0.56
ICU2	120	64	56	45270	22706	22564	1310	0.56

When we see the strategic positioning of departments, Laboratory department, Functional Diagnostic department, Cardiology departments, Surgery departments and Intensive Care Unit (ICU) departments are receiving more requests from other departments as well as sending more results to other departments (see Table 1). In other words, these departments are significant in giving support to and influencing function of other departments during the diagnosis and treatment processes of ACS patients.

Therefore, maintaining functionality, capacity and geographical location of these departments is vital so as to deliver effective and efficient care for ACS patients.

These departments are also fundamental in connecting communities of departments (five communities were identified with an average clustering coefficient of 0.62) as they have high betweenness and closeness centrality (see Table 1).

Finally, the results were reported to domain experts and they suggested that admission department, two cardiology departments, two ICU departments and two surgery departments should be selected for constructing a holistic patient flow simulation. In order to consider the impact of other departments, the rest were also modeled as one department.

2.2   Holistic Patient Flow Simulation

The holistic patient flow simulation model has sub-models such as patient inflow simulation models, in-hospital patient flow simulation model, LoS prediction model, CoT prediction model, and queuing model. As patients arrive at a hospital following a random state, Poisson distribution [28] is employed to simulate arrival of patients to a hospital based on the inter-arrival rate extracted from the Admission department, an entry point to a hospital.

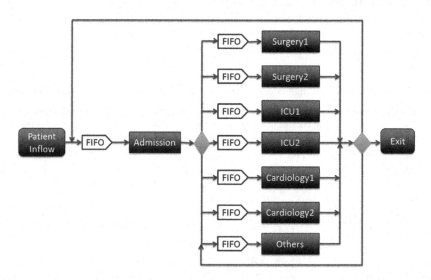

Fig. 2. The conceptual model of in-hospital patient flow simulation.

In-hospital patient flow simulation model imitates the concurrent flow of patients through eight departments (see Fig. 2). This model is built using compartmental modeling and DES method and implemented in SimPy, process-based discrete-event simulation framework based on standard Python [30].

Movement of patients from one department to another department is governed by a probability law constructed as transition matrix [29], each column summing to 1, extracted from empirical data. The relationship among departments is either one way or two ways represented by one or two directional arrows (see Fig. 2).

The in-hospital patient flow model starts simulating by accepting patients from patient inflow simulation model. The Admission department is the entry point to the hospital. Each department is attached with LoS, CoT and queueing models.

Length of Stay (LoS) has been used as a surrogate to evaluate the effectiveness of healthcare [31, 32]. But, a measure often employed to model LoS is an average LoS which does not characterize the underlying distribution as LoS data being positively skewed and multimodal [31, 32]. Because of this, density estimation methods such as the Normal, the Gamma, the Exponential and the Weibull distribution, which are mostly used statistical models to model LoS data [33–35], are not a good choice.

Hence, Gaussian Mixture Model (GMM) [36] and Kernel Density Estimation (KDE) [37] are selected for further experiment. To determine the number of individual Gaussian distributions for GMM, 10 experiments were conducted for each department. Bayesian Information Criterion (BIC) [38] is used to select the best models.

The results illustrate that modeling LoS at departments requires different number of Gaussian mixtures. This indicates that LoS at each department should be modeled separately as they provide medical treatment in different ways and procedures.

Both GMM and KDE methods have fitted the LoS data properly. Two-sample Kolmogorov-Smirnov test [39] is used to select the best fit. As a result, the KDE models fitted the LoS data better than the GMM at all departments. Hence, KDE model is selected and attached to each department to predict LoS at department level.

The amount of CoT differs from department to department as the departments provide medical treatment using different procedure and equipment. For this reason, separate CoT prediction models are developed for each department.

These models are constructed using KDE as both LoS and CoT demonstrate the same behavior. FIFO queuing technique is attached to each department so that basic statistics such as length of queue and wait time could be generated at departmental level and can be used for further analysis and decision making.

2.3   Model Validation

In six years, the hospital has admitted 9701 ACS patients, whereas the proposed model has admitted 9779 ACS patients. After patients arrive at the Admission department, they move from department to department. Figure 3 compares the number of patients visited each department in the real system and in the proposed model, whereas Fig. 4 and Fig. 5 present comparison of LoS and CoT, in that order.

The graphical presentations exhibit sub-models perform pretty well in modeling LoS and CoT. As a result, we may conclude that the patient flow simulation can be used to assess different case studies to demonstrate benefits of system approach in constructing patient flow simulation.

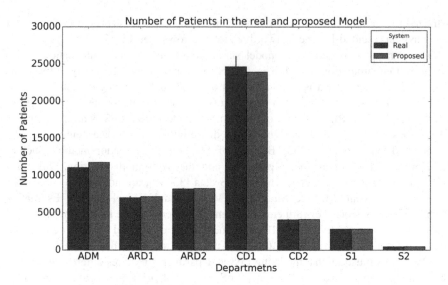

Fig. 3. Comparison of patients' movement in the proposed and real system.

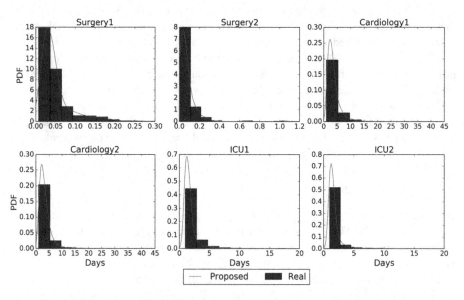

Fig. 4. Comparison of LoS at each department in the proposed and real system.

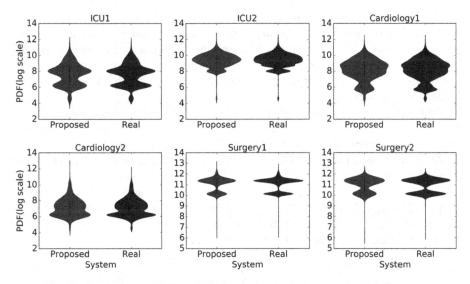

Fig. 5. Comparison of CoT at each department in the proposed and real system.

3 Case Studies Demonstrating Benefits of System Approach

The main purpose of this study is to conduct "what if" analysis and demonstrate system approach as a solution to model and analyze patient flow. The departments under consideration provide medical treatments with limited capacity (Admission = unlimited, ICU1 = 10, ICU2 = 10, CD1 = 40, CD2, 40, S1 = 2, S2 = 1, and Other = unlimited). The ACS patients mean inter-arrival rate over six years is 325 min. However, the rate varies over years showing a downward trend (485, 365, 352, 219, 267, 210 representing the years from 2010–2015). Hence, simulation of load of departments as ACS patients' inter-arrival rate varies (500, 400, 300, 200, 100, 50, 5 min) is discussed here below. The run time for all experiments is one year.

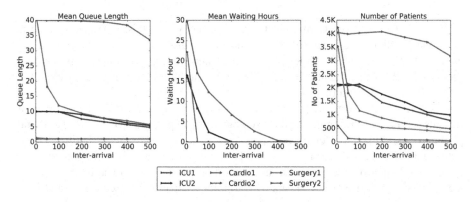

Fig. 6. Load of departments as the inter-arrival rate changes.

The number of patients increases as the inter-arrival rate decreases in all departments (see Fig. 6). However, the number of patients at ICU1 and ICU2 decreases as the inter-arrival reaches 100 min. This is because other departments are not sending the expected number of patients to both ICU departments due to overcrowding. This illustrates congestion of one or more departments affect the smooth flow of patients in the healthcare process.

On top of that, as the inter-arrival rate decreases, Cardiology1 becomes congested than Cardiology2, ICU2 becomes overcrowded than ICU1, and Surgery1 also becomes packed than Surgery2 (see Fig. 6). To reduce load of departments, two possible solutions are: 1) increasing the capacity of highly loaded department and/or 2) Pooling or merging the same departments so that they share their resources.

First, let us increase the capacity of CD1 from 40 to 50 and see how the system reacts. Increasing capacity of CD1 reduces the mean waiting hour until the inter-arrival reaches 100 min. However, it affects ICU1 department by increasing the number of patients, the mean waiting hour and the mean queue length.

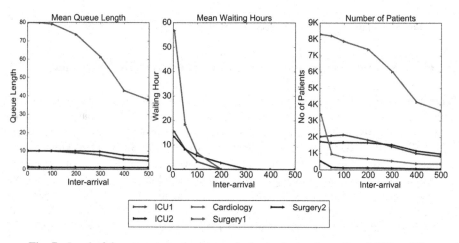

Fig. 7. Load of departments as the inter-arrival rate change (capacity of CD1 = 50).

Second, let us merge Cardiology1 and Cardiology2. In this experiment, there is only one Cardiology department and all flows to Cardiology1 and Cardiology2 are directed to the merged Cardiology department and the transition matrix is adjusted accordingly. Six experiments were conducted by varying inter-arrival rate (500, 400, 300, 200, 100, 50, 5 min).

As a result, the combined Cardiology department has served similar number of patients served by Cardiology1 and Cardiology2 before pooling (see Fig. 7 and 8) and pooling reduces the mean length of queue and the mean waiting hours significantly by 100% until inter-arrival reaches 100 min. However, as the inter-arrival approaches to zero, mean length of queue and mean waiting hour of the pooled Cardiology department become greater than the separated Cardiology departments (see Fig. 7 and 8).

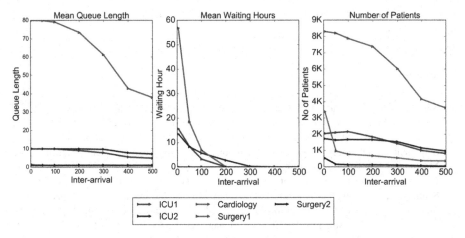

Fig. 8. Load of departments after pooling CD1 and CD2.

4 Conclusion

All over the world, healthcare systems are under pressure because of large share of aging population, pandemic (e.g., Covid-19), scarcity of resources and poor healthcare planning, organization and management. As a well-coordinated and collaborative care improves patient outcomes and decreases medical costs, there is a need for effective organization of healthcare processes.

Modeling and analyzing healthcare processes based on patient flow to and in a hospital is essential as patient flow demonstrates organizational structures, resource demand and utilization patterns, clinical and operational pathways, bottlenecks, prospect activities and "what if" scenarios. Patient flow can be investigated from clinical or operational perspectives. From operational perspective, analysis of patient flow in a single department such as ambulatory care unit, Intensive Care Unit (ICU), emergency department or surgery department was investigated in detail. On the other hand, patients' interaction with limited resources such as doctor, endoscopy or bed was also studied.

However, patient flow often described as a systemic issue requiring a systemic approach. Studies on holistic patient flow simulation are limited and/or poorly understood. Hence, this article proposes construction of patient flow simulation using system approach. To this end, first, network of hospital departments is investigated to identify the most important departments in the diagnosis and treatment process of ACS patients. Second, the result is used as an input to construct a holistic patient flow simulation using system approach and DES method. Finally, case studies are conducted to demonstrate benefits of system approach in constructing patient flow simulation.

The case studies indicate that healthcare systems must be modeled and investigated as a complex system of interconnected processes so that the real impact of operational as well as parametric change on the entire system or parts of the system could be observed.

Acknowledgment. This research is financially supported by The Russian Science Foundation, Agreement #19-11-00326.

References

1. Soulakis, N.D., et al.: Visualizing collaborative electronic health record usage for hospitalized patients with heart failure. J. Am. Med. Inform. Assoc. **22**(2), 299–311 (2015)
2. Chand, S., Moskowitz, H., Norris, J.B., Shade, S., Willis, D.R.: Improving patient flow at an outpatient clinic: Study of sources of variability and improvement factors. Health Care Manag. Sci. **12**(3), 325–340 (2009)
3. Côté, M.J.: Understanding patient flow. Decis. Line **31**, 8–13 (2000)
4. Santibáñez, P., Chow, V.S., French, J., Puterman, M.L., Tyldesley, S.: Reducing patient wait times and improving resource utilization at British Columbia Cancer Agency's ambulatory care unit through simulation. Health Care Manag. Sci. **12**(4), 392–407 (2009)
5. Christensen, B.A.: Improving ICU patient flow through discrete-event simulation. Massachusetts Institute of Technology (2012)
6. Konrad, R., et al.: Modeling the impact of changing patient flow processes in an emergency department: insights from a computer simulation study. Oper. Res. Heal. Care **2**(4), 66–74 (2013)
7. Cocke, S., et al.: UVA emergency department patient flow simulation and analysis. In: 2016 IEEE Systems and Information Engineering Design Symposium, pp. 118–123 (2016)
8. Hurwitz, J.E., et al.: A flexible simulation platform to quantify and manage emergency department crowding. BMC Med. Inform. Decis. Mak. **14**(1), 50 (2014)
9. Antonelli, D., Bruno, G., Taurino, T.: Simulation-based analysis of patient flow in elective surgery. In: Matta, A., Li, J., Sahin, E., Lanzarone, E., Fowler, J. (eds.) Proceedings of the International Conference on Health Care Systems Engineering. SPMS, vol. 61, pp. 87–97. Springer, Cham (2014). https://doi.org/10.1007/978-3-319-01848-5_7
10. Azari-Rad, S., Yontef, A., Aleman, D.M., Urbach, D.R.: A simulation model for perioperative process improvement. Oper. Res. Heal. Care **3**, 22–30 (2014)
11. Swisher, J.R., Jacobson, S.H.: Evaluating the design of a family practice healthcare clinic using discrete-event simulation. Health Care Manag. Sci. **5**(2), 75–88 (2002)
12. Almeida, R., Paterson, W.G., Craig, N., Hookey, L.: A patient flow analysis: identification of process inefficiencies and workflow metrics at an ambulatory endoscopy unit. Can. J. Gastroenterol. Hepatol. **2016**, 1–7 (2016)
13. Monks, T., et al.: A modelling tool for capacity planning in acute and community stroke services. BMC Health Serv. Res. **16**, 1–8 (2016)
14. Rebuge, Á., Ferreira, D.R.: Business process analysis in healthcare environments: a methodology based on process mining. Inf. Syst. **37**(2), 99–116 (2012)
15. Rojas, E., Munoz-Gama, J., Sepúlveda, M., Capurro, D.: Process mining in healthcare: a literature review. J. Biomed. Inform. **61**, 224–236 (2016)
16. Gunal, M.M.: A guide for building hospital simulation models. Health Syst. **1**(1), 17–25 (2012)
17. Anatoli Djanatliev, F.M.: Hospital processes within an integrated system view: a hybrid simulation approach. In: Proceedings of the 2016 Winter Simulation Conference, pp. 1364–1375 (2016)
18. Kannampallil, T.G., Schauer, G.F., Cohen, T., Patel, V.L.: Considering complexity in healthcare systems. J. Biomed. Inform. **44**(6), 943–947 (2011)

19. Kreindler, S.A.: The three paradoxes of patient flow: an explanatory case study. BMC Health Serv. Res. **17**(1), 481 (2017)
20. Vanberkel, P.T., Boucherie, R.J., Hans, E.W., Hurink, J.L., Litvak, N.: A survey of health care models that encompass multiple departments. University of Twente, Faculty of Mathematical Sciences (2009)
21. Abuhay, T.M., Krikunov, A.V., Bolgova, E.V., Ratova, L.G., Kovalchuk, S.V.: Simulation of patient flow and load of departments in a specialized medical center. Procedia Comput. Sci. **101**, 143–151 (2016)
22. Kovalchuk, S.V., Funkner, A.A., Metsker, O.G., Yakovlev, A.N.: Simulation of patient flow in multiple healthcare units using process and data mining techniques for model identification. J. Biomed. Inform. **82**, 128–142 (2018)
23. Suhaimi, N., Vahdat, V., Griffin, J.: Building a flexible simulation model for modeling multiple outpatient orthopedic clinics. In: 2018 Winter Simulation Conference (WSC), pp. 2612–2623 (2018)
24. Tabassum, S., Pereira, F.S.F., Fernandes, S., Gama, J.: Social network analysis: an overview. Wiley Interdiscip. Rev. Data Min. Knowl. Discov. **8**(5), 1–21 (2018)
25. Dunn, A.G., Westbrook, J.I.: Interpreting social network metrics in healthcare organisations: a review and guide to validating small networks. Soc. Sci. Med. **72**(7), 1064–1068 (2011)
26. Benhiba, L., Loutfi, A., Abdou, M., Idrissi, J.: A classification of healthcare social network analysis applications. In: HEALTHINF 2017-10th International Conference on Health Informatics, pp. 147–158 (2017)
27. Gephi-The Open Graph Viz Platform https://gephi.org/. Accessed 23 Jan 2019
28. Banks, J.: Discrete-event System Simulation. International Series in Industrial and Systems Engineering, vol. Fourth. Prentice-Hall, Upper Saddle River (2005)
29. Chapter 8: Markov Chains. https://www.stat.auckland.ac.nz/~fewster/325/notes/ch8.pdf. Accessed 24 Oct 2018
30. scipy.stats.rv_discrete — SciPy v0.19.0 Reference Guide (2017). https://docs.scipy.org/doc/scipy-0.19.0/reference/generated/scipy.stats.rv_discrete.html. Accessed 30 May 2017
31. Papi, M., Pontecorvi, L., Setola, R.: A new model for the length of stay of hospital patients. Health Care Manag. Sci. **19**(1), 58–65 (2014). https://doi.org/10.1007/s10729-014-9288-9
32. Marshall, A., Vasilakis, C., El-Darzi, E.: Length of stay-based patient flow models: recent developments and future directions. Health Care Manag. Sci. **8**, 213–220 (2005)
33. Ickowicz, A., Sparks, R., Wiley, J.: Modelling hospital length of stay using convolutive mixtures distributions. Stat. Med. **36**(1), 122–135 (2016)
34. Lee, A.H., Ng, A.S., Yau, K.K.: Determinants of maternity length of stay: a Gamma mixture risk-adjusted model. Health Care Manag. Sci. **4**(4), 249–55 (2001)
35. Houthooft, R., et al.: Predictive modelling of survival and length of stay in critically ill patients using sequential organ failure scores. Artif. Intell. Med. **63**, 191–207 (2015)
36. Reynolds, D.: Gaussian Mixture Models. https://pdfs.semanticscholar.org/734b/07b53c23f74a3b004d7fe341ae4fce462fc6.pdf. Accessed 19 Oct 2018
37. Chen, Y.-C.: A Tutorial on Kernel Density Estimation and Recent Advances (2017)
38. Vrieze, S.I.: Model selection and psychological theory: a discussion of the differences between the Akaike information criterion (AIC) and the Bayesian information criterion (BIC). Psychol. Methods **17**(2), 228–243 (2012)
39. Simard, R., L'Ecuyer, P.: Computing the two-sided Kolmogorov-Smirnov distribution. J. Stat. Softw. **39**(11), 1–18 (2011)

Investigating Coordination of Hospital Departments in Delivering Healthcare for Acute Coronary Syndrome Patients Using Data-Driven Network Analysis

Tesfamariam M. Abuhay[1,3]([✉]), Yemisrach G. Nigatie[3],
Oleg G. Metsker[1], Aleksey N. Yakovlev[1,2],
and Sergey V. Kovalchuk[1]

[1] ITMO University, Saint Petersburg, Russia
tesfamariam.m.abuhay@gmail.com,
olegmetsker@gmail.com, yakovlev_an@almazovcentre.ru,
sergey.v.kovalchuk@gmail.com
[2] Almazov National Medical Research Centre, Saint Petersburg, Russia
[3] University of Gondar, Gondar, Ethiopia
yemisrach.getinet@uog.edu.et

Abstract. Healthcare systems are challenged to deliver high-quality and efficient care. Studying patient flow in a hospital is particularly fundamental as it demonstrates effectiveness and efficiency of a hospital. Since hospital is a collection of physically nearby services under one administration, its performance and outcome are shaped by the interaction of its discrete components. Coordination of processes at different levels of organizational structure of a hospital can be studied using network analysis. Hence, *this article presents a data-driven static and temporal network of departments*. Both networks are directed and weighted and constructed using seven years' (2010–2016) empirical data of 24902 Acute Coronary Syndrome (ACS) patients. The ties reflect an episode-based transfer of ACS patients from department to department in a hospital. The weight represents the number of patients transferred among departments. As a result, the underlying structure of network of departments that deliver healthcare for ACS patients is described, the main departments and their role in the diagnosis and treatment process of ACS patients are identified, the role of departments over seven years is analyzed and communities of departments are discovered. The results of this study may help hospital administration to effectively organize and manage the coordination of departments based on their significance, strategic positioning and role in the diagnosis and treatment process which, in-turn, nurtures value-based and precision healthcare.

Keywords: Healthcare operations management · Network analysis · Graph theory · Data-Driven modeling · Complex system

© Springer Nature Switzerland AG 2020
V. V. Krzhizhanovskaya et al. (Eds.): ICCS 2020, LNCS 12140, pp. 430–440, 2020.
https://doi.org/10.1007/978-3-030-50423-6_32

1 Introduction

Provision of healthcare is one of the fundamental expenditures and political agendas of every government. Norway, Switzerland and the United States are the world's three biggest healthcare spenders – paying per person $9,715 (9.6% of GDP), $9,276 (11.5% of GDP), and $9,146 (17.1% of GDP), respectively [1]. However, healthcare systems are challenged to deliver high-quality and efficient care because of aging population, epidemic and/pandemic (e.g., COVID-19), scarcity of resources and poor planning, organization and management of healthcare processes [2–4].

As a well-coordinated and collaborative care improves patient outcomes and decreases medical costs [5], there is a need for effective organization of operational processes in a hospital. Besides to this, recent healthcare delivery system reforms such as value-based healthcare, accountable care and patient-centered medical homes require a fundamental change in providers' relationship to improve care coordination [6].

Acute Coronary Syndrome (ACS) is an umbrella term for an emergency situations where the blood supplied to the heart muscle is suddenly blocked [7, 8]. According to World Health Organization (WHO) [9], an estimated 17.9 million people died from Cardiovascular Diseases (CVDs) in 2016, representing 31% of all global deaths. Of these deaths, an estimated 7.4 million were due to ACS.

Whenever people feel symptoms of ACS, they visit a hospital as emergency or planned patient. During their stay in a hospital, ACS patients may move from one department to another department to get medical treatment or patients' laboratory samples and/or medical equipment may move from department to department.

Since hospital is a system that combines inter-connected and physically nearby services [10], its behavior and outcome are shaped by the interactions of its discrete components [11, 12]. In other words, departments in a hospital deliver different but interdependent services. An output of one department's operation can be an input and/or precondition for one or many departments in the diagnosis and treatment processes. This makes the underlying processes of a hospital highly dynamic, interconnected, complex, ad hoc and multi-disciplinary [5, 13, 14].

Coordination of operational processes at different level of organizational structure of a hospital can be conceptualized, studied and quantified using graph/network analysis or Social Network Analysis (SNA) [6, 15].

In organizational behavior studies, SNA can be defined as a set of social entities, such as people, groups, and organizations, with some relationships or interactions between them [16, 17]. SNA allows to model, map, characterize and quantify topological properties of a network, discover patterns of relations and identify the roles of nodes and sub-groups within a network [16, 18].

A systematic reviews [6, 15, 19–22] mentioned that SNA can be applied in healthcare setting to study interaction of healthcare professionals such as physician-nurse interactions and physician-physicians communication; diffusion of innovations, including adoption of medical technology, prescribing practices, and evidence-based medicine; and professional ties among providers from different organizations, settings, or health professions.

Chambers et al. [19] mentioned that 50 of 52 studies used survey or observation to collect data for constructing and studying complex networks in healthcare setting. Nowadays, healthcare administrative data have been widely used to demonstrate SNA [5]. According to [5, 6] and [12], the standard practice to construct network of healthcare providers (healthcare professionals) is based on "patient-sharing" concept meaning; two providers are considered to be connected to one another if they both deliver care to the same patient. To do so, first, bipartite network of patient-physician should be created. Next, this bipartite network can be projected to unipartite network of physicians only, where the ties reflect patient-sharing between physicians. For instance, Soulakis et al. [5] made an attempt to visualize and describe collaborative electronic health record (EHR) usage for hospitalized patients with heart failure by creating 2 types of networks: the first is a directed bipartite network which represents interactions between providers and patient records and the second network is undirected and depicts shared patient record access between providers. In 2018, Onnela et al. [12] have compared standard methods for constructing physician networks from patient-physician encounter data with a new method based on clinical episodes of care using data on 100% of traditional Medicare beneficiaries from 51 regions for the years 2005–2010.

However, hospital is a combination of departments and each department consists both human and material resources including medical equipment. As a result, effectiveness and efficiency of a hospital depends on organization, collaboration and availability of both human and material resources. Due to this, network of human resources (healthcare professionals) only may not reflect the real or complete picture of the underlying structure of the diagnosis and treatment process in a hospital. To the best of our knowledge, no one has studied collaboration of hospital departments using network analysis.

This article, therefore, investigates data-driven network of departments to answer the following research questions: *what is the underlying structure of network of departments that deliver healthcare for ACS patients? what are the main departments and their role in the diagnosis and treatment process of ACS patients? does the role of departments change over time? can we detect communities of departments which are highly interconnected?*

Answering these questions may give insight about the underlying organizational structure of departments, the role, strategic positioning and influence of departments in the diagnosis and treatment process and the interaction among departments and subgroups of departments (communities). This would help to effectively structure and manage the collaboration among departments and sub-group of departments by maintaining the functionality, capacity and geographical proximity of departments according to their role and significance in the diagnosis and treatment process of ACS patients.

As fraction of seconds matter a lot in diagnosing and treating ACS patients, the results of this study may also help to optimize the operational processes and minimize time, cost and effort of patients and health professionals.

The rest of the paper is organized as follows: Sect. 2 outlines methods including data collection and preprocessing; Sect. 3 discusses results obtained and Sect. 4 presents conclusion and future works.

2 Methods

This study was conducted in collaboration with the Almazov National Medical Research Centre[1], a major scientific contributor and healthcare provider that delivers high-tech medical care. The study was reviewed and approved by the Institute Review Board (IRB) of the National Center of Cognitive Research (NCCR) at ITMO University.

Seven years', from 2010 to 2016, empirical data of 24902 ACS patients was collected from this hospital. The patient identifiers were excluded from the dataset to protect the privacy of patients. The event log data describes movement of patients from department to department in the center and the corresponding timestamps. Those departments visited by ACS patients from 2010 to 2016 are included in this study.

Each patient's event log data was sorted based on an event date. In the dataset, there are two IDs, patient-ID and episode-ID, which uniquely identify a patient and clinical episode of a patient, respectively. The patient-ID is constant over the life time of a patient, whereas the episode-ID changes as the clinical episode of a patient changes. Patient-ID has been commonly used to construct network of providers [6]. However, one patient may have many episodes in different time period and departments providing care to a patient in the context of one clinical episode may not be directly connected to another clinical episode [12]. In this case, the standard approach based on patient-ID would produce a network that does not correspond to the real connections. In this study, therefore, episode-ID was used and network of departments was constructed based on the chronological transfer or flow of ACS patients. The proposed approach does not require creating bipartite network of patient-provider and changing (project-ing) the resulting network to unipartite network of providers only.

Two types of unipartite network was constructed, the first one is the static network and the second one is temporal network with one-year time window from 2010 to 2016. In order to reduce potential noises, the departments with less than 10 interaction (the number of patients transferred between departments) in seven years and less than 5 interaction in one year were excluded from the static and the temporal networks, respectively.

The networks were constructed as directed graph, where nodes represent depart-ments and ties reflect the flow or transfer of patients from one department to another department. For instance, if a patient had a surgery and transferred to Intensive Care Unit (ICU), it forms a directed graph from surgery (source) to ICU (target). The number of patients transferred between departments was employed as a weight. i.e., the proposed network is both directed as well as weighted graph. Finally, Gephi 0.9.2 [23] was employed to visualize the structure of the network and generate both network and node level statistics.

---

[1] Φhttp://www.almazovcentre.ru/?lang=en.

3   Results

3.1   Static Network of Departments

The static network of departments contains 227 nodes that represent departments and 4305 edges that represent flow or transfer of patients. Both degree and weighted degree distributions (see Fig. 1) are positively skewed with a large majority of departments having a low degree and a small number of departments having a high degree. This shows that network of departments are scale free [16] meaning; there are few departments that are highly connected to other departments in the diagnosis and treatment process of ACS patients.

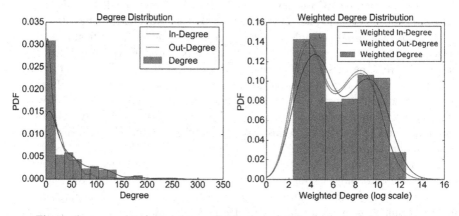

Fig. 1. Degree and weighted degree distribution of static network of departments.

The average degree and weighted degree account for 19 and 5800, in that order. This means that one department has an interaction with 19 departments on average by transferring or receiving 5800 patients on average over seven years. Even though the network of departments is sparse with density equals to 0.1, the average path length is short which accounts for 2.3. i.e., a given department may reach other departments in the network with 2.3 hops on average. Out of 227 departments in the static network, 27 departments are strongly connected meaning; 27 departments are connected to each other by at least one path and have no connections with the rest of the network.

Table 1. Top five departments with high support and influence.

Departments	Degree	In-degree	Out-degree	Weighted-degree	Weighted-in-degree	Weighted-out-degree
Regular laboratory	260	131	129	282260	141055	141205
Emergency laboratory1	240	124	116	277469	138259	139210
Functional diagnostic	219	105	114	171371	85964	85407
Cardiology one	182	93	89	101915	52381	49534
Cardiology two	182	90	92	96148	49525	46623

When we see the strategic positioning of departments in the static network, Regular Laboratory department, Emergency Laboratory1 department and Functional Diagnostic department are receiving more requests as well as sending more results to other departments (see Table 1).

In addition to this, cardiology departments, ICU departments and heart surgery departments are also vital in the diagnosis and treatment processes of ACS patients. This shows that ACS requires intensive diagnosis and treatment procedures and care. These departments are significant in terms of giving support to and influencing activities of other departments. Therefore, maintaining the functionality, capacity and geographical proximity of these departments is vital so as to deliver effective and efficient healthcare for ACS patients.

These departments are also fundamental in connecting communities of departments as they have high betweeness and closeness centrality in the network (see Table 2). In addition to bridging regions of the network, they may also facilitate the information flow across the network. For instance, these departments may serve as a hub for posting notices to patients as well as staff members of the hospital, hosting awareness creation activities, placing shared resources and propagating technology transfer projects.

Table 2. Betweeness and closeness centrality of departments.

Departments	Betweeness centrality	Departments	Closeness centrality
Regular laboratory	6449	Regular Laboratory	0.71
Emergency laboratory1	5089	Functional Diagnostic	0.68
Cardiology one	3879	Emergency Laboratory1	0.68
Cardiology two	3819	ICU	0.63

Five communities of departments (see Fig. 2) with different proportion (C1 (Violet) = 37.4%, C2 (Green) = 27.6%, C3 (Azure) = 19%, C4 (Orange) = 14% and C5 (Forest) = 4%) were identified using modularity and community extraction algorithm proposed by Blondel et al. [24].

The layout of the network of departments was produced in two steps: first, ForceAtlas2 algorithm [25] was applied to arrange nodes and edges. Second, Expansion algorithm was employed to scale up the network and make the layout more visible.

The average clustering coefficient equals to 0.62 which shows that there is strong interaction among departments within a module or community. These strongly connected departments can be reorganized and placed next to one another to maintain the geographical proximity between them. This may minimize time, cost and energy spend by the patients as well as health professionals to move from one department to another department. This may also improve the data and/or information exchange which, in turn, advances the coordination of departments as well as the effectiveness and efficiency of the diagnosis and treatment process in the center.

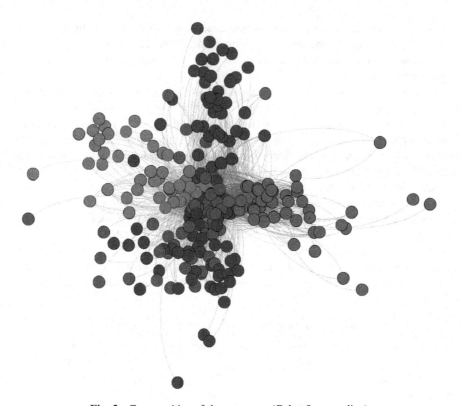

Fig. 2. Communities of departments. (Color figure online)

3.2 Temporal Network of Departments

Due to different reasons such as advancement of technology or business process reengineering, the structure of network of departments and the role of departments may change over time. In this section, analysis of the structure of network of departments and the role of departments over seven years is discussed. The node size in the network increases from year to year (see Table 3) which may indicate an introduction of new departments (e.g., emergency laboratory 2, see Fig. 4) or technologies or a change in working process that allow existing departments to engage actively in the diagnosis and treatment process.

As the node size increases, the edge size also increases with positive correlation coefficient of 0.94 (see Fig. 3). There is also positive correlation between edge size and average degree with a correlation coefficient of 0.76. Besides to this, modularity has a positive correlation with node size, edge size, Average Path Length (APL), and the number of strongly connected components.

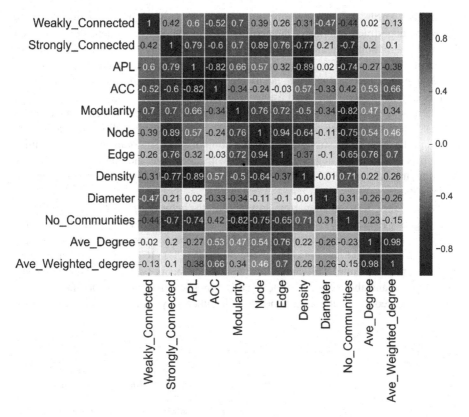

Fig. 3. Correlation coefficient of measurements of temporal network over years.

Table 3. Structure of network of departments over the course of seven years.

Year	Weakly_ Connected	Strongly_ Connected	APL	ACC	Modularity	Node	Edge	Density	Diameter	Ave_Degree	Ave_Weighted_degree
2010	1	9	2.2	0.5	0.24	70	635	0.13	5	9	729
2011	1	5	2	0.62	0.16	77	1025	0.2	5	13	1440
2012	1	9	2.1	0.64	0.23	110	1282	0.1	4	12	1387
2013	1	3	2	0.7	0.32	101	1799	0.2	4	18	2275
2014	1	20	2.2	0.6	0.35	131	1909	0.1	5	15	1754
2015	1	28	2.2	0.55	0.34	170	2773	0.1	5	16	1897
2016	2	23	2.3	0.5	0.45	148	2060	0.1	4	14	1415

The strongly connected components (which refer to a sub-graphs in which all the departments are connected to each other by at least one path and have no connections with the rest of the graph [16]) also increases over time which was 9 in 2010 and became 23 in 2016. This may indicate that the probability of forming strong connection within community or clique of departments is higher than establishing connection between departments from different communities even though both node size and edge size grow.

In other words, as a new department added to the network and got connected with one of the departments in one of the communities in the network, it tends to strengthen the local interaction instead of forming new connection with other nodes from another region of the network. This is supported with a constant average path length that does not change over time and a strong positive correlation of modularity with node size as well as edge size (see Fig. 3 and Table 3).

There is negative correlation between APL and Average Clustering Coefficient (ACC). Density of the temporal network also has negative correlation with node size, APL and the number of strongly connected components. The number of communities in the network also has negative correlation with node size and edge size which may illustrate that the node size and edge size do not contribute to the number of communities.

When we see the role of departments over time, the betweenness centrality (see Fig. 4) of Emergency Laboratory1, Regular Laboratory and Functional Diagnostic departments demonstrate growing trend, whereas all cardiology departments display random walk.

On the other hand, the betweeness centrality of ICU1, ICU2 and Outpatient clinic stay constant over time. This may show that departments such as Emergency Laboratory1, Regular Laboratory and Functional Diagnostic departments which facilitate the diagnosis procedure are the back bone of the healthcare system that deliver medical for ACS patients.

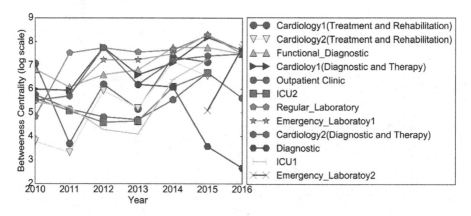

Fig. 4. Departments with high betweeness centrality over the course of seven years.

4 Conclusion and Future Work

In this article, data-driven static and temporal networks of departments are proposed to study the underlying structure of network of departments that deliver healthcare for patients, to identify the main departments and their role in the diagnosis and treatment process, to investigate evolution of role of departments over time, and to discover communities of departments.

Seven years', from 2010 to 2016, empirical data of 24902 Acute Coronary Syndrome (ACS) patients was employed to construct both static and temporal networks based on an episode-based transfer of patients.

As a result, we found out that Laboratory department, Emergency Laboratory1 department and Functional Diagnostic department are receiving more requests as well as sending more results to other departments. Five communities were discovered with an average clustering coefficient of 0.62. These departments are also fundamental in connecting communities of departments as they have high betweeness and closeness centrality in the network.

The results of this study may help hospital administration to effectively organize and manage the interaction among departments and sub-group of departments and maintain functionality, capacity and geographical proximity of departments according to their strategic positioning and role in the diagnosis and treatment process of ACS patients so that time, cost and energy could be saved and value-based healthcare could be achieved.

Finally, in the future, analysis of optimal arrangement of physical location of departments in a hospital will be conducted given degree, weighted degree, betweeness centrality, and closeness centrality measures. In addition, cost and time that take to transfer patients or information among departments will be considered. So that cost and time could be saved and patients' satisfaction could be improved.

Besides to this, how to achieve system approach in constructing holistic patient flow simulation, while maintaining the balance between the complexity and the simplicity of the model can be investigated using the results of this study.

Acknowledgement. This research is financially supported by The Russian Scientific Foundation, Agreement #17-15-01177.

References

1. Hutt, R.: Economics of healthcare: which countries are getting it right? World Economic Forum (2016). https://www.weforum.org/agenda/2016/04/which-countries-have-the-most-cost-effective-healthcare/. Accessed 30 Aug 2018
2. Haseltine, W.A.: Aging populations will challenge healthcare systems all over the world, Forbes (2018). https://www.forbes.com/sites/williamhaseltine/2018/04/02/aging-populations-will-challenge-healthcare-systems-all-over-the-world/#751ab4bd2cc3. Accessed 14 Dec 2018
3. Papi, M., Pontecorvi, L., Setola, R.: A new model for the length of stay of hospital patients. Health Care Manag. Sci. **19**(1), 58–65 (2014). https://doi.org/10.1007/s10729-014-9288-9
4. Chand, S., Moskowitz, H., Norris, J.B., Shade, S., Willis, D.R.: Improving patient flow at an outpatient clinic: Study of sources of variability and improvement factors. Health Care Manag. Sci. **12**(3), 325–340 (2009)
5. Soulakis, N.D., et al.: Visualizing collaborative electronic health record usage for hospitalized patients with heart failure. J. Am. Med. Inform. Assoc. **22**(2), 299–311 (2015)
6. DuGoff, E.H., Fernandes-Taylor, S., Weissman, G.E., Huntley, J.H., Pollack, C.E.: A scoping review of patient-sharing network studies using administrative data. Transl. Behav. Med. **8**(4), 598–625 (2018)

7. Suraj, A., Kundu, S., Norcross, W.: Diagnosis of acute coronary syndrome, American family physician (2005). https://www.aafp.org/afp/2005/0701/p119.pdf. Accessed 26 May 2018

8. "Acute coronary syndrome," American heart association (2017). http://www.heart.org/ HEARTORG/Conditions/HeartAttack/AboutHeartAttacks/Acute-Coronary-Syndrome_ UCM_428752_Article.jsp#.WwgvDi5uaUk. Accessed 25 May 2018

9. WHO, Cardiovascular diseases (CVDs) (2017). http://www.who.int/news-room/fact-sheets/ detail/cardiovascular-diseases-(cvds). Accessed 24 Aug 2018

10. Bhattacharjee, P., Kumar Ray, P.: Patient flow modelling and performance analysis of healthcare delivery processes in hospitals: a review and reflections. Comput. Ind. Eng. **78**, 299–312 (2014)

11. Bullmore, E., Sporns, O.: Complex brain networks: graph theoretical analysis of structural and functional systems. Nat. Rev. Neurosci. **10**(3), 186–198 (2009)

12. Onnela, J.-P., O'Malley, A.J., Keating, N.L., Landon, B.E.: Comparison of physician networks constructed from thresholded ties versus shared clinical episodes. Appl. Netw. Sci. **3**(1), 1–13 (2018). https://doi.org/10.1007/s41109-018-0084-1

13. Rebuge, Á., Ferreira, D.R.: Business process analysis in healthcare environments: a methodology based on process mining. Inf. Syst. **37**(2), 99–116 (2012)

14. Rojas, E., Munoz-Gama, J., Sepúlveda, M., Capurro, D.: Process mining in healthcare: a literature review. J. Biomed. Inform. **61**, 224–236 (2016)

15. Chapter 4, Emerging Trends in Care Coordination Measurement|Agency for Healthcare Research and Quality. https://www.ahrq.gov/professionals/prevention-chronic-care/improve/ coordination/atlas2014/chapter4.html#social. Accessed 28 Aug 2018

16. Tabassum, S., Pereira, F.S.F., Fernandes, S., Gama, J.: Social network analysis: an overview. Wiley Interdiscip. Rev. Data Min. Knowl. Discov. **8**(5), 1–21 (2018)

17. Dunn, A.G., Westbrook, J.I.: Interpreting social network metrics in healthcare organisations: a review and guide to validating small networks. Soc. Sci. Med. **72**(7), 1064–1068 (2011)

18. Benhiba, L., Loutfi, A., Abdou, M., Idrissi, J.: A classification of healthcare social network analysis applications. In: HEALTHINF 2017-10th International Conference on Health Informatics, pp. 147–158 (2017)

19. Chambers, D., Wilson, P., Thompson, C., Harden, M.: Social network analysis in healthcare settings: a systematic scoping review. PLoS ONE **7**(8), e41911 (2012)

20. Bae, S.-H., Nikolaev, A., Seo, J.Y., Castner, J.: Health care provider social network analysis: a systematic review. Nurs. Outlook **63**(5), 566–584 (2015)

21. Wang, F., Srinivasan, U., Uddin, S., Chawla, S.: Application of network analysis on healthcare. In: 2014 IEEE/ACM International Conference on Advances in Social Networks Analysis and Mining (ASONAM 2014), no. Asonam, pp. 596–603 (2014)

22. De Brún, A., McAuliffe, E.: Social network analysis as a methodological approach to explore health systems: a case study exploring support among senior managers/executives in a hospital network. Int. J. Environ. Res. Public Health **15**(3), 1–11 (2018)

23. Gephi-The Open Graph Viz Platform. https://gephi.org/. Accessed 23 Jan 2019

24. Iop, A., Blondel, V.D., Guillaume, J.-L., Lambiotte, R., Lefebvre, E.: Fast unfolding of communities in large networks. J. Stat. Mech. **2008**, 10008 (2008)

25. Jacomy, M., Venturini, T., Heymann, S., Bastian, M.: ForceAtlas2, a continuous graph layout algorithm for handy network visualization designed for the Gephi Software. PLoS ONE **9**(6), e98679 (2014)

A Machine Learning Approach to Short-Term Body Weight Prediction in a Dietary Intervention Program

Oladapo Babajide[1]($\boxtimes$), Tawfik Hissam[1], Palczewska Anna[1],
Gorbenko Anatoliy[1], Arne Astrup[2], J. Alfredo Martinez[3],
Jean-Michel Oppert[4], and Thorkild I. A. Sørensen[5]

[1] School of Built Environment, Engineering and Computing,
Leeds Beckett University, Leeds, UK
O.Babajide5449@student.leedsbeckett.ac.uk
[2] Department of Nutrition, Exercise and Sports,
University of Copenhagen, Copenhagen, Denmark
ast@nexs.ku.dk
[3] Centre for Nutrition Research, University of Navarra,
CIBERobn Obesity, and IMDEA Program on Precision Nutrition, Madrid, Spain
jalfmtz@unav.es
[4] Department of Nutrition Pitie-Salpetriere Hospital,
Institute of Cardiometabolism and Nutrition (ICAN),
Sorbonne University, Paris, France
jean-michel.oppert@aphp.fr
[5] Novo Nordisk Foundation Center for Basic Metabolic Research
and Department of Public Health, Faculty of Medical and Health Sciences,
University of Copenhagen, Copenhagen, Denmark
tias@sund.ku.dk

Abstract. Weight and obesity management is one of the emerging challenges in current health management. Nutrient-gene interactions in human obesity (NUGENOB) seek to find various solutions to challenges posed by obesity and over-weight. This research was based on utilising a dietary intervention method as a means of addressing the problem of managing obesity and overweight. The dietary intervention program was done for a period of ten weeks. Traditional statistical techniques have been utilised in analyzing the potential gains in weight and diet intervention programs. This work investigates the applicability of machine learning to improve on the prediction of body weight in a dietary intervention program. Models that were utilised include Dynamic model, Machine Learning models (Linear regression, Support vector machine (SVM), Random Forest (RF), Artificial Neural Networks (ANN)). The performance of these estimation models was compared based on evaluation metrics like RMSE, MAE and R2. The results indicate that the Machine learning models (ANN and RF) perform better than the other models in predicting body weight at the end of the dietary intervention program.

Keywords: Weight and obesity management · Body weight and weight-loss prediction · Supervised machine learning

© Springer Nature Switzerland AG 2020
V. V. Krzhizhanovskaya et al. (Eds.): ICCS 2020, LNCS 12140, pp. 441–455, 2020.
https://doi.org/10.1007/978-3-030-50423-6_33

1 Introduction

The main purpose of any weight-loss intervention is to ensure long-term weight loss. The success of this depends on the initial weight loss. There is, therefore, a need to track the progress of the patient in the dietary intervention program. More so, there is a need to understand and consider the end-goal of patient's adaptability and ability to reach any degrees of weight change, which is related to different approaches for weight management. Modifications in lifestyle that is aiming at weight-loss weight loss are often readily applied as a preferred treatment for overweight and obese patients. A lot of research work has gone into modelling body weight dynamics in humans [1, 2]. Some of the models aggregate observed clinical and laboratory data to make estimates of expected outcomes [3] which can be referred to as the statistical model [4]. The statistical model (1-D model) that predicts long term bodyweight seeks to require less numerous individual parameter estimates (variable inputs) which would be easier to implement in clinical practise [3]. It is known that statistical models aim to identify relationships between variables, but the predictive capabilities (in terms of their accuracies) of these statistical models are low [5].

This work aims to apply supervised machine learning methods, to predict future body weight for individuals attending the dietary intervention program at a lower margin of error. Additionally, we would like to find the best machine learning model and compare it with the statistical and dynamic model for the predictive analysis of body weight. In this study, the dietary intervention project is the Nutrient-gene interaction in human obesity (NUGENOB) [6] which was funded by the EU's 5th framework. It should be noted that since the during of the dietary intervention was in a period of 10 weeks, it is considered as a short-term dietary intervention program.

The paper is organised as follows. In the next section, we discuss related works followed by Data Collection method in Sect. 3. Section 4 presents the methodologies for predictive analysis. We discuss in Sect. 5, the regression model evaluation metrics. Section 6 presents results, and Sect. 7 presents the discussion, Sect. 8 discusses the conclusion and future work.

2 Related Works

Most of the related works in relation to weight and obesity management in terms of predicting weight relied on statistical/dynamic models. Therefore, this section will be split into two parts:

2.1 A Dynamic Model Approach for Body Weight Change Management

From the mathematical point of view, the human body obeys the laws of energy conservation which is based on the first law of thermodynamics [1]. The body is called an open system because either low or high intake of food will add energy to the process.

This energy balance equation is known to take the form of [1]:

$$R = I - E \tag{1}$$

Where I is the energy intake and E is the Energy expenditure. R is the rate of kcal/d that are stored or lost [2, 3]. This basis of energy transfer has helped to provide a better perception of how changes in any energy components can affect body weight change. Adopting this view, led to the development of compartmental equations with state variable tracking changes in energy derived from protein, fat, and carbohydrate [4, 5]. The Hall model was able to identify different state of transitions from energy intake to Energy expenditure which in the long run provides essential information into the mechanism behind human body weight change [4].

Forbes model also provides an intuition into how Fat-free fat mass (FFM) and fat mass are companions; i.e. increase/decrease in fat mass will be followed by an increase/decrease in FFM [6]. The equation for women relates to this;

$$FFM_{(t)} = 10.4ln\left(\frac{F_{(t)}}{D}\right) \tag{2}$$

Where D = 2.55 and $F_{(t)}$ is the Fat mass

$$FFM_{(t)} = 10.4ln\left(\frac{F_{(t)}}{S}\right) \tag{3}$$

Where S = 0.29 and $F_{(t)}$ is the Fat mass

Furthermore, Chow and Hall were able to create a more sophisticated method by coupling the fat-free mass (FFM) model proposed by Forbes equation through a two-dimensional dynamic model [7].

2.2   Machine Learning Approach for Body Weight Management

As an alternative to the dynamic approach for body weight change dynamics, machine learning has proven to be useful due to its ability to perform predictive analysis and drawing inference on health data [8]. This section discusses various approach by which machine learning has come to play in management overweight and obesity.

Machine learning in medicinal services can be viewed as a type of preventive healthcare. Preventive healthcare services guarantee that measures are taken to forestall disease occurrence, as opposed to disease treatment [9]. There are various degrees of preventive healthcare strategies techniques. The methodology frequently utilized through machine learning system is secondary prevention of health care. This procedure plans to identify, analyze health conditions before the development of the symptoms of complications arising in general wellbeing/health status of patients.

One of the capabilities of machine learning applications is the ability to identifying patterns in data. Such abilities can be utilized to early diagnosis of diseases and health conditions like cardiovascular diseases. Various techniques like Parameter Decreasing

Methods (PDM) and Artificial Neural Network (ANN) were applied to identify variables related to the development of obesity [10]. This strategy distinguished 32 factors, for example, individual data about the way of life containing nutritional habits and genetic profile which are credited to potential elements expanding the danger of cardiovascular illnesses. The utilization of this combined technique yielded an accuracy of 77.89% in the approval tests in characterization assignments identifying with stoutness. The BMI examinations had an accuracy of 69.15% in the forecast of a risk factor for CVD as an independent factor [10]. This method has helped in detecting weight gain at the early stage of development. Early detection of weight gain could be a signal for patients in taking positive action about their lifestyle and simultaneously minimizing health care governmental costs [10, 11].

Furthermore, in the quest to applying secondary preventive health care, a machine learning system was created (a fuzzy logic system), which aimed at predicting degrees of obesity to guide the physician's decision making [12]. The system was designed because it is perceived that BMI doesn't evaluate muscle to fat ratio precisely since it does exclude factors like age, sex, ethnicity, and bulk, giving a false diagnosis of body fatness [13, 14]. Also, another technique that was utilised in predicting overweight/obesity based on body mass index and gene polymorphisms is "Decision trees".

Decision trees were utilised in predicting early healthy and unhealthy eating habits [15].

Decision trees helped in identifying allelic variants associated with rapid body weight gain and obesity [16].

Also, neural networks have enhanced the capacity to predict long-term outcomes [17] from pre-operative data in bariatric surgery patients substantially over linear [18] and logistic regression [19]. Neural networks may have similar potential to amplify predictions of long-term weight loss success during lifestyle interventions and long-term behavioural adherence to physical activity recommendations [20].

From another dimension of using the concept of unsupervised machine learning, analysis of the patient's profile at the entry of the dietary intervention program was carried out by applying K-means clustering [21] on NUGENOB data. It resulted in better the understanding of weight-loss threshold in a dietary intervention program.

3 Data Collection

Health data analysis can enhance the efficiency of health policymakers to capture health-related issues [22]. Health data that are utilized in getting more insight into trends in diseases outbreak which can serve as the primary contact to an individual in a relevant population [23]. Concerning weight and obesity management, dietary intervention is a measure aimed at controlling body weight and obesity. Health data are usually generated during health intervention programs. An example of a health intervention program is Nutrient-gene interactions in human obesity (NUGENOB). The objective of this intervention program was, to examine if a 10-week low-fat hypo-energetic diet has a more beneficial effect on body weight, body composition and concentrations of fasting plasma lipids, glucose and insulin than a high-fat hypo-energetic diet [24, 25]. This was achieved by conducting a Randomised intervention trial with obese subjects from eight centres in seven European countries. The samples extracted from the NUGENOB database include attributes from individual subjects in the dietary intervention. These attributes are also referred to as the subject's profile.

The subject's profile is categorised as:

- Subject's Diet's Composition, e.g. High and Low-Fat food content
- Anthropometry Measurement, e.g. height and weight, waist
- Metabolic Rate Measurements
- Body composition measured by Bioimpedance analysis, e.g. fat mass and fat-free mass
- Biochemical Components Measurements, e.g. LDL - and HDL-cholesterol, and Fasting Plasma Insulin

The number of data-features is 25. It should be noted that not all features would be utilized for both machine learning and dynamic approach. Feature selection would be carried out for machine learning algorithms to select the best variable required for the predictive analysis. For better understanding of impact of diets on patients, the body weight percentage change distribution was calculated, and the distribution is shown in Fig. 1.

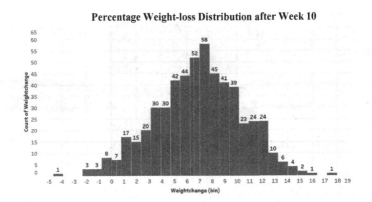

Fig. 1. Percentage body-weight-distribution after week 10

4 Methodologies

Two methods were applied in predicting body weight at the end of the dietary intervention program. They are:

4.1 Machine Learning Approach to Body Weight Change Dynamics

Machine learning (ML) are computational methods that computers use to make and improve predictions based on data. Machine Learning is a branch of artificial intelligence that can be used for predictive analytics. The use of machine learning in predictive analysis of body weight at the end of the dietary intervention program (week 10) seeks to evaluate the performance and capabilities on how it can provide augmented information for health care providers. In the long run, this is expected to improve the efficiencies of health care providers [26].

A blueprint on how input variables are structured in the machine learning models is shown in Fig. 2. Our Machine learning models are implemented using the CARET in R [27]. The process involved in the machine predictive analytics is described in Fig. 3. After the acquisition of data, the next phase is the feature selection. Feature selection in machine learning refers to techniques involved in selecting the best variables required for a predictive model.

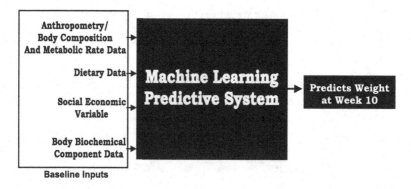

Fig. 2. Machine learning blueprint for weight-change predictive analysis

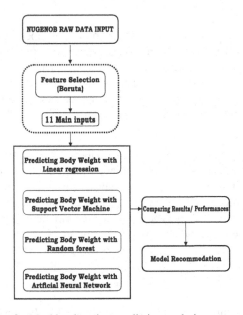

Fig. 3. Machine learning predictive analytics processes

Feature selection is usually carried out to increase the efficiency in terms of computational cost and modelling and mostly to improve the performance of the model. Feature selection for machine learning modelling can be achieved using a method called Boruta. Boruta algorithm is a wrapper built around the Random Forest classification algorithm implemented in the R package RandomForest [28]. It provides a stable and unbiased selection of essential features from an information system. Boruta wrapper is run on the NUGENOB dataset with all the attributes, and it yielded 11 attributes as essential variables. The selected variables are shown in Table 1.

Table 1. Showing features selected variables

Variable names		
1. Age	2. Body Weight@ Week 0 (Baseline)	3. Gender
4. Mean waist-hip ratio baseline	5. Fat mass baseline	6. Fasting glucose baseline
7. Basic metabolic rate baseline	8. Energy expenditure. T.0	9. Height
10. HOMA Insulin resistance(I0)	11. Fasting insulin baseline	

Machine Learning Models

The two common methods used in machine learning are supervised and unsupervised learning. Supervised learning technique is used when the historical data is available for a particular problem and this deem suitable in our case. The system is trained with the inputs (historical data) and respective responses and then used for the prediction of the response of new data. Conventional supervised approaches include an artificial neural network, Random Forest, support vector machines.

i. Multivariate Regression Analysis

In multiple linear regression, there is a many-to-one relationship, between a wide variety of independent variables (input) variables and one dependent (response) variable. Including more input variables does not always mean that regression will be better or provide better predictions. In some cases, adding more variables can make things worse as it results in overfitting. The optimal scenario for a good linear regression model is for all of the independent variables (inputs) to be correlated with the output variable, but not with each other. Linear regression can be written in the form:

$$y_i = b_0 + b_1 + b_1 x_{i^1} + \ldots b_p x_{i^p} + e_i \tag{4}$$

Where y_i represents the numeric response for the ith sample, b_0 represents the estimated intercept, b_j represents the estimated coefficient for the jth predictor, e_i represents a Random error that cannot be explained by the model. When a model can be written in the form of the equation above, it is called *linear in the parameters*.

ii. Support Vector Machine

Support Vector machines are used for both classification and regression problems. When using this approach, a hyperplane needs to be specified, which means a boundary of the decision must be defined. The hyperplane is used to separate sets of objects belonging to different classes. Support Vector Machine can handle linearly separable objects and non-linearly separable objects of classes.

Mathematically, support vector machines [29] are usually maximum-margin linear models. When there exists no loss of generality that Y = {− 1, 1} and that b = 0, support vector machines are works by solving the primal optimisation problem [30]. If there are exists non-linearly separable objects, methods such as kernels (complex mathematical functions) are utilized to separate the object which are members of different classes.

The most commonly used metric to measure the straight-line distance between two samples is defined as follows:

$$\min_{w,\xi} \left\{ \frac{1}{2} ||w||^2 + C \sum_{i=1}^{N} \xi_i \right\} \tag{5}$$

$$\text{Subject to : } y_i(w \cdot x_i) \geq 1 - \xi_i, \ \xi_i \geq 0$$

For a complete description of the random forest model, we refer the reader to [30].

iii. Random Forest

Random Forest is an ensemble learning technique in which numerous decision trees are built and consolidated to get an increasingly precise and stable prediction. The algorithm starts with a random selection of samples with replacement from the sample data. This sample is called a "bootstrapped" sample. From this random sample, 63% of the original observations occur at least once. Samples in the original data set which are not selected are called out of -bag observations. They are used in checking the error rate and used in estimating feature importance. This process is repeated many times, with each sub-sample generating a single decision tree. On the long run, it results in a forest of decision trees. The Random Forest technique is an adaptable, quick, machine learning algorithm which is a mixture of tree predictors. The Random Forest produces good outcomes more often since it deals with various kinds of data, including numerical, binary, and nominal. It has been utilized for both classification and regression. The reality behind the Random Forest is the combination of Random trees to interpret the model. Furthermore, the Random Forest is based on the utilisation of majority voting and probabilities [31]. Random Forest is also good at solving overfitting issue. A more detailed process of the random forest model is explained in [31].

iv. Artificial Neural Networks

Artificial neural network mimics the functionality of the human brain. It can be seen as a collection of nodes called artificial neurons. All of these nodes can transmit information to one another. The neurons can be represented by some state (0 or 1), and each node may also have some weight assigned to them that defines its strength or importance in the system. The structure of ANN is divided into layers of multiple nodes; the data travels from the first layer (input layer) and after passing through middle layers (hidden layers) it reaches the output layer, every layer transforms the data into some relevant information and finally gives the desired output [32]. Transfer and activation functions play an essential role in the functioning of neurons. The transfer function sums up all the weighted inputs as:

$$z = \sum_{x=1}^{n} w_i x_i + w_b$$

For a complete description of the artificial neural networks, we refer the reader to [32].

4.2 A Dynamic Modelling Approach to Body Weight Change Dynamics

The utilisation of weight-change models can enable patients to adhere to diets during a calorie restriction program. This is because weight change models generate predicted curves which is a form of diagnostic mechanism to test the difference between the actual predicted weight loss and actual weight loss [27]. There are numerous existing models, but they all require parameter estimates which on the long poses challenges for clinical implementation [27]. A new model was developed by [33], which provided a minimal amount of inputs from baseline variables (age, height, weight, gender and calorie intake. A blueprint on how input variables are structured in the dynamic models is shown in Fig. 4.

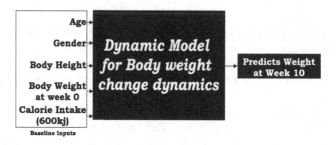

Fig. 4. Dynamic model blueprint for weight-change predictive analysis

This new model is an improvement to Forbe's model [6] which takes in a minimal amount of input in calculating the fat-free mass. We refer to the reader to [33] complete biological details on the full model development, and present here is the performance analysis of the dynamic model with other machine learning models. Dynamic modelling was implemented using the multi-subject weight change predictor simulator software [27].

5   Regression Model Evaluation Metrics

In the process of predictive analysis, errors will be generated. Such errors need to be measured to understand better the accuracies of machine learning algorithms. When the lower error is achieved, the better the predictive performance of the algorithms in term of accuracies. Dataset is split into a training set (80%) and test set (20%). The training set will be used to train the algorithm. The test set is usually used in measuring the performance of the algorithms. The training samples for this study case is 443 (train-set), while the testing samples (test-set) is 107 as depicted in Fig. 5. The test set is usually called "the unseen data". The result from the machine learning algorithms (predicted result) will be compared to the test data (Unseen data). Thus, various metrics will be utilised in measuring the degree of error between the actual and predicted results. In this work, we used the Root Mean Square Error (RMSE), the Mean Absolute Error (MAE), the fitness degree R2, A residual plots, will be utilised in measuring the performance of these algorithms. These measures were standard in the literature, generally for prediction analytics tasks. These measures includes:

- **Mean Absolute Error (MAE)**
  The MAE is used in measuring the accuracy of continuous variables. The errors generated from prediction analysis using the four selected algorithms are presented in Table 3.
- **Root Mean Square Error (RMSE)**
  The RMSE is essential to measure the prediction's accuracy because it allows the error to be the same magnitude as the quantity being predicted [34]. For best predictability, lower RMSE is needed.

- **Coefficient of multiple determination R^2 coefficients**
  R^2 coefficient also called fitness degree. Better performances are achieved when R^2 values are near 1. Ideally, if $R^2 = 1$, the original series and the predicted one would be superimposed. For best predictability, higher R^2 is needed.

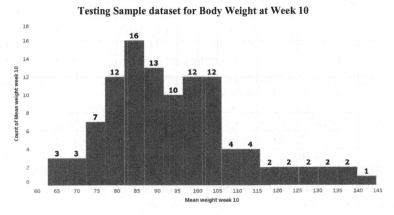

Fig. 5. Test set body-weight at week 10 distribution

Table 2. Model comparison performance based on different metrics

	LM	SVM	RF	ANN	Dynamic model
RMSE	4.309964	4.349037	3.268409	3.55828	4.629934
MAE	3.438419	3.493781	2.64141	2.763006	3.791817
R-squared	0.930191	0.930122	0.960241	0.953391	0.954028

Key: LM - Linear Regression, SVM - Support Vector machine, RF - Random Forest, ANN - Artificial Neural Network

6 Results

The predictive analysis was carried out on predicting body weight at week 10 using both Dynamic models and machine learning models. Four machine learning algorithms (Linear regression, Support Vector Machine, Random Forest and Artificial neural networks) were used for the predictive analysis of weight at the end dietary intervention program (week 10). The performance of each machine learning algorithms and dynamic model are carried out by calculating the mean absolute error and root mean square based on the test-set are presented in Table 2. The test-set sample distribution is displayed in Fig. 5.

It shows that Random Forest has the lowest error in predicting capabilities for weight at week 10. Furthermore, in terms of R-square, as illustrated in Table 2, dynamic models and machine learning models achieved R-square of over 93%. Random Forest has the highest R-square value of 96%, which is a very good fit.

7 Discussion

It should be noted that during the dietary intervention, two types of diets were administered to patients. The two types of diets are low and high hypo-energetic diets, i.e. the total calories in these diets is 600 kcal. These types of diets for obesity and weight management in this study assumes that all participants adhere strictly to 600 kcal per day in other to have a considerable weight-change or weight-loss. Machine learning and dynamic models come into play in the predictability of future change in the body weight of participants in a dietary intervention. The quest is to either predict actual body weight or to predict weight-loss at the end of the dietary intervention program. In this study, body weight-loss means quantifying by how much participants would lose weight during the diet intervention program.

The results for predicting actual body weight and computing the body weight-loss are explained later on in the text. From the above results (Table 2), Random Forest and Artificial neural networks algorithm perform best in predicting body weight, i.e. actual body weight at week 10 (last week of diet intervention). The result is evaluated by calculating their predictive error (Mean Absolute Error). The errors were found to be 2.64 and 2.76 kg. Having a lower error is one of the most important factors in predictive analysis.

Linear regression is a form of a statistical model that identify the relationships between variables. However, it comes at the cost of predictive capabilities, which will always result in a higher error [35]. This is reflected in the high mean absolute error for the linear regression model (± 3.438 kg). Dynamic models in the context of predicting body weight change, explains energy inflows and outflows in the body, based variables like age, height, weight baseline and gender in relation to time. Utilising dynamic models in predicting short term body-weight produces a high error (Mean Absolute Error) as compared to other Machine learning models with a lower error. However, in terms of achieving a good r-square (variance), both dynamic models and machine learning models are good to explain the variation of the dependent variables from the independent variables(s). Applying Boruta method of feature selection for predicting body-weight at week 10 (last week of diet intervention), variables such as initial body weight at week 0, initial fat-mass at week 0, and energy expenditure and basic metabolic rate play a significant role for predicting body weight at week 10. The technique utilized in identifying variable's relationship with the response as compared to other variables used in the model is Random Forest variable importance.

From the results obtained, a mean absolute error of ± 2.6 kg is achieved from utilizing Random Forest algorithm. In order to expatiate more on this; assuming we have a 10% reduction of actual body weight (94.23 kg) at week 10, it corresponds to 9.42 kg weight loss. The model would have predicted 84.81 kg $\pm$ 2.6 kg i.e. 3% error in prediction.

Also, when computing weight-loss, comparing the actual percentage weight-loss with the predicted percentage weight-loss, i.e. actual weight loss (9.9%) and predicted average weight-loss (7.23%), the percentage prediction error for the weight-loss would be up to 27.6%. This error incurred in weight-loss prediction analysis is high. It is quite easy from this study to predict the actual short-term body weight than short term body weight loss.

Mean absolute error (MAE) has been utilized to interpret the effect of the error on the models. In order to utilize RMSE, the underlying assumption when presenting the RMSE is that the errors are unbiased and follow normal distribution [36]. Conducting the Shapiro-Wilk normality test on the test-set. We had a p-value of 0.00276 (p-value < 0.05) that shows that the distribution of the residuals are significantly different from normal distribution which is also described in Fig. 5. This reflects that the distribution in the test set is not normally distributed. Hence, RMSE cannot be fully relied on. RMSE and the MAE are defined differently, we should expect the results to be different. Both metrics are usually used in assessing the performance of the machine learning model. Various research indicates that MAE is the most natural measure of average error magnitude, and that (unlike RMSE) it is an unambiguous measure of average error magnitude [37]. It follows that the RMSE will increase (along with the total square error) at the same rate as the variance associated with the frequency distribution of error magnitudes increases which in turn will make the RMSE always greater than MAE. Therefore, in this study, MAE would be the main metrics utilized in assessing the model.

8   Conclusion and Future Work

One of the strengths of Random Forest is its ability to perform much better than other machine learning algorithms due to its ability to handle small data sets. Since Random Forest technique is majorly on building trees, it tries to capture and identify patterns with a small dataset and is still able to generate a minimal error. In our case, the training dataset is 443. In contrast to neural networks, it needs more dataset to train with, and thus with the current dataset, it is expected that the predictive power of neural networks would be much lower than Random Forest.

Comparing the computation performance for other models, it is evident that the Random Forest performs best in the predictive analysis of body weight. Computationally, machine learning models achieve lower predictive error compared to dynamic models in predicting short-term body weight. However, from the clinical point of view, the minimum mean absolute percentage error produced from the discussed model (Random Forest) in predicting body weight-loss is still high. The research work shows the capability of both machine learning models and dynamic model in predicting body weight and weight-loss. Future work includes hybridisation of machine learning and dynamic models in predicting body weight-loss that are represented in terms of classes, i.e. High, Medium and Low weight loss. This approach will provide more solution in body weight-loss predictability. Also, further research work could address the inclusion of dietary type for the predictive analysis, which can also provide information on which diet to recommend under a specific set of conditions.

Acknowledgement. Authors are grateful to all the individuals who have enrolled in the cohort. We also acknowledge the NUGENOB team and other individuals namely: Peter Arner, Philippe Froguel, Torben Hansen, Dominique Langin, Ian Macdonald, Oluf Borbye PGedersen, Stephan Rössner, Wim H Saris, Vladimir Stich, Camilla Verdich, Abimbola Adebayo and Olayemi Babajide for their contribution in the acquisition of the subset of NUGENOB dataset.

References

1. Fogler, H.S.: Elements of Chemical Reaction Engineering, p. 876. Prentice-Hall, Englewood Cliffs (1999)
2. Jéquier, E., Tappy, L.: Regulation of body weight in humans. Physiol. Rev. **79**(2), 451–480 (1999)
3. McArdle, W.D., Katch, V.L., Katch, F.: Exercise Physiology: Energy, Nutrition, and Human Performance. Lippincott Williams & Wilkins, Philadelphia (2001)
4. Hall, K.D.: Computational model of in vivo human energy metabolism during semistarvation and refeeding. Am. J. Physiol. Endocrinol. Metab. **291**(1), 23–37 (2006)
5. Hall, K.D.: Predicting metabolic adaptation, body weight change, and energy intake in humans. Am. J. Physiol. Endocrinol. Metab. **298**(3), E449–E466 (2009)
6. Forbes, G.B.: Lean body mass-body fat interrelationships in humans. Nutr. Rev. **45**(10), 225–231 (1987)
7. Chow, C.C., Hall, K.D.: The dynamics of human body weight change. PLoS Comput. Biol. **4**(3), e1000045 (2008)
8. Mast, M.: Intel Healthcare-analytics-whitepaper
9. Leavell, H.R., Clark, E.G.: The Science and Art of Preventing Disease, Prolonging Life, and Promoting Physical and Mental Health and Efficiency. Krieger Publishing Company, New York (1979)
10. Valavanis, I.K., et al.: Gene - nutrition interactions in the onset of obesity as cardiovascular disease risk factor based on a computational intelligence method. In: IEEE International Conference on Bioinformatics and Bioengineering (2008)
11. Finkelstein, E.A., Trogdon, J.G., Brown, D.S., Allaire, B.T., Dellea, P.S., Kamal-Bahl, S.J.: The lifetime medical cost burden of overweight and obesity: implications for obesity prevention. Obesity (Silver) **16**(8), 1843–1848 (2008)
12. Heo, M., Faith, M.S., Pietrobelli, A., Heymsfield, S.B.: Percentage of body fat cutoffs by sex, age, and race-ethnicity in the US adult population from NHANES 1999-2004. Am. J. Clin. Nutr. **95**(3), 594–602 (2012)
13. Shah, N.R., Braverman, E.R.: Measuring adiposity in patients: the utility of body mass index (BMI) percent body fat, and leptin. PLoS one **7**(4), e33308 (2012)
14. Marmett, B., Carvalhoa, R.B., Fortesb, M.S., Cazellab, S.C.: Artificial intelligence technologies to manage obesity. Vittalle – Revista de Ciências da Saúde **30**(2), 73–79 (2018)
15. Spanakis, G., Weiss, G., Boh, B., Kerkhofs, V., Roefs, A.: Utilizing longitudinal data to build decision trees for profile building and predicting eating behavior. Procedia Comput. Sci. **100**, 782–789 (2016)
16. Rodríguez-Pardo, C., et al.: Decision tree learning to predict overweight/obesity based on body mass index and gene polymporphisms. Gene **699**, 88–93 (2019)
17. Thomas, D.M., Kuiper, P., Zaveri, H., Surve, A., Cottam, D.R.: Neural networks to predict long-term bariatric surgery outcomes. Bariatr. Times **14**(12), 14–17 (2017)
18. Courcoulas, A.P., et al.: Preoperative factors and 3-year weight change in the longitudinal assessment of bariatric surgery (LABS) consortium. Surg. Obes. Relat. Dis. **11**(5), 1109–1118 (2015)
19. Hatoum, I.J., et al.: Clinical factors associated with remission of obesity-related comorbidities after bariatric surgery. JAMA Surg. **151**(2), 130–137 (2016)
20. Degregory, K.W., et al.: Obesity/Data Management A review of machine learning in obesity. Obes. Rev. **19**(5), 668–685 (2018)

21. Babajide, O., et al.: Application of unsupervised learning in weight-loss categorisation for weight management programs. In: 10th International Conference on Dependable Systems, Services and Technologies (DESSERT) (2019)
22. Segen, J.C.: McGraw-Hill Concise Dictionary of Modern Medicine. McGraw-Hill, New York (2002)
23. Safran, C.: Toward a national framework for the secondary use of health data: an American medical informatics association white paper. J. Am. Med. Inf. Assoc. **14**(1), 1–9 (2007)
24. Petersen, M., et al.: Randomized, multi-center trial of two hypo-energetic diets in obese subjects: high- versus low-fat content. Int. J. Obes. **30**, 552–560 (2006)
25. Srensen, T.I.A., et al.: Genetic polymorphisms and weight loss in obesity: a randomised trial of hypo-energetic high- versus low-fat diets. PLoS Clin. Trials **1**(2), e12 (2006)
26. Bartley, A.: Predictive Analytics in Healthcare: A data-driven approach to transforming care delivery. Intel
27. Thomas, D.M., Martin, C.K., Heymsfield, S., Redman, L.M., Schoeller, D.A., Levine, J.A.: A simple model predicting individual weight change in humans. J. Biol. Dyn. **5**(6), 579–599 (2011)
28. Liaw, A., Wiener, M.: Classification and regression by RandomForest (2002)
29. Cortes, C.: Support-vector networks. Mach. Learn. **20**, 273–297 (1995). https://doi.org/10.1007/BF00994018
30. Louppe, G.: Understanding random forests: from theory to practice (2014)
31. Kohonen, T.: An introduction to neural computing. Neural Netw. **1**(1), 3–16 (1988)
32. Thomas, D.M., Ciesla, A., Levine, J.A., Stevens, J.G., Martin, C.K.: A mathematical model of weight change with adaptation. Math. Biosci. Eng. MBE **6**(4), 873–887 (2009)
33. Gropper, S.S., Smith, J.L., Groff, J.L.: Advanced Nutrition and Human Metabolism. Thomson Wadsworth, Belmont (2005)
34. Han, J., Kamber, M., Pei, J.: Data Mining. Concepts and Techniques (The Morgan Kaufmann Series in Data Management Systems), 3rd edn. Morgan Kaufmann, Waltham (2011)
35. Mastanduno, M.: Machine Learning Versus Statistics: When to use each. healthcatalyst, 23 May 2018. https://healthcare.ai/machine-learning-versus-statistics-use/. Accessed 27 Dec 2019
36. Christiansen, E., Garby, L.: Prediction of body weight changes caused by changes in energy balance. Eur. J. Clin. Invest. **32**(11), 826–830 (2002)
37. Hidalgo-Muñoz, A.R., López, M.M., Santos, I.M., Vázquez-Marrufo, M., Lang, E.W., Tomé, A.M.: Affective valence detection from EEG signals using wrapper methods. In: Emotion and Attention Recognition Based on Biological Signals and Images. InTech (2017)
38. Chai, T., Draxler, R.R.: Root mean square error (RMSE) or mean absolute error (MAE)? - arguments against avoiding RMSE in the literature. Geosci. Model Dev. **7**(3), 1247–1250 (2014)
39. Willmott, C.J., Matsuura, K.: Advantages of the mean absolute error (MAE) over the root mean square error (RMSE) in assessing average model performance. Research **30**(1), 79–82 (2005)

An Analysis of Demographic Data in Irish Healthcare Domain to Support Semantic Uplift

Kris McGlinn[1]([⊠])[iD] and Pamela Hussey[2][iD]

[1] ADAPT, Trinity College Dublin, Dublin, Ireland
kris.mcglinn@adaptcentre.ie
[2] Health Informatics, Dublin College University, Dublin, Ireland
pamela.hussey@dcu.ie

Abstract. Healthcare data in Ireland is often fragmented and siloed making it difficult to access and use, and of the data that is digitized, it is rarely standardised from the perspective of data interoperability. The Web of Data (WoD) is an initiative to make data open and interconnected, stored and shared across the World Wide Web. Once a data schema is described using an ontology and published, it resides on the web, and any data described using Linked Data can be associated with this ontology so that the semantics of the data are open and freely available to a global audience. In this article we explore the semantic uplift of demographic data in the Irish context through an analysis of Irish data catalogues, and explore how demographic data is represented in health standards internationally. Through this analysis we identify the Fast Healthcare Interoperability Resources (FHIR) ontology as a basis for managing demographic health care data in Ireland.

Keywords: Knowledge engineering · Linked data · Interoperability

1 Introduction

It is estimated that up to 30% of the total health care budget in Ireland is spent managing the large amounts of data generated, i.e. collecting data, querying data and storing data. Reports indicate that much of the healthcare data in Ireland, like the US, often exists in silos, fragmented processes and is only accessible to disparate stakeholders [11,21]. Of the data that is digitized, it is rarely standardised from the perspective of data interoperability, meaning it does not adhere to any particular standardised terminologies, schema or syntax. This impacts on data quality and data value in the longer term [14].

The Web of Data is an initiative to make data open and interconnected, stored and shared across the World Wide Web[1] using a well established architecture, the semantic web stack [23]. Central to this is the concept of Linked Data

---

[1] https://www.w3.org/2013/data/.

© Springer Nature Switzerland AG 2020
V. V. Krzhizhanovskaya et al. (Eds.): ICCS 2020, LNCS 12140, pp. 456–467, 2020.
https://doi.org/10.1007/978-3-030-50423-6_34

(LD), a way of structuring and sharing data on the web based on the Resource Description Framework (RDF). By defining data models using semantic web technologies, it becomes possible to make data schemas available using standard web access mechanisms, e.g. HTTP. Once a data schema is described using an ontology (or as an RDF vocabulary) and published, it resides on the web, and any data described using LD can be associated with this ontology (or vocabulary) so that the semantics of the data are open and freely available to a global audience. Combined with SPARQL, an RDF query language, this is a powerful tool for sharing and re-using data and has the potential to provide improved support for interoperability in health and social care in Ireland. Already within the health care domain, organisations such as HL7[2] are exploring the use of resource based data models as evident in the Fast Healthcare Interoperability Resources (FHIR) [1,5]. FHIR provides an architecture which aligns with the semantic web and can provide a sound basis for supporting data interoperability in the health domain.

In this article we present an analysis of health care data in the Irish context [12], with a particular focus on demographic data as this is well established within the health domain and already has several standard approaches to supporting its exchange, e.g. the ISO 13606 demographics package [6], OpenEHR [4] and FHIR (captured under Person and Patient) [1]. This analysis forms part of a methodology for supporting the uplift of data into the Web of Data. The article is structured as follows, first we provide a description of our methodology for supporting semantic uplift, next a brief overview of some of the relevant health care standards for capturing demographic data, then an analysis of health care data in the Irish context with a focus on demographic data. Finally, some recommendations for managing health care data that can be applied within the Irish context for sharing demographic data in accordance with national planned eHealth Strategy. It is also one of two articles, the second providing an overview of where this work fits into the wider requirements for interoperability in health care in Ireland and internationally [15].

2   Methodology for Semantic Uplift

Semantic uplift is the conversion of structured or semi-structured data into Linked Data based upon semantic-web technologies. Our process for supporting semantic uplift within the healthcare domain is based on a standard methodology for ontology development[3], which consists of defining the scope, reuse of existing ontologies and vocabularies, enumeration of terms, definition of classes, properties and constraints, and finally the creation of instances (Fig. 1). Ontology development (chevron 3–6 in Fig. 1) is only required where analysis determines no existing vocabulary can be found to satisfy the data exchange requirements defined within the scoping stage, or to support the interlinking process where

---

[2] https://www.hl7.org/.

[3] https://protege.stanford.edu/publications/ontology_development/ontology101.pdf.

multiple ontologies have been found. This article addressed specifically the analysis of existing standards and ontologies (Sect. 3), and existing data schemas (Sect. 4).

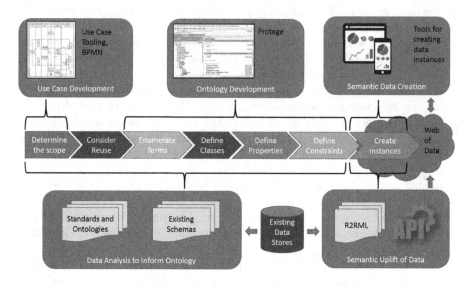

Fig. 1. Overview of methodology for developing ontology

3   Analysis of Demographic Data in Electronic Health Care Standards

Electronic Health Record (EHR) systems in Europe and internationally are digitized records of patient and population health information that can be shared between health care settings [19]. Here we examine how several key international standards model demographic data.

3.1   ISO 13606

In the European Union CEN's TC 251 Technical Committee is a decision making body for Standardization. They develop standards in the field of Health Information and Communications Technology (ICT). The goal is interoperability between independent EHR systems and towards this goal CEN TC 251 have generated a large number of standards related to health care for harmonisation of systems and data [9], of particular relevance is ISO 13606-1 [2] which specifies a generic information model of part or all of the EHR of a single identified subject of care between EHR systems, or between EHR systems and a centralized EHR data repository. ISO/EN 13606 is based on a dual model: a Reference

Model (RM) for information, and an Archetype Object Model (AOM) for defining knowledge, i.e. the concepts of the clinical domain by means of Archetypes. Archetypes are patterns that represent the specific characteristics of the clinical data, and aim to support domain experts to create and change archetypes, giving them control over how EHRs are built and to reflect their knowledge.

Several classes are available to represent how a clinical system will be delivered to a recipient. These are EHR_EXTRACT - a top level container for a transaction between an EHR Provider and an EHR Recipient, FOLDER - a means for compartmentalizing within EHR, COMPOSITION - information committed to an EHR by some agent, SECTION within a composition, ENTRY - a result from a clinical action, i.e an observation or test result, CLUSTER - a means to organise multiple entries, e.g. in a time series and ELEMENT - containing a single data value. Built upon these classes, the ISO 13606 demographics package was developed to solve three scenarios related to demographic data: 1) minimum identification to permit demographic matching between two systems; 2) a rich enough descriptor set to populate a recipient's demographic system with enough to identify and contact persons or organisations, and 3) for the whole thing to be optional if the exchange is occurring inside a shared demographics realm [3]. It enables the modelling of values such as a persons name, postal address, gender, birth date, birth order and deceased time.

The main shortcomings with 13606-1 are related to its licensing, which is not open, inhibiting re-use. There are also issues with its definition of interfaces as these are minimal and would benefit from expansion [3]. A good candidate for this is SPARQL, and more recent work has looked specifically at converting ISO 13606-1 into an OWL ontology, called OntoCR [16], although work on OntoCR appears to have ceased as of 2016 and the ontology is not publicly available.

3.2   OpenEHR

OpenEHR[4] is another standard for electronic health records which unlike 13606-1 is open source, maintained by the openEHR foundation and HL7 (see Sect. 3.3). ISO 13606-1 forms the backbone of the openEHR reference model, which can be viewed as a super-set of the 13606 RM, and the archetype model in ISO "13606 Part 2: Archetypes" is similar to that published by openEHR. OpenEHR provides specifications for the management, storage and retrieval of an Electronic Patient Record (EPR), and not just the communication of EHR data [22]. Like ISO 13606, OpenEHR has two distinct levels of models; a high level standardised RM with generic concepts; and a lower level more specific (clinical) model based upon archetypes, the archetype model (AM). Using this two-level approach, openEHR accounts for changes in clinical concepts by only requiring modifications to the archetypes, without having to change the reference model [18]. The types of archetypes are listed as follows; demographic, composition, section, entry (these include observation, instruction, action, evaluation and admin_entry), cluster and element.

---

[4] https://www.openehr.org/.

Demographic data[5] is influenced by clinical adaptations including the HL7v3 Reference Information Model (RIM) (Sect. 3.3). At the top most level is the concept of "Party". This is a superclass of Actor and Role. Actor in turn is a superclass of Agent, Group, Organisation and Person. A class "Party_Relationship" provides for the definition of relationships between parties. In addition to these classes, there also exist "Party_Identity", Contact and Address.

To ensure interoperability, consistent use of archetypes between different openEHR systems is required. To this end, the openEHR community has provided the Clinical Knowledge Manager (CKM), an international repository[6], where clinicians can freely develop, manage, publish and use archetypes, and several countries have established Electronic Patient Record strategies involving openEHR (such as UK, Norway and Australia) [22]. A criticism of openEHR is that the complexity of the data models, for example the CKM has over 500 Archetypes available, designed to cover all possible data elements, understanding these requires an investment of time and effort. Also the modelling of new archetypes is a complex task, where careful weighing of benefits against costs must be considered [7]. To support greater re-use and continuous improvement of archetypes, work has been conducted on converting openEHR to OWL, and OWL versions of the standards are available. These include an OWL description of the openEHR demographic model[7]. Having contacted the author of this ontology it is no longer being actively developed, and has not been applied to a specific use case.

3.3   HL7

HL7 is developed by HL7 international, concerned with interoperability standards for health informatics. Within the Irish context, the Health Service Executive (HSE) Healthlink is a key resource for GP messaging standards using message files specified in HL7 [13]. HL7 consists of several different standards as well as a framework to develop these standards, the HL7 Development Framework (HDF). This is a framework of modeling and administrative processes, policies, and deliverables used to produce specifications that are used by the healthcare information management community to overcome challenges and barriers to interoperability among computerized healthcare-related information systems. /Standards developed within HL7 include the Clinical Document Architecture (CDA) based on the RIM for the generation of EHR document, HL7 v2 to support hospital workflows and HL7 v3 which aims to support all healthcare workflows, and unlike HL7 v2 is based upon object oriented principles.

A promising new standard is the HL7 Fast Healthcare Interoperability Resource (FHIR), developed by the HL7 FHIR working group [17]. While openEHRs focus is on the data model and complete data models, FHIR is more concerned with information exchange and the description of the APIs,

---

[5] https://specifications.openehr.org/releases/RM/latest/demographic.html.
[6] https://www.openehr.org/ckm/.
[7] http://trajano.us.es/~isabel/EHR.

providing the option to extend information models as required. FHIR is a web-based standard and uses the REpresentational State Transfer (REST). RESTful web services typically communicate over HTTP, and thus provide interoperability between computer systems on the Internet. Combined with a Linked Data module utilizing Semantic Web technologies, i.e. RDF and OWL, the semantic expression capability of FHIR can be expanded and facilitate inference and data linkage across datasets. Resources can be combined using PROFILES to identify packages of data to address clinical and administrative needs. Profiles constrain what a particular application needs to communicate based on Resources and Extensions (data elements, self-defined, that are not part of the core set), i.e. you only send data that is required for specific purposes. Examples of Profiles are for referral of a patient; for populating registries; adverse event reporting; ordering a medication; and providing data to a clinical decision support algorithm such as a risk assessment calculation. FHIR offers a promising method for supporting interoperability based upon RDF and ontologies [20]. The FHIR ontology[8] has well defined and detailed descriptions related to demographics, such as the concept of a Person, with data properties; name, address, telecom, photo, birthDate, gender, etc. It therefore provides a strong basis to support data interoperability within the Irish health services. In the next section we explore the representation of demographic data within the Irish Health domain.

4   Analysis of Healthcare Collections in Ireland

In this section we analyse healthcare collections in Ireland with a focus on demographic data, as it is the most well represented data category, with 28 of the 75 catalogues explicitly mentioning they cover demographic data, the largest of any specific category. The consistent representation of demographic data is therefore essential to support interoperability between health services.

The analysis was done based upon the Health Information and Quality Authority (HIQA) Catalogue of national health and social care data collections [12]. HIQA is an independent authority established to drive high-quality and safe care for people using health and social care services in Ireland. HIQA's role is to develop standards, inspect and review these services and support informed decisions on how services are delivered. Towards this goal, HIQA has published the "Catalogue of national health and social care data collections" (version 3). The aim of this third version of the Catalogue is to enable all stakeholders (including the general public, patients and service users, clinicians, researchers, and healthcare providers) to readily access information about health and social care data collections in Ireland. The catalogue consists of a comprehensive list of national health and social care data collections. These are national repositories of routinely collected health and social care data in the Republic of Ireland.

The catalogue lists 75 of these collections, and for each collection provides data in terms of title, managing organisation, description/summary, data providers, available data dictionaries, data content (i.e. a breakdown of the type

---

[8] http://build.fhir.org/fhir.ttl.

of data collected) etc. In order to structure the analysis, HIQAs National standard demographic dataset and guidance for use in health and social care settings in Ireland was used [10]. This provides guidelines on a set of concepts and properties for describing demographic data, such as related to name, date of birth, contact details, address, etc. (see Fig. 2).

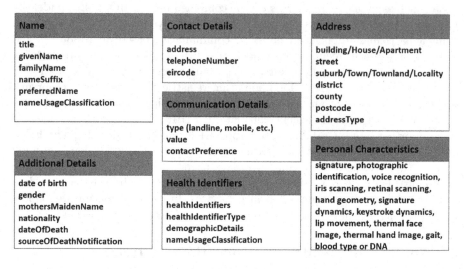

Fig. 2. HIQA Overview of Demographic Data

4.1   Methodology for Analysis of Health Catalogue

The methodology consisted of three main phases. In the first phase the data dictionaries given in the data dictionary field were analysed. The second phase the different named concepts were extracted from the data field. The third phase consisted of a harmonisation process, to identify a set of classifications for the different concepts identified.

4.2   Results of Analysis

The analysis began with the collections with associated data dictionaries, 39 had "no", "not available", or "not available online". Of the remaining 36, 15 provided links (such as www.noca.ie) with no obvious way to access the data dictionary or required a password, 3 had broken links, 5 mentioned resources that could not be located (e.g. Under revision as part of HRB LINK project) and so these were discounted. The remaining 13 data dictionaries were mostly pdf documents, such as the Ambulatory Care Report (ACR), Cardiac First Responder (CFR),

Patient Care Report (PCR), Patient Treatment Register (PTR) standards, as well as EUROCAT[9], and heartwatch. The Irish Mental Health Care provides an excel file.

Secondly all 75 collections "data content" field was analysed. Typical examples of this type of data (without a corresponding data dictionary) is "Name, address, date of birth, gender, District Electoral Division (DED), HSE area, Local Health Office (LHO) area, task force area, date commenced on methadone, type of methadone treatment, prescribing doctor, dispensing clinic, date and reason for discontinuation of methadone, client photograph and client signature.", although the range of data concepts covered is highly varied, reflecting the nature of the health services. From this analysis a matrix of collections against listed data concepts was created (such as name, HSE area, so on) and a tick was given for a data concept if it is present in a collection. Due to the wide range of data concepts, a process of harmonisation took place to identify classes for data concepts, either taken the name of the data concept directly, i.e. "Name", or deriving an appropriate class for a set of data concepts, e.g. "Patient" and "Person", based on our analysis of openEHR and FHIR, and also the HIQA schema.

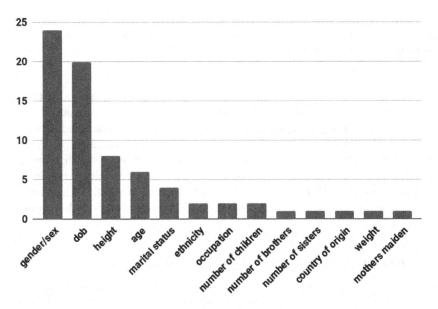

Fig. 3. Number of "Person" related data properties identified across analysed health care catalogues.

Figure 3 gives the count for each concept explicitly referenced in the data content field, with gender and date of birth (dob) being the most represented.

---

[9] https://www.hse.ie/eng/about/who/healthwellbeing/knowledge-management/health-intelligence-work/eurocat.html.

It should be noted that it is expected that more data related to the person class is included in the data collections, as often they refer to "demographic data, for example..." listing then one or two examples of the type of demographic data (this explains the high number of gender and dob). Figure 4 shows number of properties for both Name and Contact Details, both are classes directly related to Person. There are three Object Properties (associations) that relate Person to these classes.

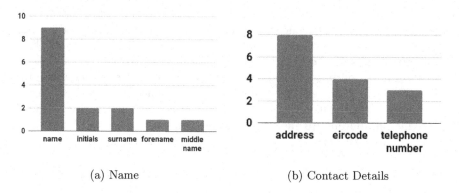

(a) Name   (b) Contact Details

Fig. 4. Number of "Name" and "Contact Details" related data properties identified across analysed health care catalogues

Through this process over 50 potential classes have been identified within the Irish health domain, ranging from (not an exhaustive list): person, name, contact details, patient, patient infant, patient pregnant, disabled person, address, location, medical/clinical information/assessment, treatment, therapy, prescription, observations, test, results, diagnosis, event, incident, injury, death, paediatric mortality, service, procedure, operation, product, device, vehicle,vaccination/ immunisations, disease, infection, staff, practitioner, admission, child admission unit, legal status, approved centre, etc. Each class has two or more data concepts taken from the analysis.

4.3   Demographic Data Alignments with Standards

Table 1 gives a very high level overview of some of the data concepts and properties identified in the above standards with respect to demographics. As can be seen, the 13606 demographics package includes concepts such as Person, Postal Address, gender, name, gender and birth time. It does not explicitly have a concept Patient, although it does have an "entity role relationship", so the potential exists to model a patient relationship between person entities in a similar fashion as OpenEHRs demographic information model, in which a Patient can be defined using the "Party_Relationship" held between a Person and an Organisation. The OpenEHR Person concept is more detailed than 13606-1 allowing a person to have a contact, and also a "Party_Identity" which can be broken down into elements for first name, last name, etc. The OpenEHR CKM provides more detailed archetypes for defining demographics (e.g. the DEMOGRAPHIC-CLUSTER archetypes), these cover the concepts found in our analysis of the Irish Health domain. Nonetheless, the concepts are scattered across the different archetypes. The FHIR ontology on the other hand covers all the required concepts and each can be found explicitly defined in the ontology. Given the focus of FHIR on information exchange, its adherence to semantic web principles, we therefore believe FHIR is the most suitable approach for managing Irish healthcare data going forward.

Table 1. Occurrence of demographic related concepts in standards

	ISO 13606-1	OpenEHR	FHIR
Person	x	x	x
Patient		x	x
Postal address	x	x	x
Contact		x	x
Name	x	x	x
First, last name		x	x
Birth date	x	x	x
Gender	x	x	x
Height		x	x

5   Conclusion

In this article we presented a review of standards relevant to the definition of demographic data in the health care domain in Europe and internationally and identified the need for harmonisation to ensure data interoperability. Semantic web technologies such as RDF, OWL and SPARQL have been identified as a prime candidate for supporting greater data interoperability, as demonstrated by the research efforts to convert existing resources and the move towards these technologies in HL7 FHIR.

An analysis was conducted over the Irish health catalogue provided by the Irish Health Information and Quality Authority (HIQA). This gives an overview of 75 collections (data sets maintained by different health services in Ireland) and provides information on each in terms of the types of data being collected. From this analysis, typical data concepts, i.e. those related to demographic data on patients (i.e. age, gender etc.) have been identified and these can each be directly mapped to Patient and Person concepts modelled within the FHIR ontology. We therefore believe that FHIR is currently explicit enough to support interoperability of demographic data within the Irish health context. By extending vocabularies such as FHIR, additional data properties required within the Irish context can be provided while maintaining interoperability with the wider international community. This is an important step towards greater data interoperability for health services in Ireland.

This work is being conducted within the greater context of eHealth Ireland with coordination and collaboration of various stakeholders such as the eHealth ecosystem [8], which includes patients, providers, software vendors, legislators, and health information technology (IT) professionals. This is important to foster ownership ensuring that health care data is not viewed solely as a commodity for profit, rather than a means to improve health care in Ireland. The next steps for this work are to examine a wider range of data concepts, beyond demographic data, and determine if FHIR is suitable for managing these data resources, particularly within the context of the FAIRVASC project and the management of data related to the rare disease vasculitis.

Acknowledgements. This research was conducted in the Centre of eIntegrated Care (CeIC) in DCU and has received funding from the ADAPT Centre, funded under the SFI Re- search Centres Programme (Grant 13/RC/2106) and co-funded by the European Regional Development Fund.

References

1. Fast Healthcare Interoperability Resources - FHIR v4.0.1. http://hl7.org/fhir/
2. BS EN 13606–1:2007 Health informatics - Electronic health record communication - Part 1: Reference Model (2007). https://www.iso.org/standard/40784.html. Accessed 13 May 2019
3. Austin, T., Sun, S., Hassan, T., Kalra, D.: Evaluation of ISO EN 13606 as a result of its implementation in XML. Health Inform. J. **19**(4), 264–280 (2013). https://doi.org/10.1177/1460458212473993, http://www.ncbi.nlm.nih.gov/pubmed/23995217, http://www.pubmedcentral.nih.gov/articlerender.fcgi?artid=PMC4107818
4. Beale, T., Heard, S., Kalra, D., Lloyd, D.: Demographic information model. Rev. Lit. Arts Am. 1–25 (2008). https://specifications.openehr.org/releases/RM/latest/demographic.html
5. Bender, D., Sartipi, K.: HL7 FHIR: an agile and RESTful approach to healthcare information exchange. In: Proceedings - IEEE Symposium on Computer-Based Medical Systems, pp. 326–331. IEEE, June 2013. https://doi.org/10.1109/CBMS.2013.6627810, http://ieeexplore.ieee.org/document/6627810/
6. CEN: Health informatics - Electronic health record communication - Part 1: Reference model (ISO 13606–1:2008). Technical report (2012). https://www.iso.org/obp/ui/#iso:std:iso:13606:-1:ed-2:v1:en
7. Christensen, B., Ellingsen, G.: Evaluating model-driven development for large-scale EHRs through the openEHR approach. Int. J. Med. Inform. **89**, 43–54 (2016). https://doi.org/10.1016/j.ijmedinf.2016.02.004
8. EHealth Ireland Ecosystem: Ecosystem Network - Council of Clinical Information Officers (2015). http://www.ehealthireland.ie/Stakeholder-Engagement/eHealthIrelandEcoSystem/. Accessed 13 May 2019
9. European Committee for Standardization (CEN): CEN - Technical Bodies - CEN/TC 251 (2016). https://tinyurl.com/yao5ctay. Accessed 21 Jan 2020

10. Health Information and Quality Authority (HIQA): National Standard Demographic Dataset and Guidance for use in health and social care settings in Ireland (2013). https://www.hiqa.ie/system/files/National-Standard-Demographic-Dataset-2013.pdf. Accessed 13 May 2019
11. HIQA: Guidance on Classification and Terminology Standards for Ireland (December), pp. 1–47 (2013). https://www.hiqa.ie/sites/default/files/2017-07/Guidance-on-terminology-standards-for-Ireland.pdf
12. HIQA: Guiding Principles for National Health and Social Care Data Collections. Technical report (2017)
13. HSE: eHealth Strategy for Ireland, p. 75 (2013). https://doi.org/10.1017/CBO9781107415324.004, https://www.ehealthireland.ie/Knowledge-Information-Plan/eHealth-Strategy-for-Ireland.pdf
14. Hussey, P., Tully, M.: National Data Dictionary Metadata Registry Framework Briefing Paper (2017). https://www.ehealthireland.ie/Our-Team/Enterprise-Architecture/HSE-National-Data-Dictionary-Briefing-Paper.pdf. Accessed 13 May 2019
15. Hussey, P., McGlinn, K.: The role of academia in reorientation models of care - insights on eHealth from the front line. Informatics 6, 37 (2019)
16. Lozano-Rubí, R., Muñoz Carrero, A., Serrano Balazote, P., Pastor, X.: OntoCR: A CEN/ISO-13606 clinical repository based on ontologies. J. Biomed. Inform. 60, 224–233 (2016). https://doi.org/10.1016/j.jbi.2016.02.007, https://www.sciencedirect.com/science/article/pii/S1532046416000290
17. Luz, M.P., Nogueira, J.R.D.M., Cavalini, L.T., Cook, T.W.: Providing full semantic interoperability for the fast healthcare interoperability resources schemas with resource description framework. In: 2015 International Conference on Healthcare Informatics, pp. 463–466, October 2015. https://doi.org/10.1109/ICHI.2015.74
18. Marcos, M., Maldonado, J.A., Martínez-Salvador, B., Boscá, D., Robles, M.: Interoperability of clinical decision-support systems and electronic health records using archetypes: a case study in clinical trial eligibility. J. Biomed. Inform. 46(4), 676–689 (2013). https://doi.org/10.1016/j.jbi.2013.05.004, http://www.sciencedirect.com/science/article/pii/S1532046413000701
19. Muñoz, P., Trigo, J., Martínez, I., Muñoz, A., Escayola, J., García, J.: The ISO/EN 13606 standard for the interoperable exchange of electronic health records. J. Healthc. Eng. 2(1), 1–24 (2011). https://doi.org/10.1260/2040-2295.2.1.1, http://www.hindawi.com/journals/jhe/2011/316579/
20. Peng, C., Goswami, P., Bai, G.: An ontological approach to integrate health resources from different categories of services. In: HEALTHINFO 2018, The Third International Conference on Informatics and Assistive Technologies for Health-Care, Medical Support and Wellbeing, pp. 48–54 (2018)
21. Reisman, M.: EHRs: the challenge of making electronic data usable and interoperable. Pharm. Ther. 42(9), 572–575 (2017)
22. Ulriksen, G.H., Pedersen, R., Ellingsen, G.: Infrastructuring in healthcare through the openEHR architecture. Comput. Support. Coop. Work (CSCW) 26(1), 33–69 (2017). https://doi.org/10.1007/s10606-017-9269-x
23. Yadagiri, N., Ramesh, P.: Semantic web and the libraries: an overview. Int. J. Libr. Sci.^{TM} 7(1), 80–94 (2013). http://www.ceserp.com/cp-jourwww.ceser.in/ceserp/index.php/ijls/article/view/2989

From Population to Subject-Specific Reference Intervals

Murih Pusparum[1,2(✉)] ⓘ, Gökhan Ertaylan[2] ⓘ, and Olivier Thas[1,3,4] ⓘ

[1] Hasselt University, 3500 Hasselt, Belgium
murih.pusparum@uhasselt.be
[2] Flemish Institute for Technological Research (VITO), 2400 Mol, Belgium
murih.pusparum@vito.be
[3] Department of Data Analysis and Mathematical Modelling,
Ghent University, 9000 Ghent, Belgium
[4] National Institute for Applied Statistics Research Australia (NIASRA),
Wollongong, NSW 2500, Australia

Abstract. In clinical practice, normal values or reference intervals are the main point of reference for interpreting a wide array of measurements, including biochemical laboratory tests, anthropometrical measurements, physiological or physical ability tests. They are historically defined to separate a healthy population from unhealthy and therefore serve a diagnostic purpose. Numerous cross-sectional studies use various classical parametric and nonparametric approaches to calculate reference intervals. Based on a large cross-sectional study (N = 60,799), we compute reference intervals for subpopulations (e.g. males and females) which illustrate that subpopulations may have their own specific and more narrow reference intervals. We further argue that each healthy subject may actually have its own reference interval (subject-specific reference intervals or SSRIs). However, for estimating such SSRIs longitudinal data are required, for which the traditional reference interval estimating methods cannot be used. In this study, a linear quantile mixed model (LQMM) is proposed for estimating SSRIs from longitudinal data. The SSRIs can help clinicians to give a more accurate diagnosis as they provide an interval for each individual patient. We conclude that it is worthwhile to develop a dedicated methodology to bring the idea of subject-specific reference intervals to the preventive healthcare landscape.

Keywords: Clinical statistics · Clinical biochemistry · Reference intervals · Longitudinal data · Quantile mixed models

1 Introduction

In the era of personalized medicine, we are surrounded by sensors and methodologies to capture and store data from a single individual at an unprecedented scale. These data later can be associated with Linked Data technologies [1]. Furthermore, artificial intelligence, and more specifically various machine learning (ML) techniques, are making their way to daily practice. We are presented with

© Springer Nature Switzerland AG 2020
V. V. Krzhizhanovskaya et al. (Eds.): ICCS 2020, LNCS 12140, pp. 468–482, 2020.
https://doi.org/10.1007/978-3-030-50423-6_35

algorithms that can outperform pathologists [2] for diagnosing and staging of diseases [2,3]. However, one disadvantage of AI and ML approaches is the lack of transparency in the decision making process. Although there is progress to make these algorithms transparent, these efforts are in their infancy. Hence, statistical methods and reasoning are essential to bridge the gap towards their implementation in clinical practice. Specifically, in the preventive healthcare landscape, there is a clear lack of methodological development to render these technologies viable in practice.

In the field of medicine, interpreting clinical laboratory results can be done in several ways. The most common way is by comparing them to a standard value or range that has been calculated from a reference population of healthy individuals. Such intervals are known as normal values or reference intervals, but for further purposes we will refer to them as *population reference intervals* (PRI). In clinical practise, a patient will be considered healthy when the laboratory results show values within this PRI. In the preventive healthcare landscape, however, a PRI only gives little advantage since it is only designed for diagnostic purposes. Moreover, it is assumed to be constant over time and space.

To obtain the PRI, a cross-sectional prospective or retrospective study is typically considered. In this type of studies, the data of a particular physiological or clinical parameters will be collected from a large number of healthy subjects. The participants must be as similar as possible with the target population in which the PRIs will be used. For example, a study for estimating PRIs of BNP (brain natriuretic peptide) using only university students will be inappropriate as this BNP test is normally run for elderly people [4].

The classical definition of a PRI is the central 95% of the reference population of the parameter of interest. This central 95% is located between the 2.5 and 97.5 percentiles of the reference population. Various methods for estimating the PRIs have been proposed. Parametric methods start from the assumption that the distribution of the parameter of interest can be described by a particular distribution (usually Gaussian). The percentiles can then be directly computed from this distribution when its parameters (mean and variance for the Gaussian distribution) are estimated from a dataset. In general, when the distributional assumption holds, the parametric methods will be better in the sense they provide more precise estimates of the PRI than the nonparametric methods for the same sample size. However, without any distributional assumption, the nonparametric methods are more suitable as they can still produce unbiased estimators of the PRI, whereas the parametric methods may give biased results when the wrong distribution is used. With nonparametric methods, a minimum number of 120 participants is proposed for calculating the intervals [4]. Statistically speaking, larger sample sizes will result in better estimates in terms of bias and precision.

These classical estimation methods typically require a cross-sectional dataset, containing a single measurement of the parameter for each subject in the study. The application of these methods can be seen in many studies using a large number of cross-sectional samples for estimating PRIs for common clinical markers [5–9]. Longitudinal studies, on the other hand, are characterised by multiple (repeated) measurements of the parameter for each subject in the study.

Several studies involving longitudinal dataset for calculating PRIs have been performed [10,11]. However, instead of using the classical methods, a simple random effects model and a semi-parametric method were used in these two studies, respectively, to produce pointwise PRIs i.e. PRIs that only have a valid probabilistic interpretation for each time point separately.

In this paper, we demonstrate the use of classical methods for reference interval calculation on a large cross-sectional study. We also apply the methods to subpopulations (e.g. males and females, for illustration purpose) and we will argue that reference intervals can be made more specific (i.e. more informative) when applied to such subpopulations. By extending this reasoning to every individual, we end up with reference intervals for each subject (subject-specific reference intervals, SSRI). The assumption under SSRI is that each parameter measured from any subject has a biological variation that is specific to this individual and has potential upper and lower boundaries that can be inferred from data, allowing a better interpretation of this parameter range taking into account subject specific variation. For the estimation of these SSRI, data on single subject level are required, and hence data from longitudinal studies are needed. We will propose to estimate subject-specific reference intervals with linear quantile mixed models (LQMM). Data and methods are described in Sect. 2, while the results are discussed in Sect. 3. A conclusion and some suggestions for future research will be given in Sect. 3.3.

2   Materials and Methods

2.1   Data Description

There are two different types of datasets that are used in this paper. The first dataset comes from a cross-sectional study conducted in 2012–2016 and consisting of 60,799 participants from the Balearic Island, Spain, with ages ranging from 19 to 70 years [12]. This dataset will be referred to as the *Balearic data*. The measurements fall into three categories: a personal and health habits category (e.g. gender, age and smoking status), an anthropometric or a physiological measurements category (e.g. BMI and body fat percentage), and a clinical category (e.g. HDL and LDL cholesterol level). To reduce the scope of the research, only 5 parameters from each of the physiological and the clinical category were considered. Physiological parameters consist of a body shape index (ABSI), body mass index (BMI), waist circumference, systolic blood pressure, and diastolic blood pressure. The clinical parameters include total cholesterol, HDL cholesterol, LDL cholesterol, triglycerides, and glucose level.

The second dataset comes from our in-house ongoing longitudinal cohort study of 30 individuals with monthly physiological and clinical measurements over a period of 9 months and with age ranging from 45 to 60 years at the time of recruitment. This dataset will be referred to as the *IAM Frontier data* [13]. The same 10 physiological and clinical measurements as for the Balearic data were assessed. Characteristics of the two datasets are presented in Table 1. Due to privacy reasons and confidentiality, the order of the individuals presented in each graph in Sect. 3 is randomised.

Table 1. Data characteristics of the Balearic and the IAM Frontier datasets.

	Balearic data	I AM Frontier data
Type	Cross-sectional	Longitudinal
Number of subjects	60,799	30
Time points	Single (1 time point)	Multiple (9 time points)

Category	Parameter
Physiological	A body shape index (ABSI)
	Body mass index (BMI)
	Waist circumference
	Systolic blood pressure (Systolic BP)
	Diastolic blood pressure (Diastolic BP)
Clinical	Total cholesterol
	HDL cholesterol
	LDL cholesterol
	Triglycerides
	Glucose

2.2 Classical Parametric and Nonparametric Reference Intervals

Let n denote the total number of sample observations. For the Balearic data, the RIs were estimated using a classical parametric method and two nonparametric methods. The parametric method estimates the 2.5 and 97.5 percentiles as

$$\bar{x} \pm z_{0.975} s_x \tag{1}$$

where $\bar{x}$ and s_x indicate the sample mean and the sample standard deviation, and $z_{0.975}$ is the 97.5 percentile of a standard normal distribution [14].

With the nonparametric methods the bounds of the reference interval are computed as the sample 2.5 percentile and the sample 97.5 percentile. These are estimated from the order statistics, which is the ordered set of sample observations. In particular, for a sample if n observations, the order statistics can be denoted by $y_{[1]} \leq y_{[2]} \leq \cdots \leq y_{[n]}$. We consider two nonparametric methods. The first estimates the 2.5 and the 97.5 percentile as the $0.025(n+1)$-th and $0.975(n+1)$-th order statistics. The second method estimates these percentiles as the $[(0.025 \times n) + 0.5]$-th and $[(0.975 \times n) + 0.5]$-th order statistics. If any of these numbers is not an integer then it is rounded to the nearest value, for example a value of 12.3 is rounded to 12 and 12.6 is rounded to 13. For rounding off a .5 decimal, it follows the 'round-to-even' rule, therefore 12.5 equals 12 and 13.5 equals 14. These two nonparametric methods will be referred to as NP1 and NP2. The bootstrap or resampling technique was also applied in combination with these methods [15]. We will call the PRIs obtained from these five approaches the *classical reference intervals* (CRIs) and the summary is presented in Table 2.

For some clinical parameters, one-sided reference intervals are needed. For example, for LDL cholesterol only an upper bound is used in clinical practice. In such cases, the PRIs still refer to 95% of the reference population, but now the lower bound is fixed at the minimal value of 0, and the upper bound is given by the 95 percentile of the distribution. The methods described in the previous paragraphs can still be used, but with the 97.5 percentile replaced with the 95 percentile.

Table 2. Summary of classical parametric and nonparametric methods

Method	Formula	Characteristics
Parametric	$\bar{x} \pm z_{0.975}s_x$	Rely on distributional assumptions
Nonparametric		Without distributional assumptions, based on order statistics
Nonparametric 1 (NP1)	[0.025(n + 1), 0.975(n + 1)]	
Bootstrapped NP1	[0.025(n + 1), 0.975(n + 1)] in combination with bootstrapped samples	
Nonparametric 2 (NP2)	[0.025n + 0.5, 0.975n + 0.5)]	
Bootstrapped NP1	[0.025n + 0.5, 0.975n + 0.5)] in combination with bootstrapped samples	

2.3   Linear Quantile Mixed Models Longitudinal Data

For the IAM Frontier data, linear quantile mixed models (LQMM) were fitted to obtain the RIs estimates. Linear quantile regression models [16] are a class of statistical models that express a particular quantile or percentile (e.g. quantile $\tau \in (0,1)$) of the outcome distribution as a linear function of one or more regressors. In our setting, we do not have regressors, but we do have repeated measurements on multiple subjects. This can be formulated as a simple special case of a linear quantile mixed model (LQMM), which extend the class of linear quantile regression models by the inclusion of random effects. In particular, we propose a LQMM which only includes one fixed-intercept and one random-intercept model the between-subject variability of the reference intervals. With Y the outcome variable (i.e. clinical parameter of interest) of subject $i = 1, \ldots, n$, with random effect u_i and with $Q(\tau \mid u_i)$ the subject-specific quantile function of outcome Y evaluated in the $100 \times \tau$ percentile, the model can be written as

$$Q(\tau \mid u_i) = \beta_0^{(\tau)} + u_i, \tag{2}$$

in which $\beta_0^{(\tau)}$ represents the fixed intercept. The model is completed by specifying the distribution of the random effects; in this paper this is restricted to the zero-mean normal distribution with variance Ψ_u^2. Note that the intercept parameter $\beta_0^{(\tau)}$ has the interpretation of the $100 \times \tau$ percentile bound of the PRI, whereas

$\beta_0^{(\tau)} + u_i$ has the interpretation of the SSRI for subject i. We need this model with $\tau = 0.025$ (lower bound) and with $\tau = 0.975$ (upper bound).

This class of models were first described in a study in 2007 and the authors gave details on how the model parameters can be estimated from longitudinal data [17]. They also proposed a method for predicting the subject-specific random effects u_i. Their models and methods were further generalised and improved in [18]. The methods are implemented in the *lqmm* package [19,20] of the statistical software R [21].

An important characteristic of the LQMM and its parameter estimation procedure, is that it can give subject-specific RIs with only few repeated measurements for each subject. This is a typical feature of random effects models: the random effects distribution allows for information-sharing between subjects.

3   Results and Discussions

3.1   Population Reference Intervals for the Balearic Data

For each parameter a boxplot was produced with reference lines corresponding to the lower and upper bounds of PRIs that have been previously published [22–26]. Figure 1 shows boxplots for two physiological and two clinical parameters, split by gender. The boxplots for the other parameters can be found in Appendix. The graphs illustrate that for some parameters there may be difference between males and females. To the contrary, the published PRIs used in clinical practice often do not have gender-specific intervals. The example of this case can be seen in systolic blood pressure, body mass index, diastolic blood pressure, and triglycerides level (Fig. 5 in Appendix). This suggests that, for these parameters, it may be better to work with PRIs for subpopulations.

The numerical results for the published PRIs for all ten parameters are shown in Table 3. Figure 1 also shows a fairly long tail in the distributions of almost all parameters his is an indication of a skewed distribution and hence the parametric methods based on the normal assumption may not be appropriate here. The nonparametric methods may thus be advised.

Figure 2 shows the published PRI and PRIs computed by applying the parametric and nonparametric methods (CRIs) to the Balearic dataset. The PRIs for all parameters can be found in Table 4. From Fig. 2, it can be seen that the CRIs computed by the five parametric and nonparametric methods give wider intervals than the published PRIs. Only for waist circumference the published PRI is very close to the CRIs. Among the CRIs, the nonparametric methods generally give similar intervals as the parametric method. However, for some of the parameters such as HDL, BMI and glucose level (see Fig. 6 in Appendix), the intervals calculated by the parametric method are quite different as compared to the nonparametric. In these parameters, we observed deviations from the Normal distribution. Since the parametric method relies on distributional assumptions (usually Gaussian), a departure from this assumption may result in different estimates of intervals of the nonparametric methods.

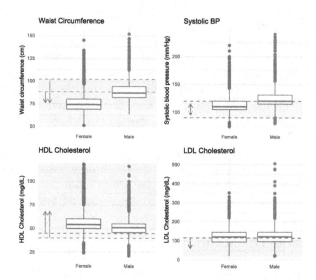

Fig. 1. Boxplots for 4 parameters in the physiological (top) and clinical (bottom) categories. The grey transparent area and the dashed lines correspond to the published PRIs while the arrows indicate their directions. The red and green dashed lines represent the lower/upper bounds of the published PRIs for males and females, respectively, and the grey dashed lines represent the published PRI for males and females together. (Color figure online)

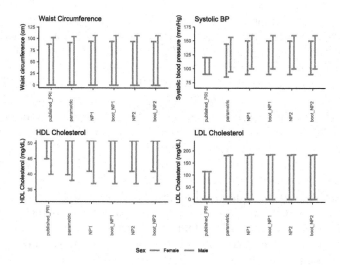

Fig. 2. For 4 parameters in the physiological (top) and clinical (bottom) categories, gender-specific PRIs are shown, estimated with various methods. Red and green lines represent the RIs for males and females, respectively. For waist circumference and LDL cholesterol, only the upper bounds were computed, and for HDL cholesterol only the lower bounds. For systolic blood pressure (BP) both lower and upper bounds were computed. For all calculations the Balearic dataset was used. (Color figure online)

Table 3. Published PRIs for all ten parameters. Only for waist circumference and HDL cholesterol level, gender-specific PRIs are reported.

Parameter	Unit	Lower limit	Upper limit
Physiological category			
Waist circumference[a]	cm	0	F < 88, M < 102*
A body shape index[b]	Z-score	$-\infty$	0.229
BMI[c]	kg/m^2	18.5	24.9
Systolic BP[a]	mm/Hg	90	120
Diastolic BP[a]	mm/Hg	60	80
Clinical category[d]			
Total cholesterol	mg/dL	0	190
HDL cholesterol	mg/dL	F ≥ 45, M ≥ 40*	∞
LDL cholesterol	mg/dL	0	115
Triglycerides	mg/dL	0	150
Glucose	mg/dL	70	110

*F for females and M for males. [a]National Heart, Lung, and Blood Institute (NHLBI), [b]Krakauer & Krakauer (2012), [c]World Health Organization (WHO), [d]Laposata (2019).

Table 4. Reference intervals calculated by various methods for all parameters in the Balearic dataset.

Parameter	Sex*	Published PRI	Parametric RI	NP1	Boot. NP1	NP2	Boot. NP2
Physiological category							
Waist circumference	M	(0, 102)	(0, 104.51)	(0, 107)	(0, 106.81)	(0, 107)	(0, 106.79)
	F	(0, 88)	(0, 91.55)	(0, 94)	(0, 94.04)	(0, 94)	(0, 94.04)
A body shape index	M	(−∞, 0.229)	(−∞, 1.34)	(−∞, 1.29)	(−∞, 1.29)	(−∞, 1.29)	(−∞, 1.29)
	F	(−∞, 0.229)	(−∞, 0.77)	(−∞, 0.99)	(−∞, 0.99)	(−∞, 0.99)	(−∞, 0.99)
BMI	M	(18.5, 24.9)	(18.59, 35.17)	(20.02, 36.82)	(20.01, 36.82)	(20.02, 36.82)	(20.02, 36.83)
	F	(18.5, 24.9)	(15.40, 34.65)	(18.36, 37.29)	(18.36, 37.31)	(18.36, 37.29)	(18.36, 37.30)
Systolic BP	M	(90, 120)	(94.61, 156.08)	(100, 160)	(100, 160)	(100, 160)	(100, 160)
	F	(90, 120)	(84.97, 144.21)	(90, 150)	(90, 150)	(90, 150)	(90, 150)
Diastolic BP	M	(60, 80)	(54.75, 97.24)	(60, 100)	(60, 100)	(60, 100)	(60, 100)
	F	(60, 80)	(50.02, 90.83)	(53, 93)	(52.65, 92.63)	(53, 92)	(52.72, 92.61)
Clinical category							
Total cholesterol	M	(0, 190)	(0, 260.53)	(0, 263)	(0, 263.00)	(0, 263)	(0, 263.02)
	F	(0, 190)	(0, 252.82)	(0, 256)	(0, 256.22)	(0, 256)	(0, 256.25)
HDL cholesterol	M	(40, ∞)	(37.98, ∞)	(37, ∞)	(36.99, ∞)	(37, ∞)	(36.99, ∞)
	F	(45, ∞)	(39.88, ∞)	(41, ∞)	(41.00, ∞)	(41, ∞)	(41.00, ∞)
LDL cholesterol	M	(0, 115)	(0, 183.05)	(0, 185)	(0, 185.09)	(0, 185)	(0, 185.08)
	F	(0, 115)	(0, 181.16)	(0, 183)	(0, 183.47)	(0, 183)	(0, 183.43)
Triglycerides	M	(0, 150)	(0, 271.30)	(0, 275)	(0, 275.03)	(0, 275)	(0, 275.01)
	F	(0, 150)	(0, 163.46)	(0, 167)	(0, 166.55)	(0, 167)	(0, 166.49)
Glucose	M	(70, 110)	(49.06, 132.18)	(65, 135)	(65.37, 135.40)	(65, 135)	(65.40, 135.44)
	F	(70, 110)	(55.66, 114.75)	(64, 113)	(63.99, 113.14)	(64, 113)	(63.99, 113.11)

*F for females and M for males.

When separately computing the PRIs for the males and females, we see that the bounds may be quite different. This is a first argument in favour of refining the PRIs towards smaller sub-populations. For example, for systolic blood pressure there are no gender-specific reference intervals published, but when estimated from the Balearic data we observe a clear difference between males and females. Similar findings were also observed in the other parameters, which are displayed in Fig. 6 in Appendix.

3.2   Subject-Specific Reference Intervals for the IAM Frontier Data

The IAM Frontier dataset contains data of 30 individuals that were measured at nine time-points. The individual profiles are shown in Fig. 3. They show for all subjects how the measurements evolve over time. The plot indicates differences between the two genders: males generally have larger waist circumference, higher systolic BP and higher LDL cholesterol than females, but they have lower HDL cholesterol. Females have higher HDL than males at least until the age of 50 [27,28] and the difference on the sex hormones between males and females can explain this phenomenon [29]. The plot also suggests a large between-subject variability and small within-subject variability, which is a common characteristic of repeated measurements. This phenomenon can be quantified by the intra-class

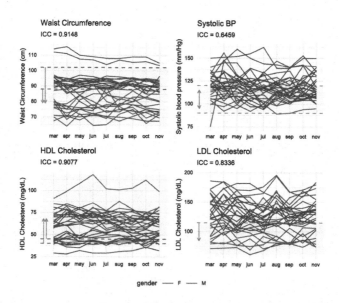

Fig. 3. Subject-specific profiles for all individuals in the IAM Frontier dataset. The grey transparent area and the dashed lines correspond to the published PRIs. The red and green dashed lines represent the lower/upper bounds of the published PRIs for males and females, respectively, and the grey dashed lines represent the published PRI for males and females together (no distinction between genders). The arrows indicate the directions of the intervals. (Color figure online)

correlation (ICC). Figure 3 also shows the ICC for each parameter. A large ICC is an indication that the within-subject variance is small as compared to the between-subject variance, or, equivalently, that the correlation between observations of the same individual is large. Large ICCs are observed for waist circumference and HDL cholesterol levels. Systolic blood pressure, on the other hand, has an ICC of only 65%.

We argue that for parameters with a large ICC, a subject-specific RI (SSRI) would be preferred over a population RI (PRI). The former can be calculated from with quantile mixed models (LQMM). The results of this approach are displayed in Fig. 4. The graph also shows the PRIs that were computed with the classical nonparametric method, using all observations. Since these classical methods are not valid with longitudinal data, these PRIs are only shown for illustration purposes. The results for the other parameters can be consulted in Appendix. Figure 4 shows that the SSRIs vary between subjects. For the two-sided intervals, the SSRIs are generally smaller than the PRIs computed from the same data. For the one-sided intervals, we see that the SSRI bounds vary about the PRI bound; this variation follows the subject-specific observations. Our results suggests that SSRIs may be more informative than PRIs.

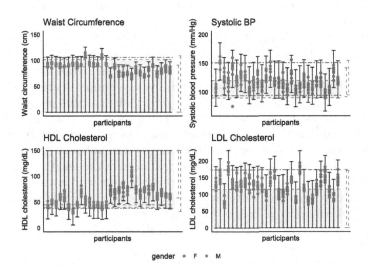

Fig. 4. SSRI for all subjects, estimated using LQMM with the IAM Frontier dataset. Red and green points refer to males and females observations. The grey area together with the red and green dashed lines indicate the published PRI for males and females, respectively, and the grey dashed lines indicate the published PRI for males and females together (no distinction between genders). The blue and the vertical red and green dashed lines indicate the estimated PRI and PRIs for males and females computed from the same data (the order of the individuals is randomised in each graph). (Color figure online)

3.3   Discussion and Conclusion

We have applied conventional methods for estimating reference intervals for many parameters in the Balearic dataset, which comes from a cross-sectional study with 60,799 participants. Since such reference intervals are computed from a large cross-sectional sample from a reference population of healthy individuals, they are referred to as population reference intervals (PRI). **Our analyses demonstrated that parametric and nonparametric methods do not always give the same results, from which we conclude that it is better to rely on the nonparametric methods for they do not rely on distributional assumptions.** By computing reference intervals for subgroups of participants (e.g. males and females), we demonstrated that reference intervals for subpopulations may be different. This pleas for not using a single PRI for all subjects, but rather work with PRI for subpopulations.

In this paper, we considered reference intervals for individuals, referred to as Subject-Specific Reference Intervals (SSRI). Our motivation came from *the perspective of personalised medicine, which starts from the supposition that each person is unique, and from the observation in longitudinal data that often the within-subject variability of a clinical parameter over time is small as compared to the between-subject variability.* However, since longitudinal data often do not include a very large number of observations for each individual, the conventional nonparametric methods for RI calculation cannot be used for individual subjects. We have proposed to use linear quantile mixed models (LQMM) for the calculation of the SSRIs. This method makes use of the assumption that the upper (and lower) SSRI bounds vary between subjects as a normal distribution, allowing for the calculation of SSRIs even with only 9 observations per subject. We have applied the method to several parameters in the longitudinal IAM Frontier dataset. **The results show, as expected, that there is variability between the SSRI, which is an indication for the need of subject-specific intervals.** The results also show that for some parameters the lengths of the SSRI are smaller than those of the PRI. If such intervals were used in clinical practice then a deviation from the healthy status may be sooner detected. Similarly, for one-sided intervals, the bounds of the SSRI vary about the PRI, following the distribution of the repeated measurements of the individual.

Despite our first positive findings of the use of LQMM for the calculation of SSRI, more research is needed. The LQMM relies on the distribution assumption that quantiles vary between subjects according to a normal distribution. This assumption need to be assessed, and the consequences of deviations from this assumption need to be evaluated. Moreover, the theory behind the LQMM is asymptotic in nature, which does not guarantee that the SSRIs are unbiased when only limited numbers of time-points are available. Future research could focus on a thorough evaluation of the LQMM for SSRI calculation and on further improving the methods so as to give reliable SSRIs even if model assumptions are not satisfied.

We believe that when SSRIs are widely used in clinical practice, they will allow for more precise diagnoses and hence they will be beneficial both for the patients and clinicians. We anticipate that in the future, the collaboration with artificial intelligence (AI) and machine learning (ML) algorithms could produce SSRIs for subjects for which even no longitudinal data is available. The well understood statistical methods produced from this research can perhaps eventually overcome the lack of algorithm transparency that is often criticised in the AI and ML approaches.

Appendix

See Fig. 7.

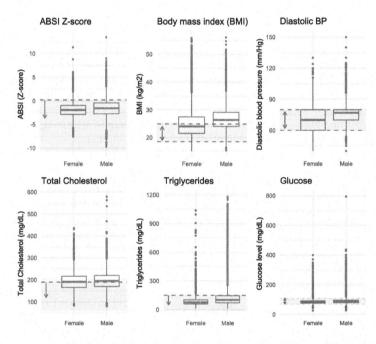

Fig. 5. Boxplots for 4 parameters in the physiological (top) and clinical (bottom) categories. The grey transparent area and the dashed lines represent the lower and upper bounds of the published PRIs for males and females together (no distinction between genders). The arrows indicate the directions of the intervals.

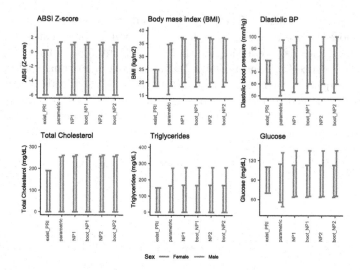

Fig. 6. For 4 parameters in the physiological (top) and clinical (bottom) categories, gender-specific PRIs are shown, estimated with various methods. Red and green lines represent the RIs for males and females, respectively. For ABSI Z-score, total cholesterol, and triglycerides level, only the upper bounds were computed. For diastolic blood pressure (BP) and glucose level both lower and upper bounds were computed. For all calculations the Balearic dataset was used. (Color figure online)

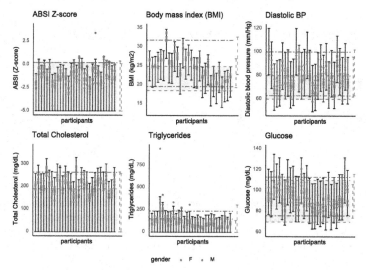

Fig. 7. Subject-specific reference intervals (SSRI) for all subjects, estimated using LQMM with the IAM Frontier dataset. Red and green dotted points refer to males and females observations. The grey transparent area together with the dashed lines indicate the published PRI while the blue and the vertical red and green dashed lines indicate the estimated PRI and PRI for males and females computed from the same data (the order of the individuals is randomised in each graph). (Color figure online)

References

1. Zaveri, A., Ertaylan, G.: Linked data for life sciences. Algorithms **10**(4), 126 (2017)
2. Ščupáková, K., et al.: Cellular resolution in clinical MALDI mass spectrometry imaging: the latest advancements and current challenges. Clin. Chem. Lab. Med. (CCLM) (2019). https://doi.org/10.1515/cclm-2019-0858
3. de Kok, T.M., et al.: Deep learning methods to translate gene expression changes induced in vitro in rat hepatocytes to human in vivo. Toxicol. Lett. **314**, S170–S170 (2019)
4. Graham, J., Barker, A.: Reference intervals. Clin. Biochem. Rev. **29**(i), 93–97 (2008)
5. Rustad, P., et al.: The Nordic reference interval project 2000, recommended reference intervals for 25 common biochemical properties. Scand. J. Clin. Lab. Invest. **64**, 271–284 (2004)
6. Katayev, A., Balciza, C., Seccombe, D.W.: Establishing reference intervals for clinical laboratory test results, is there a better way? Am. J. Clin. Pathol. **133**(2), 180–186 (2010)
7. Ichihara, K., et al.: Collaborative derivation of reference intervals for major clinical laboratory tests in Japan. Ann. Clin. Biochem. **53**(3), 347–356 (2016)
8. Adeli, K., Higgins, V., Trajcevski, K., White-Al Habeeb, N.: The Canadian laboratory initiative on pediatric reference intervals: a CALIPER white paper. Crit. Rev. Clin. Lab. Sci. **54**(6), 358–413 (2017)
9. Cheneke, W., et al.: Reference interval for clinical chemistry test parameters from apparently healthy individuals in Southwest Ethiopia. Ethiop. J. Lab. Med. **5**(5), 62–69 (2018)
10. Royston, P.: Calculation of unconditional and conditional reference intervals for foetal size and growth from longitudinal measurements. Stat. Med. **14**, 1417–1436 (1995)
11. Vogel, M., Kirsten, T., Kratzsch, J., Engel, C., Kiess, W.: A combined approach to generate laboratory reference intervals using unbalanced longitudinal data. J. Pediatr. Endocrinol. Metab. **30**(7), 767–773 (2017)
12. Romero-Saldaña, M., et al.: Validation of a non-invasive method for the early detection of metabolic syndrome: a diagnostic accuracy test in a working population. BMJ Open **8**(10), 1–11 (2018)
13. I AM Frontier study - VITO, Belgium. http://https://iammyhealth.eu/en/i-am-frontier. Accessed 8 Jan 2020
14. Solberg, H.E.: Approved recommendation (1987) on the theory of reference values. Part 5: statistical treatment of collected reference value. Determination of reference limit. Clin. Chim. Acta **170**, S13–S32 (1987)
15. Linnet, K.: Nonparametric estimation of reference intervals by simple and bootstrap-based procedures. Clin. Chem. **46**(6), 867–869 (2000)
16. Koenker, R.: Quantile Regression. Cambridge University Press, Cambridge (2005)
17. Geraci, M., Bottai, M.: Quantile regression for longitudinal data using the asymmetric Laplace distribution. Biostatistics **8**(1), 140–154 (2007)
18. Geraci, M., Bottai, M.: Linear quantile mixed models. Stat. Comput. **24**(3), 461–479 (2013). https://doi.org/10.1007/s11222-013-9381-9
19. Geraci, M.: Linear quantile mixed models: the lqmm package for laplace quantile regression. J. Stat. Softw. **57**(13), 1–29 (2014)
20. Geraci, M.: lqmm: Linear Quantile Mixed Models. R package version 1.5. http://CRAN.R-project.org/package=lqmm. Accessed 22 Jan 2020

21. R Core Team: R: A language and environment for statistical computing. R Foundation for Statistical Computing, Vienna, Austria. http://www.R-project.org/. Accessed 22 Jan 2020
22. World Health Organization: Mean Body Mass Index (BMI). https://www.who.int/gho/ncd/risk-factors/bmi-text/en/. Accessed 22 Jan 2020
23. Krakauer, N.Y., Krakauer, J.C.: A new body shape index predicts mortality hazard independently of body mass index. PLoS ONE **7**(7), e39504 (2012)
24. National Heart, Lung, and Blood Institute (NHLBI): Clinical Guidelines on the Identification, Evaluation, and Treatment of Overweight and Obesity in Adults. NIH Publication, Maryland USA (1998)
25. National Heart, Lung, and Blood Institute (NHLBI): Low Blood Pressure. https://www.nhlbi.nih.gov/health-topics/low-blood-pressure. Accessed 22 Jan 2020
26. Laposata, M.: Laposata's Laboratory Medicine: Diagnosis of Disease in the Clinical Laboratory, 3rd edn. McGraw-Hill Education, Ohio (2019)
27. Kim, H.K., et al.: Gender difference in the level of HDL cholesterol in Korean adults. Korean J. Family Med. **32**(3), 173–181 (2011)
28. Davis, C.E., et al.: Sex difference in high density lipoprotein cholesterol in six countries. Am. J. Epidemiol. **143**(11), 1100–1106 (1996)
29. Rossouw, J.E.: Hormones, genetic factors, and gender differences in cardiovascular disease. Cardiovasc. Res. **53**(3), 550–557 (2002)

Analyzing the Spatial Distribution of Acute Coronary Syndrome Cases Using Synthesized Data on Arterial Hypertension Prevalence

Vasiliy N. Leonenko[1,2]([envelope]) [ORCID]

[1] ITMO University, 49 Kronverksky Pr., St. Petersburg 197101, Russia
vnleonenko@yandex.ru
[2] Almazov National Medical Research Centre,
2 Akkuratova st., St. Petersburg 197341, Russia

Abstract. In the current study, the authors demonstrate the method aimed at analyzing the distribution of acute coronary syndrome (ACS) cases in Saint Petersburg. The employed approach utilizes a synthetic population of Saint Petersburg and a statistical model for arterial hypertension prevalence. The number of ACS–related emergency services calls in an area is matched with the population density and the prospected number of individuals with arterial hypertension, which makes it possible to find locations with excessive ACS incidence. Three categories of locations, depending on the joint distribution of the above-mentioned indicators, are proposed as a result of data analysis. The method is implemented in Python programming language, the visualization is made using QGIS open software. The proposed method can be used to assess the prevalence of certain health conditions in the population and to match them with the corresponding severe health outcomes.

Keywords: Acute coronary syndrome · Arterial hypertension · Synthetic populations · Statistical modeling · Python

1 Introduction

Acute coronary syndrome (ACS) is a range of health conditions associated with a sudden reduced blood flow to the heart. This condition is treatable if diagnosed quickly, but since the fast diagnostics is not always possible, the death toll of ACS in the world population is dramatic [6]. The modeling approach for forecasting the distribution of ACS cases would allow the healthcare specialists to be better prepared for the ACS cases, both in emergency services and in stationary healthcare facilities [3]. One of the simple forecasting methods is related to the application of statistical analysis to the retrospective EMS calls data associated with acute heart conditions. However, if the corresponding time series data set is not long, the accurate prediction is impossible without using additional data

This research is financially supported by The Russian Science Foundation, Agreement #19-11-00326.

related to the possible prerequisites for acute coronary syndrome calls, such as health conditions that increase the risk of ACS.

One of the factors in the population which might raise the probability of acute coronary syndrome is arterial hypertension (or, shortly, AH)—a medical condition associated with elevated blood pressure [14]. Arterial hypertension is one of the main factors leading to atherogenesis and the development of vulnerable plaques, which in turn might be responsible for the development of acute coronary syndromes [10]. Thus, we might assume that the urban area populated predominantly by individuals with AH might demonstrate higher rates of ACS. Based on that assumption, it might be possible to use spatially explicit AH data as an additional predictor of prospective ACS cases. Unfortunately, the data on AH prevalence with the geographical matching are rarely found, and for Russian settings, they are virtually non–existent. Nevertheless, they could be generated synthetically, which adds uncertainty to the analysis but, on the other hand, makes possible the analysis itself.

In this paper, we describe methods and algorithms to analyze the distribution of ACS–associated emergency medical service calls (shortly, EMS calls) by matching them with synthesized data on arterial hypertension prevalence. Using Saint Petersburg as a case study, we address the following question: may the synthesized AH data combined with EMS calls dataset provide additional information connected with ACS distribution in the population, compared to absolute data and relative data on EMS calls alone?

2   Data

2.1   EMS Calls

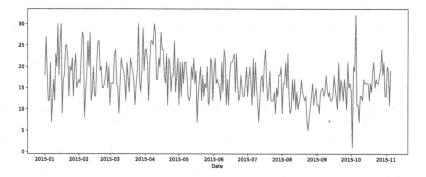

Fig. 1. The daily dynamics of emergency service calls connected with acute coronary syndrome (Jan – Nov, 2015)

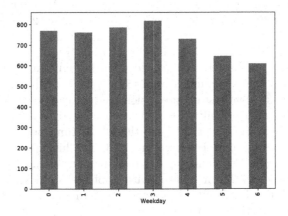

Fig. 2. The cumulative number of ACS emergency service calls in different weekdays.

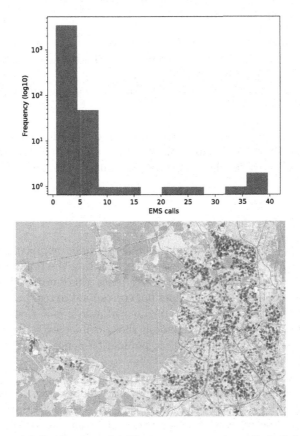

Fig. 3. The spatial distribution of ACS emergency service calls in Saint Petersburg

The EMS data we used in the research contain 5125 ACS–related EMS calls registered in Saint Petersburg from January to November 2015 [7]. The back-of-the-envelope analysis of the time series corresponding to daily number of calls (Fig. 1) and the weekly EMS calls distribution (Fig. 2) did not reveal any statistically significant patterns connected with distribution of calls over time, although it is clear that the number of EMS calls has a decline in the weekends. Thus, there is no straightforward prediction method to forecast fluctuations of the cumulative number of daily EMS calls connected with ACS.

The spatial distribution of calls for the whole time period based on the addresses from the database is shown in Fig. 3. The histogram for cumulative distribution was built by calculating the total number of EMS calls in a given spatial cell with the size 250 m × 250 m, with empty cells (0 EMS calls) excluded from the distribution. It was established that the form of the histogram does not change significantly if the cell sizes vary (up to 2 km × 2 km). It can be seen that the predominant majority of the spatial cells had 1 to 5 EMS calls, and only for single cells this number exceeds 8. Based on general knowledge, we assumed that the increased concentration of the EMS calls within particular cells may be caused by one of the following reasons:

- The cell has higher population density compared to the other cells;
- The cell has higher concentration of people with arterial hypertension, which might cause higher ACS probability;
- The cell includes people who are more prone to acute coronary syndrome due to unknown reasons.

To distinguish these cases and thus to be able to perform a more meaningful analysis of EMS calls distribution, we assess the spatial distribution of city dwellers and people with high blood pressure using the synthetic population approach.

2.2   Synthetic Population

A "synthetic population" is a synthesized, spatially explicit human agent database (essentially, a simulated census) representing the population of a city, region or country. By its cumulative characteristics, this database is equivalent to the real population, but its records does not correspond to real people. Statistical and mechanistic models built on top of the synthetic populations helped tackle a variety of research problems, including those connected with public health. In this study, we have used a synthetic population generated according to the standard of RTI International [13].

According to the standard of RTI International, the principal data for any given synthetic population is stored in four files: people.txt (each record contains id, age, gender, household id, workplace id, school id), households.txt (contains id and coordinates), workplaces.txt (contains id, coordinates and capacity of the workplaces), and schools.txt (contains id, coordinates, capacity). Our synthetic population is based on 2010 data from "Edinaya sistema ucheta naseleniya Sankt Peterburga" ("Unified population accounting system of Saint Petersburg") [4], which was checked for errors and complemented by the coordinates of the given locations. The schools records were based on the school list from the official web–site of the Government of Saint Petersburg [5]. The distribution of working places for adults and their coordinates were derived from the data obtained with the help of Yandex.Auditorii API [15]. The detailed description of the population generation can be found in [8].

2.3   Assessing AH Risk and Individual AH Status

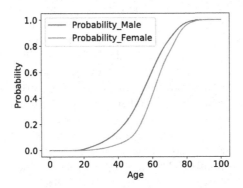

Fig. 4. The cumulative distribution function used to define the AH status of an individual, based on data from [12].

Further on we assess the probability for an individual in the synthetic population to have arterial hypertension. Each individual receives two additional characteristics [9]:

- The AH risk (the probability of having arterial hypertension). Based on [12], we assumed that the mentioned probability depends on age and gender of an individual. The corresponding cumulative distribution function was found using the data of 4521 patients during 2010–2015 and is shown in Fig. 4.
- The actual AH status (positive or negative). The corresponding value (0 or 1) is generated by the Monte Carlo algorithm according to the AH risk calculated in the previous step. The AH status might be used in simulation models which include demographic processes and population-wide simulation of the onset and development of AH.

The proportion of the synthetic population affected by arterial hypertension is found to be 26.6% which roughly correlates with the AH prevalence data in the USA according to American Heart Association Statistical Fact Sheet 2013 Update (1 out of every 3) [1] and is lower than the estimate for the urban population in Russia (47.5%) [11]. The cumulative and spatial distributions of

AH+ individuals in Saint Petersburg are shown in Fig. 5. It can be seen that the non–uniformity in ages and genders of the citizens potentially causes an uneven distribution of individuals exposed to arterial hypertension.

Further in the paper we match the number of AH+ dwellers of every cell with the number of EMS calls within this same cell and propose an indicator to analyze the relation between them.

2.4   Calculating the Indicators Related to EMS Calls

We convert the coordinates of EMS calls location from degrees to meters using Mercator projection. After this, we form a grid with a fixed cell size (250 m × 250 m) which covers the urban territory under consideration. Finally, using the EMS calls dataset, we calculate the overall number of EMS calls which was made within each cell of the grid. In the same way, we calculate the overall number of dwellers and AH+ individuals for the cells. This algorithm was implemented as a collection of scripts written in Python 3.7 with the libraries numpy, matplotlib, and pandas. The output of the algorithm is a .txt file with the coordinates of the cells and the cell statistics (overall number of individuals, number of AH+ individuals, overall number of EMS calls).

In order to understand the relationship between the numbers of AH+ users and the number of EMS calls, we follow our earlier research [2], where the ratio r_1 between the overdose–related EMS calls and the assessed number of opioid drug users was studied. In this paper, we compare r_1 with the alternative indicator r_2 which depends on the cumulative number of people in the cell under study instead of the assessed quantity of AH+ individuals. The formulas to calculate the mentioned ratios are the following:

$$r_1 = \frac{n_{ems} + 1}{n_{ah} + 1} \quad \text{and} \quad r_2 = \frac{n_{ems} + 1}{n_p + 1}$$

where n_{ems} is the number of registered EMS calls in a cell, n_{ah} is the assessed number of AH+ users in a cell, and n_p is the number of dwellers in a cell based on the synthetic population data. These quantities represent the number of calls per AH+ individual and calls per dweller, respectively. By adding 1 to the numerator and denominator we are able to avoid a divide by zero error, and although it provides a small skew in the data, its consistent application across all cells leaves the results and their interpretations unhindered. We use the ratio r_1 to understand which cells have large differences in the orders of magnitude compared to other cells. The ratio r_2 is introduced to compare its distribution with r_1 and thus decide whether the statistical model for AH+ probability assessment helps more accurately detect the anomalies connected with EMS calls distribution.

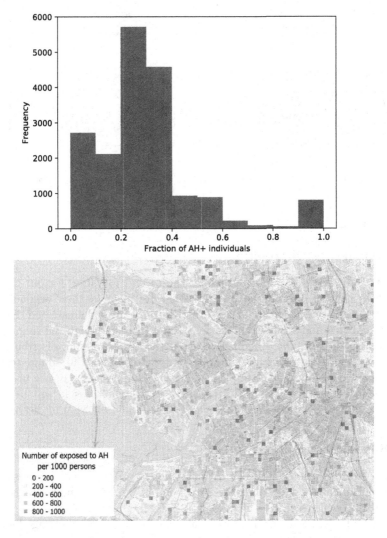

Fig. 5. The aggregated and geospatial distributions of AH+ individuals in Saint Petersburg

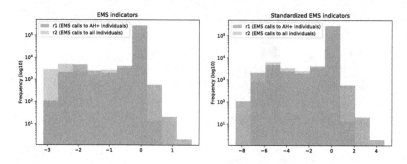

Fig. 6. The distributions of r_1 and r_2 (original and standardized).

3   Results

3.1   Cumulative Distribution

In Fig. 6, the aggregated distributions of the r_1 and r_2 values for our data are shown. On the left graph, the distributions are given in their original form, and in the right one the standardized distributions are demonstrated, i.e. with means equal to 0 and standard deviations equal to 1. Although the shape of the histograms is similar, the difference between the corresponding distributions is statistically significant, which is supported by the results of Chi–square test performed for the standardized samples. The crucial difference is in the histogram tails, i.e. in the extreme values of the indicators, which, as it will be shown further in the paper, is also accompanied by their different spatial distribution.

3.2   Spatial Distribution

In Fig. 7, a distribution of 20 cells with the highest values of r_1 and r_2 is shown (shades of blue and shades of green correspondingly). The lighter shades corresponds to the bigger cell side lengths (250, 500, 1000 and 2000 m).

The results demonstrate that the locations of high r_1 values change less with the change of cell side length, compared to r_2 (it is demonstrated on the map by several points with different shades of blue situated one near another). Also it is notable that the high r_2 values were found in lined up adjacent cells (see left and right edges of the map). This peculiarity of r_2 distribution requires further investigation, because it hampers the meaningful usage of the indicator.

The locations marked with three blue points represent concentration of high EMS calls in the isolated neighborhood with few assessed number of AH+ individuals. Most of these locations happen to be near the places connected with tourism and entertainment (1 – Gazprom Arena football stadium, 2 – Peterhof historical park) or industrial facilities (3 – bus park, trolleybus park, train depot; 4 – Izhora factory, Kolpino bus park). Location 5 corresponds to Pulkovo airport, a major transport hub (it is marked by only two blue points though). Location 6 is the one which cannot be easily connected with excessive EMS

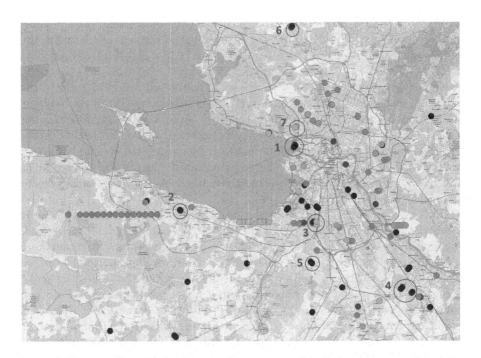

Fig. 7. Points of high r_1 and r_2 (Color figure online)

Fig. 8. Heatmap of EMS calls matched against high r_1 and r_2 locations (Color figure online)

calls—it is situated in a small suburb with plenty of housing. The possible interpretation of why it demonstrates high r_1 is the discrepancy between the actual number of dwellers for 2015 (a year for EMS calls data) compared to the 2010 information (a year for populational data). This zone was a rapidly developing construction site and subsequently witnessed a fast increase in the number of dwellers. Location 7 is also an expectational one – it is the only one which is marked by three green points (high r_2). Additionally, this zone was not marked by high r_1, although it is easily interpreted as yet another industrial district (Lenpoligraphmash printing factory). Increasing the number of points in a distribution to 100 does not change significantly the results: isolated areas with meaningful interpretation are mostly marked by the blue points, except Lenpoligraphmash at location 7.

Whereas the exceptional values of r_1 indicate isolated non–residential areas (industrial objects and places of mass concentration of people) which might be connected with the increased risk of ACS and thus require attention from healthcare services, the extreme values of r_2 indicator might be useful when we need to assess the excess of EMS calls in the densely populated residential areas. In Fig. 8, where r_1 and r_2 values are plotted against a heatmap of EMS call numbers, we see that there are two types of peak concentrations of EMS calls (bright red color). Ones are not marked with green dots (the r_2 values are not high) and thus might be explained by high concentration of dwellers in general. Others, marked with green dots, show the locations with high number of EMS calls relative to population. In case the locations does not demonstrate high r_1 values (no blue dots in the same place), they might correspond to the category of neighborhoods with ACS risk factors not associated with arterial hypertension (to be more precise, not associated with the old age of dwellers, since it is the main parameter of the statistical model for AH prevalence used in this study).

4 Discussion

In this paper, we demonstrated a statistical approach which uses synthetic populations and statistical models of arterial hypertension prevalence to distinguish several cases of ACS–associated EMS call concentration in the urban areas:

- High r_1 values for any corresponding number of EMS calls (Fig. 7) might indicate locations where acute coronary syndrome cases happen despite the low AH+ population density (for instance, particular industrial zones).
- Average to low r_2 values for high number of EMS calls (Fig. 8, red spots without green points) correspond to areas with high population density.
- High r_2 values and low r_1 values for high number of EMS calls (Fig. 8, red spots with green points) might indicate areas where the excessive number of ACS cases cannot be explained neither by the high population density, nor by AH prevalence, thus they might indicate neighborhoods with unknown negative factors.

It is worth noting that due to the properties of our EMS dataset (see Sect. 2.1 and Fig. 3) most of the locations with extremely high r_1 and r_2 correspond to the number of EMS calls in a grid cell equal to 1. Ascribing EMS calls to one or another property of the area based on such a small number of observations is definitely premature, and thus our interpretations given earlier in the text should be continuously tested using the new data on EMS calls. Despite the fact that we cannot draw any definite and final conclusions, in the author's opinion, the study successfully introduces the application of the concept of using synthesized data for health conditions of unknown prevalence (arterial hypertension) to categorize spatial distribution of their acute repercussions (acute coronary syndrome). As it was demonstrated by the authors before [2], the same approach can be successfully used in case of opioid drug usage, and we expect to broaden the scope of its application by applying it in other domains.

As to the current research, we plan the following directions of its further development:

- Currently, the time periods of the EMS calls information and synthetic population data do not match, which might cause the bias in the estimated values of the indicators. We plan to reproduce the results of this study using the actualized data sets.
- The enhanced statistical model for AH is considered to make the calculation of the number of AH+ individuals more accurate.
- The values of r_1 are almost the same for the cases of (a) 1 EMS call in presence of 0 AH+ individuals, and (b) $2n$ calls in presence of n AH+ individuals, so those cases cannot be distinguished by using indicators such as r_1, although they are essentially different. We want to explore the possibility of using a yet another indicator which will take into account the absolute number of dwellers in the neighborhood and will have a meaningful interpretation.
- We have access to a number of health records of the people hospitalized with ACS in a human–readable format, which contains information about their AH status. Using natural language processing tools, we plan to obtain a digital version of this data set and consequently to assess numerically the connection between AH and ACS cases in Saint Petersburg. This result will help reduce uncertainty in the results of the current study connected with analyzing the distribution of r_1.

References

1. AHA: American heart association statistical fact sheet 2013 update. https:// www.heart.org/idc/groups/heart-public/@wcm/@sop/@smd/documents/ downloadable/ucm_319587.pdf. Accessed 10 Apr 2020
2. Bates, S., Leonenko, V., Rineer, J., Bobashev, G.: Using synthetic populations to understand geospatial patterns in opioid related overdose and predicted opioid misuse. Comput. Math. Organ. Theory **25**(1), 36–47 (2019). https://doi.org/10. 1007/s10588-018-09281-2

3. Derevitskiy, I., Krotov, E., Voloshin, D., Yakovlev, A., Kovalchuk, S.V., Karbovskii, V.: Simulation of emergency care for patients with ACS in Saint Petersburg for ambulance decision making. Procedia Comput. Sci. **108**, 2210–2219 (2017)
4. Government of Saint Petersburg: Labor and employment committee. Information on economical and social progress. http://rspb.ru/analiticheskaya-informaciya/razvitie-ekonomiki-i-socialnoj-sfery-sankt-peterburga/. (in Russian). Accessed 13 Apr 2020
5. Government of Saint Petersburg: Official web-site. https://www.gov.spb.ru/. Accessed 13 Apr 2020
6. Jan, S., et al.: Catastrophic health expenditure on acute coronary events in asia: a prospective study. Bull. World Health Organ. **94**(3), 193 (2016)
7. Kovalchuk, S.V., Moskalenko, M.A., Yakovlev, A.N.: Towards model-based policy elaboration on city scale using game theory: application to ambulance dispatching. In: Shi, Y., et al. (eds.) ICCS 2018. LNCS, vol. 10860, pp. 404–417. Springer, Cham (2018). https://doi.org/10.1007/978-3-319-93698-7_31
8. Leonenko, V., Lobachev, A., Bobashev, G.: Spatial modeling of influenza outbreaks in Saint Petersburg using synthetic populations. In: Rodrigues, J.M.F., et al. (eds.) ICCS 2019. LNCS, vol. 11536, pp. 492–505. Springer, Cham (2019). https://doi.org/10.1007/978-3-030-22734-0_36
9. Leonenko, V.N., Kovalchuk, S.V.: Analyzing the spatial distribution of individuals predisposed to arterial hypertension in Saint Petersburg using synthetic populations. In: ITM Web of Conferences, vol. 31, p. 03002 (2020)
10. Picariello, C., Lazzeri, C., Attanà, P., Chiostri, M., Gensini, G.F., Valente, S.: The impact of hypertension on patients with acute coronary syndromes. Int. J. Hypertens. **2011** (2011). https://doi.org/10.4061/2011/563657. Article no. 563657
11. Boytsov, S.A., et al.: Arterial hypertension among individuals of 25–64 years old: prevalence, awareness, treatment and control. by the data from ECCD. Cardiovasc. Ther. Prev. **13**(4), 4–14 (2014). (in Russian)
12. Semakova, A., Zvartau, N.: Data-driven identification of hypertensive patient profiles for patient population simulation. Procedia Comput. Sci. **136**, 433–442 (2018)
13. Wheaton, W.D., et al.: Synthesized population databases: a US geospatial database for agent-based models. Methods report (RTI Press) **2009**(10), 905 (2009)
14. WHO: Hypertension. Fact sheet. https://www.who.int/news-room/fact-sheets/detail/hypertension. Accessed 13 Apr 2020
15. Yandex: Auditorii. https://audience.yandex.ru/. Accessed 13 Apr 2020

The Atrial Fibrillation Risk Score for Hyperthyroidism Patients

Ilya V. Derevitskii[1(✉)], Daria A. Savitskaya[2], Alina Y. Babenko[2], and Sergey V. Kovalchuk[1]

[1] ITMO University, Saint Petersburg, Russia
ivderevitckii@itmo.ru
[2] Almazov National Medical Research Centre, Institute of Endocrinology, St. Petersburg, Russia

Abstract. Thyrotoxicosis (TT) is associated with an increase in both total and cardiovascular mortality. One of the main thyrotoxicosis complications is Atrial Fibrillation (AF). Right AF predictors help medical personal prescribe the select patients with high risk of TAF for a closest follow-up or for an early radical treatment of thyrotoxicosis. The main goal of this study is creating a method for practical treatment and diagnostic AF. This study proposes a new method for assessing the risk of occurrence atrial fibrillation for patients with TT. This method considers both the features of the complication and the specifics of the chronic disease. A model is created based on case histories of patients with thyrotoxicosis. We used Machine Learning methods for creating several models. Each model has advantages and disadvantages depending on the diagnostic and medical purposes. The resulting models show high results in the different metrics of the prediction of a thyrotoxic AF. These models are interpreted and simple for use. Therefore, models can be used as part of the support and decision-making system (DSS) by medical specialists in the treatment AF.

Keywords: Atrial fibrillation risk · Thyrotoxicosis · Machine Learning · Risk scale · Thyrotoxicosis complications · Chronic disease

1 Introduction

Thyrotoxicosis (TT) is associated with an increase in both total and cardiovascular mortality. The majority of patients with TT are working-age individuals. Consequently, the negative social impact of TT remains incredibly significant [1]. Due to the risk of thromboembolic events and heart failure, one of the severest complications of TT is atrial fibrillation (AF). The thyrotoxic AF (TAF) incidence is as follows: 7–8% among middle-aged patients, 10–20% in seniors and 20–35% for those having ischemic heart disease or valvular disease [2–5]. High incidence, severity and also a significant adverse impact on life quality – all these facts make TAF prevention a crucial problem. Regrettably, there is no established TAF risk calculation scale at the moment. Nevertheless, devising one would allow selecting patients with a high risk of TAF for the closest follow-up or for early radical treatment of TT – surgical or radioiodine therapy instead of long-term medical treatment. This will ultimately lead to a decrease in TAF frequency.

© Springer Nature Switzerland AG 2020
V. V. Krzhizhanovskaya et al. (Eds.): ICCS 2020, LNCS 12140, pp. 495–508, 2020.
https://doi.org/10.1007/978-3-030-50423-6_37

2   Existing Methods Overview

In the literature, the topic of assessing the risk of AF development is described in sufficient detail. There are several approaches to solving this problem. In the first approach, the authors study risk factors for atrial fibrillation, predicting the probability of AF development for a specific period using regression analysis. For example, the 10-year risk of AF development is estimated [1]; multivariate Cox regression is used as tools. The results obtained with this approach are quite reliable, but such models do not include many factors that presumably affect the risk of AF. Also, these models do not take into account the specifics of AF development in patients with certain chronic diseases (in contrast to the models of the second approach). In the second approach, the risk of AF is estimated for patients with a certain chronic disease, such as diabetes [5], and it is revealed that patients with diabetes have a 49% higher risk of developing AF. In the third approach, meta-studies of populations are conducted to analyze the risk factors for AF, for example [6]. In this article, the authors described a cross-sectional study of California residents. Those results are interesting and can be used to calibrate the models obtained by MO methods; however, within using these works we cannot assess the probability of AF development in a particular person. Many factors can influence the probability of AF. It is necessary to consider both the specifics of this complication and the specifics of a chronic disease when we create a model for assessing the risk of AF. Therefore, special methods are needed for creating such a model. This study presents an approach to the estimator's risk of atrial fibrillation for patients who suffer from thyrotoxicosis. This approach considers the specifics of the influence of the thyrotoxicosis course for a particular patient on the probability of AF.

To date, a lot of data have been obtained about thyrotoxic atrial fibrillation risk factors. The majority of investigations have demonstrated that the most significant predictors of this severe arrhythmia are advancing age, male gender, prolonged thyrotoxicosis duration, high level of thyroid hormones and concomitant cardiovascular diseases [2, 7–11]. Findings from other researches have shown that factors like female gender, obesity, heart rate more than 80 beats per minute, the presence of left ventricular hypertrophy, big left atrium diameter, chronic renal disease and proteinuria, elevated liver transaminase and C-reactive protein levels predispose to thyrotoxic atrial fibrillation [12–14]. Thus, based on the comprehensive literature review, we can conclude that available data on thyrotoxic atrial fibrillation predictors are contradictory and, furthermore, a lot of studies have an insufficient evidence base. Moreover, for an individual's absolute risk evaluation, integrating multiple risk factors and identify the most significant of them, are required. Accordingly, it is necessary to develop a system for ranking risk factors by their degree of impact and models for calculating the risk of thyrotoxic AF. Using the system for ranking we need to create an easy-to-use and interpretable model for assessing the risk of developing atrial fibrillation in patients with thyrotoxicosis. All of the above explain relevance of this study.

3 Case Study

3.1 Data

The study included 420 patients with a history of overt thyrotoxicosis. The study sample was divided into two subgroups: one includes 127 (30.2%) individuals with thyrotoxic atrial fibrillation and other - 293 (69.8%) patients without. All participants were treated previously or at the time of the study for thyrotoxicosis in Almazov National Medical Research Centre or in Pavlov First Saint Petersburg State Medical University, St. Petersburg, Russia in 2000–2019.

Entry criteria:

- Men and women with overt thyrotoxicosis, associated with Graves' disease, toxic adenoma or multinodular toxic goiter.
- Age between 18 and 80 years.

Exclusionary criteria:

- Subclinical hyperthyroidism.
- A history of atrial fibrillation developed before the onset of thyrotoxicosis.
- Concomitant diseases that may affect structure and function of the cardiovascular system: hemodynamically significant valve disease, nonthyrotoxic cardiomyopathy, severe obstructive lung diseases.
- Chronic intoxication (alcohol, toxicomania).
- Pregnancy at the time of thyrotoxicosis.

The data includes 4 groups of features (abbreviations are indicated in brackets):

- Physiological: gender, age, height, weight, body mass index (BMI), the genesis of thyrotoxicosis «Genesis of TT», smoking status «SS», total cholesterol «TC», triglycerides «TG», low density lipoprotein cholesterol «LDL» level high density lipoprotein cholesterol «HDL», creatinine «CR», potassium «K» and hemoglobin «HB» levels during TT.
- Initial cardiological status before the development of TT: arterial hypertension «AH before TT», coronary heart disease before TT «CHD before TT», various heart rhythm disorders before TT«HRD before TT», congestive heart failure «CHF before TT», heart rate reducing therapy antiarrhythmic drugs, ivabradine, beta-blockers, and other types of therapy «HRRT before TT».
- Characteristics of thyrotoxicosis: free tetraiodothyronine «FT4», triiodothyronine «FT3», thyroid-stimulating hormone receptor antibodies level «TSHRA», duration of TT, for patients with AF - duration of TT before AF «DTT», subclinical TT duration (more less than 12 months) «DSTT», the number of relapses of TT «RTT», the presence of episodes of hypothyroidism «EHT», therapy of thyrotoxicosis «TTT», the duration of thyrostatic therapy «DTST», the time after TT onset when thyroidectomy or radioiodine therapy was provided «TTRTP».
- Cardiological status during the TT: arterial hypertension «AH during the TT», heart rate during the first examination «HR during the TT» and maximum recorded during TT heart rate «MHR during the TT», supraventricular extrasystole «SVE

during the TT», ventricular extrasystole «VE during the TT», other heart rhythm disorders «OHRD heart rhythm disorders», congestive heart failure «CHF during the TT», heart rate reducing therapy «HRRT during the TT».

3.2 Models Creation

The main objective of the study is to create a practical method for assessing the risk of developing atrial fibrillation in a patient with thyrotoxicosis. The model should include both features unique to the disease and physiological parameters.

At the initial stage, we have a table with observations (rows) and features (columns). The preprocessing of this dataset included five-stage. This stage shows in Fig. 1.

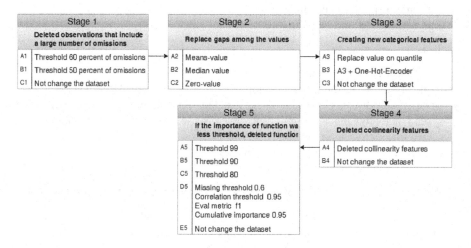

Fig. 1. Preprocessing stages

In the first stage, we deleted observations that include a large number of omissions.

In the second stage of preprocessing we replace gaps among the values. In the third stage, we calculated quantiles for each discrete feature. Next, we divided the values of all discrete features into three intervals. We use quantiles as intervals threshold. Using the B3 method, we first use the A1 method. Then we use One-Hot-Encoder methods for coding categorical values. On the fourth stage, we tested each featured couple on collinearity. On the next stage, we selected features for each dataset.

Finally, we created 120 different datasets (2 * 2 * 3 * 2 * 5). Each dataset contains a unique sequence of processing and selecting variables for prediction.

Next, the sample was divided into test and training, classes were balanced by pre-sampling features. The model was selected from the models KNeighborsClassifier, RandomForestClassifier, LogisticRegression, DecisionTreeClassifier, XGBClassifier, SVC, GaussianNB, MultinomialNB, BernoulliNB. For each model, metrics f1, accuracy, recall, precision for cross-validation were calculated for each of the datasets. Table 1 is a fragment of a table with the results of cross-validation.

Table 1. Models comparison

Model	Accuracy average	F1 average	Recall average	Precision average	Preprocessing
XGB	**0.876**	**0.793**	0.788	**0.801**	B1 A2 C3 B4 B5
BernoulliNB	0.834	0.742	0.788	0.701	A1 C2 B3 B4 A5
GaussianNB	0.819	0.73	**0.8**	0.677	B1 C2 B3 B4 C5
LR	0.835	0.726	0.715	0.742	A1 C2 B3 B4 B5
Decision Tree	0.806	0.703	0.675	0.666	B1 A2 A3 B4 B5
Random Forest	0.819	0.687	0.609	0.753	C1 B2 B3 A4 C5
MultinomialNB	0.777	0.653	0.693	0.618	A1 C2 B3 B4 C5
K Neighbors	0.796	0.614	0.537	0.720	A1 C2 B3 A4 B5
SVC	0.765	0.452	0.327	0.778	A1 C2 B3 B4 D5

This table includes the best combination of dataset preprocessing steps and model type for each model. The table is ranked by the f1 average. The best result was shown by the XGBClassifier model - 0.79 f1 average, 0.88 accuracies, and 0.8 precision. The method of processing the dataset is to remove features with more than 50% gaps, close gaps with averages, do not encode variables with naive coding and do not use quantiles, do not delete collinear features, the threshold for selecting functions is 0.9. Given the initial balance of classes 3 to 7 with a given quality, it makes sense to apply this model in practice. In terms of recall, the best result was shown by the GaussianNB model - 0.8. For reasons of interpretability, it is also advisable to use the DecisionTreeClassifier model (0.81 accuracies and 0.70 f1). Also, a good result was shown by the LR model (0.835 accuracies). This model has high interpretability. Also, for the convenience of using this model, it can be translated into a simple scale. To each attribute to associate a certain score. Given the high sensitivity and specificity, it also makes sense to use it in practice.

The main study objective is the creation of a set of models convenient for their application in practice. Today, there are many predictive models, but fewer of them are applied in practice. This is due to many factors. For example, the reluctance of doctors to work with the new software on which the model is based. Or the model has high accuracy, but it is not clear how to interpret the results. The decision-making system in the form of a scale is understandable and convenient for medical specialists. Each factor has a specific score. This score indicates the contribution of the factor to prognosis. Scales are widespread in medical practice, for example, the FINDRISK scale [15].

On base Logistic Regression model, we can create a scale for estimated atrial fibrillation risk. The preprocessing process is shown in Fig. 1. Out of 176 signs, the model showed the best quality at 60. The grid search method found model parameters.

The features were awarded points based on the regression coefficient. If the absolute value of the regression coefficient lay in the half-interval (0.0.5]–0.5 points, (0.5.1]–1 point, (1.2]–2 points, (2.3]–3 points, (3.4]–4 points.

The following values most significantly reduce the risk of atrial fibrillation: supraventricular extrasystole, which appeared against the background of thyrotoxicosis, duration of thyrotoxicosis is less than 6 months (coefficient −4); atrial pacemaker migration, Infiltrative ophthalmopathy and pretibial myxedema, duration of thyrotoxicosis of more than 6 and less than 12 months (coefficient −3); lack of arterial hypertension in the presence of thyrotoxicosis, presence of hypertension with target blood pressure values during treatment, hemoglobin more than 120 g/L and less than 132 (coefficient −2).

The most significant increase in the risk of Atrial Fibrillation is as follows: Diabetes mellitus (+4 coefficient); Heart rate on the background of TT is greater than 89 and less than 95 (coefficient 3); thyroidectomy thyrostatic therapy of thyrotoxicosis, male, lack of heart rate-reducing therapy against thyrotoxicosis, impaired glucose tolerance, ventricular extrasystole, which appeared against the background of thyrotoxicosis, duration of thyrostatic therapy, months >40.0, time of radical treatment (how many years after the debut thyrotoxicosis performed thyroidectomy or radioiodine therapy, months) less than 77(+2 coefficient).

The developed scale was designed in the form of a questionnaire, which allows, without medical tests, to assess the risk of developing Atrial Fibrillation for a particular patient. You can see a fragment of this questionnaire in Fig. 2.

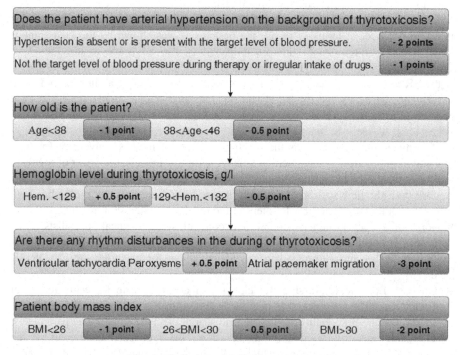

Fig. 2. Fragment of AF risk questionnaire

The final score was calculated for each patient, patients were divided into 4 risk groups. Table 2 shows the incidence of AF. Groups 2 and 3 include the most severe patients, the indicator of getting into this group shows very high values of sensitivity and specificity. This table shows the accuracy of the created scale. This scale is a standard scale for assessing the risk of complication, created using the same tools as the well-known FINDRISK scale [15] and was validated similarly [16, 17].

Table 2. Risk scale metrics for the development of AF.

Risk level	Frequency_FP0	Frequency_FP1	Score
Low	1	0	<= −5
Average	0.454545	0.545455	(−5,1]
Tall	0.0980392	0.901961	(1,5.5]
Very tall	0.0526316	0.947368	>= 5.5

3.3   Results Analysis

In Figs. 3 and 4, we visualized the 1st tree in gradient boosting.

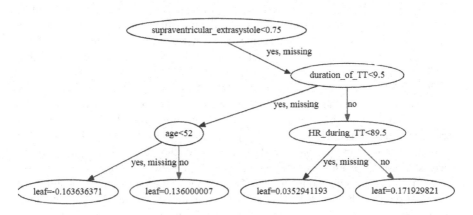

Fig. 3. Fragment of decisions tree for patients with supraventricular extrasystole

Figure 3 we visualized a fragment of decision tree for patients with supraventricular extrasystole. Presumably, the most important symptom is the presence of supraventricular extrasystole in the patient. Also, this decision tree confirms the a priori significance of the signs describing the specificity of the course of thyroid disease in a particular patient. Presumably, the longer the duration of thyrotoxicosis, the higher the probability of AF. The age and heart rate detected at the first examination also increase the probability of atrial fibrillation in patients with thyrotoxicosis and supraventricular extrasystole.

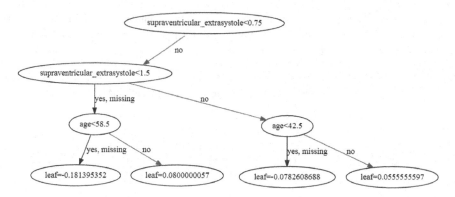

Fig. 4. Fragment of decisions tree for patients with supraventricular extrasystole

Figure 4 we visualized a fragment of decision tree for patients without supraventricular extrasystole. From the first tree, it follows that the most important criterion for the presence of AF is age. However, the specifics of constructing trees in gradient boosting does not always allow us to determine the importance of a particular trait for a predicted variable. We try to analyze the decision tree created on the basis of the DesissionTreeClassifier model. Visualization of 2 fragments of this tree is presented in Figs. 5 and 6.

Figure 5 describes the branch for patients who do not have supraventricular extrasystole. Based on the tree structure, supraventricular extrasystole is one of the most important signs that contribute to the probability of AF. The absence of supraventricular extrasystole with an 89% probability means that AF will not develop in this patient. Further, patients are divided by age, threshold 58.5. Age above 58 indicates a high probability of developing AF if in these patients the FDT3 level exceeds 1.6. For patients younger than 58 without extrasystole, the probability of developing AF is high only if thyrostatic therapy duration is above 225 or the duration of thyrotoxicosis is above 65 months. The risk of developing thyrotoxicosis is also affected by body mass index (the higher the variable, the higher the risk of AF) and heart rate-reducing therapy. If therapy was carried out, the risk of atrial fibrillation is significantly higher than in the absence of therapy.

We analyzed the second fragment - the course of the disease for patients with supraventricular extrasystole. For this group of patients, obesity of the 2nd and 3rd degrees indicates a high probability of AF even in the absence of problems with carbohydrate metabolism. If the patient does not have obesity, does not have a violation of carbohydrate metabolism, and does not undergo heart rate reducing therapy, the probability of AF is close to 0. If there are problems with MA without analyzing other factors, the patient has a 60% chance of developing AF. Further risk assessment depends on age (the younger the patient, the lower the risk), potassium level, FT3, the presence of other heart rhythm disturbances, FT3 and the duration of hypertension. These findings are confirmed by a priori knowledge and expert opinion of doctors. This decision tree is a fairly simple and well-interpreted method that can be useful as part of a support and decision-making system for medical professionals working with patients

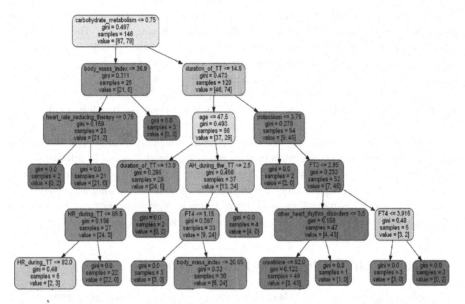

Fig. 5. Fragment of decisions tree for patients without supraventricular extrasystole

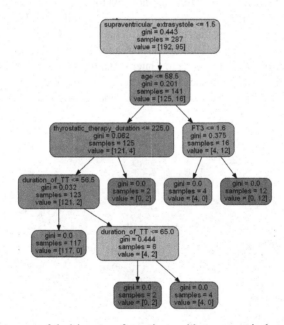

Fig. 6. Fragment of decisions tree for patients with supraventricular extrasystole

with thyrotoxicosis. Similarly, the complete structure of the tree was analyzed. The findings were interpreted by experts in the field of thyrotoxicosis. The knowledge gained about the relationship between the risk of AF and the specific course of thyrotoxicosis can be used to create a support and decision-making system.

Confusion matrices were constructed and analyzed. In terms of f1, XGBoostClassifier has the best predictive power. However, Logistic Regression has fewer false-negative errors. Given the balance of classes, DecisionTreeClassifier has good predictive power, despite the simplicity of the model (sensitivity 0.8 and specificity 0.63). DecisionTreeClassifier is the most convenient model for interpretation. The choice of a specific model depends on the goals that the medical specialist wants to achieve. If it is more important to find all patients at risk of developing AF, XGBoost is better, because the model has the most specificity. If it is important to find all patients whose AF does not develop, then it is more logical to apply Logistic regression. For ease of use and interpretation of the result, it makes sense to use DecisionTreeClassifier or scale.

4 Analysis Atrial Fibrillation Paths in Dynamic

We analyzed at the way atrial fibrillation occurs. We used data from 1450 patients suffering from thyrotoxicosis of any form. We divided the patient data into 2 groups. Group 1 includes 450 patients who have developed atrial fibrillation. Group 2 includes 1000 patients in whom atrial fibrillation did not develop in the observed period. Next, we present the patient's course of diseases in the form of a pathway. We use graphs representation. The nodes were the codes of "mkb10" (international statistical classification of diseases) with that the patient was admitted for treatment to the Almazov Medical Center. We created edges from an earlier node to a later node. In Figs. 7 and 8 showed graphs for the two groups. The colors show the clusters. These clusters were obtained by the method of maximizing modularity. The size of the node depends on its degree.

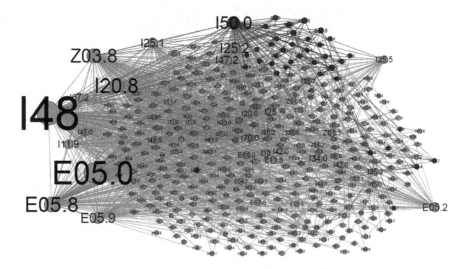

Fig. 7. Pathways of thyrotoxicosis for 450 patients with atrial fibrillation (Color figure online)

This graph contains 211 nodes "mkb10" and 1035 edges. Largest nodes in terms of degree are observation for suspected other diseases or conditions (Z03.8, purple cluster), congestive heart failure (I50.0, black cluster), past myocardial infarction (I25.2 purple cluster), other forms of angina pectoris (I20. 8, purple cluster), hypertension (I10, green cluster), atherosclerotic heart disease (I25.1, purple cluster), past myocardial infarction (I25.1, orange cluster). A diagnosis of AF is associated with each of these conditions. Therefore, when one of these conditions occurs, the patient is at risk of atrial fibrillation. It is a signal for the use of the above models. We include all large nodes in the model features.

Heart failure diagnosis is the center of gray clusters. This disease relates TT with another diagnosis in gray clusters. This cluster includes several types of anemia, heart disease, and limb disease. We assume that these diseases influence on risk AF only together with heart failure diagnosis. Observation with suspected other diseases or conditions, other forms of angina pectoris, acute transmural infarction of the lower myocardial wall, ventricular tachycardia are the largest nodes in the purple cluster. This cluster in heart disease clusters. All these diseases were included in model train features. The pathways of AF for patients of 2 types of thyrotoxicosis are described in the green cluster. These species are thyrotoxicosis with diffuse goiter and thyrotoxicosis with toxic multinodular goiter. If patient suffers from these diseases, his risk-AF-indicators include codes E66.0 (obesity) I10 (diabetes mellitus), cardiomyopathies (I42.0, I42.8). Therefore, for these patients group right diets is very important. Blue cluster included patients with other forms of thyrotoxicosis and thyrotoxicosis unspecified. For patients in this pathway group, indicators of AF include hypertension, cerebral atherosclerosis. Since 2 types of atrial fibrillation belong to this cluster, patients from this group must be automatically spent in the risk group.

The following types of atrial fibrillation fall into the blue cluster: Paroxysmal atrial fibrillation, typical atrial flutter. These forms of AF are most closely associated with a diagnosis of E05.8 (unspecified forms of thyrotoxicosis), E05.9 (unspecified forms of thyrotoxicosis). All clusters were analyzed. And we identified indicators of increased risk for each particular type of Atrial Fibrillation. Those pieces of knowledge used in predictive models of atrial fibrillation.

Based on Table 3, the graph structure is different. The average degree is higher in the trajectories of patients without AF. This can be explained by different group sizes. The different structure of the course of thyrotoxicosis for these groups of patients follows from a different number of components. We can calculate the distance of similarity for a couple of particular pathways. Therefore, we propose new features - distance from particular pathways to pathways from these graphs. We plan to study this hypothesis in future work. The obtained knowledge will be used in future work to create a dynamic model that predicts the risks of other complications.

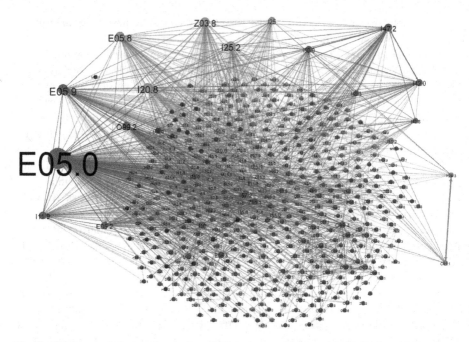

Fig. 8. Pathways of thyrotoxicosis for 1000 patients without atrial fibrillation (Color figure online)

Table 3. Graph statistics.

Statistic	Graph for first group	Graph for second group
Average degree	4,52	5.355
Graph diameter	10	9
Graph density	0.009	0.014
Components	9	4
The average length of the path	3.217	3.09
Average clustering factor	0.308	0.33

5 Conclusion and Future Work

As a result of the study, we created a practical method for assessing the risk of AF. This method includes 4 models. The choice of a particular model by a medical specialist depends on the specific treatments-diagnostic goal. The best combination of specificity and completeness was shown by a model based on Gradient Boosting. This model takes into account both the specifics of the course of thyrotoxicosis in a particular patient and many other features that affect the risk of AF. Given the high-quality indicators of the constructed models, these models may be applicable for early prediction of atrial fibrillation in a patient. The development will allow medical staff to select patients with a high risk of TAF for immediate follow-up or for early radical treatment for TT - surgical treatment or treatment with radioactive iodine instead of long-term medical treatment. We create a scale for assessing the risk of developing Atrial Fibrillation. The main advantage of the scale is its ease of use and good interpretability. Medical staff can use this scale without special software. All constructed models simplify the process of working with patients suffering from thyrotoxicosis. Further work includes improving the quality of the models used by increasing the data features and the use of neural networks, and analysis of atrial fibrillation pathways in dynamics using graphs.

Acknowledgments. This research is financially supported by The Russian Scientific Foundation, Agreement #17-15-01177.

References

1. Schnabel, R.B., et al.: Development of a risk score for atrial fibrillation (Framingham Heart Study): a community-based cohort study. Lancet **373**, 739–745 (2009). https://doi.org/10.1016/S0140-6736(09)60443-8
2. Frost, L., Vestergaard, P., Mosekilde, L.: Hyperthyroidism and risk of atrial fibrillation or flutter: a population-based study. Arch. Intern. Med. **164**, 1675–1678 (2004). https://doi.org/10.1001/archinte.164.15.1675
3. Biondi, B., Kahaly, G.J.: Cardiovascular involvement in patients with different causes of hyperthyroidism. Nat. Rev. Endocrinol. **6**, 431 (2010)
4. Klimanska, N., Klimanskaya, N., Kiselev, Yu.: Features of antiarrhythmic therapy in patients with thyrotoxic heart. In: Кліманська, Н., Климанская, Н., Киселев, Ю.: Особенности антиаритмической терапии у больных с тиреотоксическим сердцем (2005)
5. Xiong, Z., et al.: A machine learning aided systematic review and meta-analysis of the relative risk of atrial fibrillation in patients with diabetes mellitus. Front. Physiol. **9** (2018). https://doi.org/10.3389/fphys.2018.00835
6. Go, A.S., et al.: Prevalence of diagnosed atrial fibrillation in adults: National implications for rhythm management and stroke prevention: The anticoagulation and risk factors in atrial fibrillation (ATRIA) study. J. Am. Med. Assoc. **285**, 2370–2375 (2001). https://doi.org/10.1001/jama.285.18.2370
7. Klein, I., Danzi, S.: Thyroid disease and the heart. Circulation **116**, 1725–1735 (2007). https://doi.org/10.1161/CIRCULATIONAHA.106.678326
8. Babenko A.Y.: Interrelation of remodeling of the heart and blood vessels with thyrotoxicosis, vol. 12. Бабенко А. Ю. Взаимосвязь ремоделирования сердца и сосудов при тиреотоксикозе (2011)
9. Sawin, C.T., et al.: Low serum thyrotropin concentrations as a risk factor for atrial fibrillation in older persons. N. Engl. J. Med. **331**, 1249–1252 (1994). https://doi.org/10.1056/NEJM199411103311901
10. Nakazawa, H.: Is there a place for the late cardioversion of atrial fibrillation? A long-term follow-up study of patients with post-thyrotoxic atrial fibrillation. Eur. Heart J. **21**, 327–333 (2000). https://doi.org/10.1053/euhj.1999.1956

11. Gammage, M.D., et al.: Association between serum free thyroxine concentration and atrial fibrillation. Arch. Intern. Med. **167**, 928–934 (2007). https://doi.org/10.1001/archinte.167.9. 928

12. Iwasaki, T., et al.: Echocardiographic studies on the relationship between atrial fibrillation and atrial enlargement in patients with hyperthyroidism of Graves&rsquo. Disease. Cardiol. **76**, 10–17 (1989). https://doi.org/10.1159/000174467

13. Tănase, D.M., Ionescu, S.D., Ouatu, A., Ambăruș, V., Arsenescu-Georgescu, C.: Risk assessment in the development of atrial fibrillation at patients with associate thyroid dysfunctions. Rev. medico-chirurgicală a Soc. Medici și Nat. din Iași. (2013)

14. Metab Synd, E., Hernando, V.U., Eliana, M.S.: Citation: Hernando VU, Eliana MS (2015) Role of Thyroid Hormones in Different Aspects of Cardiovascular System. Endocrinol Metab Synd. **4**, 166 (2015). https://doi.org/10.4172/2161-1017.1000166

15. Lindström, J., Lindström, L., Tuomilehto, J.: The diabetes risk score a practical tool to predict type 2 diabetes risk. Diabetes care **26**, 725–731 (2003)

16. Makrilakis, K., et al.: Validation of the Finnish diabetes risk score (FINDRISC) questionnaire for screening for undiagnosed type 2 diabetes, dysglycaemia and the metabolic syndrome in Greece. Diabetes Metab. (2011). https://doi.org/10.1016/j.diabet. 2010.09.006

17. Zhang, L., Zhang, Z., Zhang, Y., Hu, G., Chen, L.: Evaluation of Finnish diabetes risk score in screening undiagnosed diabetes and prediabetes among U.S. adults by gender and race: NHANES 1999–2010. PLoS One **9**, (2014). https://doi.org/10.1371/journal.pone.0097865

Applicability of Machine Learning Methods to Multi-label Medical Text Classification

Iuliia Lenivtceva[✉], Evgenia Slasten, Mariya Kashina,
and Georgy Kopanitsa

ITMO University, 49 Kronverkskiy Prospect, 197101 Saint Petersburg,
Russian Federation
lenivezzki@gmail.com, slastenevgenia@gmail.com,
k.mariya1997@gmail.com, georgy.kopanitsa@gmail.com

Abstract. Structuring medical text using international standards allows to improve interoperability and quality of predictive modelling. Medical text classification task facilitates information extraction. In this work we investigate the applicability of several machine learning models and classifier chains (CC) to medical unstructured text classification. The experimental study was performed on a corpus of 11671 manually labeled Russian medical notes. The results showed that using CC strategy allows to improve classification performance. Ensemble of classifier chains based on linear SVC showed the best result: 0.924 micro F-measure, 0.872 micro precision and 0.927 micro recall.

Keywords: Multi-label learning · Medical text classification · Interoperability · FHIR · Data structuring

1 Introduction

Medical data standardization is crucial in terms of data exchange and integration as data formats vary greatly from one healthcare provider to another. Many international standards for terminologies (SNOMED CT [1], LOINC [2]) and data exchange (openEHR [3], ISO13606 [4], HL7 standards [5]) are successfully implemented and perform well in practice. The most developing and perspective standard for medical information today is FHIR-HL7 [6].

The data are usually stored in structured, semi-structured or unstructured form in medical databases. Structured and semi-structured data can be mapped to standards with minimum losses of information [7]. However, a big part of Electronic Health Record (EHR) is in free text [8]. Unstructured medical records are more complicated to process, however, they usually contain detailed information on patients which is valuable in modeling and research [9].

The extraction of useful knowledge becomes more challenging as medical databases become more available and contain a wide range of texts [10]. Sorting documents and searching concepts and entities in texts manually is time-consuming. Text classification

© Springer Nature Switzerland AG 2020
V. V. Krzhizhanovskaya et al. (Eds.): ICCS 2020, LNCS 12140, pp. 509–522, 2020.
https://doi.org/10.1007/978-3-030-50423-6_38

is an important task which aims to sort documents or notes according to the predefined classes [11] which facilitates entities extraction such as symptoms [12], drug names [13], dosage [14], drug reactions [15], etc. The task of information extraction (IE) is domain specific and requires considering its specificity in practice. Thus, high performance in IE can be achieved through free text classification to a particular domain [16].

The developed applications and methods for processing free texts are language specific [17]. Russian medical free text processing is challenging mostly because there is no open source medical corpora [18]. Moreover, each medical team develops their own storage format, which makes it difficult to standardize, exchange and integrate Russian medical data.

Our long-term goal is to develop methods for data extraction from Russian unstructured clinical notes and mapping these data on FHIR for better interoperability and personalized medicine. The purpose of the article is to investigate the applicability of machine learning algorithms to classify Russian unstructured and semi-structured allergy anamnesis to facilitate entities extraction.

2   Related Work

Studies on text classification using machine learning methods are widely represented in literature.

Jain et al. [16] describes classifiers based on Multinomial Naïve Bayes (MNB), k-Nearest Neighbors (k-NN) and Support Vector Machine (SVM) as the most popular models for multi-label classification. Logistic regression (LR) is also a widespread model for the task [19].

Binary relevance (BR) approach suggests to train N independent binary classifiers for multi-label classification with N labels. This approach has a linear complexity; however, it does not consider interdependences between labels [19]. Classifier Chains (CC) is a popular and representative algorithm for multi-label classification. CC suggests to link N binary classifiers in a chain with random ordering as it shows better predictive performance of the classification. The set of predicted labels is treated as extra features for the next classifiers in a chain. CC and ensembles [20] are known to solve over-fitting problem. CC are more computationally demanding than simple binary classifiers [21].

The performance metrics of multi-label classifiers applied to medical text are represented in Table 1. The literature review showed that there is no a single concept on which metrics to use when evaluating multi-label classifiers.

Table 1. Performance of medical multi-label classifiers

Classifier	#labels	Data and tools	F1		PRC		REC		Citation
			micro	macro	micro	macro	micro	macro	
BR	10	Real data	0.78		0.84		0.80		Zhao et al. [22]
CC			0.79		0.89		0.75		
Binary	45	Open dataset Medical WEKA	–	0.38	–	–	–	–	Read et al. [23]
CC			–	0.39	–	–	–	–	
kNN			–	–	–	–	–	–	
LR			–	–	–	–	–	–	
Rule-based	7	Real data	0.95		0.96		0.94		Baghdadi et al. [24]
SVM			0.99		0.97		0.98		
SVM	6	Open dataset cTAKES	0.83		–		0.934		Weng et al. [25]
NB	8	Real data WEKA	0.82		0.77		0.89		Spat et al. [26]
1-NN			0.86		0.87		0.86		
J48			0.88		0.90		0.87		
SVM	45	Real data Manual labeling	0.823	–	0.823	–	0.831	–	Argaw et al. [10]
SVM	2618	Real data	0.683	0.652	–	0.535	–	0.868	Lita et al. [27]
SVM	78	Open dataset	0.530	–	–	–	–	–	Baumel et al. [28]
BR	420	Real data	0.720	0.706	0.818	0.812	0.643	0.659	Kaur et al. [8]

3 Methods

3.1 Data Description

Clinical documents (written in Russian) of more than 250 thousand patients were provided by Almazov National Medical Research Centre (St. Petersburg, Russia) for the research. The patients' personal information was discarded. We searched for different forms of the words «allergy» and «(in)tolerance» (Russian equivalents «аллергия», «(не)переносимость») using regular expressions to find all the notes containing any information on allergy and intolerances. The corpus of 269 thousand notes was created after the search and duplicates removal. We classified allergy notes according to four labels which are described in Table 2.

Table 2. Classes description

Label	Classes description	Example in Russian	Example in English
AL	A note contains information about allergen or intolerance. It might be the name of a drug or a drug's group (nitrates). A note also might only mention that allergy or intolerance takes place	Аллергологический анамнез аллергия на укус насекомых Назначена терапия метотрексан 10 мг, отменена в связи с плохой переносимостью препарата Аллергологический анамнез аллергия на не помнит	Allergy anamnesis allergy to a bite of an insect Methotrexate 10 mg treatment was started, but due to the poor tolerance the drug was canceled Allergy does not remember exactly
R	A note contains information about the reaction to some allergen. The allergen might be specified or not	Аллергологический анамнез аллергия на атопический дерматит. Аллергия на медикаменты пенициллин крапивница йод нет	Allergy anamnesis allergy atopic dermatitis. Allergy to medications penicillin urticaria, iodine no
NN	A note declares that there is no allergy or intolerance	Аллергия нет	No allergy
N	A note does not contain information about allergy or intolerance	План лечения введение препаратов переносит удовлетворительно	Treatment plan drug administration tolerates satisfactorily

Two experts assigned an appropriate label to each note. In case of disagreement the decision was made by consensus.

The final corpus contains 11671 labeled notes.

3.2 Task Description

AllergyIntorence is one of the FHIR resources, it contains structured information on patient's allergies, intolerances and symptoms. The task of mapping this data to FHIR involves machine learning methods as it is stored in unstructured form. Figure 1 represents the main blocks of information that can be mapped to FHIR. Bold blocks denote information that is mentioned in the processed corpus.

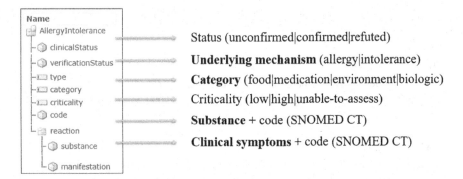

Fig. 1. Blocks of information to be mapped to FHIR

Underlying mechanism can be extracted by searching keywords «allergy» and «intolerance» in the corpus. Category refers to an exact substance type. The most sophisticated task is to extract exact substances and clinical symptoms written in Russian and to bind corresponding codes from international terminological systems to ensure interoperability. To facilitate this task classification of multi-topic clinical notes is required.

3.3   Preprocessing

The steps of preprocessing are:

1. Clean medical notes from symbols and extra spaces. Full stops are left as they play an important role in sentence tokenization.
2. Reduce notes to minimize noise during classification as the original note might contain up to 9239 words. Only 2 meaningful sentences before and after regular expression («аллергия», «(не)переносимость») are left.
3. Correct syntactic, case and spaces errors using regular expressions.
4. Dictionary-based spelling correction with Levenshtein distance calculation.
5. Tokenize and normalize words.
6. Train-test split, training set contains 7819 notes and test set – 3852.
7. Vectorize both train and test sets using Bag of Words (BOW) representation. The dictionary size for BOW is 8000 words.

3.4   Classification

We applied four shallow machine learning models: MNB, LR, SVM, k-NN and two ensembles of classifier chains: ECCLR, ECCSVM. The optimal parameters of the shallow models were adjusted by grid search. Optimal parameters of the models are introduced in Table 3.

Table 3. Parameters of classifiers

Model	Parameters
Shallow classifiers	
MNB	Alpha: 0.5
LR	Solver: saga, penalty: l2, C = 3, max_iter = 4000
Linear SVM	Loss: squared hinge, penalty: l2, max_iter = 4000, C = 1.3684
k-NN	Algorithm: brute, n_neighbors = 1, weights: uniform
Ensembles of classifier chains	
ECCLR	Ensemble of 10 logistic regression classifier chains with random ordering of labels
ECCSVM	Ensemble of 10 linear SVM classifier chains with random ordering of labels

The pipeline was built using python version 3.7.1. For lexical normalization «pymorphy2» was used. All the preprocessing steps were realized with custom skripts. «scikit-learn» package was used to implement supervised learning algorithms, evaluate models and to perform t-SNE. «Bokeh», «matplotlib» and «plotly» were used for visualization.

3.5   Evaluation Metrics

According to [21] macro and micro averaging precision, recall and F-measure are often used to evaluate multi-label classification performance. So, we used these metrics to evaluate the performance of the classification.

Micro-averaging:

$$B_{micro}(h) = B\left(\sum\nolimits_{j=1}^{q} TP_j, \sum\nolimits_{j=1}^{q} FP_j, \sum\nolimits_{j=1}^{q} TN_j, \sum\nolimits_{j=1}^{q} FN_j\right) \tag{1}$$

Macro-averaging:

$$B_{macro}(h) = \frac{1}{q}\sum\nolimits_{j=1}^{q} B(TP_j, FP_j, TN_j, FN_j) \tag{2}$$

B ∈ {Precision, Recall, F$^{\beta}$}, q – number of class labels.

Precision (positive predictive value) is the fraction of correctly identified examples of the class among all the examples identified as this class.

$$Precision(TP_j, FP_j, TN_j, FN_j) = \frac{TP_j}{TP_j + FP_j} \tag{3}$$

Recall evaluates the fraction of identified examples from the class among all the examples of this class.

$$Recall\big(TP_j, FP_j, TN_j, FN_j\big) = \frac{TP_j}{TP_j + FN_j} \tag{4}$$

F-measure is harmonic mean ($\beta = 1$) of precision and recall.

$$F^\beta\big(TP_j, FP_j, TN_j, FN_j\big) = \frac{(1+\beta^2)TP_j}{(1+\beta^2)TP_j + FP_j + \beta^2 FN_j} \tag{5}$$

TP – true positive examples, TN – true negative examples, FP – false positive examples, FN – false negative examples, $\beta = 1$.

t-SNE was performed using predicted probabilities for each label. The perplexity equals 30 according to recommendations of G.E. van der Maaten et al. [29].

4   Results

After text cleaning still there were notes which contained neither allergies nor intolerances.

Figure 2 illustrates the distribution of classes in the corpus. The classes are imbalanced.

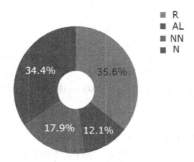

Fig. 2. Classes distribution in the corpus

Performances of different classifiers are represented in Table 4. LR and linear SVM showed the best results among shallow classifiers. However, the use of CC with LR and linear SVM as base classifiers improved performance metrics and showed best results.

Table 4. Performance of the applied classifiers

Model	Precision		Recall		F-measure	
	Micro	Macro	Micro	Macro	Micro	Macro
Shallow classifiers						
MNB	0.781	0.764	0.864	0.873	0.864	0.852
LR	**0.866**	**0.850**	**0.920**	0.915	**0.920**	**0.910**
Linear SVM	**0.865**	**0.849**	**0.919**	0.916	**0.919**	**0.909**
k-NN	0.694	0.715	0.803	0.827	0.803	0.809
Ensembles of classifier chains						
ECCLR	**0.867**	**0.852**	**0.925**	**0.921**	**0.922**	**0.912**
ECCSVM	**0.872**	**0.855**	**0.927**	**0.922**	**0.924**	**0.914**

Classification report for the best classifier is represented in Table 5.

Table 5. Classification report for ECCSVM

	Precision	Recall	F1-score	Support
AL	0.93	0.94	0.94	1317
R	0.95	0.92	0.93	1388
NN	0.92	0.93	0.93	690
N	0.83	0.89	0.86	457
Micro avg	0.92	0.93	0.92	3852
Macro avg	0.91	0.92	0.91	3852
Weighted avg	0.92	0.93	0.92	3852
Samples avg	0.92	0.93	0.92	3852

Figure 3 illustrates t-SNE representation classes.

Figure 4, Fig. 5, Fig. 6, Fig. 7 represent 10 most important keywords in the corpus which indicate that the note belongs to the corresponding class. The diagrams show how often each word can be met in the corpus (word counts) and how important this word is for classification (weights of classifier). The diagram is plotted using LR weights.

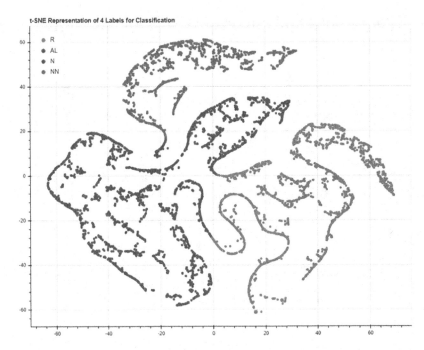

Fig. 3. t-SNE representation of classes

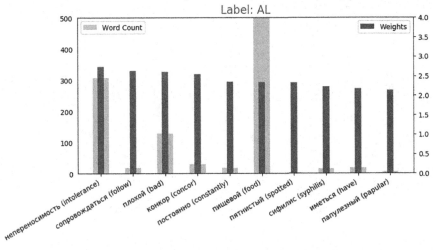

Fig. 4. Top 10 positive keywords for label AL

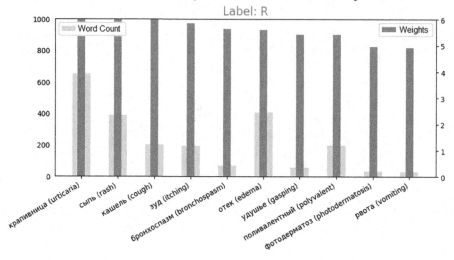

Fig. 5. Top 10 positive keywords for label R

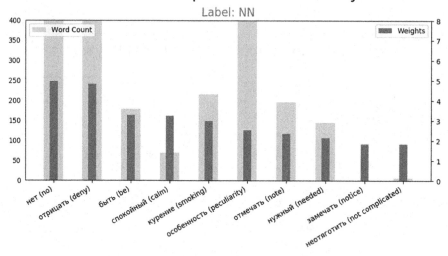

Fig. 6. Top 10 positive keywords for label NN

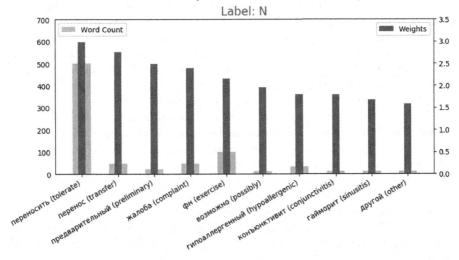

Fig. 7. Top 10 positive keywords for label N

5   Discussion

Regarding previous studies on multi-label medical text classification many authors use applications for entities extraction and algorithms implementation (Table 1). However, there is no open source applications for medical purposes developed for the Russian case such as MetaMap [30], for instance. Thus, all the steps were realized manually and with custom scripts.

In the medical text multi-label classification task with limited labeled data we concentrated on improving F-measure as it enforces a better balance between performing on relevant and irrelevant labels and, thus, suitable for multi-label task evaluation [31]. Also, precision, recall and F-measure are not sensitive to classes imbalance.

Two of the proposed shallow classifiers LR and linear SVM performed well on real unstructured labeled data. Using CC strategy allowed to improve the results of basic classifiers and the best performance was shown by ensemble of classifier chains based on linear SVC. Classification report for this classifier (Table 5) has shown that three most important labels for mapping AL, R and NN are well separated from each other and from the fourth class N. The fourth class showed lower performance which can be caused by the least number of labeled data in the corpus and the variety of topics covered in it.

Recall is higher than precision for all classifiers and for both averaging strategies. It means that classifiers are good at identifying classes and differentiating them from each other. The number of false negatives is low, which means that classifiers do not intend to lose important notes. This result is satisfying from the point of mapping task as it is important to find as many class representatives as possible.

The obtained result of 0.924 μ F-measure, 0.872 μ Precision and 0.927 μ Recall by ECCSVC outperformed almost all the represented in Table 1 results. Baghdadi et al. [24] reported high overall performance of implemented classifiers and the data were previously standardized. W.-H. Weng et al. [25] used additional tools for clinical text processing and information extraction. The closest task was solved by Argaw et al. [10] in terms of real data manual labeling. All the obtained metrics of our ECCSVC are higher, however, the number of labels in the classification task is lower.

t-SNE representation shows that classes are well separated.

Figure 4 shows 10 most important words associated with allergens and substances. The list of keywords for this task contain such entities as «intolerance» which indicates the presence of patient's intolerance in the text of anamnesis; «food» which is associated with the category of allergy in the FHIR resource; medications such as «concor» which might be associated with a substance in the FHIR resource; number of verbs indicating the presence of allergy such as «follow», «have». The words «intolerance» and «food» are also most frequent words of this class in a corpus.

Figure 5 shows 10 most important words associated with clinical symptoms in FHIR resources and reactions. All the most frequent keywords of this class are symptoms.

Figure 6 shows 10 most important words associated with the situation when no allergy was detected. This class keywords contain many negative words such as «no», «deny», «not complicated» and general purpose normalized words, which are usually met in calm allergy anamnesis: «calm», «be», «notice». The keywords of this group are not frequent in a corpus because of low number of labeled notes for this class. The NN notes would be marked as «no allergy» and would not be considered during information extraction and mappings.

Figure 7 shows 10 most important words associated with class N, which indicates that the exact note is not connected with allergy or intolerance. The most important and frequently met keyword in this class is «tolerate (переносить)». This word has one root with the word «intolerance (непереносимость)». Thus, this word frequent due to the initial mechanism of search. Other keywords represent different topics not connected with allergy and intolerance. Thus, the notes from this class would not be considered during information extraction and mappings.

6 Conclusion

In this study we investigated the applicability of several classifiers to the task of clinical free-text allergy anamnesis classification for filtering multi-topic data.

The research showed that LR, linear SVC, ECCLR and ECCSVC performed well and can be applied to the task of clinical free-text allergy anamnesis classification. The use of chaining strategy improved the performance of shallow classifiers.

In the future we plan to apply a model for Named Entity Recognition (NER) to extract named entities such as allergies and symptoms from medical free text and map them to FHIR. Also, we plan to develop a model to ICD-10 Russian codes and terms identification in medical free-text allergy anamnesis.

Acknowledgements. This work financially supported by the government of the Russian Federation through the ITMO fellowship and professorship program. This work was supported by a Russian Fund for Basic research 18-37-20002. This work is financially supported by National Center for Cognitive Research of ITMO University.

References

1. Fung, K.W., Xu, J., Rosenbloom, S.T., Campbell, J.R.: Using SNOMED CT-encoded problems to improve ICD-10-CM coding—a randomized controlled experiment. Int J Med Inform **126**, 19–25 (2019). https://doi.org/10.1016/j.ijmedinf.2019.03.002
2. Fiebeck, J., Gietzelt, M., Ballout, S., et al.: Implementing LOINC: current status and ongoing work at the Hannover Medical School. In: Studies in Health Technology and Informatics, pp. 247–248. IOS Press (2019)
3. Mascia, C., Uva, P., Leo, S., Zanetti, G.: OpenEHR modeling for genomics in clinical practice. Int. J. Med. Inform. **120**, 147–156 (2018). https://doi.org/10.1016/j.ijmedinf.2018.10.007
4. Santos, M.R., Bax, M.P., Kalra, D.: Building a logical EHR architecture based on ISO 13606 standard and semantic web technologies. In: Studies in Health Technology and Informatics (2010)
5. Ulrich, H., Kock, A.K., Duhm-Harbeck, P., et al.: Metadata repository for improved data sharing and reuse based on HL7 FHIR. In: Studies in Health Technology and Informatics (2017)
6. Hong, N., Wen, A., Mojarad, M.R., et al.: Standardizing heterogeneous annotation corpora using HL7 FHIR for facilitating their reuse and integration in clinical NLP. In: AMIA Annual Symposium Proceedings AMIA Symposium, pp. 574–583 (2018)
7. Lenivtseva, Y., Kopanitsa, G.: Investigation of content overlap in proprietary medical mappings. Stud. Health Technol. Inform. **258**, 41–45 (2019). https://doi.org/10.3233/978-1-61499-959-1-41
8. Kaur, R., Ginige, J.A.: Analysing effectiveness of multi-label classification in clinical coding. In: ACM International Conference Proceeding Series. Association for Computing Machinery (2019)
9. Wang, Y., Wang, L., Rastegar-Mojarad, M., et al.: Clinical information extraction applications: a literature review. J. Biomed. Inform. **77**, 34–49 (2018)
10. Alemu, A., Hulth, A., Megyesi, B.: General-purpose text categorization applied to the medical domain. Comput. Sci. **16** (2007)
11. Onan, A., Korukoğlu, S., Bulut, H.: Ensemble of keyword extraction methods and classifiers in text classification. Expert Syst. Appl. **57**, 232–247 (2016). https://doi.org/10.1016/j.eswa.2016.03.045
12. Métivier, J.-P., Serrano, L., Charnois, T., Cuissart, B., Widlöcher, A.: Automatic symptom extraction from texts to enhance knowledge discovery on rare diseases. In: Holmes, J.H., Bellazzi, R., Sacchi, L., Peek, N. (eds.) AIME 2015. LNCS (LNAI), vol. 9105, pp. 249–254. Springer, Cham (2015). https://doi.org/10.1007/978-3-319-19551-3_33
13. Levin, M.A., Krol, M., Doshi, A.M., Reich, D.L.: Extraction and mapping of drug names from free text to a standardized nomenclature. In: AMIA Annual Symposium Proceedings, pp. 438–442 (2007)
14. Xu, H., Jiang, M., Oetjens, M., et al.: Facilitating pharmacogenetic studies using electronic health records and natural-language processing: a case study of warfarin. J. Am. Med. Inform. Assoc. **18**, 387–391 (2011). https://doi.org/10.1136/amiajnl-2011-000208

15. Wang, X., Hripcsak, G., Markatou, M., Friedman, C.: Active computerized pharmacovigilance using natural language processing, statistics, and electronic health records: a feasibility study. J. Am. Med. Inform. Assoc. **16**, 328–337 (2009). https://doi.org/10.1197/jamia.M3028

16. Jain, A., Mandowara, J.: Text classification by combining text classifiers to improve the efficiency of classification. Int. J. Comput. Appl. **6**, 1797–2250 (2016)

17. Ali, A.R., Ijaz, M.: Urdu text classification. In: Proceedings of the 6th International Conference on Frontiers of Information Technology, FIT 2009 (2009)

18. Toldova, S., Lyashevskaya, O., Bonch-Osmolovskaya, A., Ionov, M.: Evaluation for morphologically rich language: Russian NLP. In: Proceedings on the International Conference on Artificial Intelligence (ICAI), pp. 300–306. CSREA Press, Las Vegas (2015)

19. Cheng, W., Hüllermeier, E.: Combining instance-based learning and logistic regression for multilabel classification. Mach. Learn. **76**, 211–225 (2009). https://doi.org/10.1007/s10994-009-5127-5

20. Tahir, M.A., Kittler, J., Bouridane, A.: Multilabel classification using heterogeneous ensemble of multi-label classifiers. Pattern Recogn. Lett. **33**, 513–523 (2012). https://doi.org/10.1016/j.patrec.2011.10.019

21. Zhang, M.L., Zhou, Z.H.: A review on multi-label learning algorithms. IEEE Trans. Knowl. Data Eng. **26**, 1819–1837 (2014)

22. Zhao, R.W., Li, G.Z., Liu, J.M., Wang, X.: Clinical multi-label free text classification by exploiting disease label relation. In: Proceedings - 2013 IEEE International Conference on Bioinformatics and Biomedicine, IEEE BIBM 2013, pp 311–315 (2013)

23. Read, J., Pfahringer, B., Holmes, G., Frank, E.: Classifier chains for multi-label classification. In: Buntine, W., Grobelnik, M., Mladenić, D., Shawe-Taylor, J. (eds.) ECML PKDD 2009. LNCS (LNAI), vol. 5782, pp. 254–269. Springer, Heidelberg (2009). https://doi.org/10.1007/978-3-642-04174-7_17

24. Baghdadi, Y., Bourrée, A., Robert, A., et al.: Automatic classification of free-text medical causes from death certificates for reactive mortality surveillance in France. Int. J. Med. Inform. **131**. https://doi.org/10.1016/j.ijmedinf.2019.06.022

25. Weng, W.-H., Wagholikar, K.B., McCray, A.T., et al.: Medical subdomain classification of clinical notes using a machine learning-based natural language processing approach. BMC Med. Inform. Decis. Mak. **17**, 155 (2017). https://doi.org/10.1186/s12911-017-0556-8

26. Spat, S., et al.: Multi-label classification of clinical text documents considering the impact of text pre-processing and training size. In: 23rd International Conference of the European Federation for Medical Informatics (2011)

27. Lita, L.V., Yu, S., Niculescu, S., Bi, J.: Large scale diagnostic code classification for medical patient records. In: IJCNLP, pp. 877–882 (2008)

28. Baumel, T., Nassour-Kassis, J., Cohen, R., et al.: Multi-label classification of patient notes a case study on ICD code assignment. In: AAAI Conference on Artificial Intelligence. pp. 409–416 (2017)

29. van der Maaten, L.J.P., Hinton, G.E.: Visualizing high-dimensional data using t-SNE. J. Mach. Learn. Res. **9**, 2579–2605 (2008)

30. Aronson, A.R., Lang, F.M.: An overview of MetaMap: historical perspective and recent advances. J. Am. Med. Inform. Assoc. **17**, 229–236 (2010). https://doi.org/10.1136/jamia.2009.002733

31. Dembczynski, K., Jachnik, A., Kotłowski, W., et al. Optimizing the F-measure in multi-label classification: plug-in rule approach versus structured loss minimization. In: ICML 2013: Proceedings of the 30th International Conference on International Conference on Machine Learning, pp. 1130–1138 (2013)

Machine Learning Approach for the Early Prediction of the Risk of Overweight and Obesity in Young People

Balbir Singh[(✉)] and Hissam Tawfik

Leeds Beckett University, Calverley Street, Leeds LS1 3HE, UK
{b.singh, h.tawfik}@leedsbeckett.ac.uk

Abstract. Obesity is a major global concern with more than 2.1 billion people overweight or obese worldwide which amounts to almost 30% of the global population. If the current trend continues, the overweight and obese population is likely to increase to 41% by 2030. Individuals developing signs of weight gain or obesity are also at a risk of developing serious illnesses such as type 2 diabetes, respiratory problems, heart disease and stroke. Some intervention measures such as physical activity and healthy eating can be a fundamental component to maintain a healthy lifestyle. Therefore, it is absolutely essential to detect childhood obesity as early as possible. This paper utilises the vast amount of data available via UK's millennium cohort study in order to construct a machine learning driven model to predict young people at the risk of becoming overweight or obese. The childhood BMI values from the ages 3, 5, 7 and 11 are used to predict adolescents of age 14 at the risk of becoming overweight or obese. There is an inherent imbalance in the dataset of individuals with normal BMI and the ones at risk. The results obtained are encouraging and a prediction accuracy of over 90% for the target class has been achieved. Various issues relating to data preprocessing and prediction accuracy are addressed and discussed.

Keywords: Obesity · Data driven model · Young people · Classification · Machine learning · Ensemble learning

1 Introduction

Obesity in general is a growing epidemic that affects every age group. In the UK alone, the economic impact of dealing with obesity and related illnesses is almost same as the other major issues affecting the economy such as smoking and armed conflict as reported in a publication by McKinsey Global Institute on Overcoming Obesity [1]. The report discusses that more than 30% of the world population is either overweight or obese. If this growth rate prevails, the proportion of people in overweight and obese category is likely to rise up to 41% by the year 2030. The report concludes by recommending behaviour change interventions which can result in saving money in the long run as a result of reduced healthcare and increase in productivity levels. Implementation of this type of intervention could save almost £1 billion for the National

© Springer Nature Switzerland AG 2020
V. V. Krzhizhanovskaya et al. (Eds.): ICCS 2020, LNCS 12140, pp. 523–535, 2020.
https://doi.org/10.1007/978-3-030-50423-6_39

Health Service (NHS) in the United Kingdom alone. The evidence suggests that the behavioural change interventions to combat obesity need further investigation to find workable solutions rather than waiting for a perfect solution. In the United Kingdom, the data presented in Health Survey for England reports that the percentage of obese children between the age of 2 and 15 has increased significantly since 1995 [2]. 16% of boys and 15% of girls in this age group were classed as obese. 14% of both genders were classed as overweight. This results in 30% of boys and 29% of girls being either overweight or obese. In 1995, 11% of boys and 12% of girls of 2–15 years of age were obese. A range of studies have however been monitoring the trend in obesity growth. One in specific included data from review included data from 467,294 children from Australia, China, England, France, Netherlands, New Zealand, Sweden, Switzerland and USA [3]. It was reported that the prevalence of childhood obesity may be plateauing worldwide. Another survey identified 52 obesity studies worldwide from 25 countries and reported some stability in obesity [4]. Similar claims were made by [5] and [6]. Some reports even suggest that childhood obesity may well be declining as a cumulative result of increased physical activity, television viewing decline and reduction in sugary drink consumption [7]. However, [4] also reported that this stability should be observed with caution since previous stable phases were followed by further increases in prevalence of obesity.

The Health Survey England (HSE) series was established to keep an eye on changes in nation's health [8]. The survey was designed to acquire information about certain health conditions and other risk factors affecting them. It was reported that in 2013, 26% of men and 24% of women were obese. 41% of men and 33% of women were overweight but not obese. Looking at these combined figures gives cause for concern since 67% of men and 57% of women are above their normal weight for their height.

Childhood obesity is of great public concern as up to 90% of the overweight and obese childhood population will continue to be obese as adults [9]. Obesity is strongly linked with other negative health conditions such as type 2 diabetes, cardiovascular diseases, cancers and even death [10–13]. Considering all this, it is obvious that there is a pressing need to identify individuals at a risk of developing obesity as early as possible so that some preventative measures can be put in place as early as possible. The purpose of this study is to carry out apply, analyse and evaluate machine learning algorithms to classify adolescents at a risk of becoming overweight or obese using early childhood BMI as input features to machine learning. The BMI in adults is defined as a ratio of body mass in kilograms to the square of individual's height in meters. In adults, BMIs of over 25 kg/m^2 and 30 kg/m^2 are classed as overweight and obese respectively. This formula cannot be applied to children and adolescents since their body mass index changes significantly with age [14]. It varies between 13 kg/m^2 and 17 kg/m^2 from birth to an age of 1. It then decreases to 15.5 kg/m^2 at the age of 6 and increases to 21 kg/m^2 at the age of 20. Therefore, normal overweight and obesity thresholds can't be applied to children and adolescents. For this reason, this study uses overweight and obesity cut-off points recommended by International Obesity Task Force (IOTF). This paper is organised as follows:

- Related Work: This section discusses the application of machine learning techniques in health and approaches tackling obesity.
- Methods: This section deals with the data preprocessing, application of ML algorithms for the classification of imbalanced data, techniques used for the treatment of data imbalance and the evaluation of suitable algorithms for improved prediction accuracy.
- Discussion: Performance of ML algorithms and their effect on prediction accuracy are discussed.
- Conclusion and further work: The paper concludes the work carried out and the challenges experienced. Further work to build on this research is discussed.

2 Related Work

2.1 Machine Learning Applications in Health

A range of machine learning algorithms have been applied in health domains to predict the presence certain health conditions using a number of characteristic features. Artificial Neural Networks (ANNs) are a commonly used branch of Machine Learning (ML) methods that are used to correlate input parameters to corresponding output data. Numerous examples of application of ANNs in medical applications and engineering systems have been reported with a varying degree of success. A survey of over 300 recent contributions in applying Deep Learning (DL) and Convolutional Neural Networks (CNNs) to predict health risks/disease a set of medical images highlights the scale of ML applications in medical field [15]. A large number of examples of automated disease diagnosis using computational intelligence have been reported over the last decade or so [16–18]. ANNs have also been applied in various non-linear problem-solving scenarios including robotic decision making, swarm intelligence, aviation application and Artificial Intelligence (AI) in games [19]. Identification of diseases associated with the brain activity such as Parkinson's, Schizophrenia, and Huntington's disease from the Contingent Negative Variation (CNV) response in electroencephalograph has also been successfully implemented using ANNs [20]. Multilayer perceptron (MLP) and probabilistic neural network (PNN) have been utilized for the prediction of osteoporosis with bone densitometry [21]. A combination of several ML algorithms has also been successfully implemented to predict accurate amount of medication dosage required for patients suffering from sickle cell disease [22]. Administering accurate amount of medication based on patient's condition is of paramount importance. In this study, a range of combined machine learning techniques were investigated to accurately predict the correct amount of medication required. The algorithms investigated included Random Forests, Support Vector Machines and a few variants of Neural Networks. It was reported that a combination of Multilayer Perceptron Neural Networks trained with Levenberg-Marquardt algorithm and Random Forests provided the best results. This work builds on authors' earlier research where regression was used to predict age 14 BMI using the BMI values from earlier childhood [23].

2.2   Machine Learning Techniques for Tackling Obesity

Only a small number of studies conducted to predict childhood obesity were identified through literature searches. This is possibly due to the complex nature of the problem. Hence, it is important to utilise some complex prediction algorithm using machine learning methods or even ensemble of ensemble predictors to implement a robust and accurate prediction system rather than using simpler techniques such as linear regression or other statistical methods [24]. It is proposed that such a system should be able to:

- at least match human decision maker's behaviour
- be sufficiently general and handle a wide range of variability
- be applicable in practical situations with a proven degree of success

Reference [25] applied machine learning techniques to data collected from children before the age of 2 to predict future obesity. In this study, they used data collected on children prior to the second birthday using a clinical support system. They reported an accuracy of 85%, sensitivity of 89%, positive predictive value of 84% and negative predictive value of 88% using the ID3 algorithm for decision trees without pruning. The other algorithms tested were Naïve Bayes, Random Trees, Random Forests, C4.5 decision trees with pruning and Bayes Net. Several other studies also reported the use of machine learning algorithms used for predicting obesity. Reference [26] suggested that Radial Basis ANNs (RBANNs) are far more efficient than classical Back Propagation ANNs (BPANNs) but very large datasets would be required to train such systems. This study discussed algorithms only; results were not reported. Reference [27] discussed several algorithms for predicting childhood obesity. They recommended the suitability of ANNs, Naïve Bayes and Decision Trees. Several optimisation techniques have also been applied to achieve better prediction accuracy. For example, Genetic Algorithms were employed by [28] to improve the prediction accuracy to 92%. However, it must be noted that they used a very small sample size of 12 subjects. A comprehensive study, possibly the best one identified so far, was carried out by [29] to apply machine learning techniques to predict childhood obesity. They compared the performance metrics of several machine learning prediction algorithms. They compared logistic regression with six data mining techniques: Decision Trees, Association Rules, Neural Networks, Naïve Bayes, Bayesian Networks and Support Vector Machines. They considered prediction sensitivity the most important element in predicting obesity for their study. The highest reported sensitivity for their work was 62% in the case of Naïve Bayes and Bayesian Networks. This research group used a limited range of demographics (gender) and biometrics (weight, height and BMI) and the subjects were 2 year old children. It is envisaged that the prediction accuracy can be further improved by using a different set of parameters, using big data and other machine learning techniques such as deep learning to handle big data. Reference [30] applied machine learning techniques to measure and monitor physical activity in children. They evaluated Multilayer Perceptrons (MLPs), Support Vector Machines, Decision Trees, Naïve Bayes and K = 3 Nearest Neighbour algorithms. It was reported that MLPs outperformed all the other algorithms yielding an overall accuracy of 96%, sensitivity of 95% and specificity of 99%. It should be noted that the sample size in this case was also relatively small (22 participants). The investigation of Deep Learning techniques for future work was proposed.

3   Methods

3.1   Data

The data for this study are used from UK's Millennium Cohort Study (MCS) [31]. The MCS is an ESRC funded research project that followed every child born in year 2000 and 2001. This study is the most recent of Britain's well-known study that followed national longitudinal birth cohorts. The very first MCS sweep surveyed 18,818 babies and 18,552 families related to those children. This wave was conducted in 2001 and 2002 when the babies were 9 months old. Subsequent sweeps were carried out at the age of 3 years, 5 years, 7 years, 11 years and 14 years to examine each child's growth as shown in Table 1:

Table 1. Millennium Cohort Study survey sweeps.

Millennium Cohort Study profiles						
Survey	MCS1	MCS2	MCS3	MCS4	MCS5	MCS6
Age	9 months	3 years	5 year	7 years	11 years	14 years

3.2   Classification Challenge

As discussed in the Introduction section, the BMI in adults is defined as a ratio of body mass in kilograms to the square of individual's height in meters. There are predetermined BMI values setting the boundaries for normal, overweight and obese levels. In children and adolescents, these levels are calculated differently and vary with their age and gender. The data for each survey sweep were collected over the course of approximately 12 months. This means that the children in a given sweep having exact same BMI will have different obesity levels depending upon their age and gender. This makes the classification problem significantly challenging since the target obesity levels at the age of 14 don't have linear relationship with the earlier measurements.

3.3   Data Preprocessing

A vast amount of data about children, their raising families and physical surroundings was collected including weight, height, obesity flags, socio economic conditions and indices of multiple deprivations. For the purpose of this study, body mass index values from 3 years, 5 years, 7 years and 11 years of age were used as input features to the machine learning algorithm and obesity flags for age 14 such as Normal and At risk (Overweight and Obese) were used as target variables as show in Fig. 1. The obesity flags for 14 year old subjects are determined by the following factors:

- Age at the time of survey
- Weight in kilograms and height in meters and
- Gender

These obesity flags are available in the dataset and are determined using growth charts published by the IOTF.

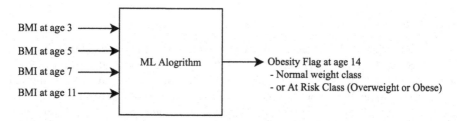

Fig. 1. Machine learning based obesity prediction model

The dataset has a large number of missing values and outliers. There are 11110 non missing values for the obesity flag for age 14 (MCS6) survey. Since the obesity flag for the age 14 survey is being used as a target variable, imputing the missing values for this variable are not going to be useful for this study. A reliable predictor variable is required to implement a robust machine learning algorithm. Therefore, to be able to use reliable data, instances of missing values for the age 14 obesity flag are deleted. As it can be seen from the count field of Table 2, a large number of data values are still missing for the input variables.

Table 2. Missing values

BMI	Non null values	Missing values
BMI2 (Age 3)	9525	1585
BMI3 (Age 5)	10382	728
BMI4 (Age 7)	10147	963
BMI5 (Age 11)	10398	712

The dataset analysis also indicates that the minimum and maximum data values are significantly higher than the first and third quartiles, highlighting the presence of outlier data points as shown in Table 3.

Table 3. Description of uncleaned data

	BMI2	BMI3	BMI4	BMI5
Count	9525	10382	10147	10398
Mean	16.78	16.25	16.51	19.08
Std	2.01	1.83	2.24	3.55
Min	9.1	10.19	10.11	9.48
25%	15.7	15.16	15.09	16.51
50%	16.6	16	16.08	18.3
75%	17.5	17.01	17.38	20.95
Max	63.6	42.8	35.36	61.72

As highlighted earlier, the dataset has a considerably large number of outliers which need to be identified and dealt with in order to produce meaningful results. One way of visualising and estimating outliers is to use boxplots for all of the input variables. By choosing the boxplot whisker at an appropriate level, the data points that fall beyond these points can be identified as outliers. A boxplot diagram for the input variables BMI2, BMI3, BMI4 and BMI5 is shown in Fig. 2.

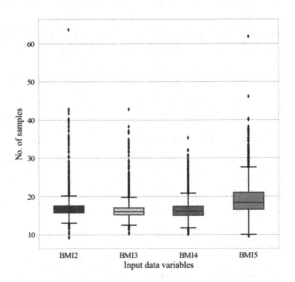

Fig. 2. Boxplot of uncleaned dataset

All data values beyond first and third quartile are identified and imputed using the mean value of each of the input variable. The boxplot for the cleaned data is shown in Fig. 3:

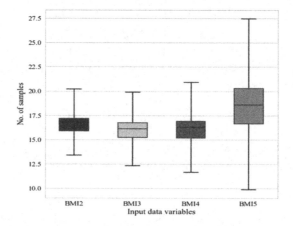

Fig. 3. Boxplot of cleaned dataset

3.4   Machine Learning Algorithm Implementation

The dataset is composed of 11110 instances and has three classes based on the obesity
label. There are 8160 normal cases, 2126 overweight and 824 obese. Majority of the
instances falling under the normal category, 19% under overweight and only just over
7% belonging to the obese category, makes the dataset highly imbalanced. As overweight and obese categories are both at risk, therefore instances belonging to these two
classes have been combined and have been labeled as 'At risk'. Although this reduces
the imbalance a little, but majority of the algorithms are only able to classify the
majority class with a high degree of accuracy.

Table 4. Classification results of imbalanced data.

Metric	Algorithm	Normal	At Risk	Weighted average
Precision	KNN	0.87	0.71	0.83
	J48 pruned tree	0.87	0.78	0.85
	Random forest	0.88	0.72	0.84
	Bagging	0.89	0.73	0.85
	SVM	0.87	0.74	0.84
	MLP	0.85	0.75	0.82
	Voting	0.87	0.74	0.84
Recall	KNN	0.91	0.63	0.83
	J48 pruned tree	**0.94**	**0.62**	**0.85**
	Random forest	0.91	0.66	0.84
	Bagging	0.91	0.68	0.85
	SVM	0.92	0.63	0.84
	MLP	0.93	0.54	0.83
	Voting	0.92	0.63	0.84

A range of popular classification algorithms were employed to classify adolescents of age 14 at a risk of becoming overweight or obese using BMIs measured at ages 3, 5, 7 and 11 years. The results from the best performing algorithms are tabulated in Table 4. Although the overall accuracy is well over 80% for all of the algorithms used, because of the class and BMI imbalance issues, the individual class accuracy for the At risk class is still very poor as indicated by the recall metric shown in Table 4. For example, the weighted average for recall in the case of J48 decision tree algorithm is 85% and the normal class has a significantly higher accuracy of 94%, the individual class accuracy for the 'At risk' class is only 62%.

Apparently, higher sensitivity or recall score for the minority class is of paramount importance in order to detect individuals at the risk of becoming overweight or obese as early as possible.

3.5 Proposed Approach

As discussed above, it is important to classify the minority class with a high degree of accuracy. To deal with the this class imbalance issue, it was proposed to synthesise new class instances for the minority class using SMOTE [32] and train ML algorithms on synthetically increased data. SMOTE generates additional data points using KNN approach by inserting synthetic data points on lines joining K nearest neighbours as shown in Fig. 4.

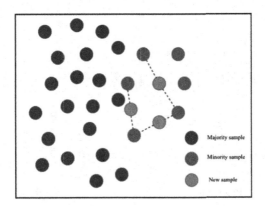

Fig. 4. Generating synthetic data points using SMOTE

A SMOTE oversampling of 138% was applied to the minority class to make the number of cases in the minority class equal to the number of cases in the majority class. However, for robustness, a trained network must be tested on an original, imbalanced dataset. For this reason, 30% sample of the original imbalanced data was kept for testing the ML algorithm accuracy. The same machine learning algorithms used in the case of imbalanced data classification were used to classify synthetically balanced dataset. The precision and recall score metrics are shown in Table 5 and plotted in Fig. 5.

Table 5. Classification results of balanced data.

Metric	Algorithm	Normal	At Risk	Weighted average
Precision	KNN	0.93	0.57	0.83
	J48 pruned tree	0.95	0.57	0.85
	Random forest	0.91	0.63	0.84
	Bagging	0.92	0.62	0.84
	SVM	0.95	0.56	0.85
	MLP	**0.96**	0.51	0.84
	Voting	0.96	0.56	0.85
Recall	KNN	0.78	0.83	0.79
	J48 pruned tree	0.76	0.88	0.79
	Random forest	0.84	0.76	0.82
	Bagging	0.82	0.80	0.82
	SVM	0.74	0.89	0.78
	MLP	0.68	**0.92**	0.74
	Voting	0.74	0.91	0.78

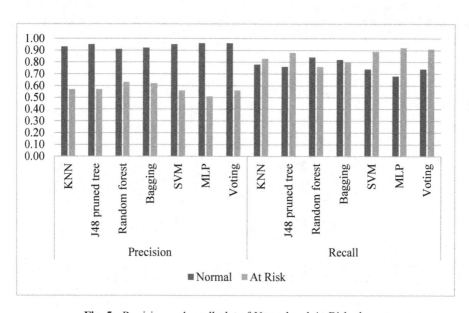

Fig. 5. Precision and recall plot of Normal and At Risk classes

The recall metric shows a significant increase for the 'At risk' class accuracy. Although the normal class accuracy has decreased somewhat, but the increase in the minority class accuracy outweighs this loss since detecting individuals at a risk of becoming overweight or obese is far more important than misclassifying a normal individual.

The recall metric is comparison is plotted in Fig. 6 as a bar graph where the output of each algorithm is plotted before and after oversampling the minority class.

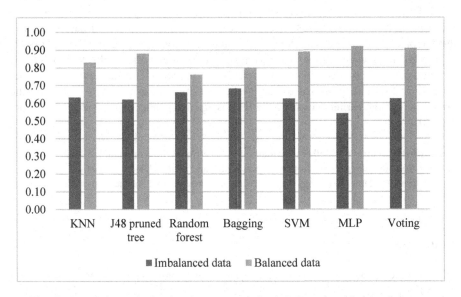

Fig. 6. Comparison of minority class accuracy for imbalanced and balanced datasets

4 Discussion

It is clear from Table 5 and Figs. 5 and 6 that synthetically enhancing the minority class significantly increases the sensitivity or recall metric for the minority class. Some of the algorithms that didn't perform satisfactorily in the case of imbalanced data, their performance is also improved significantly when the dataset is synthetically balanced. For example, the Multi-Layer Perceptron (MLP) resulted in a minority class accuracy of 54% when the dataset was imbalanced but jumped to 92% in the case of balanced data. It should be noted that the performance of all the algorithms was tested on a sample of unaltered, imbalanced dataset. This dataset had 2448 in the normal class and 885 in the At Risk class. The precision value for the normal class for the MLP algorithm is 96%. This means, out of all those predicted as 'Normal', only 4% were predicted as 'At Risk'. A recall value of 92% means that out of all those labeled as 'At Risk', only 8% were predicted as 'Normal'. This results in an F1 score of 93.96%.

5 Conclusion and Further Work

Obesity in general but more importantly childhood obesity is of great global concern since a majority of obese children grow up to obese adults. Many studies have employed statistical methods to predict the probabilities of children growing into adults as obese. In this paper we evaluated several popular machine learning algorithms to accurately predict adolescents at risk of becoming overweight or obese at the teenage stage. Issues relating to low prediction accuracy because of data imbalance have been considered and dealt with using Synthetic Minority Oversampling Technique (SMOTE). An ensemble of classifiers capitalises on the prediction accuracies of individual classifier algorithms. For robustness, a sample of unaltered data is used to test the prediction accuracies. The results obtained are encouraging and the authors are continuing to develop further on this research. Future work is planned to include longitudinal and cross sectional data relating to participants to achieve even higher accuracies. It is also proposed to use predict obesity state at the age of 14 using the obesity flags from earlier ages since the BMI is age and gender dependent for ages from 2 to 20.

Acknowledgements. The authors are grateful to 'The Centre for Longitudinal Studies, Institute of Education' for the use of data used in this study and to the 'UK Data Archive Service' for making them available. However, they are in no way responsible for the analysis and interpretation of these data.

References

1. Dobbs, R., et al.: Overcoming obesity : an initial economic analysis, McKinsey Global Institute (2014)
2. Fat, L.N.: Children's body mass index, overweight and obesity. Heal. Surv. Engl. (2014). Chapter 10
3. Olds, T., et al.: Evidence that the prevalence of childhood overweight is plateauing: data from nine countries. Int. J. Pediatr. Obes. 6(5–6), 342–360 (2011)
4. Rokholm, B., Baker, J.L., Sørensen, T.I.A.: The levelling off of the obesity epidemic since the year 1999 - a review of evidence and perspectives. Obes. Rev. 11, 835–846 (2010)
5. Blüher, S., et al.: Age-specific stabilization in obesity prevalence in German children: a cross-sectional study from 1999 to 2008. Int. J. Pediatr. Obes. 6(sup3), e199–e206 (2011)
6. Moss, A., Klenk, J., Simon, K., Thaiss, H., Reinehr, T., Wabitsch, M.: Declining prevalence rates for overweight and obesity in German children starting school. Eur. J. Pediatr. 171, 289–299 (2011). https://doi.org/10.1007/s00431-011-1531-5
7. Wabitsch, M., Moss, A., Kromeyer-Hauschild, K.: Unexpected plateauing of childhood obesity rates in developed countries. BMC Med. 12, 17 (2014)
8. Moody, A.: Adult anthropometric measures, overweight and obesity. In: Health and Social Care Information Centre (2013)
9. Singh, A.S., Mulder, C., Twisk, J.W.R., Van Mechelen, W., Chinapaw, M.J.M.: Tracking of childhood overweight into adulthood: a systematic review of the literature. Obes. Rev. 9, 474–488 (2008)
10. Dietz, W.H.: Health consequences of obesity in youth: childhood predictors of adult disease. Pediatrics 101, 518–525 (1998)
11. Engeland, A., Bjørge, T., Søgaard, A.J., Tverdal, A.: Body mass index in adolescence in relation to total mortality: 32-year follow-up of 227,000 Norwegian boys and girls. Am. J. Epidemiol. 157, 517–523 (2003)
12. Butland, B., et al.: Tackling Obesities: Future Choices – Project Report, 2nd edn. (2007)

13. Freedman, D.S., Mei, Z., Srinivasan, S.R., Berenson, G.S., Dietz, W.H.: Cardiovascular risk factors and excess adiposity among overweight children and adolescents: the Bogalusa heart study. J. Pediatr. **150**, 12–17 (2007)

14. Cole, T.J., Bellizzi, M.C., Flegal, K.M., Dietz, W.H.: Establishing a standard definition for child overweight and obesity worldwide. BMJ **320**, 1–6 (2000)

15. Litjens, G., et al.: A survey on deep learning in medical image analysis. Med. Image Anal. **42**, 60–88 (2017)

16. Szolovits, P., Patil, R.S., Schwartz, W.B.: Artificial intelligence in medical diagnosis. Ann. Intern. Med. **108**(1), 80–87 (1988)

17. Ishak, W.H.W., Siraj, F.: Artificial intelligence in medical application: an exploration. Health Inform. Eur. J. **16**, 1–9 (2008)

18. Jarvis-Selinger, S., Bates, J., Araki, Y., Lear, S.A.: Internet-based support for cardiovascular disease management. Int. J. Telemed. Appl. **2011**, 9 (2011)

19. Kumar, K., Thakur, G.S.M.: Advanced applications of neural networks and artificial intelligence: a review. Int. J. Inf. Technol. Comput. Sci. **4**, 57 (2012)

20. Jervis, B.W., et al.: Artificial neural network and spectrum analysis methods for detecting brain diseases from the CNV response in the electroencephalogram. IEE Proc. - Sci. Meas. Technol. **141**, 432–440 (1994)

21. Mantzaris, D.H., Anastassopoulos, G.C., Lymberopoulos, D.K.: Medical disease prediction using artificial neural networks. In: 8th IEEE International Conference on BioInformatics and BioEngineering, BIBE 2008 (2008)

22. Khalaf, M., et al.: Machine learning approaches to the application of disease modifying therapy for sickle cell using classification models. Neurocomputing **228**, 154–164 (2017)

23. Singh, B., Tawfik, H.: A machine learning approach for predicting weight gain risks in young adults. In: Conference Proceedings of 2019 10th International Conference on Dependable Systems, Services and Technologies, DESSERT 2019 (2019)

24. Michie, D., Spiegelhalter, D.J., Taylor, C.C., Campbell, J. (eds.): Machine Learning, Neural and Statistical Classification. Ellis Horwood, Upper Saddle River (1994)

25. Dugan, T.M., Mukhopadhyay, S., Carroll, A., Downs, S.: Machine learning techniques for prediction of early childhood obesity. Appl. Clin. Inform. **6**, 506–520 (2015)

26. Novak, B., Bigec, M.: Application of artificial neural networks for childhood obesity prediction. In: Proceedings - 1995 2nd New Zealand International Two-Stream Conference on Artificial Neural Networks and Expert Systems, ANNES 1995 (1995)

27. Adnan, M.H.B.M., Husain, W., Damanhoori, F.: A survey on utilization of data mining for childhood obesity prediction. In: 8th Asia-Pacific Symposium Information Telecommunication Technologies (2010)

28. Hariz, M., Adnan, B.M., Husain, W., Aini, N., Rashid, A.: Parameter identification and selection for childhood obesity prediction using data mining. In: 2nd International Conference on Management and Artificial Intelligence (2012)

29. Zhang, S., Tjortjis, C., Zeng, X., Qiao, H., Buchan, I., Keane, J.: Comparing data mining methods with logistic regression in childhood obesity prediction. Inf. Syst. Front. **11**, 449–460 (2009)

30. Fergus, P., et al.: A machine learning approach to measure and monitor physical activity in children. Neurocomputing **228**, 220–230 (2017)

31. Smith, K., Joshi, H.: The Millennium Cohort Study. Popul. Trends **107**, 30–34 (2002)

32. Chawla, N.V., Bowyer, K.W., Hall, L.O., Kegelmeyer, W.P.: SMOTE: synthetic minority over-sampling technique. J. Artif. Intell. Res. **16**, 321–357 (2002)

Gait Abnormality Detection in People with Cerebral Palsy Using an Uncertainty-Based State-Space Model

Saikat Chakraborty[(✉)], Noble Thomas, and Anup Nandy

Machine Intelligence and Bio-motion Research Lab, Department of Computer Science and Engineering, National Institute of Technology, Rourkela, Odisha, India
saikat.sc@gmail.com

Abstract. Assessment and quantification of feature uncertainty in modeling gait pattern is crucial in clinical decision making. Automatic diagnostic systems for Cerebral Palsy gait often ignored the uncertainty factor while recognizing the gait pattern. In addition, they also suffer from limited clinical interpretability. This study establishes a low-cost data acquisition set up and proposes a state-space model where the temporal evolution of gait pattern was recognized by analyzing the feature uncertainty using Dempster-Shafer theory of evidence. An attempt was also made to quantify the degree of abnormality by proposing gait deviation indexes. Results indicate that our proposed model outperformed state-of-the-art with an overall 87.5% of detection accuracy (sensitivity 80.00%, and specificity 100%). In a gait cycle of a Cerebral Palsy patient, first double limb support and left single limb support were observed to be affected mainly. Incorporation of feature uncertainty in quantifying the degree of abnormality is demonstrated to be promising. Larger value of feature uncertainty was observed for the patients having higher degree of abnormality. Sub-phase wise assessment of gait pattern improves the interpretability of the results which is crucial in clinical decision making.

Keywords: Dempster-Shafer theory · Uncertainty · Kinect · Cerebral Palsy

1 Introduction

Cerebral Palsy (CP) is a neurological disorder attributed to non-progressive damage of fetal or infant brain, causing limited movement and postural instability [1]. Around the world, more than 4 per 1000 children suffer from CP [2]. In developing countries the prevalence of CP is alarming. For example, in India, more than 15–20% of physically disabled children are suffering from CP [3]. Proper therapeutic intervention can improve the quality of gait in children with CP [4]. Hence, the diagnosis of gait pattern is crucial in investigating the efficacy of an intervention [5].

Approaches for machine learning (ML)-based automated gait diagnosis can be broadly divided into two groups: feature-based classification technique and cycle segmentation-based state-space modeling technique. Feature-based techniques require extraction of a suitable set of features which is highly depended

© Springer Nature Switzerland AG 2020
V. V. Krzhizhanovskaya et al. (Eds.): ICCS 2020, LNCS 12140, pp. 536–549, 2020.
https://doi.org/10.1007/978-3-030-50423-6_40

on expert knowledge or network architecture type (in deep learning). Despite providing satisfactory performances this strategy of diagnosis suffers from low clinical interpretability of the outcome [6]. In clinical diagnosis, the interpretability of the results carries more importance than just reporting classification accuracy [7]. On the other hand, state-space models recognize the gait pattern by temporal segmentation of a gait cycle. It provides sub-phase wise comprehensive gait analysis with relevant clinical interpretation [8]. Characterization of gait signal in each sub-phase unfolds the spatio-temporal evolution of the system dynamics over a period of time.

In coordinative movement, the uncertainty in prediction of the initial state of system dynamics impacts the measurement of gait variables (i.e. features) [9]. Decision relating to the assignment of a class level (i.e. normal or abnormal) using such a feature set associates a level of uncertainty [10]. Quantification of this uncertainty is crucial in clinical decision making [10]. In literature, different studies have demonstrated automated gait diagnosis system for people with CP [5,11–13]. However, the analysis of feature uncertainty in decision making was often ignored. Dempster-Shafer theory of evidence (DST) [14,15] has been successfully used to map the uncertainty in the classification process [16–18] in different problem domains. It assign a mass probability to particular class considering the uncertainty factor associated with the evidence. Hence, the use of DST in classifying CP gait seems to be promising and clinically viable. Again, an investigation of the impact of each of gait sub-phases on the overall abnormal gait pattern of CP patients seems warranted. Another fact is that, the existing automated CP gait diagnostic systems associates high-cost sensors which make the overall system expensive. Most of the clinics, specially in developing countries, are not able to afford those costly systems. Thus, a low-cost automated diagnosis system is also needed for people with CP.

This study aimed to construct a novel state-space based automated gait diagnosis system for children and adolescent with CP (CAwCP) by quantifying uncertainty of the selected feature set. First, a low-cost sensor-based architecture was proposed from which velocity of ankle joints were extracted. Second, a state-space model was build where state duration and transition was modeled using DST. Variability of state duration was used to recognize the gait pattern. The use of DST allows the presence of uncertainty in gait velocity. Quantification and incorporation of this uncertainty facilitated subject specific gait modeling. Third, an index for quantification of the degree of abnormality was also proposed. Finally, state-of-the-art feature-based gait classification approaches were compared with the proposed system. The contributions of this work are as follows:

- Establishing a low-cost multi-sensor-based architecture for data acquisition.
- Developing a state-space model for automated gait diagnosis for CAwCP that incorporate uncertainty of features to estimate the temporal evolution of signal.
- Proposing degree of abnormality indexes based on dynamic stability and feature uncertainty.

The rest of the paper is organized as follows: Sect. 2 describes elaborately the state-of-the-art methods to detect CP gait abnormality. Construction of a low-cost architecture for data acquisition, description of the proposed state-space model and degree of abnormality indexes are demonstrated in Sect. 3. Section 4 presents the results with an elaborate discussion. The paper concludes in Sect. 5 by providing a future research direction.

2   Related Work

Several attempts have been made to construct automated gait diagnostic system for CP patients.

Kamruzzaman et al. [5] proposed an automated gait diagnosis system for children with CP (CwCP) using support vector machine (SVM). Normalized stride length and cadence were used as an input features for the classifier. They reported SVM as comparatively better classifier with an overall abnormality detection accuracy of 96.80%. Wolf et al. [13] accumulated gait data from different sensors and constructed a large feature pool by gait sub-phases analysis. Features were ranked using mutual information and then used for fuzzy rule-based classification. A gait deviation index was also introduced in their study. Zhang et al. [11] investigated the significance of Bayesian classifier to diagnosis CP patients. They computed normalized stride length and cadence and reported Bayesian classifier as better than other popular classifiers. Gestel et al. [19] used Bayesian networks (BN) in combination with expert knowledge to form an semi-automated diagnostic system. They reported a promising detection accuracy (88.4%) using sagittal plane gait dynamics. Laet et al. [12] obtained joint motion pattern from Delphi-consensus study [20] and used it to detect CP gait using Logistic Regression and naïve Bayes classifiers. They recommended the inclusion of expert knowledge in feature selection and discretization of continuous features to detect gait abnormality. Zhang et al. [21] extracted a set of crucial kinematic parameters from sagittal plane gait pattern of CwCP and given it as input to seven popular supervised learning algorithms. They reported artificial neural network (ANN) as the best model followed by SVM, decision tree (DT), and random forest (RF). Krzak et al. [22] collected kinematic gait parameters of people with CP using Milwaukee Foot Model (MFM). They applied k-means clustering and obtained five distinct groups having similar gait pattern. Dobson et al. [23] and recently Papageorgiou et al. [24] have thoroughly analyzed the automated classification systems available for CP patients and reported cluster-based algorithms as the most used technique for gait detection in CP patients. But, cluster-based models may construct clinically irrelevant artificial group and suffer from limited clinical interpretations [21]. Though the above mentioned models have obtained satisfactory results, they have ignored the uncertainty factor associated with the feature set which is crucial in deciding the class label. Again, those systems are highly expensive also.

Table 1. Demographic information of the subjects*

Participants (N = 30)	CP type	GMFCS levels	Age (years)	Height (cm)	Gender
CAwCP (15)	Diplegic:5	I, II	11.71 ± 4.46	124.00 ± 16.92	M:3, F:2
	Hemiplegic (RS):5	II	10.84 ± 3.6	114.66 ± 15.28	M:2, F:3
	Hemiplegic (LS):3	I, II	12.00 ± 6.38	132.25 ± 20.17	M:2, F:1
	Athetoid:2	I, II	15.66 ± 2.52	149.67 ± 5.51	M:1, F:1
			12.55 ± 2.13	130.15 ± 14.87	M:8, F:7
TDCA (15)			12.45 ± 3.51	132.06 ± 14.09	M:9, F:6
P-value			0.78	0.27	0.71

*GMFCS: Gross Motor Function Classification System, RS: Right side, LS: Left side, M: male, F: female

3 Methods

3.1 Participants

Fifteen CAwCP patients, without having any other disease or surgical history that can affect the gait pattern, and who can walk without any aid, were recruited from the Indian Institute of Cerebral Palsy (IICP), Kolkata. Along with that, fifteen typically developed children and adolescent (TDCA) without having any type of musculoskeletal or neurological disease that can affect gait pattern, also were recruited. This work was approved by the ethics board of local institution. The objectives and protocols of our experiment were explained to the participants and an informed consent form was signed from each of them. Table 1 shows the demographic information of the subjects which demonstrates that there was no significant difference ($p > 0.05$) between the two groups in terms of anthropometric parameters.

3.2 Experimental Setup and Data Acquisition

We have used low-cost Kinect (v2) sensor for data acquisition which make our proposed system cost-effective. Inspired by the work of Geerse et al. [25] we have constructed a three Kinect-based client-server architecture (see Fig. 1) which covers 10 m walking distance. The Kinects were placed on tripods (Slik F153) sequentially at 35° angle with the walking direction. The arrows emerging from Kinects cover the horizontal field of view (FoV) of Kinect i.e. 70° [26]. The distance of the Kinects from the left border of the track was 2 m. Height and tilt angle of the Kinects were 0.8 m and 0° respectively, while the distance between the Kinects was 3.5 m. The width of the track was 0.84 m. This setup allowed ≈0.5 m of overlapped tracking volumes (between two successive sensors) which was empirically estimated to be sufficient (for both the groups) for the next sensor to recognize a person's body and start tracking. A computer was set as the server which controlled each of the Kinect connected to separate computers. System clocks of the computers were synchronized using Greyware's DomainTime II (Greyware Automation Products, Inc.), which follows PTP protocol. A training session was provided to the participants before the experiment. Subjects were asked to start walk at self selected speed from $4\,\mathrm{m}$ distance from the 1st Kinect and after walking $1\,\mathrm{m}$ distance (marked by line) data were started to capture. Subjects were asked to walk upto the 'End' line. The total distance of the path was $12\,\mathrm{m}$ out of which $10\,\mathrm{m}$ distance was considered for data collection. The extra distances (at start and ending points) were given to reduce the effect of acceleration and deceleration on gait variables. Five trials for each participants were taken with $2\,\mathrm{min}$ of resting gap. The client-server architecture followed the same protocol like [25, 26] to record, combine, remove noise, and process body point data series from multiple Kinects.

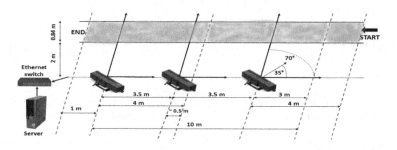

Fig. 1. Data acquisition in multi-Kinect environment

3.3 Dempster-Shafer Classifier

Our proposed state-space model was constructed by estimating uncertainty of feature using DST. This section describes the theoretical ground of DST for using it as a classifier, often termed as *Cardiff Classifier* [17].

The Cardiff classifier is constructed on the basis of some elementary and mutually exclusive hypotheses which constitute the frame of discernment (FOD) [16]. If there are two elementary hypotheses, for e.g. h: belongs to class i, and $\neg h$: not belongs to class i, then FOD is called binary frame of discernment (BFOD), denoted as Θ. There are 4 number of possible hypotheses derived from the elementary hypotheses: $\{\{\varnothing\}, \{h\}, \{\neg h\}, \{h, \neg h\}\}$. On the basis of existing evidence, each hypothesis is assigned a probability mass value $m(.)$ which is called degree of belief (DOB) [16]. It refers the strength of support for a classification. Each of the hypotheses $\{\theta_i\}$, $(i = 1, ..., 2^{|\Theta|})$ having $m(\theta_i) > 0$ constitute body of evidence (BOE) [10, 16]. The basic property of $m(.)$ is given by [27]:

$$0 \leq m(.) \leq 1, m(\varnothing) = 0, \sum_{i=1}^{2^{|\Theta|}} m(\theta_i) = 1 \tag{1}$$

In DST, the probability masses are assigned to each subset of Θ. This property allows uncertainty in classification [16].

At the first step during the construction of the classifier, each source of evidence i.e. input variable (v) is mapped to a confidence factor $cf(v)$ (0–1 scale) using some predefined function derived from the pattern of data distribution. $cf(v)$ is then transformed to BOE as [27]:

$$m(\{h\}) = \frac{B}{(1-A)}cf(v) - \frac{AB}{(1-A)},$$

$$m(\{\neg h\}) = \frac{-B}{(1-A)}cf(v) + B, \tag{2}$$

$$m(\{h, \neg h\}) = 1 - m(\{h\}) - m(\{\neg h\})$$

In Eq. 2 the value $m(\{h, \neg h\})$ refers the DOB of either belong to a class or not belong to a class. Hence, this value quantify the uncertainty associated with a particular feature (i.e. input variable or evidence). The control variables A and B define the dependence of $m(\{h\})$ on $cf(v)$ and the maximal support for $m(\{h\})$ or $m(\{\neg h\})$. In case of more than one source of evidence for a hypothesis, Dempster's rule of combination is used to combine the individual BOEs to get the final BOE (FBOE) (see [16] for more details). The highest FBOE determine the class label of a sample.

3.4   State-Space Model Construction

Human gait cycle can be broadly classified into four dominant sub-phases: first double limb support (FDLS), single limb support for the left limb (LSL), second double limb support (SDLS), and single limb support for the right limb (RSL) (see Fig. 2). Based on this, a hypothetical state-space model was proposed (Fig. 3) which quantify the temporal evolution of gait signal.

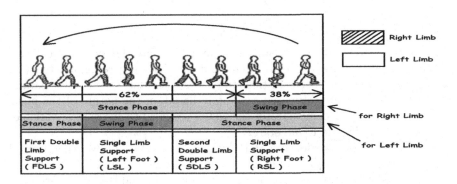

Fig. 2. Sub-phases of a normal gait cycle

For CAwCP, gait velocity was reported as one of the most discriminative feature [3]. Hence, velocity of both ankle joints (anterior-posterior (A-P) and vertical (V) direction) were selected as the sources of evidence. DST was used

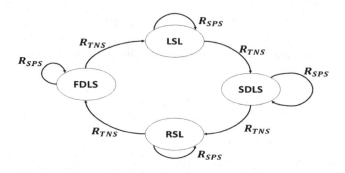

Fig. 3. Hypothetical state-space model

as a binary classifier for assigning a frame to a particular state out of two subsequent states. Twenty-one subjects (11 TDCA, 10 CAwCP) were selected for in-sample group (i.e. training set), and remaining (4 TDCA, 5 CAwCP) were selected for out-of-sample (i.e. test set) group. State transition rules was constructed on the basis of elementary hypotheses. The BFOD was defined as: {present state (PS) i.e. hypothesis h, next state (NS) i.e. hypothesis $\neg h$}. Thus the hypothesis space constitutes: $\{\{\varnothing\}, \{PS\}, \{NS\}, \{PS, NS\}\}$. We assumed that the temporal sequence of the states (see Fig. 3) should not deviate for any subject (both CAwCP and TDCA). Gait cycle was segmented into 4 states (i.e. FDLS, LSL, SDLS, RSL) using the gait event detection algorithm described by Zeni et al. [28]. In order to convert the input variables into confidence factor, we found a suitable probability distribution that fits the best to our input data. Kolmogorov-Smirnov test was performed to investigate the distribution pattern of the input data. Fréchet distribution (see Eq. 3) was found to fit the best for our input data:

$$cf(v) = e^{-\left(\dfrac{v - m}{s}\right)^{-\alpha}} \qquad if \quad v > m \tag{3}$$

In Eq. 3, $\alpha > 0$ is the shape parameter, m and $s > 0$ are the location and scale parameters respectively. Hence, for each input variable 5 control parameters (i.e. α, m, s, A, B) were considered in this study. For each state, the control parameters, α, m and s, were estimated separately based on the nature of the input variables [10]. Table 2 demonstrates the values assigned to the control variables (i.e. α, m and s) for the FDLS state.

Following the work of Jones et al. [10], the values of A and B were determined using expert knowledge. For individual input variable, the limit of uncertainty was estimated which was used to compute the value of A and B. If for a particular variable v, the upper and lower limit of uncertainty are ϕ_U and ϕ_L respectively, then A and B can be written as [10]:

$$A = \frac{(\phi_U - \phi_L)}{(1 + \phi_U - 2\phi_L)}$$
$$B = 1 - \phi_L \tag{4}$$

Table 2. Values of the control variables α, m and s for FDLS state

Input variable	α	m	s
Left ankle (A-P)	5.60	−0.98	1.72
Left ankle (V)	6.04	0.90	1.02
Right ankle (A-P)	7.05	−1.70	2.12
Right ankle (V)	36.91	−8.45	8.72

Inspired by the work of Safranek et al. [27] and Jones et al. [10] which considered the input variables as low-level measurements (i.e. assessment which associate substantial level of uncertainty), we assigned 0.98 and 0.2 to ϕ_U and ϕ_L respectively. Hence, in our experiment the value of A and B were 0.49 and 0.80 respectively. Assignment of values to the control variables was performed using in-sample data. BOE of each input variable was then combined to get the FBOE which quantified the support for a frame to assign to a particular state.

It was assumed that a gait cycle can start from any state (out of 4 states). Hence, for the initial time frame of a gait cycle, the FBOE corresponding to all four states were computed. The state having the highest FBOE value was selected as the initial state of the gait cycle. For the subsequent phases the transition of a frame was govern by the following rules:

- R_{SPS}: if FBOE($\{PS\}$) > FBOE($\{NS\}$) + FBOE($\{PS, NS\}$), OR if FBOE ($\{PS\}$) > FBOE($\{NS\}$) but FBOE($\{PS\}$) < FBOE($\{NS\}$) + FBOE ($\{PS, NS\}$), then the corresponding frame will stay at present state;
- R_{TNS}: if FBOE($\{NS\}$) > FBOE($\{PS\}$) but FBOE($\{NS\}$) < FBOE($\{PS\}$) + FBOE($\{PS, NS\}$), then the corresponding frame will transit to the immediate next state.

3.5 Abnormality Detection

Dynamic stability, generally quantified by the value of coefficient of variation (CoV) of a feature [29], was reported to be comparatively low for CAwCP population [3]. In order to detect gait abnormality, we computed mean CoV (ω_m) of the time duration for each state for the in-sample group (both CAwCP (ω_m^{ab}) and TDCA (ω_m^n)). The out-sample subjects, used the values of the control variables learned from the in-sample group to compute the corresponding FBOEs. Then the CoV (ω_t) vector for a test subject was estimated. Class label (CL) for an out-sample subject was determined based on $L2$ norm distance as follows:

$$
\begin{aligned}
\omega_m^n &= \{\omega_{mFDLS}^n, \omega_{mLSL}^n, \omega_{mSDLS}^n, \omega_{mRSL}^n\} \\
\omega_m^{ab} &= \{\omega_{mFDLS}^{ab}, \omega_{mLSL}^{ab}, \omega_{mSDLS}^{ab}, \omega_{mRSL}^{ab}\} \\
\omega_t &= \{\omega_{FDLS}^t, \omega_{LSL}^t, \omega_{SDLS}^t, \omega_{RSL}^t\} \\
CL &= min(L2(\omega_t, \omega_m^{ab}), L2(\omega_t, \omega_m^n))
\end{aligned}
\tag{5}
$$

Performance of the proposed system was assessed using some classical metrics i.e. accuracy, sensitivity, and specificity. An attempt was also made to quantify the degree of gait abnormality for individual subject using the proposed model. In CP patients, GMFCS level indicates the degree of abnormality [30]. We hypothesized that higher value of CoV of the time duration for each state indicates higher degree of abnormality. Hence, the first abnormality index ($A1$) was computed by taking average of CoV values throughout the 4 states:

$$A1 = \frac{\omega_{FDLS} + \omega_{LSL} + \omega_{SDLS} + \omega_{RSL}}{4} \tag{6}$$

Again, we assumed that more abnormal CP should have higher uncertainty in their gait pattern. On the basis of that we proposed the second abnormality index ($A2$):

$$A2 = \frac{\sum_{i=1}^{N} FBOE(PS, NS)_i}{N} \tag{7}$$

In Eq. 7, N refers total number of frames.

4   Results and Discussion

In this study a state-space based automated gait diagnosis system for CAwCP patients was proposed which perform sub-phase wise gait assessment by quantifying feature uncertainty in temporal evolution of gait cycle. Person specific degree of abnormality was also analyzed.

Our proposed model charecterize gait cycle with an average ± 3 frame difference from the corresponding ground truth (for both TDCA and CAwCP). For comparison, we have implemented ML models (i.e. ANN, SVM, and Bayesian classifier) which were used for CP gait abnormality detection in state-of-the-art. The same feature set, used for our state-space construction, was given as input to those models. ANN was implemented with 3-layer architecture (i/p layer nodes: 4, o/p layer nodes: 1), while for SVM, three kernel functions (i.e. Radial basis function (RBF), polynomial, and linear) were tested. Hyperparameters (ANN: learning rate and hidden nodes; SVM (RBF, polynomial, linear): regularization parameters; SVM (RBF): gamma; SVM (polynomial): degree) was tuned using grid search. In Bayesian classifier, Gaussian mixture model (GMM) was used to approximate the class conditional distribution which was trained using expectation-maximization (EM) algorithm [31]. Leave-one-out cross validation was used to reduce the generalization error. Following the work of Beynon et al. [16] and Jones et al. [10], the belief values obtained from the Cardiff classifier were used for classification also. In that case the FOD was: {normal, abnormal} and FBOE values were used for class labeling.

Our proposed model outperformed state-of-the-art with an overall 87.5% detection accuracy (see Table 3). Figure 4 demonstrates the normalized confusion matrix for our proposed model and Cardiff classifier. It can be seen that the false positive value is substantially high for the Cardiff classifier which caused to degrade its performance compared to other classifiers. Deviation of gait pattern of a CAwCP patient from TDCA was not uniform for all phases of a gait cycle. The input feature for the Cardiff classifier characterized an entire gait cycle from where the control parameters were estimated. Hence, the FBOEs, computed using those control parameters, failed to classify correctly some subjects. This might be a cause of lower performance of this classifier. Our state-space model avoids this problem by analyzing the gait sub-phase wise. Specificity is the highest for our model (see Table 3), but sensitivity is lower than SVM (RBF) and SVM (linear). This might be due to the varied CP gait pattern, where our model incorrectly performed some state transitions for some of the subjects.

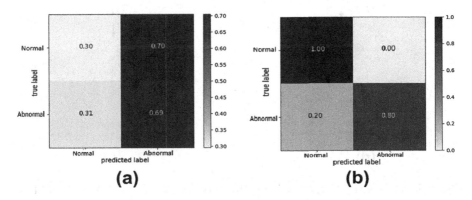

Fig. 4. Normalized confusion matrix. (a) Cardiff classifier, (b) Proposed model

Table 3. Comparative analysis of the proposed model with state-of-the-art

Models	State-of-the-art	Accuracy (%)	Sensitivity (%)	Specificity (%)
SVM (RBF)	Kamruzzaman et al. [5]	86.27	92.06	70.67
SVM (Linear)	Zhang et al. [21]	67.12	94.31	40.41
SVM (Polynomial)	Kamruzzaman et al. [5]	61.11	43.26	75.48
ANN	Zhang et al. [21]	67.23	59.43	78.23
Bayesian classifier	Zhang et al. [11]	80.27	72.71	89.32
Cardiff classifier		49.50	49.60	49.20
Proposed		**87.50**	80.00	**100**

In terms of clinical significance, our model also outperformed others by providing a clear interpretation of the results sub-phase wise, whereas, the computation for the other ML models (i.e. ANN,SVM) is basically a 'black-box'. The simplex plots (Fig. 5(a), 5(b), 5(c), and 5(d)) demonstrate the allocation of data frames in subsequent states during a gait cycle. It can be observed that the transition frame (marked by 'red dot') constitutes high uncertainty which causes it to move to the immediate next state. Uncertainty value then again decreases for which the subsequent frames maps to the next state (which become the current state now). It was noticed that, on an average, the transition frame uncertainty for CAwCP patients is higher than TDCA,

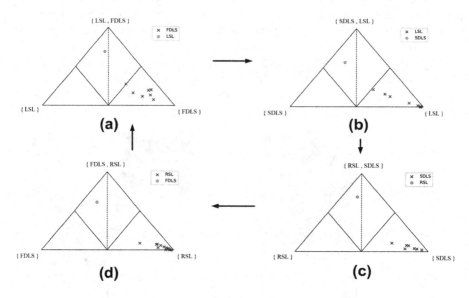

Fig. 5. Simplex plots for a gait cycle of Subj.1 (TDCA). (a) FDLS to LSL, (b) LSL to SDLS, (c) SDLS to RSL, (d) RSL to FDLS

Figure 6 demonstrates sub-phase wise gait distortion of CAwCP compared to TDCA. It shows that CoV of CAwCP differs in FDLS and LSL phases significantly ($\approx 37.5\%$ and $\approx 50\%$) from the corresponding values of TDCA. This information implies that the dynamic stability of CAwCP mainly reduces in this two phases, hence, clinicians should take special attention for these two states during an intervention.

Results in Table 4 shows that both $A1$ and $A2$ and consequently the total score are higher for GMFCS II patients than GMFCS I. This implies that dynamic stability decreases in CP patients having higher degree of abnormality.

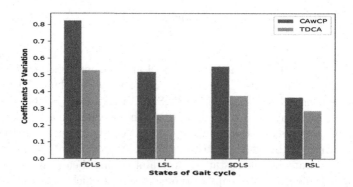

Fig. 6. Sub-phase wise gait assessment

It also demonstrates that CP patients of higher degree associate higher uncertainty value in their gait pattern which may be a consequence to compensate the effect of lower muscle control [3]. The proposed abnormality indexes exhibit promising aspect to quantify the degree of abnormality.

Table 4. Degree of gait abnormality assessment

GMFCS level of CP	Average A1	Average A2	Total score
I	0.5807	0.1394	0.7201
II	0.7997	0.3772	1.1769

5   Conclusion

Automatic gait diagnosis with clinical interpretation is crucial for CAwCP patients. Feature uncertainty takes an important part in abnormal gait detection. The prime contribution of this study is to propose a state-space model which recognize temporal evolution of gait pattern by quantifying feature uncertainty. Data were captured using a low-cost multi-Kinect setup which provided the standard 10 m walking path. An attempt was also made to quantify the degree of abnormality by proposing abnormality indexes. Results shows that the performance of our model is comparable with state-of-the-art. As a future research aspect, some data-driven constraints may be imposed on the state transition rules to further reduce the frame offset (frame difference from the ground truth) of our model. Investigating the performance of the proposed system in detecting CAwCP gait in more challenged environment (i.e. duel task walking, walking on uneven ground) seems warrented. The proposed model performs sub-phase wise gait assessment and provides succinct interpretability of the results which will help the clinicians in decision making during rehabilitation treatment.

Acknowledgment. We would like to be extremely thankful to Science and Engineering Research Board (SERB), DST, Govt. of India (FILE NO: ECR/2017/000408) to partially support this research work. We also like to thank IICP, Kolkata for providing Cerebral Palsy patients and necessary arrangements to collect data.

References

1. Richards, C.L., Malouin, F.: Cerebral palsy: definition, assessment and rehabilitation. In: Handbook of Clinical Neurology, vol. 111, pp. 183–195. Elsevier (2013)
2. Stavsky, M., Mor, O., Mastrolia, S.A., Greenbaum, S., Than, N.G., Erez, O.: Cerebral palsy–trends in epidemiology and recent development in prenatal mechanisms of disease, treatment, and prevention. Front. Pediatr. **5**, 21 (2017)
3. Chakraborty, S., Nandy, A., Kesar, T.M.: Gait deficits and dynamic stability in children and adolescents with cerebral palsy: a systematic review and meta-analysis. Clin. Biomech. **71**, 11–23 (2020)

4. Smania, N., et al.: Improved gait after repetitive locomotor training in children with cerebral palsy. Am. J. Phys. Med. Rehabil. **90**(2), 137–149 (2011)

5. Kamruzzaman, J., Begg, R.K.: Support vector machines and other pattern recognition approaches to the diagnosis of cerebral palsy gait. IEEE Trans. Biomed. Eng. **53**(12), 2479–2490 (2006)

6. Gilpin, L.H., Bau, D., Yuan, B.Z., Bajwa, A., Specter, M., Kagal, L.: Explaining explanations: an overview of interpretability of machine learning. In: 2018 IEEE 5th International Conference on Data Science and Advanced Analytics (DSAA), pp. 80–89. IEEE (2018)

7. Breiman, L.: Statistical modeling: the two cultures (with comments and a rejoinder by the author). Stat. Sci. **16**(3), 199–231 (2001)

8. Ma, H., Liao, W.H.: Human gait modeling and analysis using a semi-Markov process with ground reaction forces. IEEE Trans. Neural Syst. Rehabil. Eng. **25**(6), 597–607 (2017)

9. van Emmerik, R.E.A., Ducharme, S.W., Amado, A.C., Hamill, J.: Comparing dynamical systems concepts and techniques for biomechanical analysis. J. Sport Health Sci. **5**(1), 3–13 (2016)

10. Jones, L., Beynon, M.J., Holt, C.A., Roy, S.: An application of the Dempster-Shafer theory of evidence to the classification of knee function and detection of improvement due to total knee replacement surgery. J. Biomech. **39**(13), 2512–2520 (2006)

11. Zhang, B., Zhang, Y., Begg, R.K.: Gait classification in children with cerebral palsy by Bayesian approach. Pattern Recogn. **42**(4), 581–586 (2009)

12. De Laet, T., Papageorgiou, E., Nieuwenhuys, A., Desloovere, K.: Does expert knowledge improve automatic probabilistic classification of gait joint motion patterns in children with cerebral palsy? PLoS ONE **12**(6), e0178378 (2017)

13. Wolf, S., et al.: Automated feature assessment in instrumented gait analysis. Gait Posture **23**(3), 331–338 (2006)

14. Dempster, A.P.: A generalization of Bayesian inference. J. Roy. Stat. Soc.: Ser. B (Methodol.) **30**(2), 205–232 (1968)

15. Shafer, G.: A Mathematical Theory of Evidence, vol. 42. Princeton University Press, Princeton (1976)

16. Beynon, M.J., Jones, L., Holt, C.A.: Classification of osteoarthritic and normal knee function using three-dimensional motion analysis and the Dempster-Shafer theory of evidence. IEEE Trans. Syst. Man Cybern.-Part A: Syst. Hum. **36**(1), 173–186 (2005)

17. Biggs, P.R., Whatling, G.M., Wilson, C., Holt, C.A.: Correlations between patient-perceived outcome and objectively-measured biomechanical change following total knee replacement. Gait Posture **70**, 65–70 (2019)

18. Liu, M., Zhang, F., Datseris, P., Huang, H.H.: Improving finite state impedance control of active-transfemoral prosthesis using Dempster-Shafer based state transition rules. J. Intell. Rob. Syst. **76**(3–4), 461–474 (2014)

19. Van Gestel, L., et al.: Probabilistic gait classification in children with cerebral palsy: a Bayesian approach. Res. Dev. Disabil. **32**(6), 2542–2552 (2011)

20. Nieuwenhuys, A., et al.: Identification of joint patterns during gait in children with cerebral palsy: a delphi consensus study. Dev. Med. Child Neurol. **58**(3), 306–313 (2016)

21. Zhang, Y., Ma, Y.: Application of supervised machine learning algorithms in the classification of sagittal gait patterns of cerebral palsy children with spastic diplegia. Comput. Biol. Med. **106**, 33–39 (2019)

22. Krzak, J.J., et al.: Kinematic foot types in youth with equinovarus secondary to hemiplegia. Gait Posture **41**(2), 402–408 (2015)
23. Dobson, F., Morris, M.E., Baker, R., Graham, H.K.: Gait classification in children with cerebral palsy: a systematic review. Gait Posture **25**(1), 140–152 (2007)
24. Papageorgiou, E., Nieuwenhuys, A., Vandekerckhove, I., Van Campenhout, A., Ortibus, E., Desloovere, K.: Systematic review on gait classifications in children with cerebral palsy: an update. Gait Posture **69**, 209–223 (2019)
25. Geerse, D.J., Coolen, B.H., Roerdink, M.: Kinematic validation of a multi-kinect v2 instrumented 10-meter walkway for quantitative gait assessments. PLoS ONE **10**(10), e0139913 (2015)
26. Müller, B., Ilg, W., Giese, M.A., Ludolph, N.: Validation of enhanced kinect sensor based motion capturing for gait assessment. PLoS ONE **12**(4), e0175813 (2017)
27. Safranek, R.J., Gottschlich, S., Kak, A.C.: Evidence accumulation using binary frames of discernment for verification vision. IEEE Trans. Robot. Autom. **6**(4), 405–417 (1990)
28. Zeni Jr., J.A., Richards, J.G., Higginson, J.S.: Two simple methods for determining gait events during treadmill and overground walking using kinematic data. Gait Posture **27**(4), 710–714 (2008)
29. Hausdorff, J.M.: Gait dynamics, fractals and falls: finding meaning in the stride-to-stride fluctuations of human walking. Hum. Mov. Sci. **26**(4), 555–589 (2007)
30. Rethlefsen, S.A., Ryan, D.D., Kay, R.M.: Classification systems in cerebral palsy. Orthop. Clin. **41**(4), 457–467 (2010)
31. Duda, R.O., Hart, P.E., Stork, D.G.: Pattern Classification. Wiley, Hoboken (2012)

Analyses of Public Health Databases via Clinical Pathway Modelling: TBWEB

Anderson C. Apunike[1(✉)] [ID], Lívia Oliveira-Ciabati[1] [ID],
Tiago L. M. Sanches[1] [ID], Lariza L. de Oliveira[1] [ID], Mauro N. Sanchez[2] [ID],
Rafael M. Galliez[3] [ID], and Domingos Alves[1] [ID]

[1] University of São Paulo, Av. Bandeirantes n°3900,
Ribeirão Preto, SP 14040-900, Brazil
anderson.apunike@gmail.com
[2] University of Brasilia, Distrito Federal, Brasília, DF 70910-900, Brazil
[3] Federal University of Rio de Janeiro, Av. Pedro Calmon n°550,
Rio de Janeiro, RJ 21941-901, Brazil

Abstract. One of the purposes of public health databases is to serve as
repositories for storing information regarding the treatment of patients.
TBWEB (TuBerculose WEB) is an epidemiological surveillance system
for tuberculosis cases in the state of São Paulo, Brazil. This paper proposes an analysis of the TBWEB database with the use of clinical pathways modelling. Firstly, the database was analysed in order to find the
interventions registered on the database. The clinical pathways were
obtained from the database by the use of process mining techniques. Similar pathways were grouped into clusters in order to find the most common treatment sequences. Each cluster was characterised and the risk
of bad outcomes associated with each cluster was discovered. Some clusters had an association with the risk of negative outcomes. This method
can be applied to other databases, serve as a base for decision-making
systems and can be used to monitor public health databases.

Keywords: Clinical pathways · Process mining · Public health ·
Clustering

1 Introduction

Over the years, informatics has changed the way data is stored and retrieved.
With the progress of informatics, electronic repositories known as databases
came into existence. A database is a set of related data that is organised and
stored in a way that facilitates access, manipulation and control [9]. Interventions carried out on patients are stored in electronic health records and these
records are stored in databases. Each patient follows a treatment sequence over
the course of treatment. The interventions that make up the treatment sequence
forms the clinical pathway. A clinical pathway is defined as a temporal sequence
of clinical interventions established by a specialist or by a multidisciplinary team

© Springer Nature Switzerland AG 2020
V. V. Krzhizhanovskaya et al. (Eds.): ICCS 2020, LNCS 12140, pp. 550–562, 2020.
https://doi.org/10.1007/978-3-030-50423-6_41

to treat certain patients or reach certain objectives [12]. The development of clinical pathways involves setting up goals, clinical practice revision and the establishment, application and analysis of the protocol created by the multidisciplinary team [5]. Clinical pathways serve as a way of documenting treatment, improving teamwork and facilitating communication [8].

It is possible to extract clinical pathways from electronic health records with the use of process mining techniques [6]. Process mining is defined as the use of event logs to discover, monitor, and improve the processes of an establishment [20]. This technique originated from business management and it has been increasingly applied to health [22]. Examples of the application of process mining can be found in chemotherapy [4] and in patients with Acute Coronary Syndrome [10]. To the best of our knowledge, there are no studies where this technique was applied to study tuberculosis care. The modelling process of the clinical pathways begins with discovering the chronological sequence of the events of interest. After finding the correct sequence of events, each event is represented by a character or symbol. The order of events are followed and their representative characters or symbols are put together to form a string (or a sequence of symbols) that represents the clinical pathway. To visualize the clinical pathway, diagrams such as networks, flow charts or Petri nets can be used [20].

With the representation of clinical pathways with visual elements, it is possible to have a generalised view of the treatment sequences that exist in the database. Furthermore, risk assessment techniques are able to detect which treatment sequences that are associated with the risk of bad outcomes. Pathways with such characteristics can be identified and avoided in order to improve the outcome of the treatment. In clinical practice, it is common to follow guidelines, protocols or recommendations during treatment. The pathways obtained from process mining can be audited to verify if the treatment sequences follow such recommendations.

The primary objective of this work was to perform secondary analysis on the TBWEB database, which is the database of the system used for tuberculosis epidemiological surveillance in the state of São Paulo, Brazil. This analysis was carried out with the use of the clinical pathways extracted from TBWEB database with process mining. The other objectives were to characterize the clinical pathways and perform a risk assessment on the pathways and detect which pathways are associated with the risk of negative outcomes. The results obtained from this work can serve as a basis for clinical and programmatic decision making regarding the treatment of tuberculosis patients.

2   Methods

2.1   Study Dataset

The TBWEB system originated in 2004 and it belongs to the State Health Secretariat of São Paulo State [11]. The system serves as a platform for registering and monitoring tuberculosis cases in São Paulo State [1]. The dataset that was analysed was comprised of tuberculosis cases whose treatment began from 2006 to 2016. There were no age restrictions. The exclusion criteria was latent tuber-culosis cases because such cases are monitored in another system [3]. In total, the dataset had 212,569 cases over a ten-year period. The TBWEB dataset was obtained from the Centro de Informação e Informática (CIIS) of Ribeirão Preto Medical School (FMRP) of the University of São Paulo, Ribeirão Preto, Brazil.

2.2   Statistical Software

The statistical software developed for clinical pathway modelling and subse-quent analyses was written in R language. RStudio version 1.0.136 [19] served as the Integrated Development Environment for writing the codes and executing commands. The following R packages were used over the course of the analy-ses: Stringr [21] for carrying out operations on strings, openxlsx for saving files in spreadsheets and igraph [7] for drawing networks. Risk assessment was per-formed with the use of epiR package [18] in order to calculate the relative risk associated with the pathways.

2.3   Analysis Plan

The analyses performed with the TBWEB dataset followed a plan of four phases: study and preparation of the data, setup and filtering of the clinical pathways, classification and clustering of the clinical pathways, characterization and anal-ysis of the clinical pathways.

Study and Preparation of the Data. This phase consists of studying the dataset, understanding its structure and the variables contained in the dataset. The preparation of the data involved choosing the variables of interest and imple-menting the exclusion criteria on the dataset, if there are any.

In the TBWEB dataset, variables related to treatment regimen, bacilloscopy results, the states of the patient (registered on a monthly basis) and treatment outcomes were chosen to set up the clinical pathways. Records with missing data or with values out of a plausible range were not considered for analyses. More details on the chosen variables are shown on the table below (see Table 1).

Setup and Filtering of the Clinical Pathways. In this step, the exact sequence of events and interventions was followed, concatenating characters or symbols in order to get the string that represents the clinical pathway. If there are certain patterns of pathways that fall into the exclusion criteria, such pathways were removed from the analysis, thus filtered out from other pathways.

Table 1. Description of the selected TBWEB variables.

Type	Variable	Description
Treatment regimen	RHZ	rifampicin + isoniazid + pyrazinamide
Treatment regimen	RHZE	rifampicin + isoniazid + pyrazinamide + ethambutol
Treatment regimen	SZEEt	reptomycin + pyrazinamide + ethambutol + ethionamide
Treatment regimen	MR	Treatment regimen for Multi-Drug Resistant tuberculosis
Treatment regimen	OTHERS	None of the above
Treatment regimen	No Info	No information
Bacilloscopy results/states (BAC)	Positive	Positive BAC result
Bacilloscopy results/states (BAC)	Negative	Negative BAC result
Bacilloscopy results/states (BAC)	Progress	BAC exam in progress
Bacilloscopy results/states (BAC)	No	No BAC test
Bacilloscopy results/states (BAC)	No Info	No information on Bac test
Patient states or Treatment outcome	Default	Patient took medicine for more than 30 days and interrupted treatment for more than 30 consecutive days
Patient states or Treatment outcome	Primary Default	Patient took medicine for less than 30 days and interrupted treatment for more than 30 consecutive days or the diagnosed patient did not start treatment at all
Patient states or Treatment outcome	Inpatient treatment	Patient stays in hospital, receiving 24-h care
Patient states or Treatment outcome	Outpatient treatment	Patient comes to the hospital, receives treatment and leaves the hospital
Patient states or Treatment outcome	Change diagnosis	Change in treatment regimen due to intolerance or toxicity
Patient states or Treatment outcome	TB Death	Death tuberculosis
Patient states or Treatment outcome	NTB Death	Death by other causes
Patient states or Treatment outcome	Transfer	Patient was transferred to another hospital, state or country
Patient states or Treatment outcome	Others	Other states (none of the previously mentioned states or outcomes)
Patient states or Treatment outcome	No info	No information (only for patient states)

Classification and Clustering of the Clinical Pathways. After discovering the clinical pathways, the next step is to find groups of related pathways. If there is any classification system to classify patients based on their clinical conditions, such a system can be applied to the pathways in order to obtain groups of pathways with related conditions. If there is no classification system based on clinical conditions, the pathways can be clustered directly.

To cluster the pathways, hierarchical clustering methods were taken into consideration because of the way the results are displayed in the form of a tree (or dendrogram). There are other methods of clustering but it was decided to limit the clustering of the pathways to hierarchical clustering methods. The objective of the clustering process was to form subgroups of pathways based on their level of similarity. Hierarchical clustering requires a distance metric in order to group the input data based on their similarity. Due to the fact that the clinical pathways are in the form of strings, two dimensional distance methods like Euclidean or Manhattan distances do not apply. Therefore, the Levenshtein distance [14] is applied to the pathways as a distance metric in this case. This distance method calculates the total number of transformations (insertions, deletions, substitutions) to transform one string to another. For example, the Levenshtein distance between "wafer" and "water" is one because only one transformation is required to switch from "wafer" to "water". After finding the distances between the pathways, a hierarchical clustering method is applied to the pathways. Clustering methods like average-linkage, complete-linkage or mcquitty can be used. Average-linkage method is a clustering method where the distance between clusters is the mean distance between all the pairs of objects from each cluster [17]. In complete-linkage method, the distance between clusters is the maximum distance between two objects in each cluster [13]. The Mcquitty method is slightly different, as it calculates the distance between clusters based on the average distance of the newly formed cluster with one previously formed [16].

Hierarchical clustering generates a dendrogram that shows the clusters of pathways. In order to find the subgroups (or clusters) of pathways, the optimal number of clusters has to be found and depending on the configuration of the dendrogram, the optimal number of clusters can be hard to discover. The Elbow method [2] was used to find the optimal number of clusters. This method is comprised of four steps:

- Form clusters based on values. This process involves clustering data according to values known as "k". The value k can vary from 1 to N clusters for example. In the dendrogram, clusters can be obtained by cutting the dendrogram at different heights. Therefore, in order to obtain different clusters according to the value k, the dendrogram was cut in various sections.
- Calculate the within-cluster sum of squares (wss) for each value. The within-cluster sum of squares involves calculating the sum of squares of all the pathways within each cluster. There are N clusters for every value of k. Therefore, Twss value which is the total within-cluster sum of squares for each k is calculated by adding the sum of squares of all the pathways within each cluster (wss). The formula below (1) illustrates the wss formula for k, where k is the number of clusters, N is the maximum number of clusters related to k, Mi and Mj are the number of pathways in clusters Pi and Pj respectively.

$$Twss(k) = \sum_{k=1}^{N} \left(\sum_{i=1}^{Mi} \sum_{j=1}^{Mj} Levenshtein_D istance(Pi, Pj)^2, i \neq j \right) \quad (1)$$

After calculating the within-cluster sum of squares for different numbers of clusters, the next step is to plot a graph of the Twss values against the number of clusters. The optimal number of clusters corresponds to the value that has the "elbow" which is the point where the within-cluster sum of values decreases in a lesser rate compared to previous values.

Characterization and Analysis of the Clinical Pathways. After determining the optimal number of clusters, the final step is to characterize and analyze the clusters of pathways. This is done by calculating the relative risk of each cluster, performing descriptive analysis on all the clusters and finding the representative pathway of each cluster. In this work, relative risk is used to find an association between the clusters of clinical pathways and bad outcomes. Here, bad outcomes are referred to the occurrence of events such as death or default of treatment. In each case, the exposed group is the group that belongs to the cluster in question. The unexposed group refers to all the pathways that belong to other clusters Knowing the risk associated with each cluster, the next step of this phase is to perform a descriptive analysis of the clusters to know the characteristics of the patients that belong to each cluster. This descriptive analysis is centered around demographic variables, comorbidities and variables that describe the clinical condition of the patients. After describing the clusters, the final step is to discover the representative pathways of each cluster. In order to find the representative pathway of a cluster, the adjacency matrix of the cluster has to be found. An adjacency matrix is a matrix which rows and columns are the states or events that make up the pathways and is filled with the total number of transitions from one state or event to another. With the cluster's adjacency matrix, a graph can be plotted to show all the pathways that form the cluster and the number of times the patients moved from one event to another. The representative pathway is the path with the highest transitions from one state to another, starting from the initial nodes (treatment regimens) to one of the final nodes (treatment outcome).

3   Results and Discussions

This work is a pilot study on TBWEB and the analysis plan was applied to the dataset of tuberculosis cases in 2011. The choice to focus on one year and not on the entire dataset was a technical decision, as dealing with more than ten thousand pathways lead to memory issues when calculating the distances between the pathways. So the way out was to choose the yar with the highest number of pathways less than ten thousand. The 2011 dataset was chosen and it had 7321 pathways. The pathways were discovered and clustered with three methods; average-linkage, complete-linkage and mcquitty. Out of the three clustering methods, complete-linkage was chosen to define the clusters because when the elbow method was applied on all the methods, complete-linkage returned the highest number of clusters. This study focused on a hierarchical clustering method that has the highest number of clusters. The figure below (see Fig. 1) shows the dendrogram generated using complete-linkage method. Since complete-method was chosen to define the clusters, the dendrograms obtained from other hierarchical clustering methods are available at the link: https://tinyurl.com/all-clusterings-tbweb.

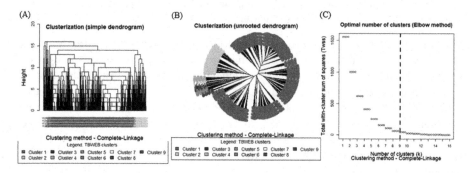

Fig. 1. Clusterization of TBWEB pathways with Complete-Linkage Method. (A) Simple dendrogram. (B) Unrooted dendrogram. (C) Application of Elbow Method to determine the optimal number of clusters.

The figure above illustrates the results obtained while clustering the pathways. According to the elbow method, 9 clusters was the optimal number of clusters. In Fig. 1B, the total within-cluster sum of squares (Twss) decreases as the number of clusters increases. In this case, the bend is located at 9 because as from that point the value of Twss drops in a lesser rate compared to previous values. The results of the risk analysis performed on all the clusters are illustrated in the table below (see Table 2).

Table 2. Risk assessment of the clusters.

Clusters	Number of pathways	Good outcomes	Bad outcomes	Relative risk value	95% confidence interval
Cluster 1	6037	79 (1.3%)	5958 (98.7%)	0.1	[0.08, 0.13]
Cluster 2	846	45 (5.3%)	801 (94.7%)	0.7	[1.22, 2.29]
Cluster 3	89	7 (7.9%)	82 (92.1%)	2.3	[1.13, 4.80]
Cluster 4	149	5 (3.4%)	144 (96.6%)	1.0	[0.41, 2.34]
Cluster 5	106	98 (92.5%)	8 (7.5%)	25.6	[36.93, 51.47]
Cluster 6	29	0 (0%)	29 (100%)	-	-
Cluster 7	24	2 (8.3%)	22 (91.7%)	2.4	[0.64, 9.26]
Cluster 8	13	11 (84.6%)	2 (15.4%)	25.8	[19.81, 33.52]
Cluster 9	23	4 (14.3%)	24 (85.7%)	4.2	[1.69, 10.54]

Based on the risk analysis results, cluster 1 can be considered as protective, since the pathways contained in it poses a low risk of bad outcomes. On the other hand, the other clusters are associated with varying degrees of clusters, with clusters 5 and 8 having the highest risks of bad outcomes. The table below shows the results of the descriptive analysis performed on the clusters 1, 5 and 8 (see Table 3). The full table with all the clusters can be found at https://tinyurl.com/full-analysis-clusters-tbweb. Table 3 is divided into 14 sections and each section represents the proportions of a group of variables.

Table 3. Descriptive analysis of the clusters.

Attributes	Cluster 1	Cluster 5	Cluster 8
Number of pathways	6037	106	13
Section 1: Sex			
Male	4196 (69.5%)	89 (84%)	10 (76.9%)
Female	1841 (30.5%)	17 (16%)	3 (23.1%)
Section 2: Age group			
Child (0–9 years)	101 (1.7%)	0 (0%)	0 (0%)
Teenager (10–19 years)	468 (7.8%)	7 (6.6%)	1 (7.7%)
Adult (20–59 years)	4885 (80.9%)	96 (90.5%)	12 (92.3%)
Senior adult (60 years and above)	580 (9.6%)	3 (2.8%)	0 (0%)
No information	3	0	0
Section 3: Education			
No years in school	219 (4.1%)	2 (2.2%)	0 (0%)
1 to 3 years in school	630 (11.9%)	11 (12.1%)	2 (20%)
4 to 7 years in school	1943 (36.7%)	39 (42.9%)	5 (50%)
8 to 11 years in school	2079 (39.3%)	33 (36.3%)	3 (30%)
12 to 14 years in school	275 (5.2%)	4 (4.4%)	0 (0%)
15 years in school and above	147 (2.8%)	2 (2.2%)	0 (0%)
No information	744	15	3
Section 4: Occupation			
Employed	3497 (62.1%)	67 (68.4%)	8 (61.5%)
Housekeeper	495 (8.8%)	4 (4.1%)	1 (7.7%)
Unemployed	593 (10.5%)	21 (21.4%)	1 (7.7%)
Retired	351 (6.2%)	2 (2%)	0 (0%)
Imprisoned	691 (12.3%)	4 (4.1%)	3 (23.1%)
No information	410	8	0
Section 5: HIV Test			
Positive	408 (6.9%)	15 (14.2%)	0 (0%)
Negative	4900 (82.6%)	76 (71.7%)	12 (92.3%)
HIV test in progress	33 (0.6%)	1 (0.9%)	0 (0%)
No HIV test	591 (10%)	14 (13.2%)	1 (7.7%)
No information	105	0	0

(*continued*)

Table 3. (*continued*)

Attributes	Cluster 1	Cluster 5	Cluster 8
Section 6: AIDS			
AIDS (present)	366 (6.1%)	12 (11.3%)	0 (0%)
AIDS (absent)	5671 (93.9%)	94 (88.7%)	13 (100%)
No information	0	0	0
Section 7: Diabetes			
Diabetic	333 (5.5%)	5 (4.7%)	0 (0%)
Non-diabetic	5704 (94.5%)	101 (95.3%)	13 (100%)
No information	0	0	0
Section 8: Alcoholism			
Alcoholic	743 (12.3%)	27 (25.5%)	2 (15.4%)
Nonalcoholic	5294 (87.7%)	79 (74.5%)	11 (84.6%)
No information	0	0	0
Section 9: Mental disorders			
Mental disorder (present)	89 (1.5%)	0 (0%)	0 (0%)
Mental disorder (absent)	5948 (98.5%)	106 (100%)	13 (100%)
No information	0	0	0
Section 10: Drug addiction			
Drug addict	495 (8.2%)	21 (19.8%)	4 (30.8%)
Non-drug addict	5542 (91.8%)	85 (80.2%)	9 (69.2%)
No information	0	0	0
Section 11: Smoking			
Smoker	33 (0.5%)	0 (0%)	0 (0%)
Non-smoker	6004 (99.5%)	106 (100%)	13 (100%)
No information	0	0	0
Section 12: Case Type			
New case	5464 (90.5%)	85 (80.2%)	7 (53.8%)
Relapse	393 (6.5%)	7 (6.6%)	2 (15.4%)
Retreatment	180 (3%)	14 (13.2%)	4 (30.8%)
Section 13: Type of tuberculosis			
Pulmonary tuberculosis	5036 (83.4%)	89 (84%)	13 (100%)
Extrapulmonary tuberculosis	848 (14%)	13 (12.3%)	0 (0%)
Pulmonary and Extrapulmonary tuberculosis	142 (2.4%)	3 (2.8%)	0 (0%)
Disseminated tuberculosis	11 (0.2%)	1 (0.9%)	0 (0%)
Section 14: Pregnancy			
Pregnant	24 (1.3%)	2 (11.8%)	0 (0%)
Not pregnant	1829 (98.7%)	15 (88.2%)	3 (100%)
No information	4184	89	10
Section 15: Outcomes			
Cure	5958 (98.7%)	8 (7.5%)	2 (15.4%)
Default (Abandoned treatment)	57 (0.9%)	97 (91.5%)	11 (84.6%)
Death by tuberculosis	1 (0%)	1 (0.9%)	0 (0%)
Death by other causes	21 (0.3%)	0 (0%)	0 (0%)
No information	0	0	0

The descriptive analysis adds more information about the clusters and enables comparisons between clusters. From the diagram above, it is observed that clusters 5 and 8 have high default rates (Table 3, Section 15) and this reflected on Table 2 where both clusters had the highest risk of bad outcomes (death or default). Cluster 5 has the highest proportion of pregnant women (Table 3, Section 14) and cluster 8 has the highest proportion of retreatment and relapse (Table 3, Section 12). This shows that factors such as pregnancy, relapse or retreatment can be considered as risk factors. Drug addiction (Table 3, Section 10) might be considered as a risk factor due to the that cluster 1 has is among the clusters with the lowest drug addiction proportions. Since clusters 1 and 5 are in stark contrast to one another in terms of risk, the representative pathways of both clusters are shown in the figure below (see Figs. 2 and 3) to have an idea of the most frequent procedures and patient states in the two clusters. Here the representative pathway shows the most travelled path in the cluster of clinical pathways. The represenative pathways with the highest risks

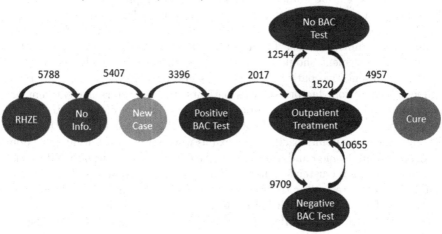

Fig. 2. Representative pathways - Cluster 1, lowest risk.

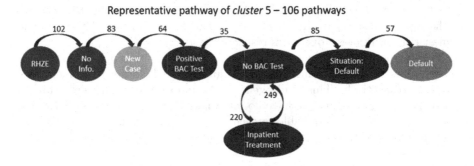

Fig. 3. Representative pathways - Cluster 5, highest risk.

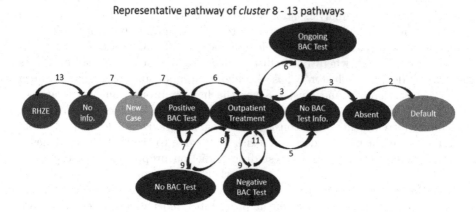

Fig. 4. Representative pathways - Cluster 8, second-highest risk.

of bad outcomes can be found on Figs. 3 and 4. The representative pathways of all clusters can be found at https://tinyurl.com/all-pathways-tbweb.

The figure above brings some insights on the clinical pathways, for example, on health recommendations. The Brazilian Health Ministry recommends that tuberculosis treatment should commence with the use of RHZE [15] and over the nine representative pathways of the clusters, RHZE was the first treatment regimen. In addition, factors such as patients' presence during treatment and the conduction of bacilloscopy tests can interfere on risk of bad outcomes. In cluster 1 (no risk), there is no patient absence in the representative pathway and although at certain points bacilloscopy exams were not carried out, at some point the exam was performed and it gave a negative result and eventually led to cure. On the other hand, cluster 5 indicates that the most travelled pathway did not have bacilloscopy test after the first positive result and there was absence over the course of treatment which led to default. Cluster 8 which had the second-highest risk also had no baciloscopy tests or no information regarding the test at some points and the representative pathway ended in default. This leads to the belief that carrying out of regular tests, the availability of test results and the patient's continued presence during the course of treatment has an impact in avoiding negative outcomes. This can be confirmed by comparing the structure of the representative pathway of clusters 1 and 5 which poses the lowest and highest risks of negative outcomes respectively.

This can be confirmed by comparing the structure of the representative pathway of clusters 1 and 5 which poses the lowest and highest risks of negative outcomes respectively. This study generated lots of data which were attached to this paper via links and we have no information regarding the validity of the links. In case the images or tables attached to the links are no longer available, enter in contact with the authors.

4   Conclusions

This work has shown that clinical pathway modelling can be used as a method of analyzing public health databases. It was possible to discover the treatments that were registered on the database and their respective outcomes. This was a pilot study and a specific subset of the entire database was selected to perform the analyses. In this work, the whole analyses were conducted on tuberculosis treatments that were registered in 2011. In other words, the whole TBWEB database was not analyzed. The idea of classifying and clustering pathways and finding the representative pathways gives a general view of the different treatment regimens that exist in a public health database. Also, risk analysis helps in measuring the risk associated with a cluster of clinical pathways.

Future applications of this analysis can be in the form of extending this method of analysis to the entire TBWEB database or extend the analyses to other public health databases. Also, this study was limited to hierarchical clustering methods, so future works can focus on repeating this analyses with other clustering methods. In addition, the discovered pathways can be checked if they are in accordance to the real treatment procedure of tuberculosis. This serves as a conformance checking process which is one of the core techniques in process mining. Furthermore, another way of continuing this study is to verify of the pathways influence outcomes. This can be done by excluding the final values regarding the treatment outcome in the pathway strings and repeating the same analyses (clustering and analyzing death rate and/or default rate among clusters). Such a procedure can prove if the clinical pathways influence outcomes. Moreover, depending on the volume of data, alternatives to calculate the distance between the pathways before clustering need be discovered. A divide and conquer strategy, calculating parts of the distance matrix and joining the distances or running the distance calculations on a server instead of a personal computer can be ways to overcome potential Big-Data problems.

Finally, predictive or decision-support systems can be created by combining clinical pathway modelling and artificial intelligence. In this way, the future treatment processes of a patient's treatment can be predicted or the medical team can be alerted if a specific treatment pathway poses a high risk of bad outcome to the patient, allowing the medical team to change the treatment regimen and take new decisions to improve the treatment outcome. Another application of analyzing public health databases with clinical pathway modelling is to perform this analysis periodically on a public health database to monitor the evolution of the risk of bad outcomes on an individual basis, thus adding more value to precision-medicine.

References

1. Cve prof. alexandre vranjac - manual de utilização do tbweb versão 1.6. http://www.saude.sp.gov.br/resources/cve-centro-de-vigilancia-epidemiologica/areas-de-vigilancia/tuberculose/manuais-tecnicos/dvtbc_tbweb_2008.pdf. Accessed 07 Feb 2020

2. Datanovia - determining the optimal number of clusters: 3 must know methods. https://www.datanovia.com/en/lessons/determining-the-optimal-number-of-clusters-3-must-know-methods. Accessed 07 Feb 2020

3. Il-tb - sistema de informação para notificação das pessoas em tratamento de iltb. http://sitetb.saude.gov.br/iltb

4. Baker, K.E.A.: Process mining routinely collected electronic health records to define real-life clinical pathways during chemotherapy. Int. J. Med. Inf. **103**, 32–41 (2017)

5. Campbell, H.E.A.: Integrated care pathways. Br. Med. J. **316**, 133–137 (1998)

6. Caron, F.E.A.: A process mining-based investigation of adverse events in care processes. Health Inf. Manag. J. **43**, 16–25 (2014)

7. Csardi, G., Nepusz, T.: The igraph software package for complex network research. Int. J. (Complex Syst.) **1695**, 1–9 (2005)

8. Deneckere, S.E.A.: Care pathways lead to better teamwork: results of a systematic review. Soc. Sci. Med. **75**, 264–268 (2012)

9. Elmasri, R., Navathe, S.: Sistemas de Banco de Dados. Pearson-Addison-Wesley, São Paulo (2005)

10. Funkner, A.E.A.: Data-driven modeling of clinical pathways using electronic health records. Procedia Comput. Sci. **121**, 835–842 (2017)

11. Galesi, V.: Dados de tuberculose de estado de são paulo. Revista de Saúde Pública **41**, 121 (2007)

12. Hunter, B., Segrott, J.: Using a clinical pathway to support normal birth: impact on practitioner roles and working practices. Birth **37**, 227–36 (2010)

13. Johnson, S.: Hierarchical clustering schemes. Psychometrika **2**, 241–254 (1967)

14. Levenshtein, V.I.: Binary codes capable of correcting deletions, insertions, and reversals. Sov. Phys. Dokl. **10**, 707–710 (1966)

15. Maciel, E.: Efeitos adversos causados pelo novo esquema de tratamento da tuberculose preconizado pelo ministério da saúde do brasil. J. Bras. Pneumol. **36**, 232–238 (2010)

16. McQuitty, L.: Similarity analysis by reciprocal pairs for discrete and continuous data. Educ. Psychol. Measur. **26**, 1695 (1966)

17. Sokal, R., Michener, C.: A statistical method for evaluating systematic relationships. Univ. Kansas Sci. Bull. **28**, 1409–1438 (1958)

18. Stevenson, M.E.A.: epir: tools for the analysis of epidemiological data (2020). https://CRAN.R-project.org/package=epiR. Accessed 7 Feb 2020

19. RStudio Team: Rstudio manual: integrated development environment for R (2015). http://www.rstudio.com/

20. Van Der Aalst, W.: Process Mining: Discovery, Conformance and Enhancement of Business Processes. Springer, London (2011). https://doi.org/10.1007/978-3-642-19345-3

21. Wickham, H.: stringr: Simple, consistent wrappers for common string operations (2020). https://CRAN.R-project.org/package=stringr. Accessed 7 Feb 2020

22. Williams, R.E.A.: Process mining in primary care: a literature review. Stud. Health Technol. Inf. **247**, 37–380 (2018)

Preliminary Results on Pulmonary Tuberculosis Detection in Chest X-Ray Using Convolutional Neural Networks

Márcio Eloi Colombo Filho[1]([✉]) [ID], Rafael Mello Galliez[2] [ID],
Filipe Andrade Bernardi[1] [ID], Lariza Laura de Oliveira[3] [ID],
Afrânio Kritski[2] [ID], Marcel Koenigkam Santos[3] [ID],
and Domingos Alves[3] [ID]

[1] Interunit Postgraduate Program in Bioengineering,
University of São Paulo, São Carlos, SP, Brazil
{marcioeloicf, filipepaulista12}@usp.br
[2] School of Medicine, Federal University of Rio de Janeiro,
Rio de Janeiro, Brazil
galliez77@gmail.com, kritskia@gmail.com
[3] Ribeirão Preto Medical School,
University of São Paulo, Ribeirão Preto, São Paulo, Brazil
larizalaura@gmail.com, {marcelk46, quiron}@fmrp.usp.br

Abstract. Tuberculosis (TB), is an ancient disease that probably affects humans since pre-hominids. This disease is caused by bacteria belonging to the mycobacterium tuberculosis complex and usually affects the lungs in up to 67% of cases. In 2019, there were estimated to be over 10 million tuberculosis cases in the world, in the same year TB was between the ten leading causes of death, and the deadliest from a single infectious agent. Chest X-ray (CXR) has recently been promoted by the WHO as a tool possibly placed early in screening and triaging algorithms for TB detection. Numerous TB prevalence surveys have demonstrated that CXR is the most sensitive screening tool for pulmonary TB and that a significant proportion of people with TB are asymptomatic in the early stages of the disease. This study presents experimentation of classic convolutional neural network architectures on public CRX databases in order to create a tool applied to the diagnostic aid of TB in chest X-ray images. As result the study has an AUC ranging from 0.78 to 0.84, sensitivity from 0.76 to 0.86 and specificity from 0.58 to 0.74 depending on the network architecture. The observed performance by these metrics alone are within the range of metrics found in the literature, although there is much room for metrics improvement and bias avoiding. Also, the usage of the model in a triage use-case could be used to validate the efficiency of the model in the future.

Keywords: Tuberculosis · Chest X-ray · Convolutional neural networks

© Springer Nature Switzerland AG 2020
V. V. Krzhizhanovskaya et al. (Eds.): ICCS 2020, LNCS 12140, pp. 563–576, 2020.
https://doi.org/10.1007/978-3-030-50423-6_42

1   Introduction

1.1   Tuberculosis

Tuberculosis (TB), is an ancient disease that affects humans and probably existed in pre-hominids, and still is nowadays an important cause of death worldwide. This disease is caused by bacteria belonging to the mycobacterium tuberculosis complex and usually affects the lungs, although other organs are affected in up to 33% of cases [1]. When properly treated, tuberculosis caused by drug-sensitive strains is curable in almost all cases. If left untreated, the disease can be fatal in 5 years in 50 to 65% of the cases. Transmission usually occurs by the aerial spread of droplets produced by patients with infectious pulmonary tuberculosis [1].

Despite the progress achieved in TB control over the past two and a half decades, with more than 50 million deaths averted globally, it is still the leading cause of death in people living with HIV, accounting for one in five deaths in the world [2]. In 2019, there were estimated to be over 10 million TB cases in the world, in the same year TB was between the ten leading causes of death, and the deadliest cause from a single infectious agent [3].

Most people who develop TB can be cured, with early diagnosis and appropriate drug treatment. Still, for many countries, the end of the disease as an epidemic and major public health problem is far from the reality. Twenty-five years ago, in 1993, WHO declared TB a global health emergency [4]. In response, the End TB Strategy has the overall goal of ending the global TB epidemic, to achieve that goal it defines the targets (2030, 2035) and milestones (2020, 2025) for the needed reductions in tuberculosis cases and deaths. The sustainable development goals include a target to end the epidemic by 2030 [4].

One of these efforts supports the continued collation of the evidence and best practices for various digital health endeavors in TB prevention and care. This will make a stronger 'investment case' for innovative development and the essential implementation of digital health initiatives at scale [5]. Adequate triage and diagnosis are a prerequisite for the prognosis and success of any treatment and may involve several professionals and specialties. In this context, the choice of the essentially clinical mechanism that allows the measurement of the evaluator's impression becomes an important tool in helping a more accurate diagnosis [6].

1.2   Chest X-Ray for Detecting TB

Chest X-ray (CXR) has recently been promoted by the WHO as a tool that can be placed early in screening and triaging algorithms (see Fig. 1). A great number of prevalence surveys on TB demonstrated that CXR is the most sensitive screening tool for pulmonary TB and that TB is asymptomatic on a significant proportion of people while still in the early course of the disease [7]. When used as a triage test, CXR should be followed by further diagnostic evaluation to establish a diagnosis, it is important that any CXR abnormality consistent with TB be further evaluated with a bacteriological test [8].

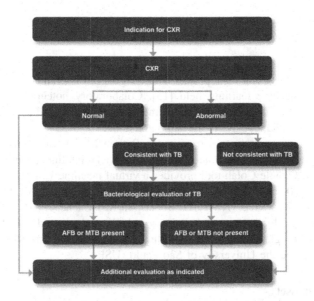

Fig. 1. Using chest radiography as a triage tool (Source: Chest radiography in tuberculosis detection – summary of current WHO recommendations and guidance on programmatic approaches, p. 11).

More than identifying active TB disease, CXR also identifies those who have inactive TB or fibrotic lesions without a history of TB treatment. Once active TB has been excluded, patients with fibrotic lesions should be followed-up, given this population is at the highest risk of developing active TB disease and/or other respiratory complications [9].

In Porto Alegre city, located in the southern region of Brazil, a projection of the impact on case detection and health system costs of alternative triage approaches for tuberculosis found that most of the triage approaches modeled without X-ray were predicted to provide no significant benefit [10]. The same study also found that adding X-ray as a triage tool for HIV-negative and unknown HIV status cases combined with appropriate triage approaches could substantially save costs over using an automated molecular test without triage, while identifying approximately the same number of cases [10].

In many high TB burden countries, it has been reported a relative lack of radiology interpretation expertise [11], this condition could result in impaired screening efficacy. There has been interest in the use of computer-aided diagnosis for the detection of pulmonary TB at chest radiography, once automated detection of disease is a cost-effective technique aiding screening evaluation [12, 13]. There are already studies in the medical literature using artificial intelligence tools to evaluate CXR images, but the developed tool's availability is limited, especially in high burden countries [14].

This study presents experimentation of classic convolutional neural network architectures on public CRX databases in order to create a tool applied to the diagnostic aid of TB in chest X-ray images.

2   Methods

2.1   Google Colab

Google Colaboratory (Colab) is a project with the goal of disseminating education and research in the machine learning field [15]. Colaboratory notebooks are based on Jupyter and work as a Google Docs object, thus they can be shared as such and many users can work simultaneously on the same notebook.

Colab provides pre-configured runtimes in Python 2 or 3 with TensorFlow, Matplotlib, and Keras, essential machine learning and artificial intelligence libraries. To run the experiments, Google Colab tool provides a virtual machine with 25.51 GB RAM, GPU 1xTesla K80, having 2496 CUDA cores, CPU 1x single-core hyperthreaded Xeon Processors @2.3 GHz (No Turbo Boost), 45 MB Cache.

The images were loaded into Google Drive, which can be linked to Colab for direct access to files. Google Drive provides unlimited storage disk due to a partnership between Google and the University of São Paulo (USP).

2.2   Image Datasets

PadChest. PadChest is a dataset of labeled chest X-ray images along with their associated reports. This dataset includes more than 160,000 large-scale, high-resolution images from 67,000 patients that were interpreted and reported by radiologists at Hospital San Juan (Spain) from 2009 to 2017. The images have additional data attached containing information on image acquisition and patient demography [16]. This dataset contains a total of 152 images classified with TB label.

National Institutes of Health. In 2017 the National Institutes of Health (NIH), a component of the U.S. Department of Health and Human Services, released over 100,000 anonymized chest X-ray images and their corresponding data from more than 30,000 patients, including many with advanced lung disease [17]. Other two notorious and vastly used publicly available datasets maintained by the National Institutes of Health, are from Montgomery County, Maryland, and Shenzhen, China [18] which contains respectively 58 and 336 images labeled as TB.

2.3   Network Architectures

AlexNet. In the ILSVRC classification and the localization challenge of 2012, the AlexNet architecture came on top as the winner [19]. The network architecture has 60 million parameters and 650 thousand neurons [20]. The standard settings were employed in this study: Convolutional, maxpooling, and fully connected layers, ReLU activations and the SGD optimization algorithm with a batch size of 128, momentum of 0.9, step learning annealing starting at 0.01 and reduced three times, weight decay of 0.0005, dropout layers with p = 0.5 design patterns, as in the original architecture article [20].

GoogLeNet. In the year of 2014, GoogLeNet was the winner of the ILSVRC detection challenge, and also came in second place on the localization challenge [21]. The standard settings were employed in this study: Convolutional, maxpooling, and fully connected layers are used, in addition, has a layer called an inception module that runs the inputs through four separate pipelines and joins them after that [22]. Also, ReLU activations, asynchronous SGD, momentum of 0.9, step learning annealing decreasing with 4% every eight epochs, a dropout layer with p = 0.7, as in the original architecture article [22].

ResNet. In 2015 ResNet won the classification challenge with only provided training data [23]. ResNet enables backpropagation through a shortcut between layers, this allows weights to be calculated more efficiently [24]. ResNet has a high number of layers but can be fast due to that mechanism [24]. The standard settings were employed in this study: momentum of 0.9, reduction of step learning annealing by a factor of ten every time the rate of change in error stagnates, weight decay of 0.0001 and batch normalization, as in the original architecture article [25].

2.4   Auxiliary Tools

HDF5 Dataset Generator. Functions responsible for generating the training, validation and test sets. It begins by taking a set of images and converting them to NumPy arrays, then utilizing the sklearn train_test_split function [26] to randomly split the images into the sets. Each set is then written to HDF5 format. HDF5 is a binary data format created by the HDF5 group [27] to store on disk numerical datasets too large to be stored in memory while facilitating easy access and computation on the rows of the datasets.

Data Augmentation Tool. Set of functions responsible for zooming, rotating and flipping the images in order to higher the generalization of the model. The Keras ImageDataGenerator class function [28] was utilized, with parameters set as: rotation range of 90°, max zoom range of 15%, max width and height shift range of 20% and horizontal flip set as true.

2.5   Results Validation

In the experiments presented in this analysis, we choose a set of metrics based on confusion matrix [29]. Table 1 shows a confusion matrix 2 × 2 for a binary classifier.

Table 1. Confusion Matrix

	Actual class negative	Actual class positive
Predicted class negative	True negative (TN)	False negative (FN)
Predicted class positive	False positive (FP)	True positive (TP)

The used metrics are described in the equations below:

- Sensibility (also called true positive rate, or recall)

$$Sens = TP/(TP + FN) \tag{1}$$

- Specificity (or true negative rate)

$$Spec = TN/(TN + FP) \tag{2}$$

- Precision

$$Pr = TP/(TP + FP) \tag{3}$$

- F1-Score

$$F1 = 2 * ((Pr * Sens)/(Pr + Sens)) \tag{4}$$

Another metric used is the AUC - The area under the Receiver Operating Characteristic Curve (ROC) which plots the TPR (true positive rate) versus the FPR (false positive rate) [29].

3   Results

The first step was creating a new database merging the images from the Montgomery, Shenzhen, and PadChest datasets. The number of images from each dataset is described in Table 2.

Table 2. Number of images by dataset and class.

Dataset	Number of "no_tb" images	Number of "yes_tb" images	Total
Montgomery	80	58	138
Shinzen	326	336	662
PadChest	140	152	292
Total	546	546	1092

The image data set was split into 3 sets of images: training, validation and test sets using the HDF5 Dataset Generator. During each training the data augmentation function was responsible for preprocessing the images in order to higher the generalization of the model. The datasets sizes are described in Table 3.

Table 3. Number of images by set.

Set	Number of "no_tb" images	Number of "yes_tb" images
Training	446	446
Validation	50	50
Test	50	50

The first model used the AlexNet architecture by Krizhevsky et al. implemented in Keras, the images were resized to $227 \times 227 \times 3$ pixels utilizing the CV2 library, in order to fit the net architecture input size. The training and validation loss/accuracy over the 75 epochs is shown in Fig. 2.

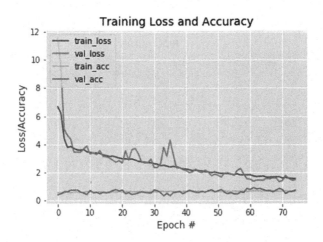

Fig. 2. AlexNet Loss and Accuracy history per Epoch.

The test confusion matrix, test performance, classification metrics and area under the ROC curve are shown in Table 4, Table 5, Table 6 and Fig. 3, respectively.

Table 4. AlexNet main classification metrics.

	Precision	Recall	F1-score	Support
Class: no_tb	0.68	0.86	0.76	50
Class: yes_tb	0.81	0.60	0.69	50
Accuracy	–	–	0.73	100
Macro avg	0.75	0.73	0.73	100
Weighted avg	0.75	0.73	0.73	100

Following the first experiment, the GoogLeNet architecture was implemented, the input size for this CNN model is $224 \times 224 \times 3$. The same database and datasets were

Table 5. AlexNet confusion Matrix.

	Actual no_tb	Actual yes_tb
Predicted no_tb	43	7
Predicted yes_tb	20	30

Table 6. AlexNet model performance.

Metric	Value
Accuracy	0.73
Sensitivity	0.86
Specificity	0.60

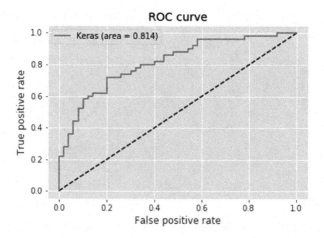

Fig. 3. AlexNet ROC curve and Area Under the Curve.

utilized in the training. The training loss/accuracy over the 100 epochs is shown in Fig. 4 and Fig. 5.

The test confusion matrix, test performance, classification metrics and area under the ROC curve are shown in Table 7, Table 8, Table 9 and Fig. 6, respectively.

The ResNet architecture was also implemented, trained and validated utilizing the same parameter as the other experiments. The input size for this CNN model is 224 × 224 × 3. Training and validation loss/accuracy over the 100 epochs is shown in Fig. 7.

The test confusion matrix, test performance, classification metrics and area under the ROC curve are shown in Table 10, Table 11, Table 12 and Fig. 8, respectively.

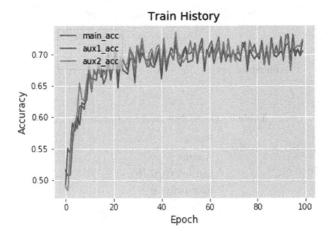

Fig. 4. Accuracy history per Epoch.

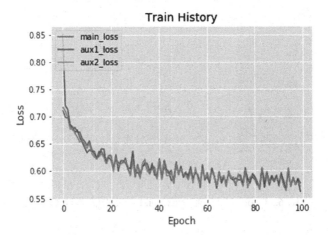

Fig. 5. GoogleNet Loss history per Epoch.

Table 7. GoogleNet main classification metrics.

	Precision	Recall	F1-score	Support
Class: no_tb	0.75	0.76	0.75	50
Class: yes_tb	0.76	0.74	0.75	50
Accuracy	–	–	0.75	100
Macro avg	0.75	0.73	0.75	100
Weighted avg	0.75	0.73	0.75	100

Table 8. GoogleNet confusion Matrix.

	Actual no_tb	Actual yes_tb
Predicted no_tb	38	12
Predicted yes_tb	13	37

Table 9. GoogleNet model performance.

Metric	Value
Accuracy	0.75
Sensitivity	0.76
Specificity	0.74

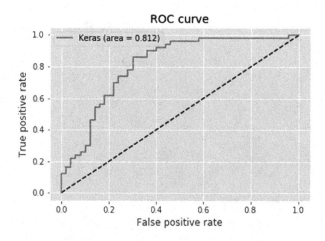

Fig. 6. GoogleNet ROC curve and Area Under the Curve.

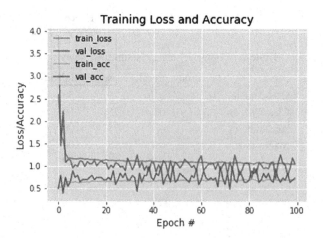

Fig. 7. ResNet Loss and Accuracy history per Epoch.

Table 10. ResNet main classification metrics.

	Precision	Recall	F1-score	Support
Class: no_tb	0.64	0.76	0.70	50
Class: yes_tb	0.71	0.58	0.64	50
Accuracy	–	–	0.67	100
Macro avg	0.68	0.67	0.67	100
Weighted avg	0.68	0.67	0.67	100

Table 11. ResNet confusion Matrix.

	Actual no_tb	Actual yes_tb
Predicted no_tb	38	12
Predicted yes_tb	21	29

Table 12. ResNet model performance.

Metric	Value
Accuracy	0.67
Sensitivity	0.76
Specificity	0.58

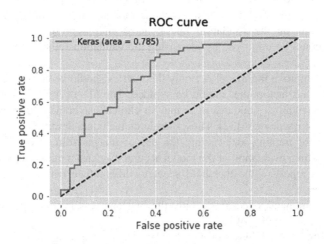

Fig. 8. ResNet ROC curve and Area Under the Curve.

4   Discussion

To gather information about a higher number of articles, as well as to evaluate the quality and possible biases of each, a systematic review was chosen as a base for comparison and discussion. According to the definition presented in Harris et al. [30] this study is classified as a Development study, that focuses on reporting methods for creating a CAD program for pulmonary TB, and includes an assessment of diagnostic accuracy.

Of the 40 studies evaluated by Harris's review, 33 reported measures of accuracy assessments, these studies had the AUC ranged from 0.78 to 0.99, sensitivity from 0.56 to 0.97, and specificity from 0.36 to 0.95. The WHO states that it is necessary for screening tests to have sensitivity greater than 0.9 and specificity greater than 0.7 [31].

This study had an AUC ranging from 0.78 to 0.84, sensitivity from 0.76 to 0.86 and specificity from 0.58 to 0.74 depending on the network architecture. The observed performance by these metrics alone are within the range of metrics found in the literature, although still far from the highest metrics obtained and did not meet the WHO standards.

The reason for such results can be speculated. One of the main possible reasons is the size of the image dataset as well as the number of datasets from different sources. A higher number of images, with high quality diagnosis, could improve the model's performance. A higher number of datasets from different sources could increase the model's generalizability.

Observing the three architecture's loss curve during training (Fig. 2, Fig. 5 and Fig. 7) they seem to indicate a good fit, once training and validation loss curves both decrease to a point of stability maintaining a minimal gap between them. Although, the noisy movements on the validation line indicate an unrepresentative dataset. The performance could be improved by increasing the validation set size compared to the training set.

To further improve the model not only the metrics should be considered, but there are also many bias factors that should be avoided. The FDA (The US Food and Drug Administration) requires standards to be met for clinical use in their guidelines of CAD applied to radiology devices [32]. Those standards include a description of how CXRs were selected for training and testing, the use of images from distinct datasets for training and testing, evaluation of the model accuracy against a microbiologic reference standard and a report of the threshold score to differentiate between a positive and negative classes.

Of the requirements cited above only the first one was met by this study, which leaves open the possibility of bias.

The potential risk of bias can also be detected by applying tools for systematic reviews of diagnostic accuracy studies, the Quality Assessment of Diagnostic Accuracy Studies (QUADAS)-2 [33] is one of the approaches.

5 Conclusion

The preliminary results are between the range of metrics presented in the literature, although there is much room for improvement of metrics and bias avoiding. Also, the usage of the model in a triage use-case could be used to validate the efficiency of the model.

References

1. Fauci, A.S., Braunwald, E., Kasper, D.L., Harrison, T.R.: Princípios de medicina interna, vol. I. McGraw-Hill Interamericana, Madrid (2009)
2. World Health Organization (WHO). Guidelines for treatment of drug-susceptible tuberculosis and patient care, 2017 update
3. World Health Organization. Global tuberculosis report 2019. World Health Organization (2019)
4. World Health Organization. Global tuberculosis report 2018. World Health Organization (2018)
5. World Health Organization. Digital health for the End TB Strategy: an agenda for action. World Health Organization (2015)
6. Zhou, S.-M., et al.: Defining disease phenotypes in primary care electronic health records by a machine learning approach: a case study in identifying rheumatoid arthritis. PloS one, **11** (5), e0154515. 2016 World Health Organization. Tuberculosis Prevalence Surveys: A Handbook. World Health Organization, Geneva (WHO/HTM/TB/2010.17) (2011). http://www.who.int/tb/advisory_bodies/impact_measurement_taskforce/resources_documents/thelimebook/en/. Accessed 5 Oct 2016
7. TB Care I. International standards for tuberculosis care, third edition. The Hague: TB CARE I (2014) http://www.who.int/tb/publications/ISTC_3rdEd.pdf. Accessed 5 Oct 2016
8. World Health Organization. Systematic screening for active tuberculosis: Principles and recommendations. Geneva: World Health Organization (WHO/HTM/TB/2013.04) (2013). http://apps.who.int/iris/bitstream/10665/84971/1/9789241548601_eng.pdf?ua=1. Accessed 27 Sept 2016
9. Melendez, J., Sánchez, C.I., Philipsen, R.H., et al.: An automated tuberculosis screening strategy combining X-ray-based computer-aided detection and clinical information. Sci. Rep. **6**, 25265 (2016)
10. Rahman, A.A.S., et al.: Modelling the impact of chest X-ray and alternative triage approaches prior to seeking a tuberculosis diagnosis. BMC Infect. Dis. **19**(1), 93 (2019)
11. Antani, S.: Automated detection of lung diseases in chest X-rays. a report to the board of scientific counselors. US National Library of Medicine. Published April 2015. https://lhncbc.nlm.nih.gov/system/files/pub9126.pdf. Accessed 20 Sept 2016
12. Jaeger, S., Karargyris, A., Candemir, S., et al.: Automatic screening for tuberculosis in chest radiographs: a survey. Quant. Imaging Med. Surg. **3**(2), 89–99 (2013)
13. McAdams, H.P., Samei, E., Dobbins III, J., et al.: Recent advances in chest radiography. Radiology **241**, 663–683 (2006)
14. Lakhani, P., Sundaram, B.: Deep learning at chest radiography: automated classification of pulmonary tuberculosis by using convolutional neural networks. Radiology **284**, 574–582 (2017)
15. Colaboratory: Frequently Asked Questions, Dez. (2019). https://research.google.com/colaboratory/faq.html

16. Bustos, A., Pertusa, A., Salinas, J.M., de la Iglesia-Vayá, M.: PadChest: a large chest X-ray image dataset with multi-label annotated reports. arXiv preprint arXiv:1901.07441. Accessed 22 Jan 2019

17. Wang, X., Peng, Y., Lu, L., Lu, Z., Bagheri, M., Summers, R.M.: Chestx-ray8: hospital-scale chest X-ray database and benchmarks on weakly-supervised classification and localization of common thorax diseases. In: Proceedings of the IEEE Conference on Computer Vision and Pattern Recognition, pp. 2097–2106 (2017)

18. Jaeger, S., et al.: Two public chest X-ray datasets for computer-aided screening of pulmonary diseases. Quant. Imaging Med. Surg. 4(6), 475 (2014)

19. Imagenet large scale visual recognition challenge 2012 (ilsvrc2012) (2012). http://image-net.org/challenges/LSVRC/2012/results.html

20. Krizhevsky, A., et al.: ImageNet classification with deep convolutional neural networks, June 2017

21. Imagenet large scale visual recognition challenge 2014 (ilsvrc2014). (2014). http://image-net.org/challenges/LSVRC/2014/results

22. Szegedy, C., et al.: Going deeper with convolutions, September 2014

23. Imagenet large scale visual recognition challenge 2015 (ilsvrc2015) (2015). http://image-net.org/challenges/LSVRC/2015/results

24. Grimnes, Ø.K.: End-to-end steering angle prediction and object detection using convolutional neural networks, June 2017

25. He, K., et al.: Deep residual learning for image recognition, December 2015

26. Sklearn Documentation. https://scikit-learn.org/stable/modules/generated/sklearn.model_selection.train_test_split.html. Accessed 03 Feb 2020

27. The HDF Group. Hierarchical data format version 5. http://www.hdfgroup.org/HDF5. (cited on page 33)

28. Keras Documentation. https://keras.io/preprocessing/image/. Accessed 03 Feb 2020

29. Maratea, A., Petrosino, A., Manzo, M.: Adjusted f-measure and kernel scaling for imbalanced data learning. Inf. Sci. 257, 331–341 (2014)

30. Harris, M., et al.: A systematic review of the diagnostic accuracy of artificial intelligence-based computer programs to analyze chest X-rays for pulmonary tuberculosis. PLoS One 14 (9), 19 (2019)

31. Denkinger, C.M., Kik, S.V., Cirillo, D.M., et al.: Defining the needs for next-generation assays for tuberculosis. J. Infect. Dis. 211(Suppl 2), S29–S38 (2015)

32. Muyoyeta, M., et al.: Implementation research to inform the use of Xpert MTB/RIF in primary health care facilities in high TB and HIV settings in resource constrained settings. PLoS One 10(6), e0126376 (2015). Epub 2015/06/02. pmid:26030301; PubMed Central PMCID: PMC4451006

33. Whiting, P.F., et al.: QUADAS-2: a revised tool for the quality assessment of diagnostic accuracy studies. Ann. Intern. Med. 155(8), 529–536 (2011)

Risk-Based AED Placement - Singapore Case

Ivan Derevitskii[1], Nikita Kogtikov[3], Michael H. Lees[2], Wentong Cai[3(✉)], and Marcus E. H. Ong[4]

[1] ITMO University, Saint-Petersburg, Russia
[2] University of Amsterdam, Amsterdam, The Netherlands
[3] School of Computer Science and Engineering,
Nanyang Technological University, Singapore 639798, Singapore
`aswtcai@ntu.edu.sg`
[4] Singapore General Hospital, Singapore, Singapore

Abstract. This paper presents a novel risk-based method for Automated External Defibrillator (AED) placement. In sudden cardiac events, availability of a nearby AED is crucial for the surviving of cardiac arrest patients. The common method uses historical Out-of-Hospital Cardiac Arrest (OHCA) data for AED placement optimization. But historical data often do not cover the entire area of investigation. The goal of this work is to develop an approach to improve the method based on historical data for AED placement. To this end, we have developed a risk-based method which generates artificial OHCAs based on a risk model. We compare our risk-based method with the one based on historical data using real Singapore OHCA occurrences from Pan-Asian Resuscitation Outcome Study (PAROS). Results show that to deploy a large number of AEDs the risk-based method outperforms the method purely using historical data on the testing dataset. This paper describes our risk-based AED placement method, discusses experimental results, and outlines future work.

Keywords: AED placement · Cardiac arrest · Optimization

1 Introduction

Cardiovascular disease accounts for near 18 million deaths a year worldwide, nearly 30% of all deaths [1]. Around 50% of these deaths are caused by sudden cardiac events and without intervention only 1% of people survive such an event. Automated External Defibrillators (AEDs) are one of the keyways to reduce mortality rates due to cardiac arrest. To impact survival rate, the time between the arrest and application of the AED is crucial. For this reason, governments worldwide have initiated AED deployment strategies in an effort to reduce deaths caused by sudden cardiac events. However, the current strategies are often driven by best practice, heuristics and guidelines (see [2]).

© Springer Nature Switzerland AG 2020
V. V. Krzhizhanovskaya et al. (Eds.): ICCS 2020, LNCS 12140, pp. 577–590, 2020.
https://doi.org/10.1007/978-3-030-50423-6_43

More recently there have been efforts to attempt more analytical approaches, where historical data and projected cardiac events are used to optimize the deployment of AEDs. In these optimization approaches there are a number of key questions or challenges that need to be addressed: firstly, how to make the projection of future cardiac events; secondly, how to determine the suitability of the placement (i.e., the likelihood of an AED successfully preventing death); and finally a method or technique to optimize the deployment.

In this paper we propose a novel risk-based method and compare it with the existing method that uses purely historical data. Our method is based on an artificial cardiac arrest model that uses demography data and cardiac arrest odd ratio for different race groups. We study the case of Singapore, a small but densely populated country that has a highly varied demography and an ageing population [3]. In Singapore cardiac arrests are one of the most common causes of death, causing over 2,000 deaths annually [4]. In 2015, the Out-of-Hospital Cardiac Arrest (OHCA) survival rate in Singapore was 23.4% [4]; whereas in other countries (e.g., the Netherlands) this can be as high as 40% [5]. Through our novel optimization approach, we aim to improve survivability for the current population, but also to identify an AED deployment that can deal with future demographic changes for the country.

For our analysis we used data that capture all OHCAs in Singapore from 2010 to 2016, which contains a total of 11,861 occurrences. For the method based on historical data, we optimized the deployment using different amount of data. For all the approaches, we tested how the deployment would have performed for the OHCAs taking place in 2016. We say that the AED deployment successfully "covers" an OHCA if there is at least one AED within 100 m – a standard rule used by many other researchers (e.g., [6]). The results demonstrate that the risk-based approach provides better coverage than the method just using historical data especially when a large number of AEDs need to be deployed. Moreover, the artificial population approach can be used to generate arbitrary number of future cardiac arrests. This allows for scenario planning where population wide exogenous effects may increase OHCA rates (e.g., national trends in diet, national trends in exercise, etc.).

The remainder of this paper is organized as follows. Section 2 describes the existing service point optimization methods. Datasets used in the paper are presented in Sect. 3. The research methodology and the proposed risk-based method are briefly described in Sect. 4. Section 5 discusses optimization results. Finally, Sect. 6 concludes the paper with directions of future work.

2   Related Works

The area of AED placement optimization is already an active field of research, and it fits within a broader research area of service point optimization. In the following two subsections we provide an overview of the existing work in service point optimization as well as a comprehensive overview of the existing approaches for AED placement optimization.

2.1   Service Point Relocation Methods

Optimization of service points is used in medicine [7–9], finance [10,11], as well as in the placement of commercial service points [12]. For spatial optimization, greedy algorithms [13], tabu-search [14], binary optimization [15], and genetic algorithms [16–18] are used. In [19], optimization of adding service points to an existing service network was proposed. However, access to a service point is often not round-the-clock and depends on time, therefore it is necessary to consider not only the spatial but also the temporal nature of service requests [20].

Often the nature of service requests depends on external data (for example, demographic data when optimizing commercial service points, or data on pedestrian and transport activity when optimizing the location of emergency medical facilities). Therefore, in this work, in addition to a method based on historical data (service requests), a method based on the demography data to determine OHCA subzone risk is proposed.

The complexity of the optimization problem may also depend on the characteristics of a particular locality and with it the features of the distribution of service requests [21]. Therefore, even the use of the existing methods is of interest for AED placement in Singapore.

2.2   AED Relocation Works

Chan et al. proposed a method based on historical data as well as a demography-based approach for AED placement in [19]. They used data of OHCA in Toronto. The dataset was divided into training and testing datasets. The training dataset was used to generate the input for the linear programming optimization and the testing dataset was used for evaluating optimization results. For the demography-based approach, people's daily activities were also considered during the optimization.

Dahan et al. proposed a method based on the historical OHCA data in [22]. They studied spatiotemporal optimization, which considers not only spatial, but also temporal availability of AEDs. Tsai et al. compared spatiotemporal approach and temporal only approach in terms of OHCA coverage in [23]. Sun et al. also studied spatiotemporal AED placement optimization with historical OHCA data in [20]. A cost-based optimization method for AED placement was proposed in [24], where cost effectiveness of the demography-based and the historical-data based method was compared. Instead of using the standard 100 m rule to determine the OHCA coverage, walking time was used in the fitness function for OHCA coverage estimation in [25].

The main disadvantage of the historical-data based method is the limited area covered by the dataset; whereas the demography-based method mentioned above does not consider cardiac arrest risk for different race groups in the population. Existing studies show that ethnicity is an important factor that affects OHCA risk. For example Malay race group have higher OHCA risk comparing to Chinese [26]. In this paper, we propose a risk-based method which considers race groups in estimating cardiac arrest risk and generates artificial OHCAs as the training dataset. A modified version of the optimization algorithm proposed in [19] is used in our work.

3   Data

Three datasets are used in this paper. The first dataset contains historical OHCAs, the second one describes the spatial demography of Singapore, and the third one provides the current (as of 6 May 2019) AED locations within Singapore. The historical OHCA dataset was obtained from Pan-Asian Resuscitation Outcome Study (PAROS). It covers all OHCA cases in Singapore from April 2010 to September 2016. This includes a total of 11,861 cases. Along with the location (given as location postcode) and the time at which symptom first occurred, associated with each OHCA there is basic demographic information about the patient such as age, ethnicity, and gender. Table 1 shows some details about this dataset, including cases per year as well as the division of cases among different race groups.

Table 1. OHCA dataset

Year	2010	2011	2012	2013	2014	2015	2016
Chinese	742	893	984	1225	1349	1587	1253
Malay	165	201	222	265	362	363	280
Indian	99	185	146	161	209	277	200
Others	75	98	88	85	117	145	85
All	1081	1377	1440	1736	2037	2372	1818

Table 2. Similarity between testing and training datasets

2010–2015	2011–2015	2012–2015	2013–2015	2014–2015	2015
48.86%	45.89%	40.58%	35.1%	28.14%	18.32%

For the results reported in Sect. 5, to test the effectiveness of the method based on historical data we group historical data into six training datasets: OHCA occurrences in 2015–2015, 2014–2015, 2013–2015, 2012–2015, 2011–2015, and 2010–2015. These training datasets are used respectively by the optimization algorithm to obtain AED placements. To test optimization results, OHCA occurrences in 2016 are used. Table 2 shows similarity between the testing and training datasets based on postcodes. Note that similar to [20], OHCAs occurring in the healthcare facilities (e.g., nursing home) are excluded in both training and testing datasets.

Table 3. Demography of town center subzone in Ang Mo Kio (2015)

Chinese	Malay	Indian	Others
4260	210	360	190

In order to generate potential future OHCA cases (and to make forecast), we use publicly available spatial demographic dataset from Singapore Data Mall (see link [27] for race-gender and [28] for age-gender data). This dataset divides Singapore into a total of 323 subzones [29] (see Fig. 1). Within each subzone the dataset provides statistics on the number of people and distributions of age, race, and gender. Table 3 illustrates an example of demography of town center subzone in Ang Mo Kio (a township in Singapore) in 2015.

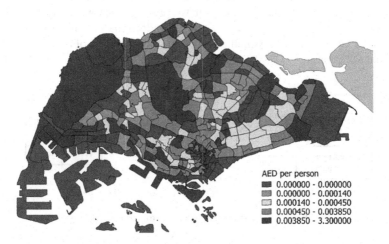

Fig. 1. A Map of Singapore divided into sub regions (each region shows the number of AEDs placed in that region per individual)

The third dataset includes the locations (longitude-latitude) of current 9,200 AEDs in Singapore (as of May 2019). Figure 1 shows the subzone boundaries as well as the AED density (AED per person) heatmap according to the dataset. It shows that the AED density varies across the entire Singapore island. However, the figure does not indicate the land use of each of the areas. For example, the large areas to the south-west (Jurong Island and Tuas) and to the east (Changi Airport) are non-residential areas. Within some other subzones there are also areas with very few residents.

There are a number of important caveats that we must mention regarding assumptions and inaccuracies about the data we use. Firstly, Singapore currently provides a number of mobile AEDs via the SMRT taxi company [30], the AEDs are placed within 1000 vehicles in the fleet. This accounts for an extra 12% of

AEDs that are currently available and are not captured in our dataset. However, these AEDs were placed at the end of 2015 so their impact on the results should be minimal. Secondly, during our optimization procedure we assume that all building locations on the map are potential AED sites and that these sites are available for 24 h. In reality, some locations may be inaccessible to the public, or only accessible during office hours. Building data were obtained via Open Street Map [31] and this dataset includes 141,821 locations of buildings in Singapore.

4   Research Methodology

In this paper we develop a risk-based method for AED placement, and to analyze its effectiveness we compare the result we obtain with alternative AED placement strategies. These include the current real-world AED placement and a traditional approach that simply uses historical occurrences of cardiac arrests to optimize the placement of AEDs. In this section we describe historical-data based and risk-based methods in detail, along with the underlying optimization technique and the evaluation criterion which assesses the effectiveness of a particular AED placement solution.

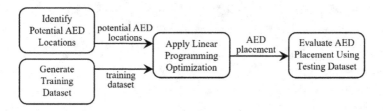

Fig. 2. Overview of research methodology

Figure 2 shows an overview of the research methodology. Initially the dataset regarding potential AED locations is created. Potential locations (buildings) are identified using Open Street Map. This means that the AED optimization is considered at the resolution of a building, and the exact placement of the AED within the building is not explicitly modelled. All methods use the same set of potential locations. The training dataset used by the optimization algorithm differs in the case of the historical-data based approach and the risk-based method. In the historical-data based approach, as explained in Sect. 3 we group OHCA occurrences in 2010–2015 into six training datasets; whereas in the risk-based method artificial OHCAs are generated based on the demographics of the population in 2015.

The potential location data and the training data are then combined by creating a weighted adjacency matrix, where the weights of the edges are calculated as the straight-line distances from each potential AED location to each OHCA location in the training data. This matrix is used as the input data in the linear programming optimization algorithm (described below and based on [19]).

Finally, an AED placement generated is evaluated by testing how many of the 2016 OHCA cases would be covered by the placement.

4.1   Optimization Approach and Evaluation Criterion

The input to the optimization algorithm takes the form of an adjacency matrix as described above. In this paper we use a modified version of the linear programming algorithm in [19] for finding optimal location of AEDs. This is a binary optimization problem (also known as the Maximal Covering Location Problem [15]), defined as follows:

$$\text{maximize} \sum_{j=1}^{J} x_j; \text{ subject to} \sum_{i=1}^{I} y_i = N \text{ and } x_j \leq \sum_{i=1}^{I} a_{ij} y_i, \forall j = 1, ..., J \quad (1)$$

where x_j is a binary parameter indicating whether cardiac arrest j is covered or not; y_i is a binary decision variable indicating whether an AED is placed in location i or not; and a_{ij} is binary parameter indicating whether cardiac arrest j is coverable by location i (i.e., within 100 m). N is the number of locations in which AEDs are placed. I is the number of potential locations where AEDs can be placed. J is the number of cardiac arrests in the training dataset.

Each of the methods will produce a placement of AEDs, that is a mapping of N AEDs to potential locations. From this placement an assessment is made to see what fraction of OHCAs recorded in 2016 would have been covered. As mentioned above, in this paper we use a simple measure of coverage, that is, an OHCA is considered covered by the placement if it is within 100 m from the nearest AED. This does not include the time it takes for an individual to locate the AED, or any vertical distance that must be covered.

4.2   Historical-Data Based Method

The historical-data based method tries to optimize the AED placement under the assumption that previous OHCA occurrences are a good predictor for future OHCA cases. We have data from a six-year period 2010–2016. In this approach, we use different combinations of historical data as a predictor for future OHCAs. In particular, we use 2016 as our testing dataset and see if the data from 2015–2015, 2014–2015, 2013–2015, 2012–2015, 2011–2015, and 2010–2015 are good predictors for the OHCAs in 2016. The effectiveness of this approach is determined by the demographic stability in Singapore. If the occurrence of OHCAs are clustered and stable over a period of years then this approach should be effective on some time scale.

4.3   Risk-Based Method

The risk-based method tries to capture various factors that might predict future OHCA cases, including population density and distribution of population in terms of ethnicity. Using the current population data, we can identify high risk subzones of the Singapore island. From this a set of potential or artificial OHCAs can be generated, which will be used as the training dataset to optimize the placement of AEDs.

Singapore is a multi-cultural city with most of the country's inhabitants coming from four major ethnic groups: Chinese (74.3%), Malay (13.4%), Indians (9%) and others (3.2%), this distribution has remained remarkably stable since Singapore's independence (more than 50 years) [32]. As in other countries [33] the risk of heart disease and cardiac arrest does vary by race. Using the OHCA data we can derive an associated risk of OHCA for a specific race group, r_{race}, as follows:

$$r_{race} = \frac{\sum_{year=2010}^{2015} n_{race}^{year}}{\sum_{year=2010}^{2015} p_{race}^{year}} \times \frac{\sum_{year=2010}^{2015} p_{Chinese}^{year}}{\sum_{year=2010}^{2015} n_{Chinese}^{year}} \tag{2}$$

where $race \in \{Chinese, Malay, Indian, Others\}$; n_{race}^{year} is the number OHCA cases for a specific race in a specific year; p_{race}^{year} is the number of people in Singapore for a specific race in a specific year. Note that race group risk is estimated using historical OHCA occurrences in 2010–2015 and that the value is normalized with the OHCA risk of the Chinese race group. The calculated risk values for various race groups are shown in Table 4.

Table 4. Risks for different races

Chinese	Malay	Indian	Others
1.000	1.181	1.635	4.833

Given an associated risk for each race group, we then calculate an associated risk for each subzone of Singapore. To do this we use the data about the total population size and race distributions in each of Singapore's subzones in 2015 (see Sect. 3 for detail). Subzone OHCA risk, R_s, is calculated as follows:

$$R_s = \frac{\sum_{i \in \{Chinese, Indian, Malay, Others\}} r_i \times dem_{i,s}}{dem_s} \tag{3}$$

where r_i is the OHCA risk for race group i; $dem_{i,s}$ is the number of people in race group i who live in subzone s in 2015; and dem_s is the total number of people living in subzone s in 2015.

Based on subzone OHCA risks, a zonal risk map can be generated. As shown in Fig. 3, we use a simple algorithm to generate a set of potential/artificial OHCAs. To generate a location for a potential OHCA, a subzone is first chosen based on subzone OHCA risks. The higher the subzone OHCA risk is, the more often a subzone will be chosen. After that a random location (building) will be selected within the subzone for the location of OHCA. This procedure terminates when the required

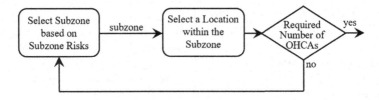

Fig. 3. Artificial OHCA generation

number of artificial OHCAs are generated. These artificial OHCAs then serve as training dataset to the optimization algorithm (see Fig. 2). With this method we can in principle generate any number of OHCAs. The number of artificial OHCAs is currently set as a parameter in our experiments.

5   AED Optimization Result

In this section we demonstrate the results of different optimization methods presented in Sect. 4 and compare their performance against the current deployment of AEDs. Recall that in all scenarios the performance of the deployment is based on the fraction of OHCAs occurring in 2016 (a total of 1,818 cases) that would be covered by at least one AED (that is, there is at least one AED existing within 100 m).

5.1   Historical-Data Based Method

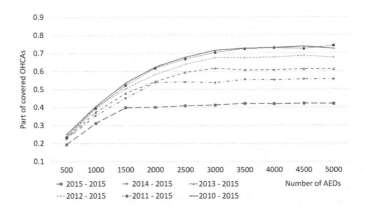

Fig. 4. Results of historical-data based method for different training datasets

Figure 4 shows the result of historical-data based method using 6 different training datasets (from 1-year data 2015–2015 to 6-year dataset 2010–2015). It clearly shows the performance depends on the training dataset. Table 1 presents the

number of OHCAs for each year, which gives a clear indication of the size of
each training dataset. Due to the increase of similarity between training and
testing datasets (see Table 2), for a given number of AEDs to be deployed, there
is an increasing improvement on the OHCA coverage when more data are added
into the training dataset. However, the improvement diminishes as more historical data are added to the training dataset since the increase of similarity becomes
smaller (see Table 2). In principle however, more data do not necessarily imply
better performance. This depends on the dynamics of the population and how
uniform the aging process is across the country. Due to the housing policy it is
likely that the most at risk subzones may move as the population of particular
housing developments age.

All scenarios shown in Fig. 4 have similar performance when a small number
of AEDs are deployed because the optimization algorithm tends to put AEDs
first in the areas that have frequent OHCA occurrences over the years. When
a large number of AEDs are deployed, the larger datasets obviously work better because they cover larger area and have higher similarity with the training
dataset (as showed in Table 2). Note that all the curves become flat after the
number of AEDs increases to a certain level. This is because even the largest
6-year training dataset does not cover all areas in Singapore and its similarity
with the testing dataset is only 48.86%. The figure also shows that as the amount
of training data increases, the limit of maximal coverage will also increase (from
43% for the 1-year data to 74% for 6-year data).

5.2 Risk-Based Method

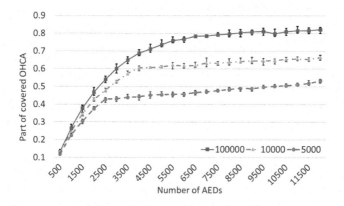

Fig. 5. Results of risk-based method for different training dataset

Figure 5 shows the results of the risk-based method. Recall that the risk-based
method uses artificial OHCAs generated from subzone risks as training dataset
for the optimization algorithm (see Sect. 4). In our experiments, the number of
generated artificial OHCAs in the training dataset varies from 5,000, 10,000, to

100,000. Obviously, for a given number AEDs to be deployed, better coverage is achieved with larger training dataset. Interestingly, we see a saturation effect similar to the historical-data based method for the training dataset with 5,000 and 10,000 artificial OHCAs. The advantage of using the risk-based method is that we can generate the training dataset with arbitrarily large number of artificial OHCAs. Note that the largest training dataset used in the historical-data based method (i.e., the 6-year dataset) contains only 10,043 OHCA cases.

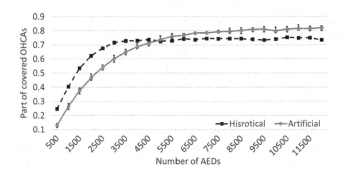

Fig. 6. Comparison between historical-data based and risk-based methods

Figure 6 compares the best results obtained using the historical-data based and risk-based methods. When the number of AEDs to be deployed is small, historical-data based method performs better than the risk-based method. However, when a large number of AEDs needs to be deployed, the risk-based method would be a better choice. In order words, the training dataset used in the risk-based method is able to provide more meaningful information to guide the deployment of large number of AEDs. The performance crossover happens at around 6,500 AEDs. For a reference, there are currently 9,203 AEDs deployed in Singapore (as of 6 May 2019).

6   Conclusion

We studied the problem of optimizing AED placement in Singapore. Both the historical-data based and the risk-based methods were used in our investigation. OHCA occurrences from 2010 to 2015 in Singapore were used as the training datasets in the historical-data based method. In the risk-based method, historical OHCAs and demography data were used to derive subzone OHCA risks in Singapore, which in turn were used to generate potential or artificial OHCAs in the training dataset. The performance of both methods was evaluated using OHCA occurrences in 2016.

The results demonstrate a number of interesting findings. Firstly, irrespective of the training datasets used in the historical-data based method, the first 2,000 AEDs placed have an equally effective coverage of OHCAs. Secondly, for all the training dataset used in the historical-data based method, the OHCA coverage does not increase further after the number of AEDs reaches a certain level. Finally, for a small number of AEDs, the historical-data based method gives a better OHCA coverage than the risk-based method. However, when a large number of AEDs needs to be deployed, the risk-based method would be a better choice.

As for our future work, to combine the two methods, the risk model can be used to generate artificial OHCAs to complement historical data. Most of the existing work determines OHCA coverage either in terms of distance or time, to better define OHCA coverage we will be developing a probabilistic model using historical OHCA data and AED location data. Countries around the world are dealing with ageing populations. The dynamics of this ageing can introduce very unique risks, which may be geographically correlated. We are also planning to investigate how these changing risk maps for cardiac arrests may impact the placement of AEDs.

Acknowledgments. This work was supported by the National Research Foundation (NRF) of Singapore, GovTech, under its Virtual Singapore Program Grant No. NRF2017VSG-AT3DCM001-031. Ivan Derevitskii is also financially supported by The Russian Science Foundation, Agreement №17-71-30029 with co-financing from Bank Saint Petersburg.

References

1. World Health Organization. https://www.who.int/news-room/fact-sheets/detail/cardiovascular-diseases-(cvds). Accessed Apr 2020
2. Resuscitation Council (UK). https://www.resus.org.uk/publications/a-guide-to-a. Accessed Apr 2020
3. Department of Statistics Singapore. https://www.singstat.gov.sg/-/media/files/visualising_data/infographics/population/population-trends2019.pdf. Accessed Apr 2020
4. Singapore Heart Foundation. https://www.myheart.org.sg/programmes/save-a-life-initiative/. Accessed Apr 2020
5. Boyce, L.W., et al.: High survival rate of 43% in out-of-hospital cardiac arrest patients in an optimised chain of survival. Netherlands Heart J. **23**(1), 20–25 (2015)
6. Chan, T.C.Y., Demirtas, D., Kwon, R.H.: Optimizing the deployment of public access defibrillators. Manage. Sci. **62**(12), 3617–3635 (2016)
7. Brotcorne, L., Laporte, G., Semet, F.: Ambulance location and relocation models. Eur. J. Oper. Res. **147**, 451–463 (2003)
8. Baray, J.Ô., Cliquet, G.: Optimizing locations through a maximum covering/p-median hierarchical model: maternity hospitals in France. J. Bus. Res. **66**(1), 127–132 (2013)
9. Li, X., Zhao, Z., Zhu, X., Wyatt, T.: Covering models and optimization techniques for emergency response facility location and planning: a review. Math. Methods Oper. Res. **74**(3), 291–310 (2011)
10. Allahi, S., Mobin, M., Vafadarnikjoo, A., Salmon, C.: An integrated AHP-GIS-MCLP method to locate bank branches. In: Industrial and Systems Engineering Research Conference (ISERC), Nashville, Tennessee, USA, June 2015

11. Miliotis, P., Dimopoulou, M., Giannikos, I.: A hierarchical location model for locating bank branches in a competitive environment. Int. Trans. Oper. Res. **9**(5), 549–565 (2002)

12. Wu, S., Kuang, H., Lo, S.-M.: Modeling shopping center location choice: shopper preference-based competitive location model. J. Urban Plan. Dev. **145**(1), 04018047 (2018)

13. Tong, D., Murray, A., Xiao, N.: Heuristics in spatial analysis: a genetic algorithm for coverage maximization. Ann. Assoc. Am. Geogr. **99**(4), 698–711 (2009)

14. Basar, A., Kabak, Ö., Topcu, Y.I.: A tabu search algorithm for multi-period bank branch location problem: a case study in a Turkish bank. Scientia Iranica **26**, 3728–3746 (2018)

15. Church, R., ReVelle, C.: The maximal covering location problem. Papers of the Regional Science Association, vol. 32, no. 1, pp. 101–118 (1974)

16. Beasley, J.E., Chu, P.C.: A genetic algorithm for the set covering problem. Eur. J. Oper. Res. **94**(2), 392–404 (1996)

17. Aickelin, U.: An indirect genetic algorithm for set covering problems. J. Oper. Res. Soc. **53**(10), 1118–1126 (2002)

18. Solar, M., Parada, V., Urrutia, R.: A parallel genetic algorithm to solve the set-covering problem. Comput. Oper. Res. **29**, 1221–1235 (2002)

19. Chan, T.C.Y., et al.: Identifying locations for public access defibrillators using mathematical optimization. Circulation **127**(17), 1801–1809 (2013)

20. Sun, C.L.F., Demirtas, D., Brooks, S.C., Morrison, L.J., Chan, T.C.: Overcoming spatial and temporal barriers to public access defibrillators via optimization. J. Am. Coll. Cardiol. **68**(8), 83–845 (2016)

21. Hansen, C.M., et al.: Automated external defibrillators inaccessible to more than half of nearby cardiac arrests in public locations during evening, nighttime, and weekends. Circulation **128**(20), 2224–2231 (2013)

22. Dahan, B., et al.: Optimization of automated external defibrillator deployment outdoors: an evidence-based approach. Resuscitation **108**, 68–74 (2016)

23. Tsai, Y.S., Ko, P.C.-I., Huang, C.-Y., Wen, T.-H.: Optimizing locations for the installation of automated external defibrillators (AEDs) in urban public streets through the use of spatial and temporal weighting schemes. Aplli. Geogr. **35**(1–2), 394–404 (2012)

24. Tierney, N.J., et al.: Novel relocation methods for automatic external defibrillator improve out-of-hospital cardiac arrest coverage under limited resources. Resuscitation **125**, 83–89 (2018)

25. Bonnet, B., Dessavre, D.G., Kraus, K., Ramirez-Marquez, J.E.: Optimal placement of public-access AEDs in urban environments. Comput. Ind. Eng. **90**, 269–280 (2015)

26. Rakun, A., et al.: Ethnic and neighborhood socioeconomic differences in incidence and survival from out of hospital cardiac arrest in Singapore. Prehosp. Emer. Care **23**(5), 619–630 (2019)

27. Singapore government data. https://data.gov.sg/dataset/resident-population-by-planning-area-subzone-ethnic-group-and-sex-2015. Accessed Apr 2020

28. Singapore government data. data.gov.sg/dataset/resident-population-by-planning-area-subzone-age-group-and-sex-2015. Accessed Apr 2020

29. Singapore subzones master plan. https://data.gov.sg/dataset/master-plan-2014-subzone-boundary-web. Accessed Apr 2020

30. HeartSine.   https://heartsine.com/2015/12/singapore-taxi-drivers-will-now-protect-public-with-on-board-defibrillators/

31. OpenStreetMap. https://www.openstreetmap.org/. Accessed Apr 2020
32. Teo, T.-A.: Civic Multiculturalism in Singapore - Revisiting Citizenship. Rights and Recognition. Palgrave, London (2019)
33. Becker, B., et al.: Racial differences in the incidence of cardiac arrest and subsequent survival. N. Engl. J. Med. **329**(9), 600–606 (1993)

Time Expressions Identification Without Human-Labeled Corpus for Clinical Text Mining in Russian

Anastasia A. Funkner⑩ and Sergey V. Kovalchuk$^{(\boxtimes)}$⑩

ITMO University, Saint Petersburg, Russia
{funkner.anastasia,kovalchuk}@itmo.ru

Abstract. To obtain accurate predictive models in medicine, it is necessary to use complete relevant information about the patient. We propose an approach for extracting temporary expressions from unlabeled natural language texts. This approach can be used for the first analysis of the corpus, for data labeling as the first stage, or for obtaining linguistic constructions that can be used for a rule-based approach to retrieve information. Our method includes the sequential use of several machine learning and natural language processing methods: classification of sentences, the transformation of word bag frequencies, clustering of sentences with time expressions, classification of new data into clusters and construction of sentence profiles using feature importances. With this method, we derive the list of the most frequent time expressions and extract events and/or time events for 9801 sentences of anamnesis in Russian. The proposed approach is independent of the corpus language and can be used for other tasks, for example, extracting an experiencer of a disease.

Keywords: Time expression recognition · Natural language processing · Corpus labeling · Clinical text mining · Machine learning

1   Introduction

Since hospitals began to use medical information systems, a large amount of data has accumulated in electronic form. Most often, information is stored in the form of an electronic medical record (EMR), which relates to a particular patient. Data can be stored in a structured form (tables with lab results), in semi-structured (filled fields in some form, for example, an operation protocol) and unstructured (free texts in natural language). Structured data usually requires a little processing and can be easily used to model medical and healthcare processes. In semi-structured data, it is necessary to extract features from natural language texts. However, the researcher has an idea about the topic or even the structure of the content in a field of the completed form and can extract features using regular expressions or linguistic patterns [1]. To work with unstructured data, it is necessary to use many methods and tools to find out specific facts about the patient.

V. V. Krzhizhanovskaya et al. (Eds.): ICCS 2020, LNCS 12140, pp. 591–602, 2020.
https://doi.org/10.1007/978-3-030-50423-6_44

The purpose of this study is to develop methods for extracting time expressions and related events. At the same time, the proposed methods should work with an unlabeled or minimally labeled input text corpus.

Processing of medical texts includes tasks which are often encountered for other texts: morphological and syntactical analysis, negation detection, temporal processing, etc. But there are specific tasks such as recovering family history from free texts [2]. The problem of time expression recognition relates to information retrieval from unstructured texts. Two approaches can be used to extract information [2]. The first approach includes rule-based methods. In this case, the necessary information is searched using regular expressions or linguistic patterns which are known in advance or are collected iteratively during multiple scanning of the text corpus [3, 4].

The second approach is based on machine learning methods. This approach does not require manually searching for the necessary constructions in the texts or building a knowledge base for each new corpus. The developed models are trained and detect patterns in the data automatically. However, for the training of any model, a labeled corpus is required [5]. However, for the Russian language, there is almost no labeled corpus of medical texts [6]. In [7, 8] only 120 EMRs were labeled for a specific hospital unit. State-of-the-art models for clinical temporal relation extraction are trained on thousands of labeled sentences [9]. Since we can not to mark up a sufficient amount of data to use the above methods, we try to develop methods for detecting time expressions and events using only 1k sentences, which are labelled as containing and not containing a time expression (TE).

The tasks of recognizing time expressions and extracting events can be solved simultaneously or in parallel using different models. The second one is called the clinical temporal relation extraction task. In this case, researchers usually train deep neural networks using a labeled corpus, which consists of sentences where events and temporal expressions are defined in advance [9, 10]. In this paper, we do not have an aim to extract both the event and the time expressions. However, Sect. 3 includes examples when the event is found with the timestamp using our approach.

In 2019, we started developing an application to process medical texts. Figure 1 shows seven modules of this application. We have already developed modules for basic text processing (extracting abbreviations, searching for numerical data, lemmatisation, if necessary), correcting typos in texts [11], detection and removing negations [12], and topic segmentation of texts. Currently, two modules are being developed for determining the experiencer of a disease and a module for determining events with a timestamp [13]. This paper is about mining temporal data using machine learning. In the future, we planned to run these modules sequentially and automatically retrieve a set of patient's features. This application can be helpful for hospital staff as a navigation system in medical history because it allows quickly to find out what and when occurred with the patient in the past. Also, the application can be useful for scientists to build more accurate predictive models. As far as we know, there is no other application or system for the Russian language that could process the text with above-mentioned methods. Existing libraries and modules are trained with general Russian corpus and do not work accurately enough for medical texts [14, 15].

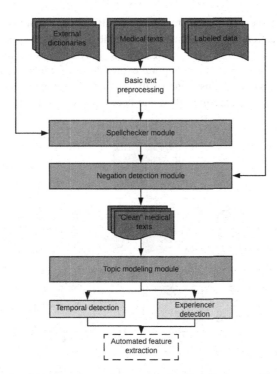

Fig. 1. An application for processing and extracting from medical texts. The green modules are finished and can be used [11, 12]. Pink modules are being developed now, and dotted ones will be created in the future. (Color figure online)

2   Method

This section describes which and in what order machine learning methods are used to extract common expressions, including temporal ones. Figure 2 shows the general scheme of the method with the flow of data and trained models between stages.

First, it is necessary to divide the data corpus into three groups, each of which is represented by a set of sentences. A group X consists of the smallest number of sentences and its size is defined by researchers so that they can label sentences in this group without spending too much time. A group Y contains a larger number of sentences and its size is limited by available computing power since $X + Y$ groups are used for clustering, in which a distance matrix is usually calculated (depends on the method). All remaining sentences form a group Z.

2.1   Manual Labeling of the Group X

The sentences of group X must be labeled as having and not having a time expression (TE and noTE). The temporal expression refers to any expression that indicates a period or a specific moment in time. According to the TimeML annotation system, temporal expressions can be divided into five classes: date (in 2010), time (at 4 pm), duration (over the past 10 years), recurring events (once a month) and others [16]. Using this classification, we label sentences with the first three classes. Other classes of expressions were rare in our corpus. Besides, in [14] there is more specific for the Russian language classification of temporal expressions, of which we also used only a few classes for markup. At this stage of the method, the researchers determine what temporal structures which are important and necessary for their corpus (see Fig. 2(1)).

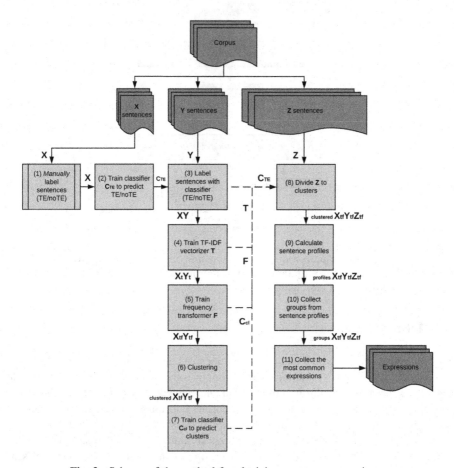

Fig. 2. Scheme of the method for obtaining common expressions.

2.2   Classifier to Label the Group Y

At this stage, it is necessary to select a classification model, train and test it with the labeled group X. The classification model helps to mark up an additional set of sentences (the group Y) without additional time. We use bag of words to present sentences for the classifier, but other representations are applied here (word embeddings, TF-IDF vectors, etc.). After training and testing, it is necessary to pay attention to sentences that turn out to be false negative or false positive after testing. There may be sentences incorrectly marked manually. As a result of this stage, we have C_{TE} classifier and joined labeled groups X and Y (see Fig. 2(2,3)).

2.3 Frequency Transformations

At stage (4) and (5), it is necessary to collect a bag of words with n-grams, if this has not been done before. The number n depends on the language of the corpus. The Russian language is characterized by long temporal expressions with compound prepositions, so we recommend using $n = 4$. Next, the transformer T is trained to obtain the term frequency (TF) and inverse document frequency (IDF). The transformer T remembers IDF of the training set and can be used later with new data. TF-IDF transformation helps to reduce the weight of frequent and non-specific words and words, which are rare and found in a small number of documents (see Fig. 2(4)). After calculating TF-IDF vectors and obtaining $X_t Y_t$, we compare the frequencies of sets with and without temporal constructions (S_{TE} and S_{noTE}: $X_t \cup Y_t = S_{TE} \cup S_{noTE}$). We assume that using frequencies of the S_{noTE} will reduce the weights of words that are not related to time, but also frequent in documents of S_{TE} (medical terms in our case). We simply subtract from each TF-ID vector the average vector of all TF-ID vectors from S_{noTE} and zero the negative components for S_{TE}. At the end, the function with this subtraction is saved in the transformer F and transformed vectors are called $X_{tf} Y_{tf}$ (see Fig. 2(5)).

2.4 Clustering and Clusters' Classifier

Obtained at the previous stage vectors are used for clustering. We assume that similar sentences can join in one cluster. Then, using the obtained clusters, another classifier is trained, and new sentences of the group Z are appended to the clusters. If the initial corpus is small and the group Z was not collected, then we can skip stages (7) and (8) (see Fig. 2). At the stage (8) all above mentioned trained models are used.

2.5 Expression Retrieval from Sentence Profile

At the next stages of the method, we try to find the expressions inside the sentence that characterize it and determine its membership in the cluster. Using the classifier from the stage (7), we obtain feature importances for each cluster and summarize them for all components of the considered sentence. We propose using the classification one-vs-all strategy and tree-based models (decision trees, random forest, tree boosting) as basic classifiers. Such models make it easy to obtain feature importances. If the basic models are not interpretable, then model-agnostic interpretation methods can be used [17]. As a result, the sum vector is called a sentence profile and includes the groups of consequent words with positive importance (expressions). We can also adjust the threshold for more stringent selection of expressions. The resulting expressions are grouped by their syntax structure to define the most common ones (see Fig. 2(9)–(10)).

3   Results

In this research, we used a set of anonymized 3434 EMRs of patients with the acute coronary syndrome (ACS) who admitted to Almazov National Medical Research Centre (Almazov Centre) during 2010–2015. Disease anamneses are one of the most unstructured records (free text without any tags; each physician writes as he/she prefers), and therefore we used them to demonstrate our approach. Totally, there are 34241 sentences in the set.

We divide the data set into 1k sentences (group X), 3k sentences (group Y) and the rest (group Z). We spent only 20 min to mark the group X. To label the group Y with TE/noTE, we train a decision tree (F1 score equals 0.95). Before TF-IDF transformation, enumerated temporal pointers were replaced by tags: years (2002, 1993 -> "#year"), months (January -> "#month"), dates (01/23/2010 -> "#fulldate"), etc. This helps not to divide temporal pointers to several components of the vector when forming a bag of words. Figure 3 shows the difference in the average TF-IDF values for a set with and without TEs, as well as the top 10 words with positive and negative differences. Specific components to a set with TE are much greater than zero and are associated with temporal pointers ("year", "#year", "#month"). The typical words for a set without TE are much less than zero. These words are associated with negations of diseases ("not"), patient's well-being ("normal", "satisfactory"), medications ("take") or a general description of the patient's condition ("disease", "chronic", "age"). These topics are usually not accompanied by temporal pointers.

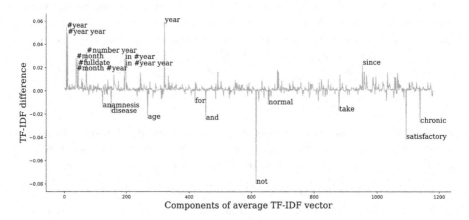

Fig. 3. The difference of average TF-IDF vectors for sets of sentences with TEs and without ones.

Hierarchical clustering is used to divide sentences with TEs. However, we perform clustering three times to divide large clusters into smaller ones. After the first clustering, 13 groups are identified so that 11 small ones include sentences with an almost identical structure. We suppose that the doctors of Almazov Centre used almost the same formulations when they write about an exact disease. For example, one of the clusters consists of almost identical phrases: "in #month #year, the patient had coronary angiography, no pathology." Large clusters gather unique sentences. Figure 4 shows the cluster identified after the first clustering and shows the thresholds with which the big clusters were then divided. As a result, 1145 sentences with TEs are divided into 23 clusters ranging in size from 14 to 237.

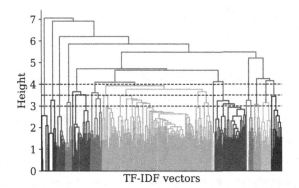

Fig. 4. Multiple hierarchy clustering of TF-IDF vectors (sentences) with TE.

Then, to divide the group Z into clusters, classifiers of tree boosting were trained using the classification one-vs-all strategy (F1 score varies from 0.83 to 0.96), and profiles were calculated for each sentence as described in Sect. 2.5. Figure 5, 6 and 7 show the profiles of sentences and defined groups of TF-IDF values. In Fig. 5 a simple sentence has two groups: temporal one ("#year year") and one with the event ("appear pressing pain behind the sternum"). In Fig. 6 we show the profile of a compound sentence, which should be divided into three simpler sentences. The event and TE of the first part were extracted: "treatment in the neurology department", "emergency treatment", and "in #year". Figure 7 depicts an unsuccessful example of separation into expressions since the event and time remain in the same group.

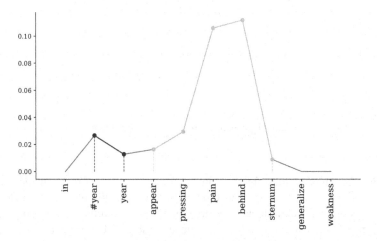

Fig. 5. A sentence profile for "In 2007 pressing pain behind the sternum and generalized weakness appeared" (the word order in x-labels is defined by Russian).

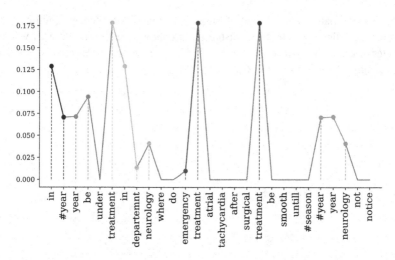

Fig. 6. A sentence profile for "In 2012 [the patient] was under treatment in the department of neurology where emergency treatment [because of] atrial tachycardia was done, after surgical treatment [the recover] was smooth, until the autumn of 2013 [the patient] did not notice neurology [problems]" (the word order in x-labels is defined by Russian).

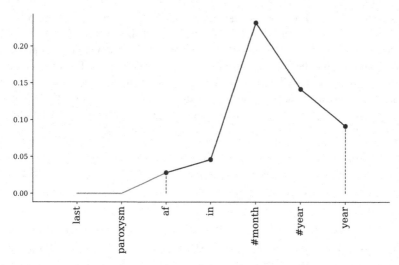

Fig. 7. A sentence profile for "Last paroxysm [of] atrial fibrillation (AF) [was] in March of 2008".

Table 1 contains the most frequent time expressions derived from sentence profiles. Almost all time expressions contain time tags ("#year", "#fulldate", etc.). However, there is a group in which time adverbs are contained. In addition to grouping by tags, we use morphological constructions obtained with the TreeTagger [18, 19]. The combination of morphological constructions and numerical regular expressions can be used to find temporal expressions in new sentences. Also, for each type of time expression, we will specify a normalization function to define the exact date or period of the event [20, 21].

Table 1. The most popular time expressions among 9801 sentences with TE.

Time expression	Treetagger construction for Russian	Examples in Russian	Examples in English	Number
in #year year	Sp-l Mc---d Ncmsnnl	в 1983 году	in 1983	1827
from #year years	Sp-l Mc---d Ncmpgn	с 45 лет	from 45 years [old]	1428
#fulldate	Mc---d	4.03.2008	4.03.2008, 16.04.10	1030
about #number years	Sp-g Mc--g Ncmpgn	около 3-ёх лет	about 3 years	1030
in #month #year years	Sp-l Ncmsln Mc---d Ncmsgn	в апреле 2010 года	In April 2010 [years]	930
#year years	Mc---d Ncmsgn	2009 года	2009 [years]	830
adverb	R	неоднократно, после, около, вплоть, вновь, впервые, затем	repeatedly, after, about, until, again, for the first time, then	830
#year	Mc---d	2008	2008	565
#number years	Mc---d Ncmpgn	10 лет	10 years	565
about #number years ago	Sp-l Mc---d Ncmpgn R	около 10 лет назад	about 10 years ago	266
#number-#number	Mc---d	35-40, 2006-2009	35-40, 2006-2009	232
during #number	Sp-l Ncnsln Mc---d	в течение 5	during 5	199
from #fulldate to #fulldate	Sp-l Mc---d Sp-l Mc---d	с 2.03.2010 по 5.04.2010	from 2.03.2010 to 5.04.2010	133
from #fulldata	Sp-l Mc---d	от 07.05.09	from 07.05.09	100
#season #year year	Ncfsin Mc---d Ncmsgn	весной 2010 года	spring 2010 [year]	100
#time	Mc---d	10:30, 23-00	10:30, 23-00	66

Table 2 and Table 3 show the example of patient anamnesis with defined expressions in Russian and English. The anamnesis consists of seven sentences and includes different types of time expressions: years, dates, periods, a start of a period, repeated actions, etc. This example shows how complex a free clinical text can be: many abbreviations and partially written words, grammatical and syntax mistakes, compound sentences. However, time expressions are found in five sentences and events in four sentences. We hope that the recovery of syntax structure and abbreviations can help to improve the quality of extracted expressions.

Table 2. The example of anamnesis with extracted expressions in Russian.

Anamnesis in Russian
[Перенёс] 2 [Инфаркта миокарда: 2001] и [2006г].
Сегодня повторно вызвал бригаду [СМП,] которой был [доставлен в НИИ Кардиологии] для решения вопроса об имплантации ЭКС.
Считает себя больным в течение [20 лет,] когда впервые стал отмечает начало подъёма [АД до 190\110 мм.рт.ст.,] [адаптирован] к [цифрам АД 140\80 мм.рт.ст].
[03.10.10] вызывал СМП, однако в транспортировке в стационар больному было отказано.
[С 2001г появились давящие] загрудинные [боли,] возникающие [при физической нагрузке] выше [обычной,] купирующиеся после её прекращения [в течение 5-7] минут или приёмом нитроглицерина.
Нынешнее ухудшение началось примерно неделю назад, когда появилось урежение [ЧСС] до 34-36 в мин., слабость, [повторные эпизоды потери] сознания.
Последняя госпитализация [с 17.09.10 по 29.09.10г] [с диагнозом:] ИБС: прогрессирующая стенокардия ПИКС [2006г].

Table 3. The example of anamnesis with extracted expressions in English.

Anamnesis in English
[Have] 2 [Myocardial infarction: 2001] and [2006].
Today, he again called the [ambulance,] which [delivered to the Cardiology Institute] to resolve the issue of implantation of ECS.
He considers himself sick for [20 years], when he first began to note the beginning of the rise [blood pressure to 190\110 mm Hg,] [adapted] to [numbers of blood pressure 140\80 mm Hg].
[03.10.10] called the ambulance, but the patient was refused transportation to the hospital.
[Since 2001 there were pressing] sternal [pains] arising [during physical exertion] above [normal,] stopping after its cessation [within 5-7] minutes or by taking nitroglycerin.
The current deterioration began about a week ago, when there was a decrease in [heart rate] to 34-36 per minute, weakness, [repeated episodes of loss] of consciousness.
The last hospitalization [from 17.09.10 to 09.29.10] [with a diagnosis of] IHD: progressive angina pectoris PIKS [2006].

4   Discussion

The method described in Sect. 2 helps to get an idea about the set of texts in a natural language quickly. As a result, the list of final constructions turns out to be quite large and contains many unnecessary or incorrectly extracted expressions (see Fig. 7). We suppose these actions can help to improve quality of extracted expressions: complex sentences need to be divided into simple ones, acronyms and abbreviations of medical terms need to be deciphered, and important constructions inside sentences need to be found with shallow parsing [22]. However, to solve these problems, a model of syntax parsing needs to be trained with a specific medical corpus in Russian. So far, we did not find such an open access model. Also, in this paper, we work with anamnesis texts that describe the patient's events in the past. However, other texts in a natural language may contain information about future events and to solve this problem we can mark sentences as past, future and without TE.

In the future, to improve the quality and create more accurate models, we will have to mark up the words in sentences, for example, the word can belong to an event, TE, or none of this. However, there are many labeling systems: TimeML, BIO, TOMN, etc. [10]. We plan to compare what a system is best for our corpus and tasks.

5   Conclusion

In this paper, we propose a way to work with an unlabeled corpus of texts. This approach can be used for the first analysis of the corpus, for data labeling as the first stage, or for obtaining linguistic constructions that can be used for a rule-based approach to retrieve information [3, 4]. The proposed approach is independent of the corpus type and can be used for other tasks, for example, extracting an experiencer of disease [13].

To solve the problem of TEs and events recognition, we plan to train the model for syntax parsing sentences and label the corpus according to one of the known labeling systems, using the already obtained expressions [10].

When the development of the modules is completed (see Fig. 1), we plan to improve the already developed models for predicting diseases [23–25] and integrate the modules into the medical system of the Almazov Centre. It will help hospital staff to navigate the patient history and allows doctors to reduce the time when they need to get acquainted with patient data during the appointment.

Acknowledgements. This work is financially supported by National Center for Cognitive Research of ITMO University.

References

1. Jackson, P., Moulinier, I.: Natural Language Processing for Online Applications: Text Retrieval, Extraction, and Categorization. John Benjamins Publishing Company, Amsterdam (2002)
2. Dalianis, H.: Clinical Text Mining: Secondary Use of Electronic Patient Records. Springer, Cham (2018). https://doi.org/10.1007/978-3-319-78503-5
3. Riloff, E.: Automatically constructing a dictionary for information extraction tasks. In: Proceedings of National Conference on Artificial Intelligence, pp. 811–816 (1993)
4. Riloff, E., Jones, R.: Learning dictionaries for information bootstrapping extraction by multi-level. In: Proceeding AAAI 1999/IAAI 1999 Proceedings of the Sixteenth National Conference on Artificial Intelligence and Eleventh Conference on Innovative Applications of Artificial Intelligence, pp. 474–479 (1999)
5. Shickel, B., Tighe, P.J., Bihorac, A., Rashidi, P.: Deep EHR: a survey of recent advances in deep learning techniques for electronic health record (EHR) analysis
6. Kudinov M.S., Romanenko A.A., Piontkovskaja I.I.: Conditional random field in segmentation and noun phrase inclination tasks for Russian. Компьютерная лингвистика и интеллектуальные технологии, pp. 297–306 (2014)
7. Shelmanov, A.O., Smirnov, I.V., Vishneva, E.A.: Information extraction from clinical texts in Russian. Komp'juternaja Lingvistika i Intellektual'nye Tehnol. **1**, 560–572 (2015)

8. Baranov, A., et al.: Technologies for complex intelligent clinical data analysis. Vestn. Ross. Akad. meditsinskikh Nauk. **71**, 160–171 (2016). https://doi.org/10.15690/vramn663

9. Lin, C., Miller, T., Dligach, D., Bethard, S., Savova, G.: A BERT-based universal model for both within- and cross-sentence clinical temporal relation extraction. In: Proceedings of the 2nd Clinical Natural Language Processing Workshop, vol. 2, pp. 65–71 (2019)

10. Lin, C., Miller, T., Dligach, D., Bethard, S., Savova, G.: Representations of time expressions for temporal relation extraction with convolutional neural networks 322–327 (2017). https://doi.org/10.18653/v1/w17-2341

11. Balabaeva, K., Funkner, A., Kovalchuk, S.: Automated spelling correction for clinical text mining in Russian (2020)

12. Funkner, A., Balabaeva, K., Kovalchuk, S.: Negation detection for clinical text mining in Russian (2020)

13. Harkema, H., Dowling, J.N., Thornblade, T., Chapman, W.W.: ConText: an algorithm for determining negation, experiencer, and temporal status from clinical reports. J. Biomed. Inform. **42**, 839–851 (2009). https://doi.org/10.1016/j.jbi.2009.05.002

14. Korobov, M.: Morphological analyzer and generator for Russian and Ukrainian languages. In: Khachay, M.Yu., Konstantinova, N., Panchenko, A., Ignatov, D.I., Labunets, V.G. (eds.) AIST 2015. CCIS, vol. 542, pp. 320–332. Springer, Cham (2015). https://doi.org/10.1007/978-3-319-26123-2_31

15. Sorokin, A.A., Shavrina, T.O.: Automatic spelling correction for Russian social media texts. In: Proceedings of the International Conference "Dialog", Moscow. pp. 688–701 (2016)

16. Ingria, R., et al.: TimeML: robust specification of event and temporal expressions in text. New Dir. Quest. Ans. **3**, 28–34 (2003)

17. Molnar, C.: Interpretable Machine Learning. Lulu, Morrisville (2019)

18. Schmid, H.: Probabilistic part-of-speech tagging using decision trees. In: Proceedings of the International Conference on New Methods in Language Processing (1994)

19. Russian statistical taggers and parsers. http://corpus.leeds.ac.uk/mocky/

20. Negri, M., Marseglia, L.: Recognition and normalization of time expressions: ITC-irst at TERN 2004. Rapp. interne, ITC-irst, Trento (2004)

21. Zhao, X., Jin, P., Yue, L.: Automatic temporal expression normalization with reference time dynamic-choosing. In: Coling 2010 - 23rd International Conference on Computational Linguistics, Proceedings of the Conference (2010)

22. Korobkin, D.M., Vasiliev, S.S., Fomenkov, S.A., Lobeyko, V.I.: Extraction of structural elements of inventions from Russian-language patents. In: Multi Conference on Computer Science and Information Systems, MCCSIS 2019 - Proceedings of the International Conferences on Big Data Analytics, Data Mining and Computational Intelligence 2019 and Theory and Practice in Modern Computing 2019 (2019)

23. Funkner, A.A., Yakovlev, A.N., Kovalchuk, S.V.: Data-driven modeling of clinical pathways using electronic health records. Procedia Comput. Sci. **121**, 835–842 (2017). https://doi.org/10.1016/j.procs.2017.11.108

24. Derevitskii, I., Funkner, A., Metsker, O., Kovalchuk, S.: Graph-based predictive modelling of chronic disease development: type 2 DM case study. Stud. Health Technol. Inform. **261**, 150–155 (2019). https://doi.org/10.3233/978-1-61499-975-1-150

25. Balabaeva, K., Kovalchuk, S., Metsker, O.: Dynamic features impact on the quality of chronic heart failure predictive modelling. Stud. Health Technol. Inform. **261**, 179–184 (2019)

Experiencer Detection and Automated Extraction of a Family Disease Tree from Medical Texts in Russian Language

Ksenia Balabaeva$^{(\boxtimes)}$ and Sergey Kovalchuk

ITMO University, Saint-Petersburg, Russia
kyubalabaeva@gmail.com, sergey.v.kovalchuk@gmail.com

Abstract. Text descriptions in natural language are an essential part of electronic health records (EHRs). Such descriptions usually contain facts about patient's life, events, diseases and other relevant information. Sometimes it may also include facts about their family members. In order to find the facts about the right person (experiencer) and convert the unstructured medical text into structured information, we developed a module of experiencer detection. We compared different vector representations and machine learning models to get the highest quality of 0.96 f-score for binary classification and 0.93 f-score for multi-classification. Additionally, we present the results plotting the family disease tree.

Keywords: EHRs · Natural language processing · Family disease tree · Word embeddings · Text classification · Language models · Information retrieval · Family history

1 Introduction

In Russia electronic health records (EHRs) consist of structured, half-structured and unstructured text data. In order to process texts and get valuable information, medical experts spend a lot of time on reading. Although, in the era of automatization and big data analysis, researchers and engineers have developed automated tools for such tasks [1, 2]. Even though in most cases text processing is a relatively simple task when it comes to structured or half-structured texts, unstructured text mining is still a challenge. It requires more intelligent and computationally complex methods for natural language processing.

This work in broad terms is dedicated to unstructured medical text processing and information retrieval. One of the widely used parts of EHRs free text is a patient's life and medical anamnesis that describes facts, events, and diseases. However, there is also information about relatives and hereditary diseases, which a patient may not have. In order to segment text parts describing family members, we have developed a module of binary sentence classification. On the one hand, it filters non-patient information for further processing, while on the other hand it helps to get structured relatives features that can be used in further analysis and modelling. The second block of our module includes retrieval of family disease tree from texts in natural language. We have

© Springer Nature Switzerland AG 2020
V. V. Krzhizhanovskaya et al. (Eds.): ICCS 2020, LNCS 12140, pp. 603–612, 2020.
https://doi.org/10.1007/978-3-030-50423-6_45

developed a multiclassification algorithm that identifies class of the relatives and maps their information to the family disease tree. To clarify the task setting, the input and output data on each step are displayed on Fig. 1.

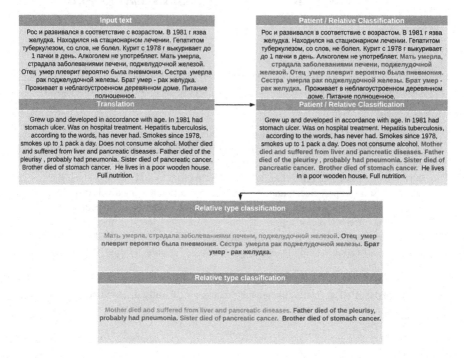

Fig. 1. Main steps of text processing in experiencer detection module. Grammar and spelling of the original text are preserved

It is also necessary to mention that we worked with texts in Russian language, therefore, our work also contributes to the research of languages with complex systems.

2 Literature Review

There are several works in the field of medical 'free text' processing, such as keyword extraction, and disease classification [3, 4]. Examples of similar task to the family disease tree extraction are also featured in literature, for instance, in [5], researchers extract family history from clinical notes, based on the word-wise labeled dataset and applying machine learning (ML) methods. Another work uses algorithms based on syntax structure of the English language [6]. One of the latest works [7] applies deep learning algorithms to extract family history. However, they are all applied to English language texts. While some modules of retrieving family tree structure from text are based on rules [8], alt, they work only with well-structured texts not observing medical context.

In general, the task of sentence classification is not new. It has several specific features. First of all, there is a problem of word representation: the length of the sentence varies, but the feature vector must be fixed. There is a classical approach for this problem, bag of words (BOW) vector representations [9], when sentences are represented as collections of words using their text frequency. However, it is a simplification of real interactions between words since BOW disregards the order.

There is also a big contribution to the word representation task from deep learning which is used to get continuous bag of words representation (CBOW) [10]. Another approach is to use skip grams (SG) that enrich representations with word's n-grams [11].

There are also end-to-end solutions in deep learning. For instance, such architectures as convolutional neural networks, LSTM and attention-based models are widely used in sentence classification [12–14].

However, the efficient application of deep learning methods usually require big volumes of data. Moreover, the majority of works uses English language for modelling, and the efficiency of the aforementioned methods on other languages is not fully researched, especially for the Russian language.

Another important issue of the family disease tree extraction is the relation bond uncertainty. For example, it is not always possible to say to which exact parent a grandmother belongs to or whether a patient's brother is a half-brother or a sibling. That is why, our goal is to map texts to the risk probability model, concerning the uncertainty in data, uncertainty in a model and uncertainty of the disease inheritance process.

3 Previous Work

The module of experiencer detection is a part of our research group text mining project (Fig. 2). It is dedicated to the medical language corpora with a high share of specific terminology. The project consists of several functional parts. The first part solves the problem of misspelling correction [15]. The second module is a negation detector, that helps to deal with rejections of events [16]. The third is the experiencer detection module, described in this paper.

The module of temporal structure processing, that helps to assign time stamps to the facts in a sentence is still under research. The green blocks represent the modules that are already developed. The blue ones are currently under development. As a result, all modules will be implemented as a Python 3 package.

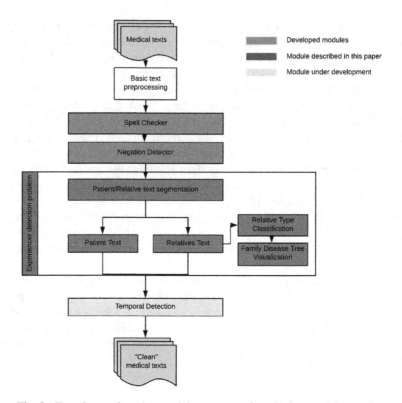

Fig. 2. Experiencer detection module as a part of medical text mining project

4 Data

We use a dataset of electronic health records (EHRs) from Almazov National Medical Research Centre, Saint Petersburg Russia. For the task of patient/non-patient classification, we selected 1376 sentences from 696 patients, with 672 sentences containing information about relatives and 704 sentences describing patients. All sentences were labeled by the authors.

For the second task, the number of unique labels was much higher, and 672 sentences were not enough. Therefore, we decided to enrich our dataset and labeled additional data. Finally, we got 1204 sentences. The labels included 9 most common classes: mother, father, sister, brother, daughter, son, grandmother, grandfather, aunt, uncle. The distribution of classes is displayed on Fig. 3.

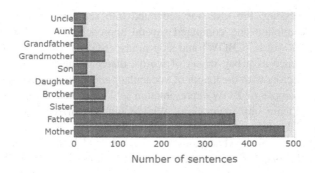

Fig. 3. Relatives' types distribution

The information about parents and their diseases in anamnesis is the most frequent. Sisters, brothers and grandmothers are discussed with approximately equal frequency. Less attention is payed to children, aunts and uncles' diseases.

We also have to mention the restrictions we implied on the data used in the paper. First of all, we only used sentences with one relative, as most sentences (78%) in fact contained information about only one relative. Secondly, multiple subjects in a sentence require more time-consuming labeling process, different task setting and a syntax trees parser. Unfortunately, we have not found satisfying syntax tree parser for Russian language yet. However, we have plans to work on that problem in future.

5 Methods

We have decomposed the module into two separate tasks: patient/relative binary classification and patient type multiclassification. All the experiments were implemented in Python programming language, version 3.7. We started our experiments with baseline development. For each task, an algorithm based on key words search was developed. As key words we used stems of common words for the names of family members. For instance, 'бабушк' for the word 'бабушка' (grandmother) or 'матер', 'мам' for the word 'матери', 'мама' (mother). We used lemmas to find different forms of words regardless of case and number and assigned the class to the sentence according to the key word found in it. This procedure reflects the most common and easiest way to get features from the initial medical texts, although, it is not accurate enough. More details on performance are provided in the results section.

The whole natural language text preprocessing was reduced to removal of punctuation and misspelling correction with the spell checker module already developed by our team [15, 16]. We deliberately avoided normalization and stop words removal, since anamnesis contain a lot of specific terms that cannot be normalized correctly using the existing open-source libraries. The reason to keep stop words lies in complex wording in Russian where they can be helpful.

After preprocessing, the text was separated into sentences and vectorized. For vectorization we applied and compared several approaches: bag of words (BOW), continuous bag of words (CBOW) and skip-grams (SG). BOW is a model that represents each sentence as a bag or set of words, disregarding order and grammar. It provides sparse vectors with the length of vocabulary and calculates the scoring of term frequency. An example of BOW representation is TFIDF vectorizer and CountVectorizer (Python sklearn implementation). While CountVectorizer simply counts word frequencies, TFIDF also uses inverse document frequency. However, bag of words approaches have several drawbacks, such as sparsity and lack of sentence meaning. That is why we decided to compare them with word embeddings that build a language model and learn text features. Word embeddings have a huge advantage of word order and context consideration. Therefore, such methods to a certain extent can catch the meaning of the sentence.

In this paper we compare the continuous bag of words (CBOW) and skip-grams, using the FastText model [9]. The difference between these approaches lies in the learning process. While CBOW aims to predict a word by the context, SG is learning to predict context by a word [10].

After that, vectors are input to the ML algorithms to start supervised learning. As models, we use logistic regression, k-nearest neighbors and random forest (python, sklearn library) and also gradient boosting (python, xgboost library) since they have different structure, suit for both binary and multiclassification, and show good performance in many tasks. The hyperparameters were tuned using GridSearch algorithm.

All models are evaluated using 5-fold cross-validation on the quality and stability. As a metric, f-score is used with macro-averaging.

6 Results

6.1 Binary Classification for Patient and Relative Sentences

Concerning the results of sentence classification on patient and relatives classes, the baseline model based on key words search had 0.6337 f-score. Fortunately, other approaches performed with a higher quality (Table 1). In terms of vectorization, bag of words methods (such as CV, TFIDF), on average, performed almost equal results to the skip-gram approach, achieving ~ 0.95 f-score. The continuous bag of words models show a much lower performance for all ML models. However, the highest score in most trials was still achieved by the SG word embedding approach (f-score 0.96).

Comparing different ML models, there is little difference in their best performances, and the f-score varies in the range of 0.95 to 0.96, even for simpler models such as LogReg or KNN. In terms of stability, cross-validation scores have, on average, standard deviation from one to two hundredth and the scores are stable enough.

Table 1. Task 1: Patient-relative classification. F1-scores on 5-fold cross-validation ± std.

Vectorization method\model	Logistic regression	XGB	Random forest	KNN
CountVectorizer (CV)	**0.9592** (± 0.0107)	0.9344 (± 0.0024)	0.9519 (± 0.0082)	0.9236 (± 0.0117)
TfIdf vectorizer (TFIDF)	0.9460 (± 0.0107)	0.9337 (± 0.0167)	0.9534 (± 0.0071)	0.9207 (± 0.0128)
CBOW	0.7211 (± 0.0095)	0.8637 (± 0.0300)	0.8542 (± 0.0229)	0.8200 (± 0.0270)
SG	0.9468 (± 0.0292)	**0.9621** (± 0.0178)	**0.9563** (± 0.0201)	**0.9614** (± 0.0151)

6.2 Patient Type Classification

The second task required building a multiclassification algorithm, for 9 types of relatives. The baseline model based on the key word search achieved f-score of 0.59278. The metrics for other approaches are displayed in Table 2. The best vectorization method for this task was bag of words (CountVectorizer). Word embeddings performed worse than even the baseline model. The reason of such results may lay in the size of our dataset. It was only 1204 samples and for several rare classes (Fig. 3), such as aunts, uncles or sons contained only 20–30 examples, which may not be sufficient for learning a language model based on neural networks.

Table 2. Task 2: Relative type multi-classification. F1-scores on 5-fold cross-validation ± std.

Vectorization method\model	Log Reg	XGBoost	Random forest	KNN
CountVectorizer	**0.9235** (± 0.0181)	**0.9305** (± 0.0285)	0.9123 (± 0.0189)	**0.6895** (± 0.0274)
TfIdf vectorizer	0.7107 (± 0.0369)	0.9297 (± 0.0324)	**0.9145** (± 0.0145)	0.6207 (± 0.0458)
CBOW	0.0914 (± 0.0056)	0.2054 (± 0.0317)	0.1607 (± 0.0174)	0.1200 (± 0.0150)
SG	0.2203 (± 0.006)	0.4781 (± 0.0509)	0.4236 (± 0.04327)	0.38754 (± 0.0426)

Unlike the previous task, not all ML model performed equally good this time. The worst results (0.6895) was shown by K-nearest neighbors algorithm. The best performance was demonstrated by the XGBoost, that achieved 0.93 f-score. Logistic Regression and Random forest also showed quite close and accurate results.

In general, current results are satisfactory, on the plot (Fig. 4) the results of XGBoost predictions made on vectors from count vectorizer are presented on the left and skip-grams are on the right. The count vectorizer representation provides almost 100% accuracy for the majority of relatives. However, it struggles to correctly predict sentences about brother, son and uncle. It usually confuses them with mother or father which have a higher share in sentence samples (Fig. 3). Thus, according to the confusion matrix for the skip gram vectorization, the performance of the embeddings strongly depends on the sample size. It is clear, that the quality of predictions is much higher for sentences about mother and father and other classes suffer from the lack of examples. In order to improve the embedding results, the data may be enriched with more instances for the other types of family members and make the dataset more balanced. According to the results, at least 3–5 hundreds of samples for each relative type are required.

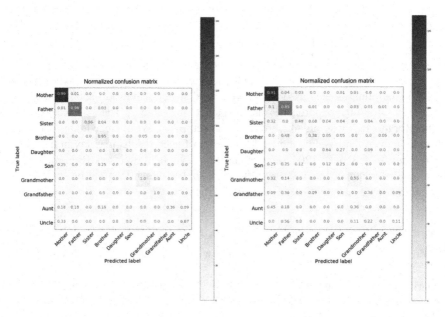

Fig. 4. Confusion Matrix for predictions of XGBoost + CountVectorizer (on the left) and Skip-Gram (on the right)

In addition to the model's prediction, we have implemented visualization of family disease tree as an output of the module (Fig. 5). As an input it takes sentences about relatives with the assigned relative type from the predictive models.

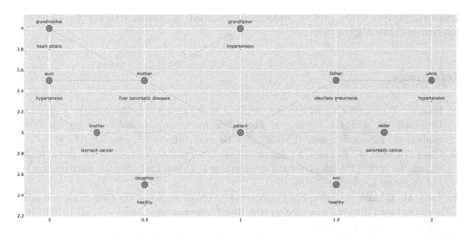

Fig. 5. Family disease tree visualization

Then, it extracts the disease or events that a relative experienced with simple syntax rules (for instance, removal of verbs and words, and a subject from the sentence). After that it plots a tree-type graph using Plotly Python library.

7 Conclusion

In conclusion, set goal is achieved and the module of experiencer detection for medical texts is developed. To reach this goal, the literature was analyzed, a dataset was collected and labeled, and experiments on the methods of vectorization and ML models were performed. A model that scored 0.96 f-metrics for patient/relative classification and 0.93 f-score on relative type classification problem was built. It was also shown that naïve methods based on the key words search can hardly solve the task and the developed models perform better.

The findings can be used for hereditary disease modelling. The module may be applied not only to medicine, but also to other spheres, where it is required to determine the subject in a sentence, for instance, in social networks or advertisements. Moreover, this approach can be applied to other languages, that do not have a language corpus or a syntax tree parser.

However, there are still things to improve. For instance, we need to expand the functionality to process the sentences with multiple subjects, for example, when there is information concerning several relatives in one sentence. For this task, the dataset should be labelled on the word level and the syntax-tree parser is to be implemented. Moreover, we need to expand the dataset especially for the task of relative type classification and try to generate and use synthetic data.

As this module is a part of the medical text mining project, infrastructure for communication with other modules should be provided. As a result, a python 3 package will be developed in future.

Acknowledgement. This work is financially supported by National Center for Cognitive Research of ITMO University.

References

1. Hanauer, D.: Supporting information retrieval from electronic health records: a report of University of Michigan's nine-year experience in developing and using the electronic medical record search engine (EMERSE). J. Biomed. Inform. **55**, 290–300 (2015)
2. Cesar dos Reis, J., Perciani, E.: Intention-based information retrieval of electronic health records. In: 25th IEEE International Conference on Enabling Technologies: Infrastructure for Collaborative Enterprise (2016)
3. Tang, M., Gandhi, P., Kabir, M.: Progress notes classification and keyword extraction using attention based deep learning models with BERT. Preprint https://arxiv.org/pdf/1910.05786.pdf. Accessed 07 Feb 2020
4. Zhang, X., Henao, R., Gan, Z.: Multi-label learning from medical plain text with convolutional residual models. https://arxiv.org/pdf/1801.05062.pdf. Accessed 07 Feb 2020
5. Bill, R., Pakhomov, S., Chen, E.: Automated extraction of family history information from clinical notes. In: AMIA Annual Symposium Proceedings, pp. 1709–1717 (2014)
6. Azab, M., Dadian, S., Nastase, V.: Towards extracting medical family history from natural language interactions: a new dataset and baselines. In: EMNLP/IJCNLP (2019)
7. Lewis, N., Gruhl, D., Yang, H.: Extracting family history diagnoses from clinical texts. In: BICoB (2011)
8. Family tree maker, github project: https://github.com/adrienverge/familytreemaker/blob/master/familytreemaker.py. Accessed 07 Feb 2020
9. Zhang, Y., Jin, R., Zhou, Z.: Understanding bag-of-words model: a statistical framework. Int. J. Mach. Learn. Cybern. **1**(11), 43–52 (2010). https://doi.org/10.1007/s13042-010-0001-0
10. Mikolov, T., Chen, K., Corrado, G., Dean, J.: Efficient estimation of word representations in vector space. In: CONFERENCE 2013, Proceedings of the International Conference on Learning Representations, ICLR (2013)
11. Mikolov, T., Sutskever, I., Chen, K., Corado, J., Dean, J.: Distributed representations of words and phrases and their compositionality. In: NIPS 2013: Proceedings of the 26th International Conference on Neural Information Processing Systems, vol.2, pp. 3111–3119 (2013)
12. Kim, Y.: Convolutional neural networks for sentence classification. In: Proceedings of the 2014 Conference on Empirical Methods in Natural Language Processing (EMNLP), pp. 1746–1751 (2014)
13. Ganda, R., Mahmood, A.: Deep learning for sentence classification. In: Conference: 2017 IEEE Long Island Systems, Applications and Technology Conference (LISAT) (2017)
14. Zhou, Q., Wang, X.: Differentiated attentive representation learning for sentence classification. In: Proceedings of the Twenty-Seventh International Joint Conference on Artificial Intelligence (IJCAI 2018) (2018)
15. Balabaeva, K., Funkner, A., Kovalchuk, S.: Automated spelling correction for clinical text mining in Russian. Preprint https://arxiv.org/abs/2004.04987 (2020). Accessed 15 Apr 2020
16. Funkner, A., Balabaeva, K., Kovalchuk, S.: Automated negation detection for medical texts in Russian Language. Preprint, https://arxiv.org/pdf/2004.04980.pdf (2020). Accessed 15 Apr 2020

Computational Methods for Emerging Problems in (dis-)Information Analysis

Machine Learning – The Results Are Not the only Thing that Matters! What About Security, Explainability and Fairness?

Michał Choraś[1,2,3], Marek Pawlicki[1,2]([⊠]), Damian Puchalski[1], and Rafał Kozik[1,2]

[1] ITTI Sp. z o.o., Poznań, Poland
[2] UTP University of Science and Technology, Bydgoszcz, Poland
{chorasm,marek.pawlicki}@utp.edu.pl
[3] FernUniversitat in Hagen (FUH), Hagen, Germany

Abstract. Recent advances in machine learning (ML) and the surge in computational power have opened the way to the proliferation of ML and Artificial Intelligence (AI) in many domains and applications. Still, apart from achieving good accuracy and results, there are many challenges that need to be discussed in order to effectively apply ML algorithms in critical applications for the good of societies. The aspects that can hinder practical and trustful ML and AI are: lack of security of ML algorithms as well as lack of fairness and explainability. In this paper we discuss those aspects and provide current state of the art analysis of the relevant works in the mentioned domains.

Keywords: Machine Learning (ML) · AI · Secure ML · Explainable ML · Fairness

1 Introduction

Recent advances in machine learning (ML) and the surge in computational power have opened the way to the proliferation of Artificial Intelligence (AI) in many domains and applications.

Still, many of the ML algorithms offered by researchers, scientists and R&D departments focus only on the numerical quality of results, high efficiency and low error rates (such as low false positives or low false negatives). But even when such goals are met, those solutions cannot (or should not) be realistically implemented in many domains, especially in critical fields or in the aspects of life that can impact whole societies, without other crucial criteria and requirements, namely: security, explainability and fairness. Moreover, frequently the outstanding results are achieved on data that is well-prepared, crafted in laboratory conditions, and are only achievable when implemented in laboratory environments.

However, when large scale applications of AI became reality, the realization came that the security of machine learning requires immediate attention. Malicious users, called 'Adversaries' in the AI world, can skilfully influence the inputs

© Springer Nature Switzerland AG 2020
V. V. Krzhizhanovskaya et al. (Eds.): ICCS 2020, LNCS 12140, pp. 615–628, 2020.
https://doi.org/10.1007/978-3-030-50423-6_46

fed to the AI algorithms in a way that changes the classification or regression results. Regardless of the ubiquity of machine learning, the awareness of the security threats and ML's susceptibility to adversarial attacks used to be fairly uncommon and the subject has received significant attention only recently.

Apart from security, another aspect that requires attention is the explainability of ML and ML-based decision systems. Many researchers and systems architects are now using deep-learning capabilities (and other black-box ML methods) to solve detection or prediction tasks. However, in most cases, the results are provided by algorithms without any justification. Some solutions are offered as if it was magic and the Truth provider, while for decision-makers in a realistic setting the question why (the system arrives at certain answers) is crucial and has to be answered.

Therefore, in this paper an overview of aspects and recent works on security, explainability, and fairness of AI/ML systems is presented, the depiction of those concerns can be found in Fig. 1. The major contributions of the paper are: current analysis of challenges in machine learning (other than only having good numeric results) as well as state of the art analysis of works in secure ML, explainable ML and fairness.

The paper is structured as follows: in Sect. 2 security of machine learning is discussed and an overview of recent works is provided. Several types of adversarial attacks are mentioned, such as evasion attacks, data poisoning, exploratory attacks (an example of deep learning use for exploratory attacks can be found in [1]) etc. In Sect. 3 a survey of fairness in ML is presented, while in Sect. 4 the focus is on related works in explainable machine learning. Conclusions are given thereafter.

2 Security and Adversarial Machine Learning in Disinformation

Recently it has come to attention that skilfully crafted inputs can affect artificial intelligence algorithms to sway the classification results in the fashion tailored to the adversary needs [2]. This new disturbance in the proliferation of Machine Learning has been a subject riveting attention of the researches very recently, and at the time of writing this paper a variety of vulnerabilities have been uncovered [2].

With the recent spike of interest in the field of securing ML algorithms, a myriad of different attack and defence methods have been discovered; no truly safe system has been developed however, and no genuinely field-proven solutions exist [3].

The solutions known at this point seem to work for certain kinds of attacks, but do not assure safety against all kinds of adversarial attacks. In certain situations, implementing those solutions could lead to the deterioration of ML performance [2].

The adversaries behaviour is affected by the extent of the knowledge the agent possesses of the target algorithm's architecture. In literature, this level of acquaintance is categorised as black box and white box [4].

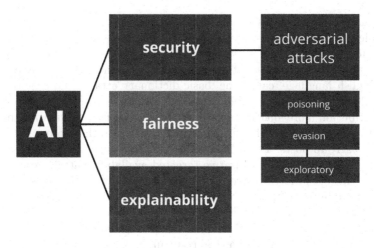

Fig. 1. Machine Learning concerns

While white box attacks presuppose full knowledge of the attacked algorithm; black box strikes are performed with no preceding knowledge of the model [4].

- Targeted Poisoning
- transferable clean-label attack (convex polytope attack)
- Feature Collision Attack
- one-shot kill Attack
- single poison instance attack
- watermarking
- multiple poison instance attack
- Non-Targeted Poisoning

There are a couple of known poisoning attacks featured in the literature. In [5] a method utilising the intrinsic properties of Support Vector Machines is introduced. The overarching idea is that an adversary can craft a data point that significantly deteriorates the performance of the classifier. The formulation of that data point can be, as demonstrated by the authors, defined as the solution of an optimisation problem with regard to a performance measure. The authors note that the challenge of finding the worst possible mix of label flips is not a straightforward one. The classes chosen for the flips are the ones classified with a high confidence, this should result in a significant impact on SVM accuracy [5].

The authors of [6] express in their paper an illustration of how an attacker could manipulate ML in spam filtering by meddling with the data to either subject the user to an ad, or stop the user from receiving genuine communication. The efficiency of those attacks is illustrated.

The authors use an algorithm called SpamBayes for their research. Spam-Bayes takes the head and body of a message, tokenizes them and scores the spam to classify it as spam, ham or unsure. With this established, the paper presents a dictionary attack, in which the algorithm is subjected to an array of spam e-mails containing a set of words that are likely to be present in genuine communication. When those are marked as spam, the algorithm will be more likely to flag legitimate mail as spam. This particular attack comes in two variations: a procedure where the attack mail simply contains the whole dictionary of the English language, called the 'basic dictionary attack' and a more refined approach, where the attack is performed with the use of a message containing word distribution more alike the users message distribution, along with the colloquialisms, misspellings etc. In this particular case the authors propose a pool of Usenet newsgroup posts. The other evaluated attack is geared towards blocking a specific kind of e-mail - a causative targeted availability attack, or a focused attack. In this scenario the adversary spam the user with messages containing words that are likely to appear in a specific message. With SpamBayes retrained on these messages it is then predisposed to filtering a distinct, genuine communication as spam. This could eliminate a competing bid of a rival company, for example. Including the name of a rival company in spam e-mails, their products or the names of their employees could achieve that objective. The authors indicate that using the dictionary attacks can neglect the feasibility of a spam filter with only 1% of retraining dataset controlled, and a masterfully crafted focused attack can put a specific message in the spam box 90% of the time.

In [7] a framework for the evaluation of security of feature selection algorithms is proposed. The framework follows the outline defined by [8] in which the authors evaluate the attacker's goal, the extent of the adversary's knowledge of the workings of the algorithm, and their capability in data manipulation. The goal of the malicious user is either targeted or indiscriminate, and it aims to infringe on one or more of the well-known infosec triad items: availability, integrity or privacy. The specific acquaintance of the adversary with the workings of the system can be one of the following:

- knowledge of the training data (partial or full)
- knowledge of the feature representation (partial or full)
- knowledge of the feature selection algorithm
- Perfect Knowledge (worst case scenario)
- Limited Knowledge

With regard to the attacker's capability, in the case of causative (poisoning) attacks, the attacker can usually influence just a subsection of the training set. The adversary has to bear in mind that the labeling process varies in different use cases, with the use of honeypots and anti-virus software giving setting the constraints of the datapoints that have to be crafted in the malware detection example.

The authors evaluate the robustness of 3 widely-used feature selection algorithms: Lasso, Ridge Regression and the Elastic Net against poisoning attacks with regard to the percentage of injected poisoned data points. The results show that poisoning 20% of the data inflates the classification error 10-fold. In addition to the influence on classification, the authors notice that even with a minute amount of poisoning samples the stability index drops to zero. This means that the attacker can influence feature selection.

In [9] the authors investigate a poisoning attack geared towards targeting specific test instances with the ability to fool a labelling authority, which they name 'clean-label' attacks. Their work does not assume knowledge of the training data, but does require the knowledge of the model. It is an optimisation-based attack for both the transfer-learning and end-to-end DNN training cases. The overall procedure of the attacks, called 'Poison Frogs' by the authors, is as follows: the basic version of this attack starts with choosing the target datapoint, then making alterations to that datapoint to make it seem like it belongs to the base class. A poison crafted that way is then inserted into the dataset. The objective is met if the target datapoint is classified as the base class at test time. Arriving at a poisonous datapoint to be inserted into the training set comes as a result of a process called 'feature collision'. It is a process that exploits the nonlinear complexity of the function propagating the input through the second-to-last layer of the neural network to find a datapoint which 'collides' with the target datapoint, but is also close to the base class in the feature space. This allows the poisoned datapoint to bypass the scrutiny of any labelling authority, and also remain in the target class distribution. The optimisation is performed with a forward-backward-splitting iterative procedure.

The [10] paper evaluates a poisoning procedure geared towards poisoning multi-class gradient-decent based classifiers. To this end the authors utilize the recently proposed back-gradient optimization. This approach allows for a replacement of one of the optimisation problems with a set of iterations of updating the parameters.

The authors introduce an attack procedure to poison deep neural networks taking into consideration the weight updates, rather than training a surrogate model trained on deep feature representations. They demonstrate the method on a convolutional neural network (CNN) trained on the well-known MNIST digit dataset, a task which requires the optimisation of over 450000 parameters. They find that deep networks seem more resilient to poisoning attacks than regular ML algorithms, at least in conditions of poisoning under 1% of the data. The authors also conduct a transferability experiment in which they conclude that poisons crafted against linear regression (LR) algorithm are ineffective against a CNN, and poisons crafted against a CNN have a similar effect on LR as random label flips. A more comprehensive assessment of the effects of poisoning attacks crafted against deep neural networks with the use of the back-gradient algorithm is necessary. The notion of transferability of adversarial attacks is explored in depth in [11], where the authors conclude that, for evasion attacks in their case, many attacks could be effectively used across models trained with different techniques, and prove their findings by attacking classifiers hosted by Amazon and Google without any knowledge of the attacked models. An in-depth analysis of the transferability of both evasion and poisoning attacks is performed in [12]. [13] investigates poisoning attacks carried out by an attacker with full knowledge of the algorithm. The assumption is that the adversary aims to poison the model with the minimum amount of poisoning examples. The attacker function is defined as a bilevel optimisation problem. The authors notice that this function is similar to machine teaching, where the objective is to have maximum possible influence over the subject by carefully crafting the training dataset. The authors point to the mapping of a teacher to the attacker and from a student to the AI algorithm. The paper thus offers economical solutions to the bilevel optimisation problem present in both fields. Essentially, the authors suggest that, under certain regulatory conditions, the problem can be reduced with the use of Karush-Kuhn-Tucker theorem (KKT) to a single-level constrained optimization problem. Thus, a formal framework for optimal attacks is introduced, which is then applied in 3 different cases - SVM, Linear Regression and logistic regression. In [14] the authors propose a way of bypassing the gradient calculation by partially utilising the concept of a Generative Adversarial Network (GAN). In this approach an autoencoder is applied to craft the poisoned datapoints, with the loss function deciding the rewards. The data is fed to a neural network, and the gradients are sent back to the generator. The effectiveness of their method is tested thoroughly on the well-known MNIST and CIFAR-10 datasets. The chosen architecture is a two-layer feed forward neural network with recognition accuracy of 96.82% on the MNIST dataset, and for CIFAR-10 a convolutional neural network with two convolutional layers and two fully-connected layers, with the accuracy of 71.2%. For demonstrative purposes, one poisoned datapoint is injected at a time. The authors conclude that the generative attack method shows improvement over the direct gradient methods and stipulate that it is viable for attacking deep learning and its datasets, although more research is required. A targeted backdoor attack is proposed in [15]. The premise of the method is to create a backdoor to an authentication system based on artificial intelligence, allowing the adversary to pass the authentication process by deceiving it. The poisoning datapoints are created specifically to force an algorithm to classify a specific instance as a label of the attacker's choice. The authors propose a method that works with relatively small poison samples and with the adversary possessing no knowledge of the algorithm utilised. This claim is backed up by a demonstration of how inserting just 50 samples gets a 90% success rate.

3   Fairness in Machine Learning

Fairness in Machine Learning (or Artificial Intelligence in a broader sense) is a concept which is getting an increased amount of attention with the growing popularity of AI in different society-impacting applications. Fairness in AI is mainly ethically and legally motivated. It is also a fertile ground to spread politically motivated disinformation, when used maliciously.

The background of the fairness concept in AI results from the misinformed, but widespread perception of Artificial Intelligence and AI/ML-based decision making as fully objective. In practice, the fairness of AI-driven decisions depends highly on the data provided as the input to learning algorithms. This data can be (and often is) biased due to several reasons: 1) bias of human operators providing this data as input, resulting e.g. in biased labeling of samples, 2) data unbalance/misrepresentation of e.g. specific minority groups, 3) historical bias (discrimination) [16].

In addition, in [17] there is a list of potential causes of bias in training datasets leading to unfairness in AI:

- Skewed sample – misrepresentation of training data in some areas that evolves over time. In that way the future observations confirm biased prediction and misrepresented data samples give less chances for contradicting observations.
- Tainted examples - the bias existing in the old data caused by human bias is replicated by the system trained on this data.
- Limited features - reliability of some labels from a minority group (e.g. unreliably collected or less informative) impacts the system and may cause the lower accuracy for the predictions related the minority group.
- Sample size disparity – causing difficulties in building a reliable model of the group described by an insufficient data sample.
- Proxies – correlation of sensitive biased attributes (even if not used to train an ML system, encrypted, etc.) with other features preserves a bias in predicted output.

All those reasons result in inaccurate decisions based on (or related to) sensitive attributes such as gender, race or others. A decision-making process is affected by disparate treatment if its decisions are based on these sensitive attributes. In summary, the definition of unfair machine learning process can be formulated as a situation in which an output tends to be disproportionately benefitting (or unfair) towards the group characterized by certain sensitive attribute values. However, this commonly used definition is too abstract to reach a consensus on the mathematical formulations of fairness. The majority of definitions of fairness in ML include the following elements [16, 18, 19]: group fairness (including demographic parity, equalized odds and predictive rate parity), unawareness, individual fairness and counterfactual fairness.

The author of [16] discusses some solutions to address each of these elements:

- Not including sensitive attributes' values in a training dataset (addressing the unawareness and disparate treatment of subjects). The challenge here is in the existence of the proxies, i.e. non-sensitive attributes used to train ML system highly correlated with eliminated sensitive attributes.
- Statistical parity of different groups in the training sample – e.g. application of the so-called 80% rule - the size of the sample belonging to the group with the lowest selection rate should be at least 80% in comparison to the mostly represented group (proportion should be higher than 4/5). This could prevent from extreme misrepresentation of minor groups.
- Optimal adjustment of learned predictor to reduce discrimination against a specified sensitive attribute in supervised learning according to the equalized odds definition [20].
- Replacing the original value of the sensitive attribute by the counterfactual value propagated "downstream" in the causal graph. This addresses counterfactual fairness and provides a way to check and explain the possible impact of bias via a causal graph [21]. As pointed in [16], in practice in many applications it is hard to build a causal graph and the elimination of correlated attributes can result in significantly decreased accuracy of prediction.

State of the Art on Algorithm for Fair ML. As the fairness in ML/AI is a trending topic, there are many algorithms focusing on improving fairness in ML described in the literature. Most of them fall into three categories: preprocessing, optimization at training time, and post-processing [16]. In general, algorithms belonging to the same category are characterized by common advantages and flaws.

Preprocessing. The idea is based on building a new representation of input data by removing the information correlated to the sensitive attribute and at the same time preserving the remaining input information as much as possible. The downstream task (e.g. classification, regression, ranking) can thus use the "cleaned" data representation and produce results that preserve demographic parity and individual fairness. In [22] authors use the optimal transport theory to remove disparate impact of input data. They also provide numerical analysis of the database fair correction. In [23] authors propose a learning algorithm for fair classification addressing both group fairness and individual fairness by obfuscating information about membership in the protected group. Authors of [24] propose a model based on a variational autoencoding architecture with priors that encourage independence between sensitive and latent factors of data variation. To remove any remaining dependencies an additional penalty term based on the "Maximum Mean Discrepancy" (MMD) measure is additionally introduced. A statistical framework for removing information about a protected variable from a dataset is presented in [25], along with the practical application to a real-life dataset of recidivism, proving successful predictions independent of the protected variable, with the predictive accuracy preserved. [26] proposes a convex optimization for learning a data transformation with three goals: controlling discrimination, limiting distortion in individual data samples, and preserving utility.

The authors of [27, 28] evaluate the use of dataset balancing methods for data augmentation to counter fairness issues. Both papers report positive results.

Optimization at Training Time. Data processing at training time provides good performance on accuracy and fairness measure and ensures higher flexibility in optimizing the trade-off between these factors. The author of [16] describes that the common idea that can be found in the state-of-the-art works falling into this category of algorithms is to add a constraint or a regularization term to the existing optimization objective. Recent works considering algorithms to ensure ML fairness applied at the training time include: [29] where the problem of learning a non-discriminatory predictor from a finite training set is studied to preserve "equalized odds" fairness, [21, 30] where a flexible mechanism to design fair classifiers by leveraging a novel intuitive measure of the decision boundary (un)fairness is introduced, and [31] that addresses the problem of reducing the fair classification to a sequence of cost-sensitive classification problems, whose solutions provide a randomized classifier with the lowest (empirical) error subject to the desired constraints.

The disadvantages of abovementioned approaches include the fact that these methods are highly task-specific and they require a modification of the classifier, which can be problematic in most applications/cases. **Post-processing** The post-processing algorithms are focused on editing the posteriors to satisfy the fairness constraints and can be applied to optimize most of fairness definitions except the counterfactual fairness. The basic idea is to find a proper threshold using the original score function for each group. An exemplary recent work that falls into this category is the publication [20], in which the authors show how to optimally adjust any learned predictor to remove the discrimination according to the "equal opportunity" definition of fairness, with the assumption that data about the predictor, target, and membership in the protected group are available.

The advantage of post-processing mechanisms is that retraining/changes are not needed for the classifier (the algorithm can be applied after any classifier) [16]. **Summary** The author of [31] provides an experimental comparison of the selected algorithms applied to reduce unfairness using four real-life datasets with one or two protected sensitive attributes (gender or/and race). The selected methods include preprocessing, optimization at training time and post-processing approaches. The methods that achieve the best trade-off between accuracy and fairness are those falling into the optimization at training time category, while the advantage related to the implementation of preprocessing and post-processing methods is the preservation of fairness without modifying the classifiers. In general, experimental results prove the ability to significantly reduce or remove the disparity, in general not impacting the classifier's accuracy for all the methods [16,31].

4 AI Explainability and Interpretability

The aspects of explainability and interpretability are trending topics in the area of Machine Learning and Artificial Intelligence in general as well. As discussed in [32,33] these two terms – explainability and interpretability tend to be used (also in literature) interchangeably, however despite the fact that they are related concepts, there are some minor differences in their meanings. Interpretability addresses the aspects related to observation of AI system outputs. Interpretability of AI system is higher, if the changes of the systems outputs in result of changing algorithmic parameters are more predictable. In other words, system interpretability is related to the extent to which a human can predict the results of AI systems based on different inputs. On the contrary, explainability is related to the extent to which a human can understand and explain (literally) the internal mechanics of an AI/machine learning system. In its simplest form, the definition of explainability refers to an attempt to provide insights into the predictor's behavior [34]. According to [33], nowadays, attempts to define these concepts are not enough to form a common and monolithic definition of explainability and interpretability and to enable their formalization. It is also worth mentioning, that the "right to explanation" in the context of AI systems directly affecting individuals by their decisions, especially legally and financially is one of the subjects of the GDPR [35].

Different scientific and literary sources focus on surveying and categorisation of methods and techniques addressing explainability and interpretability of decisions resulting from AI systems use. [32] discusses the most common practical approaches, techniques and methods used to improve ML interpretability and enable more explainable AI. They include, among others, algorithmic generalization, i.e. shifting attitude from case-specific models to more general ones. Another approach is paying attention to feature importance, described also in [34] as the most popular technique addressing ML explainability, also known as feature-level interpretations, feature attributions or saliency maps. Some of feature importance-based methods found in the literature are perturbation-based methods based on Shapley values adapted from the cooperative game theory. In the explainability case, Shapley values are used to attain fair distribution of gains between players, where a cooperative game is defined between the features. In addition, some recent works [32,36] show that adversarially trained models can be characterised by increased robustness but also provide clearer feature importance scores, contributing to improved prediction explainability. Similar to the feature importance way, counterfactual explanations [34] is a technique applied in the financial and healthcare domains. Explanations using this method are based on providing point(s) and values that are close to the input values for which the decision of the classifier possibly changes (case-specific threshold values). Another method used for increasing explainability of AI-based predictions is LIME (Local Interpretable Model-Agnostic Explanations) based on approximation of the model by testing it, then applying changes to the model and analysis of the output. DeepLIFT (Deep Learning Important Features) model is used for the challenge-based analysis of deep learning/neural networks. As described in [32] DeepLift method is based on backpropagation, i.e. digging back into the feature selection inside the algorithm and "reading" neurons at subsequent layers of network. The authors of [37] evaluate the use of influence functions as a way of selecting specific training samples responsible for a given prediction, a paradigm known as prototype-based explanation.

In the literature one can find different attempts of categorization of the methods aimed at increased explainability of AI. Integrated/Intrinsic and post-hoc explainability methods [33,38] is one of such categorization. Intrinsic explainability in its simplest form is applicable to some basic variants of the low complexity models, where the explanation of a simple model is the model itself. On the other hand, more complex models are explainable in a post-hoc way, providing explanations after the decision and using techniques such as feature importance, layer-wise relevance propagation, or the mentioned Shapley values.

Similar categorization is given in [38] where in-model (integrated/intrinsic) and post-model (post-hoc) methods exist alongside additional pre-model interpretability methods. Pre-model methods are applicable before building (or selection) of the ML model and are strictly related to the input data interpretability. They use mainly classic descriptive statistical methods, such as PCA (Principal Component Analysis), t-SNE (t-Distributed Stochastic Neighbor Embedding), and clustering methods such as k-means. Another criterion described in [38] is the differentiation into model-specific and model-agnostic explanation methods. In the majority of cases model-specific explanation methods are applicable to the intrinsically interpretable models (for example analysis and interpretation of weights in a linear model), while model-agnostic methods can be applied after the model and include all post-hoc methods relying on the analysis of pairs of feature input and output. Alternative criterion based on explanation methods is described in [39]. In such differentiation, methods are categorized based on type of explanation that the given method provides, including: feature summary (providing statistic summary for each feature with their possible visualization), model internals (for intrinsic explainable or self-explainable models), data point (example-based models) and a surrogate intrinsically interpretable model - that is trained to approximate the predictions of a black box model.

According to [38,40] explanation models can be evaluated and compared using qualitative and quantitative metrics, as well as by comparison of the explanation method's properties, including its expressive power, translucency (model-specific vs. model-agnostic), portability (range of applications) and computational complexity. On the other hand, individual explanations can be characterized by accuracy, fidelity, consistency (similarity of explanations provided by different models), stability, comprehensibility, certainty, to list most relevant ones. According to the literature we can also distinguish qualitative and quantitative indicators to assess the explanation models. Factors related to quality of explainability are: form of the explanation, number of the basic units of explanation that it contains, compositionality (organization and structure of the explanation), interactions between the basic explanation units (i.e. intuitiveness of relation between them), uncertainty and stochasticity. Quantitative indicators are presented in some works (e.g. [38,41,42]). The most common metrics used to quantify the interpretation of ML models are identity, separability and stability. These three factors provide the information on to what extent identical, non-identical and similar instances of predictions are explained in identical, non-identical and similar way, respectively. In addition, according to [41] the explanation should be characterized by high completeness (coverage of the explanation), correctness and compactness. However, these indicators are applicable only to simple models (rule-based, example-based).

5   Conclusions

In this paper recent research in secure, explainable and fair machine learning was surveyed. The high number of related works shows that those aspects are becoming crucial. At the same time an increasing number of researchers are aware that in machine learning the numeric results are not the only thing that matters. This work is a part of the SAFAIR Programme (Secure and Fair AI Systems for Citizens) of the H2020 SPARTA project that focuses on security, explainability, and fairness of AI/ML systems, especially in the cybersecurity domain. Moreover, the same aspects (secure, fair and explainable ML) are a part of the project SIMARGL focusing on detection on malware by advanced

ML techniques. We believe that even more projects will contain the work on secure and explainable machine learning, and that this survey will be helpful and might inspire more researchers in ML community to seriously consider those aspects.

Acknowledgement. This work is funded under the SPARTA project, which has received funding from the European Union's Horizon 2020 research and innovation programme under grant agreement No. 830892. This work is partially funded under SIMARGL project, which has received funding from the European Union's Horizon 2020 research and innovation programme under grant agreement No. 833042.

References

1. Choraś, M., Pawlicki, M., Kozik, R.: The feasibility of deep learning use for adversarial model extraction in the cybersecurity domain. In: Yin, H., Camacho, D., Tino, P., Tallón-Ballesteros, A.J., Menezes, R., Allmendinger, R. (eds.) IDEAL 2019. LNCS, vol. 11872, pp. 353–360. Springer, Cham (2019). https://doi.org/10.1007/978-3-030-33617-2_36

2. Chakraborty, A., Alam, M., Dey, V., Chattopadhyay, A., Mukhopadhyay, D.: Adversarial attacks and defences: a survey. arXiv preprint arXiv:1810.00069 (2018)

3. Liao, X., Ding, L., Wang, Y.: Secure machine learning, a brief overview. In: 2011 Fifth International Conference on Secure Software Integration and Reliability Improvement-Companion, pp. 26–29. IEEE (2011)

4. Papernot, N., McDaniel, P., Sinha, A., Wellman, M.P.: SoK: security and privacy in machine learning. In: 2018 IEEE European Symposium on Security and Privacy (EuroS&P), pp. 399–414. IEEE (2018)

5. Biggio, B., Nelson, B., Laskov, P.: Poisoning attacks against support vector machines. arXiv preprint arXiv:1206.6389 (2012)

6. Nelson, B., et al.: Exploiting machine learning to subvert your spam filter. LEET **8**, 1–9 (2008)

7. Xiao, H., Biggio, B., Brown, G., Fumera, G., Eckert, C., Roli, F.: Is feature selection secure against training data poisoning? In: International Conference on Machine Learning, pp. 1689–1698 (2015)

8. Biggio, B., Fumera, G., Roli, F.: Pattern recognition systems under attack: design issues and research challenges. Int. J. Pattern Recognit. Artif. Intell. **28**(07), 1460002 (2014)

9. Shafahi, A., et al.: Poison frogs! Targeted clean-label poisoning attacks on neural networks. In: Advances in Neural Information Processing Systems, pp. 6103–6113 (2018)

10. Muñoz-González, L., et al.: Towards poisoning of deep learning algorithms with back-gradient optimization. In: Proceedings of the 10th ACM Workshop on Artificial Intelligence and Security - AISec 17. ACM Press (2017)

11. Papernot, N., McDaniel, P., Goodfellow, I.: Transferability in machine learning: from phenomena to black-box attacks using adversarial samples. arXiv preprint arXiv:1605.07277 (2016)

12. Demontis, A., et al.: Why do adversarial attacks transfer? Explaining transferability of evasion and poisoning attacks. In: 28th {USENIX} Security Symposium ({USENIX} Security 19), pp. 321–338 (2019)

13. Mei, S., Zhu, X.: Using machine teaching to identify optimal training-set attacks on machine learners. In: Twenty-Ninth AAAI Conference on Artificial Intelligence (2015)
14. Yang, C., Wu, Q., Li, H., Chen, Y.: Generative poisoning attack method against neural networks. arXiv preprint arXiv:1703.01340 (2017)
15. Chen, X., Liu, C., Li, B., Lu, K., Song, D.: Targeted backdoor attacks on deep learning systems using data poisoning. arXiv preprint arXiv:1712.05526 (2017)
16. Zhong, Z.: A tutorial on fairness in machine learning, July 2019
17. Barocas, S., Selbst, A.D.: Big data's disparate impact. SSRN Electron. J. **104**, 671 (2016)
18. Gajane,P., Pechenizkiy, M.: On Formalizing Fairness in Prediction with Machine Learning. arXiv e-prints, page arXiv:1710.03184, October 2017
19. Verma, S., Rubin, J.: Fairness definitions explained. In: Proceedings of the International Workshop on Software Fairness - FairWare 18. ACM Press (2018)
20. Hardt, M., Price, E., Srebro, N., et al.: Equality of opportunity in supervised learning. In: Advances in Neural Information Processing Systems, pp. 3315–3323 (2016)
21. Zafar, M.B., Valera, I., Rodriguez, M.G., Gummadi, K.P.: Fairness beyond disparate treatment & disparate impact. In: Proceedings of the 26th International Conference on World Wide Web - WWW 17. ACM Press (2017)
22. Del Barrio, E., Gamboa, F., Gordaliza, P., Loubes, J.-M.: Obtaining fairness using optimal transport theory. arXiv preprint arXiv:1806.03195 (2018)
23. Zemel, R., Wu, Y., Swersky, K., Pitassi, T., Dwork, C.: Learning fair representations. In: International Conference on Machine Learning, pp. 325–333 (2013)
24. Louizos, C., Swersky, K., Li, Y., Welling, M., Zemel, R.: The variational fair autoencoder. arXiv preprint arXiv:1511.00830 (2015)
25. Lum, K., Johndrow, J.: A statistical framework for fair predictive algorithms. arXiv preprint arXiv:1610.08077 (2016)
26. Calmon, F., Wei, D., Vinzamuri, B., Ramamurthy, K.N., Varshney, K.R.: Optimized pre-processing for discrimination prevention. In: Advances in Neural Information Processing Systems, pp. 3992–4001 (2017)
27. Iosifidis, V., Ntoutsi, E.: Dealing with bias via data augmentation in supervised learning scenarios. Jo Bates Paul D. Clough Robert Jäschke, p. 24 (2018)
28. Sharma, S., Zhang, Y., Aliaga, J.M.R., Bouneffouf, D., Muthusamy, V., Varshney, K.R.: Data augmentation for discrimination prevention and bias disambiguation. In: Proceedings of the AAAI/ACM Conference on AI, Ethics, and Society, pp. 358–364 (2020)
29. Woodworth, B., Gunasekar, S., Ohannessian, M.I., Srebro,N.: Learning non-discriminatory predictors. arXiv preprint arXiv:1702.06081 (2017)
30. Zafar, M.B., Valera, I., Rodriguez, M.G., Gummadi, K.P.: Fairness constraints: mechanisms for fair classification. arXiv preprint arXiv:1507.05259 (2015)
31. Agarwal, A., Beygelzimer, A., Dudík, M., Langford, J., Wallach, H.: A reductions approach to fair classification. arXiv preprint arXiv:1803.02453 (2018)
32. Gall, R.: Machine learning explainability vs interpretability: two concepts that could help restore trust in AI. KDnuggets News **19**(1) (2019)
33. Dosilovic, F.K., Brcic, M., Hlupic, N.: Explainable artificial intelligence: a survey. In 2018 41st International Convention on Information and Communication Technology, Electronics and Microelectronics (MIPRO). IEEE, May 2018
34. Bhatt, U., et al.: Explainable machine learning in deployment. arXiv preprint arXiv:1909.06342 (2019)

35. Albert, C.: We are ready for machine learning explainability?, June 2019. https://towardsdatascience.com/we-are-ready-to-ml-explainability-2e7960cb950d. Accessed 31 Mar 2020

36. Etmann, C., Lunz, S., Maass, P., Schönlieb, C.-B.: On the connection between adversarial robustness and saliency map interpretability. arXiv preprint arXiv:1905.04172 (2019)

37. Koh, P.W., Liang, P.: Understanding black-box predictions via influence functions. In: Proceedings of the 34th International Conference on Machine Learning, vol. 70, pp. 1885–1894. JMLR.org (2017)

38. Carvalho, D.V., Pereira, E.M., Cardoso, J.S.: Machine learning interpretability: a survey on methods and metrics. Electronics 8(8), 832 (2019)

39. Molnar, C.: A guide for making black box models explainable (2018). https://christophm.github.io/interpretable-ml-book/. Accessed 28 Mar 2019

40. Robnik-Šikonja, M., Bohanec, M.: Perturbation-based explanations of prediction models. In: Zhou, J., Chen, F. (eds.) Human and Machine Learning. HIS, pp. 159–175. Springer, Cham (2018). https://doi.org/10.1007/978-3-319-90403-0_9

41. Silva, W., Fernandes, K., Cardoso, M.J., Cardoso, J.S.: Towards complementary explanations using deep neural networks. In: Stoyanov, D., et al. (eds.) MLCN/DLF/IMIMIC -2018. LNCS, vol. 11038, pp. 133–140. Springer, Cham (2018). https://doi.org/10.1007/978-3-030-02628-8_15

42. Honegger, M.: Shedding light on black box machine learning algorithms: development of an axiomatic framework to assess the quality of methods that explain individual predictions. arXiv preprint arXiv:1808.05054 (2018)

Syntactic and Semantic Bias Detection and Countermeasures

Roman Englert[1,2(✉)] and Jörg Muschiol[1]

[1] FOM University of Applied Sciences,
Herkulesstraße 32, 45127 Essen, Germany
roman.englert@fom-net.de
[2] Faculty III, New Media and Information Systems, Siegen University,
Kohlbettstraße 15, 57072 Siegen, Germany

Abstract. Applied Artificial Intelligence (AAI) and, especially Machine Learning (ML), both had recently a breakthrough with high-performant hardware for Deep Learning [1]. Additionally, big companies like Huawei and Google are adapting their product philosophy to AAI and ML [2–4]. Using ML-based systems require always a training data set to achieve a usable, i.e. trained, AAI system. The quality of the training data set determines the quality of the predictions. One important quality factor is that the training data are unbiased. Bias may lead in the worst case to incorrect and unusable predictions. This paper investigates the most important types of bias, namely syntactic and semantic bias. Countermeasures and methods to detect these biases are provided to diminish the deficiencies.

Keywords: Bias detection · Training samples · Multivariate regression · Root-out-bias

1 Introduction

The term bias has several meanings: in AI and Machine Learning (ML) "any preference for one hypothesis over another, beyond mere consistency with the examples, is called a bias" [5]. In other words a (declarative) bias helps to understand how prior knowledge can be used to identify the hypothesis space within which to search. The bias is independent of the applied ML technique, i.e. probability theory or (inductive) logic programming. In psychology is the omission bias "the tendency to judge harmful actions as worse, or less moral than equally harmful omissions (inactions) because actions are more obvious than inactions" [6], and in mathematics is a bias a systematic error.

An applied understanding of bias is prejudice by morally incomplete data, where the application is in training ML algorithms/models with data sets. As an example serves an American computer scientist discovering that Google's facial recognition Software only spotted his face, if he wore a white mask (see Fig. 1) [7]. Since ML models provide predictions based on the used training data sets, potential biases need to be recognized and defined before training data sets are being generated.

© Springer Nature Switzerland AG 2020
V. V. Krzhizhanovskaya et al. (Eds.): ICCS 2020, LNCS 12140, pp. 629–638, 2020.
https://doi.org/10.1007/978-3-030-50423-6_47

Fig. 1. An American computer scientist, found his computer system recognized the white mask, but not his face.

Bias detection requires an investigation of the training sample. In the case of the American computer scientist, the bias is called semantic bias, since a feature is missing in the training data. If the training sample contains mathematical computable biases like features are dependent on each other, or heteroscedasticity [8], then the bias is called syntactic. Both types of bias require dedicated methods to achieve a bias mitigation, the former the root-out-bias method with an interrogation by a human expert [9], and the latter a pre-processing of the training data [10–12]. For the mitigation of bias we focus on data transformation techniques.

The paper is structured as follows: The importance of AAI and training data is described in Sect. 2. Then, the computation of syntactic bias including the state-of-the art of research is described in Sect. 3. Subsequently, Sect. 4 contains an example for training data that became insignificant after bias inspection. And Sect. 5 contains the detection of missing features with the root-out-bias method and an example based on an interrogation by a human expert. Section 6 concludes the paper with a summary and an outlook discussing further research for the automatization of syntactic and semantic bias detection.

2   AAI Strategies of Google and Huawei and the Importance of Proper Training Samples

After a long period of research, AI reached a maturity level and had 2007 the breakthrough with high-performant Deep Learning chips developed by Huawei and Google [1]. These two leading companies announced to focus their R&D completely onto AI [2–4]. Their AI strategies are described in the following, since they influence the applied data science world and, thus, highlight the importance of proper training data. An example for a famous improper training sample, the Wooldridge data set "affairs" [13], is shown in the subsequent section.

Huawei started focusing on AI already in the eighties with research on AI algorithms for classification (e.g. regression), social analytics (e.g., PageRank), dimensionality reduction (e.g., KPCA), anomaly detection (e.g., local outlier factor), and clustering of samples (e.g., K-means, GMM), to name a few important ones. Huawei's AI strategy is based on AI research with the following focus fields [4]:

- Image processing and interpretation
- Natural language understanding
- Decision making and inferences from knowledge domains

Accompanying goals are an optimization of required data, less computational effort and less energy consumption. Additionally, Machine Learning is aimed to be secure and trustworthy, and should be processed autonomously and fully automated. In order to reach this comprehensive AI goal Huawei implements four measures:

1. A full-stack AI portfolio consisting of distributed cloud, devices, algorithms and applications for multi-purposes.
2. The development of a talent hub through the cooperation with universities and industry.
3. Huawei's portfolio is completely based on AI.
4. All processes are based on AI and an efficiency increase is expected.

An example for a widely known process is email spam filtering. This strategy will influence research at universities and development.

Google has a different AI strategy [3, 14]: As hitherto they had a focus onto the search engine and data collection based on services like email, Google Scholar and Maps/Earth. Today Google offers cloud-based ML applications [3]:

- Contact Center AI: A trainable customer support system for the automated communication.
- Document Understanding AI: ML-based understanding of documents using text extraction and information tagging.
- Cloud Talent Solution: Matching of job offers and applicants.
- Recommendations AI: Personalized product recommendations with real-time adaptations to customer behavior.

Thus, the former focus fields are becoming less important, and data are in future collected through cloud-based ML applications. Google focuses on research in the areas of Deep Learning (Neural Networks with several hidden layers), document analysis, Pattern Recognition, feature (text) extraction, recommendations, and (probabilistic) matching. Adaptive AI systems require huge training samples without bias. The aforementioned AI applications reason the importance for proper unbiased training samples.

3 Syntactic Bias and Its Mitigation

The preprocessing of training samples to mitigate bias is described in varies papers [10–12]. Some frequent occurring anomalies are discussed in the following (the mathematical representation is based on [8]):

1. Normal distribution: If features are normally-distributed and the sample size is big enough, i.e., the central limit theorem holds, then varies tests like the t-test can be applied. The t-test determines, whether the average of a random sample deviates more than a given value p from the population mean.
2. Significance: Features (variables) of training samples that are not significant according to, e.g., the 5% level, can be dropped. Assume that the considered feature of the training sample is normally-distributed. Then, the f- and the t-test can be combined used to check the significance of the feature: Assume p is 5%, then the variable can be dropped, if the p-level of the t-test is less equal 0.05, and the f-test shows that the regressors coefficients are equal (to zero). The latter one investigates the joint significance of the features.
3. Dummy traps: This effect holds for regression, when the model suffers from (multi) collinearity. In this case can one variable be (linearly) predicted of the others. As example consider the variables x, y, and z, then these variables are collinear, iff

$$x = a \cdot y + b \cdot z. \tag{1}$$

with a and b are real-valued vectors. In this case, the estimate of the impact of variable x onto another dependent variable is less precise compared to the situation without collinearity, and the variable x should be dropped.
4. Independent and identical distributed: This property for distributions of variables is important for the central limit theorem, stating that the probability distribution of the sum of independent and identical distributed variables (with finite variance) approximates the normal distribution. As a consequence, e.g., the significance of features can be investigated.

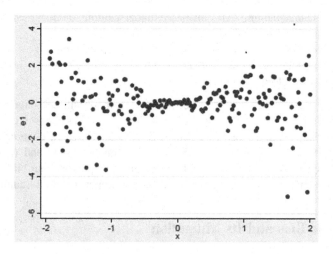

Fig. 2. An example for heteroscedasticity [15].

5. Heteroscedasticity: In this case the dispersion of sub-populations differs (see Fig. 2), and statistical hypothesis tests for their significance can become invalid. Heteroscedasticity can be detected by applying the f-test, which identifies how good a model fits to the training sample, since the f-test compares the variance of two sub-samples. Especially for non-linear models may occur severe impacts, where models can become inconsistent. As an example consider the speed measurement of a moving object. The measurement close to the object is more precise than if the object has a greater distance. Thus, the measurement data are expected to contain heteroscedasticity, which is a general challenge of measurements [8]. Heteroscedasticity can be mitigated using various approaches. An overview of detecting heteroscedasticity and mitigating its effects is provided in [22]. The mitigation is based on an analysis and transformation of the residuals that cause the heteroscedasticity effect.

The above discussed anomalies can also be mitigated by applying mappings of random samples to new distributions with constraints [11]. In the following the mitigation of bias using an inspection tool is described. As exemplary tool, the What-if tool from Google is used [16].

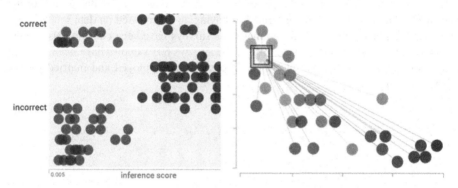

Fig. 3. What-if tool: visualization of inference results and arrangement of data points by similarity [16].

The What-if tool supports model inspection with the visualization of inference results with correct/incorrect data points (false/positives) and to compare two models (Fig. 3, left). Furthermore, data points can be arranged by similarity, whereat the user can create distance features and apply them to the model for inspection (Fig. 3, right).

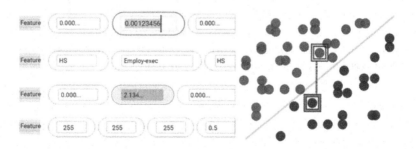

Fig. 4. Re-weighing of data points and comparison of counterfactuals to data points [16].

The mitigation of bias is supported by re-weighing of features (Fig. 4, left). However, this may require some iterations and trials, and may become a tedious task. Analogously, the possibility to compare data points and their counterfactuals to gain insight into the portion of data points to model, resp. the statistical distribution of features. Nevertheless, this manual inspection may be tedious and requires expert knowledge and experience, but automated tools like Fairness 360° [18] require that an expert chooses a re-weighing measure among more than 70 choices and inspects the result of its application to the training sample. Bias mitigation and syntactic bias detection, both, require a tool and a human expert.

4 An Example for Training Data that Became Insignificant After Bias Inspection

Social contacts are an important component of an individual's life and many adults follow the wish to marry and start a family [13]. However, it is still an unsolved question whether the human being suits monogamy, since the "happily ever after" feeling does not always last and a partner may engage in an extramarital affair. For the evaluation the probability is investigated of having an affair based on data set "affairs".

In 1969 and 1974 data from two magazine surveys have been collected about men and women time spent with beaus. From these surveys 601 samples have been selected based on the criteria that the interrogated people were employed and married for the first time (Fig. 5).

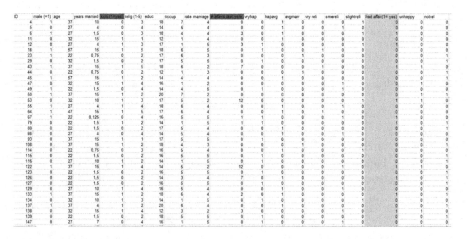

Fig. 5. 601 samples from Wooldridge's data set "affairs" (excerpt). Dummy traps are highlighted red and the regressand yellow. (Color figure online)

4.1 Data Preprocessing

The data set is provided as an EXCEL file that consists of 601 rows containing the answers of the interrogated persons and nineteen columns with features, i.e., nineteen candidates for variables. Since the probability of having an affair has to be investigated, the column "id" that identifies an individual will be ignored for the regression computation, and column "affair" (=1, if had an affair) is taken as regressand (the dependent variable that is explained), the dependent random variable for the regression model. First, the data are sharpened by dropping those variables that are not significant according to the 5% level, i.e., p-level of t-test $\leq$ 0.05 (see also Sect. 3).

Furthermore, the data set contains twice a dummy trap, i.e., the model suffers from collinearity (cf. Fig. 5, red highlighted). First, columns twelve till fifteen contain a zero or one ("vryhap", "hapavg", "avgmarr", "unhap") and are perfect collinear, i.e., each row sums up to one for these four (binary) variables. Hence, the column "unhap" is left out and the interpretation for the remaining three variables is "relative to having an unhappy marriage". Analogous, columns sixteen till nineteen contain a zero or one ("vryrel", "smerel", "slghtrel", "notrel") and are perfect collinear, i.e., each row sums up to one for these four variables. Hence, the column "notrel" is left out and the interpretation for the remaining three variables is "relative to being not religious".

After computing the regression including the t-test (significance test of variables), two variables are determined with a p-value $\leq$ 0.05, and thus, these variables are not significant and dropped: "kids" and "naffairs". The check for heteroscedasticity was negative. To summarize, thirteen variables remain as regressors (independent variables): "male", "age", "yrsmarr", "relig", "educ", "occup", "ratemarr", "vryhap", "hapavg", "avgmarr", "vryrel", "smerel" and "slghtrel" (cf. Fig. 5).

4.2 Interpretation of the Parameter Estimates of the Significant Variables in the Model

The parameter estimates lead to the following model for the regressand

$$\begin{aligned}
\Pr(\hat{y} = 1) = {} & 0.8342 + 0.0511 * male - 0.0073 * age + 0.0181 * yrsmarr \\
& - 0.1856 * relig + 0.0032 * educ + 0.0045 * occup + 0.0102 * ratemarr \\
& - 0.3541 * vryhap - 0.2698 * hapavg - 0.2078 * avgmarr + 0.4884 * vryrel \\
& + 0.2879 * smerel + 0.2590 * slghtrel
\end{aligned} \tag{2}$$

The regressand "affair" is used to estimate the linear probability model. Note, the intercept is 0.8342 (expected mean value, if all variables are 0, however "age" cannot be 0). First, the variables "age", "relig", "vryhap", hapavg", and "avgmarr" have negative effects on the probability of a person to engage in an affair. The marginal effect of the variable "age" is, that when a person turns one year older, it reduces the probability of having an affair by 0.0073. The more religious a person is ("relig": 1...5 (high)), the probability to have an affair will be reduced by 0.1856 times "relig". The dummy variables "vryhap", "hapavg" and "avgmarr" each have a negative effect on the likelihood of a person having a paramour in relation to an unhappy marriage, namely:

−0.3541, −0.2698, −0.2078. Second, the five variables "male", "yrsmarr", "educ", "occup" and "ratemarr" have a small positive coefficient (<0.06): The average effect on the likelihood to have an affair is for a male with a factor of 0.0511 higher than for a woman ("male"). Third, the dummy variables "vryrel", "smerel" and "slghtrel" have a relatively large positive contribution (0.4884, 0.2879, 0.2590) on the likelihood to have an affair (in relation to being not religious).

The R-squared value is low with 0,1261, meaning that the data explain poorly the regressand. Hence, a RESET test [18] with quadratic and cubic order is recommended in order to test for a more precise fit.

5   Semantic Bias and the Root-Out-Bias Method

Semantic bias is in contrast to syntactic bias not computable and may have different causes that are difficult to detect. As an example consider the American computer scientist who test a face recognition Software and found his computer system recognized the white mask, but not his face (Fig. 1, Sect. 1). The training sample for the utilized Neural Network missed at least one feature, and thus, is biased. The omitted and missing feature is ambiguous, since it might be for instance "skin type" describing characteristics of the skin, or the feature "culture" depicting the cultural background, assuming that the cultural background influences the appearance of someone. This type of bias is called semantic bias.

Another example is the Chabot Tay from Microsoft [20, 21]: The goal with Tay was to perform research and to gain deeper insights into conversational understanding. Tay was automatically trained by its conversations with unknown persons via the Internet. Unfortunately, Tay was trained with biased statements according to gender, and additionally, to racists statements. Microsoft had to take off Tay immediately. This demonstrates that speech could be biased, and must be inspected before presenting the statements to the underlying Neural Network.

A method to identify semantic bias is the so-called "root-out-bias" [9]: Since semantic bias is not computable, a human expert must be involved. The root-out-bias consists of two steps: The first step is to openly question what preconceptions could currently exist in a domain that is aimed to be modeled. This important step requires experience, and must be done by experts. As outcome potential biases are identified, and then, in step two requirements to the training data set must be defined.

In order to illustrate the root-out-bias method it is applied to the Wooldridge data set from Sect. 4. The data set contains 16 features to explain the probability that someone has an extramarital affair. One feature is the sex of an interrogated person, where is assumed that an affair only takes place between different sexes [13]. This assumption is outdated, since 2015 the same-gender marriage was invented in the US [19]. Thus, nowadays the data set is semantically-biased and this assumption should be revised. The new requirement for this data set can be that people are interrogated despite the gender of their marriage partner.

This example demonstrates that semantic bias has sweeping facets and is not only culturally-dependent (also law-dependent), and may change in societies and cultures by time.

6 Summary and Outlook

This paper analyses syntactic and semantic biases within training samples. Syntactic bias can be detected by computation. But the mitigation requires a human expert who can be supported by tools for data inspection and visualization. In contrast, semantic bias cannot be computed and must be detected by a structured procedure, e.g., root-out-bias method by an experienced human interrogator. Varies examples demonstrate the importance of proper training samples to avoid bias. Biased training samples can lead to social and morally unacceptable AI systems that need to be taken off.

As an outlook the investigation for the automatization of the detection and mitigation of syntactic and semantic bias is aimed. For syntactic bias provides the mathematical modeling of biases the "parts" of the training sample that needs to be improved. And semantic bias may be analyzed using semantic networks that brings knowledge items semantically into relation to each other.

References

1. Groth, O., Nitzberg, M.: Solomon's Code. Humanity in a World of Thinking Machines. Pegasus Books (2018)
2. Artificial Intelligence Technology Scan, Teqmine Technology Analysts. https://teqmine.com/google-ai-strategy/. Accessed 28 Oct 2019
3. Google's AI-based Cloud Services. https://cloud.google.com/solutions/ai/?hl=de. Accessed 28 Oct 2019
4. Huawei Releases AI Strategy und Full-Stack, All-Scenario AI Portfolio. https://www.huawei.com/en/press-events/news/2018/10/Huawei-HC-2018-Eric-Xu-AI. Accessed 28 Oct 2019
5. Russell, S., Norvig, P.: AI a Modern Approach. Prentice-Hall, Upper Saddle River (1995)
6. Psychology omission bias definition. https://en.wikipedia.org/wiki/Omission_bias. Accessed 28 Oct 2019
7. Bias in the face recognition Software of Google. https://www.bbc.com/news/technology-45561955. Accessed 28 Oct 2019
8. Casella, G., Berger, R.: Statistical Inference. Duxbury Advanced Series, 2nd edn. Thomson Learning Inc., Boston (2002)
9. Root-out-bias method. https://sloanreview.mit.edu/article/the-risk-of-machine-learning-bias-and-how-to-prevent-it. Accessed 28 Oct 2019
10. Kamiran, F., Calder, T.: Data preprocessing techniques for classification without discrimination. Knowl. Inf. Syst. 33(1), 1–33 (2012). https://doi.org/10.1007/s10115-011-0463-8
11. Calmon, F, Wei, D., Vinzamuri, B., Ramamurthy, K., Varshney, K.: Optimized preprocessing for discrimination prevention. In: Conference on Neural Information Processing Systems. Advances in Neural Information Processing Systems, vol. 30, pp. 3992–4001 (2017)
12. Zemel, R., Wu, Y., Swersky, K., Pitassi, T., Dwork, C.: Learning fair representations. international conference on machine learning. In: Proceedings of Machine Learning Research, pp. 325–333 (2013)
13. Fair, R.: A theory of extramarital affairs. J. Polit. Econ. 86(1), 45–62 (1978)
14. Google's AI. https://ai.google. Accessed 10 Nov 2019

15. Richard, W.: Heteroskedasticity, University of Notre Dame. https://www3.nd.edu/~rwilliam. Accessed 11 Nov 2019
16. Google's What-If-Tool. https://pair-code.github.io/what-if-tool. Accessed 11 Nov 2019
17. Fairness 360°. https://aif360.mybluemix.net. Accessed 11 Nov 2019
18. Sapra, S.: A regression error specification test (RESET) for generalized linear models. Econ. Bull. **3**(1), 1–6 (2005)
19. Same-gender marriage US. https://www.nytimes.com/2015/06/27/us/supreme-court-same-sex-marriage.html. Accessed 16 Nov 2019
20. Microsoft is deleting its AI chatbot's incredibly racist tweets. https://img.sauf.ca/pictures/2016-03-24/d360716e3199095063ebd4749b78fc4c.pdf. Accessed 16 Nov 2019
21. Davis, E.: AI amusements: the tragic tale of tay the chatbot. ACM AI Matters **2**(4), 20–24 (2016)
22. Rosopa, P., Schaffer, M., Schroeder, A.: Managing heteroscedasticity in general linear models. Psychol. Methods **18**(3), 335–351 (2013)

Detecting Rumours in Disasters: An Imbalanced Learning Approach

Amir Ebrahimi Fard$^{(\boxtimes)}$, Majid Mohammadi$^{(\boxtimes)}$, and Bartel van de Walle$^{(\boxtimes)}$

Delft University of Technology, Delft, The Netherlands
{A.EbrahimiFard,M.Mohammadi,B.A.vandeWalle}@tudelft.nl

Abstract. The online spread of rumours in disasters can create panic and anxiety and disrupt crisis operations. Hence, it is crucial to take measure against such a distressing phenomenon since it can turn into a crisis by itself. In this work, the automatic rumour detection in natural disasters is addressed from an imbalanced learning perspective due to the rumour dearth versus non-rumour abundance in social networks.

We first provide two datasets by collecting and annotating tweets regarding the Hurricane Florence and Kerala flood. We then capture the properties of rumours and non-rumours in those disasters using 83 theory-based and early-available features, 47 of which are proposed for the first time. The proposed features show a high discrimination power that help us distinguish rumours from non-rumours more reliably. Next, We build the rumour identification models using imbalanced learning to address the scarcity of rumours compared to non-rumour. Additionally, to replicate the rumour detection in the real-world situation, we practice cross-incident learning by training the classifier with the samples of one incident and test it with the other one. In the end we measure the impact of imbalanced learning using Bayesian Wilcoxon Signed-rank test and observe a significant improvement in the classifiers performance.

Keywords: Rumour detection · Imbalanced learning · Building dataset · Feature engineering · Twitter

1 Introduction

Rumours are unverified information circulating about the topics that people perceive important and are used as sense-making or risk management mechanism [4]. Rumours tend to thrive in situations that are ambiguous and/or pose a threat in which meanings are uncertain; questions are unsettled, information is missing, and/or lines of communications are absent. One of the contexts that satisfies all those conditions is the crisis. Due to the lack of information and mistrust towards the available sources of formal channels at the time of disaster being happening, people feel frustrated and seek information from informal channels. Eventually, if no information is available, people engage in affirmative rumouring, which means speculation based on whatever evidence and framework of understanding

© Springer Nature Switzerland AG 2020
V. V. Krzhizhanovskaya et al. (Eds.): ICCS 2020, LNCS 12140, pp. 639–652, 2020.
https://doi.org/10.1007/978-3-030-50423-6_48

they posses [4]. Traditionally, rumours were propagated by means of word of mouth. However, in recent years, the rapid growth of social networks escalated this problem and turned it into a serious issue by accelerating, widening, and deepening the circulation of misinformation among people [21,25]. Thus, it is of the essence to develop a solution for this problem, because it can otherwise turn into a potential threat to the main societal institutions such as peace and democracy.

One of the promising approaches to quell online rumours is artificial intelligence (AI) which can tackle misinformation at scale and across languages and time zones. Among the AI approaches, binary classification dominates the literature of rumour detection [33]. In this approach the classifiers are trained by historical samples form two classes of rumour and non-rumour. Recent studies have shown theoretically and empirically that compared to all different kinds of information circulating in social networks, rumours are in minority [5,6,15]. In other word, the fraction of non-rumours has the majority in the flow of information in social networks. From a data collection point of view, this leads to an imbalanced dataset containing uneven volume of rumour and non-rumour samples. This imbalance gives rise to the "class imbalance" problem or "curse of imbalanced dataset", which is the problem of learning a concept from a class with a small number of samples [16]. In machine learning, the problem of imbalanced data has been addressed by a learning paradigm called imbalanced learning.

In this study, we address the rumour dearth versus non-rumour abundance by imbalanced learning. As Fig. 1 illustrates, we first collected more than 200,000 tweets regarding Kerala flood and Hurricane Florence. For the annotation, we used a large-scale labeling technique based on signal words of each incident.

We then extract 83 theory-based and early available features from every data point. Out of those features, 47 of which are proposed for the first time. Our feature selection method represents a high level of effectiveness for the newly proposed features. In the next step, we conduct a series of experiments using imbalanced learning. For the experiments, we use cross-topic learning in which a classifier is trained on one incident and is tested on the other one, instead of training and test on the same dataset. In the end, using Bayesian Wilcoxon Signed-rank test, we measure the effectiveness of imbalanced learning.

Given this, the main contributions of this article are summarised as follows:

1. Improving the models performance using imbalanced learning due to the imbalanced nature of rumour against non-rumour in social media.
2. Building two annotated datasets comprising more than 200,000 tweets regarding Kerala flood and the Hurricane Florence.
3. Proposing 47 features for computational rumour detection and evaluating them using five different learning methods.
4. Designing a set of cross-topic novel experiments which replicate the real-world situation in rumour detection through training with one dataset and testing with the other one.

This paper is organised as follows. Section 2, provides an overview on the related studies in computational rumour detection. Section 3 explains data

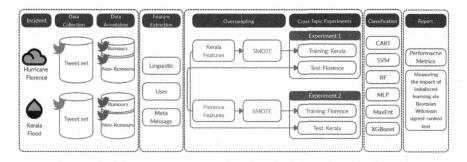

Fig. 1. Research flow of rumour detection with imbalanced learning.

collection, annotation, and feature extraction. In Sect. 4 we report and discuss the experiments results and features importance. Finally in Sect. 5 we conclude this study and give some future research directions.

2 Related Work

Computational rumour detection is a classification problem that aims to distinguish rumours from non-rumours. Similar to other classification problem, data collection, feature extraction, and model training constitute the pillars of a rumour detection model. In fact a dataset comprising rumours and non-rumours, a set of relevant features, and algorithm for the context of rumour spreading are prerequisite conditions for building a rumour detection model.

Since almost a decade ago that first articles in rumour detection with the computational approach were published, most of the studies meet this set of conditions and contribute in at least one of those areas. Several studies provided annotated datasets. They are mostly collected from Twitter [12,19,24,32] and Sina Weibo [17,29] as they provide a fairly easy API access. Some other studies, proposed new features according to specification of rumour propagation in predefined categories of content, user, network, and temporal features [2,12]. For instance due to the importance of rumour identification as early as possible, some of scholars proposed features to capture early appearance of rumours. For instance, Zhao et al. [31] propose a novel technique for the early rumour detection using signature text phrases. In another research, Wang et al. [26] show a feature pattern consisting of both user's attitude and information diffusion for the early detection of rumours in social networks. In the same vein, Kwon et al. [13] showed that some features are more informative in the early stages of rumour diffusion. Furthermore, some other studies proposed new algorithms for identification of rumours. For instance, Ma et al. [18] used the recurrent neural network (RNN) for the rumour detection, Lozano et al. [8] used the combination of convolutional neural networks (CNN) with both automatic rule mining and manually written rules, Chen et al. [3] utilized a CNN for short text categorization using multiple filter sizes, or Zubiaga et al. [34] model the rumour tweets as sequences by extracting their features in the course of time and applying the conditional random field (CRF) for classification. New techniques do not necessarily mean designing a completely new algorithm, but it mostly means applying an algorithm which is developed before to the context of rumour detection. For example, LSTM (as an RNN technique) and CRF techniques were introduced in 1997 [10] and 2001 [14], respectively, but they were used later for rumour detection [18,34], and were truly counted as significant contributions.

The other area of importance is rumour detection in different subject domains. As we discussed earlier, any domain that falls into 3C's categories can be a hotbed for rumour emergence. There are several studies [2,12] that highlight the importance of domain in rumour detection, but they do not analyse the impact of domain in rumour detection [20]. To the best of our knowledge, there is one single research on computational rumour detection in a specific domain which is recently published by Sicilia et al. [20]. They study the rumour detection in the health domain, in particular, the case of Zika virus. In this work, they take the context into account by collecting data and proposing novel features. However for the third element namely algorithm, they leave it without contribution by using classifiers from different learning paradigms.

3   Data and Features

For this study, we prepared two datasets regarding the Kerala flood and Hurricane Florence. The 2018 Kerala flood was the worst monsoon flooding in a century in Southern India with 400 fatalities and $2.7bn worth of damages. The Hurricane Florence was a category four hurricane hit Carolinas in the south-east of the United States. The hurricane caused more than 50 fatalities and up to $22bn damages. We set up a streaming API of Twitter to collect all tweets, retweets, replies, and mentions that included hashtags, text strings, and/or handles related to Kerala flood and Hurricane Florence. We could collect 100,880 tweets regarding Kerala flood and 101,844 tweets for Hurricane florence.

To select the rumour cases we searched several credible news outlets and fact-checking websites (e.g. Snopes, The Washington Post, The Hindu, and Indian Express). We identified in total three rumours in Kerala flood and four rumours in Hurricane Florence with a high level of consistency among different news outlets and fact-checking sources. Then we extracted the rumour-related tweets corresponding to these events if the tweet contains the keyword relevant to the rumour [13]. The tweets without explicit keywords were assigned a non-rumour label. After the data annotation the Kerala dataset consists of 2,000 rumour-related and 98,880 non-rumour-related tweets. Florence dataset also comprises 2,382 rumours and 99,462 non-rumours[1]. The imbalance between the number of rumours and non-rumours in an incident aligns with the previous findings [6,15].

For feature extraction, we use 83 features to represent rumour and non-rumour tweets. The features are either taken from the literature of computational

---

[1] The datasets are publicly available: https://bit.ly/2WxVhY0.

rumour detection or introduced in this work[2]. Tweet features are classified into three groups: linguistic & content, user, and meta-message. The linguistic & content category contains all the features about syntactic and semantic aspects of tweets. The features related to the account holders and their social networks fall into the user category, and all the features about tweets metadata belong to the meta-message class. In this study, we want to detect rumours in the disasters as early as possible, we can thus rely solely on the features that are available during the initial phases of rumour diffusion. In this regard, we deliberately ignore propagation and temporal features, as they are not available in the early diffusion phases [13]. We also skip features related to likes, retweets, and comments since they have a high volatility and varying values in the initial stages of rumour diffusion. Table 1 demonstrates all the features in three categories of linguistic & content, user, and meta-message. In this table, dagger (†) and diamond (◇) symbols specify features inspired by social bot detection literature and proposed ones, respectively.

4   Experimental Results

In this section we first report the models performance and measure the impact of imbalanced learning in the identification of rumours, then we evaluate the discrimination power of proposed features.

4.1   Models Performance

In this section, we evaluate the features weights, and report the results of our models and compare their performance in cross-incidents experiments subsequently. Then we measure the impact of imbalanced learning on rumour identification. We also conduct a baseline analysis by comparing the performance of our models with state of the art.

For feature evaluation, our goal is to assign a weight to each feature. This allows us to rank features ordinaly as well as knowing to what degree each feature is informative for classification algorithms. The conventional methods such as χ^2 test and recursive feature elimination are either time-consuming or not suitable for our datasets due to their size[3]. Therefore, we define a score using random forest, XGBoost, adaptive boosting, regression tree, and extremely randomised trees as classification algorithms with an embedded feature selection mechanism. By summing up the features weights we obtain the degree of significance for each feature. To find the significant features we determine a threshold for the features score. If the score of a feature in each dataset is less than the given threshold, it is assumed to be *insignificant*. To determine the threshold, we choose

---

[2] The newly introduced features are explained and their relevance are discussed in the first section of the supplementary materials (available at: https://bit.ly/2PJ3FmR).

[3] The further explanations regarding the issues of the conventional feature selection methods can be found in the second section of the supplementary materials (available at: https://bit.ly/2PJ3FmR).

Table 1. List of all the features in three categories: linguistic & content, user and meta-message. New features are marked with ◇. The features inspired by social bot detection literature that, to the best of our knowledge, have not been used in rumour detection literature yet are marked with †.

Feature class	Features
Linguistic & content	Number of exclamation marks in a tweet [2,32]
	Number of question marks in a tweet [2,32]
	Number of characters in a tweet [2]
	Number of words in a tweet [2]
	Number of uppercase letters in a tweet [2,32]
	Number of lowercase letters in a tweet [2]
	Number of first person pronoun in a tweet [2]
	Number of second person pronoun in a tweet [2]
	Number of third person pronoun in a tweet [2]
	Number of capital words in a tweet [2]
	Average word complexity in a tweet [24]
	Number of vulgar words in a tweet [24]
	Number of abbreviations in a tweet [24]
	Number of emojis in a tweet [24]
	Polarity of a tweet [30]
	Subjectivity of a tweet [27]
	Tone of a tweet [30]
	Positive words score of a tweet [29]
	Negative words score of a tweet [29]
	† Frequency of Part of Speech (POS) tags in a tweet (19 features) [23]
	◇ Frequency of Name Entity Recognition (NER) tags in a tweet (17 features)
	Opinion and insight score [24]
	Anxiety score [22]
	Tentativeness score [24]
	◇ Certainty score
	Sentence complexity [24]
User	Profile description (binary) [2,28–30]
	Verified account (binary) [2,28,29,32]
	Number of Statuses [2,13,28,29]
	Influence [2,13,28–30]
	Number of following [13,28–30]
	User role [24]
	◇ Attention
	Account age (day) [2]
	◇ Openness (binary)
	Profile location (binary) [29]
	† Profile picture (binary) [23]
	Profile URL (binary) [30]
	◇ Average follow speed
	◇ Average being followed speed
	◇ Average like speed
	◇ Average tweet speed
	† Screen name length [23]
	† Number of digits in screen name [23]
Meta-message	Number of hashtags in a tweet [2,19]
	Number of mentions in a tweet [2]
	Tweet URL (Binary) [2]
	Number of multimedia in a tweet [28–30]
	◇ Location sharing (binary)

a very small number of 0.001 in order to select features with minimum level of informativeness. The selection of this number as the significance threshold is inspired by the significance level in null hypothesis testing. Additionally, we call a feature *consistently significant* if its score is higher than the threshold in both datasets.

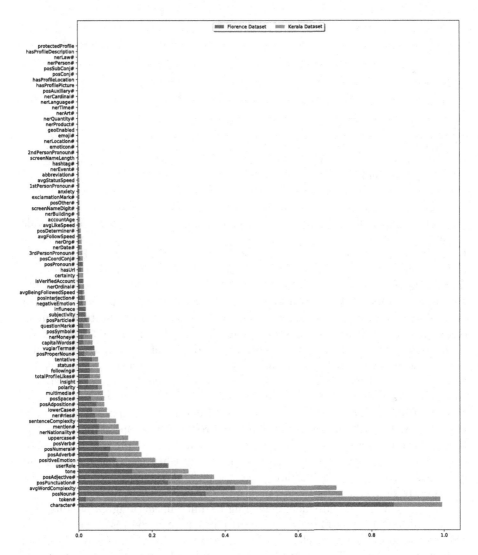

Fig. 2. Evaluation of the feature significance in the Florence and Kerala datasets. Blue and orange bars indicate the significance of the features in Florence and Kerala datasets, respectively. The longer a bar is, the more important the corresponding feature will be. (Color figure online)

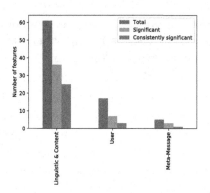 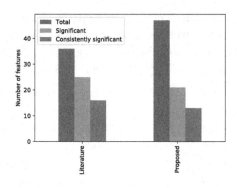

(a) Comparison of features based on their types.

(b) Comparison of features based on their source.

Fig. 3. Comparison of the feature significance. The blue, orange, and green bars denote the total, significant, and consistently significant number of features, respectively. (Color figure online)

Figure 2 illustrates the feature significance in the Florence and Kerala datasets with blue and orange bars, respectively. In this figure, the length of each bar is proportionate to the importance of the corresponding feature. Figure 3 displays significant and consistently significant features based on their types and sources. In this figure, the abscissa represents the source or type, and the ordinate is the number of features in each category. Figure 3a illustrates the number of significant and consistently significant features as well as the total number of features for each feature category. According to this figure, 60% of linguistic & content features, 41% of user features, and 60% of meta-message features are significant. Among the significant features, 70% of linguistic & content features, 43% of user features, and 33% of meta-message features are considered as consistently significant. This means that linguistic & content, user, and meta-message features have the highest fraction of consistently significant features, respectively. Figure 3b shows the same measures for the proposed features and the ones which were extracted from the literature. Based on this figure, 70% of the literature features and 45% of the proposed features are significant. Out of the significant ones, 64% of literature and 62% of the proposed features are consistently significant.

In this study seven classifiers belonging to different learning paradigms were considered for the experiments [20]: multi-Layer perceptron (MLP) as a neural network, support vector machine (SVM) as a kernel machine, classification and regression trees (CART) as a decision tree, random forest (RF) as an ensemble of trees, XGBoost as a boosting approach, and maximum entropy (MaxEnt) as an exponential model. We tried to use entirely distinct algorithms in order to verify their effectiveness in the early rumour detection problem. We conduct four experiments to address two crucial challenges in computational rumour detection. The first challenge is the imbalanced nature of rumours to non-rumours [6,15]. This is what we are also observing in this study as the ratio of rumour to non-rumour in datasets is approximately 1:50. To tackle this issue, in addition to training the classifier with imbalanced dataset we use an oversampling technique to train the classifiers with the balanced form of the training sets. For oversampling, we use synthetic minority oversampling technique (SMOTE) as a powerful imbalanced learning technique that has shown a great deal of success in various applications. The SMOTE algorithm creates artificial data based on the feature space similarities between existing minority examples [9]. It is worth emphasising that we use this algorithm only for training data, in other words, the test dataset is still intact and preserves its imbalanced shape.

The second challenge is the common practice of the field namely training and testing on the same dataset. In a real rumour propagation case, there is almost no time for data collection, feature extraction, and model training; thus, such an approach for rumour detection cannot be operationalised. The other approach is using historical rumours for identification of upcoming rumours. In this regard, we use one of our datasets for training and the other one for testing; then we switch the training and test set to assess the robustness of the proposed approach. In other words, once we use Kerala as training set and Florence as the test set (which is indicated by Kerala $\Rightarrow$ Florence notation), then we switch the datasets and use Florence as training set and Kerala as test set (Florence $\Rightarrow$ Kerala). To evaluate the classifiers performacne in different experimental settings we use precision-recall (PR) curve as it can provide an informative representation of performance assessment [9].

In the upper panel of Table 2, the performance of each classifier is demonstrated in four figures using PR Curve. Each figure corresponds to an experiment, and each curve shows the classifier's performance regarding various threshold. By comparing the upper-left figure (a) to the upper-right one (b) and the bottom-left figure (c) to the bottom-right one (d), it is readily seen that oversampling had a positive impact and has improved classifiers performance in most cases. However, the degree of improvement is not the same for all classifiers. The improvement for classifiers with acceptable performance is insignificant, while classifiers with less impressive performance experience more salient enhancement. There are also classifiers which are insensitive to oversampling or the ones which receive slight negative influence from oversampling.

To be able to compare the classifiers' performance in a more concrete way, we have measured the area under the PR curve (AUPRC) for each classifier. The bottom panel of Table 2 shows the AUPRC regarding each classifier. As the table shows, for classifiers with the high score in Kerala $\Rightarrow$ Florence such as SVM, MaxEnt, and XGBoost oversampling leads to slight improvement, but for classifiers with lower performance such as RF or MLP, oversampling results in a much higher improvement. Despite the poor performance of CART, oversampling does not improve it that much. Similarly, in the Florence $\Rightarrow$ Kerala experiment, CART has the lowest performance and oversampling cannot improve it significantly. MLP with very high performance receives marginal improvement.

For the other two classifiers with high performance score oversampling shows a slight negative effect which is quite insignificant. For XGBoost and RF as classifiers with satisfactory AUPRC, oversampling fairly improves their performance. We use Bayesian Wilcoxon Singed-rank test [1] to verify the effectiveness of oversampling on the used classifiers. Based on this test, the probability of oversampling being effective is 0.99, which shows that oversampling significantly enhances the performance of classifiers.

For the baseline analysis, we use PHEME dataset [34] which is publicly available for computational rumour detection. We extract 83 features from each tweet and train the classifiers that we introduced before. We repeat the baseline experimental setup by using 5-fold cross-validation in the experiment. Table 3 demonstrates the precision, recall, and F-measure of classifiers along with the results of CRF in [34]. Based on this table, the proposed features and the canonical machine learning classifiers outperform significantly CRF in spite of the fact that the proposed experiment settings are much simpler than that in [34].

4.2   Features Performance

In order to assess the impact of the proposed features on the rumour detection, we also eliminated the proposed features and conducted the same experiments. Then, we subtracted the scores corresponding to each classifier to obtain the discriminant power of the proposed significant features. Let f be the intersections of the proposed and significant features, then

$$DP(f) = PF_{Significant\ features} - PF_{leave\ f\ out\ set} \tag{1}$$

where DP is the discriminant power of the feature set f, $PF_{Significant\ features}$ is the performance of the classifier with all significant features, and $PF_{leave\ f\ out\ set}$ is the performance of the classifier after removing the feature set f. The higher positive values for the discriminant power means that the feature set f is more significant, and zero or negative values is an indicator that shows the feature set f is not effective in increasing the classifier performance.

Table 4 tabulates the discriminant power of each classifier on two experiments. Interestingly, the average performance of all classifiers will increase if the proposed features are used. In particular, AUPRC of the first experiment has increased the most when the classification model is XGBoost. In the second experiment CART and XGBoost receive the highest improvement.

When we look at the classifiers in each experiment one by one, the performance often declines when we remove the proposed significant features. However, for MLP the performance improves after the removal of those features. This problem, namely performance improvement after feature reduction, is a non-trivial machine learning problem which has been revisited in the literature before [7,11]. One of the few insights about this problem that has been discussed in the literature is that classification performance achieved with different set of features is highly sensitive to the type of data and type of the classifier [11]; therefore getting the average of the classifiers performance may give a more reliable and broader picture of the discrimination power of chosen features.

Table 2. Assessing the performance of the classifiers using PR curve (upper panel) and AUPRC (bottom panel).

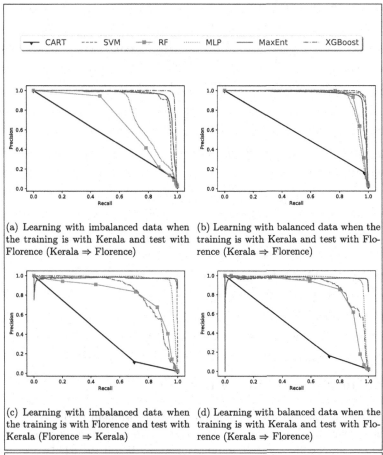

(a) Learning with imbalanced data when the training is with Kerala and test with Florence (Kerala ⇒ Florence)

(b) Learning with balanced data when the training is with Kerala and test with Florence (Kerala ⇒ Florence)

(c) Learning with imbalanced data when the training is with Florence and test with Kerala (Florence ⇒ Kerala)

(d) Learning with balanced data when the training is with Kerala and test with Florence (Kerala ⇒ Florence)

	Classifiers	Training with imbalanced data (AUPRC)	Training with balanced data (AUPRC)
	CART	54.2%	56.8%
	SVM	94.6%	97.8%
Kerala ⇒	RF	71.5%	94.3%
Florence	MLP	80.9%	93.3%
	MaxEnt	95.9%	96.5%
	XGBoost	98.7%	98.8%
	CART	41.4%	44.6%
	SVM	97.7%	97%
Florence ⇒	RF	81.3%	85.5%
Kerala	MLP	96.3%	96.4%
	MaxEnt	97.6%	97.2%
	XGBoost	83.9%	89.2%

Table 3. Baseline analysis on PHEME dataset [34]

Classifiers	PR	RE	F1-Score
CART	90.5%	91.7%	91.1%
SVM	92.6%	55.5%	69.4%
RF	93.9%	92.4%	93.1%
MLP	92.6%	90.6%	91.6%
MaxEnt	91.2%	79%	84.7%
XGBoost	95%	94.4%	94.7%
CRF [34]	66.7%	55.6%	60.7%

Table 4. The discriminant power of features for the classifiers with respect to AUPRC metric on two experiments

Classifiers	Florence ⇒ Kerala	Kerala ⇒ Florence
CART	10.4%	29.6%
SVM	0.3%	6%
RF	8.3%	3.1%
MLP	0.2%	−3.4%
MaxEnt	7.6%	4.1%
XGBoost	16.3%	25%
Average	7.2%	8.4%

5   Conclusion and Future Directions

In this work, we tackled the rumour propagation problem in disasters through imbalanced learning. We provided two datasets regarding Kerala flood and Hurricane Florence by collecting and annotating 100,880 and 101,844 tweets, respectively. We then used 83 features for creating a learning model by compiling existing features in the literature and introducing several new ones. In order to identify rumours in a timely manner, we focused solely on the early available features. We evaluated all features and observed that the proposed features could enhance the performance of the subsequent learning model for the rumour detection. For model building, we conducted a series of experiments using imbalanced learning. For the experiments, we used cross-topic learning which instead of training and test on the same dataset, it is trained on one event and is tested on the other one. Then we measure the impact of imbalanced learning using Bayesian Wilcoxon Signed-rank test. Our results show improvement in almost all the classifiers when the oversampling technique is applied on the training sets.

One of the possible future direction in this domain is data collection and annotation. This field needs more publicly available datasets from different social media and in different kinds of disasters. Such datasets will enable scholars to validate their results in broader contexts and investigate the impacts of contextual factors on their results. Another avenue could be more domain specific studies. In order to have a universal system for rumour detection, we need to know the behaviour of rumours in different subject domains. Performing domain-specific rumour analysis on various subject domains would be a practical step toward discovering the behaviour of rumours.

References

1. Benavoli, A., Corani, C., Demšar, J., Zaffalon, M.: Time for a change: a tutorial for comparing multiple classifiers through Bayesian analysis. J. Mach. Learn. Res. **18**(1), 2653–2688 (2017)
2. Castillo, C., Mendoza, M., Poblete, B.: Information credibility on Twitter. In: Proceedings of the 20th International Conference on World Wide Web - WWW 2011, p. 675. ACM Press, New York (2011)
3. Chen, Y.C., Liu, Z.Y., Kao, H.Y.: IKM at SemEval-2017 Task 8: convolutional neural networks for stance detection and rumor verification. In: The 11th International Workshop on Semantic Evaluation (SemEval-2017), pp. 465–469 (2017)
4. DiFonzo, N., Bordia, P.: Rumor psychology: social and organizational approaches. American Psychological Association (2007)
5. Dunbar, R.I.: Gossip in evolutionary perspective. Rev. Gener. Psychol. **8**(2), 100–110 (2004)
6. Fard, A.E., et al.: Rumour as an anomaly: rumour detection with one-class classification. In: 2019 IEEE International Conference on Engineering, Technology and Innovation (ICE/ITMC), pp. 1–9. IEEE (2019)
7. Forman, G.: An extensive empirical study of feature selection metrics for text classification. J. Mach. Learn. Res. **3**(Mar), 1289–1305 (2003)
8. García, L.M., Lilja, H., Tjörnhammar, E., Karasalo, M.: Mama Edha at SemEval-2017 task 8: stance classification with CNN and rules. In: The 11th International Workshop on Semantic Evaluation (SemEval-2017), pp. 481–485 (2017)
9. He, H., Garcia, E.: Learning from imbalanced data. IEEE Trans. Knowl. Data Eng. **21**(9), 1263–1284 (2009)
10. Hochreiter, S., Schmidhuber, J.: Long short-term memory. Neural Comput. **9**(8), 1735–1780 (1997)
11. Janecek, A.G.K., Gansterer, W.N., Demel, M.A., Ecker, G.F.: On the relationship between feature selection and classification accuracy. In: Proceedings of the 2008 International Conference on New Challenges for Feature Selection in Data Mining and Knowledge Discovery, pp. 90–105. JMLR.org (2008)
12. Kwon, S., Cha, M., Jung, K., On, W.C.: Prominent features of rumor propagation in online social media. In: International Conference on Data Mining. IEEE (2013)
13. Kwon, S., Cha, M., Jung, K.: Rumor detection over varying time windows. PLOS One **12**(1), e0168344 (2017)
14. Lafferty, J., McCallum, A., Pereira, F.C.: Conditional random fields: probabilistic models for segmenting and labeling sequence data (2001)
15. Lazer, D.M.J., et al.: The science of fake news. Science (New York, N.Y.) **359**(6380), 1094–1096 (2018)
16. Lemaître, G., Nogueira, F., Aridas, C.K.: Imbalanced-learn: a python toolbox to tackle the curse of imbalanced datasets in machine learning. J. Mach. Learn. Res. **18**(1), 559–563 (2017)

17. Liang, G., He, W., Xu, C., Chen, L., Zeng, J.: Rumor identification in microblogging systems based on users' behavior. IEEE Trans. Comput. Soc. Syst. **2**(3), 99–108 (2015)

18. Ma, J., Gao, W., Mitra, P., Kwon, S., Jansen, B., Wong, K.: Detecting rumors from microblogs with recurrent neural networks. In: IJCAI, pp. 3818–3824 (2016)

19. Qazvinian, V., Rosengren, E., Radev, D.R., Mei, Q.: Rumor has it: identifying misinformation in microblogs. In: Proceedings of the Conference on Empirical Methods in Natural Language Processing, pp. 1589–1599 (2011)

20. Sicilia, R., Lo Giudice, S., Pei, Y., Pechenizkiy, M.: Twitter rumour detection in the health domain. Expert Syst. Appl. **110**, 33–40 (2018)

21. Starbird, K., Maddock, J., Orand, M., Achterman, P., Mason, R.: Rumors, false flags, and digital vigilantes: misinformation on Twitter after the 2013 Boston marathon bombing. In: IConference (2014)

22. Turenne, N.: The rumour spectrum. PLOS One **13**(1), e0189080 (2018)

23. Varol, O., et al.: Feature engineering for social bot detection. In: Feature Engineering for Social Bot Detection, pp. 311–334. CRC Press, March 2018

24. Vosoughi, S., Mohsenvand, M.N., Roy, D.: Rumor Gauge. ACM Trans. Knowl. Discov. Data **11**(4), 1–36 (2017)

25. Vosoughi, S., Roy, D., Aral, S.: The spread of true and false news online. Science (New York, N.Y.) **359**(6380), 1146–1151 (2018)

26. Wang, S., Moise, I., Helbing, D.: Early signals of trending rumor event in streaming social media. In: Computer Software and Applications Conference (COMPSAC), pp. 654–659. IEEE (2017)

27. Wijeratne, S., et al.: Feature engineering for Twitter-based applications. In: Feature Engineering for Machine Learning and Data Analytics, pp. 359–384 (2017)

28. Wu, K., Yang, S., Zhu, K.Q.: False rumors detection on Sina Weibo by propagation structures. In: 2015 IEEE 31st International Conference on Data Engineering, pp. 651–662. IEEE, April 2015

29. Yang, F., Liu, Y., Yu, X., Yang, M.: Automatic detection of rumor on Sina Weibo. In: Proceedings of the ACM SIGKDD Workshop on Mining Data Semantics (2012)

30. Zhang, Q., Zhang, S., Dong, J., Xiong, J., Cheng, X.: Automatic detection of rumor on social network. In: Li, J., Ji, H., Zhao, D., Feng, Y. (eds.) NLPCC -2015. LNCS (LNAI), vol. 9362, pp. 113–122. Springer, Cham (2015). https://doi.org/10.1007/978-3-319-25207-0_10

31. Zhao, Z., Resnick, P., Mei, Q.: Enquiring minds. In: Proceedings of the 24th International Conference on World Wide Web - WWW 2015, pp. 1395–1405. ACM Press, New York (2015)

32. Zubiaga, A., Liakata, M., Procter, R.: Exploiting context for rumour detection in social media. In: Ciampaglia, G.L., Mashhadi, A., Yasseri, T. (eds.) SocInfo 2017. LNCS, vol. 10539, pp. 109–123. Springer, Cham (2017). https://doi.org/10.1007/978-3-319-67217-5_8

33. Zubiaga, A., Aker, A., Bontcheva, K., Liakata, M., Procter, R.: Detection and resolution of rumours in social media. ACM Comput. Surv. **51**(2), 1–36 (2018)

34. Zubiaga, A., Liakata, M., Procter, R.: Learning reporting dynamics during breaking news for rumour detection in social media (2016)

Sentiment Analysis for Fake News Detection by Means of Neural Networks

Sebastian Kula[1,2], Michał Choraś[1], Rafał Kozik[1(✉)], Paweł Ksieniewicz[1,3], and Michał Woźniak[1,3]

[1] UTP University of Science and Technology, Bydgoszcz, Poland
rkozik@utp.edu.pl
[2] Kazimierz Wielki University, Bydgoszcz, Poland
skula@ukw.edu.pl
[3] Wrocław University of Science and Technology, Wrocław, Poland

Abstract. The problem of fake news has become one of the most challenging issues having an impact on societies. Nowadays, false information may spread quickly through social media. In that regard, fake news needs to be detected as fast as possible to avoid negative influence on people who may rely on such information while making important decisions (e.g., presidential elections). In this paper, we present an innovative solution for fake news detection that utilizes deep learning methods. Our experiments prove that the proposed approach allows us to achieve promising results.

Keywords: Online disinformation · Fake news · Neural networks · Deep learning · Sentiment analysis

1 Introduction

Fake news is often defined as a hoax or false information that is spread employing the news media, either printed or online social networks. This phenomenon is not new in human history, and one can find examples of fake news originating in the nineteenth century (e.g., Great Moon Hoax [1]). However, due to the increasing popularity of social media widely used for political purposes, the problem of fake news has gained more importance in recent years. It also imposes a great detection challenge. Manual fact-checking in many cases, is difficult, time-consuming, and expensive. Therefore, the community has been looking for various automated detection solutions that would speed up this process. In recent years, different NLP (Natural Language Processing) methods have been proposed to solve the fake news detection problem.

The main contribution of this paper is the proposition and evaluation of neural network-based approach to text analysis and fake news detection. The contribution includes the application of the remote, cloud computing platform, GPU cards, state-of-the-art Machine Learning and Deep Learning libraries; all the above-mentioned works allowed to create a working model for fake news detection in a relatively short time and with the use of open-source solutions.

© Springer Nature Switzerland AG 2020
V. V. Krzhizhanovskaya et al. (Eds.): ICCS 2020, LNCS 12140, pp. 653–666, 2020.
https://doi.org/10.1007/978-3-030-50423-6_49

The paper is structured as follows: after the introduction, in Sect. 2 the proposed approach is presented in detail. In Sect. 3 the used datasets are overviewed. Experimental setup and results are described in Sect. 4, whereas conclusions are given after that.

2   Related Work

There are various machine learning approaches for fake news detection and the main effort is put into efficient feature extraction, as well as an appropriate choice of the classification model.

In [2], authors have adopted Naïve Bayes to recognize fake and legitimate news. On the other hand, Shu et al. in [3], explicitly elaborated and listed a set of attributes that may help to indicate fake news. These attributes include the source (author or the publisher of the news), headline (a sort of title that is intended to draw readers' attention), body (the main text describing the news), image or video that is intentionally used for spotting the fake news.

In the literature, some approaches utilise computer vision for fake news detection. An interesting method, falling into that category, for image-based fake photos detection has been presented in [4].

Recently, with the emergence of deep learning, a significant number of researchers have started applying this type of model to solve various classification and regression problems. The deep learning methods are capable of autonomously computing the hierarchical representation of the data and allow achieving results that surpass other state-of-the-art approaches.

Zang et al. [5] have proposed a deep recurrent diffusive neural network to address the problem of fake news detection. On the other hand, in contrast to the traditional RNN model, in [6] authors adapted a pre-trained BERT model (Bidirectional Encoder Representations from Transformers), that consists of several stacked transformer-encoder blocks.

3   The Proposed Approach

To tackle the challenge of fake news, the application of the Flair library is proposed [7], which offers outstanding features in terms of neural network design, includes many state-of-the-art methods, among them numerous methods based on the deep learning, also enabling GPU-based training. Flair is a Natural Language Processing library designed for all word embeddings as well as arbitrary combinations of embeddings [7]. The crucial elements of creating the fake news detection model were carried out with the support of the Flair library. The training process was carried out based on deep learning methods afterword embeddings had been carried out using the modern and effective procedures in this area. As shown in Fig. 1, in our work, we chose to use various types of neural networks to solve the problem of text-based fake news detection.

3.1   Text Pre-processing Using NLP

The goal of text pre-processing is to obtain the text, which is a reduced representation of the raw text. The reduced text enables the detection of specific patterns of the raw text simultaneously. A reduction strategy was adopted, consisting of the elimination of unnecessary elements and, through this step, achieving a higher generalization of the text. To detect unnecessary items and overrepresentation of words, statistical analysis of their occurrences in datasets was used. The clean text was obtained by creating a separate code, unrelated to the Flair library.

Fig. 1. The processing pipeline of the proposed solution

Due to Flair use of embedding layers, it is not necessary to run the usual pre-processing steps such as constructing a vocabulary of words in the dataset or encoding words as one-hot vectors [7]. In the Flair, each embedding layer implements either the TokenEmbedding or the DocumentEmbedding interface for word and document embeddings respectively [7]. In our approach, we treated the content of articles as documents and we applied the DocumentEmbedding interface.

3.2   Word Embeddings in Flair

Neural networks used in NLP tasks do not operate directly on texts, sentences, or words, but on their representation in the numerical form. This process of converting them into numbers is called word embeddings and it is one of the key elements enabling sentiment analysis and fake news detection.

The main methods of word embeddings are 'word2vec', 'glove', and 'FastText', which are classified as canonical methods. In addition to the listed above, the Flair library supports a growing list of embeddings such as hierarchical character features, ELMo embeddings, ELMo transformer embeddings, BERT embeddings, byte pair embeddings, Flair embeddings and Pooled Flair embeddings [7].

In this work, the 'glove' method was used. For comparative purposes, the 'twitter' word embeddings, 'news' word embeddings and 'crawl' word embeddings were used as well. The synthesis of the methods used is summarized below.

The 'glove' is an open-source project at Stanford University; its code is freely available [8]. The 'glove' overcomes the disadvantages of the models focusing only on local statistics and the models focusing only on global statistics. For example, methods like latent semantic analysis (LSA) efficiently leverage statistical information, but they do relatively poorly on the word analogy task [8]. The other example, skip-gram methods, may do better on the analogy task, but they poorly utilize the statistics of the corpus [8]. The 'glove' is a specific weighted least squares model that trains on global word-word co-occurrence counts and thus makes efficient use of statistics [8]. The 'glove' is a global log-bilinear regression model for the unsupervised learning of word representations that outperforms other models on word analogy and word similarity [8].

To launch word embeddings in the Flair with the use of 'glove', the user enters the following WordEmbeddings ('glove') command in the code.

The FastText method was created by the Facebook AI Research lab based on the models contained in the article [9]. The FastText method is based on the bag of n-grams and subword units. Each word is represented as a bag of character n-grams [10]. The method indicates better results than the state-of-the-art methods in word similarity and word analogy experiments [10].

Both methods are available in the Flair library as pre-trained databases of word embeddings. The 'glove' and the FastText methods were created based on data obtained from Wikipedia. Pre-trained models used in this paper, like 'news', were created using FastText embeddings over news and Wikipedia data; the 'crawl' was created using the FastText embeddings over web crawls; 'twitter' was created using two billion tweets.

3.3 Recurrent Neural Network

Currently, the text classification methods most often use methods based on Deep Neural Networks (DNN), which have better performance in Natural Language Processing (NLP) tasks solving than other neural networks [19,20]. DNNs are characterized by high complexity and a large number of hidden layers, which is their distinguishing feature in comparison with standard Artificial Neural Networks (ANN).

Deep Neural Networks have already been extensively used in many areas of artificial intelligence, such as speech recognition, image recognition, text translation, sentiment analysis, and spam detection. There is a whole range of DNN methods used in NLP. In this article, we focus on Recurrent Neural Network (RNN) as well as Gated Recurrent Unit (GRU) and Long-Short Term Memory (LSTM) methods that are classified as RNN methods (networks).

The feature distinguishing RNN networks from other ANN networks is their recurrency, referring to the flow of signals between input and output of the network. This type of networks has a kind of feedback loop, which means that the output is also the input for the next state and affects its output value. Such a network architecture results in the fact that the network has a kind of memory that theoretically allows for information storage. Apart from the difference mentioned above, the RNN network works like a regular, one-way ANN network, that is during the training weights and propagation errors are calculated.

The disadvantage of the RNN network is the phenomenon of the vanishing gradient, which makes it impossible to remember and search for the bindings between data that occur after a more extended period. There are several methods that overcome this undesirable phenomenon; they include GRU and LSTM networks. Both networks are described in the next two subsections.

The GRU is a network with recursive architecture, improved by introducing special types of gates: the reset gate r_t and the update gate z_t. They together control how information is updated to the state [11]. In mathematical terms, the z_t gate is as follows [11]:

$$z_t = \sigma(W_z x_t + U_z h_{t-1})$$

where σ is sigmoid function, W_z and U_z are weights, h_{t-1} is previous state, x_t is the sequence vector at time t. The formula for r_t gate is as follows [11]:

$$r_t = \sigma(W_r x_t + U_r h_{t-1})$$

The formula is similar to the formula for the update gate, with the noticeable different weights W_r and U_r. The candidate activation h'_t is computed similarly to that of the traditional recurrent unit [11,12]:

$$h'_t = tanh(W_h x_t + r_t \odot U_h h_{t-1})$$

where r_t is a set of reset gates and $\odot$ is an element-wise multiplication [12]. When the reset gate is zero, the previous computed state is erased from the network. The activation h_t of the GRU at time t is a linear interpolation between the previous activation h_{t-1} and the candidate activation h'_t [11,12]:

$$h_t = (1 - z_t) \odot h_{t-1} + z_t \odot h'_t$$

The neural network constructed in this way allows to control and collect the data and thus to tackle the issue of the vanishing gradient.

The LSTM is a neural network that is similar to the GRU, except that it is more complex and requires more computing power during training. The LSTM contains the following gates: input, output, forget, memory cell and new memory cell content [12]. The forget gate determines whether the memory cell content will be preserved or erased; the input gate determines whether the new memory cell content will be added to the memory cell; the output gate decides what content from the memory cell will be on the output. There are many versions of the LSTM implementation; the original one was presented in the article [13].

4   Experimental Evaluation

The purpose of the experiment is to conduct the training with the data and then to validate the model using the Flair library. The application of the Flair library to create the fake news detection model has not been reported in the literature yet.

4.1   Experimental Setup

Finding the right dataset is fundamental to create an efficient, reliable fake news detection model. Simultaneously, the access to such datasets is limited and creates a challenge to acquire current, ready-to-learn databases. In the article, we applied freely available datasets, which are accessible on the websites of Kaggle and the Information Security and Object Technology (ISOT) research lab. Two different sets of data were applied, one called "ISOT Fake News Dataset" [14], the other called "Getting real about fake news" (GRaFN) [15].

Two models were taught, the first one is based on the application of the ISOT dataset for training, and the second model, because the collection acquired from the Kaggle contains mostly fake news, was taught with the use of both collections, through attaching the real news collection from ISOT to the collection downloaded from the Kaggle webpage.

Information in both datasets contains news published on websites. The ISOT collection is dominated by the vast majority of political information and news from around the world. The dataset contains two files (true and fake) in csv file format. The real information database was created based on the websites of a reliable Reuters news agency, and the fake information was collected from the pages marked as unreliable by Politifact [16]. The dataset contains a total of 44898 items, 21417 of which are real items and 23481 are fake items. Each file contains four columns: article title, text, article publication date and the subject which can relate to one of six types of information (world-news, politics-news, government-news, middle-east, US news, left-news) [14]. In order to prepare the data for pre-processing in a proper way, an analysis of the occurrence of e-mail addresses, social media addresses, website addresses (https and www) was conducted, the results are presented in Table 1.

Table 1. The analysis of the occurrence of email addresses, social media addresses, website addresses (https and www) in the ISOT dataset

Type of address	ISOT dataset (True)	ISOT dataset (Fake)
email and social media addresses	803	27888
https addresses	0	94
www addresses	48	726

In the file with true data, 28909 instances of Reuters information agency name were observed; such a large number of instances can become a special feature of the pattern, hence we decided to eliminate this property from the data. The number of occurrences of city names was also analysed to determine the nature of the information in geographical terms. Table 2 presents the 10 most common city names in the items.

The second dataset contains texts classified into eight categories (bias with 443 occurrences, conspiracy - 430 occurrences, fake - 19 occurrences, bs - 11492

Table 2. The 10 most common city names occurrence in the ISOT dataset

Name of the city	ISOT dataset (True)	ISOT dataset (Fake)
WASHINGTON	6921	107
NEW YORK	949	16
LONDON	761	3
MOSCOW	673	1
BERLIN	530	2
BEIJING	503	0
PARIS	331	13
ANKARA	275	0
MEXICO CITY	237	4
TOKYO	228	0

occurrences, hate - 246 occurrences, junksci - 102 occurrences, satire - 146 occurrences and state - 121 occurrences); in this paper, all these categories but satire can be considered to be various forms of fake news.

The dataset contains 20 columns, 12999 rows containing the same number of articles in the text column. The dataset includes the text from 244 websites, collected within 30 days [15]. Most of the articles were written in English (12403 articles), but there are also articles in Russian (203 articles), Spanish (172 articles), German (111 articles), French (38 articles), Arabic (22 articles), Portuguese (11 articles), Turkish (10 articles), Italian (9 articles) or Chinese (1 article). Only the English-language articles were used to create the model. Similarly to the first dataset for the correct pre-processing, the number of e-mail addresses occurrences, social media addresses occurrences, website addresses (https and www) occurrences, emoticons occurrences were detected; data in Table 3.

Table 3. The analysis of the occurrence of email addresses, social media addresses, website addresses (https and www) and emoticons in the "Getting real about fake news" dataset

Type of string	Number of occurrences in GRaFN dataset
email and social media addresses	3512
https addresses	25
www addresses	1499
emoticons	93

The described datasets were used not only for the training process but also for validation and testing of neural network models. The initial step to create the model was to prepare the data by pre-processing them. The data underwent many handlings consisting of the elimination of unnecessary elements or the ones disturbing the training process and simultaneously occurring in significant amounts in the data. First, datasets were limited to two key columns, the *label* column, determining if the information is true or false, and the *text* column, containing the contents of the articles. Then, all the items beyond standard text were eliminated from the articles in the *text* column. Social media addresses, e-mail addresses, website addresses (https and www), emoticons and even punctuation and periods were removed.

Two data collections were used to conduct the experiments. Collection 1 is based entirely on the ISOT dataset and, according to the cross-validation procedure implemented in the Flair, contains three elements: data for the training, for the testing, and the validation. To ensure adequate representativeness of the data in the collection 1, the ISOT dataset was divided into three parts, in the following proportion: 80% allocated for the training, 10% for the testing and 10% for the validation. Collection 2 was created from the combination of the extracted part of the ISOT dataset, containing real news, and the GRaFN dataset. As in the collection 1, data representativeness in collection 2 is ensured by dividing into three parts in the same proportion as for the collection 1.

Pre-processing and training operations followed by post-processing were carried out in the cloud service environment called the Colaboratory. The pandas library version 0.25.3, the Flair library version 0.4.3, Jupyter notebook and the Python programming language were used. In the hardware scope, the hardware resources available on the Colaboratory platform were used, in the form of the GPU card P100 PCIE-16 GB, cuda version 10.1, 12.72 GB RAM memory, 68.4 GB HDD memory.

Before the training, in addition to preparing the data and hardware, the neural network architecture needs to be designed and its hyperparameters need to be specified. The Flair library allows the user to specify the values of many hyperparameters; some selected ones are presented in Table 4. All created models had exactly the same hyperparameters values set before the training.

Table 4. Hyperparameters values of RNN GRU

Name of the hyperparameter	Hyperparameter value
Learning rate	0.1
Batch size	32
Anneal factor	0.5
Patience	5
Max number of epochs	5
Hidden states size	512

A crucial element in machine learning is the correct selection of metrics. The classification tasks focus most often on the following metrics: accuracy, f1 score, precision and recall. The paper presents the analysis of accuracy, precision and recall during the tests and f1 score during the validation processes. The accuracy was defined as a number correct predictions divided by the total number of predictions, multiplied by 100%. The f1-score was defined according to the following formula:

$$f1score = \frac{2 * precision * recall}{precision + recall}$$

where precision and recall are validation values.

4.2 Plan of the Experiment

The experiment involved three different routines (training and validation) for generating models and tests of the created models. The routine 1 consisted of using the collection 2, the LSTM neural network and word embeddings 'glove', the routine 2 consisted of using the collection 1, the neural network GRU and the word embeddings 'glove', the routine 3 was the extension of the routine 2 by adding parameterization, consisting of changing the word embedding methods from 'glove' to 'news', 'twitter' and 'crawl'. The goal of the tests was to demonstrate the usability of the created models in practice.

4.3 Results and Discussion

When analyzing the results, the focus was on examining the training loss, the validation loss, the f1 score parameter, the accuracy and the computation time needed to train the neural network. The results are shown in Figs. 2, 3, 4, 5, 6, 7 and 8. Based on the analysis of losses, it can be concluded that, apart from the case presented in Fig. 5 for the 'twitter' method, there is no risk of overfitting the network, because the training loss curve usually has lower values or relatively slightly higher values than the validation loss curve. For all methods, a high f1 score was obtained - above 0.9 for the epoch 5. In Flair, the f1 score metric is crucial in the validation process. On its basis, the best model is automatically selected for tests. It is observed that not always the best result and the best model is obtained for the last epoch. For example, for the 'twitter' method the highest f1 score was obtained for the fourth epoch.

All trained models have obtained sufficient testing accuracy to be applied in the practical tasks of fake news detection. Based on a dataset with similar topics as the ISOT dataset, which is politics-related topics, in [14] for the fake news detection challenge, the authors obtained a maximum accuracy of 92%. In all our experiments, the maximum obtained accuracy outperformed this value (Table 5).

As a result of the training, models with calculated parameters, i.e., weights were obtained. These models were tested on the remote Colaboratory platform. The first stage of testing consisted of entering selected texts contained in the ISOT dataset. In the second stage of testing, the Independent [17], which is a British online newspaper, and the 11 Sci-Fi Short Stories website [18] were used. Articles in the Independent were treated as a source of credible texts

Table 5. Resulted metrics for testing of models for the label fake (the comparison between word embeddings techniques, 'glove', 'news', 'twitter', 'crawl')

Metric	'glove'	'news'	'twitter'	'crawl'
True positives (TP)	2275	2080	2269	2217
True negatives (TN)	2139	2121	2064	2100
False positives (FP)	4	22	79	43
False negatives (FN)	2	197	8	60
Precision	0.9982	0.9895	0.9664	0.9810
Recall	0.9991	0.9135	0.9965	0.9736
Accuracy	99.86%	95.04%	98.03%	97.67%

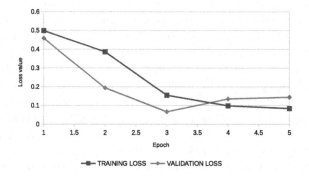

Fig. 2. Relations between the number of epochs versus the training and validation losses; the results obtained during the training of the model with the use of both dataset collections, and the word embeddings technique 'glove'

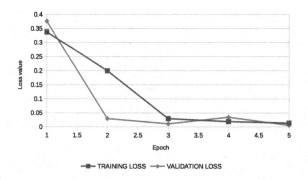

Fig. 3. Relations between the number of epochs versus the training and validation losses; the results obtained during the training of the model with the use of the ISOT dataset collection, and the word embeddings technique 'glove'

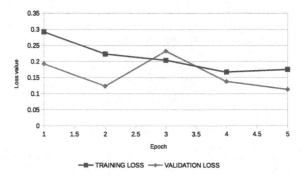

Fig. 4. Relations between the number of epochs versus the training and validation losses; the results obtained during the training of the model with the use of the ISOT dataset collection, and the word embeddings technique 'news'

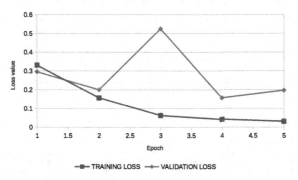

Fig. 5. Relations between the number of epochs versus the training and validation losses; the results obtained during the training of the model with the use of the ISOT dataset collection, and the word embeddings technique 'twitter'

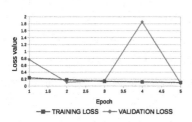

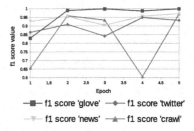

Fig. 6. Relations between the number of epochs versus the training and validation losses; the results obtained during the training of the model with the use of the ISOT dataset collection, and the word embeddings technique 'crawl'

Fig. 7. Analysis of f1 score; the comparison between word embeddings techniques

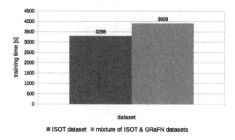

Fig. 8. Computation time needed for models training; the comparison between datasets applied for the training

(true), and science fiction stories as a source of unreliable materials (false). For both the above-described tests, the model worked correctly and detected true and false information. Tests were carried out repeatedly, confirming the validity, robustness and credibility of the model. Examples of correct model operation are shown in Fig. 9. All entered texts were subjected to the procedure of eliminating irrelevant elements from texts, before submitting them to the model. The procedure is identical to the one carried out in the pre-processing stage on the raw data.

```
# create example sentence
sentence = Sentence('The space station staff liked her when they interviewed her she seemed
polite and quiet and incurious. That was important. One of the astronauts, a bearded
Russian with kind eyes, asked her a question: Will you be lonely in space? She looked at
the faint lines scrawled around his eyes and forehead, and she supposed he had a family
somewhere, maybe small children. Yes, she said, but I have always been lonely. The
astronaut nodded, and she could see he understood. She could see his aquiline profile as he
turned to someone off screen, and she knew she would get the job.')


# predict tags and print
classifier.predict(sentence)

print(sentence.labels)

    __label__fake (0.86463862657547)
```

Fig. 9. The screenshot of the launched model, which recognizes that the information from the 11 Sci-Fi Short Stories website is fake

The model can be used in real-time solutions because its execution time is relatively short. The results indicated the impact of word embedding techniques on the accuracy and the f1 score. The highest results were obtained for the 'glove' method. In the process of creating the model and training the neural network, it was observed that one of the crucial elements to obtain robust results is the correctly performed pre-processing on the raw data. The relatively short time needed to train the neural network was achievable by applying for the GPU card.

4.4 Threats to Validity

The proposed method should be tested on other datasets. The critical and most desired scenario would be to have fake and true news obtained from the same source/news agency. In our experiments, true news is not from the same source as fake news. However, such datasets are still to be offered by reliable news agencies (which most often claim not to have fake news at all).

Another aspect is the lack of a clear definition of what is exactly meant by fake news. For example, conspiracy theories type of information is not always considered as fake news. In such definitions, the motive of the source is taken into account, but it cannot be determined only by analyzing the text as we do in this work.

5 Conclusions

The paper presents the stages of creating the model applied for fake news detection. The model is based on DNN networks trained with the Flair library. The pre-processing, training and post-processing phases are described in detail for the obtained models. The novelty of the paper is the application of the Flair library for detecting true and false information, as well as the application of the cloud solution called the Collaboratory. The model fulfills its tasks and allows for the analysis of texts with high accuracy. During the training process, the accuracy was up to 99.8%.

The current work concerned the distinction between label fake and label true. However, there are many additional subcategories under the fake news category; future work will concern the creation of a model to distinguish those sub-categories.

Acknowledgement. This work is funded under SocialTruth project, which has received funding from the European Union's Horizon 2020 research and innovation programme under grant agreement No. 825477.

References

1. Goodman, M.: The Sun and the Moon: The Remarkable True Account of Hoaxers, Showmen, Dueling Journalists, and Lunar Man-Bats in Nineteenth-Century New York. Basic Books, New York (2008)
2. Jain, A., Kasbe, A.: Fake News Detection. In: 2018 IEEE International Students' Conference on Electrical, Electronics and Computer Science (SCEECS), Bhopal, pp. 1–5 (2018)
3. Shu, K., Sliva, A., Wang, S., Tang, J., Liu, H.: Fake news detection on social media: a data mining perspective. CoRR, abs/1708.01967 (2017)
4. Choraś, M., Giełczyk, A., Demestichas, K., Puchalski, D., Kozik, R.: Pattern recognition solutions for fake news detection. In: Saeed, K., Homenda, W. (eds.) CISIM 2018. LNCS, vol. 11127, pp. 130–139. Springer, Cham (2018). https://doi.org/10.1007/978-3-319-99954-8_12

5. Zhang, J., Cui, L., Fu, Y., Gouza, F.B.: Fake news detection with deep diffusive network model. CoRR, abs/1805.08751 (2018)
6. Devlin, J., Chang, M.-W., Lee, K., Toutanova, K.: BERT: pre-training of deep bidirectional transformers for language understanding. arXiv preprint arXiv:1810.04805 (2018)
7. Akbik, A., Bergmann, T., Blythe, D., Rasul, K., Schweter, S., Vollgraf, R.: FLAIR: an easy-to-use framework for state-of-the-art NLP. In: Ammar, W., Louis, A., Mostafazadeh, N. (eds.) NAACL-HLT (Demonstrations), pp. 54–59. Association for Computational Linguistics (2019)
8. Pennington, J., Socher, R., Manning, C.D.: GloVe: global vectors for word representation. In: Empirical Methods in Natural Language Processing (EMNLP), pp. 1532–1543 (2014)
9. Joulin, A., Grave, E., Bojanowski, P., Mikolov, T.: Bag of Tricks for efficient text classification. In: Proceedings of the 15th Conference of the European Chapter of the Association for Computational Linguistics: Volume 2, Short Papers, pp. 427–431 (2017)
10. Bojanowski, P., Grave, E., Joulin, A., Mikolov, T.: Enriching word vectors with subword information. Trans. Assoc. Comput. Linguist. 5, 135–146 (2017)
11. Yang, Z., Yang, D., Dyer, C., He, X., Smola, A.J., Hovy, E.H.: Hierarchical attention networks for document classification. In: Proceedings of the 2016 Conference of the North American Chapter of the Association for Computational Linguistics: Human Language Technologies, pp. 1480–1489 (2016)
12. Chung, J., Gülcehre, C., Cho, K., Bengio, Y.: Empirical evaluation of gated recurrent neural networks on sequence modeling. In: NIPS 2014 Workshop on Deep Learning and Representation Learning, Montréal, Canada (2014)
13. Hochreiter, S., Schmidhuber, J.: Long short-term memory. Neural Comput. 9, 1735–1780 (1997)
14. Ahmed, H., Traore, I., Saad, S.: Detecting opinion spams and fake news using text classification. J. Secur. Priv. 1(1), e9 (2018)
15. Getting real about fake news. https://www.kaggle.com/mrisdal/fake-news. Accessed 25 Nov 2019
16. Ahmed, H., Traoré, I., Saad, S.: Detection of online fake news using N-gram analysis and machine learning techniques. In: Traoré, I., Woungang, I., Awad, A. (eds.) ISDDC 2017. LNCS, vol. 10618, pp. 127–138. Springer, Cham (2017). https://doi.org/10.1007/978-3-319-69155-8_9
17. The Independent webpage. https://www.independent.co.uk/news/world/middle-east/raqqa-isis-terror-fears-europe-a7401511.html. Accessed 28 Dec 2019
18. 11 Sci-Fi Short Stories, the Janitor in Space - American Short Fiction website. http://americanshortfiction.org/2014/07/01/janitor-space/. Accessed 28 Dec 2019
19. Ksieniewicz, P., Choraś, M., Kozik, R., Woźniak, M.: Machine learning methods for fake news classification. In: Yin, H., Camacho, D., Tino, P., Tallón-Ballesteros, A.J., Menezes, R., Allmendinger, R. (eds.) IDEAL 2019. LNCS, vol. 11872, pp. 332–339. Springer, Cham (2019). https://doi.org/10.1007/978-3-030-33617-2_34
20. Choraś, M., Pawlicki, M., Kozik, R., Demestichas, K.P., Kosmides, P., Gupta, M.: SocialTruth project approach to online disinformation (fake news) detection and mitigation. In: Proceedings of ARES, Canterbury, UK, pp. 68:1–68:10 (2019)

Author Index

Abuhay, Tesfamariam M. 418, 430
Achermann, Guillem 252
Alfredo Martinez, J. 441
Alves, Domingos 363, 550, 563
Anatoliy, Gorbenko 441
Andrade Bernardi, Filipe 563
Andrysiak, Tomasz 170
Anna, Palczewska 441
Apunike, Anderson C. 550
Ashikaga, Hiroshi 334
Astrup, Arne 441

Babajide, Oladapo 441
Babenko, Alina Y. 495
Balabaeva, Ksenia 603
Balatsko, Maksym 225
Bardina, Mariia 305
Bartkow, Piotr 291
Bera, Aneta 18
Bernardi, Filipe Andrade 363
Bernus, Olivier 334
Bielecka, Marzena 406
Bielecki, Andrzej 406
Bortko, Kamil 291
Brabec, Jan 74
Bruggeman, Jeroen 243
Burduk, Robert 128

Cai, Wentong 577
Chakraborty, Saikat 536
Choraś, Michał 196, 615, 653
Colombo Filho, Márcio Eloi 563

de Lima, Inácia Bezerra 363
De Luca, Gabriele 252
de Oliveira, Lariza Laura 550, 563
de Walle, Bartel van 639
Derevitskii, Ilya V. 495
Derevitskii, Ivan 577
Dubois, Remi 334

Englert, Roman 629
Ertaylan, Gökhan 468

Fard, Amir Ebrahimi 639
Franc, Vojtěch 74
Friedjungová, Magda 225
Funkner, Anastasia A. 591

Galliez, Rafael M. 550
Goścień, Róża 211
Guleva, Valentina 305

Haïssaguerre, Michel 334
Hissam, Tawfik 441
Hocini, Mélèze 334
Holotyak, Taras 184
Hołubowicz, Witold 196
Hussey, Pamela 456

Igor, Vozniuk 390

Jankowski, Jarosław 277, 291
Jiang, Zhengwei 156
Jiřina, Marcel 225

Kamath, S. Sowmya 321
Kashina, Mariya 509
Kitlas Golińska, Agnieszka 376
Klęsk, Przemysław 18
Klikowski, Jakub 117
Klinkowski, Mirosław 211
Koenigkam Santos, Marcel 563
Kogtikov, Nikita 577
Komárek, Tomáš 74
Kopanitsa, Georgy 390, 509
Korytkowski, Marcin 184
Kovalchuk, Sergey V. 418, 430, 495, 591,
    603
Kozik, Rafał 196, 615, 653
Krishnan, Gokul S. 321

Kritski, Afrânio 563
Ksieniewicz, Paweł 103, 128, 211, 653
Kula, Sebastian 653
Kurzynski, Marek 88

Łazęcka, Małgorzata 3
Lees, Michael H. 577
Lenivtceva, Iuliia 509
Leonenko, Vasiliy N. 483
Lesiński, Wojciech 376
Liu, Jian 156
Lu, Feng 266

Machlica, Lukáš 74
Maria, Prohorova 390
McGlinn, Kris 456
Mello Galliez, Rafael 563
Metsker, Oleg G. 390, 418, 430
Michalski, Radosław 277
Mielniczuk, Jan 3
Miyoshi, Newton Shydeo Brandão 363
Mohammadi, Majid 639
Morozova, Elena 390
Muschiol, Jörg 629

Nandy, Anup 536
Nannes, Berend 334
Nigatie, Yemisrach G. 430
Nowak, Jakub 184
Nowakowski, Arkadiusz 45

Obuchowicz, Rafał 406
Oliveira-Ciabati, Lívia 550
Ong, Marcus E. H. 577
Oppert, Jean-Michel 441

Pawlicki, Marek 196, 615
Pazura, Patryk 277, 291
Peng, Peng 266
Piórkowski, Adam 406
Przybylski, Andrzej 376
Puchalski, Damian 615
Pusparum, Murih 468

Quax, Rick 334

Rudnicki, Witold R. 376

Saganowski, Łukasz 170
Sanches, Tiago L. M. 550
Sanchez, Mauro N. 550
Savitskaya, Daria A. 495
Scherer, Rafał 184
Simoni, Michele 252
Singh, Balbir 523
Slasten, Evgenia 509
Śliwka, Piotr 348
Socha, Leslaw 348
Sørensen, Thorkild I. A. 441
Sprik, Rudolf 243
Strąk, Łukasz 45
Sychel, Dariusz 18

Tawfik, Hissam 523
Teisseyre, Paweł 3
Thas, Olivier 468
Thomas, Noble 536
Topolski, Mariusz 35
Trajdos, Pawel 88

Unold, Olgierd 45

Vaganov, Danila 305
Vašata, Daniel 225
Vinci, André Luiz Teixeira 363
Voloshynovskiy, Slava 184

Walkowiak, Krzysztof 211
Wang, Quiyun 156
Wang, Xuren 156
Wieczorek, Wojciech 45
Woźniak, Michał 59, 117, 141, 653

Xiao, Qingsai 156

Yakovlev, Aleksey N. 418, 430
Yamada, Diego Bettiol 363
Yao, Yepeng 156
Yoshiura, Vinicius Tohoru 363

Żak, Michał 141
Zyblewski, Paweł 59

Printed in the United States
by Baker & Taylor Publisher Services